Lecture Notes in Computer Science 8043

Commenced Publication in 1973
Founding and Former Series Editors:
Gerhard Goos, Juris Hartmanis, and Jan van Leeuwen

Editorial Board

David Hutchison
  Lancaster University, UK

Takeo Kanade
  Carnegie Mellon University, Pittsburgh, PA, USA

Josef Kittler
  University of Surrey, Guildford, UK

Jon M. Kleinberg
  Cornell University, Ithaca, NY, USA

Alfred Kobsa
  University of California, Irvine, CA, USA

Friedemann Mattern
  ETH Zurich, Switzerland

John C. Mitchell
  Stanford University, CA, USA

Moni Naor
  Weizmann Institute of Science, Rehovot, Israel

Oscar Nierstrasz
  University of Bern, Switzerland

C. Pandu Rangan
  Indian Institute of Technology, Madras, India

Bernhard Steffen
  TU Dortmund University, Germany

Madhu Sudan
  Microsoft Research, Cambridge, MA, USA

Demetri Terzopoulos
  University of California, Los Angeles, CA, USA

Doug Tygar
  University of California, Berkeley, CA, USA

Gerhard Weikum
  Max Planck Institute for Informatics, Saarbruecken, Germany

Ran Canetti   Juan A. Garay (Eds.)

Advances in Cryptology – CRYPTO 2013

33rd Annual Cryptology Conference
Santa Barbara, CA, USA, August 18-22, 2013
Proceedings, Part II

 Springer

Volume Editors

Ran Canetti
Boston University and Tel Aviv University
111 Cummington Street
Boston, MA 02215, USA
E-mail: canetti@bu.edu

Juan A. Garay
AT&T Labs – Research
180 Park Avenue
Florham Park, NJ 07932, USA
E-mail: garay@research.att.com

ISSN 0302-9743   e-ISSN 1611-3349
ISBN 978-3-642-40083-4   e-ISBN 978-3-642-40084-1
DOI 10.1007/978-3-642-40084-1
Springer Heidelberg Dordrecht London New York

Library of Congress Control Number: 2013944216

CR Subject Classification (1998): E.3, F.2, K.6.5, G.2.1, D.4.6, C.2.0, J.1

LNCS Sublibrary: SL 4 – Security and Cryptology

Typesetting: Camera-ready by author, data conversion by Scientific Publishing Services, Chennai, India

Printed on acid-free paper

Springer is part of Springer Science+Business Media (www.springer.com)

Preface

CRYPTO 2013, the 33rd Annual International Cryptology Conference, was held August 18–22, 2013, on the campus of the University of California, Santa Barbara. The event was sponsored by the International Association for Cryptologic Research (IACR) in cooperation with the UCSB Computer Science Department and the IEEE Computer Society's Technical Committee on Security and Privacy.

The program represents the recent significant advances in all areas of cryptology. Sixty-one papers were included in the program, a record number for IACR flagship conferences. This two-volume proceedings contains the revised versions of all the papers. One pair of papers shared a single presentation slot in the program. There were also two invited talks. On Monday, Cindy Cohn from the Electronic Frontier Foundation gave a talk entitled "Crypto Wars Part 2 Have Begun." On Wednesday, Adam Langley from Google spoke about "Why the Web Still Runs on RC4," in a joint session with CHES 2013. To accommodate the increase in the number of papers, sessions were held throughout Tuesday and Thursday afternoons. The rump session took place as usual on Tuesday evening, and was chaired by Dan Bernstein and Tanja Lange.

For the Best Paper Award, the Program Committee (PC) unanimously selected the paper "On the Function Field Sieve and the Impact of Higher Splitting Probabilities" by Faruk Gologlu, Robert Granger, Gary McGuire and Jens Zumbragel.

This year we also awarded a *Best Young-Author Paper Award*. To be eligible for the award, all authors of the paper had to either be full-time students or have received their PhDs in 2011 or later. The award was given to the paper "Counter-Cryptanalysis: Reconstructing Flame's New Variant Collision Attack" by Marc Stevens.

Faced with a large number of high-quality submissions, the PC decided to significantly increase the number of papers in the program from last year's 48 papers, at the price of making the program longer and keeping the paper presentations short (20 minutes per paper, including questions and answers). Another option that was seriously considered was to move to parallel sessions on some of the days of the conference. This would have allowed for somewhat longer paper presentations, and an early adjourn on Thursday. In the end, we opted to retain the single-session format, with the hope of keeping the community more unified by allowing participants to attend all talks.

The papers were reviewed by a PC consisting of 40 leading researchers in the field, in addition to the two co-chairs. Each PC member was allowed to submit one paper, plus an additional one if co-authored with a student. PC-authored papers were held to higher standards during the review process. Papers were reviewed in a double-blind fashion. Initially, each paper was assigned to three reviewers (four for PC-authored papers). During the discussion phase, when necessary, extra reviews were solicited. As part of the paper discussion phase, we held a two-day PC meeting on May 2 and 3, at the AT&T building in downtown Manhattan.

We strived to ensure that all papers received a fair and objective evaluation by experts as well as a broader group of PC members. The final decisions were made based on the reviews and discussion, and taking other factors such as balance of the program into account.

This year we initiated an early review and rebuttal process, where authors received preliminary reviews on their submissions about midway through the review period, and were given the option to comment on the reviews within a window of several days. The authors' comments were then taken into account in the discussions within the PC and in the final reviews. This process was labor-intensive; however, we feel it was worthwhile, as it resulted in a significantly better understanding of many submissions.

We would like to sincerely thank the authors of all submissions—those whose papers made it into the program and those whose papers did not. Our sincere gratitude also goes out to the PC members, who have invested an incredible amount of work in reviewing papers, interacting with the authors via the rebuttal mechanism, and participating in so many discussions on papers, their contribution, and the state of the art in their fields of expertise. We also sympathize with the occasional frustration from seeing decisions go against personal recommendations and preferences, in spite of the hard work invested.

We are also indebted to the many external reviewers, who significantly contributed to the comprehensive evaluation of papers. A list of PC members and external reviewers appears after this note. Despite all our efforts, the list of external reviewers may have errors or omissions; we apologize for that in advance.

We would like to thank Helena Handschuh, the General Chair, for working closely with us throughout the whole process, providing the much-needed support in every step, including creating and maintaining the website, and taking care of all aspects of the conference's logistics.

Special thanks are due to Shai Halevi, who provided us with unlimited support of his *websubrev* software, which we used for the whole conference planning, paper evaluation, and interaction with PC members and authors. Josh Benaloh, was our IACR point of contact, always providing timely and informative answers to our questions. Alfred Hofmann and his colleagues at Springer provided a meticulous service for the timely production of this volume.

Finally, we would like to thank Qualcomm, Microsoft, Google, Good Technologies, and Cryptography Research Inc. for their generous support.

August 2013   Ran Canetti
Juan A. Garay

Crypto 2013
The 33rd International Cryptology Conference

Sponsored by *the International Association for Cryptologic Research*

General Chair

Helena Handschuh   Cryptography Research Inc. and K.U. Leuven

Program Co-chairs

Ran Canetti   Boston University and Tel Aviv University
Juan A. Garay   AT&T Labs — Research

Program Committee

Masayuki Abe	NTT, Japan
Mihir Bellare	UCSD, USA
Zvika Brakerski	Stanford University, USA
Jan Camenisch	IBM Research, Zürich, Switzerland
David Cash	Rutgers University, USA
Kai-Min Chung	Cornell University, USA and Academia Sinica, Taiwan
Jean-Sebastien Coron	University of Luxembourg
Dana Dachman-Soled	Microsoft Research, USA
Stefan Dziembowski	University of Warsaw, Poland and University of Rome I, Italy
Iftach Haitner	Tel Aviv University, Israel
Shai Halevi	IBM Research, USA
Goichiro Hanaoka	AIST, Japan
Dennis Hofheinz	KIT, Germany
Jonathan Katz	University of Maryland, USA
Lars R Knudsen	DTU, Denmark
Eyal Kushilevitz	Technion, Israel
Kristin Lauter	Microsoft Research, USA
Huijia Lin	MIT and Boston University, USA
Yehuda Lindell	Bar Ilan University, Israel
Vadim Lyubashevsky	ENS, France
John Mitchell	Stanford University, USA
Tal Moran	Inter-Disciplinary Center, Israel
Jesper B Nielsen	University of Aarhus, Denmark
Christof Paar	University of Bochum, Germany

Manoj M Prabhakaran   University of Illinois at Urbana-Champaign,
                           USA
Tal Rabin   IBM Research, USA
Charles Rackoff   University of Toronto, Canada
Christian Rechberger   DTU, Denmark
Thomas Ristenpart   University of Wisconsin, USA
Guy Rothblum   Microsoft Research, USA
Rei Safavi   University of Calgary, Canada
                           (advisory member)
Christian Schaffner   University of Amsterdam, The Netherlands
Hovav Shacham   UCSD, USA
Vitaly Shmatikov   UT Austin, USA
Nigel Smart   University of Bristol, UK
Adam Smith   Penn State University, USA
Martijn Stam   University of Bristol, UK
John P Steinberger   Tsinghua University, China
Frederik Vercauteren   K.U. Leuven, Belgium
Xiaoyun Wang   Tsinghua University, China
Daniel Wichs   Northeastern University, USA

External Reviewers

Divesh Aggarwal	Ignacio Cascudo	Eiichiro Fujisaki
Adi Akavia	Nishanth Chandran	Steven Galbraith
Martin Albrecht	Melissa Chase	Sanjam Garg
Elena Andreeva	Nathan Chenette	Ran Gelles
Benny Applebaum	Alessandro Chiesa	Rosario Gennaro
Gilad Asharov	Sherman S.M. Chow	Craig Gentry
Gilles van Assche	Craig Costello	Benedikt Gierlichs
Nuttapong Attrapadung	Scott Coull	Vipul Goyal
Paulo Barreto	Ivan Damgaard	Louis Granboulan
Timo Bartkewitz	Maria Dubovitskaya	Adam Groce
Raef Bassily	Leo Ducas	Jens Groth
Amos Beimel	Frédéric Dupuis	Kris Haralambiev
David Bernhard	Konrad Durnoga	Moritz Hardt
Dan Bernstein	Markus Drmuth	Carmit Hazay
Nir Bitansky	Keita Emura	Nadia Heninger
Andrey Bogdanov	Robert Enderlein	Jens Hermans
Joppe Bos	Sebastian Faust	Gottfried Herold
Christina Boura	Serge Fehr	Martin Hirt
Elette Boyle	Sean Hallgren	Viet Tung Hoang
Cas Cremers	Feng-Hao	Susan Hohenberger
Christophe De Cannire	Dario Fiore	Yuval Ishai
Anne Canteaut	Marc Fischlin	Tibor Jager
Angelo De Caro	Tore Kasper Frederiksen	Abhishek Jain

Thomas P. Jakobsen
Chen Jie
Charanjit Jutla
Seny Kamara
Tomasz Kazana
Aoki Kazumaro
Sriram Keelveedhi
Dmitry Khovratovich
Eike Kiltz
Ilya Kizhvatov
Markulf Kohlweiss
Venkata Koppula
Hugo Krawczyk
Stephan Krenn
Ranjit Kumaresan
Kaoru Kurosawa
Tanja Lange
Enrique Larraia
Martin M. Lauridsen
Gregor Leander
Chen-Kuei Lee
Anja Lehmann
Gaetan Leurent
Kevin Lewi
Allison Lewko
Feng-Hao Liu
Feng-hao Liu
Jake Loftus
Steve Lu
Edward Lui
Mohammad Mahmoody
Hemanta Maji
Takahiro Matsuda
Chrysanti Mavrotami
Travis Mayberry
Sarah Meiklejohn
Florian Mendel
Alexander Meurer
Daniele Micciancio
Eric Miles
Kazuhiko Minematsu
Ilya Mironov
Peter Montgomery
Amir Moradi
Paz Morillo
Kirill Morozov

Nicky Mouha
Naveed Muhammad
Michael Naehrig
Maria Naya-Plasencia
Gregory Neven
Phong Nguyen
Ryo Nishimaki
Kobbi Nissim
Adam O'Neill
Tatsuaki Okamoto
Claudio Orlandi
Rafi Ostrovsky
Omer Paneth
Bryan Parno
Anat
    Paskin-Cherniavsky
Christopher J. Peikert
Yuval Peres
Ludovic Perret
Eduoardo Persichetti
Joop van de Pol
Christopher Portmann
Emmanuel Prouff
Ananth Raghunathan
Kasper B. Rasmussen
Mariana Raykova
Oded Regev
Leonid Reyzin
Ben Riva
Matthieu Rivain
Phillip Rogaway
Mike Rosulek
Ron Rothblum
Yannis Rouselakis
Yusuke Sakai
Koichi Sakumoto
Louis Salvail
Palash Sarkar
Giannicola Scarpa
Dominique Schroeder
Peter Schwabe
Karn Seth
Ronen Shaltiel
Chih-Hao Shen
Thomas Shrimpton
Maciej Skórski

Graeme Smith
Fang Song
Damien Stehle
Ron Steinfeld
Koutarou Suzuki
Katsuyuki Takashima
Sidharth Telang
Aris Tentes
Isamu Teranishi
Seth Terashima
Stefano Tessaro
Susan Thomson
Mehdi Tibouch
Jean-Pierre Tillich
Tomas Toft
Eran Tromer
Max Tuengerthal
Madhur Tulsiani
Vinod Vaikuntanathan
Vesselin Velichkov
K. Venkata
Muthuramakrishnan
Venkitasubramaniam
Thomas Vidick
Colin Walter
Meiqin Wang
Brent Waters
Dirk Westhoff
Carolyn Whitnall
Ronald de Wolf
David Wu
Xiaodi Wu
Keita Xagawa
Shota Yamada
Scott Yilek
Kazuki Yoneyama
Hongbo Yu
Greg Zaverucha
Maciej Zdanowicz
Mark Zhandry
Colin Jia Zheng
Joe Zimmerman
Angela Zottarel

Table of Contents – Part II

Session 11: Implementation-Oriented Protocols

Invited Talk: Why the Web Still Runs on RC4

Session 12: Number-Theoretic Hardness

Session 13: MPC — Foundations

Session 14: Codes and Secret Sharing

Session 15: Signatures and Authentication

Session 16: Quantum Security

Session 17: New Primitives

Session 18: Functional Encryption I

Session 19: Functional Encryption II

Table of Contents – Part I

Session 4: Cryptanalysis II

Session 5: MPC – New Directions

Session 6: Leakage Resilience

Session 7: Symmetric Encryption and PRFs

Session 8: Key Exchange

Session 9: Multi Linear Maps

Session 10: Ideal Ciphers

Fast Cut-and-Choose Based Protocols
for Malicious and Covert Adversaries*

Yehuda Lindell

Dept. of Computer Science
Bar-Ilan University, Israel
lindell@biu.ac.il

Abstract. In the setting of secure two-party computation, two parties
wish to securely compute a joint function of their private inputs, while
revealing only the output. One of the primary techniques for achieving
efficient secure two-party computation is that of Yao's garbled circuits
(FOCS 1986). In the semi-honest model, where just one garbled circuit is constructed and evaluated, Yao's protocol has proven itself to be
very efficient. However, a malicious adversary who constructs the garbled circuit may construct a garbling of a different circuit computing a
different function, and this cannot be detected (due to the garbling). In
order to solve this problem, many circuits are sent and some of them
are opened to check that they are correct while the others are evaluated.
This methodology, called *cut-and-choose*, introduces significant overhead,
both in computation and in communication, and is mainly due to the
number of circuits that must be used in order to prevent cheating.

In this paper, we present a cut-and-choose protocol for secure computation based on garbled circuits, with security in the presence of malicious
adversaries, that vastly improves on all previous protocols of this type.
Concretely, for a cheating probability of at most 2^{-40}, the best previous
works send between 125 and 128 circuits. In contrast, in our protocol 40
circuits alone suffice (with some additional overhead). Asymptotically, we
achieve a cheating probability of 2^{-s} where s is the number of garbled
circuits, in contrast to the previous best of $2^{-0.32s}$. We achieve this by
introducing a new cut-and-choose methodology with the property that
in order to cheat, *all* of the evaluated circuits must be incorrect, and not
just the *majority* as in previous works.

1   Introduction

Background. Protocols for secure two-party computation enable a pair of parties
P_1 and P_2 with private inputs x and y, respectively, to compute a function f
of their inputs while preserving a number of security properties. The most central of these properties are *privacy* (meaning that the parties learn the output

---

* This work was funded by the European Research Council under the European
Union's Seventh Framework Programme (FP/2007-2013) / ERC Grant Agreement
n. 239868.

R. Canetti and J.A. Garay (Eds.): CRYPTO 2013, Part II, LNCS 8043, pp. 1–17, 2013.

$f(x, y)$ but nothing else), *correctness* (meaning that the output received is indeed $f(x, y)$ and not something else), and *independence of inputs* (meaning that neither party can choose its input as a function of the other party's input). The standard way of formalizing these security properties is to compare the output of a real protocol execution to an "ideal execution" in which the parties send their inputs to an incorruptible trusted party who computes the output for the parties. Informally speaking, a protocol is then secure if no real adversary attacking the real protocol can do more harm than an ideal adversary (or simulator) who interacts in the ideal model [12,10,24,2,4,11]. An important parameter when considering this problem relates to the power of the adversary. Three important models are the *semi-honest model* (where the adversary follows the protocol specification exactly but tries to learn more than it should by inspecting the protocol transcript), the *malicious model* (where the adversary can follow any arbitrary polynomial-time strategy), and the *covert model* (where the adversary may behave maliciously but is guaranteed to be caught with probability ϵ if it does [1]).

Efficient Secure Computation and Yao's Garbled Circuits. The problem of efficient secure computation has recently gained much interest. There are now a wide variety of protocols, achieving great efficiency in a variety of settings. These include protocols that require exponentiations for every gate in the circuit [28,16] (these can be reasonable for small circuits but not large ones with tens or hundreds of thousands of gates), protocols that use the "cut and choose" technique on garbled circuits [20,21,29,25], and more [27,14,15,6,18,3,26,7]. The recent protocols of [26,7] have very fast online running time. However, for the case of Boolean circuits and when counting the entire running time (and not just the online time), the method of cut-and-choose on garbled circuits is still the most efficient way of achieving security in the presence of covert and malicious adversaries.

Protocols for cut-and-choose on garbled circuits [20,21,29,25] all work in the following way. Party P_1 constructs a large number of garbled circuits and sends them to party P_2. Party P_2 then chooses a subset of the circuits which are opened and checked. If all of these circuits are correct, then the remaining circuits are evaluated as in Yao's protocol [30], and P_2 takes the *majority* output value as the output. The cut-and-choose approach forces P_1 to garble the *correct* circuit, since otherwise it will be caught cheating. However, it is important to note that even if all of the opened circuits are correct, it is not guaranteed that all of the unopened circuits are correct. This is due to the fact that if there are only a small number of incorrect circuits, then with reasonable probability these may not be chosen to be opened. For this reason, it is critical that P_2 outputs the majority output, since the probability that a majority of unopened circuits are incorrect when all opened circuits are correct is exponentially small in the number of circuits. We stress that it is not possible for P_2 to abort in case it receives different outputs in different circuits, even though in such a case it knows that P_1 cheated, because this opens the door to the following attack. A malicious P_1 can construct a single incorrect circuit that computes the following function: if the first bit of P_2's input equals 0 then output random garbage; else compute the correct function. Now, if this circuit is not opened (which happens with probability $1/2$) and if the first bit of P_2's input equals 0, then P_2 will receive a different output in this circuit and in the others. In contrast, if the first bit of P_2's input equals 1 then it always receives the same output in all circuits. Thus, if the protocol instructs P_2 to abort if it receives different outputs, then P_1 will learn the first bit of P_2's input (based on whether or not P_2 aborts). By having P_2 take the majority value as output, P_1 can only cheat if the majority of the unopened circuits are incorrect, while all the opened ones are correct. In [21] it was shown that when s circuits are sent and half of them are opened, the probability that P_1 can cheat is at most $2^{-0.311s}$. Thus, concretely, in order to obtain an error probability of 2^{-40}, it is necessary to set $s = 128$ and so use 128 circuits, which means that the approximate cost of achieving security in the presence of malicious adversaries is 128 times the cost of achieving security in the presence of semi-honest adversaries. In [29], it was shown that by opening and checking 60% of the circuits instead of 50%, then the error becomes $2^{-0.32s}$ which means that it suffices to send 125 circuits in order to obtain a concrete error of 2^{-40}. It was claimed in [29] that these parameters are "optimal for the cut-and-choose method" and that they establish "a close characterization of the limit of the cut-and-choose method". We show that these protocols are actually far from the "limit" of this method.

Our Results. In this paper, we present a novel twist on the cut-and-choose strategy used in [20,21,29,25] that enables us to achieve an error of just 2^{-s} with s circuits (and some small additional overhead). Concretely, this means that just 40 circuits are needed for error 2^{-40}. Our protocol is therefore much more efficient than previous protocols (there is some small additional overhead but this is greatly outweighed by the savings in the garbled circuits themselves unless the circuit being computed is small). We stress that the bottleneck in protocols of this type is the computation and communication of the s garbled circuits. This has been demonstrated in implementations. In [9], the cost of the circuit communication and computation for secure AES computation is approximately 80% of the work. Likewise in [17, Table 7] regarding secure AES computation, the bandwidth due to the circuits was 83% of all bandwidth and the time was over 50% of the time. On large circuits, as in the edit distance, this is even more significant with the circuit generation and evaluation taking 99.999% of the time [17, Table 9]. Thus, the reduction of this portion of the computation to a third of the cost is of great significance.

We present a high-level outline of our new technique in Section 2. For now, we remark that the cut-and-choose technique on Yao's garbled circuits introduces a number of challenges. For example, since the parties evaluate numerous circuits, it is necessary to enforce that the parties use the same input in all circuit computations. In addition, a selective input attack whereby P_1 provides correct garbled inputs only for a subset of the possible inputs of P_2 must be prevented (since otherwise P_2 will abort if its input is not in the subset because it cannot compute any circuit in this case, and thus P_1 will learn something about P_2's

input based on whether or not it aborts). There are a number of different solutions to these problems that have been presented in [20,21,29,25,9]. The full
protocol that we present here is based on the protocol of [21]. However, these
solutions are rather "modular" (although this is meant in an informal sense),
and can also be applied to our new technique; this is discussed at the end of
Section 2. Understanding which technique is best will require implementation
since they introduce tradeoffs that are not easily comparable. We leave this for
future work, and focus on the main point of this work which is that it is possible to achieve error 2^{-s} with just s circuits. In Section 3.1 we present an exact
efficiency count of our protocol.

Covert Adversaries. Although not always explicitly proven, the known protocols
for cut-and-choose on garbled circuits achieve covert security where the deterrent
probability ϵ that the adversary is caught cheating equals 1 minus the statistical
error of the protocol. That is, the protocol of [21] yields covert security of $\epsilon = 1 - 2^{-0.311s}$ (actually, a little better), and the protocol of [29] yields covert security
with $\epsilon = 1 - 2^{-0.32s}$. Our protocol achieves covert security with deterrent $\epsilon = 1 - 2^{-s+1}$ (i.e., the error is 2^{-s+1}) which is far more efficient than all previous work.
Specifically, in order to obtain $\epsilon = 0.99$, the number of circuits needed in [21] is
24. In contrast, with our protocol, it suffices to use 8 circuits. Furthermore, with
just 11 circuits, we achieve $\epsilon = 0.999$, which is a high deterrent.

2   The New Technique and Protocol Outline

The idea behind our new cut-and-choose strategy is to design a protocol with
the property that the party who constructs the circuits (P_1) can cheat if and
only if *all* of the checked circuits are correct and *all* of the evaluated circuits
are incorrect. Recall that in previous protocols, if the circuit evaluator (P_2)
aborts if the evaluated circuits don't all give the same output, then this can
reveal information about P_2's input to P_1. This results in an absurd situation:
P_2 knows that P_1 is cheating but cannot do anything about it. In our protocol,
we run an additional *small* secure computation after the cut-and-choose phase
so that if P_2 catches P_1 cheating (namely, if P_2 receives inconsistent outputs)
then in the second secure computation it learns P_1's full input x. This enables
P_2 to locally compute the correct output $f(x, y)$ once again. Thus, it is no longer
necessary for P_2 to take the majority output. Details follow.

Phase 1 – first cut-and-choose:

- Parties P_1 (with input x) and P_2 (with input y) essentially run a protocol
  based on cut-and-choose of garbled circuits, that is secure for malicious adversaries (like [21] or [29]). P_1 constructs just s circuits (for error 2^{-s}) and
  the strategy for choosing check or evaluation circuits is such that each circuit
  is *independently* chosen as a check or evaluation circuit with probability $1/2$
  (unlike all previous protocols where a fixed number of circuits are checked).
- If all of the circuits successfully evaluated by P_2 give the same output z,
  then P_2 locally stores z. Otherwise, P_2 stores a "proof" that it received two

inconsistent output values in two different circuits. Such a proof could be having a garbled value associated with 0 on an output wire in one circuit, and a garbled value associated with 1 on the same output wire in a different circuit. (This is a proof since if P_2 obtains a single consistent output then the garbled values it receives on an output wire in different circuits are all associated with the same bit.)

Phase 2 – Secure Evaluation of Cheating: P_1 and P_2 run a protocol that is secure for malicious adversaries with error 2^{-s} (e.g., they use the protocol of [21,29] with approximately $3s$ circuits), in order to compute the following:

- P_1 inputs the same input x as in the computation of phase 1 (and *proves* this).
- P_2 inputs random values if it received a single output z in phase 1, and inputs the proof of inconsistent output values otherwise.
- If P_2's input is a valid proof of inconsistent output values, then P_2 receives P_1's input x; otherwise, it receives nothing.

If this secure computation terminates with abort, then the parties abort.

Phase 3 – Output Determination: If P_2 received a single output z in phase 1 then it outputs z and halts. Otherwise, if it received inconsistent outputs then it received x in phase 2. P_2 locally computes $z = f(x, y)$ and outputs it. We stress that P_2 does not provide any indication as to whether z was received from phase 1 or locally computed.

Security. The argument for the security of the protocol is as follows. Consider first the case that P_1 is corrupted and so may not construct the garbled circuits correctly. If all of the check circuits are correct and all of the evaluation circuits are incorrect, then P_2 may receive the same incorrect output in phase 1 and will therefore output it. However, this can only happen if each incorrect circuit is an evaluation circuit and each correct circuit is a check circuit. Since each circuit is an evaluation or check circuit with probability exactly $1/2$ this happens with probability exactly 2^{-s}. Next, if all of the evaluation circuits (that yield valid output) are correct, then the correct output will be obtained by P_2. This leaves the case that there are two different evaluation circuits that give two different outputs. However, in such a case, P_2 will obtain the required "proof of cheating" and so will learn x in the 2nd phase, thereby enabling it to still output the correct value. Since P_1 cannot determine which case yielded output for P_2, this can be easily simulated.

Next consider the case that P_2 is corrupted. In this case, the only way that P_2 can cheat is if it can provide output in the second phase that enables it to receive x. However, since P_1 constructs the circuits correctly, P_2 will not obtain inconsistent outputs and so will not be able to provide such a "proof". (We remark that the number of circuits s sent is used for the case that P_1 is corrupted; for the case that P_2 is corrupted a single circuit would actually suffice. Thus, there is no need to justify the use of fewer circuits than in previous protocols for this corruption case.)

Implementing Phase 2. The main challenge in designing the protocol is phase 2. As we have hinted, we will use the knowledge of two different garbled values for a single output wire as a "proof" that P_2 received inconsistent outputs. However, it is also necessary to make sure that P_1 uses the same input in phase 1 and in phase 2; otherwise it could use x or x', respectively, and then learn whether P_2 received output via phase 1 or 2. The important observation is that all known protocols already have a mechanism for ensuring that P_1 uses the same input in all computed circuits, and this mechanism can be used for the circuits in phase 1 and 2, since it does not depend on the circuits being computed being the same.

Another issue that arises is the efficiency of the computation in phase 2. In order to make the circuit for phase 2 small, it is necessary to construct all of the output wires in all the circuits of phase 1 so that they have the same garbled values on the output wires. This in turn makes it necessary to open and check the circuits only after phase 2 (since opening a circuit to check it reveals both garbled values on an output wire which means that this knowledge can no longer be a proof that P_1 cheated). Thus, the structure of the actual protocol is more complex than previous protocols; however, this relates only to its description and not efficiency.

We remark that we use the method of [21] in order to prove the consistency of P_1's input in the different circuits and between phase 1 and phase 2. However, we believe that the methods used in [29,25], for example, would also work, but have not proven this.

3   The Protocol

Preliminaries – Modified Batch Single-Choice Cut-and-Choose OT. The cut-and-choose OT primitive was introduced in [21]. Intuitively, a cut-and-choose OT is a series of 1-out-of-2 oblivious transfers with the special property that in some of the transfers the receiver obtains a single value (as in regular oblivious transfer), while in the others the receiver obtains both values. For cut-and-choose on Yao's garbled circuits, the functionality is used for the receiver to obtain all garbled input values in the circuits that it wishes to open and check, and to obtain only the garbled input values associated with its input on the circuits to be evaluated.

In [21], the functionality defined is such that the receiver obtains both values in exactly half of the transfers; this is because in [21] exactly half of the circuits are opened. In this work, we modify the functionality so that the receiver can choose at its own will in which transfers it receives just one value and in which it receives both. We do this since we want P_2 to check each circuit with probability exactly $1/2$, independently of all other circuits. This yields an error of 2^{-s} instead of $\binom{s}{s/2}^{-1}$, which is smaller (this is especially significant in the setting of covert adversaries).

This modification introduces a problem since at a later stage in the protocol the receiver needs to prove to the sender for which transfers it received both values and for which it received only one. If it is known that the receiver obtains both values in exactly half of the transfers, or for any other known number, then the receiver can just send both values in these transfers (assuming that they are otherwise unknown, as is the case in the Yao circuit use of the functionality), and the sender knows that the receiver did *not* obtain both values in all others; this is what is done in [21]. However, here the receiver can obtain both values in an unknown number of transfers, as it desires. We therefore need to introduce a mechanism enabling the receiver to prove to the sender in which transfers it did not receive both values, in a way that it cannot cheat. We solve this by having the sender input s random "check" values, and having the receiver obtain such a value in every transfer for which it receives a single value only. Thus, at a later time, the receiver can send the appropriate check values, and this constitutes a proof that it did not receive both values in these transfers. See Figure 1 for the formal functionality definition.

FIGURE 1 (Modified Batch Single-Choice Cut-and-Choose OT $\mathcal{F}_{\mathrm{ccot}}$)

Inputs:

- S inputs ℓ *vectors* of pairs $\boldsymbol{x}_i$ of length s, for $i = 1, \ldots, \ell$. (Every vector consists of s pairs; i.e., $\boldsymbol{x}_i = \langle (x_0^{i,1}, x_1^{i,1}), (x_0^{i,2}, x_1^{i,2}), \ldots, (x_0^{i,s}, x_1^{i,s}) \rangle$. There are ℓ such vectors.) In addition, S inputs s "check values" $\chi_1, \ldots, \chi_s$. All values are in $\{0, 1\}^n$.
- R inputs $\sigma_1, \ldots \sigma_\ell \in \{0, 1\}$ and a set of indices $\mathcal{J} \subseteq [s]$.

Output: The sender receives no output. The receiver obtains the following:

- For every $i = 1, \ldots, \ell$ and for every $j \in \mathcal{J}$, the receiver R obtains the jth pair in vector $\boldsymbol{x}_i$. (I.e., for every $i = 1, \ldots, \ell$ and every $j \in \mathcal{J}$, R obtains $(x_0^{i,j}, x_1^{i,j})$.)
- For every $i = 1, \ldots, \ell$, the receiver R obtains the σ_i value in every pair of the vector $\boldsymbol{x}_i$. (I.e., for every $i = 1, \ldots, \ell$, R obtains $\langle x_{\sigma_i}^{i,1}, x_{\sigma_i}^{i,2}, \ldots, x_{\sigma_i}^{i,s} \rangle$.)
- For every $k \notin \mathcal{J}$, the receiver R obtains χ_k.

A protocol for securely computing the $\mathcal{F}_{\mathrm{ccot}}$ functionality, that is based on the protocol in [21], is provided in the full version of this paper [22]. The computational complexity of the protocol is as follows:

Operation	Exact Cost	Approximate Cost
Regular exponentiations	$1.5s\ell + 18.5s + 25$	$1.5s\ell$
Fixed-base exponentiations	$9s\ell + \ell + 2s + 1$	$9s\ell$
Bandwidth (group elements)	$5s\ell + \ell + 11s + 15$	$5s\ell$

Encoded Translation Tables. We modify the output translation tables typically used in Yao's garbled circuits as follows. Let k_i^0, k_i^1 be the garbled values on wire i, which is an output wire, and let H be a collision-resistant hash function. Then, the encoded output translation table for this wire is simply $\left[H(k_i^0), H(k_i^1) \right]$. We require that $k_i^0 \neq k_i^1$ and if this doesn't hold (which will be evident since then $H(k_i^0) = H(k_i^1)$), P_2 will automatically abort. Observe that given a garbled value

k, it is possible to determine whether k is the 0 or 1 key (or possibly neither) by just computing $H(k)$ and seeing if it equals the first or second value in the pair, or neither. However, given the encoded translation table, it is not feasible to find the actual garbled values, since this is equivalent to inverting the one-way function. This is needed in our protocol, as we will see below. We remark that both k_i^0, k_i^1 are revealed by the end of the protocol, and only need to remain secret until Step 7 has concluded (see the protocol below). Thus, they can be relatively short values.

PROTOCOL 2 (Computing $f(x, y)$)

Inputs: P_1 has input $x \in \{0, 1\}^\ell$ and P_2 has input $y \in \{0, 1\}^\ell$.

Auxiliary input: a statistical security parameter s, the description of a circuit C such that $C(x, y) = f(x, y)$, and $(\mathbb{G}, q, g)$ where $\mathbb{G}$ is a cyclic group with generator g and prime order q, and q is of length n. In addition, they hold a hash function H that is a suitable randomness extractor; see [8].

Specified output: Party P_2 receives $f(x, y)$ and party P_1 receives no output; denote the length of the output of $f(x, y)$ by m.

The protocol:

1. INPUT KEY CHOICE AND CIRCUIT PREPARATION:
   (a) P_1 chooses random values $a_1^0, a_1^1, \ldots, a_\ell^0, a_\ell^1; r_1, \ldots, r_s \in_R \mathbb{Z}_q$ and $b_1^0, b_1^1, \ldots, b_m^0, b_m^1 \in_R \{0, 1\}^n$.
   (b) Let $w_1, \ldots, w_\ell$ be the input wires corresponding to P_1's input in C, and denote by $w_{i,j}$ the instance of wire w_i in the jth garbled circuit, and by $k_{i,j}^b$ the key associated with bit b on wire $w_{i,j}$. P_1 sets the keys for its input wires to:
   $$k_{i,j}^0 = H(g^{a_i^0 \cdot r_j}) \text{ and } k_{i,j}^1 = H(g^{a_i^1 \cdot r_j}).$$
   (c) Let $w_1', \ldots, w_m'$ be the output wires in C. Then, the keys for wire w_i' in *all* garbled circuits are b_i^0 and b_i^1 (unlike all other wires in the circuit, the same values are used for the output wires in all circuits).
   (d) P_1 constructs s independent copies of a garbled circuit of C, denoted $GC_1, \ldots, GC_s$, using random keys except for wires $w_1, \ldots, w_\ell$ (P_1's input wires) and $w_1', \ldots, w_m'$ (the output wires) which are as above.

2. OBLIVIOUS TRANSFERS: P_1 and P_2 run a modified batch single-choice cut-and-choose oblivious transfer, with parameters ℓ (the number of parallel executions) and s (the number of pairs in each execution):
   (a) P_1 defines vectors $z_1, \ldots z_\ell$ so that z_i contains the s pairs of random symmetric keys associated with P_2's ith input bit y_i in all garbled circuits $GC_1, \ldots, GC_s$. P_1 also chooses random values $\chi_1, \ldots, \chi_s \in_R \{0, 1\}^n$. P_1 inputs these vectors and the $\chi_1, \ldots, \chi_s$ values.
   (b) P_2 chooses a random subset $\mathcal{J} \subset [s]$ where every $j \in \mathcal{J}$ with probability exactly $1/2$, under the constraint that $\mathcal{J} \neq [s]$. P_2 inputs the set $\mathcal{J}$ and bits $\sigma_1, \ldots, \sigma_\ell \in \{0, 1\}$, where $\sigma_i = y_i$ for every i.
   (c) P_2 receives all the keys associated with its input wires in all circuits GC_j for $j \in \mathcal{J}$, and receives the keys associated with its input y on its input wires in all other circuits.
   (d) P_2 receives χ_j for every $j \notin \mathcal{J}$.

PROTOCOL 3 (PROTOCOL 2 – continued)

3. SEND CIRCUITS AND COMMITMENTS: P_1 sends P_2 the garbled circuits (i.e., the garbled gates). In addition, P_1 sends P_2 the "seed" for the randomness extractor H, and the following displayed values (which constitute a "commitment" to the garbled values associated with P_1's input wires):

$$\left\{(i,0,g^{a_i^0}),(i,1,g^{a_i^1})\right\}_{i=1}^{\ell} \quad \text{and} \quad \left\{(j,g^{r_j})\right\}_{j=1}^{s}$$

In addition, P_1 sends P_2 the encoded output translation tables, as follows:

$$\left[\left(H(b_1^0),H(b_1^1)\right),\ldots,\left(H(b_m^0),H(b_m^1)\right)\right].$$

If $H(b_i^0) = H(b_i^1)$ for any $1 \leq i \leq m$, then P_2 aborts.

4. SEND CUT-AND-CHOOSE CHALLENGE: P_2 sends P_1 the set $\mathcal{J}$ along with the values χ_j for every $j \notin \mathcal{J}$. If the values received by P_1 are incorrect, it outputs $\bot$ and aborts. Circuits GC_j for $j \in \mathcal{J}$ are called check-circuits, and for $j \notin \mathcal{J}$ are called evaluation-circuits.

5. P_1 SENDS ITS GARBLED INPUT VALUES IN THE EVALUATION-CIRCUITS: P_1 sends the keys associated with its inputs in the evaluation circuits: For every $j \notin \mathcal{J}$ and every wire $i = 1,\ldots,\ell$, party P_1 sends the value $k'_{i,j} = g^{a_i^{x_i}\cdot r_j}$; P_2 sets $k_{i,j} = H(k'_{i,j})$.

6. CIRCUIT EVALUATION: P_2 uses the keys associated with P_1's input obtained in Step 5 and the keys associated with its own input obtained in Step 2c to evaluate the circuits GC_j for every $j \notin \mathcal{J}$. If P_2 receives only one valid output value per output wire (i.e., one of b_i^0, b_i^1, verified against the encoded output translation tables) and it does not abort in the next step, then this will be its output. If P_2 receives two valid outputs on one output wire (i.e., both b_i^0 and b_i^1 for output wire w_i') then it uses these in the next step. If there exists an output wire for which P_2 did not receive a valid value in *any* evaluation circuit (neither b_i^0 nor b_i^1), then P_2 aborts.

7. RUN SECURE COMPUTATION TO DETECT CHEATING:
   (a) P_1 defines a circuit with the values $b_1^0, b_1^1, \ldots, b_m^0, b_m^1$ hardcoded. The circuit computes the following function:
      i. P_1's input is a string $x \in \{0,1\}^{\ell}$, and it has no output.
      ii. P_2's input is a pair of values b_0, b_1.
      iii. If there exists a value i ($1 \leq i \leq m$) such that $b_0 = b_i^0$ and $b_1 = b_i^1$, then P_2's output is x; otherwise it receives no output.
   (b) P_1 and P_2 run the protocol of [21] on this circuit (except for the proof of P_1's input values), as follows:
      i. P_1 inputs its input x; If P_2 received b_i^0, b_1^i for some $1 \leq i \leq m$, then it inputs the pair b_i^0, b_i^1; otherwise it inputs garbage.
      ii. The garbled circuit constructed by P_1 uses the same a_i^0, a_i^1 values as above (i.e., the same triples $(i,0,g^{a_i^0}),(i,1,g^{a_i^1})$), but independent r_j values. In addition, regular translation tables are used, and not encoded translation tables. Finally, the parties use $3s$ copies of the circuit (and not s).
      iii. P_2 takes the majority output from the evaluation circuits, as in [21]. If any of the checked circuits are invalid, then P_2 aborts. We stress that this check includes the check that the circuit has the correct $b_1^0, b_1^1, \ldots, b_m^0, b_m^1$ values hardcoded; P_2 checks this relative to the encoded translation tables that it received.

PROTOCOL 2 – continued

7. RUN SECURE COMPUTATION TO DETECT CHEATING (CONT.): If this computation results in an abort, then both parties halt at this point and output $\perp$. (Note that in the protocol of [21] both parties must know the circuit. However, the oblivious transfers that determine P_2's input are run before the circuit is sent and checked. Thus, P_1 can send the $b_1^0, b_1^1, \ldots, b_m^0, b_m^1$ values to P_2 after the oblivious transfers are concluded; P_2 can check these values against the encoded translation tables and can then check that these are the values that are hardwired into the circuit.)

8. CHECK CIRCUITS FOR COMPUTING $f(x, y)$:
   (a) SEND ALL INPUT GARBLED VALUES IN CHECK-CIRCUITS: For every check-circuit GC_j, party P_1 sends the value r_j to P_2, and P_2 checks that these are consistent with the pairs $\{(j, g^{r_j})\}_{j \in \mathcal{J}}$ received in Step 3. If not, P_2 aborts outputting $\perp$.
   (b) CORRECTNESS OF CHECK CIRCUITS: For every $j \in \mathcal{J}$, P_2 uses the $g^{a_i^0}, g^{a_i^1}$ values it received in Step 3, and the r_j values it received in Step 8a, to compute the values $k_{i,j}^0 = H(g^{a_i^0 \cdot r_j}), k_{i,j}^1 = H(g^{a_i^1 \cdot r_j})$ associated with P_1's input in GC_j. In addition it sets the garbled values associated with its own input in GC_j to be as obtained in the cut-and-choose OT. Given all the garbled values for all input wires in GC_j, party P_2 decrypts the circuit and verifies that it is a garbled version of C, using the encoded translation tables for the output values. If there exists a circuit for which this does not hold, then P_2 aborts and outputs $\perp$.

9. VERIFY CONSISTENCY OF P_1'S INPUT: Let $\hat{\mathcal{J}}$ be the set of check circuits in the computation in Step 7, and let $\hat{r}_j$ be the value used to generate the keys associated with P_1's input in the jth circuit, just like r_j in Step 1a (i.e., $H(g^{a_i^0 \cdot \hat{r}_j})$ is the 0-key on the ith input wire of P_1 in the jth garbled circuit used in Step 7). Let $\hat{k}_{i,j}$ be the analogous value of $k_{i,j}'$ in Step 5 received by P_2 in the computation in Step 7.
   For every input wire $i = 1, \ldots, \ell$, party P_1 proves a zero-knowledge proof of knowledge that there exists a $\sigma_i \in \{0, 1\}$ such that for every $j \notin \mathcal{J}$ and every $j' \notin \mathcal{J}'$, $k_{i,j}' = g^{a_i^{\sigma_i} \cdot r_j}$ AND $\hat{k}_{i,j} = g^{a_i^{\sigma_i} \cdot \hat{r}_j}$ (note that P_2 has g^{r_j} and $g^{\hat{r}_j}$ for every j, and $g^{a_i^0}, g^{a_i^1}$ for every i; thus this is just a Diffie-Hellman tuple proof). If any of the proofs fail, then P_2 aborts and outputs $\perp$.

10. OUTPUT EVALUATION: If P_2 received no inconsistent outputs from the evaluation circuits GC_i ($i \notin \mathcal{J}$), then it decodes the outputs it received using the encoded translation tables, and outputs the string received. If P_2 received inconsistent output, then let x be the output that P_2 received from the second computation in Step 7. Then, P_2 computes $f(x, y)$ and outputs it.

The Circuit for Step 7. A naive circuit for computing the function in Step 7 can be quite large. Specifically, to compare two n bit strings requires $2n$ XORs followed by $2n$ ORs; if the output is 0 then the strings are equal. This has to be repeated m times, once for every i, and then the results have to be ORed. Thus, there are $2mn + m$ non-XOR gates. Assuming n is of size 80 (e.g., which suffices for the output values) and m is of size 128, this requires $20,480$ non-XOR gates, which is very large. An alternative is therefore to compute the following garbled circuit:

1. For every $i = 1, \ldots, m$,
   (a) Compare $b_0 \| b_1$ to $b_i^0 \| b_i^1$ (where '$\|$' denotes concatenation) by XORing bit-by-bit, and take the NOT of each bit. This is done as in a regular garbled circuit; by combining the NOT together with the XOR this has the same cost as a single XOR gate.
   (b) Compute the $2n$-wise AND of the bits from above. Instead of using $2n-1$ Boolean AND gates, this can be achieved by encrypting the 1-key on the output wire under all n keys (together with redundancy so that the circuit evaluator can know if it received the correct value). Furthermore, this encryption can be a "one-time pad" and thus is just the XOR of all of the 1-keys on the input wires together with the 1-key on the output way. The 0-key for the output can be given in the clear, since it provides no additional information, but is not needed so can just not be given (note that P_2 knows exactly which case it is in). Note that the result of this operation is 1 if and only if $b_0 \| b_1 = b_i^0 \| b_i^1$ and so P_2 had both keys on the ith output wire.
2. Compute the OR of the m bits resulting from the above loop. Instead of using $m-1$ Boolean OR gates, this can be achieved by simply setting the 1-key on all of the output wires from the n-wise ANDs above to be the 1-key on the output wire of the OR. This ensures that as soon as the 1-key is received from an n-wise AND, the 1-key is received from the OR, as required. (This reveals for which i the result of the n-wise AND was 1. However, this is fine here since P_2 knows exactly where equality should be obtained in any case.)
3. Compute the AND of the output from the previous step with all of the input bits of P_1. This requires ℓ Boolean AND gates.
4. The output wires include the output of the OR (so that P_2 can know if it received x or nothing), together with the output of all of the ANDs with the input bits of P_1.

The original and optimized circuits are depicted in the full version [22]. The number of non-XOR operations required to securely compute this circuit is just ℓ binary AND gates. Assuming $\ell = 128$ (e.g., as in the secure AES example), we have that there are only 128 non-XOR gates. When using 128 circuits as in our instantiation of Step 7 via [21], this comes to 16,384 garbled gates *overall*, which is significant but not too large. We stress that the size of this circuit is independent of the size of the circuit for the function f to be computed. Thus, this becomes less significant as the circuit becomes larger. On the other hand, for very small circuits or when the input size is large relative to the overall circuit size, our approach will not be competitive. To be exact, assume a garbled circuit approach that requires $3s$ circuits. If $3s|C| < s|C| + 3s \cdot \ell$ then our protocol will be slower (since the cost of our protocol is $s|C|$ for the main computation plus $3s\ell$ for the circuit of Step 7, in contrast to $3s|C|$ for the other protocol). This implies that our protocol will be faster as long as $|C| > \frac{3\ell}{2}$. Concretely, if $\ell = 128$ and $s = 40$, it follows that our protocol will be faster as long as $|C| > 192$. Thus, our protocol is much faster, except for the case of very small circuits.

Additional Optimizations. Observe that although the above circuit is very small, P_2's input size is $2n$ and this is quite large. Since the input size has a significant effect on the cost of the protocol (especially when using cut-and-choose oblivious transfer), it would be desirable to reduce this. This can be achieved by first having P_2 input $b_0 \oplus b_1$ instead of $b_0 \| b_1$, reducing the input length to n (this is sound since if P_2 does not have both keys on any output wire then it cannot know their XOR). Furthermore, in order to obtain a cheating probability of 2^{-40} it suffices for the circuit to check only the first 40 bits of $b_0 \oplus b_1$. (Note that b_0^i and b_1^i have to be longer since $H(b_i^0), H(b_i^1)$ are published; nevertheless, only 40 bits need to be included in the circuit. When using this optimization, the length of b_0^i, b_1^i can be 128 bits and not 80, which is preferable.) Finally, by choosing all of the b_i^0, b_i^1 values so that they have the same fixed XOR (i.e., for some Δ it holds that for all i, $b_i^0 \oplus b_i^1 = \Delta$, as in the free XOR technique), the size of the circuit is further reduced. This significantly reduces the bandwidth; a diagram of this circuit appears in the full version [22].

Security. In the full version of this paper [22], we prove the following theorem:

Theorem 4. *Assume that the Decisional Diffie-Hellman assumption holds in* $(\mathbb{G}, g, q)$, *and that H is a collision-resistant hash function. Then, Protocol 2 securely computes f in the presence of malicious adversaries (with error $2^{-s} + \mu(n)$ where $\mu(\cdot)$ is some negligible function).*

3.1   A Detailed Efficiency Count and Comparison

In this section we provide an exact efficiency count of our protocol. This will enable an exact comparison of our protocol to previous and future works, as long as they also provide an exact efficiency count. We count exponentiations, symmetric encryptions and bandwidth. We let n denote the length of a symmetric encryption, and an arbitrary string of length of the security parameter (e.g., χ_j).

Step	Fixed-base exponent.	Regular exponent.	Symmetric Encryptions	Group elms sent	Symmetric comm
1	$2s\ell$	0	$4s|C|$		
2	$9s\ell$	$1.5s\ell$		$5s\ell$	
3	$\ell + s$	0		$2\ell + s$	$4ns|C|$
4					$\frac{s}{2} \cdot n$
5					nm
6			$\frac{s}{2} \cdot |C|$		
7	$9s\ell + 5040s$	$480s$	19.5ℓ	$21s\ell$	$12sn\ell$
8	$s/2 + s\ell$		$\frac{s}{2} \cdot 4|C|$		$\frac{s}{2} \cdot n$
9		$2s\ell + 18\ell$		10	$2s\ell n$
TOTAL	$21s\ell + 5040s$	$3.5s\ell + 18\ell$ $+480s$	$6.5s|C|+$ $19.5s\ell$	$26s\ell$	$4ns|C| + 14s\ell n$

The number of symmetric encryptions is counted as follows: each circuit requires $4|C|$ symmetric encryptions to construct, $4|C|$ symmetric encryption to check, and $|C|$ encryptions to evaluate (we assume a single encryption per entry; if standard double-encryption is used this should be doubled). Since approximately half of the circuits are check and half are evaluation, the garbling, checking and evaluation of the main garbled circuit accounts for approximately $s \cdot 4|C| + \frac{s}{2} \cdot 4|C| + \frac{s}{2} \cdot |C| = 6.5s|C|$ symmetric encryptions. The garbled circuit used in Step 7 has ℓ non-XOR gates and so the same analysis applies on this size. However, the number of circuits sent in this step is $3s$ and thus we obtain an additional $3 \times 6.5 \cdot s \cdot \ell = 19.5s\ell$.

The bandwidth count for Step 7 is computed based on the counts provided in [21], using $3s$ circuits. The cost of the exponentiations is based on the fact that in [21], if P_1 has input of length ℓ_1 and P_2 has input of length ℓ_2, and s' circuits are used, then there are $3.5s'\ell_1 + 10.5s'\ell_2$ fixed-base exponentiations and $s'\ell_2$ regular exponentiations. However, $0.5s'\ell_1$ of the fixed-base exponentiations are for the proof of consistency and these are counted in Step 9 instead. Now, in Step 7, P_1's input length is ℓ (it is the same x as for the entire protocol) and P_2's input is comprised of two garbled values for the output wires. Since these must remain secret for only a short amount of time, it is possible to take 80-bit values only and so P_2's input length is 160 bits (this is irrespective of P_2's input length to the function f). Taking $s' = 3s$ and plugging these lengths this into the above, we obtain the count appearing in the table.

The proof of consistency of P_1's input is carried out ℓ times (once for each bit of P_1's input) and over $s + 3s = 4s$ circuits (since there are s circuits for the main computation of C, plus another $3s$ circuits for the computation in Step 7). By the count in [21], this proof therefore costs $\frac{4s\ell}{2} + 18\ell$ exponentiations, and bandwidth of 10 group elements and another $8s\ell$ short strings (this can therefore be counted as $2s\ell n$.

A Comparison to [21]. In order to to get a concrete understanding of the efficiency improvement, we will compare the cost to [21] for the AES circuit of size 6,800 gates [31], and input and output sizes of 128. Now, as we have mentioned, the overall cost of the protocol of [21] is $3.5s'\ell_1 + 10.5s\ell_2$ fixed-base

Protocol	Fixed-base exp.	Regular exp.	Symmetric encryptions	Bandwidth
[21]	224,000	16,000	5,525,000	449,640,000
Here	309,120	37,120	1,874,800	177,725,440

Fig. 1. Comparison of protocols for secure computation of AES

exponentiations, $s'\ell_2$ regular exponentiations and $6.5s'|C|$ symmetric encryptions. In this case, $\ell_1 = \ell_2 = 128$, $s' = 125$ ($s' = 125$ was shown to suffice for 2^{-40} security in [17]), and so we have that the cost is 224,000 fixed-base exponentiations, 16,000 regular exponentiations, and $812.5|C| =$5,525,000 symmetric encryptions. In contrast, taking $\ell = 128$ and $s = 40$ we obtain here 309,120 fixed-base exponentiations, $37,120$ regular exponentiations, and 1,874,800 symmetric encryptions. In addition, the bandwidth of [21] is approximately $112,000$ group elements and 3,400,000 symmetric ciphertexts. At the minimal cost of 220 bits per group element (e.g., using point compression) and 128 bits per ciphertext, we have that this would come to approximately 449,640,000 bits, or close to half a gigabyte (in practice, it would be significantly larger due to communication overheads). In contrast, the bandwidth of our protocol for this circuit would be 133,120 group elements and 1,159,680 ciphertexts. With the same parameters as above, this would be approximately 177,725,440 bits, which is under 40% of the cost of [21]. This is very significant since bandwidth is turning out to be the bottleneck in many cases.

We stress that in larger circuits the difference would be even more striking.

4   Variants – Universal Composability and Covert Adversaries

Universal Composability [5]. As in [21], by instantiating the cut-and-choose oblivious transfer and the zero-knowledge proofs with variants that universally composable, the result is that Protocol 2 is universally composable.

Covert Adversaries [1]. Observe that in the case that P_2 is corrupted, the protocol is fully secure irrespective of the value of s used. In contrast, when P_1 is corrupted, then the cheating probability is $2^{-s} + \mu(n)$. However, this cheating probability is *independent* of the input used by the P_2 (as shown in the proof of Theorem 4). Thus, Protocol 2 is suitable for the model of security in the presence of *covert adversaries*. Intuitively, since the adversary can cheat with probability only 2^{-s} and otherwise it is caught cheating, the protocol achieves covert security with deterrent $\epsilon = 1 - 2^{-s}$. However, on closer inspection, this is incorrect. Specifically, as we have discussed above, if P_2 catches P_1 cheating due to the fact that two different circuits yield two different outputs, then it is not allowed to reveal this fact to P_1. Thus, P_2 cannot declare that P_1 is a cheat in this case, as is required in the model of covert adversaries. However, if P_2 detects even a single bad circuit in the check phase, then it can declare that P_1 is cheating, and this happens with probability at least $1/2$ (even if only a single circuit is bad).

We can use this to show that for every s, Protocol 2 securely computes f in the presence of covert adversaries **with deterrent** $\epsilon = 1 - 2^{-s+1}$. In actuality, we need to make a slight change to Protocol 2, in order to achieve this. See the full version of this paper for details [22].

As discussed in the introduction, this yields a huge efficiency improvement over previous results, especially for small values of s. For example, 100 circuits are needed to obtain $\epsilon = 0.99$ in [1], 24 circuits are needed to obtain $\epsilon = 0.99$ in [21], and here **8 circuits alone** suffice to obtain $\epsilon = 0.99$. Observe that when covert security is desired, the number of circuits sent in Step 7 needs to match the level of covert security. For example, in order to obtain $\epsilon = 0.99$, 8 circuits are used in our main protocol and 24 circuits are used in Step 7.

We remark that our protocol would be a little simpler if P_2 always asked to open exactly half the circuits (especially in the cut-and-choose oblivious transfer). In this case, the error would be $\binom{s}{s/2}^{-1}$ instead of 2^{-s}. In order to achieve an error of 2^{-40} this would require 44 circuits which is a 10% increase in complexity, and reason enough to use our strategy of opening each circuit independently with probability $1/2$. However, when considering covert security, the difference is huge. For example, with $s = 8$ we have that $\binom{8}{4}^{-1} = 1/70$ whereas $2^{-8} = 1/256$. This is a very big difference.

Acknowledgements. We thank Benny Pinkas and Ben Riva for helpful discussions.

References

1. Aumann, Y., Lindell, Y.: Security Against Covert Adversaries: Efficient Protocols for Realistic Adversaries. Journal of Cryptology 23(2), 281–343 (2010), extended abstract at In: Vadhan, S.P. (ed.) TCC 2007. LNCS, vol. 4392, pp. 137–156. Springer, Heidelberg (2007)
2. Beaver, D.: Foundations of Secure Interactive Computing. In: Feigenbaum, J. (ed.) CRYPTO 1991. LNCS, vol. 576, pp. 377–391. Springer, Heidelberg (1992)
3. Bendlin, R., Damgård, I., Orlandi, C., Zakarias, S.: Semi-homomorphic Encryption and Multiparty Computation. In: Paterson, K.G. (ed.) EUROCRYPT 2011. LNCS, vol. 6632, pp. 169–188. Springer, Heidelberg (2011)
4. Canetti, R.: Security and Composition of Multiparty Cryptographic Protocols. Journal of Cryptology 13(1), 143–202 (2000)
5. Canetti, R.: Universally Composable Security: A New Paradigm for Cryptographic Protocols. In: 42nd FOCS, pp. 136–145 (2001), http://eprint.iacr.org/2000/067
6. Damgård, I., Orlandi, C.: Multiparty Computation for Dishonest Majority: From Passive to Active Security at Low Cost. In: Rabin, T. (ed.) CRYPTO 2010. LNCS, vol. 6223, pp. 558–576. Springer, Heidelberg (2010)
7. Damgård, I., Pastro, V., Smart, N.P., Zakarias, S.: Multiparty Computation from Somewhat Homomorphic Encryption. In: Safavi-Naini, R. (ed.) CRYPTO 2012. LNCS, vol. 7417, pp. 643–662. Springer, Heidelberg (2012)

8. Dodis, Y., Gennaro, R., Håstad, J., Krawczyk, H., Rabin, T.: Randomness Extraction and Key Derivation Using the CBC, Cascade and HMAC Modes. In: Franklin, M. (ed.) CRYPTO 2004. LNCS, vol. 3152, pp. 494–510. Springer, Heidelberg (2004)
9. Frederiksen, T.K., Nielsen, J.B.: Fast and Maliciously Secure Two-Party Computation Using the GPU. Cryptology ePrint Archive: Report 2013/046 (2013)
10. Goldwasser, S., Levin, L.: Fair Computation of General Functions in Presence of Immoral Majority. In: Menezes, A., Vanstone, S.A. (eds.) CRYPTO 1990. LNCS, vol. 537, pp. 77–93. Springer, Heidelberg (1991)
11. Goldreich, O.: Foundations of Cryptography. Basic Applications, vol. 2. Cambridge University Press (2004)
12. Goldreich, O., Micali, S., Wigderson, A.: How to Play any Mental Game – A Completeness Theorem for Protocols with Honest Majority. In: 19th STOC, pp. 218–229 (1987)
13. Hazay, C., Lindell, Y.: Efficient Secure Two-Party Protocols: Techniques and Constructions. Springer (November 2010)
14. Ishai, Y., Prabhakaran, M., Sahai, A.: Founding Cryptography on Oblivious Transfer – Efficiently. In: Wagner, D. (ed.) CRYPTO 2008. LNCS, vol. 5157, pp. 572–591. Springer, Heidelberg (2008)
15. Ishai, Y., Prabhakaran, M., Sahai, A.: Secure Arithmetic Computation with No Honest Majority. In: Reingold, O. (ed.) TCC 2009. LNCS, vol. 5444, pp. 294–314. Springer, Heidelberg (2009)
16. Jarecki, S., Shmatikov, V.: Efficient Two-Party Secure Computation on Committed Inputs. In: Naor, M. (ed.) EUROCRYPT 2007. LNCS, vol. 4515, pp. 97–114. Springer, Heidelberg (2007)
17. Kreuter, B., Shelat, A., Shen, C.: Billion-Gate Secure Computation with Malicious Adversaries. In: The 21st USENIX Security Symposium (2012)
18. Lindell, Y., Oxman, E., Pinkas, B.: The IPS Compiler: Optimizations, Variants and Concrete Efficiency. In: Rogaway, P. (ed.) CRYPTO 2011. LNCS, vol. 6841, pp. 259–276. Springer, Heidelberg (2011)
19. Lindell, Y., Pinkas, B.: A Proof of Yao's Protocol for Secure Two-Party Computation. Journal of Cryptology 22(2), 161–188 (2009)
20. Lindell, Y., Pinkas, B.: An Efficient Protocol for Secure Two-Party Computation in the Presence of Malicious Adversaries. In: Naor, M. (ed.) EUROCRYPT 2007. LNCS, vol. 4515, pp. 52–78. Springer, Heidelberg (2007)
21. Lindell, Y., Pinkas, B.: Secure Two-Party Computation via Cut-and-Choose Oblivious Transfer. In: Ishai, Y. (ed.) TCC 2011. LNCS, vol. 6597, pp. 329–346. Springer, Heidelberg (2011)
22. Lindell, Y.: Fast Cut-and-Choose Based Protocols for Malicious and Covert Adversaries. Cryptology ePrint Archive: Report 2013/079 (2013)
23. Menezes, A.J., van Oorschot, P.C., Vanstone, S.A.: Handbook of Applied Cryptography. CRC Press (2001)
24. Micali, S., Rogaway, P.: Secure Computation. In: Feigenbaum, J. (ed.) CRYPTO 1991. LNCS, vol. 576, pp. 392–404. Springer, Heidelberg (1992)
25. Mohassel, P., Riva, B.: Garbled Circuits Checking Garbled Circuits: More Efficient and Secure Two-Party Computation. Cryptology ePrint Archive, Report 2013/051 (2013)
26. Nielsen, J.B., Nordholt, P.S., Orlandi, C., Sheshank Burra, S.: A New Approach to Practical Active-Secure Two-Party Computation. In: Safavi-Naini, R. (ed.) CRYPTO 2012. LNCS, vol. 7417, pp. 681–700. Springer, Heidelberg (2012)

27. Nielsen, J.B., Orlandi, C.: LEGO for Two-Party Secure Computation. In: Reingold, O. (ed.) TCC 2009. LNCS, vol. 5444, pp. 368–386. Springer, Heidelberg (2009)
28. Schoenmakers, B., Tuyls, P.: Practical Two-Party Computation Based on the Conditional Gate. In: Lee, P.J. (ed.) ASIACRYPT 2004. LNCS, vol. 3329, pp. 119–136. Springer, Heidelberg (2004)
29. Shelat, A., Shen, C.-H.: Two-Output Secure Computation with Malicious Adversaries. In: Paterson, K.G. (ed.) EUROCRYPT 2011. LNCS, vol. 6632, pp. 386–405. Springer, Heidelberg (2011)
30. Yao, A.: How to Generate and Exchange Secrets. In: 27th FOCS, pp. 162–167 (1986), See [19] for details
31. Bristol Cryptography Group. Circuits of Basic Functions Suitable For MPC and FHE, http://www.cs.bris.ac.uk/Research/CryptographySecurity/MPC/

Efficient Secure Two-Party Computation Using Symmetric Cut-and-Choose

Yan Huang[1], Jonathan Katz[1,*], and David Evans[2,**]

[1] Dept. of Computer Science, University of Maryland
{jkatz,yhuang}@cs.umd.edu
[2] Dept. of Computer Science, University of Virginia
evans@cs.virginia.edu

Abstract. Beginning with the work of Lindell and Pinkas, researchers have proposed several protocols for secure two-party computation based on the *cut-and-choose* paradigm. In current instantiations of this approach, one party generates κ garbled circuits; some fraction of those are "checked" by the other party, and the remaining fraction are evaluated.

We introduce here the idea of *symmetric* cut-and-choose protocols, in which *both* parties generate κ circuits to be checked by the other party. The main advantage of our technique is that κ can be reduced by a factor of 3 while attaining the same statistical security level as in prior work. Since the number of garbled circuits dominates the costs of the protocol, especially as larger circuits are evaluated, our protocol is expected to run up to 3 times faster than existing schemes. Preliminary experiments validate this claim.

1 Introduction

Secure two-party computation was shown to be feasible in the late 1980s [35,8]. But it is only in the past 10 years that the research community has devoted significant efforts toward making such protocols *practical*. Work in this direction was spurred by the Fairplay paper [25], which implemented Yao's protocol for two-party computation with security in the semi-honest setting. More recent work [10,12,11] has shown that Yao's protocol (in combination with other techniques) can be surprisingly efficient when semi-honest security is sufficient.

More desirable, of course, is to achieve security against malicious adversaries. While this is known to be feasible, in principle, using generic zero knowledge [8], a generic approach of this sort does not currently seem likely to result in efficient protocols even if specialized zero-knowledge proofs (as suggested in [15]) are used. The first technique to be explored for making efficient two-party computation protocols secure against malicious adversaries was the *cut-and-choose* paradigm. In that approach, roughly speaking, one party generates κ garbled circuits (where κ is a statistical security parameter); some fraction of those are "checked" by

* This work was supported by DARPA and NSF award #1111599.
** This work was supported by AFOSR and NSF award #1111781.

R. Canetti and J.A. Garay (Eds.): CRYPTO 2013, Part II, LNCS 8043, pp. 18–35, 2013.

the other party—who aborts if any misbehavior is detected—and the remaining fraction are evaluated with the results being used to derive the final output (we return to the exact mechanism for doing so in the next section). Cut-and-choose was used in a relatively naive way in [25] to give inverse-polynomial security. (In fact, the approach taken was later shown to be flawed [26,16].) A rigorous analysis of the cut-and-choose paradigm was first given by Lindell and Pinkas [21], and their work was followed by numerous others exploring variations of this technique and their application to (ever more) efficient secure two-party computation [34,24,30,32,23,33,18].

In parallel with the above, other efficient approaches to achieving "full" malicious security in the two-party setting have also been explored. Approaches based on the IPS compiler [14] appear to have good asymptotic complexity [20], but seem challenging to implement (indeed, we are not aware of any implementations); other approaches [29,5,4] have round complexity proportional to the depth of the circuit being evaluated. Another direction is to explore weaker security guarantees [1,26,13], still against arbitrary malicious behavior. In the remainder of this paper we restrict our attention to protocols achieving the strongest notion of malicious security.

The critical question regarding the cut-and-choose approach is: *how many garbled-circuit copies (namely, κ) are needed to ensure some desired security level?* The value of κ has the greatest impact on the efficiency of cut-and-choose protocols, especially as larger circuits C are evaluated. The computational/communication complexity of such protocols is $O(\kappa \cdot k \cdot |C|) + \mathsf{poly}(n, k, s)$, where k is a cryptographic security parameter and n is the input length. Since $|C| \gg k, n$ (typical values are $k \approx 128$ and $n < 1000$, while $|C| \approx 10^9$ in [18]), the importance of minimizing κ is clear.

1.1 Prior Work

In previous applications of the cut-and-choose paradigm, one party (say, P_1) acts as the garbled-circuit generator and the other (P_2) acts as the garbled-circuit evaluator; assume for simplicity that only P_2 gets output. If the oblivious-transfer (OT) protocol used is secure against malicious adversaries, the main issue is to ensure correctness of P_2's output. (Note, however, that correctness is closely connected with privacy, since P_1 can potentially carry out a *selective failure* attack in which the output of P_2 is correlated with P_2's input, in a way which would not be possible in an ideal evaluation of the function.) Toward that end, P_2 checks some number c of the κ circuits generated by P_1 to make sure they were constructed correctly. Assuming they were, the remaining $\kappa - c$ garbled circuits are evaluated by P_2, who then outputs the *majority* value of those circuits' results on each output wire. (This informal description omits various details, since we wish to focus on the cut-and-choose aspect of the protocols.)

From the above we see that a malicious P_1 can successfully cheat if they generate b "bad" garbled circuits and (1) none of those bad garbled circuits is among the c garbled circuits checked by P_1, and (2) of the remaining $\kappa - c$ garbled circuits being evaluated, half or more are bad. Doing the analysis, prior work [21,23] culminating in the work of Shelat and Shen [33] shows that using κ garbled circuits yields security level $2^{-0.32\kappa}$. Moreover, this bound was shown to be the best possible for a certain class of cut-and-choose approaches [33].

1.2   Our Contribution

We recast the cut-and-choose approach in a *symmetric* setting, where each party generates κ garbled circuits to be checked by the other party. In doing so, we are motivated by work of Mohassel and Franklin [26] (see also [13]) who show how symmetric creation/evaluation of garbled circuits (but without any cut-and-choose) can be used to achieve security with only one bit of "disallowed" leakage against malicious adversaries. Here we show how to extend their approach to achieve the standard (i.e., "full") notion of malicious security.

After checking each other's garbled circuits, each party in our protocol evaluates the remaining garbled circuits of the other party, and then the results of both parties' evaluations are securely "combined" to yield the final output. Informally, a party outputs a value v for some output wire of the circuit if and only if at least one of their own garbled circuits, and *at least one of the garbled circuits generated by the other party*, evaluate to v on that wire. Since an honest party always generates correct garbled circuits, our analysis shows that correctness holds as long as *at least one* of the evaluated circuits provided by the other party is correct. (This is in contrast to one-sided cut-and-choose, where a *majority* of the evaluated circuits must be correct.) Thus, a malicious party can successfully cheat only if they generate exactly $\kappa - c$ "bad" garbled circuits, and none of those is checked by the other party. Setting $c = \kappa/2$ (which minimizes the cheating probability), the probability of successful cheating is $\binom{\kappa}{\kappa/2}^{-1} = 2^{-\kappa + O(\log \kappa)}$. We can therefore achieve the same security level as previous work while reducing[1] κ by roughly a factor of 3.

As an added advantage, our protocol naturally supports having both parties receive output (an explicit concern of [33]), with no performance penalty if only one party should learn the output.

In concurrent work, Lindell [19] shows a different approach that achieves $2^{-\kappa}$ security using κ circuits generated by only one of the parties.

1.3   Outline of the Paper

In Section 2 we review the cryptographic building blocks used in our protocol. We provide an overview of the protocol in Section 3 along with some intuition for why it is secure. In Section 4 we provide a formal description of our protocol, and we prove security in Section 5. In Appendix A we give some preliminary experimental results showing that we outperform the recent work of [18].

---

[1] To be clear: in our protocol *each party* generates κ garbled circuits and so the total number of garbled circuits is 2κ. However, since this work is done in parallel by the two parties—in addition to whatever parallel processing is available on each user's own machine—and since the communication is symmetric, the "wall-clock time" of our protocol is expected to improve on previous protocols by up to a factor of 3.

2 Notation and Building Blocks

For simplicity, we describe our protocol using concrete (rather than asymptotic) notation. Nevertheless, it should be clear that our protocol can be cast in an asymptotic setting without difficulty.

Let $\mathbb{G}$ be a group of prime order q with generator g. We assume the computational Diffie-Hellman (CDH) problem is hard in $\mathbb{G}$. We let H be a hash function that will be treated in the analysis as a random oracle. We let Com be a commitment scheme.

We use the standard definitions of secure two-party computation [7].

2.1 Naor-Pinkas Oblivious Transfer

In our protocol we do not use oblivious transfer as a "black box," but instead rely on specific details of the OT protocol used. Although several candidate OT protocols could be used, for concreteness and efficiency we use an OT protocol due to Naor and Pinkas [27] which we now describe.

Say we have a sender holding inputs $x_0, x_1 \in \{0,1\}^*$, and a receiver holding input $b \in \{0,1\}$. In the first round, the sender chooses random $C \leftarrow \mathbb{G}$ and sends C to the other party. The receiver picks $k \leftarrow \mathbb{Z}_q$, defines $h^0 = g^k$ and $h^1 = C/g^k$, and sends $h = h^b$ to the sender. In turn, the sender chooses $r \leftarrow \mathbb{Z}_q$ and sends $g^r, H(h^r) \oplus x_0, H((C/h)^r) \oplus x_1$ to the other party. The receiver recovers x_b by computing $(g^r)^k$ and using the appropriate component of the sender's final message. Note that several independent OTs can share the same first message C.

This OT protocol is simulatable for a malicious receiver under the CDH assumption in the random oracle model. It achieves privacy (but is not known to be simulatable) against a malicious sender, and this suffices for our purposes. A variant of this protocol requires the receiver to give a (standard) perfect witness-indistinguishable proof of knowledge of $\log_g h$ or $\log_g(C/h)$ after sending h. We use this variant in our analysis since it simplifies the proof.

2.2 Garbled Circuits

We use a modification of standard garbled circuits [35]. Fix a function $f : \{0,1\}^n \times \{0,1\}^n \to \{0,1\}^n$. We abstract the construction/evaluation of a garbled circuit for f via algorithms GenGC, EvalGC with the following properties. GenGC is a randomized algorithm that takes as input $2n$ input-wire labels $v_1^0, v_1^1, \ldots, v_n^0, v_n^1 \in \mathbb{G}$ (corresponding to the first input of f), $2n$ input-wire labels $v_{n+1}^0, v_{n+1}^1, \ldots, v_{2n}^0, v_{2n}^1 \in \{0,1\}^n$ (corresponding to the second input of f), and $2n$ output-wire labels $w_1^0, w_1^1, \ldots, w_n^0, w_n^1 \in \mathbb{Z}_q$. It outputs a garbled circuit GC. Deterministic algorithm EvalGC takes as input GC and $2n$ input-wire labels $v_1, \ldots, v_{2n}$; it outputs n values $b_1 \| w_1, \ldots, b_n \| w_n$, with $b_1, \ldots, b_n \in \{0,1\}$. Note that EvalGC explicitly outputs wire labels in addition to bits.

Correctness requires that for any set of input/output-wire labels, any garbled circuit GC output by GenGC $\left(\{v_i^0, v_i^1\}_{i=1}^{2n}, \{w_i^0, w_i^1\}_{i=1}^{n} \right)$, and any $x, y \in \{0,1\}^n$

with $z = f(x, y)$, we have

$$\mathsf{EvalGC}\left(\mathsf{GC}, \{v_i^{x_i}\}_{i=1}^n, \{v_i^{y_i}\}_{i=n+1}^{2n}\right) = z_1 \| w_1^{z_1}, \ldots, z_n \| w_n^{z_n}.$$

Security requires a simulator SimGC such that for all x, y with $z = f(x, y)$, any $v_1^{x_1}, \ldots, v_n^{x_n} \in \mathbb{G}$, $v_{n+1}^{y_n}, \ldots, v_{2n}^{y_n} \in \{0, 1\}^n$, and $w_1^0, w_1^1, \ldots, w_n^0, w_n^1 \in \mathbb{Z}_q$, the distribution

$$\left\{ \begin{array}{l} v_1^{1-x_1}, \ldots, v_n^{1-x_n} \leftarrow \mathbb{G}; \\ v_{n+1}^{1-y_1}, \ldots, v_{2n}^{1-y_n} \leftarrow \{0,1\}^n; \quad : \left(\mathsf{GC}, \{v_i^{x_i}\}_{i=1}^n, \{v_{n+i}^{y_i}\}_{i=1}^n\right) \\ \mathsf{GC} \leftarrow \mathsf{GenGC}\left(\{v_i^0, v_i^1\}_{i=1}^{2n}, \{w_i^0, w_i^1\}_{i=1}^n\right) \end{array} \right\}$$

is computationally indistinguishable from

$$\left\{\mathsf{GC} \leftarrow \mathsf{SimGC}\left(x, z, \{v_i^{x_i}\}_{i=1}^n, \{v_{n+i}^{y_i}\}_{i=1}^n, \{w_i^{z_i}\}_{i=1}^n\right) : \left(\mathsf{GC}, \{v_i^{x_i}\}_{i=1}^n, \{v_{n+i}^{y_i}\}_{i=1}^n\right)\right\}.$$

In particular, this means (informally) that (1) given GC, $\{v_i^{x_i}\}_{i=1}^n$, and $\{v_i^{y_{n+i}}\}_{i=1}^n$, no information is leaked about $\{w_i^{1-z_i}\}_{i=1}^n$ where $z = f(x, y)$, and (2) this holds regardless of how the $\{v_i^{x_i}\}_{i=1}^n$, $\{v_{n+i}^{y_i}\}_{i=1}^n$ are chosen (as long as the other input-wire labels are random). These properties are not standard, but are easily seen to hold by modifying the construction/proof from [22].

Note: We always let input wires $1, \ldots, n$ denote the inputs of the party generating the circuit. Thus, technically, P_1 generates garbled circuits for the function f, and P_2 generates garbled circuits for the function $f'(y, x) \overset{\text{def}}{=} f(x, y)$.

2.3   Verifiable Secret Sharing

We use a notion of (non-interactive) verifiable secret sharing (VSS) that is weaker than the usual one in the literature. For our purposes, a t-out-of-κ VSS scheme comprises three algorithms $\mathsf{Share}, \mathsf{Vrfy}, \mathsf{Rec}$ with the following functionality:

- Share takes input $s \in \mathbb{Z}_q$ and outputs κ shares $w_1, \ldots, w_\kappa \in \mathbb{Z}_q$ and additional information pub.
- Vrfy takes as input the information pub, an index i, and a candidate share $w_i \in \mathbb{Z}_q$. It outputs a bit, with 1 denoting validity.
- Rec takes as input pub and t indices/shares $\{(i_j, w_{i_j})\}_{j=1}^t$. It outputs a value $s \in \mathbb{Z}_q$.

We require that for any $s \in \mathbb{Z}_q$, any $w_1, \ldots, w_\kappa, \mathsf{pub}$ output by $\mathsf{Share}(s)$, and any $i_1, \ldots, i_t \subset [\kappa]$, we have $\mathsf{Vrfy}(\mathsf{pub}, i, w_i) = 1$ and $\mathsf{Rec}(\mathsf{pub}, \{(i_j, w_{i_j})\}_{j=1}^t) = s$.

We define a secrecy requirement for an honest dealer, and a verifiability requirement for honest receivers. Secrecy requires hardness of recovering a random secret s given pub and at most $t - 1$ shares. Formally, the following should be small for all efficient algorithms $\mathcal{A}$ and any $i_1, \ldots, i_{t-1}$:

$$\Pr[s \leftarrow \mathbb{Z}_q; (\mathsf{pub}, w_1, \ldots, w_\kappa) \leftarrow \mathsf{Share}(s) : \mathcal{A}(\mathsf{pub}, w_{i_1}, \ldots, w_{i_{t-1}}) = s].$$

Verifiability requires that the dealer cannot generate pub and two different sets of valid shares that reconstruct to different secrets. Formally, the following is small for all efficient algorithms $\mathcal{A}$:

$$\Pr\left[\begin{array}{l}\left(\mathsf{pub}, \{(i_j, w_j)\}_{j=1}^t, \{(i'_j, w'_j)\}_{j=1}^t\right) \leftarrow \mathcal{A} \\ \\ \qquad\qquad\quad \forall j : \mathsf{Vrfy}(\mathsf{pub}, i_j, w_j) = 1 \\ \qquad : \quad \bigwedge \forall j : \mathsf{Vrfy}(\mathsf{pub}, i'_j, w'_j) = 1 \\ \quad \bigwedge \mathsf{Rec}(\mathsf{pub}, \{(i_j, w_j)\}_{j=1}^t) \neq \mathsf{Rec}(\mathsf{pub}, \{(i'_j, w'_j)\}_{j=1}^t)\end{array}\right].$$

Feldman VSS [6] satisfies the above properties under the discrete-logarithm assumption.

3   High-Level Description of the Protocol

At a high level, the protocol proceeds in the following stages:

1. **Generate garbled circuits:** Each party generates κ garbled circuits along with their corresponding input-wire labels.
2. **Oblivious transfer:** Each party uses the Naor-Pinkas OT protocol (cf. Section 2.1) to obtain its input-wire labels for the garbled circuits constructed by the other party. This is done in such a way that a party must use the *same* effective input across all circuits.
3. **"Cut-and-choose":** Each party sends the garbled circuits they constructed to the other party. Using coin tossing, parties choose half of each of their circuits for checking. Then:
   (a) For each of its check circuits, each party (1) sends all the input-wire labels for that circuit (to prove that the check circuit was constructed correctly) and (2) reveals all the values it used as the OT sender in step 2 (to prove that it used the correct input-wire labels in the OT execution corresponding to the check circuit).
   (b) For each of its remaining circuits (the evaluation circuits), each party sends the input-wire labels corresponding to its own input.
4. **Output determination:** Each party evaluates the garbled circuits they received from the other party, using the input-wire labels obtained in steps 2 and 3(b). For each output wire i of the circuit, each party decides on output $z_i \in \{0, 1\}$ iff at least one of the circuits they evaluated (that the other party constructed) gave output z_i *and* at least one of the circuits the other party evaluated (that they constructed) gave output z_i.

We defer the details of step 4, and for now just assume it can be done. We also assume that if a party successfully passes the cut-and-choose step, then for at least one of that party's evaluation circuits (1) the evaluation circuit is constructed correctly and (2) the correct input-wire labels were used in the corresponding OT; this assumption holds except with probability at most $\binom{\kappa}{\kappa/2}^{-1}$.

The main issue to address is to ensure that a malicious party uses the same (effective) input in step 2 (when it obtains input-wire labels for its own input from the other party using OT) and for *all* the input-wire labels it sends in step 3(b) (for the garbled circuits that it generated). We achieve this by noting that when an honest receiver obtains the input-wire labels for its ith input wire during the OT step, it sends a message h_i for which (1) it knows $\log_g h_i$ when its effective input (on the ith wire) is 0, and (2) it knows $\log_g(C/h_i)$ when its effective input (on the ith wire) is 1. We require the parties to use this same "template" for the input-wire labels corresponding to their own input in the garbled circuits they prepare. That is, for each garbled circuit and each input wire i corresponding to an input of the circuit generator, the input-wire label v_i^0 corresponding to 0 is chosen such that $\log_g v_i^0$ is known, and the input-wire label v_i^1 corresponding to 1 is chosen such that $\log_g(C/v_i^1)$ is known. Moreover, this property is verified to hold (for the check circuits) during the cut-and-choose step. When sending its ith input-wire label v_i in step 3(b), each party must then also prove[2] that it knows $\log_g(v_i/h_i)$. This is reminiscent of a similar technique used by Shelat and Shen [33] to enforce input consistency among input-wire labels sent by the circuit generator; here, we extend it to enforce consistency also to the input-wire labels *received* as a circuit evaluator.

Given this—and still assuming step 4 can be carried out—one can informally verify that the protocol is secure. Assume for concreteness that P_2 is honest. Privacy of P_2's input is easy to see. As for correctness, P_2 constructed all its garbled circuits correctly and sent input-wire labels for its own input y in all its evaluation circuits. In step 2, P_1 obtained input-wire labels for *its* own (effective) input x in all of P_2's evaluation circuits. So all of P_2's garbled circuits that were evaluated by P_1 yield output $z \overset{\text{def}}{=} f(x,y)$. In the other direction, with high probability at least one of P_1's evaluation circuits GC^* was constructed correctly, and moreover the correct input-wire labels (for P_2's input) were used in the corresponding OT; thus, P_2 obtained the correct input-wire labels for its input y in GC^*. Furthermore, from the previous paragraph we know that the input-wire labels for P_1's input in GC^* correspond to the same input x it used before. Thus, evaluation of GC^* by P_2 also yields $z = f(x,y)$, and thus z will be the final output of P_2 in the protocol.

The missing piece is to show how to implement step 4, and this is the most involved part of our protocol. The basic idea is for each party to choose a secret value s_i^b for each output wire i of the circuit and each possible value $b \in \{0,1\}$ that wire can take. Each secret is then split into κ shares $w_{1,i}^b, \ldots, w_{\kappa,i}^b$ using a $(\kappa/2 + 1)$-out-of-κ secret-sharing scheme. Share $w_{j,i}^b$ is then used as the label corresponding to b on the ith output wire of the jth garbled circuit. The net result is that for each output wire i and bit b, the other party can reconstruct s_i^b if and only if it learns $\kappa/2 + 1$ of the shares corresponding to that wire and bit.

Note that $\kappa/2$ shares of *every* wire and bit will be revealed as part of the cut-and-choose phase. Assuming again that P_2 is honest, we thus have the following:

---

[2] Actually, as in [33], the party can simply reveal $\log_g(v_i/h_i)$.

- As noted earlier, all of the garbled circuits that P_2 constructed will evaluate to the same value $z = f(x, y)$. This means that P_1 only learns shares corresponding to the secrets $s_1^{z_1}, \ldots, s_n^{z_n}$, and learns nothing about the remaining secrets $s_1^{1-z_1}, \ldots, s_n^{1-z_n}$. This gives P_2 a way to "test" whether the circuits it constructed (that were evaluated by P_1) resulted in output z by checking which of each pair of secrets P_1 knows (e.g., using a secure equality test).
- In the opposite direction, as long as *one* of the garbled circuits constructed by P_1 (and evaluated by P_2) yields z, this will give P_2 one additional share of each of $\tilde{s}_1^{z_1}, \ldots, \tilde{s}_n^{z_n}$ (where we use $\tilde{s}$ here to denote that these secrets are chosen by P_1) and hence P_2 will be able to reconstruct each of those secrets. Note that it does not matter *which* garbled circuit evaluates to z, as any correctly constructed circuit that evaluates to z reveals the requisite share.

One point omitted from the above discussion is that now it must be possible to check during the cut-and-choose phase that correct shares were used when constructing the garbled circuits. For this reason, we use *verifiable* secret sharing (see Section 2.3). We defer to the next section additional technical details of the protocol needed for the proof of security.

4 Formal Specification of the Protocol

Fix a function $f : \{0,1\}^n \times \{0,1\}^n \to \{0,1\}^n$ that parties P_1 and P_2 wish to compute over their respective inputs $x, y \in \{0,1\}^n$. We assume both parties learn the output, but it is easy to modify the protocol so that only one party learns the output. The protocol proceeds as follows.

1. P_1 chooses $C \leftarrow \mathbb{G}$ and sends it to P_2. Symmetrically, P_2 chooses $\tilde{C} \leftarrow \mathbb{G}$ and sends it to P_1.
2. P_1 generates $4n$ input-wire labels for each of κ garbled circuits in the following way. For $j = 1, \ldots, \kappa$, it chooses $a_{j,1}^0, a_{j,1}^1, \ldots, a_{j,n}^0, a_{j,n}^1 \leftarrow \mathbb{Z}_q$ and sets the first $2n$ input-wire labels of circuit j to be of the form $\{v_{j,i}^0 = g^{a_{j,i}^0}\}_{i=1}^n$ and $\{v_{j,i}^1 = \tilde{C}/g^{a_{j,i}^1}\}_{i=1}^n$. It chooses the next $2n$ input-wire labels of circuit j uniformly as $v_{j,n+1}^0, v_{j,n+1}^1, \ldots, v_{j,2n}^0, v_{j,2n}^1 \leftarrow \{0,1\}^n$.
   Symmetrically,[3] P_2 generates $4n$ input-wire labels $\tilde{v}_{j,1}^0, \tilde{v}_{j,1}^1, \ldots, \tilde{v}_{j,2n}^0, \tilde{v}_{j,2n}^1$ for $j = 1, \ldots, \kappa$.
   Each party then uses Naor-Pinkas OT to obtain the input-wire labels corresponding to its own input in the circuits generated by the other party. I.e., for $i = 1, \ldots, n$ party P_1 chooses $k_i \leftarrow \mathbb{Z}_q$, generates $(h_i^0, h_i^1) = (g^{k_i}, \tilde{C}/g^{k_i})$, and sends $h_i \overset{\text{def}}{=} h_i^{x_i}$ to P_2. Then P_2 generates κ independent responses as in the Naor-Pinkas protocol, using inputs $(\tilde{v}_{j,n+i}^0, \tilde{v}_{j,n+i}^1)$ in the jth such instance where, recall, $\tilde{v}_{j,n+i}^b$ denotes the label corresponding to bit b on the $(n+i)$th input wire in the jth garbled circuit. P_1 recovers $\tilde{v}_{1,n+i}^{x_i}, \ldots, \tilde{v}_{\kappa,n+i}^{x_i}$. P_2 acts symmetrically to obtain $v_{1,n+i}^{y_i}, \ldots, v_{\kappa,n+i}^{y_i}$ for $i = 1, \ldots, n$.

---

[3] Recall that the first n input wires always denote the inputs of the party generating the circuit.

3. For $i \in \{1, \ldots, n\}$ and $b \in \{0, 1\}$, party P_1 chooses $s_i^b \leftarrow \mathbb{Z}_q$ and generates a $(\kappa/2 + 1)$-out-of-κ secret sharing $(\mathsf{pub}_i^b, w_{1,i}^b, \ldots, w_{\kappa,i}^b) \leftarrow \mathsf{Share}(s_i^b)$. It uses $w_{j,i}^b$ as the label for bit b on the ith output wire in the jth circuit, i.e., for $j = 1, \ldots, \kappa$ it computes the garbled circuit

$$\mathsf{GC}_j = \mathsf{GenGC}\left(\{v_{j,i}^0, v_{j,i}^1\}_{i=1}^{2n}, \{w_{j,i}^0, w_{j,i}^1\}_{i=1}^{n}\right).$$

It sends $\{\mathsf{GC}_j\}_{j=1}^{\kappa}$ and $\{\mathsf{pub}_i^0, \mathsf{pub}_i^1\}_{i=1}^{n}$ to P_2.

P_2 acts symmetrically to obtain $\widetilde{s}_i^b$ and $(\widetilde{\mathsf{pub}}_i^b, \widetilde{w}_{1,i}^b, \ldots, \widetilde{w}_{\kappa,i}^b)$ and to generate $\widetilde{\mathsf{GC}}_j$; it sends $\{\widetilde{\mathsf{GC}}_j\}_{j=1}^{\kappa}$ and $\{\widetilde{\mathsf{pub}}_i^0, \widetilde{\mathsf{pub}}_i^1\}_{i=1}^{n}$ to P_1.

4. For $j = 1, \ldots, \kappa$ and $i = 1, \ldots, n$, party P_1 commits to the input-wire labels $v_{j,i}^0$ and $v_{j,i}^1$ corresponding to its own input, in random permuted order. Let $\mathsf{ComSet}_{j,i}$ denote the resulting pair of commitments. P_2 acts symmetrically.

5. The parties run secure coin-tossing protocols to generate strings $\mathcal{J}, \widetilde{\mathcal{J}} \in \{0, 1\}^{\kappa}$ that are each uniform among strings containing exactly $\kappa/2$ ones.[4] These are interpreted in the natural way as subsets of $\{1, \ldots, \kappa\}$ of size $\kappa/2$. $\widetilde{\mathcal{J}}$ is used to check the garbled circuits constructed by P_1. Specifically, for $j = 1, \ldots, \kappa$:

   (a) If $j \in \widetilde{\mathcal{J}}$ the jth circuit is a *check circuit*. Then P_1 sends $\{a_{j,i}^0, a_{j,i}^1\}_{i=1}^{n}$, $\{v_{j,i}^0, v_{j,i}^1\}_{i=2n+1}^{2n}$, $\{w_{j,i}^0, w_{j,i}^1\}_{i=1}^{n}$, and the randomness it used to generate GC_j. It also reveals the sender-randomness it used in all the OTs corresponding to the jth circuit, and opens both commitments in $\mathsf{ComSet}_{j,i}$ for $i = 1, \ldots, n$.

   P_2 sets $v_{j,i}^0 = g^{a_{j,i}^0}$ and $v_{j,i}^1 = \widetilde{C}/g^{a_{j,i}^1}$ for $i = 1, \ldots, n$. It re-generates the jth garbled circuit and verifies that it matches GC_j. It verifies that $\{v_{j,i}^0, v_{j,i}^1\}_{i=2n+1}^{2n}$ were used in the OTs for the jth circuit, and that the commitments in $\mathsf{ComSet}_{j,i}$ open to $\{v_{j,i}^0, v_{j,i}^1\}$ in some order. Finally, it checks that $\mathsf{Vrfy}(\mathsf{pub}_i^b, j, w_{j,i}^b) = 1$ for $i = 1, \ldots, n$ and $b \in \{0, 1\}$. It aborts if any of these fail.

   (b) If $j \notin \widetilde{\mathcal{J}}$ the jth circuit is an *evaluation circuit*. In this case, P_1 sends $(v_{j,1}, \ldots, v_{j,n}) \stackrel{\text{def}}{=} (v_{j,1}^{x_1}, \ldots, v_{j,n}^{x_n})$ (i.e., the wire labels corresponding to P_1's input in the jth circuit) to P_2. It also opens the commitment in $\mathsf{ComSet}_{j,i}$ that corresponds to $v_{j,i}$. Finally, it sends $\log_g(v_{j,1}/h_1)$, ..., $\log_g(v_{j,n}/h_n)$. (Recall that $h_1, \ldots, h_n$ are the values used by P_1 when acting as receiver in the Naor-Pinkas OT protocol.)

   P_2 checks that one of the commitments in $\mathsf{ComSet}_{j,i}$ opens to $v_{j,i}$, and verifies the discrete logarithms sent by P_1. It aborts if any inconsistencies are found.

   Symmetrically, the parties use $\mathcal{J}$ to check the garbled circuits constructed by P_2.

---

[4] This can be implemented easily by using a standard coin-tossing protocol to generate polynomially many uniform bits, and then using those bits as the random coins for applying a Knuth shuffle to the string $0^{\kappa/2} 1^{\kappa/2}$.

6. For each evaluation circuit j of P_2, party P_1 evaluates $\widetilde{GC}_j$ using the input-wire labels it obtained in steps 2 and 5. By doing so, it learns n values $\widetilde{b}_{j,1} \| \widetilde{w}_{j,1}, \ldots, \widetilde{b}_{j,n} \| \widetilde{w}_{j,n}$.
   For $i = 1, \ldots, n$ and $b \in \{0, 1\}$, party P_1 tries to recover[5] $\widetilde{s}_i^b$. To do so, it finds an evaluation circuit j for which $\widetilde{b}_{j,i} = b$ and $\widetilde{w}_{j,i}$ is a valid share of $\widetilde{s}_i^b$ (i.e., $\mathsf{Vrfy}(\mathsf{pub}_i^{\widetilde{b}_{j,i}}, j, \widetilde{w}_{j,i}) = 1$). If no such j exists, it chooses $t_i^b \leftarrow \mathbb{Z}_q$. Otherwise, it computes t_i^b by running Rec using $\mathsf{pub}_i^{\widetilde{b}_{j,i}}$, the $\kappa/2$ shares $\{(k, \widetilde{w}_{k,i}^{\widetilde{b}_{j,i}})\}_{k \in \mathcal{J}}$ it learned in step 5, and the additional share $(j, \widetilde{w}_{j,i})$.
   P_2 acts symmetrically to compute $\widetilde{t}_i^0, \widetilde{t}_i^1$ for $i = 1, \ldots, n$.
7. For $i = 1, \ldots, n$, the parties do the following: Run a secure equality test, with P_1 using input $s_i^0 \| t_i^0$ and P_2 using input $\widetilde{t}_i^0 \| \widetilde{s}_i^0$. If the result is 1, each party sets $z_i = 0$ and goes to the next i. Otherwise, the parties run a second equality test with P_1 using input $s_i^1 \| t_i^1$ and P_2 using input $\widetilde{t}_i^1 \| \widetilde{s}_i^1$. If the result is 1, each party sets $z_i = 1$ and goes to the next i. If neither equality test succeeds for some i then cheating is detected and the parties abort.
   Assuming no abort has occurred, each party then outputs $z = z_1 \cdots z_n$.

4.1 Optimizations

For simplicity in our proof of security in the following section, we analyze the protocol as presented above. However, we observe that the following optimizations can be applied to the protocol (and the reader can verify that the proof can be suitably modified for each of these).

Naor-Pinkas OT. We assume a variant of Naor-Pinkas OT is used in which the receiver gives a witness-indistinguishable (WI) proof of knowledge that its message was computed correctly (see Section 2.1). This is used in our proof to extract the receiver's input. In fact, as shown in [27], such WI proofs are not necessary and extraction can be done using the random-oracle queries of the receiver. The same is true in our setting, though it complicates the presentation of the proof.

Secure Coin Tossing. In the (programmable) random-oracle model, very efficient coin tossing is possible since it is trivial to construct an equivocal and extractable commitment scheme.

Secure Equality Testing. In our proof, we assume a hybrid world in which the parties have access to an ideal functionality for equality testing; equivalently (relying on standard composition theorems [3]), we assume that the equality test is done using a fully secure protocols for this task.

In fact, using a fully secure equality test is overkill for our purposes. Instead, we can use a different approach that is very efficient in the random-oracle model. First, assume the VSS scheme has the stronger property of indistinguishability, i.e., given pub and $t - 1$ shares of a uniform secret $s \in \{0, 1\}^n$, it is hard to

---

[5] In an honest execution, only one of $\widetilde{s}_i^0$ or $\widetilde{s}_i^1$ will be recovered.

distinguish s from an independent uniform value $s' \in \{0,1\}^n$. (Any VSS scheme satisfying the unpredictability requirement from Section 2.3 can be modified to achieve this stronger notion in the random-oracle model by simply hashing the secret.) Then, rather than performing an equality test using values $s_i^0 \| t_i^0$ and $\tilde{t}_i^0 \| \tilde{s}_i^0$ (resp., $s_i^1 \| t_i^1$ and $\tilde{t}_i^1 \| \tilde{s}_i^1$) as before, the parties now carry out an equality test on values $s_i^0 \oplus t_i^0$ and $\tilde{t}_i^0 \oplus \tilde{s}_i^0$ (resp., $s_i^1 \oplus t_i^1$ and $\tilde{t}_i^1 \oplus \tilde{s}_i^1$). At this point, we observe that a full-fledged equality test is not needed since (1) the honest party's input to the equality test is either known to the malicious party or is (indistinguishable from) uniform, and (2) in either case, it is ok if the honest party's input to the equality test is leaked to the other party after equality is checked. Thus, it suffices to use a "cheap" equality test in which P_1 (resp., P_2) commits to, e.g., $s_i^0 \oplus t_i^0$ (resp., to $\tilde{t}_i^0 \oplus \tilde{s}_i^0$) using an extractable and equivocal commitment scheme (which is easily constructed in the random-oracle model), and then each party decommits and checks equality of the decommitted results in the clear.

Saving Bandwidth. Following an observation in [9], we can modify the way we do cut-and-choose as follows: Parties construct their jth garbled circuit by choosing a random seed seed_j and using that seed to generate certain (pseudo)random choices they need for constructing that circuit. (In our case, this would mean using seed_j to generate $\{a_{j,i}^0, a_{j,i}^1\}_{i=1}^n$, $\{v_{j,i}^0, v_{j,i}^1\}_{i=n+1}^{2n}$, and the randomness used to generate GC_j.) Then, in step 3, the parties send the *hash* $\mathsf{hGC}_j = H(\mathsf{GC}_j)$ in place of GC_j. If j is a check circuit then seed_j is sent; the other party re-generates GC_j and verifies that $H(\mathsf{GC}_j) = \mathsf{hGC}_j$. If j is an evaluation circuit then GC_j is sent and the other party checks that $H(\mathsf{GC}_j) = \mathsf{hGC}_j$. Since $|\mathsf{seed}_j| + |\mathsf{hGC}_j| \ll |\mathsf{GC}_j|$, this has the effect of reducing the bandwidth in steps 3 and 5 (which dominate the bandwidth of the entire protocol) by roughly half.

Batch Verification. We can use batch verification [2] when simultaneously verifying validity of shares in step 5(a) (assuming Feldman VSS is used) and the discrete logarithms in step 5(b).

Efficient Garbled Circuits. Our protocol is fully compatible with existing optimizations for garbled circuits such as garbled-row reduction [28] and the free-XOR technique [17].[6]

5   Proof of Security

Theorem 1. *Under the assumptions outlined in Section 2, and modeling H as a random oracle, the protocol in the previous section securely computes f in the presence of malicious adversaries.*

---

[6] We cannot apply the free-XOR optimization at first-level gates because of the way the circuit generator chooses the input-wire labels. However, the free-XOR method can be used at all lower levels of the circuit.

Since we are not in an asymptotic setting, technically speaking "secure" is not well-defined. In the proof below, all steps introduce a computational security factor (which can be set as small as desired by setting the cryptographic security parameter large enough) except for one step which introduces a statistical security factor of $\binom{\kappa}{\kappa/2}^{-1} = 2^{-\kappa + O(\log \kappa)}$.

All our assumptions are standard, and can be based on the CDH assumption in $\mathbb{G}$. We remark that the only place the random oracle is used is for the Naor-Pinkas OT. It would be possible to remove the random oracle by switching, e.g., to the OT protocol of [31] (and modifying the rest of the protocol accordingly). Although this would impact the efficiency, the effect would be proportional to the input length and not the size of the circuit being computed.

Proof. We analyze the protocol in a hybrid world in which the parties have access to ideal functionalities for coin tossing and equality testing. Using standard composition theorems [3], this implies security when those sub-routines are instantiated using secure protocols for those tasks.

Since the protocol is symmetric, we assume without loss of generality that P_1 is malicious. Let y denote the input of P_2. We define a sequence of experiments, beginning with the real execution of the protocol between P_1 and P_2 (in the hybrid world discussed above) and ending with an ideal execution involving a simulator $\mathcal{S}$ playing the role of the first party and interacting with a trusted party computing f. We show that each experiment is indistinguishable from the one before it, taking into account both the view/output of the malicious party and the output of P_2.

Experiment 0. This is the real execution of the protocol (in the hybrid world discussed above) between P_1 and the honest P_2 holding input y.

Experiment 1. Here we change the way P_2 behaves when acting as OT sender in step 2 and when sending commitments in step 4. First of all, we now pick $\mathcal{J}$ at the outset of the experiment. This defines the check circuits and evaluation circuits for P_2. Next, in each instance i in which P_1 acts as OT receiver in step 2 and sends message h_i, we extract (using the WI proof of knowledge) either $\log_g h_i$ or $\log_g(\widetilde{C}/h_i)$. In the former case we set $x_i = 0$ and in the latter case we set $x_i = 1$. Then, when computing the κ responses for the ith OT, in each response that corresponds to an evaluation circuit j of P_2 we continue to use $\widetilde{v}^{x_i}_{j,n+i}$ but we replace $\widetilde{v}^{1-x_i}_{j,n+i}$ with the all-0 string. (Responses that correspond to check circuits of P_2 are treated exactly as before.)

In addition, for each evaluation circuit j of P_2 and $i = 1, \ldots, n$, we now set $\mathsf{ComSet}_{j,i} = \{\mathsf{Com}(\widetilde{v}^{y_i}_{j,i}), \mathsf{Com}(g)\}$, in random permuted order.

Indistinguishability of Experiments 0 and 1 follows easily from the security of Naor-Pinkas OT (based on the CDH assumption in the random-oracle model) and computational hiding of Com.

Experiment 2. Now we generate all the evaluation circuits of P_2 using the garbled-circuit simulator SimGC. In more detail: after extracting P_1's effective input x as in the previous experiment, we compute $z = f(x, y)$. In step 3, once the $\{\widetilde{w}_{j,i}^b\}$ have been determined we compute for every evaluation circuit j the simulated garbled circuit[7] $\widetilde{\mathsf{GC}}_j \leftarrow \mathsf{SimGC}\left(x, z, \{\widetilde{v}_{j,i}^{y_i}\}_{i=1}^n, \{\widetilde{v}_{j,n+i}^{x_i}\}_{i=1}^n, \{\widetilde{w}_{j,i}^{z_i}\}_{i=1}^n\right)$. The remainder of the experiment is exactly as in Experiment 1.

Indistinguishability of Experiments 1 and 2 follows from security of the garbled-circuit simulation algorithm as defined in Section 2.2.

Note that in Experiment 2, we no longer use $\{\widetilde{v}_{j,i}^{1-y_i}\}_{i=1}^n$, $\{\widetilde{v}_{j,n+i}^{1-x_i}\}_{i=1}^n$, or $\{\widetilde{w}_{j,i}^{1-z_i}\}_{i=1}^n$ for any evaluation circuit j of P_2.

Experiment 3. This is the same as the previous experiment, except that now when performing the ith pair of equality tests we proceed as follows: if $z_i = 1$, we return 0 to both parties in the first equality test; if $z_i = 0$, we return 0 to both parties in the second equality test (if run).

Indistinguishability of this experiment from Experiment 2 follows from secrecy of VSS. Specifically, for $i = 1, \ldots, n$ only $\widetilde{\mathsf{pub}}_i^{1-z_i}$ and $\kappa/2$ shares of the secret $\widetilde{s}_i^{1-z_i}$ are used throughout the entire experiment before the equality tests. Thus, the probability (in Experiment 2) that P_1 can make any of the equality tests corresponding to $1 - z_i$ return 1 is negligible.

Experiment 4. If P_1 successfully responds to the "challenge" $\widetilde{\mathcal{J}}$ chosen during the cut-and-choose step, we repeatedly rewind P_1 in an attempt to find a $\widetilde{\mathcal{J}}' \neq \widetilde{\mathcal{J}}$ for which P_1 also responds correctly.[8] If no such $\widetilde{\mathcal{J}}'$ is found, output fail. Otherwise, re-send the original challenge $\widetilde{\mathcal{J}}$ and continue as in the previous experiment.

The only difference between this experiment and the previous one occurs in case P_1 responds correctly to only a single challenge $\widetilde{\mathcal{J}}$ and that challenge happens to be the one chosen during the experiment. This can occur with probability at most $1/\binom{\kappa}{\kappa/2}$.

Experiment 5. We now change how we compute $\widetilde{t}_i^{z_i}$ for all i. (Recall that $\widetilde{t}_i^{z_i}$ represents P_2's guess for P_1's secret $s_i^{z_i}$.) Assuming P_1 answers two different challenges $\widetilde{\mathcal{J}}, \widetilde{\mathcal{J}}'$ correctly, there is some $j^* \in \{1, \ldots, \kappa\}$ such that j^* is an evaluation circuit with respect to $\widetilde{\mathcal{J}}$ but a check circuit with respect to $\widetilde{\mathcal{J}}'$. For any such j^*, we reconstruct $s_i^{z_i}$ using the share $w_{j^*,i}^{z_i}$ sent by P_1 when answering challenge $\widetilde{\mathcal{J}}'$, along with the $\kappa/2$ other shares of $s_i^{z_i}$ that were sent by P_1 when answering challenge $\widetilde{\mathcal{J}}$. We then set $\widetilde{t}_i^{z_i} = s_i^{z_i}$ and use that value in the relevant equality test later.

We claim that this experiment is indistinguishable from the previous one; this is the crux of the proof. To prove this, we show that the shares $\{w_{j^*,i}^{z_i}\}_{i=1}^n$ computed in Experiment 5 are, except with negligible probability, the same

---

[7] Recall that the first n input wires always denote the inputs of the party generating the circuit, so in this case correspond to input y.

[8] We use standard techniques in order to ensure that the experiment runs in expected polynomial time. Specifically, in parallel with rewinding P_1 and sending a random challenge $\widetilde{\mathcal{J}}' \neq \widetilde{\mathcal{J}}$ we also enumerate over all possible $\widetilde{\mathcal{J}}'$; we output fail if we find that $\widetilde{\mathcal{J}}$ is the only challenge to which P_1 responds correctly.

shares that P_2 obtains by evaluating circuit GC_{j*} in Experiment 4. Verifiability of the secret-sharing scheme then implies that, except with negligible probability, the same values $\{\tilde{t}_i^{z_i}\}_{i=1}^n$ are computed in both experiments (namely, even if in Experiment 4 a valid share from an evaluation circuit other than $j*$ is used by P_2 to reconstruct some $s_i^{z_i}$).

Fix i. To see that the same share $w_{j*,i}^{z_i}$ is computed in each experiment, observe that in Experiment 4 the share $w_{j*,i}^{z_i}$ is computed by evaluating garbled circuit GC_{j*} using input-wire labels for P_2's input that P_2 obtains from the OTs corresponding to circuit $j*$, and input-wire labels for P_1's input that were sent by P_1 in step 5. Because P_1 responds correctly to challenge $\tilde{\mathcal{J}}'$, in which $j*$ is a check circuit, we know that: (1) GC_{j*} is correctly constructed; (2) the input-wire labels that P_2 obtained from the OTs are correct labels for GC_{j*} that correspond to P_2's input y; (3) the input-wire labels for its own input that P_1 sends must be correct labels for GC_{j*} (this follows from binding of the commitments in $\{\mathsf{ComSet}_{j*,i}\}_{i=1}^n$) and moreover must correspond to the same effective input x defined by P_1's execution as OT receiver (otherwise we obtain a discrete logarithm of the random group element $\tilde{C}$). Since GC_{j*}, when evaluated on input-wire labels corresponding to x and y, yields the share $w_{j*,i}^{z_i}$ on the ith output wire, we are done.

In Experiment 5 none of P_1's evaluation circuits need to be evaluated by P_2. Moreover, P_2 no longer needs to compute its output in any of the OTs in which it acts as receiver.

Experiment 6. In the previous experiment, when P_2 acts as OT receiver it sends $\tilde{h}_i$ with either $\log_g \tilde{h}_i$ or $\log_g(C/\tilde{h}_i)$ known (depending on y_i). The input-wire labels $\{\tilde{v}_{j,i}^{y_i}\}_{i=1}^n$ (when j is an evaluation circuit) are chosen in a similar way. In this experiment, for $i = 1, \ldots, n$ we choose $\tilde{h}_i$ uniform with $\log_g \tilde{h}_i$ known so that we are simply running the OT execution honestly using input 0. Similarly, choose $\tilde{v}_{j,i}^{y_i}$ uniform with $\log_g \tilde{v}_{j,i}^{y_i}$ known for every evaluation circuit j. (Note that this allows P_2 to reveal $\log_g(\tilde{v}_{j,i}/\tilde{h}_i)$ in step 5 for every evaluation circuit j.)

This experiment is distributed *identically* to the previous experiment, since g^k and C/g^k (where k is uniform in each case) have the same distribution. (P_2 also gives a WI proof of knowledge of either $\log_g \tilde{h}_i$ or $\log_g(C/\tilde{h}_i)$, but we assume a *perfect* WI proof is used.)

To conclude, we observe that Experiment 6 can equivalently be described in terms of an ideal-world execution in which the honest P_2 and a simulator $\mathcal{S}$ (playing the role of the first party, and running P_1 as a subroutine) interact with a trusted party computing f. Namely, $\mathcal{S}$ works as follows:

1. Choose $\mathcal{J}$ in advance; this defines the check circuits and the evaluation circuits for the simulated P_2. Choose $\tilde{C} \leftarrow \mathbb{G}$ and send it to P_1. Receive in return $C \in \mathbb{G}$.

2. For each check circuit j, generate input-wire labels as in the real protocol. For each evaluation circuit j, choose $\tilde{a}_{j,1}, \ldots, \tilde{a}_{j,n} \leftarrow \mathbb{Z}_q$ and set $\tilde{v}_{j,i} = g^{\tilde{a}_{j,i}}$ for $i = 1, \ldots, n$. Also choose $\tilde{v}_{j,n+i} \leftarrow \{0,1\}^n$ for $i = 1, \ldots, n$. When P_2 acts as OT receiver, run the OT protocol honestly using input 0.

In each instance i in which P_2 acts as OT sender, extract from P_1 (by rewinding the WI proof of knowledge) either $\log_g h_i$ or $\log_g(\widetilde{C}/h_i)$. In the former case set $x_i = 0$ and in the latter case set $x_i = 1$. Then, for check circuits send the final OT message exactly as in the real protocol, and for any evaluation circuit j send the final OT message using $\widetilde{v}_{j,n+i}$ as the x_i-input, and the 0-string as the $(1 - x_i)$-input.

3. Send x to the trusted party, and receive in return an output z.
   Generate $\{\widetilde{\mathsf{pub}}_i^b, \widetilde{w}_{j,i}^b\}$ as in the real protocol. Then for each evaluation circuit j, compute $\widetilde{\mathsf{GC}}_j \leftarrow \mathsf{SimGC}\left(x, z, \{\widetilde{v}_{j,i}\}_{i=1}^{2n}, \{\widetilde{w}_{j,i}^{z_i}\}_{i=1}^{n}\right)$; for each check circuit j, compute $\widetilde{\mathsf{GC}}_j$ as in the real protocol. Send $\{\widetilde{\mathsf{GC}}_j\}_{j=1}^{\kappa}$ and $\{\widetilde{\mathsf{pub}}_i^0, \widetilde{\mathsf{pub}}_i^1\}_{i=1}^{n}$ to P_1.
   Receive in return $\{\mathsf{GC}_j\}_{j=1}^{\kappa}$ and $\{\mathsf{pub}_i^0, \mathsf{pub}_i^1\}_{i=1}^{n}$ from P_1.

4. For each check circuit j, compute $\{\widetilde{\mathsf{ComSet}}_{j,i}\}_{i=1}^{n}$ as in the real protocol. For each evaluation circuit j, set $\widetilde{\mathsf{ComSet}}_{j,i} = \{\mathsf{Com}(\widetilde{v}_{j,i}), \mathsf{Com}(g)\}$ in random permuted order. Send all these pairs of commitments to P_1, and receive in return all the pairs of commitments from P_1.

5. Give P_1 the value $\mathcal{J}$ as the output of the appropriate coin-tossing protocol. Respond for all check circuits as in the real protocol. For each evaluation circuit j, send $\{\widetilde{v}_{j,i}\}_{i=1}^{n}$, open the appropriate commitment from $\{\widetilde{\mathsf{ComSet}}_{j,i}\}_{i=1}^{n}$, and send $\{\log_g(\widetilde{v}_{j,i}/\widetilde{h}_i)\}_{i=1}^{n}$, where $\widetilde{h}_i$ is the message sent by P_2 in the ith OT when P_2 is receiver.
   Choose $\widetilde{\mathcal{J}}$ at random as in the real protocol, and give it to P_1. If P_1 responds correctly, then repeatedly rewind to find $\widetilde{\mathcal{J}}' \neq \widetilde{\mathcal{J}}$ for which P_1 responds correctly. (If none is found, $\mathcal{S}$ aborts with output fail.) Rewind again and continue the interaction using $\widetilde{\mathcal{J}}$.

6. Let j^* be a circuit which is an evaluation circuit in $\widetilde{\mathcal{J}}$, but a check circuit in $\widetilde{\mathcal{J}}'$. For $i = 1, \ldots, n$, use the $\kappa/2$ shares of $s_i^{z_i}$ from P_1's check circuits (with respect to $\mathcal{J}$) plus the additional share of $s_i^{z_i}$ from circuit j^* (that was a check circuit with respect to $\widetilde{\mathcal{J}}'$) to reconstruct $s_i^{z_i}$. Set $\widetilde{t}_i^{z_i} = s_i^{z_i}$.

7. For $i = 1, \ldots, n$, do the following.
   – If $z_i = 0$, obtain P_1's input $s_i^0 \| t_i^0$ to the first equality test. If $s_i^0 \| t_i^0 = \widetilde{t}_i^0 \| \widetilde{s}_i^0$, return 1; else return 0. Return 0 to the second equality test (if run).
   – If $z_i = 1$, return 0 to the first equality test. Then obtain P_1's input $s_i^1 \| t_i^1$ to the second equality test. If $s_i^1 \| t_i^1 = \widetilde{t}_i^1 \| \widetilde{s}_i^1$, return 1; else return 0.

   If for some i both equality tests return 0, abort. If an abort occurred, send abort to the trusted party; otherwise, send continue.

This completes the proof.

Acknowledgments. Yan Huang would like to thank Ivan Damgård and Jesper Nielsen for their helpful comments on this work.

References

1. Aumann, Y., Lindell, Y.: Security against covert adversaries: Efficient protocols for realistic adversaries. Journal of Cryptology 23(2), 281–343 (2010)
2. Bellare, M., Garay, J.A., Rabin, T.: Fast batch verification for modular exponentiation and digital signatures. In: Nyberg, K. (ed.) EUROCRYPT 1998. LNCS, vol. 1403, pp. 236–250. Springer, Heidelberg (1998)
3. Canetti, R.: Security and composition of multiparty cryptographic protocols. Journal of Cryptology 13(1), 143–202 (2000)
4. Damgård, I., Keller, M., Larraia, E., Miles, C., Smart, N.P.: Implementing AES via an actively/Covertly secure dishonest-majority MPC protocol. In: Visconti, I., De Prisco, R. (eds.) SCN 2012. LNCS, vol. 7485, pp. 241–263. Springer, Heidelberg (2012)
5. Damgård, I., Pastro, V., Smart, N.P., Zakarias, S.: Multiparty computation from somewhat homomorphic encryption. In: Safavi-Naini, R. (ed.) CRYPTO 2012. LNCS, vol. 7417, pp. 643–662. Springer, Heidelberg (2012)
6. Feldman, P.: A practical scheme for non-interactive verifiable secret sharing. In: 28th Annual Symposium on Foundations of Computer Science (FOCS), pp. 427–437. IEEE (1987)
7. Goldreich, O.: Foundations of Cryptography. Basic Applications, vol. 2. Cambridge University Press, Cambridge (2004)
8. Goldreich, O., Micali, S., Wigderson, A.: How to play any mental game, or a completeness theorem for protocols with honest majority. In: 19th Annual ACM Symposium on Theory of Computing (STOC), pp. 218–229. ACM Press (1987)
9. Goyal, V., Mohassel, P., Smith, A.: Efficient two party and multi party computation against covert adversaries. In: Smart, N.P. (ed.) EUROCRYPT 2008. LNCS, vol. 4965, pp. 289–306. Springer, Heidelberg (2008)
10. Henecka, W., Kögl, S., Sadeghi, A.-R., Schneider, T., Wehrenberg, I.: TASTY: tool for automating secure two-party computations. In: 17th ACM Conf. on Computer and Communications Security (CCS), pp. 451–462. ACM Press (2010)
11. Huang, Y., Evans, D., Katz, J.: Private set intersection: Are garbled circuits better than custom protocols? In: Network and Distributed System Security Symposium (NDSS). The Internet Society (2012)
12. Huang, Y., Evans, D., Katz, J., Malka, L.: Faster secure two-party computation using garbled circuits. In: 20th USENIX Security Symposium. USENIX Association (2011)
13. Huang, Y., Katz, J., Evans, D.: Quid pro quo-tocols: Strengthening semi-honest protocols with dual execution. In: IEEE Symposium on Security & Privacy, pp. 239–254. IEEE (2012)
14. Ishai, Y., Prabhakaran, M., Sahai, A.: Founding cryptography on oblivious transfer – efficiently. In: Wagner, D. (ed.) CRYPTO 2008. LNCS, vol. 5157, pp. 572–591. Springer, Heidelberg (2008)
15. Jarecki, S., Shmatikov, V.: Efficient two-party secure computation on committed inputs. In: Naor, M. (ed.) EUROCRYPT 2007. LNCS, vol. 4515, pp. 97–114. Springer, Heidelberg (2007)
16. Kiraz, M., Schoenmakers, B.: A protocol issue for the malicious case of yao's garbled-circuit construction. In: Proc. 27th Symp. on Information Theory in the Benelux, pp. 283–290 (2006)

17. Kolesnikov, V., Schneider, T.: Improved garbled circuit: Free XOR gates and applications. In: Aceto, L., Damgård, I., Goldberg, L.A., Halldórsson, M.M., Ingólfsdóttir, A., Walukiewicz, I. (eds.) ICALP 2008, Part II. LNCS, vol. 5126, pp. 486–498. Springer, Heidelberg (2008)

18. Kreuter, B., Shelat, A., Shen, C.: Billion-gate secure computation with malicious adversaries. In: 21st USENIX Security Symposium. USENIX Association (2012)

19. Lindell, Y.: Fast cut-and-choose based protocols for malicious and covert adversaries. In: Canetti, R., Garay, J.A. (eds.) CRYPTO 2013, Part II. LNCS, vol. 8043, pp. 1–17. Springer, Heidelberg (2013)

20. Lindell, Y., Oxman, E., Pinkas, B.: The IPS compiler: Optimizations, variants and concrete efficiency. In: Rogaway, P. (ed.) CRYPTO 2011. LNCS, vol. 6841, pp. 259–276. Springer, Heidelberg (2011)

21. Lindell, Y., Pinkas, B.: An efficient protocol for secure two-party computation in the presence of malicious adversaries. In: Naor, M. (ed.) EUROCRYPT 2007. LNCS, vol. 4515, pp. 52–78. Springer, Heidelberg (2007)

22. Lindell, Y., Pinkas, B.: A proof of security of Yao's protocol for two-party computation. Journal of Cryptology 22(2), 161–188 (2009)

23. Lindell, Y., Pinkas, B.: Secure two-party computation via cut-and-choose oblivious transfer. Journal of Cryptology 25(4), 680–722 (2012)

24. Lindell, Y., Pinkas, B., Smart, N.P.: Implementing two-party computation efficiently with security against malicious adversaries. In: Ostrovsky, R., De Prisco, R., Visconti, I. (eds.) SCN 2008. LNCS, vol. 5229, pp. 2–20. Springer, Heidelberg (2008)

25. Malkhi, D., Nisan, N., Pinkas, B., Sella, Y.: Fairplay — a secure two-party computation system. In: Proc. 13th USENIX Security Symposium, pp. 287–302. USENIX Association (2004)

26. Mohassel, P., Franklin, M.: Efficiency tradeoffs for malicious two-party computation. In: Yung, M., Dodis, Y., Kiayias, A., Malkin, T. (eds.) PKC 2006. LNCS, vol. 3958, pp. 458–473. Springer, Heidelberg (2006)

27. Naor, M., Pinkas, B.: Efficient oblivious transfer protocols. In: 12th SODA Annual ACM-SIAM Symposium on Discrete Algorithms (SODA), pp. 448–457. ACM-SIAM (2001)

28. Naor, M., Pinkas, B., Sumner, R.: Privacy preserving auctions and mechanism design. In: Proc. 1st ACM Conf. on Electronic Commerce, pp. 129–139. ACM (1999)

29. Nielsen, J.B., Nordholt, P.S., Orlandi, C., Burra, S.S.: A new approach to practical active-secure two-party computation. In: Safavi-Naini, R. (ed.) CRYPTO 2012. LNCS, vol. 7417, pp. 681–700. Springer, Heidelberg (2012)

30. Nielsen, J.B., Orlandi, C.: LEGO for two-party secure computation. In: Reingold, O. (ed.) TCC 2009. LNCS, vol. 5444, pp. 368–386. Springer, Heidelberg (2009)

31. Peikert, C., Vaikuntanathan, V., Waters, B.: A framework for efficient and composable oblivious transfer. In: Wagner, D. (ed.) CRYPTO 2008. LNCS, vol. 5157, pp. 554–571. Springer, Heidelberg (2008)

32. Pinkas, B., Schneider, T., Smart, N., Williams, S.: Secure two-party computation is practical. In: Matsui, M. (ed.) ASIACRYPT 2009. LNCS, vol. 5912, pp. 250–267. Springer, Heidelberg (2009)

33. Shelat, A., Shen, C.-H.: Two-output secure computation with malicious adversaries. In: Paterson, K.G. (ed.) EUROCRYPT 2011. LNCS, vol. 6632, pp. 386–405. Springer, Heidelberg (2011)

34. Woodruff, D.P.: Revisiting the efficiency of malicious two-party computation. In: Naor, M. (ed.) EUROCRYPT 2007. LNCS, vol. 4515, pp. 79–96. Springer, Heidelberg (2007)
35. Yao, A.C.-C.: How to generate and exchange secrets. In: 27th Annual Symposium on Foundations of Computer Science (FOCS), pp. 162–167. IEEE (1986)

A   Experimental Results

We describe some preliminary experimentals indicating that our protocol significantly outperforms the recent work of [18].

We implemented our protocol in Java using all the optimizations of Section 4.1. We evaluated the protocol at the 80-bit security level, which means in particular that (1) each party generates 84 garbled circuits, 42 of which are checked; (2) the length of all wire labels is 80 bits; and (3) we use an order-q subgroup of $\mathbb{Z}_p^*$ where $|p| = 1024, |q| = 160$. We ran experiments over a LAN using two laptops with Intel Core i7 2.4GHz processors. Note that 80-bit security was also used in the experiments of [18].

In typical settings where the number of gates in the underlying circuit is much larger than the number of inputs/outputs, the dominant overall cost of the protocol is the generation, sending, and checking of the garbled circuits. When each side uses only a single core, our protocol evaluates circuits at the rate of 1.4 ms/gate. By comparison, the implementation of Kreuter et al. [18] evaluates circuits at the rate of about 8 ms/gate when a single thread is used.

When each side utilizes two cores, our protocol evaluates circuits at the rate of 0.8 ms/gate; by comparison, the two-threaded execution in [18] achieved a rate of roughly 4 ms/gate. We do not gain a factor of 2 in performance by leveraging a second core in part because the parties are sometimes idle, and in part because of inter-thread interference (e.g., due to cache contention and dependence on shared hardware and I/O).

Our measured performance gains relative to [18] exceed the expected factor of 3. This may be due to differences in hardware or implementation, or the complexity of managing multiple threads in the implementation of [18] regardless of how many cores are being used.

The number of public-key operations used in our protocol scales linearly with the lengths of the inputs and outputs, though we stress again that in typical scenarios the number of gates is much larger than the number of inputs/outputs and so the overall performance impact of these public-key operations is small. Nevertheless, we measured performance of this aspect of our protocol as well. When each side uses a single core, our protocol processes inputs at the rate of 0.7 s/bit (our experiments assume the lengths of the parties' inputs are the same). Output is computed at the rate of 0.1 s/bit.

Garbled Circuits Checking Garbled Circuits: More Efficient and Secure Two-Party Computation

Payman Mohassel[1] and Ben Riva[2,*]

[1] University of Calgary, Canada
[2] Tel Aviv University, Israel

Abstract. Applying cut-and-choose techniques to Yao's garbled circuit protocol has been a promising approach for designing efficient Two-Party Computation (2PC) with malicious and covert security, as is evident from various optimizations and software implementations in the recent years. We revisit the security and efficiency properties of this popular approach and propose alternative constructions and a new definition that are more suitable for use in practice.

- We design an efficient fully-secure 2PC protocol for two-output functions that only requires $O(t|C|)$ symmetric-key operations (with small constant factors, and ignoring factors that are independent of the circuit in use) in the Random Oracle Model, where $|C|$ is the circuit size and t is a statistical security parameter. This is essentially the *optimal* complexity for protocols based on cut-and-choose, resolving a main question left open by the previous work on the subject. Our protocol utilizes novel techniques for enforcing *garbler's input consistency* and handling *two-output functions* that are more efficient than all prior solutions.
- Motivated by the goal of eliminating the *all-or-nothing* nature of 2PC with covert security (that privacy and correctness are fully compromised if the adversary is not caught in the challenge phase), we propose a new security definition for 2PC that strengthens the guarantees provided by the standard covert model, and offers a smoother security vs. efficiency tradeoff to protocol designers in choosing the *right deterrence factor*. In our new notion, correctness is always guaranteed, privacy is fully guaranteed with probability $(1-\epsilon)$, and with probability ϵ (i.e. the event of undetected cheating), privacy is only "partially compromised" with at most a *single bit* of information leaked, in *case of an abort*. We present two efficient 2PC constructions achieving our new notion. Both protocols are competitive with the previous covert 2PC protocols based on cut-and-choose.

A distinct feature of the techniques we use in all our constructions is to check consistency of inputs and outputs using new gadgets that are themselves *garbled circuits*, and to verify validity of these gadgets using *multi-stage* cut-and-choose openings.

* Research supported by the Check Point Institute for Information Security and an ISF grant.

R. Canetti and J.A. Garay (Eds.): CRYPTO 2013, Part II, LNCS 8043, pp. 36–53, 2013.

1   Introduction

Informally, a secure two-party protocol for a known function $f(\cdot, \cdot)$ is a protocol between Alice and Bob with private inputs x and y that satisfies the following two requirements: (1) *Correctness*: If at least one of the players is honest then the result should be the correct output of $f(x, y)$; (2) *Privacy*: No player learns any information about the other player's input, except for the function output.

Security is defined with respect to an adversary, who is *semi-honest* if the corrupted players always follow the protocol, is *malicious* if the players can arbitrarily deviate, and is *covert* in case a cheating player has an incentive not to be caught (or more specifically, any deviation can be detected with a constant probability).

A classical solution for the case of semi-honest players (i.e., players who do not deviate from the protocol) is to use a *garbled circuit* and *oblivious transfer* [21,12]: The resulting protocol is fairly efficient since computing each gate requires a constant number of symmetric-key encryptions. Furthermore, recent results show how to improve both the computation and communication cost of the garbling process (e.g., getting XOR gates for free [9], reducing communication [4,18], and designing tailored circuits [5]).

The case of malicious players is more complicated and less efficient. A classical solution is to use zero-knowledge proofs to verify that the players follow the protocol. However, the proofs in this case are rather inefficient. [8,16] show how to garble a circuit in such a way that these proofs can be instantiated more efficiently. Still, these constructions require a constant number of exponentiations per gate, making them inefficient for large circuits.

The Cut-and-Choose Approach. A slightly more explored direction is based on using the cut-and-choose method for checking the garbled circuit. (E.g., see implementations by [18,19,10].) Instead of sending only one (and possibly not properly constructed) garbled circuit, Alice sends t garbled circuits. Then, Bob asks her to *open* a constant fraction of them. For those circuits, Alice sends all the randomness she used in the garbling process. Bob can check that the opened circuits were indeed correctly garbled. If that is not the case, Bob knows that Alice has cheated and aborts. Otherwise, Bob evaluates the remaining garbled circuits and computes the majority output. It is shown in [13,19] that with high probability the majority of the evaluated garbled circuits are properly constructed.

However, the above cut-and-choose of the circuits is not sufficient to obtain a fully-secure 2PC. There are three well-known issues to resolve: (1) *Garbler's input consistency:* Since Bob evaluates many circuits, he needs assurance that Alice uses the same input in all of them. (2) *Evaluator's input consistency:* Alice can use different input labels in the oblivious transfers and in creation of the garbled circuits, in such a way that reveals Bob's input. (E.g., she can use invalid labels for the input bit 0 in the oblivious transfer, but valid ones for 1, causing Bob to abort if his input bit is 0.) (3) *Two-output functions:* There are cases in which the players want to securely compute two different functions f_1, f_2 where each party only learns his *own* output and is assured he has obtained the correct result.

When addressing these issues, the deciding efficiency factors are both the number and the type of additional cryptographic operations required. By *expensive operations*, we refer to cryptographic primitives that require exponentiations (e.g. oblivious transfer, or public-key encryption), and by *inexpensive operations* we mean the use of primitives that do not require exponentiations (e.g. symmetric-key encryption, commitments, or hashing). To simplify the exposition, from now on we omit small constants and complexities that are independent of the computation size or input length, unless said otherwise.

To address the first issue, i.e. how to make sure Alice is using the same input in all circuits, [14,11] present two methods that require $O(t^2 \cdot n_1)$ inexpensive cryptographic operations (commitments), where n_1 is the length of Alice's input, and t is the number of circuits we use in the cut-and-choose. ([20] shows how to reduce this asymptotic overhead, but with large constants even for small security parameters.) [14,13,19] show alternative methods that require $O(t \cdot n_1)$ expensive cryptographic operations (i.e. exponentiations). These consistency-checking mechanisms can lead to significant overhead. Recall that garbling of a single gate requires a constant number of symmetric encryptions, where the constant is 4 in most implementations. Thus, e.g. for $t = 130$, the price of checking consistency for a single input bit is roughly equivalent to the price of garbling several tens of additional gates in each circuit in the first method, and even more in the second. Moreover, the first method has a large communication overhead (e.g., for input size $n_1 = 500$ and $t = 130$, it requires several millions of commitments, with a total communication overhead of hundreds of megabytes).

To address the second issue, i.e. making sure Alice is using the same labels in her OT answers and the garbled circuits, [11] presents a method that requires $O(t \cdot \max(4n_2, 8t))$ expensive cryptographic operations (specifically, oblivious transfers), where n_2 is the length of Bob's input. [13,19] introduce alternative methods that require $O(t \cdot n_2)$ expensive cryptographic operations.

To address the last issue, of verifying the computation output, [11] proposes to apply a *one time MAC* to the output and XOR the result with a random input to hide the outcome (both are done as part of the circuit). However, this solution increases Alice's input with additional $q_1 + 2t$ input bits and increases the circuit size by $O(t \cdot q_1)$ gates, where q_1 is Alice's output length (i.e. overall overhead of $O(t^2 \cdot q_1)$ inexpensive operations). [19] suggests a solution that requires the use of digital signatures and a witness-indistinguishable proof, resulting in a total overhead of $O(t \cdot q_1)$ expensive operations.

In the covert setting [1] the techniques are similar, although the issue of the garbler's input consistency is not always relevant [4,1].

All-or-Nothing Security vs. Security with Input-Dependent Abort. All the cut-and-choose protocols discussed above provide an *all-or-nothing* guarantee, which means that both correctness and privacy are preserved with the same probability (the probability of getting caught in case of cheating), and are completely compromised if cheating is not detected. For example, in case of a protocol with covert security and deterrence factor of $1/2$, there is a 50% chance that the protocol reveals the honest party's input and provides him with an incorrect output.

This can become an obstacle to using covert security, in some practical scenarios. For example, the participants of an MPC protocol may not be able to afford the lack of correctness or privacy (even if only with a constant probability), due to the high financial/legal cost, or the loss of reputation.

[14] suggests an alternative to the all-or-nothing approach and designs a secure two-party protocol that always guarantees correctness but may leak one bit of information to a malicious party. While this security guarantee is weaker than the standard definition of security against covert/malicious adversaries, it ensures correctness and "partial privacy" even in case of successful cheating, making it a reasonable relaxation in some scenarios.

The idea behind the protocols of [14] is as follows: Alice garbles a circuit gc_1 and sends it to Bob, along with the labels of Alice's input-wires. They execute a fully-secure oblivious transfer protocol in which Bob learns the labels for his input-wires. Then, they run the same steps in the other direction, where Bob garbles gc_2 and Alice is the receiver. Next, each player evaluates the garbled circuit he or she received, resulting in output-wire label out_i (we require that the output-wire labels are the actual outputs concatenated with random labels). Last, each player computes the *supposed to be* concatenation $out_1 \circ out_2$. (Alice gets out_1 from her evaluation, and can determine the value of out_2 by herself. Bob does the same.) Now they run a protocol for securely testing whether their values $out_1 \circ out_2$ are the same. If they are indeed the same, they output b. Otherwise, they abort.

The resulting protocol is highly efficient, requiring only two garbled circuits and the associated oblivious transfers. (See [6] for an optimized variant of the protocol and its performance.) Since one of the players is honest, the result from his garbled circuit will be correct. Thus, if the honest party does not abort, the output is indeed correct. On the other hand, if one of the players is malicious, he can *always* learn one bit of information by observing whether the honest party aborts or not in the final equality test. We call this scenario *Input-Dependent Abort* (IDA) (following [7]).

1.1   Our Contributions

Given the discussion above, we put forth and answer the following two questions: (1) Can we improve on the efficiency of the existing solutions for checking input-consistency and handling two-output functions, to the extent that they are no longer considered a major computation/communication overhead? (2) Can we design cut-and-choose protocols that do not suffer from the all-or-nothing limitation of standard constructions but that provide better security guarantees than those of 2PC with input-dependent abort?

In the process of answering these questions, we introduce a set of new techniques to enforce consistency of inputs and outputs in garbled circuits. Interestingly, these techniques themselves employ *specially-designed garbled circuits* (gadgets) correctness of which is checked as part of a modified cut-and-choose process containing multiple opening stages.

Fully-Secure 2PC Based on Cut-and-Choose with Small Overheads.
Towards answering the first question, we propose new and efficient solutions
for the three problems of (1) Garbler's input consistency (2) Evaluator's input
consistency and (3) Handling two-output functions, that asymptotically and
concretely improve on all previous solutions.

First, we show how to use garbled *XOR-gates* to efficiently enforce the garbler's input consistency, while requiring only $O(t \cdot n_1)$ inexpensive operations.
This approach asymptotically improves the solutions in [14,11], and only requires
inexpensive operations in contrast to the solution of [19]. Second, we observe that
the solution of [11] to the evaluator's input consistency issue can be improved by
combining it with the OT extension of [15] and the Free-XOR technique of [9].
The resulting protocol requires only $O(t \cdot \max(4n_2, 8t))$ inexpensive operations.
Third, we show how to use garbled *identity-gates* to efficiently solve the two-
output function problem, while requiring only $O(t \cdot q_1)$ inexpensive operations,
where q_1 is the garbler's output length, improving on the recent construction
of [19] which requires the same number of expensive operations. The resulting
2PC protocol is constant round and asymptotically better than all previous constructions based on the cut-and-choose method [14,11,13,19] (except for [20],
which is impractical due to large constants). In Table 1, we compare the protocol's complexity with previous constructions. We stress that the efficiency of
our protocol relies on the efficient OT extension of [15], which allows one to
efficiently extend a small number of OTs to n OTs with the price of only $O(n)$
invocations of a hash function. The protocol of [15] is in the Random Oracle
Model (ROM) and our construction inherits the same weakness. (Besides using
ROM for the OT-extension of [15], in some of our techniques we show two alter-
natives: A more efficient instantiation in the ROM, and one without the ROM
requirement, which still is more efficient than current techniques.)

We remark that our proposed solutions can be modified to work with any of
the existing garbled-circuit optimization techniques of [9,4,18,5,10].

Furthermore, in the full version of this paper we describe how to use our
techniques to construct a fully-secure 2PC protocol for the case where y is not
private, using only a single garbled circuit. This scenario which we call *authenticated computation with private inputs* naturally arises in applications such as
anonymous credentials or targeted advertising.

Table 1. Comparison of different fully secure 2PC protocols. n_i is the length of P_i's
input, q_1 is the length of P_1's output, and t is a statistical security parameter (where
t garbled circuits are used in the cut-and-choose). The number of base OTs in the OT
extension is omitted as it is independent of the circuit and input sizes.

	P_1's input	P_2's input	Two-output Overhead
[11]	inexpensive$(t^2 n_1)$	expensive$(\max(4n_2, 8t))+$ inexpensive$(t \cdot \max(4n_2, 8t))$	inexpensive$(t^2 q_1)$
[13,19]	expensive$(t n_1)$	expensive$(t n_2)$	expensive$(t q_1)$
Our protocol	inexpensive$(t n_1)$	inexpensive$(t \cdot \max(4n_2, 8t))$	inexpensive$(t q_1)$

Our main contributions are the new techniques we use for solving the Garbler's input consistency issue and handling two-output functions. Next, to give a flavor of our techniques, we present the ideas behind our solutions.

Multi-Stage Cut-and-Choose and Handling Two-Output Functions. From now on we denote by P_1 the garbler (Alice), and by P_2 the evaluator (Bob). Note that the main difficulty here is to convince the garbler, P_1, that the output he receives is correct. (Privacy of the output is easily achieved by xoring the output with a random string.)

A common method for authenticating the output of a garbled circuit is to send the random labels resulted from the evaluation of the garbled circuit. However, when we use the cut-and-choose method, many circuits are being evaluated, and sending the labels for all the garbled circuits can leak secret information (e.g., P_1 can create a single bad circuit that simply outputs P_2's input, and not get caught with high probability). We can fix this issue by using the same output-wire labels in all the garbled circuits, but then we would lose our authenticity guarantee since P_2 learns all the output-wire labels from the opened circuits and can use that information to tamper with the output of the evaluated circuits.

We propose a workaround that allows us to simultaneously use the same output-wire labels in all circuits, and preserve the authenticity guarantee, in cut-and-choose 2PC. We separate the "cut" step from the "opening" step (this is a recurring idea in all our constructions). After P_1 sends the t garbled circuits, P_2 picks a random subset S which he wants to check and sends it to P_1. Then, instead of opening the garbled circuits in S, they proceed to the evaluation of the rest of the garbled circuits. I.e., P_1 sends the labels of his input-wires for the garbled circuits not in S; P_2 evaluates all of them and takes the majority; he then commits to the output along with the corresponding output-wire labels. (Note that since the opening step is not performed yet, P_2 cannot guess the unknown output-wire labels and commit to the wrong output). Now, they complete the cut-and-choose and do the opening step: P_1 sends the randomness he used for all the garbled circuits in S, and P_2 verifies that everything was done correctly. If so, P_2 decommits the output and reveals to P_1 the actual output and its output-wire labels. To summarize, since P_1 learns the output only after P_2 has verified the garbled circuits, he cannot cheat in this new cut-and-choose strategy, any differently than he could in regular cut-and-choose. On the other hand, since P_2 is committed to his output before the opening, he cannot change the output after he sees the opened circuits.[1]

---

[1] We note that the above solution is not enough. First, the commitment in use must be non-malleable with respect to the garbled circuits being opened. E.g., consider a garbling scheme that outputs also commitments of the possible output-wire labels; P_2 could use one of those commitments as his commitment and later use the information he learned from the opening to decommit successfully. Second, the commitment has to be equivocal to allow us to later simulate P_2's message. Both requirements can be solved in the plain model by using trapdoor commitments [3] and efficient Zero-Knowledge Proof of Knowledge (ZKPoK), or in the Random Oracle Model, by committing using a hash function. The first solution requires $O(q_1)$ expensive operations while the second requires only one call to the hash function.

The above solution can be applied to most previous 2PC protocols based on cut-and-choose to obtain their two-output variants. But, since the circuit checking is done after the circuit evaluation, the above solution falls short when combined with circuit streaming or parallelized garbling techniques [5,10]. In the full version of this paper we describe a second variant of this protocol that is compatible with those techniques. The cost of this variant is only additional $t \cdot q_1$ commitments.

XOR-Gadgets and Garbler's Input Consistency. Here, our goal is to make sure P_1 uses the same input in all (or at least most of) the evaluated garbled circuits. Observe that we do not have the same issue with P_2's input since for each specific input bit, P_2 learns the t corresponding input-wire labels using a single OT. But, since P_1 does not use OT to learn the labels for his input-wires, the same approach does not work here.

First, we augment the circuit C being computed with a small circuit we call an *XOR-gadget.* Say we want to compute the circuit $C(x, y)$ where x is P_1's input, and y is P_2's. Instead of working with C, the players work with a circuit that computes $C_1(x, y, r) = (C(x, y), x \oplus r)$, where r is a random input string of length $|x|$ generated by P_1. Note that x is kept private from P_2 if r is chosen randomly. Denote P_1's inputs to the t garbled circuits of C_1 by $x_1^1, x_2^1, \ldots, x_t^1$ and $r_1^1, r_2^1, \ldots, r_t^1$. If P_1 is honest, the r_i^1-s are chosen independently at random while all the x_i^1-s are equal to x.

Let $C_2(x, r) = x \oplus r$, where x and r are P_1's inputs of the same length. (Note that y is not an input here.) In addition to P_1's garbled circuits, P_2 also generates t XOR-gadgets, which are garbled circuits of C_2. These garbled XOR-gadgets will be evaluated by P_1 and on his own inputs. (For simplicity, we assume for now that P_2 is semi-honest.) Denote P_1's inputs to these t garbled circuits by $x_1^2, x_2^2, \ldots, x_t^2$ and $r_1^2, r_2^2, \ldots, r_t^2$. If P_1 is honest, then $r_i^1 = r_i^2$ for all i, and all the x_i^2-s are equal to P_1's actual input x.

We enforce that x_i^1-s are the same in the majority of the evaluated circuits, using a combination of *three different checks*: (1) Check that P_1 uses the same value x' for all x_i^2-s. We can easily enforce this since P_1 learns the input-wire label for each bit using a single OT. (E.g., if the first bit of x' is zero, P_1 will learn t concatenated labels that correspond to the bit zero in the t XOR-gadgets P_2 prepared.) (2) Check that $(x_i^2 + r_i^2) = (x_i^1 + r_i^1)$ in all the evaluated circuits. We enforce this by evaluating the two XOR-gadgets corresponding to the i-th garbled circuit (one created by P_1 and one created by P_2), and checking the equality of their outputs (see Section 3 for subtleties that need to be addressed when doing so). (3) Check that $r_i^1 = r_i^2$ in the majority of the evaluated circuits. We enforce this as part of the cut-and-choose: When P_1 sends his garbled circuits, he also sends the labels that correspond to all r_i^1-s. After P_1 learns the labels for r_i^2-s (from the OTs), they do the opening phase and P_1 opens the subset of garbled circuits. In addition, for each opened circuit, P_1 reveals the labels of the r_i^2-s he learned, and P_2 verifies that $r_i^1 = r_i^2$. (Note that once P_1 sends the labels of r_i^1 and the garbled circuit, he cannot change r_i^1. On the other hand, P_1 cannot fake a valid label for r_i^2 that is different from the one he learned in the

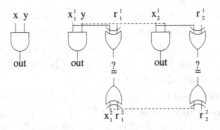

Fig. 1. Example of garbling the simple AND circuit on the left that computes the AND between P_1's bit x and P_2's bit y. P_1 garbles the upper circuits and P_2 the lower ones. Specifically, P_1 garbles two AND circuits (i.e., $t = 2$) and two XOR-gates, and P_2 garbles two XOR-gates. P_2's input is the same for all garbled circuits because of the OT (the top dashed line). Recall that the first input P_1 learns in all of P_2's XOR-gates is the same since P_1 learns the corresponding input-wire labels from the OT (the lower dashed line). Also, that the equality of r_i^1 and r_i^2, $i = 1, 2$, is checked in the cut-and-choose (e.g., by P_1 revealing the labels of r_1^1 and r_1^2 if P_2 picked to check the first set) and hence holds with high probability. Combining these two observations with the fact that P_2 compares the outputs of the XOR-gates, P_2 gets the assurance that $x_1^1 = x_2^1$.

OTs.) As a result, P_2 knows that with high probability (in terms of t) $r_i^1 = r_i^2$ in the majority of the evaluated circuits.

It is easy to see that the above three checks imply (with high probability) that x_i^1-s are the same in the majority of the evaluated circuits. Since P_2 outputs the majority result, this is sufficient for our needs.

Figure 1 shows an example of the above technique for the circuit that computes AND and $t = 2$. We stress that the above is only part of our techniques, and in particular, does not guarantee protection against a malicious P_2.

Security with Input-Dependent Abort in Presence of Covert Adversaries. We propose a new security notion that naturally combines security with input-dependent abort of [7] (alternatively, security with limited leakage of [14,6]), with security against covert adversaries [1]. The resulting security guarantee, denoted by ϵ-CovIDA, is a *strict strengthening of covert security*: In covert security, with probability ϵ both correctness and privacy are gone! Our definition always guarantees correctness, and with probability ϵ, privacy is only "slightly compromised", i.e. only a *single* bit of information may be leaked in case of an abort.

We stress that simply combining the protocols of [14,6] with the cut-and-choose method is *not* secure under our definition. Say that instead of garbling a single circuit, each player P_i garbles t circuits $gc_1^i, \ldots, gc_t^i$ and sends them to the other player. Players pick a random value $e \in [t]$, *open* all the circuits $gc_{j \neq e}^i$ (i.e., reveal the randomness used to generate them), and verify that they were constructed properly. This assures that with probability $1 - 1/t$, the remaining two circuits (one circuit from each player) is properly constructed. Parties then engage in the dual-execution protocol discussed above using these two garbled circuits. Although this protocol guarantees correctness similar to [14,6], it does not satisfy our security definition. One problem is that a malicious player can use different inputs for the two evaluated circuits, and learn whether their outputs are the same or not based on the final outcome. This attack is successful even if *all* the circuits are constructed correctly.

We show two constructions that do achieve our definition. Both constructions require a constant number of rounds. In our first construction, each player garbles only $\frac{1}{\epsilon}$ circuits and $\frac{n+2q}{\epsilon}$ additional XOR gates, where n is the length of the input and q is the length of the output. We emphasize that compared to the protocols of [14,6], where the adversary can *always* learn one bit of information, our protocol leaks one bit only with probability ϵ.

The first construction is sufficient for large values of ϵ but fails to scale for the smaller ones. For example, if one aims for a probability of leakage of less than 2^{-10}, the first protocol would require the exchange of a thousand garbled circuits. A more desirable goal is a protocol with a cost that grows only logarithmically in $\frac{1}{\epsilon}$. We achieve this in our second protocol.

The costs of both constructions are roughly the costs of running their covert counterparts in both directions. E.g. the second protocol requires $O(2\log(\frac{1}{\epsilon})(|C| + n + q))$ inexpensive operations and $O(\log(\frac{1}{\epsilon})(n + q))$ expensive ones, while the covert protocol of [13] requires $O(\log(\frac{1}{\epsilon})|C|)$ inexpensive operations and $O(\log(\frac{1}{\epsilon})n)$ expensive ones.

2 Preliminaries

Throughout this work we denote by t a statistical security parameter and by s a computational security parameter. For a fixed circuit in use, we denote by INP_i the set of indexes of P_i's input-wires to the circuit, by INP the set $\text{INP}_1 \cup \text{INP}_2$, by OUT_i the set of indexes of P_i's output-wires, and by OUT the set $\text{OUT}_1 \cup \text{OUT}_2$. For shortening, we sometimes refer to $|\text{INP}_i|$ by n_i, to $|\text{OUT}_i|$ by q_i, and set $n = n_1 + n_2$ and $q = q_1 + q_2$.

Denote by $\text{Enc}(sk, m)$ the encryption of message m under secret key sk, by $\text{PRG}(s, l)$ the l-bit string generated by a pseudo-random generator with seed s, and by $\text{Com}(m, r)$ the commitment on message m using randomness r. The decommitment of $\text{Com}(m, r)$ is m and r. (In some cases we use the abbreviations $\text{PRG}(s)$ and $\text{Com}(m)$.)

We also use the following notation for the next cryptographic primitives and functionalities.

Yao's Garbling. For the sake of simplicity and generality, we do not go into the details of the garbling mechanism and only introduce the notations we need to describe our protocols. We refer the reader to [12,2] for different approaches to creating the garbled circuits.

Given a garbled circuit gc, we denote by $\text{label}(gc, j, b)$ the label of wire j corresponding to bit value b. Also, we denote by $\text{Garb}(C, r)$ the (deterministic) garbling of circuit C using randomness r. (In practice, r would be a short seed for a pseudo-random function). For simplicity, we assume that the labels of the circuit's output-wires include also the actual output bits (thus, allowing the evaluator to learn the output).

We require the garbling scheme to be *private* and *authenticated*, meaning that given a garbled circuit and input labels of a specific input, nothing is revealed except for the output of the circuit, and, that the output-wire labels authenticate the actual output (thus, the actual output cannot be forged). Also, we require that given a garbled circuit and an input label, one can verify whether the input label is a valid input label.

Batch Committing Oblivious Transfer (BCOT). Here, sender S has n sets, each of m pairs of inputs, $\{(x_0^{j,z}, x_1^{j,z})\}_{j=1...n, z=1...m}$, and receiver R has a vector of input bits $\bar{b} = (b_1, \cdots, b_n)$. The receiver R learns the outputs according to his input bits, $x_{b_j}^{j,z}$ for all j and z. In addition, R learns commitments on *all* the sender's inputs.

[19] shows an implementation of BCOT with a cost of $O(mn)$ expensive operations. Combining their protocol with the OT-extension of [15] in the Random Oracle Model results in an alternative protocol that requires only $O(s)$ expensive operations and $O(nm)$ inexpensive ones. However, in the latter construction, the commitments on the sender's inputs cannot be opened separately and one needs to decommit all the inputs at once (we use both instantiations in our protocols). See the full version for more details. We denote the first protocol by BCOT1 and the second by BCOT2.

Two-Stage Equality Testing. In this protocol, player P_1 has input x_1 and player P_2 has input x_2. They want to test whether $x_1 = x_2$. The functionality is split into two stages in order to emulate a commitment on the inputs before revealing the result (we will use this property in one of our constructions). I.e., in the first stage players submit their inputs and learn nothing, and in the second stage, only if they both ask for the output, they receive the result. This functionality can be realized using ElGamal encryption and ZKPoKs.

3 An Efficient 2PC for Two-Output Functions with Full Security

In this section, we review the main ideas behind our efficient 2PC protocol with full security against malicious adversaries, considering the case where only P_2 needs to learn the output. In the full version we show how to extend the ideas in order to handle two-output functions. A detailed description of the protocol and the proof of security appear in full version as well.

Consistency of the evaluator's input is taken care of by combining the technique of [11] with the OT-extension of [15] and the Free-XOR technique [9]. In a nutshell, P_2's input is encoded using $\max(4n_2, 8t)$ bits in a way that any leakage of less than t of the bits does not reveal meaningful information about P_2's input. During the cut-and-choose, P_2 asks P_1 to reveal all his inputs to the OTs. If some of the inputs are not consistent with the one P_2 has learned from the OTs, P_2 aborts. This abort leaks information only in case P_1 guessed successfully more than t bits in P_2's encoded input. However, this can happen with only a negligible probability by the way the encoding is done.

As we discussed in Section 1.1, consistency of the garbler's input is addressed using the XOR-gadgets. In the following we describe the main steps of that part.

Garbling stage and the XOR-gadgets. Say the players want to compute $C(x, y)$, where x is P_1's input and y is P_2's input. Based on C, we define the following two circuits: (1) $C_1(x, y, r)$, which computes $(C(x, y), x \oplus r)$ where r is a random input string of length $|x|$ selected by P_1; (2) $C_2(x, r)$, which computes $x \oplus r$, where x and r are P_1's inputs and are of the same length. In both circuits we assume the indexes of the input-wires are the same as in C and we define the function $\alpha(k)$ to be the function that given $k \in \mathrm{INP}_1$ returns the index of the input-wire of the random bit that is xored with input-wire k. (For simplicity, we assume the same function is applicable for both C_1 and C_2.)

P_1 picks a random string z_i and generates a garbled circuit $gc_i = \mathsf{Garb}(C_1, z_i)$, for $i = 1 \ldots t$. In addition, P_2 picks a random string z_i' and generates a garbled circuit $xg_i = \mathsf{Garb}(C_2, z_i')$, for $i = 1 \ldots t$. Both players send the garbled circuits they created to each other. Next, P_1 picks r_j at random for $j \in [t]$ and sends to P_2 the labels that correspond to r_j in gc_j.

OTs for Input Labels. Parties execute OTs and BCOTs in order for each to learn the input-wire labels for his inputs in the circuits/gadgets created by his counterpart. More specifically, first they run any simulatable OT protocol with the OT-extension of [15], where P_1 acts as the sender and P_2 acts as the receiver. They use the technique of [11] for protecting against inconsistent inputs as described earlier. P_1's inputs are the labels of P_2's input-wires in all gc_j (i.e., the inputs are $\mathsf{label}(gc_j, k, 0)$ and $\mathsf{label}(gc_j, k, 1)$ for $k \in \mathrm{INP}_2$ and $j \in [t]$). P_2's input is his actual input. (We ignore here the details of encoding P_2's input.) Second, they execute BCOT2 twice where P_2 acts as the sender and P_1 acts as the receiver: (1) P_2's inputs are the labels of the input-wires in his XOR-gadgets xg_j, and P_1's inputs are his random input and actual input to the gadget (i.e., P_2 inputs are $\mathsf{label}(xg_j, k, 0)$ and $\mathsf{label}(xg_j, k, 1)$ while P_1's inputs are his actual input bits, and (2) P_2's inputs are $\mathsf{label}(xg_j, \alpha(k), 0)$ and $\mathsf{label}(xg_j, \alpha(k), 1)$ while P_1's inputs are the bits of r_j. Note that in the first BCOT2, P_1 inputs a single bit for each input bit and receives t input-wire labels. That restricts him to use the same input in all the XOR-gadgets.).

(In the detailed protocol, the players execute the OTs before sending the garbled circuits. Still, the intuition is similar.)

We stress that P_1 is yet to send the labels for his input wires in the circuits he garbled himself, i.e. gc_i-s.

Cut-and-Choose (first stage). After the OTs/BCOTs, P_1 opens a constant fraction of his garbled circuits/gadgets. In particular, P_1 opens the garbled circuit gc_j for all $j \notin E$, where E is chosen randomly using a joint coin-tossing protocol. (A joint coin-tossing protocol is needed for the simulation to work.) Moreover, P_1 reveals the random strings r_j-s he used in the opened circuits (by showing the labels he learned from BCOT2), and all his inputs to the OTs for the opened circuits. P_2 checks the correctness of the opened circuits and verifies that the same r_j was used in both gc_j and xg_j for all $j \notin E$. (He also verifies that the values he has received in the OTs for his inputs are consistent with what P_1 revealed, following the technique of [11].)

Cut-and-Choose (second stage). P_1 evaluates all the XOR-gadgets he received from P_2, and sends a commitment on all the output-wire labels he obtained to P_2. P_2 answers with opening *all* the XOR-gadgets xg_j for $j \in E$, and by decommitting all his inputs to BCOT2. P_1 checks that all the XOR-gadgets he received were properly constructed, and that the labels are consistent with the decommitments. If so, P_1 decommits the output-wire labels of the XOR-gadgets to P_2.

In general, the last step is not sound for all commitments since P_1 can send a commitment for which he does not know the corresponding message and later be able to decommit once P_2 opens the XOR-gadgets (see Footnote 1). There are several ways to overcome this issue. One option is to require P_1 to *prove* that he *knows* how to construct this commitment, or more formally, P_1 commits on the output labels with $\mathsf{Com}(labels, r)$ and proves using a ZKPoK that he knows *labels* and r. This step can be implemented efficiently for Pedersen's commitment [17], requiring only a small constant number of exponentiations. (When *labels* is longer than the commitment input length, P_1 picks a random seed *seed*, sends $\mathsf{Com}(seed, r)$, $\mathsf{PRG}(seed) \oplus labels$ and ZKPoK that he knows *seed* and r.) A more efficient option is to implement $\mathsf{Com}(labels, r)$ in the Random Oracle model using $\mathsf{H}(key \circ labels \circ r)$, where the commitment key *key* is chosen at random by the receiver (i.e., P_2 in our case). The complexity in this case is only a single call to the random oracle.

Circuit Evaluation. P_1 sends to P_2 the labels of his inputs for the remaining garbled circuits and XOR-gadgets. P_2 uses them to evaluate all his remaining circuits and gadgets. He checks that the output-wires of the XOR-gadgets are the same as the values P_1 sent him. If so, he takes the majority of the outputs to be his output.

Summary. Note that now, with high probability, not only do we know that the majority of the circuits being evaluated are correct, but also that P_1 used the same r_j-s in the XOR-gadget pairs (Check 3 from introduction). Also, recall that in the BCOT for XOR-gadgets created by P_2, P_1 can learn the labels for exactly one possible value of x. Thus, his x is the same for all the t XOR-gadgets P_2 generated (Check 1). Combined with the fact that P_2 checks equality of the output of the XOR-gadget pairs (Check 2), he is ensured that the same input bits are being used in majority of the gc_j-s. See Figure 1 for a diagram explaining the above intuition.

4 Security with Input-Dependent Abort in the Presence of Covert Adversaries

4.1 The Model

Following [11,1,6], we use the ideal/real paradigm for our security definition.

Real-Model Execution. The real-model execution of protocol Π takes place between players (P_1, P_2), at most one of whom is corrupted by a probabilistic polynomial-time machine adversary $\mathcal{A}$. At the beginning of the execution, each party P_i receives its input x_i. The adversary $\mathcal{A}$ receives an auxiliary information aux and an index that indicates which party it corrupts. For that party, $\mathcal{A}$ receives its input and sends messages on its behalf. Honest parties follow the protocol.

Let $\text{REAL}_{\Pi, \mathcal{A}(aux)}(x_1, x_2)$ be the output vector of the honest party and the adversary $\mathcal{A}$ from the real execution of Π, where aux is an auxiliary information and x_i is player P_i's input.

Ideal-Model Execution. Let $f : (\{0,1\}^*)^2 \to \{0,1\}^*$ be a two-party functionality. In the ideal-model execution, all the parties interact with a trusted party that evaluates f. As in the real-model execution, the ideal execution begins with each party P_i receiving its input x_i, and $\mathcal{A}$ receives the auxiliary information aux. The ideal execution proceeds as follows:

Send inputs to trusted party: Each party P_1, P_2 sends x_i' to the trusted party, where $x_i' = x_i$ if P_i is honest and x_i' is an arbitrary value if P_i is controlled by $\mathcal{A}$.

Abort option: If any $x_i' = \text{abort}$, then the trusted party returns abort to all parties and halts.

Attempted cheat option: If P_i sends $\text{cheat}_i(\epsilon')$, then:

- If $\epsilon' > \epsilon$, the trusted party sends corrupted_i to all parties and the adversary $\mathcal{A}$, and halts.
- Else, with probability $1 - \epsilon'$ the trusted party sends corrupted_i to all parties and the adversary $\mathcal{A}$ and halts.
- With probability ϵ',
  - The trusted party sends undetected and $f(x_1', x_2')$ to the adversary $\mathcal{A}$.
  - $\mathcal{A}$ responds with an arbitrary boolean function g.
  - The trusted party computes $g(x_1', x_2')$. If the result is 0 then the trusted party sends abort to all parties and the adversary $\mathcal{A}$ and halts. (i.e. $\mathcal{A}$ can learn $g(x_1', x_2')$ by observing whether the trusted party aborts or not.)

Otherwise, the trusted party sends $f(x_1', x_2')$ to the adversary.

Second abort option: The adversary sends either abort or continue. In the first case, the trusted party sends abort to all parties. Else, it sends $f(x_1', x_2')$.

Outputs: The honest parties output whatever they are sent by the trusted party. $\mathcal{A}$ outputs an arbitrary function of its view.

Let $\text{IDEAL}^{\epsilon}_{f,\mathcal{A}(aux)}(x_1, x_2)$ be the output vector of the honest party and the adversary $\mathcal{A}$ from the execution in the ideal model.

Definition 1. *A two-party protocol Π is secure with input-dependent abort in the presence of covert adversaries with ϵ-deterrent (ϵ-CovIDA) if for any probabilistic polynomial-time adversary $\mathcal{A}$ in the real model, there exists a probabilistic polynomial time adversary $\mathcal{S}$ in the ideal model such that*

$$\left\{ \text{REAL}_{\Pi, \mathcal{A}(aux)}(x_1, x_2) \right\}_{x_1, x_2, aux \in \{0,1\}^*} \stackrel{c}{\approx} \left\{ \text{IDEAL}^{\epsilon}_{f, \mathcal{S}(aux)}(x_1, x_2) \right\}_{x_1, x_2, aux \in \{0,1\}^*}$$

for all $|x_1| = |x_2|$ and aux.

Comparison with Covert Security. When we let $\epsilon = 1/t$ for any constant t, the above definition is strictly stronger than the standard definition of security against covert adversaries. In covert security, in case of undetected cheating which happens with probability ϵ, the adversary learns *all* the honest parties' private inputs and is able to change the outcome of computation to *whatever* value it wishes (i.e. no privacy or correctness guarantee). In our definition, however, the adversary can learn at most a single bit of information (from the abort), and under no condition is able to change the output (full correctness).

In the above definition, in contrast to the standard covert security, the adversary can choose the exact probability he gets caught (i.e. $1 - \epsilon'$) as long as this probability is larger than $1 - \epsilon$ (where ϵ is the deterrence factor). Note that this is not a relaxation in security since the adversary can only increase the probability of itself getting caught. We believe that this variant of the definition where the adversary can choose $\epsilon' > \epsilon$ with which it can get caught is of independent interest. Specifically, it yields an alternative definition for covert security that is more convenient to use in simulation-based proofs. (To obtain this alternative definition for covert security, replace the steps that are done with probability ϵ' with the following: (1) The trusted party sends x_1', x_2' to $\mathcal{A}$; (2) $\mathcal{A}$ sends the value y to the trusted party, and the trusted party sends it to all parties as their output.)

A Remark on Adaptiveness of Leakage Function. In the above definition, the leakage function g can be chosen adaptively after seeing $f(x_1', x_2')$. Somewhat surprisingly, this does not give any extra power to the adversary compared to the non-adaptive case since even in the non-adaptive case, g can be chosen to be a function that computes $f(x_1', x_2')$, emulates the adversary's computation given that value and evaluates the leakage function he would have chosen in the adaptive case.

4.2   An Efficient Protocol with $\frac{2}{\epsilon}$ Circuits

In this section, we review the main steps of our ϵ-CovIDA protocol and highlight the new techniques. A detailed description of the protocol and how to reduce the number of circuits (from linear in $\frac{1}{\epsilon}$ to logarithmic) appear in the full version.

As discussed in the introduction, in the dual-execution protocol of [14,6] parties engage in two different executions of the semi-honest Yao's garbled circuit protocol, and then run an equality testing protocol to confirm that the outputs of the two executions are the same before revealing the actual output values. We show how to extend this protocol to work in the presence of covert adversaries using the ideas presented in Section 3. For simplicity of the description, from now on we work with $t = \frac{1}{\epsilon}$ (a statistical security parameter) instead of ϵ since t would be the number of circuits each party garbles.

Dual-Execution & Cut-and-Choose. Our first step is to combine the dual-execution protocol with a standard cut-and-choose protocol for covert players. Each player P_i garbles t circuits $gc_1^i, \ldots, gc_t^i$ and sends them to the other player. Parties pick a random value $e \in [t]$, *open* all the circuits $gc_{j \neq e}^i$ and verify that they were constructed properly. This assures that with probability $1 - 1/t$, the remaining *circuit-pair* (gc_j^1, gc_j^2) is properly constructed. As before, they send the garbler's input-wire labels for the e-th circuit, execute OTs for the respective evaluators to learn their input-wire labels, evaluate the circuits, call the Equality Testing functionality and output accordingly.

The above protocol would guarantee correctness similar to the dual-execution protocol, and it would ensure that the evaluated circuits are correct with probability $1 - 1/t$. However, the protocol does not satisfy our security definition. One issue is that a malicious player learns the output of the computation even if the other player catches him cheating (as a result of the equality test). We show how this can be avoided by masking the output of the computation with random strings, chosen by the two players, and revealing them at the end of the computation in order to unmask the actual output.

A more subtle attack to address is that a malicious player can learn one bit of information about an honest party's input with probability greater than $1/t$ (in fact with probability 1): a malicious player can use different inputs in each of the two evaluated circuits, and learn whether their outputs are the same or not based on the final outcome. This attack is successful even if *all* the circuits are constructed correctly. We prevent this attack using the XOR-gadget techniques discussed earlier, along with some enhancements. We discuss the details next:

XOR-Gadgets. Define $C(x \circ m_1, y \circ m_2)$ to be the circuit that receives inputs x, y and two masks m_1, m_2 and computes $f(x, y) \oplus m_1 \oplus m_2$. Based on C, let P_1's input x' be $x \circ m_1$ and P_2's input y' be $y \circ m_2$, where m_i is a random string of length q (f's output length) selected by P_i. We define the following four circuits: (1) $C_1(x', y', r_1) = (C(x', y'), x' \oplus r_1)$, where r_1 is a random input string of length $|x'|$ selected by P_1; (2) $C_2(x', y', r_2) = (C(x', y'), y' \oplus r_2)$ where r_2 is a random input string of length $|y'|$ selected by P_2; (3) $C_1'(y', r_2) = y' \oplus r_2$ evaluated by P_2 on his own inputs; (4) $C_2'(x', r_1) = x' \oplus r_1$ evaluated by P_1 on his own inputs; In all circuits we assume the indexes of the input-wires are the same as in C and we define the function $\alpha(k)$ to be the function that given $k \in$ INP returns the index of the input-wire of the random bit input-wire that is xored with input-wire k. (For simplicity, we assume the same function is applicable for all C_i-s and C_i'-s.)

Instead of garbling C, each player P_i generates and sends t garbled circuits for C_i: $gc_1^i, \ldots, gc_t^i$ and t garbled circuits of C_i': $xg_1^i, \ldots, xg_t^i$. After sending the sets of garbled circuits, for each $j \in [t]$, player P_i picks at random a string r_j^i and sends the input-wire labels that correspond to r_j^i in gc_j^i.

OTs for Input Labels. Then, they execute BCOTs in order to learn the input-wire labels for both their actual inputs and the r_j^i-s in their counterpart's circuits. More specifically, first they use BCOT1 where P_1 acts as the sender and P_2 acts as the receiver. P_1's inputs are the input-wire labels of P_2's input-wire k in all gc_j^1-s and xg_j^1-s (i.e., the input pairs are $\left(\mathsf{label}(gc_j^1, k, 0), \mathsf{label}(gc_j^1, k, 1)\right)_{j \in [t]}$ and $\mathsf{label}(xg_1^1, k, 0) \circ \cdots \circ \mathsf{label}(xg_t^1, k, 0), \mathsf{label}(xg_1^1, k, 1) \circ \cdots \circ \mathsf{label}(xg_t^1, k, 1)$ for $k \in$ INP$_2$). P_2's input is his actual input. Second, they use BCOT2 with the labels for the rest of the input-wires of xg_j^1 (i.e., $\mathsf{label}(xg_j^1, \alpha(k), 0), \mathsf{label}(xg_j^1, \alpha(k), 1)$ for $k \in$ INP$_2$ and $j \in [t]$, where P_2's inputs are the bits of r_j^2). The players run the same protocols in the opposite direction (switching roles). At the end, each player learns the labels for his input-wires of gc_j^{3-i} and of xg_j^{3-i}. But we note that P_i is yet to send the labels for his input wires in the circuits he garbled himself, i.e. gc_j^i and xg_j^i.

Cut-and-Choose Phase (first opening). Next, as before, parties agree on a random $e \in [t]$ (using a joint coin-tossing protocol), and open the rest of the garbled circuits. In particular, they open the garbled circuit-pairs (gc_j^1, gc_j^2) and the XOR-gadgets (xg_j^1, xg_j^2) for all $j \neq e$. Moreover, for $j \neq e$, they reveal to each other the random strings r_j^i-s they used in the opened circuits (by showing the labels they learned in BCOT2), and then they decommit all the inputs they used as senders in BCOT1 for the opened circuits. The players check the correctness of the circuits and verify that the same r_j^i-s were used in both gc_j^i and xg_j^{3-i}. (Note that at the end of the opening phase, the players know that with $1 - 1/t$ probability the remaining circuit-pair (gc_e^1, gc_e^2) and the XOR gadget-pair (xg_e^1, xg_e^2) are properly constructed, and, that the inputs r_e^i used by the players in both gc_e^i, and xg_e^{3-i} are the same.)

Evaluation. Each party sends to his counterpart the input-wire labels for his inputs in the unopened circuit-pair. Parties then evaluate the circuit-pair (gc_e^1, gc_e^2) and the XOR-gadgets (xg_e^1, xg_e^2). (i.e., P_i evaluates gc_e^{3-i}, and xg_e^{3-i}.) P_i sends a commitment (along with a ZKPoK, as in Section 3) on the concatenation of the output labels he obtained after evaluating xg_e^{3-i} to P_{3-i}.

Cut-and-Choose Phase (second opening). P_{3-i} now opens the remaining XOR-gadget xg_e^{3-i}, and decommits all his inputs as a sender to the BCOTs of the XOR-gates (i.e., $\mathsf{label}(xg_1^{3-i}, k, 0) \circ \cdots \circ \mathsf{label}(xg_t^{3-i}, k, 0), \mathsf{label}(xg_1^{3-i}, k, 1) \circ \cdots \circ \mathsf{label}(xg_t^{3-i}, k, 1)$ in BCOT1, and $\mathsf{label}(xg_e^{3-i}, \alpha(k), 0), \mathsf{label}(xg_e^{3-i}, \alpha(k), 1)$ in BCOT2, both for $k \in$ INP$_i$). (We stress that only the XOR-gates of wires INP$_i$

are opened, and that those were generated using random labels independently of the garbled circuits. The XOR-gadgets of wires INP_{3-i} are checked as part of the previous phase.) P_i verifies that these XOR-gates were generated properly and that the BCOTs inputs were consistent with the XOR-gates. If everything is correct he decommits his commitment, otherwise he outputs $\perp$ and aborts. (Note that P_i reveals his output only *after* he verified that all the XOR-gates P_{3-i} generated were properly constructed. Since the only secrets in these gates are P_i's inputs, revealing them does not help P_i learn any new information.) P_{3-i} verifies that the decommitted values are valid output-wire labels, and compares the actual output with their output he obtains from evaluation of xg_e^i. If either check fails, P_{3-i} outputs $\perp$.

Equality-test. If there is no abort, players call the Equality Testing functionality as before. Note that now, with probability $1 - 1/t$, not only we know that the circuits being evaluated are correct, but also that the players use the same r_e^i-s in the final XOR gadget-pair. Combined with the fact that the players check equality of the output of the final XOR gadget-pair, they are ensured (with probability $1 - 1/t$) that the same input strings are being used in gc_e^1 and gc_e^2 or else, $x \oplus r_e^i$ would be different.

Output Unmasking. If the Equality Testing functionality returns False, the players abort. Otherwise, they unmask the output. (Recall that at this stage, each player knows the value of $C(x', y') = f(x, y) \oplus m_1 \oplus m_2$.) Player P_i sends the value of m_i along with labels that correspond to m_i in gc_e^{3-i}. These labels prove that m_i is indeed the value that P_i have used in the protocol.

Putting things together, correctness is always guaranteed due to the dual execution; full-privacy is guaranteed with probability $1 - 1/t$ due to the discussion above; and privacy with 1-bit leakage is guaranteed in the case that a cheating adversary is not caught, which only happens with probability $1/t$.

Acknowledgements. We would like to thank Ran Canetti, Benny Pinkas and Yehuda Lindell for their comments and helpful discussions. We also thank Yehuda Lindell for referring us to the malleability issue discussed in Footnote 1.

References

1. Aumann, Y., Lindell, Y.: Security against covert adversaries: Efficient protocols for realistic adversaries. J. Cryptol. 23(2), 281–343 (2010)
2. Bellare, M., Hoang, V.T., Rogaway, P.: Foundations of garbled circuits. In: CCS 2012, pp. 784–796. ACM (2012)
3. Fischlin, M.: Trapdoor Commitment Schemes and Their Applications. Ph.D. Thesis (Doktorarbeit), Department of Mathematics, Goethe-University, Frankfurt, Germany (2001)
4. Goyal, V., Mohassel, P., Smith, A.: Efficient two party and multi party computation against covert adversaries. In: Smart, N.P. (ed.) EUROCRYPT 2008. LNCS, vol. 4965, pp. 289–306. Springer, Heidelberg (2008)

5. Huang, Y., Evans, D., Katz, J., Malka, L.: Faster secure two-party computation using garbled circuits. In: Security 2011, pp. 35–35. USENIX Association (2011)
6. Huang, Y., Katz, J., Evans, D.: Quid-pro-quo-tocols: Strengthening semi-honest protocols with dual execution. In: SP 2012, pp. 272–284. IEEE Computer Society (2012)
7. Ishai, Y., Kushilevitz, E., Ostrovsky, R., Prabhakaran, M., Sahai, A.: Efficient non-interactive secure computation. In: Paterson, K.G. (ed.) EUROCRYPT 2011. LNCS, vol. 6632, pp. 406–425. Springer, Heidelberg (2011)
8. Jarecki, S., Shmatikov, V.: Efficient two-party secure computation on committed inputs. In: Naor, M. (ed.) EUROCRYPT 2007. LNCS, vol. 4515, pp. 97–114. Springer, Heidelberg (2007)
9. Kolesnikov, V., Schneider, T.: Improved garbled circuit: Free XOR gates and applications. In: Aceto, L., Damgård, I., Goldberg, L.A., Halldórsson, M.M., Ingólfsdóttir, A., Walukiewicz, I. (eds.) ICALP 2008, Part II. LNCS, vol. 5126, pp. 486–498. Springer, Heidelberg (2008)
10. Kreuter, B., Shelat, A., Shen, C.H.: Billion-gate secure computation with malicious adversaries. In: Security 2012, p. 14. USENIX Association (2012)
11. Lindell, Y., Pinkas, B.: An efficient protocol for secure two-party computation in the presence of malicious adversaries. In: Naor, M. (ed.) EUROCRYPT 2007. LNCS, vol. 4515, pp. 52–78. Springer, Heidelberg (2007)
12. Lindell, Y., Pinkas, B.: A proof of security of Yao's protocol for two-party computation. J. Cryptol. 22(2), 161–188 (2009)
13. Lindell, Y., Pinkas, B.: Secure two-party computation via cut-and-choose oblivious transfer. In: Ishai, Y. (ed.) TCC 2011. LNCS, vol. 6597, pp. 329–346. Springer, Heidelberg (2011)
14. Mohassel, P., Franklin, M.: Efficiency tradeoffs for malicious two-party computation. In: Yung, M., Dodis, Y., Kiayias, A., Malkin, T. (eds.) PKC 2006. LNCS, vol. 3958, pp. 458–473. Springer, Heidelberg (2006)
15. Nielsen, J.B., Nordholt, P.S., Orlandi, C., Burra, S.S.: A new approach to practical active-secure two-party computation. In: Safavi-Naini, R. (ed.) CRYPTO 2012. LNCS, vol. 7417, pp. 681–700. Springer, Heidelberg (2012)
16. Nielsen, J.B., Orlandi, C.: LEGO for two-party secure computation. In: Reingold, O. (ed.) TCC 2009. LNCS, vol. 5444, pp. 368–386. Springer, Heidelberg (2009)
17. Pedersen, T.P.: Non-interactive and information-theoretic secure verifiable secret sharing. In: Feigenbaum, J. (ed.) CRYPTO 1991. LNCS, vol. 576, pp. 129–140. Springer, Heidelberg (1992)
18. Pinkas, B., Schneider, T., Smart, N.P., Williams, S.C.: Secure two-party computation is practical. In: Matsui, M. (ed.) ASIACRYPT 2009. LNCS, vol. 5912, pp. 250–267. Springer, Heidelberg (2009)
19. Shelat, A., Shen, C.-H.: Two-output secure computation with malicious adversaries. In: Paterson, K.G. (ed.) EUROCRYPT 2011. LNCS, vol. 6632, pp. 386–405. Springer, Heidelberg (2011)
20. Woodruff, D.P.: Revisiting the efficiency of malicious two-party computation. In: Naor, M. (ed.) EUROCRYPT 2007. LNCS, vol. 4515, pp. 79–96. Springer, Heidelberg (2007)
21. Yao, A.C.C.: How to generate and exchange secrets. In: SFCS 1986, pp. 162–167. IEEE Computer Society (1986)

Improved OT Extension for Transferring Short Secrets

Vladimir Kolesnikov[1] and Ranjit Kumaresan[2]

[1] Bell Labs, Murray Hill, NJ 07974, USA
kolesnikov@research.bell-labs.com
[2] Technion, Haifa, Israel
ranjit@cs.technion.ac.il

Abstract. We propose an optimization and generalization of OT extension of Ishai et al. of Crypto 2003. For computational security parameter k, our OT extension for short secrets offers $O(\log k)$ factor performance improvement in communication and computation, compared to prior work. In concrete terms, for today's security parameters, this means approx. factor 2-3 improvement.

This results in corresponding improvements in applications relying on such OT. In particular, for two-party semi-honest SFE, this results in $O(\log k)$ factor improvement in communication over state of the art Yao Garbled Circuit, and has the same asymptotic complexity as the recent multi-round construction of Kolesnikov and Kumaresan of SCN 2012. For multi-party semi-honest SFE, where their construction is inapplicable, our construction implies $O(\log k)$ factor communication and computation improvement over best previous constructions. As with our OT extension, for today's security parameters, this means approximately factor 2 improvement in semi-honest multi-party SFE.

Our building block of independent interest is a novel IKNP-based framework for 1-out-of-n OT extension, which offers $O(\log n)$ factor performance improvement over previous work (for $n \leq k$), and concrete factor improvement of up to 5 for today's security parameters ($n=k=128$).

Our protocol is the first practical OT with communication/computation cost sublinear in the security parameter (prior sublinear constructions Ishai et al. [15,16] are not efficient in concrete terms).

Keywords: OT extension, 1-out-of-2 OT, 1-out-of-n OT.

1 Introduction

Our main contribution is an asymptotic and concrete efficiency improvement of Oblivious Transfer (OT) extension of Ishai et al. [14]. Our improvement applies to OT transfers of short secrets. In this Introduction we first motivate the problem, and then give intuition behind our approach.

Oblivious Transfer (OT) is a fundamental cryptographic primitive that is used as a building block in a variety of cryptographic protocols. It is a critical piece in

R. Canetti and J.A. Garay (Eds.): CRYPTO 2013, Part II, LNCS 8043, pp. 54–70, 2013.

general secure computation [29,10,18], as well as in a number of tailored solutions to specific problems of interest, such as contract signing [7]. OT performance improvement directly translates into that of secure function evaluation (SFE). In turn, SFE performance is the subject of major research effort in cryptography [14,22,20,6,12,25]. Our work can be plugged into several existing candidate solutions, resulting in factor 2-3 performance improvement, which is a major step forward in the state of the art of secure computation.

1.1 Secure Computation

SFE allows two (or more) parties to evaluate any function on their respective inputs x and y, while maintaining privacy of both x and y. SFE is justifiably a subject of an immense amount of research. Efficient SFE algorithms enable a variety of electronic transactions, previously impossible due to mutual mistrust of participants. Examples include auctions, contract signing, set intersection, etc. As computation and communication resources have increased, SFE of many useful functions has become practical for common use. Still, SFE of many of today's functions of interest carries costs sufficient to deter would-be adopters, who instead choose stronger trust models, entice users to give up their privacy with incentives, or use similar crypto-workarounds. We believe that truly practical efficiency is required for SFE to see use in real-life applications.

The current state of the art of SFE research is quite sophisticated. Particularly in the semi-honest model, there have been very few asymptotic/qualitative improvements since the original protocols of Yao [29] and Goldreich et al. [9]. Possibly the most important development in the area of SFE since the 1980's was the very efficient OT extension technique of Ishai et al. [14], which allowed to evaluate an arbitrarily large number of OTs by executing a small (security parameter) number of (possibly inefficient) "bootstrapping" OT instances, and a number of symmetric key primitives. This possibility of cheap OTs made a dramatic difference for securely computing functions with large inputs relative to the size of the function, as well as for GMW-like approaches, where OTs are performed in each level of the circiut.

As secure computation moves from theory to practice, even "small" improvements can have a significant effect. Today, even small factor performance improvements to state-of-the-art algorithms are quite hard to achieve, and are most welcome. This is especially true about the semi-honest model protocols, where the space for improvement appears to be much smaller than in the malicious model.

In this work, we propose an improvement to OT extension of Ishai et al. [14], for the case of OT of short secrets. As we will describe below, this will result in a new multi-party SFE protocol, which is approximately factor 2 (asymptotically factor $O(\log k)$) more efficient than state of the art. Our constructions also improve on standard two-party garbled circuit protocols in asymptotic ($O(\log k)$) and concrete terms, and offer performance in line with the recent work of [19].

1.2   Secure Computation and OT Efficiency Considerations

The efficiency of OT plays a critical role in the overall efficiency of secure computation. It is so to the point that OT performance determines which is the most efficient approach. Until recently, in the semi-honest model, Yao's Garbled Circuit was a clear winner. With the work of [19], which can be seen as a hybrid between GMW and Yao, and our improved OT extension technique, the GMW approach will outperform Yao with a factor of ≈ 2 for today's security parameters. Asymptotically, the performance improvement is logarithmic in the security parameter, as compared to GC-based SFE.

On the Cost of SFE Rounds. One common consideration in SFE protocol design is the number of rounds. Indeed, in some scenarios the latency associated with the communication rounds can more than double the total execution time. This holds, e.g., when the evaluated circuit is small; with the GMW evaluation, where we need a round of communication per layer of the circuit, the latency may be costly for deep and narrow circuits. This may cause somewhat increased latency of an individual computation – a possible inconvenience to the user of interactive applications.

At the same time, many SFE protocols allow for significant precomputation and streaming, where message transmission may begin (and even a response may be received) before the sender completes the computation and transmission of the message. Thus, round-related latency will usually not be a wasted time and will not cause extra delays. Most importantly, with the speed of the CPU advancing faster than that of communication, the true bottleneck for SFE already is the channel transmission capacity, even for high-speed gigabit LAN.

Thus, we argue that in many scenarios, the number of communication rounds in SFE often plays an insignificant role in practice, and round-related latency either has no impact on performance, or it can be tolerated in exchange of achieving higher throughput.

1.3   Our Contributions

Our main contribution is an asymptotic and concrete efficiency improvement of Oblivious Transfer (OT) extension of Ishai et al. [14]. Our improvement applies to OT transfers of short secrets.

1-out of-2 OT Extension. For a security parameter k, our $O(\log k)$ asymptotic improvement results in concrete efficiency improvement of about factor up to 2 for today's security parameters. This yields corresponding asymptotic and concrete improvements in multi-party computation in the semi-honest setting, when applied to state of the art solutions based on GMW protocols.

Our new 1-out of-n OT extension protocol offers $O(\log n)$ factor performance improvement over previous work (for $n < k$ and constant secret length), and concrete factor improvement of up to 5 for today's security parameters.

Further, our protocol is the first OT sublinear in the security parameter other than the non-black-box construction of Ishai et al. [15], and is the only practical

OT with this property. Our resulting secure computation protocols can also be viewed as a significant improvement of the technique of [19], which offered logarithmic in k improvement over state-of-the-art Yao's GC, but, in particular, did not extend to multiparty setting. We work in the (non-programmable) RO model, but, like in [14], we can also use a variant of correlation-robust hash functions.

We also present a new simple trick for OT extension (compatible with ours as well as with[14]), which, in particular, allows to futher cut in half the cost of OT of 1-bit secrets and reduce by 25% the cost OT of k-bit secrets. This optimization is described in Section 6 and Appendix A. To clearly state the performance improvement of our main OT extension protocol, the numbers elsewhere *do not* reflect this optimization.

Applications and Practical Performance Impact. As noted above, our 1-out of-2 OT construction immediately offers approximately factor 2 improvement in nearly all multi-party protocols – GMW and its variants.

In two-party computation, a similar, but more limited in scope, improvement was recently achieved [19]. In particular, [19] didn't work well on very shallow circuits, such as inner product computation. For such circuits, we have $O(\log k)$ improvement over 2PC state of the art, including [19].

More importantly, there is growing evidence that new GMW optimizations will often allow (multiround) GMW-based SFE protocols to outperform (constant round) Yao GC based SFE in practice, despite the round-related latencies. For example, a recent work of Schneider and Zohner [28] introduces and implements several optimizations to mitigate latency impact. It demonstrates performance improvement *of factor up to 100* of GMW over a recent Yao-based implementation of secure face matching even in high-latency (100ms round-trip, intercontinental) network. We expect that future SFE research and CPU-vs-network evolution will further improve GMW relative to Yao.

In sum, our work improves state of the art of 2PC computation for a significant class of problems where GMW protocols outperform Yao.

As noted, our 1-out of-n OT gives logarithmic performance improvement in transferring one in n random secret keys. However, in some cases, where the OT of specific secrets is required, the improvement factor may be smaller due to the fact that all n secrets encrypted with the n keys need to be transferred. In this case, logarithmic improvement applies only to the offline phase, where the secrets are not available.

Another application which immediately benefits from this work is string-selection OT (SOT), a variant of 1-out of-n OT and a building block of [19]. In SOT, the receiver selects one of the sender's two secrets based on his $\log n$-bit selection string.

1.4   Related Work

OT is a critical and heavily used component in much of cryptography, and in particular in secure computation protocols. Naturally, a lot of effort went into optimizing its performance. Unfortunately, there are fundamental limits to OT efficiency. Impagliazzo and Rudich [13] showed that a black-box reduction from oblivious transfer to a one-way function or a one-way permutation would imply $\mathbf{P} \neq \mathbf{NP}$. It is further not known whether such non-black-box reductions exist.

Beaver [3] was the first to propose OT extension, a non-black-box scheme where a large number of OTs can be obtained from a small number of OTs (possibly executed by using public-key primitives) and one-way functions. Lindell and Zarosim [21] recently showed that one-way functions are in fact needed for OT extension.

Ishai, Kilian, Nissim, and Petrank [14], in their breakthrough work showed a truly practical black-box OT extension. Its cost, in addition to the security parameter number of base OTs, is only two random oracle (RO) evaluations and output transfers. By dramatically changing the cost structure of two-party SFE, especially in the semi-honest model, this work enabled greatly improved SFE for functions with large inputs, previously considered too costly due to the need of a large number of public key operations. It also started a rise in the study of GMW-based SFE protocols, where an OT is needed per multiplicative node. Indeed, recent (yet unoptimized) GMW-based and multiple-round protocols began to outperform traditional GC protocols. In particular, [25] outperforms state-of-the-art GC protocols in the malicious model, and [19] outperforms state-of-the-art GC protocols in the semi-honest model. In addition to considering the semi-honest model, [14] presents a construction secure against malicious participants. In a few follow-up works [24,11,17], the performance of the malicious setting of the IKNP OT extension was substantially improved. We present the high-level idea of the basic IKNP construction in Section 3.2.

By employing a more efficient pseudorandom generator in Beaver's non-black-box OT extension protocol, Ishai, Kushilevitz, Ostrovsky, and Sahai [15] obtained an asymptotically more efficient (but expensive in concrete terms) construction for oblivious transfer extension, and consequently for secure computation. In fact, their protocol enjoys a *constant computational/communication overhead* over an insecure evaluation of the function to be evaluated. In order to obtain these strong efficiency results, Ishai et al. [15] make strong complexity-theoretic assumptions on pseudorandom generators. Specifically, they assume that there exists an (arbitrary stretch) pseudorandom generator in $\mathbf{NC}^0$ [2,1].

In this work, we show logarithmic in the security parameter improvement for black-box OT extension transfer of short secrets. In other words, we improve efficiency of the black-box OT extension protocol of Ishai et al. [14] asymptotically by a $\log(k/\ell)$ factor when the length of the transferred secrets is ℓ. This has important practical applications for secure computation solutions in the semihonest model, such as GMW, that require precisely 1-out-of-2 OT of 1-bit secrets. We calculate both asymptotic and concrete performance of the resulting protocols. Our constructions are presented in the semi-honest model.

We stress that in contrast to the non-black-box techniques of Ishai et al. [15], our extension protocol makes only black-box use of a (non-programmable) random oracle. Also, unlike [15] who mainly focus on asymptotic complexity, we calculate also the concrete efficiency of our construction, and demonstrate a factor of approximately 2 improvement over state-of-the-art protocols [14,6].

Finally, we mention PIR work (e.g., [16]) that construct communication efficient 1-out of-n OT protocols but perform $O(n)$ computationally intensive (e.g., public-key operations) per instance. In contrast, we perform a fixed number of public-key operations independent of the number of OT instances.

2   Overview of Our Approach

We give a high-level overview of our solution prior to presenting its technical details in Section 4. We aim that the reader somewhat familiar with the IKNP construction [14] should understand the main idea of our construction from this overview.

Consider the random $m \times k$ matrix designed by [14], which is transferred column-wise via k 1-out-of-2 base OTs from the receiver $\mathcal{R}$ to the sender $\mathcal{S}$. In [14], each row of this matrix is used to implement a 1-out-of-2 OT, as it has the randomness from which a random OT can be constructed.

Our main observation is that, *for the same communication cost*, each row of this matrix can be instead used to perform a 1-out-of-n OT, but of shorter secrets. Further, a 1-out-of-n OT of $\log n$-bit long secrets can be trivially used to construct $\log n$ instances of 1-out-of-2 1-bit OTs, which is precisely the kind of OT needed in the GMW protocol and its variants. Thus, effectively, we trade the length of the OT-transferred secrets for the number of OTs, which results in significant gain for MPC applications.

The intuition for our 1-out-of-n OT is as follows. First, recall that in IKNP, for each column of the $m \times k$ matrix, $\mathcal{S}$ randomly selects (via OT), whether he receives the random column, or the random column XORed with the m-bit long input of $\mathcal{R}$. Viewed row-wise, this effectively means that for each row j, $\mathcal{S}$ either receives (via OT) the j-th row of the randomly chosen $m \times k$ matrix (if $\mathcal{R}$'s j-th selection bit is 0), or that row XORed with his k-bit selection vector to the OT (if $\mathcal{R}$'s j-th selection bit is 1). Then $\mathcal{S}$ masks each of his two j-th input secrets with (RO hashes of) vector received as output from OT and the same vector XORed with its k-bit selection vector respectively and sends both to $\mathcal{R}$, who is able to take the mask off exactly one of the two messages. The other masked message remains hidden since $\mathcal{R}$ does not learn the selection vector provided by $\mathcal{S}$.

In the following, let $\mathcal{C}$ denote a binary code, and let r_j denote the input of $\mathcal{R}$ to the j-th instance of 1-out-of-n OT. In our 1-out-of-n OT, we modify the scheme presented above such that for each row j, $\mathcal{S}$ receives (via OT) the actual j-th row of the $m \times k$ matrix XORed with a vector that is the result of the r_j-th codeword in $\mathcal{C}$ bitwise-ANDed with the k-bit selection vector. This allows $\mathcal{S}$ to generate n random pads from each row of the matrix—the i-th such pad being the j-th row it received (via OT) XORed with a vector that is the result of the i-th codeword in $\mathcal{C}$ bitwise-ANDed with the k-bit selection vector. These n random pads may then be used by $\mathcal{S}$ to carry out a 1-out-of-n OT with $\mathcal{R}$.

The security of this construction naturally depends on the underlying code. The exact property that we need is that $\mathcal{C}$ must contain at least n codewords, each of length at most k, such that the codewords in $\mathcal{C}$ are spaced as far apart as possible from each other. This, combined with the fact that $\mathcal{R}$ does not learn the selection vector provided by $\mathcal{S}$, will ensure that $\mathcal{R}$ can efficiently recover only one of the n pads used by $\mathcal{S}$. The above is presented in detail in Section 4.

Using Walsh-Hadamard code for $\mathcal{C}$ gives a 1-out-of-n OT for n equal to the security parameter k. This OT is suitable for generation of $\log n$ instances of 1-out-of-2 OTs (Section 5.1). Using a higher-rate code with high distance results in 1-out-of-n OT for any n polynomial in k (Section 5.3).

3   Preliminaries and Notation

3.1   Notation

We use the notation OT_ℓ^m to denote m instances of 1-out-of-2 string-OT where the string is ℓ bits long. Let $\mathcal{S}$ denote the sender, and let $\mathcal{R}$ denote the receiver. In 1-out-of-2 OT, the sender's input is $\{(x_{j,0}, x_{j,1})\}_{j \in [m]}$, i.e., m pairs of strings, each of length ℓ, and the receiver holds input $\{r_j\}_{j \in [m]}$, where each r_j is an integer which is either 0 or 1. Note that if $\mathcal{S}$ provides input $\{(x_{j,0}, x_{j,1})\}_{j \in [m]}$ to OT_ℓ^m, and if $\mathcal{R}$ provides input $\{r_j\}_{j \in [m]}$ to OT_ℓ^m, then $\mathcal{R}$ receives back $\{x_{j,r_j}\}_{j \in [m]}$, while $\mathcal{S}$ receives nothing.

In Section 4, we construct protocols for 1-out-of-n OT, which is a straightforward generalization of 1-out-of-2 OT. We explain this further. We use the notation $\binom{n}{1}$-OT_ℓ^m to denote m instances of 1-out-of-n string-OT where the string is ℓ bits long. In 1-out-of-n OT, the sender's input is $\{(x_{j,0}, \ldots, x_{j,n-1})\}_{j \in [m]}$, and the receiver holds input $\{r_j\}_{j \in [m]}$, where each r_j is an integer which between 0 and $n-1$. Note that if $\mathcal{S}$ provides input $\{(x_{j,0}, \ldots, x_{j,n-1})\}_{j \in [m]}$ to $\binom{n}{1}$-OT_ℓ^m, and if $\mathcal{R}$ provides input $\{r_j\}_{j \in [m]}$ to $\binom{n}{1}$-OT_ℓ^m, then $\mathcal{R}$ receives back $\{x_{j,r_j}\}_{j \in [m]}$, while $\mathcal{S}$ receives nothing.

Following the convention in IKNP, we denote vectors in bold, and matrices in capitals. For a matrix A, we let $\mathbf{a}_j$ denote the j-th row of A, and $\mathbf{a}^i$ denote the i-th column of A. If $\mathbf{a} = a_1 \| \cdots \| a_p$ and $\mathbf{b} = b_1 \| \cdots \| b_p$ are two vectors, then we define $\oplus$ and $\odot$ operations as follows. We use the notation $\mathbf{a} \oplus \mathbf{b}$ to denote the vector $(a_1 \oplus b_1) \| \cdots \| (a_p \oplus b_p)$. Similarly, the notation $\mathbf{a} \odot \mathbf{b}$ denotes the vector $(a_1 \cdot b_1) \| \cdots \| (a_p \cdot b_p)$. Finally, suppose $c \in \{0, 1\}$, then $c \cdot \mathbf{a}$ denotes the vector $(c \cdot a_1) \| \cdots \| (c \cdot a_p)$.

Our constructions assume the existence of a random oracle H. We denote the security parameter by k, and assume (without loss of generality) that it is a power of 2.

3.2   IKNP OT Extension

In this section, we present the OT extension protocol of Ishai, Kilian, Nissim, and Petrank [14]. The protocol will reduce OT_ℓ^m to OT_m^k. This implies a reduction (via use of a PRG) to OT_k^k with some additional cost. The security of the protocol holds as long as the receiver is semi-honest. (Note: the sender may be malicious.) We now describe the protocol that realizes OT_ℓ^m given ideal access to OT_m^k.

INPUT OF $\mathcal{S}$: m pairs $(x_{j,0}, x_{j,1})$ of ℓ-bit strings, $1 \leq j \leq m$.
INPUT OF $\mathcal{R}$: m selection bits $\mathbf{r} = (r_1, \dots, r_m)$.
COMMON INPUT: a security parameter k.
ORACLE: a random oracle $H : [m] \times \{0,1\}^k \to \{0,1\}^\ell$.
CRYPTOGRAPHIC PRIMITIVE: an ideal OT_m^k primitive.

1. $\mathcal{S}$ chooses $\mathbf{s} \leftarrow \{0,1\}^k$ at random. Let s_i denote the i-th bit of $\mathbf{s}$.
2. $\mathcal{R}$ forms $m \times k$ matrices T_0, T_1 in the following way:

   – Choose $\mathbf{t}_{j,0}, \mathbf{t}_{j,1} \leftarrow \{0,1\}^k$ at random such that $\mathbf{t}_{j,0} \oplus \mathbf{t}_{j,1} = (r_j \| \cdots \| r_j)$.

   Let $\mathbf{t}_0^i, \mathbf{t}_1^i$ denote the i-th column of matrices T_0, T_1 respectively.
3. $\mathcal{S}$ and $\mathcal{R}$ interact with OT_m^k in the following way:

   – $\mathcal{S}$ acts as *receiver* with input $\{s_i\}_{i \in [k]}$.
   – $\mathcal{R}$ acts as *sender* with input $\{\mathbf{t}_0^i, \mathbf{t}_1^i\}_{i \in [k]}$.
   – $\mathcal{S}$ receives output $\{\mathbf{q}^i\}_{i \in [k]}$.

   $\mathcal{S}$ forms $m \times k$ matrix Q such that the i-th column of Q is the vector $\mathbf{q}^i$. (Note $\mathbf{q}^i = \mathbf{t}_{s_i}^i$.) Let $\mathbf{q}_j$ denote the j-th row of Q. (Note $\mathbf{q}_j = ((\mathbf{t}_{j,0} \oplus \mathbf{t}_{j,1}) \odot \mathbf{s}) \oplus \mathbf{t}_{j,0}$. Simplifying, $\mathbf{q}_j \oplus \mathbf{t}_{j,0} = r_j \cdot \mathbf{s}$.)
4. For $j \in [m]$, $\mathcal{S}$ sends $y_{j,0} = x_{j,0} \oplus H(j, \mathbf{q}_j)$ and $y_{j,1} = x_{j,1} \oplus H(j, \mathbf{q}_j \oplus \mathbf{s})$.
5. For $j \in [m]$, $\mathcal{R}$ recovers $z_j = y_{j,r_j} \oplus H(j, \mathbf{t}_{j,0})$.

EFFICIENCY. The protocol makes a single call to OT_m^k. The cost of OT_m^k is the cost of OT_k^k (which is independent of m) plus a generation of $2k$ pseudorandom strings each of length m. Other than this call to OT_m^k, each party evaluates at most $2m$ times (an implementation of) a random oracle. It is easy to see that the total communication cost of OT_ℓ^m is the communication cost of implementing OT_m^k plus $2m\ell$ bits transferred between $\mathcal{S}$ and $\mathcal{R}$ in Step 4. Thus we conclude that the communication cost of OT_ℓ^m is $2mk + 2m\ell$ bits. Note that the total computational cost of the protocol is proportional to its communication cost.

3.3 Walsh-Hadamard (WH) Codes

For $\alpha \in \{0,1\}^q$, let $\text{WH}(\alpha) = (\langle \alpha, x \rangle)_{x \in \{0,1\}^q}$, where the inner product between the two vectors is taken modulo 2. That is, $\text{WH}(\alpha)$, also known as the Walsh-Hadamard encoding of α, is the 2^q-bit string consisting of inner products of each q-bit string with α. For each k, Walsh-Hadamard codes, denoted by $\mathcal{C}_{\text{WH}}^k$, are simply defined as the set $\{\text{WH}(\alpha)\}_{\alpha \in \{0,1\}^{\log k}}$. Note that $\mathcal{C}_{\text{WH}}^k$ contains k strings (or, codewords) each of length k bits. In our constructions, we will use the well-known fact that the relative distance of $\mathcal{C}_{\text{WH}}^k$ is $1/2$ when k is a power of 2.

4   Extending 1-out-of-n OT

Recall, k is a security parameter. We present a natural generalization of 1-out-of-2 OT extension protocol given in [14]. We consider 1-out-of-n OT for any $n \leq k$.[1] First, recall that it is easy to construct a 1-out-of-n OT protocol from $O(\log n)$ instances of a 1-out-of-2 OT protocol in the semi-honest setting. The communication cost of m instances of 1-out-of-n OT on ℓ-bit strings would be the cost of $\mathrm{OT}_k^{m \log n}$ plus the cost required to transmit at most mn masked secrets each of length ℓ. Thus, the communication cost of obtaining m instances of 1-out-of-n OT on ℓ-bit strings is at most $O(m(k \log n + n\ell))$ bits. Further, its computational cost is proportional to the communication cost.

Our main contribution, formally presented in this section, is showing how to generalize IKNP's technique to directly obtain (i.e., without going via a construction for 1-out-of-2 OT) an extension protocol for 1-out-of-n OT when $n \leq k$. For the same security parameter and the same size of setup matrix at IKNP, the concrete security of our construction corresponds to that provided by security parameter $k_{\mathrm{IKNP}} \approx k/2$. If exactly same concrete security as IKNP is desired, this can be achieved by setting our security parameter $k \approx 2k_{\mathrm{IKNP}}$, which results in a multiplicative factor 2 overhead compared to IKNP. However, because we do 1-out-of-n OT at this cost, our construction will still result in asymptotic and concrete performance improvement of 1-out-of-n OT.

Let $\binom{n}{1}$-OT_ℓ^m denote m instances of 1-out-of-n OT on ℓ-bit strings. As in [14], we will reduce $\binom{n}{1}$-OT_ℓ^m to OT_m^k(which can be trivially efficiently reduced to OT_k^k). As the [14] basic protocol, our protocol is secure against a malicious sender and semi-honest receiver. Our protocol will use Walsh-Hadamard codes, denoted by $\mathcal{C}_{\mathrm{WH}}^k = (\mathbf{c}_0, \dots, \mathbf{c}_{k-1})$.

We now describe our protocol that realizes $\binom{n}{1}$-OT_ℓ^m given ideal access to OT_m^k.

Construction 1 (1-out-of-n OT Extension)
INPUT OF $\mathcal{S}$: m tuples $(x_{j,0}, \dots, x_{j,n-1})$ of ℓ-bit strings, $1 \leq j \leq m$.
INPUT OF $\mathcal{R}$: m selection integers $\mathbf{r} = (r_1, \dots, r_m)$ such that $0 \leq r_j < n$ for $1 \leq j \leq m$.
COMMON INPUT: a security parameter k such that $k \geq n$, and Walsh-Hadamard codes $\mathcal{C}_{\mathrm{WH}}^k = (\mathbf{c}_0, \dots, \mathbf{c}_{k-1})$.
ORACLE: a random oracle $H : [m] \times \{0,1\}^k \to \{0,1\}^\ell$.
CRYPTOGRAPHIC PRIMITIVE: an ideal OT_m^k primitive.

1. $\mathcal{S}$ chooses $\mathbf{s} \leftarrow \{0,1\}^k$ at random. Let s_i denote the i-th bit of $\mathbf{s}$.
2. $\mathcal{R}$ forms $m \times k$ matrices T_0, T_1 in the following way:
   – Choose $\mathbf{t}_{j,0}, \mathbf{t}_{j,1} \leftarrow \{0,1\}^k$ at random such that $\mathbf{t}_{j,0} \oplus \mathbf{t}_{j,1} = \mathbf{c}_{r_j}$.
   Let $\mathbf{t}_0^i, \mathbf{t}_1^i$ denote the i-th column of matrices T_0, T_1 respectively.
3. $\mathcal{S}$ and $\mathcal{R}$ interact with OT_m^k in the following way:
   – $\mathcal{S}$ acts as receiver with input $\{s_i\}_{i \in [k]}$.

---

[1] We discuss how to extend 1-out-of-n OT for $n = \mathrm{poly}(k)$ in Section 5.3.

- $\mathcal{R}$ *acts as* sender *with input* $\{\mathbf{t}_0^i, \mathbf{t}_1^i\}_{i \in [k]}$.
- $\mathcal{S}$ *receives output* $\{\mathbf{q}^i\}_{i \in [k]}$.

$\mathcal{S}$ *forms* $m \times k$ *matrix* Q *such that the i-th column of Q is the vector* $\mathbf{q}^i$. *(Note* $\mathbf{q}^i = \mathbf{t}_{s_i}^i$.*) Let* $\mathbf{q}_j$ *denote the j-th row of* Q. *(Note* $\mathbf{q}_j = ((\mathbf{t}_{j,0} \oplus \mathbf{t}_{j,1}) \odot \mathbf{s}) \oplus \mathbf{t}_{j,0}$. *Simplifying,* $\mathbf{q}_j \oplus \mathbf{t}_{j,0} = \mathbf{c}_{r_j} \odot \mathbf{s}$.*)

4. *For* $j \in [m]$ *and for every* $0 \leq r < n$, $\mathcal{S}$ *sends* $y_{j,r} = x_{j,r} \oplus H(j, \mathbf{q}_j \oplus (\mathbf{c}_r \odot \mathbf{s}))$.
5. *For* $j \in [m]$, $\mathcal{R}$ *recovers* $z_j = y_{j,r_j} \oplus H(j, \mathbf{t}_{j,0})$.

This concludes the description of the protocol. It is easy to verify that the protocol's outputs are correct (i.e., $z_j = x_{j,r_j}$) when both parties follow the protocol.

EFFICIENCY. The protocol makes a single call to OT_m^k. The cost of OT_m^k is the cost of OT_k^k (which is independent of m) plus a generation of $2k$ pseudorandom strings each of length m. Other than this call to OT_m^k, each party evaluates at most mn times (an implementation of) a random oracle. It is easy to see that the total communication cost of OT_ℓ^m is the communication cost of implementing OT_m^k plus $mn\ell$ bits transferred between $\mathcal{S}$ and $\mathcal{R}$ in Step 4. Thus we conclude that the communication cost of OT_ℓ^m is $O(m(k + n\ell))$ bits. Note that the total computational cost of the protocol is proportional to its communication cost. Recall that $n \leq k$, and thus when $\ell = 1$, the asymptotic cost of our $\binom{n}{1}$-OT_ℓ^m protocol is $O(mk)$ which is the same as the asymptotic cost of Ishai et al.'s OT_ℓ^m protocol described in Section 3.2. In terms of concrete performance, as mentioned above, we need to use a security parameter $k \approx 2k_{\mathrm{IKNP}}$, resulting in a factor 2 overhead compared to IKNP's OT_ℓ^m execution. Because we are performing the more powerful $\binom{n}{1}$-OT_ℓ^m, this corresponds to asymptotic (and concrete!) performance improvement.

Theorem 1. *Construction 1 is a secure protocol for evaluating* $\binom{n}{1}$-OT_ℓ^m *in the semi-honest model.*

The proof of security of Theorem 1 appears in the full version.

Remarks. In Construction 1, one can replace $\mathcal{C}_{\mathrm{WH}}^k$ with an encoding map enc : $\{0,1\}^{\log n} \to \{0,1\}^k$ that has the property that for $r, r' \in \{0,1\}^{\log n}$ with $r \neq r'$, the Hamming distance between enc(r) and enc(r') is at least $\Omega(k)$. It is instructive to see that when $n = 2$ and when enc is the k-bit *repetition* encoding of the input bit, i.e., enc(r) = $(r, \ldots, r) \in \{0,1\}^k$, then we get exactly the IKNP construction. Note that for $r \neq r'$, the Hamming distance between enc(r) and enc(r') is exactly k. As we saw in Construction 1, using the encoding map enc(r) = $\mathbf{c}_r$, where $\mathbf{c}_r$ is the r-th Walsh-Hadamard codeword, gives us an $\log k$ efficiency improvement. Since the Walsh-Hadamard code is a low-rate code, the maximum value of n is restricted to be less than or equal to k. A natural question that arises is whether a code with a better rate enables us to remove this restriction. Indeed, in Section 5.3, by using more sophisticated codes (cf. Claim 5.3) we show an improvement in the (offline) communication complexity of 1-out-of-n OT extension for arbitrary $n = \mathrm{poly}(k)$.

5 Resulting Efficiency Improvements

We evaluate performance improvements of Construction 1, and corresponding two- and multi-party SFE improvements. Recall that in the semi-honest model, a single instance of 1-out-of-n OT may be used to generate $\log n$ instances of 1-out-of-2 OT over slightly shorter strings with no additional cost. More precisely, the cost of OT_ℓ^m is exactly equal to the cost of $\binom{n}{1}$-$\mathrm{OT}_{\ell \log n}^{m/\log n}$. This observation will allow us to leverage our efficient construction of $\binom{n}{1}$-OT_ℓ^m to obtain improved efficiency for 1-out-of-2 OT, and consequently for secure computation.

5.1 Efficiency Improvements for 1-out-of-2 OT

In this section, we demonstrate a $\log k$ asymptotic improvement in the efficiency of 1-out-of-2 OT when sender's secrets are just bits (i.e., length of sender's secrets, $\ell = 1$). As observed previously, we do this by constructing 1-out-of-2 OTs via 1-out-of-n OTs.

Recall that the cost of our $\binom{n}{1}$-OT_ℓ^m protocol described in Section 4 is $O(m(k + n\ell))$. Using the fact that the cost of OT_ℓ^m is exactly equal to the cost of $\binom{n}{1}$-$\mathrm{OT}_{\ell \log n}^{m/\log n}$, we conclude that OT_ℓ^m may be reduced to OT_k^k while incurring an additional cost at most $O((m/\log n) \cdot (k + n\ell \log n))$. By choosing n such that $n \log n = k/\ell$, we see that this additional cost is asymptotically $O(mk/\log(k/\ell))$. In summary, we have shown a reduction from OT_ℓ^m to OT_k^k with cost $O(mk/\log(k/\ell))$.

Contrast our result above with the result of [14], where the cost of the reduction from OT_ℓ^m to OT_k^k was $O(m(k + \ell))$. Observe that for the important case when $\ell = 1$, our construction offers a logarithmic factor improvement in the efficiency of the reduction.

As noted in Section 4, to achieve concrete security equal to that of IKNP, we need a security parameter approximately twice theirs, which results in a factor 2 overhead of our protocol. Even with this efficiency loss we have both asymptotic and concrete performance advantage over IKNP.

Concrete Efficiency. We begin with a concrete cost analysis of $\binom{n}{1}$-OT_ℓ^m. Recall that the exact cost of reduction from OT_m^k to OT_k^k involves sending $2mk$ bits. Then, in Step 4 of Construction 1, $\mathcal{S}$ transmits $mn\ell$ bits to $\mathcal{R}$. Thus, the concrete cost of $\binom{n}{1}$-OT_ℓ^m is $m(2k+n\ell)$. Using the fact that the cost of our OT_ℓ^m is exactly equal to the cost of $\binom{n}{1}$-$\mathrm{OT}_{\ell \log n}^{m/\log n}$, we conclude that OT_ℓ^m may be reduced to OT_k^k with cost $(m/\log n) \cdot (2k + n\ell \log n)$ bits. The minimum cost can then be obtained by choosing a suitable value of n.

In contrast, the concrete communication cost of IKNP's construction of OT_ℓ^m is $2m(k + \ell)$ bits. As described earlier, there's a small gap between the security guarantees between our consruction and IKNP's. We take that into account in our cost calculation, and present the results in Table 1.

Table 1. Comparison of (amortized) communication cost (measured in bits) of 1-out-of-2 bit OT for a given security level. The costs are computed assuming parties are semi-honest. The performance improvement ratio betwen our work and IKNP represents the resulting improvement factor for MPC protocols based on the GMW approach.

level of security	our cost	IKNP cost
50	74	102
112	130	226
238	227	478

5.2 Efficiency Improvements for Secure Computation

In this section, we will discuss applications of our OT_ℓ^m protocol to secure two-party and multi-party computation. As pointed out in the Introduction, efficient OT forms a criticial component of secure computation protocols, and improvements in the efficiency of OT translates to an improvement in the efficiency of secure computation protocols built on top of OT.

In the previous section, we saw how our construction asymptotically outperforms the extension protocol of [14] by a factor of $O(\log(k/\ell))$. Clearly, this improvement factor is maximized when $\ell = 1$, i.e., for 1-bit OT. Thus, our construction has maximum benefit for secure computation protocols that extensively rely on 1-bit OTs. One such example is the well known GMW protocol [9] where each AND gate of the circuit is evaluated using (two invocations of) 1-bit OTs (and negligible additional cost). Until now, efficient implementations of the GMW protocol in the semi-honest setting (e.g., [6]) relied on the OT extension protocol of [14]. Because OT costs dominate the protocol costs, simply by using our extension protocol (instead of [14]), the semi-honest GMW protocol will enjoy an asymptotic $\log k$ efficiency improvement (and improvement in concrete terms as well).

Secure Two-Party Computation. As discussed in Section 1.3, a large class of 2PC problems is solved more efficiently with GMW than Yao. For problems in this class, our OT extension improvement results in corresponding 2PC improvement. For other problems, where Yao is faster, the relative performance of the approaches is discussed next.

The concrete improvements for the specific case of two-party computation are shown in Table 2. From the table, it is evident that our protocol begins to outperform state-of-the-art constant round protocols (e.g., [26]) for reasonable levels of security. However, for practical values of the security parameter, it performs worse, in concrete terms, when compared to the best-case performance of [19], a non-constant round protocol that generalizes both Yao garbled circuits and GMW. (We note that the communication cost of our protocol is asymptotically the same as the communication cost of [19].) In more detail, the performance of [19] is highly sensitive to the topology of the circuit. Their best performance, as noted in Table 2, is for the case of constant width circuits. In contrast, our improvements are independent of the topology of the circuit being evaluated. We point out that the approach of [19] can be viewed as somewhat related to ours (but more narrow; in particular, it is not applicable to multiparty comptuation). Furthermore, our OT extension protocol can improve the performance of [19] for circuits with low depth.

Table 2. Comparison of (amortized) communication cost (measured in bits) per gate of the circuit for various semi-honest secure two-party protocols. We note that protocols of [19] do not extend to multi-party setting, while ours do.

level of security	our cost per gate	[26] cost per gate	[19] cost per gate
50	148	100	66
112	260	224	112
238	454	476	196

Secure Multi-Party Computation. Today, practical protocols for secure *multi-party* computation are based on the GMW approach (e.g., [25,6]).[2] GMW-based secure computation protocols for t parties, in the semi-honest setting, operate in almost the same way as in the two-party case except that now parties compute pairwise OTs (more precisely, a total of $2t^2$ OTs) to securely evaluate each AND gate. That is, for each AND gate of the circuit parties evaluate a total of $2t^2$ 1-bit OTs (with negligible additional cost). Therefore, simply by using our extension protocol (instead of [14]), we will improve the asymptotic complexity by a $\log k$ factor. Concrete improvements in this setting are the same as those found in Table 1. Specifically, for "50-bit security" we obtain an improvement of $102/74 = 1.378$ in the communication cost. Similarly, we obtain an improvement factor of > 2 for "238-bit security".

5.3   Efficiency Improvements for 1-out-of-n OT

Recall that the cost of extending 1-out-of-n OT from [14] is $O(m(k \log n + n\ell))$ bits. Our main construction of 1-out-of-n OT described in Section 4 reduces the cost of 1-out-of-n OT extension to $m(2k + n\ell)$ bits. As described in Section 4, for the same guarantee as in IKNP, our security parameter should be set as $k \approx 2k_{\text{IKNP}}$. Previous solutions [23,14] cost $(4mk_{\text{IKNP}} \log n + mn\ell)$ bits. Hence for $k_{\text{IKNP}} = 128$, with $n = k$ and $\ell = 1$, our solution improves upon existing solutions by a factor ≈ 5.39.

Note that the above improvement holds only when $n \leq k$. In this section, we show how to modify Construction 1 to support $n = \text{poly}(k)$. In the resulting protocol, the (offline) communication cost of the generating 1-out-of-n OT correlations will be $O(mk)$ bits, i.e., completely independent of n. This improves over the best known offline communication complexity (which was $O(mk \log n)$ bits).

---

[2] Yao GC-based approach does not seem to map naturally into the multiparty setting. This is true even for the three party semi-honest setting. A more complicated solution is possible [4], but much less practical than GMW-based approaches [6].

The total complexity (i.e., both online and offline) of our construction will asymptotically outperform existing constructions only for $n \leq ck$ where c is an arbitrary constant. For $n = \omega(k)$, the online cost of our protocol $O(mn\ell)$ dominates the total cost, but is still as efficient as existing constructions.

The main idea of our construction is to replace $\mathcal{C}_{\mathrm{WH}}^k$ with a code from a family of linear error correcting codes with the following special properties. (Our claim below is taken verbatim from [16].)

Claim ([16,5,8]). There exists a finite field $\mathbb{F}$ of characteristic 2 and an efficiently constructible family of linear error-correcting codes $C_K : \mathbb{F}^K \to \mathbb{F}^{N_K}$ with the following properties: (1) $N_K = O(K)$; (2) The dual distance of C_K is $\delta_K = \Omega(K)$; (3) The linear code C_K' spanned by all pointwise-products of pairs of codewords in C_K has minimal distance $\Delta_K = \Omega(K)$ and supports efficient decoding of up to $\mu_K = \Omega(K)$ errors. (The pointwise product of $(c_1, \ldots, c_N)$ and $(c_1', \ldots, c_N')$ is $(c_1 c_1', \ldots, c_N c_N')$.)

The last property implies that C_K also has minimal distance $d_K = \Omega(K)$.

Setting $N_K = k$ and $K \geq \log n$ is enough to provide the desired improvements stated above. The security level provided by this construction will be $\log(2^{d_K}/n^2) = \Omega(k)$ for n polynomial in k.

6 Optimizing the Reduction from $\binom{n}{1}$-OT_ℓ^m to OT_k^k

In our OT extension protocol, the OT_m^k primitive is reduced to OT_k^k. Further, the roles of $\mathcal{R}$ and $\mathcal{S}$ are reversed in our application of the reduction in our protocol. We provide an optimization that exploits this fact. This optimization was independently discovered by us and by Schneider and Zohner [27].

The main idea of the optimization is that inside the OT extension protocol of IKNP (as well as our protocol) 1-out of-2 OT of very long (m-bit long) random-looking correlated strings is executed. We cut the communication almost in half by OT-sending a PRG seed used to generate the strings. In other words, we obtain efficiency improvements by employing pseudorandom additive sharing instead of a completely random additive sharing. Because the strings need to be correlated in a specific way, a "correction" string needs to be sent so that exactly the right secret is recovered.

Note that this technique can also be applied to the IKNP construction. Such an application would reduce the IKNP cost of OT_ℓ^m from $m(2k+2\ell)$ to $m(k+2\ell)$. Observe that the reduced costs also have an impact on the oblivious key transfer phase (by constant factor 4/3) of Yao-based constructions where $\ell = k$.

For the case of 1-out-of-2 1-bit OT extension with 160-bit security, we get an improvement factor of ≈ 3.15 over the protocol of [14], and an improvement factor of ≈ 1.5 over the optimized IKNP protocol. See Appendix A for a detailed description of the protocol. We stress that Tables 1 and 2 do not take into account the optimizations described in this section.

Acknowledgments. We thank Yuval Ishai for many useful discussions, and the anonymous referees for their comments.

The first author was supported in part by the Intelligence Advanced Research Project Activity (IARPA) via Department of Interior National Business Center (DoI / NBC) contract Number D11PC20194. The U.S. Government is authorized to reproduce and distribute reprints for Governmental purposes notwithstanding any copyright annotation thereon. Disclaimer: The views and conclusions contained herein are those of the authors and should not be interpreted as necessarily representing the official policies or endorsements, either expressed or implied, of IARPA, DoI/NBC, or the U.S. Government.

References

1. Applebaum, B.: Pseudorandom generators with long stretch and low locality from random local one-way functions. In: 44th Annual ACM Symposium on Theory of Computing, pp. 805–816. ACM Press (2012)
2. Applebaum, B., Ishai, Y., Kushilevitz, E.: Cryptography in NC^0. In: 45th Annual Symposium on Foundations of Computer Science, pp. 166–175. IEEE Computer Society Press (October 2004)
3. Beaver, D.: Correlated pseudorandomness and the complexity of private computations. In: 28th Annual ACM Symposium on Theory of Computing, pp. 479–488. ACM Press (May 1996)
4. Beaver, D., Micali, S., Rogaway, P.: The round complexity of secure protocols. In: 22nd Annual ACM Symposium on Theory of Computing, pp. 503–513. ACM Press (May 1990)
5. Chen, H., Cramer, R.: Algebraic geometric secret sharing schemes and secure multiparty computations over small fields. In: Dwork, C. (ed.) CRYPTO 2006. LNCS, vol. 4117, pp. 521–536. Springer, Heidelberg (2006)
6. Choi, S.G., Hwang, K.-W., Katz, J., Malkin, T., Rubenstein, D.: Secure multiparty computation of boolean circuits with applications to privacy in on-line marketplaces. In: Dunkelman, O. (ed.) CT-RSA 2012. LNCS, vol. 7178, pp. 416–432. Springer, Heidelberg (2012)
7. Even, S., Goldreich, O., Lempel, A.: A randomized protocol for signing contracts. C. ACM 28, 637–647 (1985)
8. Garcia, A., Stichtenoth, H.: On the asymptotic behavior of some towers of function fields over finite fields. Journal of Number Theory 61(2), 248–273 (1996)
9. Goldreich, O., Micali, S., Wigderson, A.: How to play any mental game, or a completeness theorem for protocols with honest majority. In: Aho, A. (ed.) 19th Annual ACM Symposium on Theory of Computing, pp. 218–229. ACM Press (May 1987)
10. Goldreich, O., Vainish, R.: How to solve any protocol probleman efficiency improvement. In: Pomerance, C. (ed.) CRYPTO 1987. LNCS, vol. 293, pp. 73–86. Springer, Heidelberg (1988)
11. Harnik, D., Ishai, Y., Kushilevitz, E., Nielsen, J.B.: OT-combiners via secure computation. In: Canetti, R. (ed.) TCC 2008. LNCS, vol. 4948, pp. 393–411. Springer, Heidelberg (2008)
12. Huang, Y., Evans, D., Katz, J., Malka, L.: Faster secure two-party computation using garbled circuits. In: USENIX Security Symposium (2011)
13. Impagliazzo, R., Rudich, S.: Limits on the provable consequences of one-way permutations. In: 21st Annual ACM Symposium on Theory of Computing, pp. 44–61. ACM Press (May 1989)

14. Ishai, Y., Kilian, J., Nissim, K., Petrank, E.: Extending oblivious transfers efficiently. In: Boneh, D. (ed.) CRYPTO 2003. LNCS, vol. 2729, pp. 145–161. Springer, Heidelberg (2003)

15. Ishai, Y., Kushilevitz, E., Ostrovsky, R., Sahai, A.: Cryptography with constant computational overhead. In: Ladner, R.E., Dwork, C. (eds.) 40th Annual ACM Symposium on Theory of Computing, pp. 433–442. ACM Press (May 2008)

16. Ishai, Y., Kushilevitz, E., Ostrovsky, R., Sahai, A.: Extracting correlations. In: 50th Annual Symposium on Foundations of Computer Science, pp. 261–270. IEEE Computer Society Press (2009)

17. Ishai, Y., Prabhakaran, M., Sahai, A.: Founding cryptography on oblivious transfer – efficiently. In: Wagner, D. (ed.) CRYPTO 2008. LNCS, vol. 5157, pp. 572–591. Springer, Heidelberg (2008)

18. Kilian, J.: Founding cryptography on oblivious transfer. In: STOC, pp. 20–31. ACM (1988)

19. Kolesnikov, V., Kumaresan, R.: Improved secure two-party computation via information-theoretic garbled circuits. In: Visconti, I., De Prisco, R. (eds.) SCN 2012. LNCS, vol. 7485, pp. 205–221. Springer, Heidelberg (2012)

20. Kolesnikov, V., Schneider, T.: Improved garbled circuit: Free XOR gates and applications. In: Aceto, L., Damgård, I., Goldberg, L.A., Halldórsson, M.M., Ingólfsdóttir, A., Walukiewicz, I. (eds.) ICALP 2008, Part II. LNCS, vol. 5126, pp. 486–498. Springer, Heidelberg (2008)

21. Lindell, Y., Zarosim, H.: On the feasibility of extending oblivious transfer. In: Sahai, A. (ed.) TCC 2013. LNCS, vol. 7785, pp. 519–538. Springer, Heidelberg (2013)

22. Malkhi, D., Nisan, N., Pinkas, B., Sella, Y.: Fairplay - secure two-party computation system. In: USENIX Security Symposium (2004)

23. Naor, M., Pinkas, B.: Computationally secure oblivious transfer. Journal of Cryptology 18(1), 1–35 (2005)

24. Jesper Buus Nielsen. Extending oblivious transfers efficiently - how to get robustness almost for free. IACR Cryptology ePrint Archive, 2007:215 (2007)

25. Nielsen, J.B., Nordholt, P.S., Orlandi, C., Burra, S.S.: A New Approach to Practical Active-Secure Two-Party Computation. In: Safavi-Naini, R. (ed.) CRYPTO 2012. LNCS, vol. 7417, pp. 681–700. Springer, Heidelberg (2012)

26. Pinkas, B., Schneider, T., Smart, N.P., Williams, S.C.: Secure two-party computation is practical. In: Matsui, M. (ed.) ASIACRYPT 2009. LNCS, vol. 5912, pp. 250–267. Springer, Heidelberg (2009)

27. Schneider, T., Zohner, M.: Private communication (2012)

28. Schneider, T., Zohner, M.: GMW vs. Yao? efficient secure two-party computation with low depth circuits. In: FC 2013 (2013)

29. Yao, A.: How to generate and exchange secrets. In: 27th Annual Symposium on Foundations of Computer Science, pp. 162–167. IEEE Computer Society Press (October 1986)

A   Optimizing the Reduction from $\binom{n}{1}$-OT^m_ℓ to OT^k_k

Construction 2 (Optimized 1-out-of-n OT Extension)

INPUT OF $\mathcal{S}$: m tuples $(x_{j,0}, \ldots, x_{j,n-1})$ of ℓ-bit strings, $1 \le j \le m$.

INPUT OF $\mathcal{R}$: m selection integers $\mathbf{r} = (r_1, \ldots, r_m)$ such that $0 \le r_j < n$ for $1 \le j \le m$.

COMMON INPUT: a security parameter k such that $k \ge n$, and Walsh-Hadamard codes $\mathcal{C}^k_{\mathrm{WH}} = (\mathbf{c}_0, \ldots, \mathbf{c}_{k-1})$.

ORACLE: random oracles $H : [m] \times \{0,1\}^k \to \{0,1\}^\ell$, and $G : \{0,1\}^k \to \{0,1\}^m$.

CRYPTOGRAPHIC PRIMITIVE: an ideal OT^k_m primitive.

1. $\mathcal{S}$ chooses $\mathbf{s} \leftarrow \{0,1\}^k$ at random. Let s_i denote the i-th bit of $\mathbf{s}$.
2. $\mathcal{R}$ forms a $(m \times k)$ matrix D by setting $\mathbf{d}_j = \mathbf{c}_{r_j}$. $\mathcal{R}$ then forms $m \times k$ matrices T_0, T_1 in the following way:
   - Set $\mathbf{t}^i_1 = G(v_i)$ for a randomly chosen $v_i \leftarrow \{0,1\}^k$.
   - Set $\mathbf{t}^i_0 = \mathbf{d}^i \oplus \mathbf{t}^i_1$.

   In the above, $\mathbf{t}^i_0, \mathbf{t}^i_1$ denotes the i-th column of matrices T_0, T_1 respectively. (Note that T_0, T_1 form a pseudorandom sharing of the matrix D.)
3. $\mathcal{S}$ and $\mathcal{R}$ interact with OT^k_k in the following way:
   - $\mathcal{S}$ acts as receiver with input $\{s_i\}_{i \in [k]}$.
   - $\mathcal{R}$ acts as sender with inputs $\{u_i, v_i\}_{i \in [k]}$, where each u_i is chosen uniformly at random from $\{0,1\}^k$. (Note v_i was already chosen by $\mathcal{R}$ in Step 2.)
   - $\mathcal{S}$ receives output $\{\mathbf{a}^i\}_{i \in [k]}$.

   $\mathcal{S}$ forms $k \times k$ matrix A such that the i-th column of A is the vector $\mathbf{a}^i$.
4. For each $i \in [k]$, $\mathcal{R}$ sends $w_i = G(u_i) \oplus \mathbf{t}^i_0$.
5. $\mathcal{S}$ forms $m \times k$ matrix Q such that
   - if $s_i = 0$, then $\mathbf{q}^i = w_i \oplus G(\mathbf{a}^i)$,
   - else if $s_i = 1$, then $\mathbf{q}^i = G(\mathbf{a}^i)$.

   Let $\mathbf{q}_j$ denote the j-th row of Q. (Note $\mathbf{q}^i = \mathbf{t}^i_{s_i}$. Note $\mathbf{q}_j = ((\mathbf{t}_{j,0} \oplus \mathbf{t}_{j,1}) \odot \mathbf{s}) \oplus \mathbf{t}_{j,0}$. Simplifying, $\mathbf{q}_j \oplus \mathbf{t}_{j,0} = \mathbf{c}_{r_j} \odot \mathbf{s}$.)
6. For $j \in [m]$ and for every $0 \le r < n$, $\mathcal{S}$ sends $y_{j,r} = x_{j,r} \oplus H(j, \mathbf{q}_j \oplus (\mathbf{c}_r \odot \mathbf{s}))$.
7. For $j \in [m]$, $\mathcal{R}$ recovers $z_j = y_{j,r_j} \oplus H(j, \mathbf{t}_{j,0})$.

The amortized cost per instance of the $\binom{n}{1}$-OT^m_ℓ protocol above is $(k + n\ell)$. This yields a OT^m_ℓ protocol whose amortized concrete cost per instance is $n\ell + (k/\log n)$ bits.

Time-Optimal Interactive Proofs for Circuit Evaluation

Justin Thaler*

Harvard University, School of Engineering and Applied Sciences

Abstract. Several research teams have recently been working toward the development of practical general-purpose protocols for verifiable computation. These protocols enable a computationally weak *verifier* to offload computations to a powerful but untrusted *prover*, while providing the verifier with a guarantee that the prover performed the requested computations correctly. Despite substantial progress, existing implementations require further improvements before they become practical for most settings. The main bottleneck is typically the extra effort required by the prover to return an answer with a guarantee of correctness, compared to returning an answer with no guarantee.

We describe a refinement of a powerful interactive proof protocol due to Goldwasser, Kalai, and Rothblum [20]. Cormode, Mitzenmacher, and Thaler [14] show how to implement the prover in this protocol in time $O(S \log S)$, where S is the size of an arithmetic circuit computing the function of interest. Our refinements apply to circuits with sufficiently "regular" wiring patterns; for these circuits, we bring the runtime of the prover down to $O(S)$. That is, our prover can evaluate the circuit with a guarantee of correctness, with only a constant-factor blowup in work compared to evaluating the circuit with no guarantee.

We argue that our refinements capture a large class of circuits, and we complement our theoretical results with experiments on problems such as matrix multiplication and determining the number of distinct elements in a data stream. Experimentally, our refinements yield a 200x speedup for the prover over the implementation of Cormode et al., and our prover is less than 10x slower than a C++ program that simply evaluates the circuit. Along the way, we describe a special-purpose protocol for matrix multiplication that is of interest in its own right.

Our final contribution is the design of an interactive proof protocol targeted at general data parallel computation. Compared to prior work, this protocol can more efficiently verify complicated computations as long as that computation is applied independently to many different pieces of data.

1 Introduction

Protocols for verifiable computation enable a computationally weak *verifier* $\mathcal{V}$ to offload computations to a powerful but untrusted *prover* $\mathcal{P}$. These protocols aim to provide the verifier with a guarantee that the prover performed the requested computations correctly, without requiring $\mathcal{V}$ to perform the computations herself.

Surprisingly powerful protocols for verifiable computation were discovered within the computer science theory community several decades ago, in the form of interactive proofs (IPs) and their brethren, interactive arguments (IAs) and probabilistically

---

* Supported by an NSF Graduate Research Fellowship and NSF grants CNS-1011840 and CCF-0915922.

R. Canetti and J.A. Garay (Eds.): CRYPTO 2013, Part II, LNCS 8043, pp. 71–89, 2013.

checkable proofs (PCPs). In these protocols, the prover $\mathcal{P}$ solves a problem using her (possibly vast) computational resources, and tells $\mathcal{V}$ the answer. $\mathcal{P}$ and $\mathcal{V}$ then engage in a randomized protocol involving the exchange of one or more messages. During this exchange, $\mathcal{P}$'s goal is to convince $\mathcal{V}$ that the answer is correct.

Results quantifying the power of IPs, IAs, and PCPs are some of the most celebrated in all of computational complexity theory, but until recently they were mainly of theoretical interest, far too inefficient for actual deployment. In fact, the main applications of these results have traditionally been in negative applications – showing that many problems are just as hard to approximate as they are to solve exactly.

However, the surging popularity of cloud computing has brought renewed interest in positive applications of protocols for verifiable computation. A typical motivating scenario is as follows. A business processes billions or trillions of transactions a day. The volume is sufficiently high that the business cannot or will not store and process the transactions on its own. Instead, it offloads the processing to a commercial cloud computing service. The offloading of any computation raises issues of trust: the business may be concerned about relatively benign events like dropped transactions, buggy algorithms, or uncorrected hardware faults, or the business may be more paranoid and fear that the cloud operator is deliberately deceptive or has been externally compromised. Either way, each time the business poses a query to the cloud, the business may demand that the cloud also provide a guarantee that the returned answer is correct.

This is precisely what protocols for verifiable computation accomplish, with the cloud acting as the prover in the protocol, and the business acting as the verifier. In this paper, we describe a refinement of an existing general-purpose protocol originally due to Goldwasser, Kalai, and Rothblum [14, 20]. When they are applicable, our techniques achieve asymptotically optimal runtime for the prover, and we demonstrate that they yield protocols that are significantly closer to practicality than prior work.

We also make progress toward addressing another issue of existing interactive proof implementations: their applicability. The protocol of Goldwasser et al. (henceforth the GKR protocol) applies in principle to any problem computed by a small-depth arithmetic circuit, but this is not the case when more fine-grained considerations of prover and verifier efficiency are taken into account. In brief, existing implementations of interactive proof protocols for circuit evaluation require that the circuit have a highly regular wiring pattern [14, 37]. If this is not the case, then these implementations require the verifier to perform an expensive (though data-independent) preprocessing phase to pull out information about the wiring of the circuit, and they require a substantial factor blowup (logarithmic in the circuit size) in runtime for the prover relative to evaluating the circuit without a guarantee of correctness. Developing a protocol that avoids these pitfalls and applies to more general computations remains an important open question.

Our approach is the following. We do not have a magic bullet for dealing with irregular wiring patterns: if we want to avoid an expensive pre-processing phase for the verifier and minimize the blowup in runtime for the prover, we do need to make an assumption about the structure of the circuit we are verifying. Acknowledging this, we ask whether there is some general structure in real-world computations that we can leverage for efficiency gains.

To this end, we design a protocol that is highly efficient for data parallel computation. By data parallel computation, we mean any setting in which one applies the same computation independently to many pieces of data. Many outsourced computations are data parallel, with Amazon Elastic MapReduce[1] being one prominent example of a cloud computing service targeted specifically at data parallel computations. Crucially, we do not want to make significant assumptions on the sub-computation that is being applied, and in particular we want to handle sub-computations computed by circuits with highly irregular wiring patterns.

The verifier in our protocol still has to perform an offline phase to pull out information about the wiring of the circuit, but the cost of this phase is proportional to the size of a *single* instance of the sub-computation, avoiding any dependence on the number of pieces of data to which the sub-computation is applied. Similarly, the blowup in runtime suffered by the prover is the same as it would be if the prover had run the basic GKR protocol on a single instance of the sub-computation.

Our final contribution is to describe a new protocol specific to matrix multiplication that is of interest in its own right. It avoids circuit evaluation entirely, and reduces the overhead of the prover (relative to running *any* unverifiable algorithm) to an additive low-order term. A major message of our results is that the more structure that exists in a computation, the more efficiently it can be verified, and that this structure exists in many real-world computations.

1.1 Prior Work

Work on Interactive Proofs. Goldwasser, Kalai, and Rothblum described a powerful general-purpose interactive proof protocol in [20]. This protocol is framed in the context of *circuit evaluation*. Given a layered arithmetic circuit C of depth d, size $S(n)$, and fan-in 2, the GKR protocol allows a prover to evaluate C with a guarantee of correctness in time $\text{poly}(S(n))$, while the verifier runs in time $\tilde{O}(n + d \log S(n))$, where n is the length of the input and the $\tilde{O}$ notation hides polylogarithmic factors in n.

Cormode, Mitzenmacher, and Thaler showed how to bring the runtime of the prover in the GKR protocol down from $\text{poly}(S(n))$ to $O(S(n) \log S(n))$ [14]. They also built a full implementation of the protocol and ran it on benchmark problems. These results demonstrated that the protocol does indeed save the verifier significant time in practice (relative to evaluating the circuit locally); they also demonstrated surprising scalability for the prover, although the prover's runtime remained a major bottleneck. With the implementation of [14] as a baseline, Thaler et al. [35] described a parallel implementation of the GKR protocol that achieved 40x-100x speedups for the prover and 100x speedups for the (already fast) implementation of the verifier.

Vu, Setty, Blumberg, and Walfish [37] further refine and extend the implementation of Cormode et al. [14]. In particular, they combine the GKR protocol with a compiler from a high-level programming language so that programmers do not have to explicitly express computation in the form of arithmetic circuits as was the case in the implementation of [14]. This substantially extends the reach of the implementation, but it should be noted that their approach generates circuits with irregular wiring patterns, and hence

---

[1] http://aws.amazon.com/elasticmapreduce/

only works in a *batching* model, where the cost of a fairly expensive offline setup phase is amortized by verifying many instances of a single computation in batch. They also build a hybrid system that statically evaluates whether it is better to use the GKR protocol or a different, cryptography-based argument system called Zaatar (see Section 1.1), and runs the more efficient of the two protocols in an automated fashion.

A growing line of work studies protocols for verifiable computation in the context of *data streaming*. In this context, the goal is not just to save the verifier time (compared to doing the computation without a prover), but also to save the verifier space. The protocols developed in this line of work allow the client to make a single streaming pass over the input (which can occur, for example, while the client is uploading data to the cloud), keeping only a very small summary of the data set. The interactive version of this model was introduced by Cormode, Thaler, and Yi [15], who observed that many protocols from the interactive proofs literature, including the GKR protocol, can be made to work in this restrictive setting. The observations of [15] imply that all of our protocols also work with streaming verifiers. Non-interactive variants of the streaming interactive proofs model have also been studied in detail [12, 13, 22, 25].

Work on Argument Systems. There has been a lot of work on the development of efficient interactive arguments, which are essentially interactive proofs that are secure only against dishonest provers that run in polynomial time. A substantial body of work in this area has focused on the development of protocols targeted at specific problems (e.g. [2, 5, 16]). Other works have focused on the development of general-purpose argument systems. Several papers in this direction (e.g. [8, 10, 11, 18]) have used fully homomorphic encryption, which unfortunately remains impractical despite substantial recent progress. Work in this category by Chung et al. [10] focuses on streaming settings, and is therefore particularly relevant.

Several research teams have been pursuing the development of general-purpose argument systems that might be suitable for practical use. Theoretical work by Ben-Sasson et al. [4] focuses on the development of short PCPs that might be suitable for use in practice – such PCPs can be compiled into efficient interactive arguments. As short PCPs are often a bottleneck in the development of efficient argument systems, other works have focused on avoiding their use [3, 6, 7, 19]. In particular, Gennaro et al. [19] and Bitansky et al. [9] develop argument systems with a clear focus on implementation potential. Very recent work by Parno et al. [28] describes a near-practical general-purpose implementation, called Pinocchio, of an argument system based on [19]. Pinocchio is additionally non-interactive and achieves public verifiability.

Another line of implementation work focusing on general-purpose interactive argument systems is due to Setty et al. [31–33]. This line of work begins with a base argument system due to Ishai et al. [23], and substantially refines the theory to achieve an implementation that approaches practicality. The most recent system in this line of work is called Zaatar [33], and is also based on the work of Gennaro et al. [19]. An empirical comparison of the GKR-based approach and Zaatar performed by Vu et al. [37] finds the GKR approach to be significantly more efficient for quasi-straight-line computations (e.g. programs with relatively simple control flow), while Zaatar is appropriate for programs with more complicated control flow.

1.2   Our Contributions

Our primary contributions are three-fold. Our first contribution addresses one of the biggest remaining obstacles to achieving a truly practical implementation of the GKR protocol: the logarithmic factor overhead for the prover. That is, Cormode et al. show how to implement the prover in time $O(S(n) \log S(n))$, where $S(n)$ is the size of the arithmetic circuit to which the GKR protocol is applied, down from the $\Omega(S(n)^3)$ time required for a naive implementation. The hidden constant in the Big-Oh notation is at least 3, and the $\log S(n)$ factor translates to well over an order of magnitude, even for circuits with a few million gates. We remove this logarithmic factor, bringing $\mathcal{P}$'s runtime down to $O(S(n))$ for a large class of circuits. Informally, our results apply to any circuit whose wiring pattern is sufficiently "regular". We formalize the class of circuits to which our results apply in Theorem 1.

We experimentally demonstrate the generality and effectiveness of Theorem 1 via two case studies. Specifically, we apply an implementation of the protocol of Theorem 1 to a circuit computing matrix multiplication (MATMULT), as well as to a circuit computing the number of distinct items in a data stream (DISTINCT). Experimentally, our refinements yield a 200x speedup for the prover over the state of the art implementation of Cormode et al. [14]. A serial implementation of our prover is less than 10x slower than a C++ program that simply evaluates the circuit sequentially, a slowdown that is likely tolerable in realistic outsourcing scenarios where cycles are plentiful for the prover.

Our second contribution is to specify a highly efficient protocol for verifiably outsourcing arbitrary data parallel computation. Compared to prior work, this protocol can more efficiently verify complicated computations, as long as that computation is applied independently to many different pieces of data. We formalize this protocol and its efficiency guarantees in Theorem 2.

Our third contribution is to describe a new protocol specific to matrix multiplication that we believe to be of interest in its own right. This protocol is formalized in Theorem 3. Given any *unverifiable* algorithm for $n \times n$ matrix multiplication that requires time $T(n)$ using space $s(n)$, Theorem 3 allows the prover to run in time $T(n) + O(n^2)$ using space $s(n) + o(n^2)$. Note that Theorem 3, which is specific to matrix multiplication, is much less general than Theorem 1, which applies to any circuit with a sufficiently regular wiring pattern. However, Theorem 3 achieves optimal runtime and space usage for the prover up to leading constants, assuming there is no $O(n^2)$ time algorithm for matrix multiplication. While these properties are also satisfied by a classic protocol due to Freivalds [17], the protocol of Theorem 3 is significantly more amenable for use as a primitive when verifying computations that repeatedly invoke matrix multiplication. We complement Theorem 3 with experimental results demonstrating its extreme efficiency.

Do to space constraints, full proofs of are omitted from this extended abstract, and can be found in the full version of the paper.

2 Preliminaries

We begin by defining a valid interactive proof protocol for a function f.

Definition 1. *Consider a prover $\mathcal{P}$ and verifier $\mathcal{V}$ who wish to compute a function f : $\{0,1\}^n \to \mathcal{R}$ for some set $\mathcal{R}$. After the input is observed, $\mathcal{P}$ and $\mathcal{V}$ exchange a sequence of messages. Denote the output of $\mathcal{V}$ on input x, given prover $\mathcal{P}$ and $\mathcal{V}$'s random bits R, by $out(\mathcal{V},x,R,\mathcal{P})$. $\mathcal{V}$ can output $\perp$ if $\mathcal{V}$ is not convinced that $\mathcal{P}$'s claim is valid. We say $\mathcal{P}$ is a* valid prover *with respect to $\mathcal{V}$ if for all inputs x, $Pr_R[out(\mathcal{V},x,R,\mathcal{P})=f(x)]=$ 1. The property that there is at least one valid prover $\mathcal{P}$ with respect to $\mathcal{V}$ is called* completeness. *We say $\mathcal{V}$ is a* valid verifier *for f with* soundness probability δ *if there is at least one valid prover $\mathcal{P}$ with respect to $\mathcal{V}$, and for all provers $\mathcal{P}'$ and inputs x, $Pr[out(\mathcal{V},A,R,\mathcal{P}') \notin \{f(x),\perp\}] \leq \delta$. A prover-verifier pair $(\mathcal{P},\mathcal{V})$ is a* valid interactive proof protocol *for f if $\mathcal{V}$ is a valid verifier for f with soundness probability $1/3$, and $\mathcal{P}$ is a valid prover with respect to $\mathcal{V}$. If $\mathcal{P}$ and $\mathcal{V}$ exchange r messages, we say the protocol has $\lceil r/2 \rceil$ rounds.*

Informally, the completeness property guarantees that an honest prover will convince the verifier that the claimed answer is correct, while the soundness property ensures that a dishonest prover will be caught with high probability. An *interactive argument* is an interactive proof where the soundness property holds only against polynomial-time provers $\mathcal{P}'$. We remark that the constant $1/3$ used for the soundness probability in Definition 1 is chosen for consistency with the interactive proofs literature, where $1/3$ is used by convention. In our actual implementation, the soundness probability will always be less than 2^{-45}.

Cost Model. Whenever we work over a finite field $\mathbb{F}$, we assume that a single field operation can be computed in a single machine operation. For example, when we say that the prover $\mathcal{P}$ in our interactive protocols requires time $O(S(n))$, we mean that $\mathcal{P}$ must perform $O(S(n))$ additions and multiplications within the finite field over which the protocol is defined.

Input Representation. Following prior work [12, 14, 15], all of the protocols we consider can handle inputs specified in a general data stream form. Each element of the stream is a tuple (i,δ), where $i \in [n]$ and δ is an integer. The δ values may be negative, thereby modeling deletions. The data stream implicitly defines a frequency vector a, where a_i is the sum of all δ values associated with i in the stream. When checking the evaluation of a circuit C, we consider the inputs to C to be the entries of the frequency vector a. We emphasize that in all of our protocols, $\mathcal{V}$ only needs to see the raw stream and not the aggregated frequency vector a. Notice that we may interpret the frequency vector a as an object other than a vector, such as a matrix or a string. For example, in MATMULT, the data stream defines two matrices to be multiplied.

When we refer to a *streaming verifier* with space usage $s(n)$, we mean that the verifier can make a single pass over the stream of tuples defining the input, regardless of their ordering, while storing at most $s(n)$ elements in the finite field over which the protocol is defined.

Problem Definitions. To focus our discussion, we give special attention to two problems also considered in prior work [14, 28, 31–33, 35, 37]. In the MATMULT problem, the input consists of two $n \times n$ matrices $A, B \in \mathbb{Z}^{n \times n}$, and the goal is to compute the matrix product $A \cdot B$. In the DISTINCT problem, the input is a data steam consisting of m tuples (i, δ) from a universe of size n. The stream defines a frequency vector a, and the goal is to compute $|\{i : a_i \neq 0\}|$, the number of items with non-zero frequency.

Additional Notation. Let $\mathbb{F}$ be a field. For any d-variate polynomial $p(x_1, \ldots, x_d)$: $\mathbb{F}^d \to \mathbb{F}$, we use $\deg_i(p)$ to denote the degree of p in variable i. A d-variate polynomial p is said to be *multilinear* if $\deg_i(p) = 1$ for all $i \in [d]$. Given a function $V : \{0,1\}^d \to \{0,1\}$ whose domain is the d-dimensional Boolean hypercube, the *multilinear extension* (MLE) of V over $\mathbb{F}$, denoted $\tilde{V}$, is the unique multilinear polynomial $\mathbb{F}^d \to \mathbb{F}$ that agrees with V on all Boolean-valued inputs, i.e., $\tilde{V}(x) = V(x)$ for all $x \in \{0,1\}^d$.

3 Time-Optimal Protocols for Circuit Evaluation

3.1 Technical Background

Sum-Check Protocol. Our main technical tool is the well-known sum-check protocol of Lund et al. [27], and we briefly describe this protocol and summarize the properties that are most important in our analysis. Suppose we are given a v-variate polynomial g defined over a finite field $\mathbb{F}$, such that $\deg_i(g) = O(1)$ for all $i \in \{1, \ldots, v\}$. The purpose of the sum-check protocol is to compute the sum:

$$H := \sum_{b_1 \in \{0,1\}} \sum_{b_2 \in \{0,1\}} \cdots \sum_{b_v \in \{0,1\}} g(b_1, \ldots, b_v).$$

The protocol proceeds in v rounds as follows. In the first round, the prover sends a polynomial $g_1(X_1)$, and claims that $g_1(X_1) = \sum_{x_2, \ldots, x_v \in \{0,1\}^{v-1}} g(X_1, x_2, \ldots, x_v)$. Observe that if g_1 is as claimed, then $H = g_1(0) + g_1(1)$. Also observe that the polynomial $g_1(X_1)$ has degree $\deg_1(g) = O(1)$. Hence g_1 can be specified by sending the evaluation of g at each point in the $O(1)$-sized set $\{0, 1, \ldots, \deg_1(g)\}$.

Then, in round $j > 1$, $\mathcal{V}$ chooses a value r_{j-1} uniformly at random from $\mathbb{F}$ and sends r_{j-1} to $\mathcal{P}$. We refer to this step by saying that variable $j-1$ gets *bound* to value r_{j-1}. In return, the prover sends a polynomial $g_j(X_j)$, and claims that

$$g_j(X_j) = \sum_{(x_{j+1}, \ldots, x_v) \in \{0,1\}^{v-j}} g(r_1, \ldots, r_{j-1}, X_j, x_{j+1}, \ldots, x_v). \tag{1}$$

The verifier then checks that $g_{j-1}(r_{j-1}) = g_j(0) + g_j(1)$, rejecting otherwise.

In the final round, the prover has sent $g_v(X_v)$ which is claimed to be $g(r_1, \ldots, r_{v-1}, X_v)$. $\mathcal{V}$ now checks that $g_v(r_v) = g(r_1, \ldots, r_v)$. Notice that in order to perform this check, the verifier needs to be able to evaluate $g(r_1, \ldots, r_v)$ without assistance from the prover. If this test succeeds, and so do all previous tests, then the verifier accepts, and is convinced that $H = g_1(0) + g_1(1)$.

Discussion of Costs. For our purposes, the key cost of the sum-check protocol is the prover's runtime. Notice that the number of terms defining the value $g_j(i)$ in Equation (1) falls geometrically with j: in the jth message, there are only 2^{v-j} terms. The total number of terms that must be evaluated over the course of the protocol is therefore $O\left(\sum_{j=1}^{v} 2^{v-j}\right) = O(2^v)$. Consequently, if $\mathcal{P}$ is given oracle access to (evaluations of) the polynomial g, then $\mathcal{P}$ will require $O(2^v)$ time. Unfortunately, in our applications $\mathcal{P}$ will not have oracle access to g. The key to our results is to show that in our applications $\mathcal{P}$ can nonetheless evaluate g at the necessary points in $O(2^v)$ total time.

The GKR Protocol at a Glance. In the GKR protocol, $\mathcal{P}$ and $\mathcal{V}$ first agree on an arithmetic circuit C of fan-in 2 over a finite field $\mathbb{F}$ computing the function of interest (C may have multiple outputs). Each gate of C performs an addition or multiplication over $\mathbb{F}$. C is assumed to be in layered form, meaning that the circuit can be decomposed into layers, and wires only connect gates in adjacent layers. Suppose the circuit has depth d; we will number the layers from 1 to d with layer d referring to the input layer, and layer 1 referring to the output layer.

In the first message, $\mathcal{P}$ tells $\mathcal{V}$ the (claimed) output of the circuit. The protocol then works its way in iterations towards the input layer, with one iteration devoted to each layer. The purpose of iteration i is to reduce a claim about the values of the gates at layer i to a claim about the values of the gates at layer $i+1$, in the sense that it is safe for $\mathcal{V}$ to assume that the first claim is true as long as the second claim is true. This reduction is accomplished by applying the sum-check protocol to a certain polynomial $f^{(i)}$.

More concretely, the GKR protocol starts with a claim about the values of the output gates of the circuit, but $\mathcal{V}$ cannot check this claim without evaluating the circuit herself, which is precisely what we want to avoid. So the first iteration uses a sum-check protocol to reduce this claim about the outputs to a claim about the gate values at layer 2 (more specifically, to a claim about an evaluation of the multilinear extension (MLE) of the gate values at layer 2). Once again, $\mathcal{V}$ cannot check this claim herself, so the second iteration uses another sum-check protocol to reduce the latter claim to a claim about the gate values at layer 3, and so on. Eventually, $\mathcal{V}$ is left with a claim about the inputs to the circuit, and $\mathcal{V}$ can check this claim on her own.

In summary, the GKR protocol uses a sum-check protocol at each level of the circuit to enable $\mathcal{V}$ to go from verifying a randomly chosen evaluation of the MLE of the gate values at layer i to verifying a (different) evaluation of the MLE of the gate values at layer $i+1$. Importantly, apart from the input layer and output layer, $\mathcal{V}$ does not ever see all of the gate values at a layer. Instead, $\mathcal{V}$ relies on $\mathcal{P}$ to do the hard work of actually evaluating the circuit, and uses the power of the sum-check protocol to force $\mathcal{P}$ to be consistent and truthful over the course of the protocol.

Further Details. Suppose we are given a layered arithmetic circuit C of depth d and fan-in two. Let S_i denote the number of gates at layer i of the circuit C. Assume S_i is a power of 2 and let $S_i = 2^{s_i}$. To explain how each iteration of the GKR protocol proceeds, we must introduce several functions, each of which encodes certain information about the circuit. Number the gates at layer i from 0 to $S_i - 1$, and let $V_i : \{0,1\}^{s_i} \to \mathbb{F}$ denote the function that takes as input a binary gate label, and outputs the corresponding gate's value at layer i. The GKR protocol makes use of the multilinear extension $\tilde{V}_i$ of the function V_i.

The GKR protocol also makes use of the notion of a "wiring predicate" that encodes which pairs of wires from layer $i+1$ are connected to a given gate at layer i in C. We define two functions, add_i and mult_i mapping $\{0,1\}^{s_i+2s_{i+1}}$ to $\{0,1\}$, which together constitute the wiring predicate of layer i of C. Specifically, these functions take as input three gate labels (j_1, j_2, j_3), and return 1 if gate j_1 at layer i is the addition (respectively, multiplication) of gates j_2 and j_3 at layer $i+1$, and return 0 otherwise. Let $\widetilde{\text{add}}_i$ and $\widetilde{\text{mult}}_i$ denote the multilinear extensions of add_i and mult_i respectively. Finally, let $\beta_{s_i}(z,p)$ denote the function $\beta_{s_i}(z,p) = \prod_{j=1}^{s_i}((1-z_j)(1-p_j) + z_j p_j)$. It can be shown that for any $z \in \mathbb{F}^{s_i}$,

$$\tilde{V}_i(z) = \sum_{(p,\omega_1,\omega_2) \in \{0,1\}^{s_i+2s_{i+1}}} f^{(i)}(p,\omega_1,\omega_2), \text{ where}$$

$$f^{(i)}(p,\omega_1,\omega_2) = \beta_{s_i}(z,p) \cdot \left(\widetilde{\text{add}}_i(p,\omega_1,\omega_2)(\tilde{V}_{i+1}(\omega_1) + \tilde{V}_{i+1}(\omega_2)) + \widetilde{\text{mult}}_i(p,\omega_1,\omega_2)\tilde{V}_{i+1}(\omega_1) \cdot \tilde{V}_{i+1}(\omega_2) \right).$$

Iteration i begins with a claim by $\mathcal{P}$ about the value of $\tilde{V}_i(z)$ for some $z \in \mathbb{F}^{s_i}$. In order to verify this claim, the sum-check protocol is applied to the polynomial $f^{(i)}$. However, $\mathcal{V}$ can only execute her part of the sum-check protocol if she can evaluate the polynomial $f^{(i)}$ at a random point $f^{(i)}(r_1, \ldots, r_{s_i+2s_{i+1}})$. In particular, this requires evaluating $\tilde{V}_{i+1}(\omega_2^*)$, and $\tilde{V}_{i+1}(\omega_1^*)$, but $\mathcal{V}$ cannot perform these evaluations on her own without evaluating the circuit. At a high level, $\mathcal{V}$ instead asks $\mathcal{P}$ to simply tell her these two values, and uses iteration $i+1$ to verify that these values are as claimed. The full version of the paper spells out the remaining details.

3.2 Achieving Optimal Prover Runtime for Regular Circuits

In Theorem 1 below, we describe a protocol for circuit evaluation that brings $\mathcal{P}$'s runtime down to $O(S(n))$ for a large class of circuits, while maintaining the same verifier runtime as in prior implementations of the GKR protocol. Informally, Theorem 1 applies to any circuit whose wiring pattern is sufficiently "regular".

Our protocol follows the same outline as the GKR protocol, in that we proceed in iterations from the output layer of the circuit to the input layer, using a sum-check protocol at iteration i to reduce a claim about the gate values at layer i to a claim about the gate values at layer $i+1$. However, at each iteration i we apply the sum-check protocol to a carefully chosen polynomial that differs from the ones used in prior work [14, 20]. In each round j of the sum-check protocol, our choice of polynomial allows $\mathcal{P}$ to reuse work from prior rounds in order to compute the prescribed message for round j, allowing us to shave a $\log S(n)$ factor from the runtime of $\mathcal{P}$ relative to the $O(S(n) \log S(n))$-time implementation due to Cormode et al. [14].

Specifically, at iteration i, the polynomial $f^{(i)}$ that is used in the GKR protocol is defined over $\log S_i + 2\log S_{i+1}$ variables, where S_i is the number of gates at layer i. The "truth table" of $f^{(i)}$ is sparse on the Boolean hypercube, in the sense that $f^{(i)}(x)$ is non-zero for at most S_i of the $S_i \cdot S_{i+1}^2$ inputs $x \in \{0,1\}^{\log S_i + 2\log S_{i+1}}$. Cormode et al. leverage this sparsity to bring the runtime of $\mathcal{P}$ in iteration i down to $O(S_i \log S_i)$ from a naive bound of $\Omega(S_i \cdot S_{i+1}^2)$. However, this same sparsity prevents $\mathcal{P}$ from reusing work from prior iterations as we seek to do.

In contrast, we use a polynomial $g^{(i)}$ defined over only $\log S_i$ variables rather than $\log S_i + 2\log S_{i+1}$ variables. Moreover, the truth table of $g^{(i)}$ is dense on the Boolean hypercube, in the sense that $g^{(i)}(x)$ may be non-zero for all of the S_i Boolean inputs

$x \in \{0,1\}^{\log S_i}$. This density allows $\mathcal{P}$ to reuse work from prior iterations in order to speed up her computation in round i of the sum-check protocol.

In more detail, in each round j of the sum-check protocol, the prover's prescribed message is defined via a sum over a large number of terms, where the number of terms falls geometrically fast with the round number j. Moreover, it can be shown that in each round j, each gate at layer $i + 1$ contributes to exactly one term of this sum [14]. Essentially, what we do is group the gates at layer $i + 1$ by the term of the sum to which they contribute. Each such group can be treated as a single unit, ensuring that in any round of the sum-check protocol, the amount of work $\mathcal{P}$ needs to do is proportional to the number of terms in the sum rather than the number of gates S_i at layer i.

Formal Statement. Our protocol makes use of the following functions that capture the wiring structure of an arithmetic circuit C.

Definition 2. *Let C be a layered arithmetic circuit of depth $d(n)$ and size $S(n)$ over finite field $\mathbb{F}$. For every $i \in \{1, \ldots, d-1\}$, let $in_1^{(i)} : \{0,1\}^{s_i} \to \{0,1\}^{s_{i+1}}$ and $in_2^{(i)} : \{0,1\}^{s_i} \to \{0,1\}^{s_{i+1}}$ denote the functions that take as input the binary label p of a gate at layer i of C, and output the binary label of the first and second in-neighbor of gate p respectively. Similarly, let $type^{(i)} : \{0,1\}^{s_i} \to \{0,1\}$ denote the function that takes as input the binary label p of a gate at layer i of C, and outputs 0 if p is an addition gate, and 1 if p is a multiplication gate.*

Intuitively, the following two definitions capture functions whose outputs are simple bit-wise transformations of their inputs.

Definition 3. *Let f be a function mapping $\{0,1\}^v$ to $\{0,1\}^{v'}$. Number the v input bits from 1 to v, and the v' output bits from 1 to v'. We say that f is regular if f can be evaluated on any input in constant time, and there is a subset of input bits $\mathcal{S} \subseteq [v]$ with $|\mathcal{S}| = O(1)$ such that:*

1. *Each input bit in $[v] \setminus \mathcal{S}$ affects $O(1)$ of the output bits of f. Moreover, for any $j \in [v] \setminus \mathcal{S}$, the set $\mathcal{S}_j$ of output bits affected by the jth input bit can be enumerated in constant time.*
2. *Each output bit of f depends on at most one input bit.*

Definition 4. *We say that $in_1^{(i)}$ and $in_2^{(i)}$ are similar if there is a set of output bits $\mathcal{T} \subseteq [s_{i+1}]$ with $|\mathcal{T}| = O(1)$ such that for all inputs x, the jth output bit of $in_1^{(i)}$ equals the jth output bit of $in_2^{(i)}$ for all $j \in [s_{i+1}] \setminus \mathcal{T}$.*

Theorem 1. *Let C be an arithmetic circuit, and suppose that for all layers i of C, $in_1^{(i)}$, $in_2^{(i)}$, and $type^{(i)}$ are regular. Suppose moreover that $in_1^{(i)}$ is similar to $in_2^{(i)}$ for all but $O(1)$ layers i of C. Then there is a valid interactive proof protocol $(\mathcal{P}, \mathcal{V})$ for the function computed by C, with the following costs. The total communication cost is $|\mathcal{O}| + O(d(n) \log S(n))$ field elements, where $|\mathcal{O}|$ is the number of outputs of C. The time cost to $\mathcal{V}$ is $O(n \log n + d(n) \log S(n))$, and $\mathcal{V}$ can make a single streaming pass over the input, storing $O(\log(S(n)))$ field elements. The time cost to $\mathcal{P}$ is $O(S(n))$.*

The asymptotic costs of the protocol whose existence is guaranteed by Theorem 1 are identical to those of the implementation of the GKR protocol due to Cormode et al. in [14], except that in Theorem 1 $\mathcal{P}$ runs in time $O(S(n))$ rather than $O(S(n)\log S(n))$. While the conditions of Theorem 1 may appear unnatural, our techniques in fact capture a large class of circuits. Theorem 1 applies for example to circuits computing naive $n \times n$ matrix multiplication (MATMULT), computing the number of distinct items in a data stream (DISTINCT), pattern matching (which is useful, e.g., for searching email data stored in the cloud), and FFTs. To the best of our our knowledge Theorem 1 yields the fastest known prover among all interactive proof protocols for DISTINCT and for pattern matching with sublinear space and communication costs. More importantly, we will leverage the techniques underlying Theorem 1 to achieve our improved protocol for data parallel computation described in Theorem 2.

Experimental Results. We implemented the protocols implied by Theorem 1 as applied to circuits computing MATMULT and DISTINCT. The circuits are over the field $\mathbb{F}_q$ with $q = 2^{61} - 1$. The soundness probability in all cases is less than 2^{-45} (this probability is proportional to $\frac{d(n)\log S(n)}{q}$). These experiments serve as case studies to demonstrate the feasibility of Theorem 1 in practice, and to quantify the improvements over prior implementations. While Section 5 describes a specialized protocol for MATMULT that is more efficient than the protocol implied by Theorem 1, MATMULT serves as an important case study for the costs of the more general protocol described in Theorem 1, and allows for direct comparison with prior implementation work that also evaluated general-purpose protocols via their performance on the MATMULT problem [14, 28, 32, 33, 35, 37].

The main takeaways of our experiments are as follows. When Theorem 1 is applicable, the prover in the resulting protocol is 200x-250x faster than the previous state of the art implementation of the GKR protocol, and is just 5x-10x times slower than a C++ program that simply evaluates the circuit with no correctness guarantee. The communication costs and the number of rounds required by our protocols are also 2x-3x smaller than the previous state of the art. The verifier in our implementation takes essentially the same amount of time as in prior implementations of the GKR protocol; this time is much smaller than the time to perform the computation locally without a prover. See Table 1 for detailed results – in this table, our comparison point is the implementation of Cormode et al. [14], with some of the refinements of Vu et al. [37] included.

Most of the 200x speedup can be attributed directly to our improvements in protocol design over prior work: the circuit for 512x512 matrix multiplication is of size 2^{28}, and hence our $\log S(n)$ factor improvement the runtime of $\mathcal{P}$ likely accounts for at

Table 1. Experimental results for Theorem 1. For the MATMULT problem, the Total Communication column does not count the communication required to specify the answer.

Problem	Implementation	Problem Size	$\mathcal{P}$ Time	$\mathcal{V}$ Time	Rounds	Total Communication	Circuit Eval Time
MATMULT	Previous state of the art	512 x 512	9759 s	0.10 s	767	17.97 KBs	6.07 s
MATMULT	Theorem 1	512 x 512	37.85 s	0.10 s	236	5.48 KBs	6.07 s
DISTINCT	Previous state of the art	$n = 2^{20}$	3400 s	0.20 s	3916	91.3 KBs	1.88 s
DISTINCT	Theorem 1	$n = 2^{20}$	17.28 s	0.20 s	1361	40.76 KBs	1.88 s

least a 28x speedup. The 3x reduction in the number of rounds accounts for another 3x speedup. The remaining speedup factor of roughly 2x may be due to a more streamlined implementation relative to prior work, rather than improved protocol design per se.

4 Verifying General Data Parallel Computations

Theorem 1 only applies to circuits with regular wiring patterns, as do other existing implementations of interactive proof protocols for circuit evaluation [14, 37]. For circuits with irregular wiring patterns, these implementations require the verifier to perform an expensive preprocessing phase (requiring time proportional to the size of the circuit) to pull out information about the wiring of the circuit, and they require a substantial factor blowup (logarithmic in the circuit size) in runtime for the prover relative to evaluating the circuit without a guarantee of correctness.

To address these bottlenecks, we do need to make an assumption about the structure of the circuit we are verifying. Ideally our assumption will be satisfied by many real-world computations. To this end, Theorem 2 below describes a protocol that is highly efficient for any data parallel computation, by which we mean any setting in which the same sub-computation is applied independently to many pieces of data, before possibly aggregating the results. We do not want to make significant assumptions on the sub-computation that is being applied (in particular, we want to handle sub-computations computed by circuits with irregular wiring patterns), but we are willing to assume that the sub-computation is applied to many pieces of data.

For example, Theorem 2 applies to arbitrary *counting queries* on a database. In a counting query, one applies some function independently to each row of the database and sums the results. For instance, one may ask "How many people in the database satisfy Property *P*?" Our protocol allows one to verifiably outsource such a counting query with overhead that depends minimally on the size of the database, but that necessarily depends on the complexity of the property *P*.

Overview of the Protocol. Let *C* be a circuit of size *S* with an arbitrary wiring pattern, and let C^* be a "super-circuit" that applies *C* independently to *B* different inputs before possibly aggregating the results in some fashion. For example, in the case of a counting query, the aggregation phase simply sums the results of the data parallel phase. We assume that the aggregation step is sufficiently simple that the aggregation itself can be verified using existing techniques such as the basic GKR protocol or Theorem 1, and we focus on verifying the data parallel part of the computation. For instance, in the case of a counting query, the aggregation phase simply sums the outputs, and this is easily handled via Theorem 1. We stress that our protocol applies even if there is no aggregation phase; in this case $\mathcal{P}$ will begin the protocol by sending $\mathcal{V}$ all outputs of C^*, and the protocol can then be used to prove the validity of those outputs.

If we naively apply the GKR protocol to the super-circuit C^*, $\mathcal{V}$ might have to perform an expensive pre-processing phase to evaluate the wiring predicate of C^* at the necessary locations – this would require time $\Omega(B \cdot S)$. Moreover, when applying the basic GKR protocol to C^*, $\mathcal{P}$ would require time $\Theta(B \cdot S \cdot \log(B \cdot S))$. A different approach was taken by Vu et al [37], who applied the GKR protocol *B* independent times, once for each copy of C. This causes both the communication cost and $\mathcal{V}$'s online check time to grow linearly with B, the number of sub-computations.

In contrast, our protocol achieves the best of both prior approaches. We observe that although each sub-computation C can have a very complicated wiring pattern, the super-circuit C^* is maximally regular between sub-computations, as the sub-computations do not interact at all. Therefore, each time the basic GKR protocol would apply the sum-check protocol to a polynomial derived from the wiring predicate of C^*, we instead use a simpler polynomial derived only from the wiring predicate of C. By itself, this is enough to ensure that $\mathcal{V}$'s pre-processing phase requires time only $O(S)$, rather than $O(B \cdot S)$ as in a naive application of the GKR protocol to C^*. That is, the cost of $\mathcal{V}$'s pre-processing phase in our protocol is proportional to the cost of applying the basic GKR protocol only to C, not to C^*.

Furthermore, by combining this observation with the ideas underlying Theorem 1, we can bring the runtime of $\mathcal{P}$ down to $O(B \cdot S \cdot \log S)$. That is, the blowup in runtime suffered by the prover, relative to performing the computation without a guarantee of correctness, is just a factor of $\log S$ – the same as it would be if the prover had run the basic GKR protocol on a *single* instance of the sub-computation.

Notation. Let C be an arithmetic circuit over $\mathbb{F}$ of depth d and size S with an arbitrary wiring pattern, and let C^* be the circuit of depth d and size $B \cdot S$ obtained by laying B copies of C side-by-side, where $B = 2^b$. We will use the same notation as in Section 3.1, using $*$'s to denote quantities referring to C^*. For example, layer i of C has size $S_i = 2^{s_i}$ and gate values specified by the function V_i, while layer i of C^* has size $S_i^* = 2^{s_i^*}$ and gate values specified by the function V_i^*. We denote the length of the input to C^* by n^*.

We assume at the start of our protocol that $\mathcal{P}$ has made a claim about $\tilde{V}_1^*(z)$ for some $z \in \mathbb{F}^{s_1^*}$, in the sense that it is safe for $\mathcal{V}$ to believe $\mathcal{P}$ has followed the prescribed protocol as long as $\tilde{V}_1^*(z)$ is as claimed. Such a claim about $\tilde{V}_1^*(z)$ would be obtained by first applying existing verification techniques such as Theorem 1 to the aggregation phase of the data parallel computation.

Theorem 2. *For any point $z \in \mathbb{F}^{s_1^*}$, there is a valid interactive proof protocol for computing $\tilde{V}_1^*(z)$ with the following costs. $\mathcal{V}$ spends $O(S)$ time in a pre-processing phase, and $O(n^* \log n^* + d \cdot \log(B \cdot S))$ time in an online verification phase. $\mathcal{P}$ runs in total time $O(S \cdot B \cdot \log S)$. The total communication is $O(d \cdot \log(B \cdot S))$ field elements.*

Proof sketch: Consider layer i of C^*. Let $p = (p_1, p_2) \in \{0,1\}^{s_i} \times \{0,1\}^b$ be the binary label of a gate at layer i of C^*, where p_2 specifies which "copy" of C the gate is in, while p_1 designates the label of the gate within the copy. Similarly, let $\omega = (\omega_1, \omega_2) \in \{0,1\}^{s_{i+1}} \times \{0,1\}^b$ and $\gamma = (\gamma_1, \gamma_2) \in \{0,1\}^{s_{i+1}} \times \{0,1\}^b$ be the labels of two gates at layer $i+1$. It is straightforward to check that for all $(p_1, p_2) \in \{0,1\}^{s_i} \times \{0,1\}^b$,

$$V_i^*(p_1, p_2) = \sum_{\omega_1 \in \{0,1\}^{s_{i+1}}} \sum_{\gamma_1 \in \{0,1\}^{s_{i+1}}} g^{(i)}(p_1, p_2, \omega_1, \gamma_1), \text{ where } g^{(i)}(p_1, p_2, \omega_1, \gamma_1) \text{ is}$$

defined as:

$$\beta_{s_i^*}(z, (p_1, p_2)) \cdot (\widetilde{\text{add}}_i(p_1, \omega_1, \gamma_1) \, (\tilde{V}_{i+1}^*(\omega_1, p_2) + \tilde{V}_{i+1}^*(\gamma_1, p_2)) + \widetilde{\text{mult}}_i(p_1, \omega_1, \gamma_1) \, (\tilde{V}_{i+1}^*(\omega_1, p_2) \cdot \tilde{V}_{i+1}^*(\gamma_1, p_2)))$$

Essentially, the above says that a gate $p = (p_1, p_2) \in \{0,1\}^{s_i+b}$ is connected to gates $\omega = (\omega_1, \omega_2) \in \{0,1\}^{s_{i+1}+b}$ and $\gamma = (\gamma_1, \gamma_2) \in \{0,1\}^{s_{i+1}+b}$ if and only if p, ω, and γ are all in the same copy of C, and p is connected to ω and γ within the copy. The above derivation can be shown to imply that for any $z \in \mathbb{F}^{s_i^*}$,

$$\tilde{V}_i^*(z) = \sum_{(p_1, p_2, \omega_1, \gamma_1) \in \{0,1\}^{s_i} \times \{0,1\}^{b} \times \{0,1\}^{s_{i+1}} \times \{0,1\}^{s_{i+1}}} g^{(i)}(p_1, p_2, \omega_1, \gamma_1).$$

Thus, in iteration i of our protocol, we apply the sum-check protocol to $g^{(i)}$. This reduces $\mathcal{P}$'s claim about $\tilde{V}_i^*(z)$ to a claim about $\tilde{V}_{i+1}^*(z')$ for some $z' \in \mathbb{F}^{s_{i+1}}$, exactly as in the ith iteration of the GKR protocol.

Costs for $\mathcal{V}$. The bottleneck in $\mathcal{V}$'s runtime is that, in the last round of the sum-check protocol, $\mathcal{V}$ must evaluate $g^{(i)}$ at a single point. This requires evaluating $\beta_{s_i^*}$, $\tilde{\text{add}}_i$, $\tilde{\text{mult}}_i$, and $\tilde{V}_{i+1}^*$ at a constant number of points. The $\tilde{V}_{i+1}^*$ evaluations are provided by $\mathcal{P}$ in all iterations i of the protocol except the last. The bottleneck in the evaluation is the $\tilde{\text{add}}_i$ and $\tilde{\text{mult}}_i$ computations. These can be done in pre-processing in time $O(S_i)$ by enumerating the in-neighbors of each of the S_i gates at layer i [14, 37]. Adding up the pre-processing time across all iterations i of our protocol, $\mathcal{V}$'s pre-processing time is $O(\sum_i S_i) = O(S)$ as claimed.

Costs for $\mathcal{P}$. Notice $g^{(i)}$ is a polynomial in $v := s_i + 2s_{i+1} + b$ variables. We order the sum in this sum-check protocol so that the $s_i + 2s_{i+1}$ variables in p_1, ω_1, and γ_1 are bound first in arbitrary order, followed by the variables of p_2. $\mathcal{P}$ can compute the prescribed messages in the first $s_i + 2s_{i+1} = O(\log S)$ rounds exactly as in the implementation of Cormode et al. [14], who show that each gate at layers i and $i+1$ of C^* contributes to exactly one term in the sum defining $\mathcal{P}$'s message in any given round of the sum-check protocol. Hence the total time required by $\mathcal{P}$ to handle these rounds is $O(B \cdot (S_i + S_{i+1}) \cdot \log S)$.

It remains to show how $\mathcal{P}$ can compute the prescribed messages in the final b rounds of the sum-check protocol while investing $O((S_i + S_{i+1}) \cdot B)$ across all rounds of the protocol. The idea is that once the variables of p_1, ω_1, and γ_1 are bound, the truth table of $g^{(i)}$, viewed as a function of the unbound variables, is dense on the Boolean hypercube, in the sense of Section 3.2. We therefore exploit the reuse-of-work techniques underlying Theorem 1 to achieve the desired runtime for the prover.

5   Optimal Space and Time Costs for MATMULT

In Theorem 3 below, we describe a special-purpose protocol for $n \times n$ MATMULT in Theorem 3. The idea behind this protocol is as follows. The GKR protocol, as well the protocols of Theorems 1 and 2, only make use of the multilinear extension $\tilde{V}_i$ of the function V_i mapping gate labels at layer i of the circuit to their values. In some cases, there is something to be gained by using a higher-degree extension of V_i. This is precisely what we exploit here. In more detail, our special-purpose protocol can be viewed as an extension of our circuit-checking techniques applied to a circuit C performing naive matrix multiplication, but using a quadratic extension of the gate values in this circuit. This allows us to verify the computation using a single invocation of the sum-check protocol. More importantly, $\mathcal{P}$ can evaluate this higher-degree extension at the necessary points without explicitly materializing all of the gate values of C, which would not be possible if we had used the multilinear extension of the gate values of C.

In the protocol of Theorem 3, $\mathcal{P}$ just needs to compute the correct output (possibly using an algorithm that is much more sophisticated than naive matrix multiplication), and then perform $O(n^2)$ additional work to prove the output is correct. We obtain the $O(n^2)$ bound on the extra work required by $\mathcal{P}$ by exploiting the reuse-of-work technique underlying Theorems 1 and 2.

Since $\mathcal{P}$ does not have to evaluate C in full, this protocol is perhaps best viewed outside the lens of circuit evaluation. Still, the idea underlying Theorem 3 extends those underlying our circuit evaluation protocols, and we believe similar ideas may yield further improvements to general-purpose protocols in the future.

Theorem 3. *There is a valid interactive proof protocol for $n \times n$ matrix multiplication over the field $\mathbb{F}_q$ with the following costs. The communication cost is $n^2 + O(\log n)$ field elements. The runtime of the prover is $T(n) + O(n^2)$ and the space usage is $s(n) + o(n^2)$, where $T(n)$ and $s(n)$ are the time and space requirements of any (unverifiable) algorithm for $n \times n$ matrix multiplication. The verifier can make a single streaming pass over the input as well as over the claimed output in time $O(n^2 \log n)$, storing $O(\log n)$ field elements.*

5.1 Comparison to Prior Work

It is worth comparing Theorem 3 to a well-known protocol due to Freivalds [17]. Let D^* denote the claimed output matrix. In Freivalds' algorithm, the verifier stores a random vector $x \in \mathbb{F}^n$, and computes D^*x and ABx, accepting if and only if $ABx = D^*x$. Freivalds showed that this is a valid protocol. In both Freivalds' protocol and that of Theorem 3, the prover runs in time $T(n) + O(n^2)$ (in the case of Freivalds' algorithm, the $O(n^2)$ term is 0), and the verifier runs in linear or quasilinear time. We now highlight several properties of our protocol that are not achieved by prior work.

Utility as a Primitive. A major advantage of Theorem 3 relative to prior work is its utility as a primitive that can be used to verify more complicated computations. This is important as many algorithms repeatedly invoke matrix multiplication as a subroutine. For concreteness, consider the problem of computing A^{2^k} via repeated squaring. By iterating the protocol of Theorem 3 k times, we obtain a valid interactive proof protocol for computing A^{2^k} with communication cost $n^2 + O(k \log(n))$. The n^2 term is due simply to specifying the output A^{2^k}, and can often be avoided in applications – see for example the diameter protocol described two paragraphs hence. The ith iteration of the protocol for computing A^{2^k} reduces a claim about an evaluation of the multilinear extension of $A^{2^{k-i+1}}$ to an analogous claim about $A^{2^{k-i}}$. Crucially, the prover in this protocol never needs to send the verifier the intermediate matrices $A^{2^{k'}}$ for $k' < k$. In contrast, applying Freivalds' algorithm to this problem would require $O(kn^2)$ communication, as $\mathcal{P}$ must specify each of the intermediate matrices A^{2^i}.

The ability to avoid having $\mathcal{P}$ explicitly send intermediate matrices is especially important in settings in which an algorithm repeatedly invokes matrix multiplication, but the desired output of the algorithm is smaller than the size of the matrix. In these cases, it is not necessary for $\mathcal{P}$ to send *any* matrices; $\mathcal{P}$ can instead send just the desired output, and V can use Theorem 3 to check the validity of the output with only a polylogarithmic amount of additional communication. This is analogous to how the verifier in the GKR protocol can check the values of the output gates of a circuit without ever seeing the values of the interior gates of the circuit.

As a concrete example illustrating the power of our matrix multiplication protocol, consider the fundamental problem of computing the diameter of an unweighted (possibly directed) graph G on n vertices. Let A denote the adjacency matrix of G, and let I denote the $n \times n$ identity matrix. Then it is easily verified that the diameter of G is the least positive number d such that $(A+I)_{ij}^d \neq 0$ for all (i,j). We therefore obtain the following natural protocol for diameter. $\mathcal{P}$ sends the claimed output d to V, as well as an (i,j) such that $(A+I)_{ij}^{d-1} = 0$. To confirm that d is the diameter of G, it suffices for V to check two things: first, that all entries of $(A+I)^d$ are non-zero, and second that $(A+I)_{ij}^{d-1}$ is indeed non-zero.

The first task is accomplished by combining our matrix multiplication protocol of Theorem 3 with our DISTINCT protocol from Theorem 1. Indeed, let d_j denote the jth bit in the binary representation of d. Then $(A+I)^d = \prod_j^{\lceil \log d \rceil} (A+I)^{2^j}$, so computing the number of non-zero entries of $(A+I)^d$ can be treated as a sequence of $O(\log d)$ matrix multiplications, followed by a DISTINCT computation. The second task, of verifying that $(A+I)_{ij}^{d-1} = 0$, is similarly accomplished using $O(\log d)$ invocations of the matrix multiplication protocol of Theorem 3 – since V is only interested in one entry of $(A+I)^{d-1}$, $\mathcal{P}$ need not send the matrix $(A+I)^{d-1}$ in full, and the total communication here is just polylog(n).

V's runtime in this diameter protocol is $O(m \log n)$, where m is the number of edges in G. $\mathcal{P}$'s runtime in the above diameter protocol matches the best known unverifiable diameter algorithm up to a low-order additive term [30, 38], and the communication is just polylog(n). We know of no other protocol achieving this.

In many settings, practitioners will not tolerate even a 2x slowdown to achieve verifiability, so the fact that $\mathcal{P}$'s slowdown is a low-order additive term is critical. Moreover, for a graph with $n = 1$ million nodes, the total communication cost of the above protocol would be on the order of KBs – in contrast, if $\mathcal{P}$ had to send the matrices $(I+A)^d$ or $(I+A)^{d-1}$ explicitly (as required in prior work, e.g., Cormode et al. [13]), the communication cost would be at least $n^2 = 10^{12}$ words of communication, which translates to terabytes of data.

Small-Space Streaming Verifiers. In Freivalds' algorithm, V has the store the random vector x, which requires $\Omega(n)$ space. There are methods to reduce V's space usage by generating x with limited randomness: Kimbrel and Sinha [24] show how to reduce V's space to $O(\log n)$, but their solution does not work if V must make a streaming pass over arbitrarily ordered input. Chakrabarti et al. [12] extend the method of Kimbrel and Sinha to work with a streaming verifier, but this requires $\mathcal{P}$ to play back the input matrices A, B in a special order, increasing proof length to $3n^2$. Our protocol works with a streaming verifier using $O(\log n)$ space, and our proof length is $n^2 + O(\log n)$, where the n^2 term is due to specifying AB and can be avoided in applications such as the diameter example considered above.

5.2   Protocol Details

When multiplying matrices A and B such that $AB = D$, let $A(i,j)$, $B(i,j)$ and $D(i,j)$ denote functions from $\{0,1\}^{\log n} \times \{0,1\}^{\log n} \to \mathbb{F}_q$ that map input (i,j) to A_{ij}, B_{ij}, and D_{ij} respectively. Let $\tilde{A}$, $\tilde{B}$, and $\tilde{D}$ denote their multilinear extensions.

Lemma 1. *For all* $(p_1, p_2) \in \mathbb{F}^{\log n} \times \mathbb{F}^{\log n}$,

$$\tilde{D}(p_1, p_2) = \sum_{p_3 \in \{0,1\}^{\log n}} \tilde{A}(p_1, p_3) \cdot \tilde{B}(p_3, p_2)$$

Proof. For all $(p_1, p_2) \in \{0,1\}^{\log n} \times \{0,1\}^{\log n}$, the right hand side is easily seen to equal $D(p_1, p_2)$, using the fact that $D_{ij} = \sum_k A_{ik} B_{kj}$ and the fact that $\tilde{A}$ and $\tilde{B}$ agree with the functions $A(i,j)$ and $B(i,j)$ at all Boolean inputs. Moreover, the right hand side is a multilinear polynomial in the variables of (p_1, p_2). Putting these facts together implies that the right hand side is the unique multilinear extension of the function $D(i,j)$.

Lemma 1 implies the following valid interactive proof protocol for matrix multiplication: $\mathcal{P}$ sends a matrix D^* claimed to equal the product $D = AB$. $\mathcal{V}$ evaluates $\tilde{D}^*(r_1, r_2)$ at a random point $(r_1, r_2) \in \mathbb{F}^{\log n} \times \mathbb{F}^{\log n}$. It can be shown that it is safe for $\mathcal{V}$ to believe D^* is as claimed, as long as $\tilde{D}^*(r_1, r_2) = \tilde{D}(r_1, r_2)$. In order to check that $\tilde{D}^*(r_1, r_2) = \tilde{D}(r_1, r_2)$, we invoke a sum-check protocol on the polynomial $g(p_3) = \tilde{A}(r_1, p_3) \cdot \tilde{B}(p_3, r_2)$.

$\mathcal{V}$'s final check in this protocol requires her to compute $g(r_3)$ for a random point $r_3 \in \mathbb{F}^{\log n}$. $\mathcal{V}$ can do this by evaluating both of $\tilde{A}(r_1, r_3)$ and $\tilde{B}(r_3, r_2)$ with a single streaming pass over the input, and then multiplying the results. The prover can be made to run in time $T(n) + O(n^2)$ across all rounds of the sum-check protocol using the reuse-of-work technique underlying Theorem 1. Moreover, the space requirements of $\mathcal{P}$ are just $s(n) + o(n^2)$.

Implementation. We implemented the protocol of Theorem 3 over the field with $q = 2^{61} - 1$ elements. The results are shown in Table 2, where the column labelled "Additional Time for $\mathcal{P}$" denotes the time required to compute $\mathcal{P}$'s prescribed messages after $\mathcal{P}$ has already computed the correct answer. We report the naive matrix multiplication time both when the computation is done using standard multiplication of 64-bit integers, as well as when the computation is done using finite field arithmetic over $\mathbb{F}_q$. The main takeaways from Table 2 are that the verifier does indeed save substantial time relative to performing matrix multiplication locally, and that the runtime of the prover is hugely dominated by the time required simply to compute the answer.

Table 2. Experimental results for the $n \times n$ MATMULT protocol of Theorem 3

Problem Size	Naive Matrix Multiplication Time	Additional Time for $\mathcal{P}$	$\mathcal{V}$ Time	Rounds
$2^{10} \times 2^{10}$	2.17 s over $\mathbb{Z}$; 9.11 s over $\mathbb{F}_q$	0.03s	0.67 s	11
$2^{11} \times 2^{11}$	18.23 s over $\mathbb{Z}$; 73.65 s over $\mathbb{F}_q$	0.13s	2.89 s	12

References

1. Arora, S., Barak, B.: Computational Complexity:˙A Modern Approach. Cambridge University Press (2009)
2. Benabbas, S., Gennaro, R., Vahlis, Y.: Verifiable delegation of computation over large datasets. In: Rogaway, P. (ed.) CRYPTO 2011. LNCS, vol. 6841, pp. 111–131. Springer, Heidelberg (2011)
3. Ben-Sasson, E., Chiesa, A., Genkin, D., Tromer, E.: Fast reductions from RAMs to delegatable succinct constraint satisfaction problems. In: ITCS, pp. 401–414 (2013)
4. Ben-Sasson, E., Chiesa, A., Genkin, D., Tromer, E.: On the concrete-efficiency threshold of probabilistically-checkable proofs. In: STOC (2013)
5. Boneh, D., Freeman, D.M.: Homomorphic signatures for polynomial functions. In: Paterson, K.G. (ed.) EUROCRYPT 2011. LNCS, vol. 6632, pp. 149–168. Springer, Heidelberg (2011)
6. Bitansky, N., Canetti, R., Chiesa, A., Tromer, E.: From extractable collision resistance to succinct non-interactive arguments of knowledge, and back again. In: ITCS, pp. 326–349 (2012)
7. Bitansky, N., Canetti, R., Chiesa, A., Tromer, E.: Recursive composition and bootstrapping for SNARKs and proof-carrying data. In: STOC (2013)
8. Bitansky, N., Chiesa, A.: Succinct arguments from multi-prover interactive proofs and their efficiency benefits. In: Safavi-Naini, R. (ed.) CRYPTO 2012. LNCS, vol. 7417, pp. 255–272. Springer, Heidelberg (2012)
9. Bitansky, N., Chiesa, A., Ishai, Y., Paneth, O., Ostrovsky, R.: Succinct non-interactive arguments via linear interactive proofs. In: Sahai, A. (ed.) TCC 2013. LNCS, vol. 7785, pp. 315–333. Springer, Heidelberg (2013)
10. Chung, K.-M., Kalai, Y.T., Liu, F.-H., Raz, R.: Memory delegation. In: Rogaway, P. (ed.) CRYPTO 2011. LNCS, vol. 6841, pp. 151–168. Springer, Heidelberg (2011)
11. Chung, K.-M., Kalai, Y.T., Vadhan, S.P.: Improved delegation of computation using fully homomorphic encryption. In: Rabin, T. (ed.) CRYPTO 2010. LNCS, vol. 6223, pp. 483–501. Springer, Heidelberg (2010)
12. Chakrabarti, A., Cormode, G., McGregor, A.: Annotations in data streams. In: Albers, S., Marchetti-Spaccamela, A., Matias, Y., Nikoletseas, S., Thomas, W. (eds.) ICALP 2009, Part I. LNCS, vol. 5555, pp. 222–234. Springer, Heidelberg (2009)
13. Cormode, G., Mitzenmacher, M., Thaler, J.: Streaming graph computations with a helpful advisor. Algorithmica 65(2), 409–442 (2013)
14. Cormode, G., Mitzenmacher, M., Thaler, J.: Practical verified computation with streaming interactive proofs. In: ITCS, pp. 90–112 (2012)
15. Cormode, G., Thaler, J., Yi, K.: Verifying computations with streaming interactive proofs. PVLDB 5(1), 25–36 (2011)
16. Fiore, D., Gennaro, R.: Publicly verifiable delegation of large polynomials and matrix computations, with applications. In: CCS, pp. 501–512 (2012)
17. Freivalds, R.: Fast probabilistic algorithms. In: Becvar, J. (ed.) MFCS 1979. LNCS, vol. 74, pp. 57–69. Springer, Heidelberg (1979)
18. Gennaro, R., Gentry, C., Parno, B.: Non-interactive verifiable computing: Outsourcing computation to untrusted workers. In: Rabin, T. (ed.) CRYPTO 2010. LNCS, vol. 6223, pp. 465–482. Springer, Heidelberg (2010)
19. Gennaro, R., Gentry, C., Parno, B., Raykova, M.: Quadratic span programs and succint NIZKs without PCPs. In: Johansson, T., Nguyen, P.Q. (eds.) EUROCRYPT 2013. LNCS, vol. 7881, pp. 626–645. Springer, Heidelberg (2013)
20. Goldwasser, S., Kalai, Y.T., Rothblum, G.N.: Delegating computation: interactive proofs for muggles. In: STOC, pp. 113–122 (2008)

21. Groth, J.: Short pairing-based non-interactive zero-knowledge arguments. In: Abe, M. (ed.) ASIACRYPT 2010. LNCS, vol. 6477, pp. 321–340. Springer, Heidelberg (2010)

22. Gur, T., Raz, R.: Arthur-Merlin Streaming Complexity. In: Smotrovs, J., Yakaryilmaz, A. (eds.) ICALP 2013, Part I. LNCS, vol. 7965, pp. 528–539. Springer, Heidelberg (2013)

23. Ishai, Y., Kushilevitz, E., Ostrovsky, R.: Efficient arguments without short PCPs. In: CCC, pp. 278–291 (2007)

24. Kimbrel, T., Sinha, R.K.: A probabilistic algorithm for verifying matrix products Using $O(n^2)$ time and $\log_2 n + O(1)$ random bits. Inf. Process. Lett. 45(2), 107–110 (1993)

25. Klauck, H., Prakash, V.: Streaming computations with a loquacious prover. In: ITCS, pp. 305–320 (2013)

26. Lipmaa, H.: Progression-free sets and sublinear pairing-based non-interactive zero-knowledge arguments. In: Cramer, R. (ed.) TCC 2012. LNCS, vol. 7194, pp. 169–189. Springer, Heidelberg (2012)

27. Lund, C., Fortnow, L., Karloff, H., Nisan, N.: Algebraic methods for interactive proof systems. J. ACM 39, 859–868 (1992)

28. Parno, B., Gentry, C., Howell, J., Raykova, M.: Pinocchio: nearly practical verifiable computation. In: IEEE Symposium on Security and Privacy, Oakland (2013)

29. Rothblum, G.: Delegating computation reliably: paradigms and constructions. Ph.D. Thesis (2009), http://hdl.handle.net/1721.1/54637

30. Seidel, R.: On the all-pairs-shortest-path problem in unweighted undirected graphs. JCSS 51(3), 400–403 (1995)

31. Setty, S., McPherson, R., Blumberg, A.J., Walfish, M.: Making argument systems for outsourced computation practical (sometimes). In: NDSS (2012)

32. Setty, S., Vu, V., Panpalia, N., Braun, B., Blumberg, A.J., Walfish, M.: Taking proof-based verified computation a few steps closer to practicality. In: USENIX Security (2012)

33. Setty, S., Braun, B., Vu, V., Blumberg, A.J., Parno, B., Walfish, M.: Resolving the conflict between generality and plausibility in verified computation. In: EuroSys, pp. 71–84 (2013)

34. Shamir, A.: IP = PSPACE. J. ACM 39, 869–877 (1992)

35. Thaler, J., Roberts, M., Mitzenmacher, M., Pfister, H.: Verifiable computation with massively parallel interactive proofs. In: HotCloud (2012)

36. Thaler, J.: Source code, http://people.seas.harvard.edu/~jthaler/Tcode.htm

37. Vu, V., Setty, S., Blumberg, A.J., Walfish, M.: A hybrid architecture for interactive verifiable computation. In: IEEE Symposium on Security and Privacy, Oakland (May 2013) (pre-print November 2012)

38. Yuster, R.: Computing the diameter polynomially faster than APSP. CoRR, Vol. abs/1011.6181 (2010)

SNARKs for C:
Verifying Program Executions
Succinctly and in Zero Knowledge

Eli Ben-Sasson[1], Alessandro Chiesa[2], Daniel Genkin[2],
Eran Tromer[3], and Madars Virza[2]

[1] Technion
{eli,danielg3}@cs.technion.ac.il
[2] MIT
{alexch,madars}@csail.mit.edu
[3] Tel Aviv University
tromer@cs.tau.ac.il

Abstract. An argument system for NP is a proof system that allows efficient verification of NP statements, given proofs produced by an untrusted yet computationally-bounded prover. Such a system is non-interactive and publicly-verifiable if, after a trusted party publishes a proving key and a verification key, anyone can use the proving key to generate non-interactive proofs for adaptively-chosen NP statements, and proofs can be verified by anyone by using the verification key.

We present an implementation of a publicly-verifiable non-interactive argument system for NP. The system, moreover, is a zero-knowledge proof-of-knowledge. It directly proves correct executions of programs on TinyRAM, a nondeterministic random-access machine tailored for efficient verification. Given a program P and time bound T, the system allows for proving correct execution of P, on any input x, for up to T steps, after a one-time setup requiring $\tilde{O}(|P|\cdot T)$ cryptographic operations. An honest prover requires $\tilde{O}(|P|\cdot T)$ cryptographic operations to generate such a proof, while proof verification can be performed with only $O(|x|)$ cryptographic operations. This system can be used to prove the correct execution of C programs, using our TinyRAM port of the GCC compiler.

This yields a zero-knowledge Succinct Non-interactive ARgument of Knowledge (zk-SNARK) for program executions, in the preprocessing model — a powerful solution for delegating NP computations, with several features not achieved by previously-implemented primitives.

Our approach builds on recent theoretical progress in the area. We present efficiency improvements and implementations of two main ingredients:
1. Given a C program, we produce a circuit whose satisfiability encodes the correctness of execution of the program. Leveraging nondeterminism, the generated circuit's size is merely quasilinear in the size of the computation. In particular, we efficiently handle arbitrary and data-dependent loops, control flow, and memory accesses. This is in contrast with existing "circuit generators", which in the general case produce circuits of quadratic size.
2. Given a linear PCP for verifying satisfiability of circuits, we produce a corresponding SNARK. We construct such a linear PCP (which, moreover, is zero-knowledge and very efficient) by building and improving on recent work on quadratic arithmetic programs.

R. Canetti and J.A. Garay (Eds.): CRYPTO 2013, Part II, LNCS 8043, pp. 90–108, 2013.
© International Association for Cryptologic Research 2013

Keywords: computationally-sound proofs, succinct arguments, zero-knowledge, delegation of computation.

1 Introduction

Proof systems for NP let an untrusted prover convince a verifier that "$x \in L$" where L is some fixed NP-complete language. Proof systems for NP that satisfy the *zero knowledge* and *proof of knowledge* properties are a powerful tool that enables a party to prove that he or she "knows" a secret satisfying certain properties, without revealing anything about the secret itself. Such proofs are important building blocks of many cryptographic tools, including secure computation [GMW87, BGW88], group signatures [BW06, Gro06], malleable proof systems [CKLM12], anonymous credentials [BCKL08], delegatable credentials [BCC+09], electronic voting [KMO01, Gro05, Lip11], and many others. Known constructions of zero-knowledge proofs of knowledge are practical only when proving statements of special form that avoid generic NP reductions (e.g., proving pairing-product equations [Gro06]). Obtaining implementations that are both generic and efficient in practice is a long-standing goal in cryptography [BBK+09, ABB+12].

Due to differences in computational power among parties, many applications also require *succinct verification*: the verifier is able to check a nondeterministic polynomial-time computation in time that is much shorter than the time required to run the computation when given a valid NP witness. For instance, this is the case when a weak client wishes to *outsource* (or *delegate*) a computation to an untrusted worker. The additional requirement of succinct verification has still not been achieved in practice in its full generality, despite recent theoretical and practical progress.

Furthermore, a difficulty that arises when studying the efficiency of proofs for *arbitrary* NP statements is the problem of *representation*. Proof systems are typically designed for inconvenient NP-complete languages such as circuit satisfiability or algebraic constraint satisfaction problems, while in practice, many of the problem statements we are interested in proving are easiest to express via algorithms written in some high-level programming language. Modern compilers can efficiently transform these algorithms into a program to be executed on a random-access machine (RAM). Therefore, we seek proof systems that efficiently support NP statements expressed as the correct execution of a RAM program.

1.1 Succinct Verification in the Preprocessing Model

There has been a lot of work on the problem of how to enable a verifier to succinctly verify long computations. Depending on the model, the functionality, and the security notion, different constructions are known. See the extended version of this paper for a brief summary of prior theoretical work in this area.

Many constructions achieving some form of succinct verification are only computationally sound: their security is based on cryptographic assumptions, and therefore are secure only against bounded-size provers. Yet, computational soundness seems inherent in many of these cases [BHZ87, GH98, GVW02, Wee05]. Proofs (interactive or non) that are only computationally sound are also known as *arguments* [BCC88].

In this work we are interested in non-interactive succinct verification in the preprocessing model: we investigate efficient implementations of *succinct non-interactive arguments (SNARGs) in the preprocessing model*. Also, we focus on the *publicly-verifiable* case, where a non-interactive proof can be (succinctly) verified by anyone. For simplicity, we start by introducing this cryptographic primitive for circuit satisfiability: the *circuit satisfaction problem* of a circuit $C \colon \{0,1\}^n \times \{0,1\}^h \to \{0,1\}$ is the relation $\mathcal{R}_C = \{(x,a) \in \{0,1\}^n \times \{0,1\}^h : C(x,a) = 1\}$; its language is $\mathcal{L}_C = \{x \in \{0,1\}^n : \exists a \in \{0,1\}^h, C(x,a) = 1\}$.

A **publicly-verifiable preprocessing SNARG** (or, for brevity in this paper, simply **SNARG**) is a triple of algorithms (G, P, V), respectively called *key generator, prover*, and *verifier*, working as follows. The (probabilistic) key generator G, on input a security parameter λ and circuit $C \colon \{0,1\}^n \times \{0,1\}^h \to \{0,1\}$, outputs a *proving key* σ and a *verification key* τ; these are the system's public parameters, which need to be generated only once per circuit. After that, anyone can use the proving key σ to generate non-interactive proofs for the language $\mathcal{L}_C$, and anyone can use the verification key τ to check these proofs. Namely, given σ and any $(x,a) \in \mathcal{R}_C$, the honest prover $P(\sigma, x, a)$ produces a proof π attesting that $x \in \mathcal{L}_C$; the verifier $V(\tau, x, \pi)$ checks that π is a valid proof for $x \in \mathcal{L}_C$.

The efficiency requirements are as follows:

- running the generator G on input $(1^\lambda, C)$ requires $\mathrm{poly}(|C|)$ cryptographic operations;
- running the prover P on input (σ, x, a) also requires $\mathrm{poly}(|C|)$ cryptographic operations; but
- running the verifier V on input (τ, x, π) only requires $\mathrm{poly}(|x|)$ cryptographic operations; and
- an honestly-generated (publicly-verifiable non-interactive) proof has size $\mathrm{poly}(\lambda)$.

We require (adaptive) computational soundness: for every polynomial-size prover P^*, constant $c > 0$, large enough security parameter $\lambda \in \mathbb{N}$, and circuit $C \colon \{0,1\}^n \times \{0,1\}^h \to \{0,1\}$ of size λ^c, letting $(\sigma, \tau) \leftarrow G(1^\lambda, C)$, if $P^*(\sigma, \tau)$ outputs an adaptively-chosen (x, π) such that there is no a for which $(x,a) \in \mathcal{R}_C$ then $V(\tau, x, \pi)$ rejects (except with negligible probability over G's randomness).

Furthermore, if a SNARG satisfies a certain natural proof-of-knowledge property, we call it a *SNARK*. If it additionally satisfies a certain natural zero-knowledge property, we call it a *zero-knowledge SNARK* (zk-SNARK). See the extended version of this paper for definitions.

1.2   Motivation

It would be wonderful to have efficient and generic implementations of SNARGs *without* any expensive preprocessing. (That is, running the generator G only requires $\mathrm{poly}(\lambda)$ time instead of $\mathrm{poly}(|C|)$ cryptographic operations.) The two known approaches to constructing such SNARGs are Micali's "computationally-sound proofs" [Mic00], and the bootstrapping techniques of Bitansky et al. [BCCT13]. Algorithmically, both are complex (and, thus far, expensive) constructions: the former requires probabilistically-checkable proofs (PCPs) [BFLS91] and the latter the use of recursive proof-composition.

Thus, even in light of recent advances in the computational efficiency of PCPs [BS08, Din07, MR08, BCGT13b], it seems wise to first investigate efficient implementations of SNARGs in the preprocessing model, which is a less demanding model because it allows G to conduct a one-time expensive computation "as a setup phase". Despite the expensive preprocessing, this model is potentially useful for many applications: while the generator G does require a lot of work to set up the system's public parameters (which only depend on the given circuit C but not the input to C), this work can be subsequently amortized over many succinct proof verifications (where each proof is with respect to a new, adaptively-chosen, input to C).

In this work we focus on the preprocessing model, due to the simpler and tighter constructions known in it. Recent works [Gro10, Lip12, GGPR13, BCI$^+$13] constructed zk-SNARKs based on knowledge-of-exponent assumptions [Dam92, HT98, BP04] in bilinear groups, and all of these constructions achieved the attractive feature of having proofs consisting of only $O(1)$ group elements and of having verification via simple arithmetic circuits that are linear in the size of the input for the circuit.

In this vein, Bitansky et al. [BCI$^+$13] gave a general technique for constructing zk-SNARKs. First, they define a *linear PCP* to be one where the honest proof oracle is a linear function (over an underlying field), and soundness is required to hold *only* for linear proof oracles. Then, they show a transformation (also based on knowledge-of-exponent assumptions) from any linear PCP with a *low-degree verifier* to a SNARK; also, if the linear PCP is honest-verifier zero-knowledge (HVZK), then the resulting SNARK is zero knowledge.

When combining with other works the transformation of Bitansky et al. from linear PCPs, one obtains a theoretically simple and attractive route for the construction of zk-SNARKs. Specifically, the work on *quadratic-span programs* (QSPs) and *quadratic arithmetic programs* (QAPs) of Gennaro et al. [GGPR13] implies efficient constructions of (HVZK) linear PCPs with low-degree verifiers for circuit satisfiability, and the work on fast reductions of Ben-Sasson et al. [BCGT13a] implies that random-access machine computations can be efficiently reduced to circuit satisfiability.

In this paper, we study the tantalizing question of whether the aforementioned theoretical progress can be translated into efficient implementations of zk-SNARKs. As always, bringing theory to practice requires significant additional insights and improvements, and tackling these is the goal of our work.

1.3 Contributions

In this work we present an implementation of a zk-SNARK (i.e., a non-interactive argument system for NP with the properties of zero knowledge, proof of knowledge, and succinct verification in the preprocessing model). Moreover, our implementation efficiently supports NP statements expressed as the correct execution of a program on a nondeterministic random-access machine or (via a compiler we wrote) expressed as the correct execution of a C program. Our contributions can be summarized as follows:

1) Verifying circuit satisfiability via linear PCPs. We obtain an implementation of zk-SNARKs for (arithmetic) circuit satisfiability with excellent asymptotic efficiency: linear-time generator, quasilinear-time prover, and linear-time verifier. Moreover, proofs consist of only 12 group elements (a total of 780 bytes), independently of the circuit C or the input x to C. A proof provides 128 bits of security.

Our approach consists of two steps. First, we significantly optimized and implemented the transformation of Bitansky et al. [BCI+13]; our optimizations rely on multiexponentiation algorithms (see [Ber02] and references therein) and parallelism. Second, by building on the work on quadratic arithmetic programs (QAPs) of Gennaro et al. [GGPR13] and by leveraging algebraic structure of a carefully-chosen field, we give an efficient implementation of a linear PCP with a low-degree verifier. When verifying that $x \in \mathcal{L}_C$, our linear PCP has 5 queries of $2|C|$ field elements each; each query can be generated in linear time; the prover can compute the linear proof oracle via an arithmetic circuit of size $O(|C| \log |C|)$ and depth $O(\log |C|)$; the answers to the 5 queries can be verified with $O(|x|)$ field operations.

2) From correctness of program execution to circuit satisfiability. The SNARKs generated by the previous transformation are for proving the satisfiability of a given (arithmetic) circuit. However, programs are easier to write using high-level programming languages, like C, and it is often not realistic to require an arbitrary application to already provide a circuit encoding the NP statement of interest. We address this problem by providing a "circuit generator" that differs significantly and qualitatively from *all* previous implementations of circuit generators (e.g., Fairplay [MNPS04, BDNP08]): it leverages nondeterminism to reduce the size of the output circuit. Specifically, previous circuit generators produce circuits of $O(T^2)$ size for T-step computations in the worst case, whereas our generator produces circuits of only $O(T \log T)$ size. In more detail, our solution to the circuit generation problem is as follows:

(i) We design a minimalistic nondeterministic random-access machine, called TinyRAM.

(ii) We obtain a transformation, significantly more efficient than the one in [BCGT13a], that takes as input a TinyRAM program **P** and a time bound T and outputs a circuit whose satisfiability encodes the correct execution of **P** for up to T steps. Our efficiency improvements are achieved by leveraging field operations and nondeterminism in order to verify several types of crucial (boolean) computations via smaller arithmetic circuits. We implemented our transformation.

(iii) We complement the above transformation with a GCC backend, for compiling programs written in a subset of C into TinyRAM assembly. This compiler provides a convenient way to obtain TinyRAM programs for problems of interest. Crucially, we can efficiently support arbitrary and data-dependent loops, control flow, and memory accesses.

Our choice of architecture for TinyRAM strikes a balance between the ability to efficiently compile programs into TinyRAM assembly code, and the need to design small circuits for the transition function of TinyRAM.

Delegation for NP Programs. Together, our contributions yield a system to verify program executions succinctly and in zero knowledge.

In particular, our contributions provide a solution for non-interactively delegating arbitrary NP computations, also in a way that does not compromise the privacy of any input that the untrusted worker contributes to the computation. Previous implementation work did not achieve many of the features enjoyed by our implementation. (See the extended version of this paper for a comparison with prior implementation work.)

2   From Correctness of Program Execution to Circuit Satisfiability

As summarized in Section 1.3, we implemented an *efficient* transformation that reduces correctness of program execution to circuit satisfiabiliy. The following gives further design and performance details about this transformation. Concretely, in Section 2.1 we motivate and discuss our choice of architecture, TinyRAM. Then, in Section 2.2, we discuss implementation and performance of our compiler from C to TinyRAM assembly. Finally, in Section 2.3, we discuss implementation and performance of our reduction from the correctness of TinyRAM assembly to circuit satisfiability.

2.1   The TinyRAM Architecture

To reason about correctness of program executions, we first need to fix a specific random-access machine. An attractive choice is to pick the instruction set architecture (ISA) of some existing, well-supported family of CPUs (e.g., x86 or ARM). We could then reuse existing tools and software written for those CPUs. This is possible in principle.

However, the design of CPUs typically focuses on efficient ways of getting data and code, at the right time, to the different executions units of the CPU, with the goal of maximizing utilization of these units. This is achieved by complex mechanisms whose size can dwarf the *functional core* circuitry (execution units, register file, instruction decoding, and so on). Thus, modern CPUs afford, and employ, large and rich instruction sets. As explained next, the efficiency considerations are very different in our context.

Executing vs. Verifying.   CPUs and their ISAs are optimized for fast *execution* of programs. However, we are interested in fast *verification* of (alleged) past executions. In our setting, the computation *has already been executed* and we possess a trace of this execution, giving the state of the processor (registers and flags) at every time step. Our goal is to efficiently verify the correctness of the trace: that every state in the trace follows from the preceding one.

This means that values that are expensive to produce during the execution become readily available for verification in the trace. For example, in real CPUs, reading from external memory is relatively slow and a large fraction of the circuitry is dedicated to caching data. However, in the trace, the result of a load from memory is readily seen in the processor state at the end of the computation step; thus the need for caches is moot. Similarly, modern CPUs use complicated speculative-execution and branch-prediction mechanisms to keep their execution pipelines full; but a trace verifier going down the trace can "peek into the future" and readily observe control flow.

The elimination of the above mechanisms, and many others, affects the ISA. In particular, it means that the aforementioned functional core circuitry dominates cost. This leads to the next consideration.

Transition Function Complexity.   We are ultimately interested in carrying out the verification of a trace via a circuit, so we wish to optimize the circuit complexity of the *transition function* of the ISA: the size of the smallest circuit that, given two adjacent states in the trace, verifies that the transition between the two indeed respects the ISA specification.[1]

We thus seek an ISA that strikes a balance between two opposing requirements: (1) the need for a transition function of small circuit complexity and (2) the need to produce small and fast machine code, in particular when compiling from high-level programming languages. Rich architectures allow for smaller code and shorter execution trace but have transition functions of higher circuit complexity, while minimalistic architectures require longer machine code and longer execution traces, but enjoy transition functions with smaller circuit complexity.

Modern ISAs designed for general purpose CPUs (such as x86) are complex instruction set computer (CISC) machines: they support many elaborate instructions (e.g., a round of AES [Gue12]) and addressing modes. Less rich ISAs are reduced instruction set computer (RISC) machines designed for devices like smartphones (ARM) and embedded microcontrollers (Atmel AVR).

In sum, we seek a minimal ISA that enables us to design a transition function with small circuit complexity, and yet allows reasonable overheads in code size and execution time (relative to richer ISAs).

A Custom ISA. In light of the above, we designed an instruction set architecture, named TinyRAM, that is tailored for our setting. TinyRAM is a minimalistic RISC random-access machine with a Harvard architecture and word-addressable random-access memory. It has two parameters: the *word size*, denoted W, and the *number of registers*, denoted K. The global state of the machine at any time consists of:

- the *program counter*, denoted pc; it consists of W bits;
- K general-purpose *registers*, denoted $r0, r1, \ldots, r(K-1)$, each with of W bits;
- the *(condition) flag*, denoted flag; it consists of a single bit; and
- *memory*, which is a linear array of 2^W words of W bits each.

In addition, the machine has two *input tapes*, each containing a string of W-bit words. Each tape can be read sequentially in one direction only. The first input tape is for the *primary input*, denoted x; the second input tape is for the *auxiliary input*, denoted w. We treat the primary input as given, and the auxiliary input as nondeterministic advice. (See Definition 1 below.)

We carefully selected the instructions of TinyRAM so to support relatively efficient compilation from high-level programming languages (like C), as discussed in Section 2.2, and, furthermore, allow for small circuits implementing its transition function (and other checks), as discussed in Section 2.3. Briefly, the instruction set of TinyRAM includes simple load and store instructions for accessing random-access memory, as well as simple integer, shift, logical, compare, move, and jump instructions. TinyRAM can efficiently implement complex control flow, loops, subroutines, recursion, and so on. Complicated instructions, such as floating-point arithmetic, are not directly supported and can be implemented "in software" by TinyRAM programs. Supporting only

---

[1] This does not include the task of checking the correctness of values loaded from random-access memory. Memory consistency is efficiently handled separately. See Section 2.3.

fairly simple load and store operations is important for efficiently verifying consistency of random-access memory; see Section 2.3.

In keeping with the setting of verifiying computation, the only input to TinyRAM programs is via its two input tapes, and the only output is via an `accept` instruction, which also terminates execution.[2]

So far we have only informally discussed "correctness of TinyRAM program execution". This notion is formalized by defining a TinyRAM universal language.

Definition 1. *Fix the word size W and number of registers K. Let $\mathbf{P}$ be a* TinyRAM *program, let x and w be strings of W-bit words. We say that $\mathbf{P}(x, w)$ **accepts in** T* **steps** *if $\mathbf{P}$, with x on its first input tape and w on the second, executes the instruction* `accept` *in step T.*

The TinyRAM **universal language** *consists of the triples $(\mathbf{P}, x, T)$ where $\mathbf{P}$ is a* TinyRAM *program, x is a string of W-bit words, and T is a time bound, such that there exists a string w of W-bit words for which $\mathbf{P}(x, w)$ accepts in T steps.*

A specification for the TinyRAM architecture can be found in [BCG+13].

2.2 A Compiler from C to TinyRAM

The GCC compiler is a versatile framework supporting many source languages (e.g., C and Java) and many target languages (e.g., x86 and ARM assembly). Internally, the GCC compiler is partitioned into two main modules [StGDC13]. The *frontend* is responsible for converting a program written in a high-level programming language like C or Java into an intermediate representation language called *Register Transfer Language* (RTL). The *backend* is responsible for optimizing and converting RTL code into corresponding assembly code for a given architecture.

In order to automatically generate TinyRAM assembly for problems of interest, we have implemented a prototype of a GCC backend for converting RTL code to TinyRAM assembly code. Our prototype backend works with the C frontend, and can be extended to other programming languages by combining it with suitable GCC frontends (and providing the requisite standard libraries). Concretely, we have a prototype that can compile a subset of C to TinyRAM, with word size $W \in \{8, 16\}$ and number of registers $K \geq 15$.

Because TinyRAM's instruction set is quite minimal, any operation not directly supported by TinyRAM "hardware" needs to be implemented in "software". This incurs overheads in both the *code size* (the number of lines in an assembly code) and *execution time* (the number of machine steps required to execute a piece of code). By running experiments, we established that both of these overheads are not large, as discussed next.

Code Size Overhead. We first evaluate the code size produced when compiling C code examplesinto TinyRAM assembly using our GCC port, compared to the code produced by standard GCC for some common architectures: x86, ARM and AVR. (We used the $-$O1 optimization flag in all cases.) Our results show that, compared to the RISC architectures (ARM and AVR), the resulting TinyRAM code is at most twice larger than

---

[2] For ease of development, the TinyRAM simulator also supports debugging instructions that produce additional outputs. These are excluded from the execution trace and not verified.

ARM and significantly smaller than AVR. Compared to x86, which is a very rich CISC architecture, TinyRAM code is up to three times bigger. We deduce that, at least for the program styles represented by these examples, the TinyRAM architecture allows for compilation into compact programs. See the extended version of this paper for details.

Execution Time Overhead. The circuits ultimately produced by our reduction have size $O(T \log T)$, where T is the execution time (measured in machine steps). This execution time depends on the choice of architecture, and we wish to ensure that TinyRAM does not necessitate very long execution times due to deficiencies in the instruction set.

To evaluate this, we compiled examples of C code into both TinyRAM machine code and x86 machine code. Our results show that in terms of execution time measured in machine steps (i.e., clock cycles), TinyRAM is slower than x86 by a factor of merely 2 to 6, for examples that represent some realistic computations. This is despite x86 being a very rich CISC architecture that is heavily optimized for minimizing clock cycles, which is typically implemented using many millions of gates. (Recall the difference of executing vs. verifying, discussed in Section 2.1.) See the extended version for details.

These small overheads are more than compensated by the fact that TinyRAM has a very compact transition function circuit. For instance, for a wordsize $W = 16$ and number of registers $K = 16$, and for a program with 100 instructions, the transition function consists of only 708 gates.

In summary, our experiments show that, even when working with a minimalistic architecture such as TinyRAM, we do not incur large overheads in code size or execution time. In Section 2.3, we discuss the circuit complexity of TinyRAM's transition function and how to efficiently verify TinyRAM traces.

2.3   An Efficient Reduction from TinyRAM to Circuit Satisfiability

The following describes our efficient reduction from correctness of TinyRAM executions to $\mathbb{F}$-arithmetic circuit satisfiability, for any prime field $\mathbb{F}$ of sufficiently large size.

The Reduction Notion. In our setting, a *(circuit) reduction* is a triple of functions (circ, wit, wit^{-1}) working as follows. The *circuit generator* function, circ, maps a TinyRAM program $\mathbf{P}$, time bound T, and primary input size n to a corresponding $\mathbb{F}$-arithmetic circuit C that encodes the correct computation of $\mathbf{P}$ for at most T steps on primary inputs of n words. The *witness map* function, wit, maps a pair of primary and auxiliary inputs (x, w) that make $\mathbf{P}$ accept in T steps, to a satisfying assignment a for $C(x, \cdot)$. The *inverse witness map* function, wit^{-1}, maps a satisfying assignment a for $C(x, \cdot)$ to w with the property that (x, w) makes $\mathbf{P}$ accept in T steps.

Definition 2. *A reduction from* TinyRAM *(for a word size W and number of registers K) to $\mathbb{F}$-arithmetic circuit satisfiability is a triple of functions* (circ, wit, wit^{-1}) *such that, for every* TinyRAM *program $\mathbf{P}$, time bound T, and primary input size n, the following hold:*

- *$C := \mathrm{circ}(\mathbf{P}, T, n)$ is an $\mathbb{F}$-arithmetic circuit from $\mathbb{F}^{W \cdot n} \times \mathbb{F}^h$ to $\mathbb{F}^\ell$ for some h, ℓ; C's gates are bilinear;[3]*

---

[3] A gate with inputs $x_1, \ldots, x_n$ is *bilinear* if the output is $\langle \mathbf{a}, (1, x_1, \ldots, x_n) \rangle \cdot \langle \mathbf{b}, (1, x_1, \ldots, x_n) \rangle$ for some $\mathbf{a}, \mathbf{b} \in \mathbb{F}^{n+1}$.

- *for every* (x, w) *such that* $\mathbf{P}(x, w)$ *accepts in* T *steps,* $C(x, \mathrm{wit}(\mathbf{P}, T, x, w)) = 0^{\ell}$;
- *for every* (x, a) *such that* $C(x, a) = 0^{\ell}$, $\mathbf{P}(x, \mathrm{wit}^{-1}(\mathbf{P}, T, x, a))$ *accepts in* T *steps.*

The work on fast reductions of Ben-Sasson et al. [BCGT13a] implies a reduction $(\mathrm{circ}, \mathrm{wit}, \mathrm{wit}^{-1})$ where $|C| := O(T(\log T)^2)$ and $\mathrm{circ}, \mathrm{wit}, \mathrm{wit}^{-1}$ all run in $O(T(\log T)^2)$ time.[4] In our work, we optimize and implement a reduction that builds on the theoretical approach of [BCGT13a]. We shall focus our attention only on the efficiency of the circuit and witness maps (i.e., circ and wit), because these actually need to be run in practice. Before discussing our work, however, we briefly review the approach of [BCGT13a].

The Reduction in [BCGT13a]. We begin with necessary basic definitions.

- A *(local) state* of TinyRAM, denoted S, is a string of $(W + KW + 1)$ bits, encoding the values of the program counter, K registers, and condition flag at a given step.
- The *transition function* of TinyRAM, denoted Π_{TF}, is the predicate that, given a TinyRAM program $\mathbf{P}$ and two states S and S', outputs 1 if and only if the machine in state S can transition (for *some* choice of values in random-access memory) to the state S' in the next step, according to the program $\mathbf{P}$.[5]
- An *execution trace*[6] for a TinyRAM program $\mathbf{P}$, time bound T, and primary input x is a sequence of states $\mathrm{tr} = (S_1, \ldots, S_T)$. An execution trace tr is *valid* if there exists an auxiliary input w such that the sequence of states induced by $\mathbf{P}$ running with input tapes (x, w) is tr.

The goal is to design an $\mathbb{F}$-arithmetic circuit C for verifying that tr is valid that is as small as possible. This is done in three steps, as follows.

Step 1: Code Consistency. Let C_{TF} be a circuit that implements the transition function Π_{TF} of TinyRAM: namely, $C_{\mathsf{TF}}(\mathbf{P}, S, S') = 1$ if and only if $\Pi_{\mathsf{TF}}(\mathbf{P}, S, S') = 1$. By invoking C_{TF} on each pair of successive states of tr, we can verify every state transition in the trace tr, i.e., ensure that $\Pi_{\mathsf{TF}}(\mathbf{P}, S_i, S_{i+1}) = 1$ for $i = 1, \ldots, T - 1$. Doing so gives rise to a sub-circuit of C, consisting of T copies of C_{TF}, that, when given as input tr, checks that tr is *code-consistent*.

Step 2: Memory Consistency. The global state of a random-access machine, however, also includes memory. In particular, in order to verify that tr is valid, we *also* need to verify that tr is *memory-consistent*: namely, that every load operation from an address in memory actually retrieves the value of the last store to that address.

But the accesses to memory of a program $\mathbf{P}$ depend on the inputs x and w. Hence, in general, at each time step i any of the addresses in memory could be accessed by the program. The naive solution of designing the verification circuit C to maintain a

---

[4] Given a space bound S on the computation of $\mathbf{P}$ on (x, w), Ben-Sasson et al. also present a reduction where $|C|$ is only $O(T \log T \log S)$. We have so far not considered this additional, significantly more complex, optimization.

[5] Traditionally, the transition function is the function that, given the global state of a machine as input, outputs the next state. We abuse this terminology, and use it for the function that, given two local states S, S', decides whether the second can follow the first.

[6] An execution trace is also at times known as a *computation transcript* [BCGT13a].

snapshot of memory for each time step is *not* efficient: such a circuit has size that is $\Omega(T^2)$. (All previous circuit generators either adopt the naive solution or restrict a program's memory accesses to be known at compile time.)

Ben-Sasson et al. [BCGT13a] do *not* adopt the naive solution (or restrict a program's memory accesses), but instead take an approach that is more efficient; the approach builds on classical results on quasilinear-time nondeterministic reductions [Sch78, GS89, Rob91]. The high-level idea in [BCGT13a] is that memory consistency would be easier to verify if the circuit C were to also have, as additional input, the *same* trace tr but sorted according to accessed memory addresses (and breaking ties via timestamps); let us denote this sorted trace by MemSort(tr). Concretely, one can define another "local" predicate Π_{MC} such that, if Π_{MC} is satisfied by each pair of adjacent states in MemSort(tr) (and, in addition, tr is code-consistent) then tr is valid. We can then augment C with T copies of a sub-circuit C_{MC} that verifies the predicate Π_{MC} on MemSort(tr). The circuit C is thus left to verify that the auxiliary input MemSort(tr) is the result of sorting tr.

Step 3: Routing Network. The circuit C can efficiently perform this check if it is given yet another additional input: (alleged) routing decisions for a routing network which permutes tr into MemSort(tr). A T-packet *routing network* is a directed graph with T sources, T sinks, and inner nodes (switches) such that, for any permutation $\pi\colon [T] \to [T]$, there are routing decisions for the switches that cause T packets at the sources to travel to the T sinks, according to the permutation π, and without using a switch twice (i.e., with no congestion). One such a network is the Beneš network [Ben65], which has $O(\log T)$ layers of T nodes each, and each node in a layer is connected to two nodes in the next layer. The idea is to interpret the switch settings in a routing network as a coloring on the routing network. Crucially, verifying that the given switch settings (i.e., a coloring of the network) implement *some* permutation from the input nodes to the output nodes can be done via simple and local routing constraints; furthermore, given that the switches implement some permutation, verifying that they implement the sorting permutation is easy to verify too. Overall we obtain a certain graph-coloring problem all of whose constraints can be evaluated by a circuit of size $T \cdot O((\log T)^2)$, which we add to C.

In Sum. The approach from [BCGT13a] described in the above paragraphs yields a circuit C of size $T \cdot \left(|C_{TF}| + |C_{MC}| + O((\log T)^2) \right)$ for verifying a T-step trace.

Our Optimized Reduction. As mentioned, in our work we optimize and implement the theoretical approach of Ben-Sasson et al. [BCGT13a]. Despite the excellent asymptotic efficiency of the approach, getting to the point in which the verification circuit C has a manageable size in practice proved quite challenging, both theoretically and programmatically. For instance: while (as discussed in Section 2.1) we devised TinyRAM to facilitate the design of a small circuit C_{TF} for the transition function Π_{TF}, how small of a circuit can we actually design? And how well does its size scale with, say, the word size W, number of registers K, and program size $|\mathbf{P}|$?

Our Circuit Generator. At high level, our main technical contribution is leveraging (1) *"native" arithmetic* in the field $\mathbb{F}$ (which for us is a prime field) and (2) *nondeterministic advice* so to achieve highly-optimized implementations of C_{TF}, C_{MC}, and routing constraints, and ultimately obtain drastic improvements in the size of the verification circuit C output by our circuit generator circ.

To illustrate the use of (1) and (2), consider the basic task of *multiplexing bit vectors*, used numerous times in C. Given n vectors $a_1, \ldots, a_n$ of ℓ bits each and a $\lceil \log n \rceil$-bit index i, we seek a small $\mathbb{F}$-arithmetic circuit that computes the vector selected by the index. The naive multiplexer circuit requires $O(n(\ell + \log n))$ bilinear gates. In contrast, by relying on (1) and (2), we design a multiplexer circuit that needs only $O(n \lceil \frac{\ell}{|\mathbb{F}|} \rceil)$ bilinear gates. The efficiency improvement is significant because we ultimately need to work with cryptographically-large fields; for instance, in our setting, if $n = \ell = 16$, the naive implementation uses 320 gates while we only use 51.

The idea of our multiplexer construction is as follows. Suppose, first, that every input vector, as well as the index, were represented as integers, and we only had to design a $\mathbb{Z}$-arithmetic circuit to output the integer representing the selected bit vector. In this case, we could easily construct a nondeterministic $\mathbb{Z}$-arithmetic circuit of size $O(n)$ (with bilinear gates of unbounded fan-in). However, the vectors are only given to us as strings of bits, and we need to work with $\mathbb{F}$-arithmetic circuits. This gap motivates two fundamental operations: *packing* and *unpacking* of bit vectors. Packing denotes mapping a bit vector to a sequence of field elements efficiently storing these bits, and unpacking denotes the inverse operation. The packing operation is very efficient: in the prime field $\mathbb{F}_p$ with $p \geq 2^\ell$, a single gate suffices to compute $\sum_{i=1}^{\ell} 2^{i-1} a_i$ from the input $a_1, \ldots, a_\ell$. The inverse operation is much more expensive to compute directly, but we can nondeterministically *guess* the answer and verify it using a single gate. In general, $p \geq 2^\ell$ need not hold, so we use $\lceil \frac{\ell}{|\mathbb{F}|} \rceil$ field elements to store an ℓ-bit vector. Given the aforementioned efficient packing operations, our multiplexer construction works as follows: it guesses the selected ℓ-bit vector, then computes the integers corresponding to the input ℓ-bit vectors as well as the index, and then verifies the guess by selecting the correct integer according to the (integer) index.

More generally, we have found that, throughout our circuit generator, it is often advantageous to maintain, alongside certain vectors a, also the corresponding integer $\sum_i 2^{i-1} a_i$. We believe that packing and unpacking operations will be crucial for drastically decreasing the size of circuits used in future circuit generators.

With these techniques in mind, we proceed to describe the circuit generator.

- *Designing the transition function circuit C_{TF}.* The circuit C_{TF} is the most complex sub-circuit of C. The size of C_{TF} is dominated by the size of sub-circuits for *multiplexing bit strings* (for instruction fetch, register fetch, and so on) and of the *arithmetic logic unit* (ALU), which executes the architecture's non-memory operations. To obtain an efficient implementation of the ALU, we again make use of field arithmetic and nondeterministic advice. Since we work over a prime field of large characteristic, field arithmetic looks like integer arithmetic whenever there is no "wrap around". Thus, after fetching the arguments to an operation, we derive from each argument's binary representation also the corresponding integer. Then, each operation in the ALU computes on the integer representation, instead of the binary one, when it is more efficient to do so. For instance, we use this idea to compute result and overflow information for addition, subtraction, and multiplication with only $2W$,

Table 1. Number of gates in C_{TF} as a function of $|\mathbf{P}|, W, K$

$|\mathbf{P}| = 10/100/1000$	$W = 8$	$W = 16$	$W = 32$
$K = 8$	416 / 506 / 1406	520 / 610 / 1510	728 / 818 / 1718
$K = 16$	514 / 604 / 1504	618 / 708 / 1608	826 / 916 / 1816
$K = 32$	708 / 798 / 1698	812 / 902 / 1802	1020 / 1110 / 2010

$2W$, and $3W$ bilinear gates, respectively; as for division, we guess the result and verify it with a multiplication. In each case, the integer output by an operation can be "unpacked" into its binary representation, via nondeterministic advice. By carefully implementing each operation, we obtain an ALU that, e.g., when $W = 16$ only has 296 gates.

Given efficient implementations of multiplexing and ALU, it is not difficult to obtain an efficient implementation of C_{TF}. Table 1 shows the number of gates in our implementation of C_{TF} for $|\mathbf{P}| \in \{10, 10^2, 10^3\}$, $W \in \{8, 16, 32\}$ and $K \in \{8, 16, 32\}$.

- *Designing the memory consistency circuit C_{MC}.* The predicate Π_{MC} is not as complex as the transition function Π_{TF}, but it is still important to design a small circuit C_{MC} for it. The bottleneck in the computation of Π_{MC} is again multiplexing, this time for fetching the two arguments of a memory operation. Thus, the natural approach here would be to use additional copies of our efficient multiplexer circuit. Instead, we show how to avoid additional multiplexing altogether by "stealing" certain intermediate computations from C_{TF}. We thereby obtain a circuit for C_{MC} that only contains two integer comparisons and few other logical operations. For instance, when $W = 16$, C_{MC} only costs us 60 additional gates.

- *Checking routing constraints.* Asymptotically, the routing constraints on the routing network are the most expensive sub-circuit of C. It is thus crucial to compute these constraints as efficiently as possible. A first concern is to minimize the size of a packet routed through the network. Instead of setting a packet to be a local state of the machine, which consists of $(W + KW + 1)$ bits, we show that it only suffices to send a much smaller packet, consisting of about $2W$ bits, obtained from intermediate computations of C_{TF}. This optimization in fact leads to another important one: now that a packet is as small as only about $2W$ bits, we can "pack" all the bits on a *single* field element (in our setting, $\mathbb{F}$ has size at least 2^W); then, because the packets consist of single field elements, computing the routing constraints becomes particularly simple: only one bilinear gate per vertex. Concretely, the gate at a given vertex checks whether the vertex's packet is equal to at least one of the packets at the two neighbor vertices in the next layer. Overall, when T is a power of 2, all the routing constraints can be verified with only $2 \cdot T \cdot \log T$ gates. (We thus also obtain an asymptotic improvement, by a $\log T$ factor, over the circuit size in [BCGT13a], where routing constraints required $O(T(\log T)^2)$ gates.)

Of course, there are numerous additional details that go into our final construction of the verification circuit C. Overall, say that for concreteness we fix $W = 16$, $K = 16$, and $|\mathbf{P}| = 100$, then we get

$$|C| = A \cdot T \cdot \log T + B \cdot T + C \text{ , where } A = 4, B = 1116 \text{ and } C = 307.$$

In particular, for $\log T < 20$, every cycle of TinyRAM computation costs ≈ 1200 gates. Note that, while the gate count per cycle increases as T increases (as the number of routing constraints grows as $O(T \log T)$), the growth rate is slow: doubling T costs only $4 + o(1)$ additional gates per cycle.

Our Witness Map. Thus far, we have focused on achieving *soundness*: verifying the validity of an execution trace of a TinyRAM program $\mathbf{P}$ by using the circuit $C :=$ circ$(\mathbf{P}, T, n)$ output by the circuit generator circ. The circuit generator is run by the key generator when computing the public parameters. For *completeness*, we need to implement a witness map wit$(\mathbf{P}, T, x, w)$ that computes a satisfying assignment a for $C(x, \cdot)$, whenever $\mathbf{P}(x, w)$ accepts in T steps. The witness map is executed by the prover when generating a proof. See the extended version for details on this map.

3 Verifying Circuit Satisfiability via Linear PCPs

As summarized in Section 1.3, we have implemented a zk-SNARK for circuit satisfiability; see Section 1.1 for an informal definition of this cryptographic primitive, or the extended version of this paper for a formal one. In this section we describe the design and performance of this part of our system.

3.1 A Transformation from Any Linear PCP

We begin by discussing efficiency aspects of the transformation from a linear PCP to a corresponding SNARK. To do so, we first recall (at high level) the transformation itself.

Constructing a SNARK from a linear PCP. The transformation of Bitansky et al. [BCI+13] consists of an information-theoretic step followed by cryptographic step.

- *Step 1 (information-theoretic):* compile the linear PCP into a 2-message *linear interactive proof* (linear IP), i.e., one where the prover is restricted to only apply linear functions to the verifier's message.
  This is achieved by adding a *consistency-check query*, which is a random linear combination of the linear PCP queries. In more detail, if the linear PCP has k queries each with m elements from a field $\mathbb{F}$, in the resulting linear IP the verifier sends to the prover a single message q consisting of $m' = (k + 1)m$ elements in $\mathbb{F}$; the message q is the concatenation of the k linear PCP queries and the consistency-check query. A (potentially malicious) prover is restricted to only apply linear functions to q, i.e., reply with a vector $a^* \in \mathbb{F}^{k+1}$ such that $a^* = \Pi^* q + b^*$ for some $\Pi^* \in \mathbb{F}^{(k+1) \times m'}$ and $b^* \in \mathbb{F}^{k+1}$. The honest prover simply returns the vector $a = (a_1, \ldots, a_{k+1})$ where $a_i = \langle \pi, q_i \rangle$, q_i is the i-th m-element block of q, and π is the linear PCP. A prover's message a^* is verified by checking consistency of a^*_{k+1} with $a^*_1, \ldots, a^*_k$ and then invoking the linear PCP decision predicate on $a^*_1, \ldots, a^*_k$; the consistency check ensures that $a^*_i = \langle \pi^*, q_i \rangle$ for *some* linear PCP π^*.
- *Step 2 (cryptographic):* compile the linear IP into a SNARK, by forcing any polynomial-size malicious prover to act as if it were a linear function.
  This is achieved using a cryptographic encoding Enc$(\cdot)$ with the following properties.
  (i) It allows public testing of quadratic predicates on encoded elements.

(ii) It provides a certain notion of one-way security to encoded elements.

(iii) It ensures that any polynomial-size prover can only perform linear operations on the encoded elements, "up to" information leaked by the encoding.[7]

Given $\mathsf{Enc}(\cdot)$, the compilation is then conceptually simple. The SNARK generator $G(1^\lambda, C)$ samples a verifier message $q \in \mathbb{F}^{m'}$ (which depends on the circuit C but not its input) for the linear IP, and outputs, as a proving key, the encoding $\mathsf{Enc}(q) = (\mathsf{Enc}(q_i))_{i=1}^{m'}$. (We omit here the discussion of how the short verification key is generated.) Starting from $\mathsf{Enc}(q)$ and a linear PCP π, the honest SNARK prover P homomorphically evaluates the inner products $\langle \pi, q_i \rangle$ and returns as a proof the resulting encoded answers. The SNARK verifier checks a proof by running the linear IP decision predicate on the encoded answers.

For precise definitions and details, see [BCI+13].

Computational Overheads. The transformation from a linear PCP to a SNARK introduces several computational overheads. In Step 1, the only overhead is due to the consistency-check query, and is minor. However, the cryptographic overheads in Step 2 are significant, and require optimizations for practical use.

Specifically, after sampling $q \in \mathbb{F}^{m'}$, the SNARK generator G must compute $\mathsf{Enc}(q) = (\mathsf{Enc}(q_i))_{i=1}^{m'}$. In other words G needs to compute the encoding of m' field elements, where m' is on the order of the size of the circuit C. Furthermore, after computing a linear proof oracle $\pi \in \mathbb{F}^m$, the honest SNARK prover P needs to homomorphically evaluate, for $i = 1, \ldots, k+1$, the inner product $\langle \pi, q_i \rangle$ to obtain $\mathsf{Enc}(\langle \pi, q_i \rangle)$.

In our case, the encoding is $\mathsf{Enc}(\gamma) = (g^\gamma, h^\gamma)$ where $g \in G_1, h \in G_2$ and G_1, G_2 are prime-order groups; the linear homomorphism is $\mathsf{Enc}(a\gamma + b\delta) = \mathsf{Enc}(\gamma)^a \mathsf{Enc}(\delta)^b$ with coordinate-wise multiplication and exponentiation. Therefore, both G and P need to compute a large number of cryptographic exponentiations. These operations greatly affect the complexity of G and P, and must be performed efficiently.

Efficiency Optimizations. We address the cryptographic bottleneck by using multi-exponentiation algorithms and parallelization. See the extended version of this paper for the impact of these optimizations.

3.2 An Efficient Linear PCP

In the previous section we discussed how to ensure that the transformation from a linear PCP to a SNARK adds as little computational overhead as possible. In this section, we discuss the problem of implementing a linear PCP (to give as input to the transformation) that is as efficient as possible.

Our Linear PCP. Our starting point is the work on *quadratic-span programs* (QSPs) and *quadratic-arithmetic programs* (QAPs) of Gennaro et al. [GGPR13]. Indeed, Bitansky et al. [BCI+13] observed that any QSP for a relation $\mathcal{R}$ yields a corresponding 3-query linear PCP for $\mathcal{R}$, and any QAP for a relation $\mathcal{R}$ yields a corresponding 4-query linear PCP for $\mathcal{R}$. By following the QAP approach of [GGPR13], we design a

---

[7] Since the encoding cannot provide semantic security (due to the functionality requirement of allowing for evaluation of quadratic predicates on encoded elements) but only a notion of one-way security, a limited amount of information is necessarily leaked.

linear PCP that trades an increased number of 5 queries for a linear PCP that, while keeping essentially optimal asymptotics, enjoys excellent efficiency in practice.

Concretely, for checking membership in the language $\mathcal{L}_C$ for a circuit C, our linear PCP has only 5 queries of $2|C|$ field elements each (and sampling the 5 queries needs only a single random field element); generating the queries can be done in linear time. The 5 answers of the queries can be verified via 2 quadratic polynomials using only $2n + 9$ field operations, where n is the input size. The soundness error is $2|C|/|\mathbb{F}|$. Through a suitable use of FFTs, the honest prover can compute the linear proof oracle via an arithmetic circuit of size $O(|C| \log |C|)$ and depth $O(\log |C|)$ only. (In particular, the prover is highly parallelizable.)

Efficiency Optimizations. In practice, tailored FFT algorithms are more efficient than "generic" ones (i.e., ones that work over any finite field). To leverage the efficiency of tailored FFT algorithms, we further specialize our choice of elliptic curve so to ensure that G_1, G_2 are groups of a prime order p with $p - 1 = 2^\ell h$ for a large integer ℓ. This means that, in our linear PCP, we can choose the finite field $\mathbb{F} = \mathbb{F}_p$. In such a field, there is a primitive 2^ℓ-th root of unity, and multi-point evaluation/interpolation over domains consisting of roots of unity (or their multiplicative cosets) can be performed via very simple and efficient FFT algorithms. Furthermore, the choice $\mathbb{F} = \mathbb{F}_p$ also simplifies the linear-time algorithm for sampling queries.

Zero Knowledge. The transformation from a linear PCP to a SNARK is such that if the linear PCP is honest-verifier zero-knowledge (HVZK) then the SNARK is zero knowledge. (See the extended version of this paper for a definition of HVZK.) Thus, we need to ensure that our linear PCP is HVZK. Bitansky et al. [BCI+13] showed a general transformation from a linear PCP to a HVZK linear PCP of similar efficiency. We do not rely on their general transformation. Instead, our linear PCP can be made HVZK with essentially no computational overhead, via a simple modification analogous to the one used in [GGPR13] to achieve zero knowledge. With this modification, we ensure that the SNARK obtained from our linear PCP has (statistical) zero knowledge.

For more details on our linear PCP construction, see the extended version of this paper.

3.3 Performance

Plugging our linear PCP for arithmetic circuits (Section 3.2) into the transformation (Section 3.1), we thus obtain an implementation of zk-SNARKs for arithmetic circuit satisfiability with excellent asymptotic efficiency: linear-time key generator, quasilinear-time prover, and linear-time verifier. Next, we discuss concrete performance.

Our algebraic setup is as follows: we work over $E(\mathbb{F}_q)$ where E is the elliptic curve $y^2 = x^3 + x$ and q is a prime of 512 bits; the order of the group is divisible by $p = 2^{159} + 2^{107} + 1$. This curve gives 128 bits of security. Our experiments are run on a machine with eight 2.4 GHz AMD Opteron 8431 6-core processors and 16 GB of RAM.

Performance of Key Generation. Given an arithmetic circuit $C: \mathbb{F}^n \times \mathbb{F}^h \to \mathbb{F}$ as input, the SNARK key generator G outputs: a proving key σ of $(12|C| + 2n + 40)$ group elements and a verification key τ of $(n + 8)$ group elements. Each group element (when compressed) is 65 bytes. Only 8 random field elements need to be sampled for this computation. A small set of *public parameters* provides information, to both the prover and verifier, about the choice of elliptic curve; storing these public parameters only requires 310 bytes. The extended version of this paper includes performance graphs of $G(C)$ as a function of $|C|$. For instance, when $|C| \approx 2 \cdot 10^6$, G performs $\approx 4.2 \cdot 10^9$ field operations in less than 20 minutes.

Performance of Proving. Given σ and (x, a) in the relation $\mathcal{R}_C$, the SNARK prover outputs a proof consisting of 12 group elements. As before, each group element (when compressed) is 65 bytes, so the proof length in bytes is 780. The extended version of this paper includes graphs of $P(\sigma, x, a)$ as a function of $|C|$. For instance, when $|C| \approx 2 \cdot 10^6$, P performs $\approx 3.3 \cdot 10^9$ field operations in less than 15 minutes.

Performance of verifying. Given τ, an input x, and a proof π, the SNARK verifier computes the decision bit. To do so, the verifier evaluates 21 pairings and solves a multi-exponentiation problem of size $|x|$. The extended version of the paper includes performance graphs of $V(\tau, x, \pi)$ as a function of $|x|$. For instance:

- when $|x| \leq 2^6$, V performs $\approx 2.2 \cdot 10^5$ field operations in less than 50 milliseconds;
- when $|x| \leq 2^{17}$, V performs $\approx 1.3 \cdot 10^7$ field operations in less than 20 seconds.

We emphasize that the above performance holds *no matter how large is the circuit C*.

Acknowledgments. The authors gratefully thank the members of the programming team: Ohad Barta, Lior Greenblatt, Shaul Kfir, Michael Riabzev, Gil Timnat, and Arnon Yogev. We also thank Lev Pachmanov for helping out with TinyRAM programming; Dan Berstein, Tanja Lange, Peter Schwabe, and Andrew Sutherland for discussions about elliptic curves; and Ron Rivest and Nickolai Zeldovich for helpful comments and discussions. We thank Nickolai Zeldovich for the use of his group's compute nodes.

The research leading to these results — in particular, funding of the aforementioned programming team — has received funding from the European Community's Seventh Framework Programme (FP7/2007-2013) under grant agreement number 240258. This work was partially supported by the Center for Science of Information (CSoI), an NSF Science and Technology Center, under grant agreement CCF-0939370; by the Check Point Institute for Information Security; by the Israeli Ministry of Science and Technology, and by the Israeli Centers of Research Excellence I-CORE program (center 4/11).

References

[ABB+12] Almeida, J.B., Barbosa, M., Bangerter, E., Barthe, G., Krenn, S., Béguelin, S.Z.: Full proof cryptography: verifiable compilation of efficient zero-knowledge protocols. In: CCS 2012 (2012)

[BBK+09] Bangerter, E., Barzan, S., Krenn, S., Sadeghi, A.-R., Schneider, T.: Bringing zero-knowledge proofs of knowledge to practice (2009)

[BCC88] Brassard, G., Chaum, D., Crépeau, C.: Minimum disclosure proofs of knowledge. Journal of Computer and System Sciences 37(2), 156–189 (1988)

[BCC+09] Belenkiy, M., Camenisch, J., Chase, M., Kohlweiss, M., Lysyanskaya, A., Shacham, H.: Randomizable proofs and delegatable anonymous credentials. In: Halevi, S. (ed.) CRYPTO 2009. LNCS, vol. 5677, pp. 108–125. Springer, Heidelberg (2009)

[BCCT13] Bitansky, N., Canetti, R., Chiesa, A., Tromer, E.: Recursive composition and bootstrapping for SNARKs and proof-carrying data. In: STOC 2013 (2013)

[BCG⁺13]   Ben-Sasson, E., Chiesa, A., Genkin, D., Tromer, E., Virza, M.: TinyRAM architecture specification v1.00 (2013), http://scipr-lab.org/tinyram

[BCGT13a] Ben-Sasson, E., Chiesa, A., Genkin, D., Tromer, E.: Fast reductions from RAMs to delegatable succinct constraint satisfaction problems. In: ITCS (2013)

[BCGT13b] Ben-Sasson, E., Chiesa, A., Genkin, D., Tromer, E.: On the concrete efficiency of probabilistically-checkable proofs. In: STOC 2013 (2013)

[BCI⁺13]   Bitansky, N., Chiesa, A., Ishai, Y., Paneth, O., Ostrovsky, R.: Succinct non-interactive arguments via linear interactive proofs. In: Sahai, A. (ed.) TCC 2013. LNCS, vol. 7785, pp. 315–333. Springer, Heidelberg (2013)

[BCKL08]   Belenkiy, M., Chase, M., Kohlweiss, M., Lysyanskaya, A.: P-signatures and noninteractive anonymous credentials. In: Canetti, R. (ed.) TCC 2008. LNCS, vol. 4948, pp. 356–374. Springer, Heidelberg (2008)

[BDNP08]   Ben-David, A., Nisan, N., Pinkas, B.: FairplayMP: a system for secure multi-party computation. In: CCS 2008 (2008)

[Ben65]   Beneš, V.E.: Mathematical theory of connecting networks and telephone traffic. Academic Press, New York (1965)

[Ber02]   Bernstein, D.J.: Pippenger's exponentiation algorithm (2002), http://cr.yp.to/papers/pippenger.pdf

[BFLS91]   Babai, L., Fortnow, L., Levin, L.A., Szegedy, M.: Checking computations in polylogarithmic time. In: STOC 1991 (1991)

[BGW88]   Ben-Or, M., Goldwasser, S., Wigderson, A.: Completeness theorems for non-cryptographic fault-tolerant distributed computation. In: STOC 1988 (1988)

[BHZ87]   Boppana, R.B., Håstad, J., Zachos, S.: Does co-NP have short interactive proofs? Information Processing Letters 25(2), 127–132 (1987)

[BP04]   Bellare, M., Palacio, A.: The knowledge-of-exponent assumptions and 3-round zero-knowledge protocols. In: Franklin, M. (ed.) CRYPTO 2004. LNCS, vol. 3152, pp. 273–289. Springer, Heidelberg (2004)

[BS08]   Ben-Sasson, E., Sudan, M.: Short PCPs with polylog query complexity. SIAM Journal on Computing 38(2) (2008)

[BW06]   Boyen, X., Waters, B.: Compact group signatures without random oracles. In: Vaudenay, S. (ed.) EUROCRYPT 2006. LNCS, vol. 4004, pp. 427–444. Springer, Heidelberg (2006)

[CKLM12]   Chase, M., Kohlweiss, M., Lysyanskaya, A., Meiklejohn, S.: Malleable proof systems and applications. In: Pointcheval, D., Johansson, T. (eds.) EUROCRYPT 2012. LNCS, vol. 7237, pp. 281–300. Springer, Heidelberg (2012)

[Dam92]   Damgård, I.: Towards practical public key systems secure against chosen ciphertext attacks. In: Feigenbaum, J. (ed.) CRYPTO 1991. LNCS, vol. 576, pp. 445–456. Springer, Heidelberg (1992)

[Din07]   Dinur, I.: The PCP theorem by gap amplification. Journal of the ACM 54(3) (2007)

[GGPR13]   Gennaro, R., Gentry, C., Parno, B., Raykova, M.: Quadratic span programs and succinct NIZKs without PCPs. In: Johansson, T., Nguyen, P.Q. (eds.) EUROCRYPT 2013. LNCS, vol. 7881, pp. 626–645. Springer, Heidelberg (2013)

[GH98]   Goldreich, O., Håstad, J.: On the complexity of interactive proofs with bounded communication. Information Processing Letters 67(4), 205–214 (1998)

[GMW87]   Goldreich, O., Micali, S., Wigderson, A.: How to play any mental game or a completeness theorem for protocols with honest majority. In: STOC 1987 (1987)

[Gro05]   Groth, J.: Non-interactive zero-knowledge arguments for voting. In: Ioannidis, J., Keromytis, A.D., Yung, M. (eds.) ACNS 2005. LNCS, vol. 3531, pp. 467–482. Springer, Heidelberg (2005)

[Gro06]   Groth, J.: Simulation-sound NIZK proofs for a practical language and constant size group signatures. In: Lai, X., Chen, K. (eds.) ASIACRYPT 2006. LNCS, vol. 4284, pp. 444–459. Springer, Heidelberg (2006)

[Gro10]   Groth, J.: Short non-interactive zero-knowledge proofs. In: Abe, M. (ed.) ASIACRYPT 2010. LNCS, vol. 6477, pp. 341–358. Springer, Heidelberg (2010)

[GS89]   Gurevich, Y., Shelah, S.: Nearly linear time. In: Logic at Botik 1989, Symposium on Logical Foundations of Computer Science, pp. 108–118 (1989)

[Gue12]   Gueron, S.: Intel advanced encryption standard (AES) instructions set (February 2012)

[GVW02]   Goldreich, O., Vadhan, S., Wigderson, A.: On interactive proofs with a laconic prover. Computational Complexity 11(1/2), 1–53 (2002)

[HT98]   Hada, S., Tanaka, T.: On the existence of 3-round zero-knowledge protocols. In: Krawczyk, H. (ed.) CRYPTO 1998. LNCS, vol. 1462, pp. 408–423. Springer, Heidelberg (1998)

[KMO01]   Katz, J., Myers, S., Ostrovsky, R.: Cryptographic counters and applications to electronic voting. In: Pfitzmann, B. (ed.) EUROCRYPT 2001. LNCS, vol. 2045, pp. 78–92. Springer, Heidelberg (2001)

[Lip11]   Lipmaa, H.: Two simple code-verification voting protocols. Cryptology ePrint Archive, Report 2011/317 (2011)

[Lip12]   Lipmaa, H.: Progression-free sets and sublinear pairing-based non-interactive zero-knowledge arguments. In: Cramer, R. (ed.) TCC 2012. LNCS, vol. 7194, pp. 169–189. Springer, Heidelberg (2012)

[Mic00]   Micali, S.: Computationally sound proofs. SIAM Journal on Computing 30(4), 1253–1298 (2000); Preliminary version appeared in FOCS 1994 (1994)

[MNPS04]   Malkhi, D., Nisan, N., Pinkas, B., Sella, Y.: Fairplay — a secure two-party computation system. In: SSYM 2004 (2004)

[MR08]   Moshkovitz, D., Raz, R.: Two-query PCP with subconstant error. Journal of the ACM 57, 1–29 (2008); Preliminary version appeared in FOCS 2008 (2008)

[Rob91]   Robson, J.M.: An O(T log T) reduction from RAM computations to satisfiability. Theoretical Computer Science 82(1), 141–149 (1991)

[Sch78]   Schnorr, C.-P.: Satisfiability is quasilinear complete in NQL. Journal of the ACM 25, 136–145 (1978)

[StGDC13] Stallman, R.M., and the GCC Developer Community: GNU compiler collection internals (2013), http://gcc.gnu.org/onlinedocs/gccint.pdf

[Wee05]   Wee, H.: On round-efficient argument systems. In: Caires, L., Italiano, G.F., Monteiro, L., Palamidessi, C., Yung, M. (eds.) ICALP 2005. LNCS, vol. 3580, pp. 140–152. Springer, Heidelberg (2005)

On the Function Field Sieve and the Impact of Higher Splitting Probabilities*
Application to Discrete Logarithms in $\mathbb{F}_{2^{1971}}$ and $\mathbb{F}_{2^{3164}}$

Faruk Göloğlu, Robert Granger, Gary McGuire, and Jens Zumbrägel

Complex & Adaptive Systems Laboratory and
School of Mathematical Sciences
University College Dublin, Ireland
{farukgologlu,robbiegranger}@gmail.com,
{gary.mcguire,jens.zumbragel}@ucd.ie

Abstract. In this paper we propose a binary field variant of the Joux-Lercier medium-sized Function Field Sieve, which results not only in complexities as low as $L_{q^n}(1/3, (4/9)^{1/3})$ for computing arbitrary logarithms, but also in an heuristic *polynomial time* algorithm for finding the discrete logarithms of degree one and two elements when the field has a subfield of an appropriate size. To illustrate the efficiency of the method, we have successfully solved the DLP in the finite fields with 2^{1971} and 2^{3164} elements, setting a record for binary fields.

Keywords: Discrete logarithm problem, function field sieve.

1 Introduction

When it comes to selecting appropriate parameters for public-key cryptosystems, one invariably observes a trade-off between security and efficiency. At a most basic level, for example, larger keys usually mean higher security, but worse performance.

A related rule of thumb which one does well to keep in mind, is that a specialised parameter which improves efficiency, typically (or potentially) weakens security. Examples abound of such specialisations and consequent attacks: discrete logarithms modulo Mersenne (or Crandall) primes and the Special Number Field Sieve [19]; Optimal Extension Fields [2] and Weil descent for elliptic curves [8]; high-compression algebraic tori [23] and specialised index calculus [10]; quasi-cyclic or dyadic McEliece variants [21] and Gröbner basis attacks [6], and more recently elliptic curves over binary fields [7], to name just a few. In practice therefore, one should be wary of any additional structure, which may potentially weaken a system.

---

* Research supported by the Claude Shannon Institute, Science Foundation Ireland Grant 06/MI/006. The fourth author was in addition supported by SFI Grant 08/IN.1/I1950.

R. Canetti and J.A. Garay (Eds.): CRYPTO 2013, Part II, LNCS 8043, pp. 109–128, 2013.

In this paper we give a fairly extreme example of this principle in the case of
binary (or in general small characteristic) fields which possess a small to medium-
sized intermediate field. In 2006 Joux and Lercier designed a particularly efficient
variation of the Function Field Sieve (FFS) algorithm for computing discrete
logarithms [16], which at the time possessed the fastest asymptotic complexity
of all known discrete logarithm algorithms for appropriately balanced q and n,
namely $L_{q^n}(1/3, 3^{1/3}) \approx L_{q^n}(1/3, 1.442)$, where

$$L_{q^n}(a, c) = \exp\left((c + o(1)) (\log q^n)^a (\log\log q^n)^{1-a}\right),$$

and q^n is the cardinality of the finite field.

In 2012, Joux proposed a more efficient method of obtaining relations, dubbed
'pinpointing', which applies to a specialisation of the function field setup of [16].
In this approach, each relation found via classical sieving can be amplified into
many more [13], which is advantageous when sieving is the dominant phase,
rather than the linear algebra (or individual logarithm phase). The overall complexity of this technique for solving the DLP can be as low as $L_{q^n}(1/3, (8/9)^{1/3}) \approx
L_{q^n}(1/3, 0.961)$. To demonstrate the practicality of the approach, Joux solved
the DLP in two cases: in a 1175-bit field and in a 1425-bit field, setting records
for medium-sized base fields, in this case prime fields.

In this work we demonstrate that a basic assumption used in the analysis
of virtually all fast index calculus algorithms can be very wrong indeed; in the
case of binary fields possessing a subfield of an appropriate size, this leads to the
dramatic conclusion that the logarithms of degree one elements over this subfield
can be solved in *polynomial time*. As far as we are aware, no other algorithm for
the collecting of relations and the linear algebra step has beaten the $L_{q^n}(1/3)$
barrier. Our fundamental observation is that the splitting probabilities in Joux-
Lercier's variation of the FFS can be *cubic* in the reciprocal of the degree –
rather than exponential. The reason for this is the richer structure of binary
extension fields relative to prime fields, which lends weight to the argument that
such fields should be avoided in practice. We also exploit our basic observation
to efficiently compute the logarithms of degree two elements — which until now
were the bottleneck of the individual logarithm descent phase — which for a
range of binary fields results in the fastest $L_{q^n}(1/3)$ algorithm to date, namely
$L_{q^n}(1/3, (4/9)^{1/3}) \approx L_{q^n}(1/3, 0.763)$, which is precisely the square root of the
complexity of the ordinary FFS, for which $c = (32/9)^{1/3}$.

We emphasise that our relation generation method arises purely as a specialisation of [16], and is thus completely independent of [13]. However, at a high
level, our relation generation method *may* be viewed as a form of one-sided pinpointing, but with two central differences to that of [13]. Firstly, we do not need
to search for an initial splitting polynomial, since we have an explicit description
of *all* such polynomials, i.e., no sieving need take place. Secondly, as members of
this family of polynomials have arbitrarily high degree, the other 'random' side
can be made to have very small degree, which thus splits with very high probability. These two differences result in our polynomial time relation generation.

The paper is organised as follows. In §2 we recall the Joux-Lercier variant
of the FFS. In §3 we present our specialisation and our analysis of splitting

probabilities, while in §4 we present our new descent methods and analyse the complexity of the resulting algorithms. In §5 we present our implementation results and conclude in §6.

2   The Medium-Sized Base Field Function Field Sieve

In this section we briefly recall the 2006 FFS variant of Joux and Lercier [16]. Let $\mathbb{F}_{q^n}$ be the finite field in which discrete logarithms are to be solved, where q is a prime power. In order to represent $\mathbb{F}_{q^n}$, choose two univariate polynomials $g_1, g_2 \in \mathbb{F}_q[X]$ of degrees d_1 and d_2 respectively. Then whenever $X - g_1(g_2(X))$ possesses a degree n irreducible factor $F(X)$ over $\mathbb{F}_q$, one can represent $\mathbb{F}_{q^n}$ in two related ways. In particular, let $x \in \mathbb{F}_{q^n}$ be a root of $F(X) = 0$, and let $y := g_2(x)$, so that by construction $x = g_1(y)$ as well. These relations give an explicit isomorphism between $\mathbb{F}_q(x)$ and $\mathbb{F}_q(y)$, both of which represent $\mathbb{F}_{q^n}$.

In the most basic version of the algorithm (which also leads to the best complexity) one chooses $d_1 \approx d_2 \approx \sqrt{n}$, and considers elements of $\mathbb{F}_{q^n}$ represented by:

$$xy + ay + bx + c, \quad \text{with} \quad a, b, c \in \mathbb{F}_q.$$

Substituting x by $g_1(y)$, and y by $g_2(x)$, we obtain the following equality in $\mathbb{F}_{q^n}$:

$$xg_2(x) + ag_2(x) + bx + c = yg_1(y) + ay + bg_1(y) + c. \tag{1}$$

The factor base consists simply of the degree one elements of $\mathbb{F}_q(x)$ and $\mathbb{F}_q(y)$; then for every triple (a, b, c) for which both sides of (1) split over $\mathbb{F}_q$ — i.e., when all of its roots are in $\mathbb{F}_q$ — in the respective factor bases, one obtains a relation. Determining such triples can naturally be made faster using sieving techniques. Once more than $2q$ such relations have been collected, one performs a linear algebra elimination to recover the individual logarithms. To compute arbitrary discrete logarithms, one uses a 'descent' method, as we detail in §4.

In order to assess the complexity of this algorithm, throughout the paper let $Q = q^n$, let $q = L_Q(1/3, \alpha)$, and let $L_Q(1/3, c_1)$ and $L_Q(1/3, c_2)$ denote the complexity of the sieving and linear algebra phases respectively. As shown in [16], heuristically one has

$$c_1 = \alpha + \frac{2}{3\sqrt{\alpha}} \quad \text{and} \quad c_2 = 2\alpha.$$

In order to generate sufficiently many relations, α must satisfy the condition:

$$2\alpha \geq \frac{2}{3\sqrt{\alpha}}.$$

For such α's, the complexity of the entire algorithm, including the descent phase, is minimised for $\alpha = 3^{-2/3}$, with resulting complexity $L_Q(1/3, 3^{1/3})$.

3   Specialisation to Binary Fields

We now present a specialisation of the construction of [16] as presented in the previous section, and detail some interesting consequences. From now on let $\mathbb{F}_q$ denote the finite field with 2^l elements.

All of our improvements and observations arise from a rather innocent-looking choice for g_2, namely $y = x^{2^k}$. Our primary motivation for this was to automatically eliminate half of the factor base, since any linear polynomial $(y + a)$ is equal to $(x + a^{2^{-k}})^{2^k}$, and so $\log(y + a) = 2^k \log(x + a^{2^{-k}})$. However, this selection has further serendipitous consequences, the central two being:

- Whenever $k \mid l$ and $l \geq 3k$, the probability of the l.h.s. of (1) splitting over $\mathbb{F}_q$ is approximately 2^{-3k}, instead of the expected $1/(2^k + 1)!$. We show that for some asymptotic families of binary fields, this leads to a *polynomial time* algorithm to find the logarithms of all degree one elements of $\mathbb{F}_{q^n}$.
- As surprising as the above result is, for such families, the individual logarithm phase then has complexity $L_{q^n}(1/2)$. Hence one must ensure the complexity of the stages is balanced. Depending on the form of n, we show that the bottleneck of the descent changes from degree two to degree three special-$\mathfrak{q}$, since the x-side has the same form of the l.h.s. of (1), and thus enjoys the same higher splitting probability. This ensures that our claimed new $L_{q^n}(1/3)$ complexities are achieved across all the phases of the algorithm.

In the remainder of the paper we explain these advantages in more detail. In addition to the above two observations, for certain extensions which possess Galois-invariant factor bases, the use of non-prime base fields can induce extra automorphisms, which reduce its size further, see §5. Other practical speed ups arise from our choice $y = x^{2^k}$. The matrix-vector multiplications in Lanczos' algorithm consists of only cyclic rotations, i.e., shifts mod $q^n - 1$, and so no multiplications need to be performed. Furthermore, in the descent phase, one ordinarily needs to perform special-$\mathfrak{q}$ eliminations in both function fields. However, due to the simple relation between x and y, one is free to map from one side to the other in order to increase the probability of smoothness. One can also balance the degrees of both sides by utilising other auxiliary function fields arising from passing a power of 2 from the x-side to other side; this not only provides a practical speed up but is core to our new complexity results, see §4.

3.1   Higher Splitting Probabilities

Throughout this section, rather than use the field elements x, y as variables, we use X, Y to emphasise that the stated results are valid in the univariate polynomial ring over $\mathbb{F}_q$, which is implicitly either $\mathbb{F}_q[X]$ or $\mathbb{F}_q[Y]$, depending on which side of (1) is involved.

Assume $1 < k < l$. When $Y = X^{2^k}$ the l.h.s. of (1) becomes

$$X^{2^k+1} + aX^{2^k} + bX + c\,. \tag{2}$$

Assuming $c \neq ab$ and $b \neq a^{2^k}$, this polynomial may be transformed (up to a scalar factor) into the polynomial

$$f_B(\overline{X}) = \overline{X}^{2^k+1} + B\overline{X} + B, \quad \text{with} \quad B = \frac{(b+a^{2^k})^{2^k+1}}{(ab+c)^{2^k}}, \tag{3}$$

via

$$X = \frac{ab+c}{b+a^{2^k}} \overline{X} + a.$$

The polynomial f_B is related to $P_A(\overline{X}) = \overline{X}^{2^k+1} + \overline{X} + A$, which is well-studied in the literature, having arisen in several contexts including finite geometry, difference sets, as well as determining cross correlation between m-sequences; see references in [12] for further details.

We have the following theorem due to Bluher [3] (and refined in the binary case by Helleseth and Kholosha [12]), which counts the number of $B \in \mathbb{F}_q$ for which f_B splits over $\mathbb{F}_q$.

Theorem 1. *[12, Thm. 1] Let $d = \gcd(l,k)$. Then the number of $B \in \mathbb{F}_{2^l}^{\times}$ such that $f_B(\overline{X})$ has exactly $2^d + 1$ roots over $\mathbb{F}_{2^l}$ is*

$$\begin{cases} \dfrac{2^{l-d}-1}{2^{2d}-1} & \text{if } l/d \text{ odd,} \\[3mm] \dfrac{2^{l-d}-2^d}{2^{2d}-1} & \text{if } l/d \text{ even.} \end{cases}$$

Theorem 1 of [12] also states that f_B can have no more than $2^d + 1$ roots in $\mathbb{F}_q$, and so if $\gcd(l,k) < k$ then f_B can not split. Hence we must have $k \mid l$ for our application. Indeed we must also have $l \geq 3k$ in order for there to be at least one such B. Observe that under these two conditions, for B chosen uniformly at random from $\mathbb{F}_q$, the probability that f_B splits completely over $\mathbb{F}_q$ is approximately $1/2^{3k}$ – far higher than the splitting probability $1/(2^k+1)!$ for a degree $2^k + 1$ polynomial chosen uniformly at random.

Furthermore, the set S_B of all such B can be computed explicitly, without needing to perform any factorisations or smoothness tests. Indeed, the proof of Prop. 5 in [12] gives an explicit parameterisation of all such B: for $u \in G = \mathbb{F}_{2^l} \setminus \mathbb{F}_{2^{2k}}$, we have

$$S_B = \text{Im}\left(u \longrightarrow \frac{(u+u^{2^{2k}})^{2^k+1}}{(u+u^{2^k})^{2^{2k}+1}} \right).$$

By analysing the form of this map, one can avoid obtaining repeated images. However, even a naive enumeration of elements of G requires at most $\widetilde{O}(q)$ $\mathbb{F}_q$-operations, which is comparable to the complexity of relation generation, as we now show.

3.2   Relation Generation

By exploiting the above transformation of (2) to (3) and the list S_B of precomputed B's for which (3) splits, one can construct polynomials of the form (2) which always split completely over $\mathbb{F}_q$. Indeed, for any (a, b) for which $b \neq a^{2^k}$, and for each $B \in S_B$, we simply compute via (3) the corresponding unique $c \in \mathbb{F}_q$. This ensures that (2) splits and therefore requires no sieving whatsoever. In order to obtain a relation, we also require that

$$Y g_1(Y) + b g_1(Y) + aY + c \tag{4}$$

splits over $\mathbb{F}_q$, which we assume occurs with probability $1/(d_1 + 1)!$ for randomly chosen g_1. Since $|L_B| \approx q/2^{3k}$, for each (a, b) we expect to obtain

$$\frac{q}{2^{3k} (d_1 + 1)!}$$

relations. Since we need q relations, we expect to require about $2^{3k} (d_1 + 1)!$ pairs (a, b) to obtain sufficiently many. For each pair (a, b) this costs $O(q/2^{3k})$ 1-smoothness tests, or $\tilde{O}(q/2^{3k})$ $\mathbb{F}_q$-operations. Hence the total cost is $\tilde{O}(q (d_1 + 1)!)$. Finally, in order for there to be sufficiently many relations, we must have

$$\frac{q^3}{2^{3k} (d_1 + 1)!} > q, \quad \text{or} \quad q^2 > 2^{3k} (d_1 + 1)!\,.$$

Since we insist that $l \geq 3k$, having $q > (d_1 + 1)!$ is sufficient. In §4 we consider the impact of this approach on the full DLP complexity in two cases when $q = L_{q^n}(1/3, \alpha)$ and $n \approx 2^k \cdot d_1$: firstly for $2^k \approx d_1$ and secondly for $2^k \gg d_1$. However, we now consider the relation generation complexity when the base field cardinality is polynomially related to the extension degree.

3.3   Polynomial Time Relation Generation

With a view to reducing the complexity of degree one relation generation to a minimum for some example fields, we choose k as large as possible such that $k \mid l$ and $l \geq 3k$, and set d_1 to be as small as possible, assuming a g_1 can be found with $X - g_1(X^{2^k})$ possessing a degree n irreducible factor. Experimentally it seems that $d_1 = 3$ (or possibly $d_1 = 4$) is sufficient to produce an irreducible of any degree $n \leq 2^k$, for q sufficiently large. Of course, n may be as large as $2^k \cdot d_1$ in this case.

Writing $l = k \cdot k'$ with $k' \geq 3$ a constant, and $n \approx 2^k \cdot d_1$ with d_1 constant, as $l \to \infty$, the logarithms of the degree one factor base elements of $\mathbb{F}_{q^n}$ can be computed in heuristic *polynomial time*. In particular, as $n \approx 2^k \cdot d_1 = 2^{l/k'} \cdot d_1$, we have

$$Q = q^n \approx 2^{l \cdot 2^{l/k'} \cdot d_1}\,.$$

As $l \to \infty$, we therefore have

$$\frac{\log Q}{\log \log Q} = O(2^{l/k'})\,.$$

The cost of relation generation is $\widetilde{O}(q\,(d_1+1)!) = \widetilde{O}(q) = \widetilde{O}(2^l) = \widetilde{O}(\log^{k'} Q)$, whereas the cost of sparse linear algebra, using Lanczos' algorithm [18] for instance, is the product of the row weight and the square of number of variables, namely

$$(2^{l/k'} + d_1)\,\widetilde{O}(q^2) = \widetilde{O}(\log^{2k'+1} Q)\,.$$

For the optimal choice $k' = 3$ the complexity is therefore $\widetilde{O}(\log^7 Q)$. We summarise this in the following:

Heuristic Result 1. *Let $q = 2^l$ with $l = k \cdot k'$ and $k' \geq 3$ a constant, let $d_1 \geq 3$ be constant, and assume $n \approx 2^k \cdot d_1$. Assuming that $Y g_1(Y) + aY + b g_1(Y) + c$ splits over $\mathbb{F}_q$ with probability $1/(d_1+1)!$ over all triples $(a, b, c) \in (\mathbb{F}_q)^3$, the logarithms of all degree one elements of $\mathbb{F}_{q^n}$ can be computed in time $\widetilde{O}(\log^{2k'+1} Q)$.*

Note that the set of degree one elements is always defined relative to a particular representation of $\mathbb{F}_{q^n}$. As it is easy to switch between any two representations of a finite field [20], one can always map to our $\mathbb{F}_q(x)$ first. Note also that the statement of Heuristic Result 1 implicitly assumes that the factor base contains a generator of $\mathbb{F}_{q^n}^\times$. A result of Chung proves that for all prime powers s and all $r \geq 1$ such that $s > (r-1)^2$, if $\mathbb{F}_{s^r} = \mathbb{F}_s(x)$ then $\{x + a \mid a \in \mathbb{F}_s\}$ generates $\mathbb{F}_{s^r}^\times$ [4, Thm. 8]. In our context we therefore need $q^k > (n-1)^2 \approx q^2 \cdot d_1^2$ in order for our DLP algorithm to work, which is satisfied for our q and small d_1. However, the issue of whether there exists a generator in the stated factor base remains an open problem in general, see for instance [26].

3.4 An Extreme Case: $n = 2^k \pm 1$

If $n = 2^k \pm 1$ then the degree one relation generation becomes extremely fast. In particular, if $g_1(X) = \gamma X^{\mp 1}$ then as $g_2(X) = X^{2^k}$, we obtain the polynomials $X^{2^k \pm 1} + \gamma$. Furthermore, if $k \mid l$ then $X^{2^k \pm 1} + \gamma$ is irreducible whenever γ has no root of prime order $p \mid (2^k \pm 1)$. In both cases, (4) has degree two and splits with probability $1/2$.

Table 1 contains timing data for relation generation for a family of fields with $q = 2^{3k}$ and $n = 2^k - 1$, which incorporates the factor base reduction technique arising from quotienting out by the action of the k-th power of Frobenius, which has order $3n$, see §5. We used an AMD Opteron 6128 processor clocked at $2.0\,\mathrm{GHz}$. Note that the time is quasi-cubic in the bitlengh, in accordance with the discussion preceeding Heuristic Result 1.

4   Individual Logarithms and Complexity Analysis

As unexpected as Heuristic Result 1 is, it does not by itself solve the DLP. Using a descent method à la [16,5], computing individual logarithms unfortunately then has complexity $L_{q^n}(1/2)$. Hence one can not allow the extension degree n to grow as fast as Theorem 1 permits; it must be tempered relative to the base field size.

Table 1. Relation generation times for $q = 2^{3k}$ and $n = 2^k - 1$

k	$\log_2(q^n)$	#vars	time
7	2667	5506	$2.3s$
8	6120	21932	$15.0s$
9	13797	87554	$122s$
10	30690	349858	$900s$

With this in mind, we now consider the complexity of the descent, for q and n appropriately balanced so that the total complexity is $L_{q^n}(1/3)$.

For a generator $g \in \mathbb{F}_{q^n}^\times$ and a target element $h \in \langle g \rangle$, the descent proceeds by first finding an $i \in \mathbb{N}$ such that $z = h\,g^i$ is m-smooth for a suitable m, i.e., so that all of the irreducible factors of z have degrees $\leq m$. The goal of the descent is to eliminate every irreducible factor of z, by expressing each as a product of smaller degree irreducibles recursively, until only degree one elements remain, whose logarithms are known. We do so using the special-$\mathfrak{q}$ lattice approach from [16], as follows.

Let $p(x)$ be a degree d irreducible (considered as an element of $\mathbb{F}_q[X]$) which we wish to eliminate. Since $y = x^{2^k}$, we have

$$p(x)^{2^k} = \overline{p}(x^{2^k}) = \overline{p}(y)\,,$$

where the coefficients of $\overline{p}$ are those of p, powered by 2^k. Note that we also have

$$\overline{p}(y)^{2^{-k}} = p(x)\,,$$

and hence we can freely choose to eliminate p using either the x-side or the y-side of (1). For convenience we focus on the y-side. The corresponding lattice $L_{\overline{p}}$ is defined by:

$$L_{\overline{p}(Y)} = \{(w_0(Y), w_1(Y)) \in \mathbb{F}_q[Y]^2 : w_0(Y)\,g_1(Y) + w_1(Y) \equiv 0 \pmod{\overline{p}(Y)}\}\,.$$

A basis for this lattice is $(0, \overline{p}(Y)), (1, g_1(Y) \pmod{\overline{p}(Y)})$, which is clearly unbalanced. Using the extended Euclidean algorithm, we may construct a balanced basis $(u_0(Y), u_1(Y)), (v_0(Y), v_1(Y))$ for which the degrees are $\approx d/2$. Then for any $r(Y), s(Y) \in \mathbb{F}_q[Y]$ with $r(Y)$ monic we have

$$(w_0(Y), w_1(Y)) = \big(r(Y)u_0(Y) + s(Y)v_0(Y), r(Y)u_1(Y) + s(Y)v_1(Y)\big) \in L_{\overline{p}(Y)}$$

and thus RHS$(Y) \equiv 0 \pmod{\overline{p}(Y)}$, where

$$\text{RHS}(Y) = w_0(Y)\,g_1(Y) + w_1(Y)\,.$$

When RHS$(Y)/\overline{p}(Y)$ is $(d-1)$-smooth, we also check whether LHS(X) is also $(d-1)$-smooth, where

$$\text{LHS}(X) = w_0(X^{2^k})\,X + w_1(X^{2^k})\,.$$

When both sides are $(d-1)$-smooth, we may replace $\bar{p}(Y)$ with a product of irreducibles of degree at most $d-1$, and then recurse.

Let $Q = q^n$. As in [16], we assume there is a parameter α such that:

$$n = \frac{1}{\alpha}\left(\frac{\log Q}{\log\log Q}\right)^{2/3}, \qquad q = \exp\left(\alpha \sqrt[3]{\log Q \cdot \log^2\log Q}\right). \tag{5}$$

The three stages to consider are relation generation, linear algebra, and the descent, whose complexities we denote by $L_Q(1/3, c_1)$, $L_Q(1/3, c_2)$ and $L_Q(1/3, c_3)$, respectively. The total complexity is therefore $L_Q(1/3, c)$, where $c = \max\{c_1, c_2, c_3\}$. We next consider degree 2 elimination and then two special cases of field representation.

4.1 Degree 2 Elimination

We begin with degree 2 elimination as firstly it is the bottleneck in the descent, and secondly because one can exploit the higher splitting probability of the polynomials (2) as well. Let $\bar{p}(Y)$ be a degree 2 irreducible to be eliminated. A reduced basis $(u_0(Y), u_1(Y)), (v_0(Y), v_1(Y))$ for the lattice $L_{\bar{p}(Y)}$ can be found with degrees $(1, 0), (0, 1)$. Hence with r normalised to be 1 and $s \in \mathbb{F}_q$, we have

$$(w_0(Y), w_1(Y)) = \big(u_0(Y) + s\,v_0(Y), u_1(Y) + s\,v_1(Y)\big) \in L_{\bar{p}(Y)}$$

with degrees $(1, 1)$. We have thus

$$w_0(Y)\,g_1(Y) + w_1(Y) \equiv 0 \pmod{\bar{p}(Y)},$$

and so the remaining factor has degree $d_1 - 1$. The corresponding polynomial $\mathrm{LHS}(X)$ is

$$w_0(X^{2^k})\,X + w_1(X^{2^k}), \tag{6}$$

which is of the form $X^{2^k+1} + aX^{2^k} + bX + c$, and as a consequence of Theorem 1, it splits over $\mathbb{F}_q$ with probability approximately 2^{-3k}. However, as with relation generation, we can also ensure that $\mathrm{LHS}(X)$ always splits, with the following technique. Writing the basis elements explicitly as $(Y + u_{00}, u_{10}), (v_{00}, Y + v_{10})$, and with $r = 1$ and $s \in \mathbb{F}_q$ the lattice elements are $(w_0(Y), w_1(Y)) = (Y + u_{00} + sv_{00}, sY + u_{10} + sv_{10})$. Thus combining (6) and (3), for each $B \in S_B$ we find the set of roots $s \in \mathbb{F}_q$ that satisfy the $\mathbb{F}_q[S]$ polynomial

$$B \cdot (v_{00}S^2 + (u_{00} + v_{10})S + u_{10})^{2^k} + (S^{2^k} + v_{00}S + u_{00})^{2^k+1} = 0,$$

by computing its GCD with $S^q + S$. This technique extracts all such s algebraically for any B, which ensures that $\mathrm{LHS}(X)$ automatically splits.

On average one expects there to be one such $s \in \mathbb{F}_q$ for each B. Then for each such s we check whether $\mathrm{RHS}(Y)/\bar{p}(Y)$ splits, which we assume occurs with probability $1/(d_1 - 1)!$. In general we therefore need sufficiently many B's in S_B for this to occur with good probability, i.e., that $q/2^{3k} > (d_1 - 1)!$.

4.2   Case 1: $n \approx 2^k \cdot d_1$ and $2^k \approx d_1$

In this section we will show the following:

Heuristic Result 2 (i). *Let $q = 2^l$, let $k \mid l$ and let n be such that (5) holds. Then for $n \approx 2^k \cdot d_1$ where $2^k \approx d_1$, the DLP can be solved with complexity $L_Q(1/3, (8/9)^{1/3}) \approx L_Q(1/3, 0.961)$.*

This is the simplest case we present; however for the sake of completeness and ease of exposition, we explicitly tailor the derivation presented in §3.2. By our relation generation method, the l.h.s. polynomial (2) always splits, whereas the probability of (4) being smooth is approximately $1/\sqrt{n}!$. Using the standard approximation $\log n! \approx n \log n$, the logarithm of the probability P of both sides being smooth is therefore:

$$\log P \approx -\sqrt{n} \log \sqrt{n} = -\frac{1}{2}\sqrt{n} \log n \ .$$

The size of the sieving space is $q^3/2^{3k}$, and since we require q relations we must have:

$$\frac{q^3 P}{2^{3k}} \geq q \ , \quad \text{or} \quad 2 \log q \geq \left(\frac{3}{2} + \frac{\sqrt{n}}{2}\right) \log n \approx \frac{\sqrt{n}}{2} \log n \ .$$

Ignoring low order terms, by (5) this is equivalent to

$$2\alpha \geq \frac{1}{3\sqrt{\alpha}} \ , \quad \text{or} \quad \alpha \geq 6^{-2/3} \ . \tag{7}$$

Given that we require q relations, the expected time to collect these relations is

$$\frac{q}{P} = L_Q\left(1/3 , \alpha + \frac{1}{3\sqrt{\alpha}}\right) ,$$

and hence $c_1 = \alpha + \frac{1}{3\sqrt{\alpha}}$. Since the linear algebra is quadratic in the size of the factor base, we also have $c_2 = 2\alpha$.

For the descent, as in [16], let the smoothness bound be $m = \mu\sqrt{n}$. Then the probability of finding such an expression is

$$1 \, / \, L_Q\left(1/3 , \frac{1}{3\mu\sqrt{\alpha}}\right) .$$

If the descent is to be no more costly than either the relation generation or the linear algebra, then we must have

$$\frac{1}{3\mu\sqrt{\alpha}} \leq \max\left\{\alpha + \frac{1}{3\sqrt{\alpha}}, 2\alpha\right\} . \tag{8}$$

We also need to ensure three further conditions are satisfied. Firstly, that the cost of all the special-q eliminations is no more than $L_Q(1/3, \max\{c_1, c_2\})$. Secondly, that there are enough (r, s) pairs to ensure a relation is found. And thirdly, that

during the descent the degrees of the polynomials being tested for smoothness
is really descending.

By the discussion in §4.1, in order to eliminate degree 2 elements we need
$q \geq 2^{3k} (d_1 - 1)!$, or equivalently,

$$\alpha \geq \frac{1}{3\sqrt{\alpha}}, \quad \text{or} \quad \alpha \geq 3^{-2/3}.$$

Since for degree 3 special-q LHS(X) will not have the form (2), we need to
check that the smoothness probability does not impose an extra condition on α.
For $\overline{p}(Y)$ a degree 3 irreducible to be eliminated, a reduced basis $(u_0(Y), u_1(Y))$,
$(v_0(Y), v_1(Y))$ for the lattice $L_{\overline{p}(Y)}$ can be found with degrees $(1, 1), (0, 2)$. Hence
with r now allowed to be monic of degree one and $s \in \mathbb{F}_q$, we have

$$(w_0(Y), w_1(Y)) = \big((Y + r_0)u_0(Y) + s\,v_0(Y),\, (Y + r_0)u_1(Y) + s\,v_1(Y)\big) \in L_{\overline{p}(Y)},$$

with degrees $(2, 2)$. As before, we have

$$w_0(Y)\,g_1(Y) + w_1(Y) \equiv 0 \pmod{\overline{p}(Y)},$$

and the corresponding polynomial LHS(X) is

$$w_0(X^{2^k})\,X + w_1(X^{2^k}).$$

Once divided by $\overline{p}(Y)$, the degree of the Y-side is $d_1 - 1 \approx \sqrt{n}$ while the degree
of the X-side is $2^{k+1} + 1 \approx 2\sqrt{n}$. The logarithm of the probability that a degree
n polynomial over $\mathbb{F}_q$ is m-smooth, for q and n tending to infinity but m fixed,
can be estimated by $-(n/m) \log (n/m)$, as shown in [16]. Therefore the log of
the probability P of both sides being 2-smooth is:

$$\log P \approx -\frac{\sqrt{n}}{2} \log \frac{\sqrt{n}}{2} - \frac{2\sqrt{n}}{2} \log \frac{2\sqrt{n}}{2} \approx -\frac{3}{2}\sqrt{n} \log \frac{\sqrt{n}}{2} \approx -\frac{3}{4}\sqrt{n} \log n,$$

and therefore $P = 1/L_Q(1/3, \frac{1}{2\sqrt{\alpha}})$. Since the (r, s) search space has size q^2
(which is also the complexity of the linear algebra), we require that

$$2\alpha \geq \frac{1}{2\sqrt{\alpha}} \quad \text{or} \quad \alpha \geq 16^{-1/3}.$$

Since $16^{-1/3} < 3^{-2/3}$, this imposes no additional constraint on α. Hence we can
set $\alpha = 3^{-2/3}$, and one can check that in this case, $c_1 = c_2 = c_3 = 2\alpha$, giving
complexity

$$L_Q(1/3, (8/9)^{1/3}) \approx L_Q(1/3, 0.961),$$

which is precisely the complexity Joux obtained using either optimal one-sided,
or advanced pinpointing [13]. Furthermore for this α, (8) implies that $\mu \geq 1/2$.
For an upper bound, note that for special-q of degree $\mu\sqrt{n}$, the degree of RHS(Y)
is about $\sqrt{n}(1 - \mu/2)$, while the degree of LHS(X) is about $\mu n/2$, so that $\mu < 2$
ensures that the descent is effective.

4.3   Case 2: $n \approx 2^k \cdot d_1$ and $2^k \gg d_1$

In this section we will show the following:

Heuristic Result 2 (ii). *Let* $q = 2^l$, *let* $k \mid l$ *and let* n *be such that (5) holds. Then for* $n \approx 2^k \cdot d_1$ *where* $2^k \gg d_1$, *the DLP can be solved with complexity between* $L_Q(1/3, (4/9)^{1/3}) \approx L_Q(1/3, 0.763)$ *and* $L_Q(1/3, (1/2)^{1/3}) \approx L_Q(1/3, 0.794)$.

Observe that interestingly, these two complexities are precisely the square-roots of the complexities of Coppersmith' algorithm [5], for which $c = (32/9)^{1/3}$ and $4^{1/3}$, the lower of the two being the complexity of the ordinary FFS [1,14].

For n and q of the form (5), we claim that $c_1 = \alpha$, $c_2 = 2\alpha$, and that there are sufficiently many relations available. In particular, if we write $d_1 = n^\beta$ with $\beta < 1/2$ and $2^k = n^{1-\beta}$, then again by our relation generation method, the l.h.s. polynomial (2) always splits, and the log of the probability P of both sides being 1-smooth is:

$$\log P \approx -\beta n^\beta \log n.$$

By (5) we have

$$-\beta n^\beta \log n \approx -\frac{2\beta}{3\alpha^\beta} \left(\frac{\log Q}{\log \log Q} \right)^{2\beta/3} (\log \log Q)$$

$$= -\frac{2\beta}{3\alpha^\beta} (\log Q)^{2\beta/3} (\log \log Q)^{1-2\beta/3}.$$

Hence the expected time of the relation generation is

$$\frac{q}{P} = L_Q(1/3, \alpha) \cdot L_Q \left(2\beta/3, \frac{2\beta}{3\alpha^\beta} \right).$$

For $\beta < 1/2$ the second term on the right is absorbed by the $o(1)$ term in the first term, and hence $c_1 = \alpha$ and $c_2 = 2\alpha$. The size of the sieving space is $q^3/2^{3k}$, and since we require q relations we must have:

$$\frac{q^3 P}{2^{3k}} \geq q, \quad \text{or} \quad L_Q(1/3, 2\alpha) \geq L_Q \left(2\beta/3, \frac{2\beta}{3\alpha^\beta} \right),$$

which holds for any $\alpha > 0$ when $\beta < 1/2$.

For the descent (as for Case 1) the cost of finding the first $\mu\sqrt{n}$-smooth relation is $L_Q(1/3, \frac{1}{3\mu\sqrt{\alpha}})$. And as before, for degree 2 special-q, the X-side has the same form and the condition on q arising from the search space being sufficiently large is always satisfied, since

$$q \geq 2^{3k} (d_1 - 1)! = n^{3(1-\beta)} L_Q \left(2\beta/3, \frac{2\beta}{3\alpha^\beta} \right),$$

which holds for any $\alpha > 0$ when $\beta < 1/2$.

Hence degree 3 special-q are the bottleneck. As in the first case, with r allowed to be monic of degree one and $s \in \mathbb{F}_q$, the degree of RHS(Y) is $d_1 - 1$ while the degree of LHS(X) is $2^{k+1} + 1$. These degrees are clearly unbalanced. However, we can employ the following tactic to balance them.

Since $g_1(Y)^{2^k} + Y = 0$, we let $X' = g_1(Y)^{2^a}$ and thus $Y = X'^{2^{k-a}}$. We are free to choose any $1 < a < k$, as an elimination of a special-q using Y and X' can be written in terms of Y and X by powering by a power of 2. With r allowed to be monic of degree one and $s \in \mathbb{F}_q$ we have $(w_0(Y), w_1(Y)) \in L_{\bar{p}(Y)}$ with degrees $(2,2)$, and our new expressions become

$$w_0(Y) g_1(Y)^{2^a} + w_1(Y) \equiv 0 \pmod{\bar{p}(Y)}.$$

The corresponding polynomial LHS(X') is

$$w_0(X'^{2^{k-a}}) X' + w_1(X'^{2^{k-a}}).$$

Assuming the degrees are (approximately) the same, taking logs we have

$$k - a + 1 = \log_2(d_1) + a, \quad \text{or} \quad a = (k + 1 - \log_2(d_1))/2.$$

Since a must be an integer, rather than a real variable, we must choose the nearest integer to this value. In the best case, we can take a to be this exact value, and consequently both degrees are $\sqrt{2d_1}\, 2^{k/2} = \sqrt{2}\sqrt{n}$. Therefore the log of the probability P of both sides being 2-smooth is:

$$\log P \approx -\frac{\sqrt{2}}{2} \sqrt{n} \log\left(\frac{\sqrt{2}}{2}\sqrt{n}\right) - \frac{\sqrt{2}}{2}\sqrt{n}\log\left(\frac{\sqrt{2}}{2}\sqrt{n}\right) \approx -\frac{\sqrt{2}}{2}\sqrt{n}\log n,$$

and hence $P = L_Q(1/3, -\frac{\sqrt{2}}{3\sqrt{\alpha}})$. In order to have a sufficiently large search space we must therefore have

$$2\alpha \geq \frac{\sqrt{2}}{3\sqrt{\alpha}}, \quad \text{or} \quad \alpha \geq 18^{-1/3}.$$

For $\alpha = 18^{-1/3}$ the descent initiation stipulates that $\mu \geq \alpha^{-3/2}/6 = 1/\sqrt{2}$, and any $\mu \in [1/\sqrt{2}, \sqrt{n})$ suffices. We therefore have a total complexity of

$$L_Q(1/3, 2\alpha) = L_Q(1/3, (4/9)^{1/3}) \approx L_Q(1/3, 0.763).$$

On the other hand when we need to round a to the nearest integer, the degrees can become unbalanced so that the degree of one side is up to double the degree of the other. In this case a simple calculation shows that the optimal α is $16^{-1/3}$, giving a complexity of

$$L_Q(1/3, 2\alpha) = L_Q(1/3, (1/2)^{1/3}) \approx L_Q(1/3, 0.794).$$

Naturally, for a ratio of degrees in $(1/2, 2)$, we get c-values in between. This situation is redolent of Coppersmith's algorithm [5], in which precisely the same issue arises when forcing a real variable to take integer arguments only.

Note that this degree balancing technique also works for special-q of any degree, making the descent far more rapid than for Case 1.

Remark 1. Observe that the best-case complexity with $c = (4/9)^{1/3}$ is precisely the complexity of the oracle-assisted Static Diffie-Hellman Problem in finite fields of small characteristic [17, §3]. Our result may therefore seem unsurprising, since the complexity of computing the logarithms of the factor base elements is never more than the complexity of the descent, and is thus effectively free. However, this reasoning overlooks the fact that we are working with a medium-sized base field, as opposed to the traditional FFS setting with a very small base field. In contrast to the result in [17, §3], our complexities depend crucially on our degree two elimination method, in addition to the fast computation of degree one logarithms.

5   Application to the DLP in $\mathbb{F}_{2^{1971}}$ and $\mathbb{F}_{2^{3164}}$

In this section we provide details of our implementation for discrete logarithm computations in the finite fields with 2^{1971} (as announced in [9]) and 2^{3164} elements, respectively.

5.1   Discrete Logarithms in $\mathbb{F}_{2^{1971}}$

In order to represent the finite field with 2^{1971} elements we first defined $\mathbb{F}_q = \mathbb{F}_{2^{27}}$ by $\mathbb{F}_2[T]/(T^{27} + T^5 + T^2 + T + 1)$. Denoting by t a root of this irreducible in $\mathbb{F}_{2^{27}}$ we defined $\mathbb{F}_{2^{1971}} = \mathbb{F}_{q^{73}}$ by $\mathbb{F}_q[X]/(X^{73} + t)$. For x a root of $X^{73} + t$ in $\mathbb{F}_{q^{73}}$, we defined y by $y := x^8$, and we therefore also have $x = t/y^9$.

Since we use a Kummer extension, the elements of the factor base are related via the generator of the Galois group of $\mathbb{F}_{q^{73}}/\mathbb{F}_q$ [16,13], and one can therefore quotient out by the action of this automorphism to reduce the number of variables from 2^{27} to $\approx 2^{27}/73$. As stated in §3, we can take this idea even further. In fact, $x^{2^9} = c\,x$ for $c = t^7 \in \mathbb{F}_q$, so the map $\sigma : a \to a^{2^9}$ is an additional automorphism which preserves the set of degree one factor base elements. The map σ^3 equals the Frobenius $a \to a^q$ (of order 73) and hence σ generates a group G of order 219. Considering the orbits of G acting on the factor base elements, we find 612 864 orbits of full size 219, seven of size 73, and one of size 1, resulting in $N = 612\,872$ orbits, which gives the number of factor base variables.

Since the degrees of the polynomials relating x and y are nearly balanced, the complexity of our relation generation falls into Case 1 in §4.2, which matches Joux's optimal one-sided, or advanced pinpointing for Kummer extensions. However, for Kummer extensions for which the degrees are balanced — as opposed to being very skewed as in §3.4 where $2^k \gg d_1$ — the advanced pinpointing is faster in practice, and so we used it for relation generation. We computed approximately $10N$ relations in about 14 core-hours computation time. For simplicity, we keep only those relations with distinct factors; this ensures that each entry of the relation matrix is a power of two, and hence all element multiplications in the matrix-vector products consist of cyclic rotations modulo $2^{1971} - 1$.

After relation generation, we performed structured Gaussian elimination (SGE) (in a version based on [15]) to reduce the number of variables and thus to decrease the cost for the subsequent linear algebra step. During our experiments we made the observation that additional equations are indeed useful for reducing the number of variables. However, the benefit of SGE is unclear as the row weight is being increased. We therefore stopped the SGE at this point, which resulted in a $528\,812 \times 527\,766$ matrix of constant row weight 19. The running time here was about 10 minutes on a single core.

We obtained the following partial factorisation of $2^{1971} - 1$:

$7 \cdot 73^2 \cdot 439 \cdot 3943 \cdot 262657 \cdot 2298041 \cdot 10178663167 \cdot 27265714183 \cdot 9361973132609$

$\cdot 1406791071629857 \cdot 5271393791658529 \cdot 671165898617413417 \cdot 2762194134676763431$

$\cdot 4815314615204347717321 \cdot 4218592755298376314743137371 9$

$\cdot 22068362846714807160397927912339216441$

$\cdot 78133539370531820286911002468435975940517909 7 \cdot C_{338} \,,$

where C_{338} is a 338-digit composite. We took as our modulus for the linear algebra step the product of C_{338} and the six largest prime factors of the cofactor, which has 507 digits. We applied a parallel version of Lanczos' algorithm (see [18]) using OpenMP on an SGI Altix ICE 8200EX cluster using Intel (Westmere) Xeon E5650 hex-core processors and GNU Multi-Precision library [11], taking 2220 core-hours in total.

For the DLP we took as (a presumed) generator $g = x + 1 \in \mathbb{F}_{2^{1971}}^{\times}$ and the target element was set as usual to be

$$x_\pi = \sum_{i=0}^{72} \tau(\lfloor \pi \cdot q^{i+1} \rfloor \bmod q)\, x^i \,,$$

where τ takes the binary representation of an integer and maps to $\mathbb{F}_q$ via $2^i \mapsto t^i$. We first solved the target logarithm in the subgroups of order the first 11 terms in the factorisation using either linear search or Pollard's rho [22].

The descent proceeded by first finding an $i \in \mathbb{N}$ such that

$$x_\pi g^i = z_1/z_2 \,,$$

where both z_1 and z_2 were 7-smooth. We implemented the descent in such a way that at the early phase of the algorithm the expected subsequent costs are as small as possible. This means that we try to find factorisations which consist of as many small degree factors as possible. We used about 40 core-hours to find an exponent i with favourable factorisation patterns and found $i = 47\,147\,576$ to be a good choice. We then spent about 3 hours to perform the descent down to degree 3. As stated in §3 and §4, at each stage during the descent, we can eliminate a given special-$\mathfrak{q}$ on either the x-side or on the y-side, one of which may be much faster. Computing the elimination probabilities we found that eliminating on the y-side is always faster. Indeed, for degree 2 special-$\mathfrak{q}$ we *must* perform this on the y-side, as it is not possible to do so on the x-side, due to the factorisation patterns of (2).

At this point we were left with 103 special-$\mathfrak{q}$ of degree 3, as opposed to the ≈ 500 expected with a random 7-smooth split of $x_\pi g^i$. The expected cost of eliminating each of these is $2^{25.1}$ 2-smoothness tests. These special-q elements were resolved on the same SGI Altix ICE 8200EX cluster in about 850 core-hours, using Shoup's Number Theory Library [24], resulting in 1140 special-q elements of degree 2. Using the technique of §4.1, we reduced the cost of the elimination of each of these by a factor of $2^9 = 2^{3k}$, and all their logarithms were computed in 5 core-hours, completing the descent.

Thus the running time for solving an instance of the discrete logarithm problem completely in the finite field $\mathbb{F}_{2^{1971}}$ sums to $14 + 2220 + 898 = 3132$ core-hours in total. Finally, we found that $\log_g(x_\pi)$ equals

11992984215354106866091146371988855845186852755447163352368959007609021 9879 5745784008181148775933944656038305197825417423602365358899373622007711 17361 6782694231011634031353555222808041139032152735559059010822822482400219 28787 8207304028565280573096588688279004416835100344085961912427000601289864 33752 1100022143802898875460611252245879711978727508058465196231404376457393 62938 2354173616116810825627780459657892709561158924173579400674739684346062 99268 2942919573782264511826207837453495025029601399274531964897400652447954 89583 27920827882768332440907342446643941097670216203953951337767311548 3439 .

5.2   Discrete Logarithms in $\mathbb{F}_{2^{3164}}$

For this case we defined $\mathbb{F}_q = \mathbb{F}_{2^{28}} = \mathbb{F}_2[T]/(T^{28}+T+1)$. We denote by t a root of this irreducible in $\mathbb{F}_{2^{28}}$. Furthermore, let $\mathbb{F}_{q^{113}} = \mathbb{F}_q[X]/(X^{113}+t)$ and denote by x a root of $X^{113}+t$ in $\mathbb{F}_{2^{3164}}$. We defined y by $y = x^{16}$, and we therefore also have $x = t/y^7$.

As in the previous section we use the Kummer extension idea of [16,13] to reduce the size of the factor base. Again we can use a larger group than just the Galois group of $\mathbb{F}_{q^{113}}/\mathbb{F}_q$, since $x^{2^{14}} = cx$ for $c = t^9 + t^8 + t^5 + t^4 \in \mathbb{F}_q$ and thus the map $\sigma : a \to a^{2^{14}}$ is an additional factor base preserving automorphism. The map σ^2 equals the Frobenius $a \to a^q$ and hence σ generates a group G of order 226. Considering the orbits of G acting on the factor base elements, we find $N = 1\,187\,841$ orbits in total, which gives the number of factor base variables.

For relation generation, since $16 > 7$ the degrees are unbalanced and hence more favourable toward the use of our relation generation method as given in §3.2. It produces one relation in just under a second, so that more than N relations can be found in about 350 core-hours. However, thanks to our choice of g_2, Joux's pinpointing methods *also* benefit from the higher splitting probability as explained by Theorem 1, and so for this Kummer extension, it is still preferable to use Joux's advanced pinpointing method, which generates about $10N$ relations in approximately 2 hours on a single-core.

With the structured Gaussian elimination step in mind we computed approximately $10N$ relations and performed SGE on this matrix to reduce the number of variables, where we stopped again at the point when the row weight is being increased. The result was a $1\,066\,010 \times 1\,064\,991$ matrix of constant row weight 25, which constitutes a reduction of 10.3% in the number of variables.

The full factorisation of $2^{3164} - 1$ (obtained from the Cunningham tables [25]) is:

$3 \cdot 5 \cdot 29 \cdot 43 \cdot 113^2 \cdot 127 \cdot 227 \cdot 1583 \cdot 3391 \cdot 6329 \cdot 23279 \cdot 48817 \cdot 58309 \cdot 65993 \cdot 85429$

$\cdot 1868569 \cdot 2362153 \cdot 116163097 \cdot 636190001 \cdot 7920714887 \cdot 54112378027$

$\cdot 1066818132868207 \cdot 94395483835364237 \cdot 362648335437701461 \cdot 491003369344660409$

$\cdot 1507911621390132617836 9 \cdot 103845937170696551129458045825843 21$

$\cdot 16210807687504089730597044158159945072569569899134297641 53$

$\cdot 25492807273453795564805967522921896342698297652509936704025490424 22649$

$\cdot 4785290367491952770979444950472742768748481440405231269246278905154 317$

$\cdot 947326915707939568567591984149117797341195244156353967998649410983309655 6$
0269355785101434237

$\cdot 30893732435679706159469738259014519623666572271820219584074344744581789 67$
$789139446879970022670238264606111325817554799$

$\cdot 33248138195822034659908271092377125566098001373614163921550203376275101 35$
$8208879881599077605921097512410793579836318474132090869696712 1 \cdot P_{190}$,

where P_{190} is a 190-digit prime.

We then ran a parallel version of the Lanczos' algorithm on several nodes of the SGI Altix ICE 8200EX cluster, using MPI and OpenMP parallelisation techniques on 144 cores and again the GNU Multi-Precision library [11], taking 85488 core-hours in total. Note that since the nodes we used for the computation were not very "well-connected," the total running time would have been reduced to around 30000 core-hours if we had run our algorithm on 12 cores.

For the DLP we took as our (proven) generator $g = x + t + 1 \in \mathbb{F}_{2^{3164}}^\times$ and a target element set as usual to be $x_\pi = \sum_{i=0}^{113} \tau(\lfloor \pi \cdot q^{i+1} \rfloor \bmod q) \, x^i$.

As before the descent proceeded by first finding an $i \in \mathbb{N}$ such that $x_\pi \, g^i = z_1/z_2$, where both z_1 and z_2 were here 16-smooth. At each stage, we choose to sieve for the special-q on the y-side.

In this case we put even more effort in analysing and optimising the descent in the earlier stages so that the expected subsequent costs will be minimised. In fact we associated a cost k_d to each factor of degree d arising in the factorisation of the l.h.s. and r.h.s. polynomials, which we estimated by considering the distribution of factorisation pattern.

We used about 70 core-hours to find the 16-smooth initial fraction z_1/z_2, then spent 210 core-hours for the descent down to degree 4, and used 340 core-hours for processing the degree 4 polynomials. At this point we had 71 polynomials of degree 3, which needed an expected number of $2^{34.1}$ 2-smoothness tests to be resolved. These special-q elements have been processed by the same SGI Altix ICE 8200EX cluster in about 20972 core-hours, using Shoup's Number Theory Library [24], and resulted in 1239 special-q elements of degree 2. Finally, using the technique in §4.1, these elements were eliminated in about 10 core-hours, completing the descent.

The running time for solving an instance of the discrete logarithm problem completely in the finite field $\mathbb{F}_{2^{3164}}$ sums to $350 + 85488 + 20972 + 210 + 340 + 10 = 107092$ core-hours (as already indicated, this figure would be reduced to around 52000 core-hours if Lanczos' algorithm was run on 12 cores). Finally, we found that $\log_g(x_\pi)$ equals

24109586720847037799012020772616422090705143132887875333858087170248784565712688312063491036765323357553385717747797766545731784956477016880944817731731405243895025293868522646360493835468855617633181786341747893370309598402582718996263618673697554067799885512742832012390129483899153002417393400439161058228340028972042930361976940653379032557934518587736643501300307220916662531725410704479482997812210193428607010640365444303319677531146468063350633002030742348610674716684119982045443191768323538019822219249958042954261671123069707959607989886446311000373932915585804124069420045551161487903876549604900084297695444007900819088072394071341577241660482464194055035573980358979998525931969540314396297687768509998877208705617419130555318640416547078404337954037532005208916171502547565867282159415513550648407797656823989931563900000242491107399569193500692930336704230702995815576366664993721204536863038736714880164096355781178708892302786491643781333.

Observe that this computation also breaks the elliptic curve DLP for supersingular curves defined over $\mathbb{F}_{2^{791}}$, with embedding degree 4. However, since 791 is not prime, even before this break, such curves would not have been recommended, due to the potential applicability of Weil descent attacks [8].

6   Conclusion

We have presented and analysed new variants of the medium-sized base field FFS, for binary fields, which have complexities as low as $L_{q^n}(1/3, (4/9)^{1/3})$ for computing arbitrary logarithms. Furthermore, for fields possessing a subfield of an appropriate size, we have provided the first ever heuristic *polynomial time* algorithm for finding the discrete logarithms of degree one and two elements, which have both been verified experimentally. To illustrate the efficiency of the methods, we have successfully solved the DLP in the finite fields $\mathbb{F}_{2^{1971}}$ and $\mathbb{F}_{2^{3164}}$, setting a record for binary fields.

It would be interesting to know whether there are more general theorems on splitting behaviours for other polynomials arising during the descent, and also to what extent the known theorems apply to other characteristics.

Acknowledgements. The authors would like to extend their thanks to the Irish Centre for High-End Computing (ICHEC) — and Gilles Civario in particular — for their support throughout the course of our computations.

References

1. Adleman, L.M., Huang, M.-D.A.: Function field sieve method for discrete logarithms over finite fields. Inform. and Comput. 151(1-2), 5–16 (1999)
2. Bailey, D.V., Paar, C., Sarkozy, G., Hofri, M.: Computation in optimal extension fields. In: Conference on The Mathematics of Public Key Cryptography, The Fields Institute for Research in the Mathematical Sciences, pp. 12–17 (2000)
3. Bluher, A.W.: On $x^{q+1}+ax+b$. Finite Fields and Their Applications 10(3), 285–305 (2004)
4. Chung, F.R.K.: Diameters and eigenvalues. J. Amer. Math. Soc. 2(2), 187–196 (1989)
5. Coppersmith, D.: Fast evaluation of logarithms in fields of characteristic two. IEEE Trans. Inform. Theory 30(4), 587–593 (1984)
6. Faugère, J.-C., Otmani, A., Perret, L., Tillich, J.-P.: Algebraic cryptanalysis of mcEliece variants with compact keys. In: Gilbert, H. (ed.) EUROCRYPT 2010. LNCS, vol. 6110, pp. 279–298. Springer, Heidelberg (2010)
7. Faugère, J.-C., Perret, L., Petit, C., Renault, G.: Improving the complexity of index calculus algorithms in elliptic curves over binary fields. In: Pointcheval, D., Johansson, T. (eds.) EUROCRYPT 2012. LNCS, vol. 7237, pp. 27–44. Springer, Heidelberg (2012)
8. Gaudry, P., Hess, F., Smart, N.P.: Constructive and destructive facets of Weil descent on elliptic curves. J. Cryptology 15(1), 19–46 (2002)
9. Göloğlu, F., Granger, R., McGuire, G., Zumbrägel, J.: Discrete Logarithms in GF(2^{1971}). NMBRTHRY list (February 19, 2013)
10. Granger, R., Vercauteren, F.: On the discrete logarithm problem on algebraic tori. In: Shoup, V. (ed.) CRYPTO 2005. LNCS, vol. 3621, pp. 66–85. Springer, Heidelberg (2005)
11. Granlund, T.: The GMP development team: GNU MP: The GNU Multiple Precision Arithmetic Library, 5.0.5 edn (2012), http://gmplib.org/
12. Helleseth, T., Kholosha, A.: $x^{2^l+1} + x + a$ and related affine polynomials over GF(2^k). Cryptogr. Commun. 2(1), 85–109 (2010)
13. Joux, A.: Faster index calculus for the medium prime case application to 1175-bit and 1425-bit finite fields. In: Johansson, T., Nguyen, P.Q. (eds.) EUROCRYPT 2013. LNCS, vol. 7881, pp. 177–193. Springer, Heidelberg (2013)
14. Joux, A., Lercier, R.: The function field sieve is quite special. In: Fieker, C., Kohel, D.R. (eds.) ANTS 2002. LNCS, vol. 2369, pp. 431–445. Springer, Heidelberg (2002)
15. Joux, A., Lercier, R.: Improvements to the general number field sieve for discrete logarithms in prime fields: a comparison with the gaussian integer method. Math. Comput. 72(242), 953–967 (2003)
16. Joux, A., Lercier, R.: The function field sieve in the medium prime case. In: Vaudenay, S. (ed.) EUROCRYPT 2006. LNCS, vol. 4004, pp. 254–270. Springer, Heidelberg (2006)
17. Joux, A., Lercier, R., Naccache, D., Thomé, E.: Oracle-assisted static diffie-hellman is easier than discrete logarithms. In: Parker, M.G. (ed.) Cryptography and Coding 2009. LNCS, vol. 5921, pp. 351–367. Springer, Heidelberg (2009)
18. LaMacchia, B.A., Odlyzko, A.M.: Solving large sparse linear systems over finite fields. In: Menezes, A., Vanstone, S.A. (eds.) CRYPTO 1990. LNCS, vol. 537, pp. 109–133. Springer, Heidelberg (1991)
19. Lenstra, A.K., Lenstra Jr., H.W.: The development of the number field sieve. Lecture Notes in Mathematics, vol. 1554. Springer, Heidelberg (1993)

128   F. Göloğlu et al.

20. Lenstra Jr., H.W.: Finding isomorphisms between finite fields. Math. Comp. 56(193), 329–347 (1991)
21. Misoczki, R., Barreto, P.S.L.M.: Compact McEliece keys from Goppa codes. In: Jacobson Jr., M.J., Rijmen, V., Safavi-Naini, R. (eds.) SAC 2009. LNCS, vol. 5867, pp. 376–392. Springer, Heidelberg (2009)
22. Pollard, J.M.: Monte carlo methods for index computation (mod p). Math. Comp. 32(143), 918–924 (1978)
23. Rubin, K., Silverberg, A.: Torus-based cryptography. In: Boneh, D. (ed.) CRYPTO 2003. LNCS, vol. 2729, pp. 349–365. Springer, Heidelberg (2003)
24. Shoup, V.: NTL: A library for doing number theory, 5.5.2 edn. (2009), http://www.shoup.net/ntl/
25. Wagstaff, S., et al.: The Cunningham Project, http://homes.cerias.purdue.edu/~ssw/cun/index.html
26. Wan, D.: Generators and irreducible polynomials over finite fields. Math. Comp. 66(219), 1195–1212 (1997)

An Algebraic Framework
for Diffie-Hellman Assumptions

Alex Escala[1], Gottfried Herold[2], Eike Kiltz[2,*],
Carla Ràfols[2], and Jorge Villar[3,**]

[1] Universitat Autònoma de Barcelona, Spain
[2] Horst-Görtz Institute for IT Security and Faculty of Mathematics,
Ruhr-Universität Bochum, Germany
[3] Universitat Politècnica de Catalunya, Spain

Abstract. We put forward a new algebraic framework to generalize and
analyze Diffie-Hellman like Decisional Assumptions which allows us to
argue about security and applications by considering only algebraic properties. Our $\mathcal{D}_{\ell,k}$-MDDH assumption states that it is hard to decide
whether a vector in $\mathbb{G}^\ell$ is linearly dependent of the columns of some
matrix in $\mathbb{G}^{\ell \times k}$ sampled according to distribution $\mathcal{D}_{\ell,k}$. It covers known
assumptions such as DDH, 2-Lin (linear assumption), and k-Lin (the k-
linear assumption). Using our algebraic viewpoint, we can relate the
generic hardness of our assumptions in m-linear groups to the irreducibil-
ity of certain polynomials which describe the output of $\mathcal{D}_{\ell,k}$. We use the
hardness results to find new distributions for which the $\mathcal{D}_{\ell,k}$-MDDH-
Assumption holds generically in m-linear groups. In particular, our new
assumptions 2-SCasc and 2-ILin are generically hard in bilinear groups
and, compared to 2-Lin, have shorter description size, which is a relevant parameter for efficiency in many applications. These results support using our new assumptions as natural replacements for the 2-Lin
Assumption which was already used in a large number of applications.

To illustrate the conceptual advantages of our algebraic framework,
we construct several fundamental primitives based on any MDDH-
Assumption. In particular, we can give many instantiations of a primitive
in a compact way, including public-key encryption, hash-proof systems,
pseudo-random functions, and Groth-Sahai NIZK and NIWI proofs. As
an independent contribution we give more efficient NIZK and NIWI
proofs for membership in a subgroup of $\mathbb{G}^\ell$, for validity of ciphertexts
and for equality of plaintexts. The results imply very significant efficiency
improvements for a large number of schemes, most notably Naor-Yung
type of constructions.

Keywords: Diffie-Hellman Assumption, Groth-Sahai proofs, hash proof
systems, public-key encryption.

* Funded by a Sofja Kovalevskaja Award of the Alexander von Humboldt Foundation
and the German Federal Ministry for Education and Research.
** Partially supported by the Spanish Government through projects MTM2009-07694
and Consolider Ingenio 2010 CDS2007-00004 ARES.

R. Canetti and J.A. Garay (Eds.): CRYPTO 2013, Part II, LNCS 8043, pp. 129–147, 2013.

1   Introduction

Arguably, one of the most important cryptographic hardness assumptions is the Decisional Diffie-Hellman (DDH) Assumption. For a fixed additive group $\mathbb{G}$ of prime order q and a generator $\mathcal{P}$ of $\mathbb{G}$, we denote by $[a] := a\mathcal{P} \in \mathbb{G}$ the *implicit representation* of an element $a \in \mathbb{Z}_q$. The DDH Assumption states that $([a], [r], [ar]) \approx_c ([a], [r], [z]) \in \mathbb{G}^3$, where a, r, z are uniform elements in $\mathbb{Z}_q$ and $\approx_c$ denotes computationally indistinguishability of the two distributions. It has been used in numerous important applications such as secure encryption [8], key-exchange [16], hash-proof systems [9], pseudo-random functions [26], and many more.

BILINEAR GROUPS AND THE LINEAR ASSUMPTION. Bilinear groups (i.e., groups $\mathbb{G}, \mathbb{G}_T$ of prime order q equipped with a bilinear map $e : \mathbb{G} \times \mathbb{G} \to \mathbb{G}_T$) [20,3] revolutionized cryptography in recent years and and are the basis for a large number of cryptographic protocols. However, relative to a (symmetric) bilinear map, the DDH Assumption is no longer true in the group $\mathbb{G}$. (This is since $e([a], [r]) = e([1], [ar])$ and hence $[ar]$ is not longer pseudorandom given $[a]$ and $[r]$.) The need for an "alternative" decisional assumption in $\mathbb{G}$ was quickly addressed with the Linear Assumption (2-Lin) introduced by Boneh, Boyen, and Shacham [2]. It states that $([a_1], [a_2], [a_1 r_1], [a_2 r_2], [r_1 + r_2]) \approx_c ([a_1], [a_2], [a_1 r_1], [a_2 r_2], [z]) \in \mathbb{G}^5$, where $a_1, a_2, r_1, r_2, z \leftarrow \mathbb{Z}_q$. 2-Lin holds in generic bilinear groups [2] and it has virtually become the standard decisional assumption in the group $\mathbb{G}$ in the bilinear setting. It has found applications to encryption [23], signatures [2], zero-knowledge proofs [17], pseudorandom functions [4] and many more. More recently, the 2-Lin Assumption was generalized to the $(k\text{-Lin})_{k \in \mathbb{N}}$ Assumption family [19,29] (1-Lin = DDH), a family of increasingly (strictly) weaker Assumptions which are generically hard in k-linear maps.

SUBGROUP MEMBERSHIP PROBLEMS. Since the work of Cramer and Shoup [9] it has been recognized that it is useful to view the DDH Assumption as a hard subgroup membership problem in $\mathbb{G}^2$. In this formulation, the DDH Assumption states that it is hard to decide whether a given element $([r], [t]) \in \mathbb{G}^2$ is contained in the subgroup generated by $([1], [a])$. Similarly, in this language the 2-Lin Assumption says that it is hard to decide whether a given vector $([r], [s], [t]) \in \mathbb{G}^3$ is in the subgroup generated by the vectors $([a_1], [0], [1]), ([0], [a_2], [1])$. The same holds for the $(k\text{-Lin})_{k \in \mathbb{N}}$ Assumption family: for each k, the k-Lin assumption can be naturally written as a hard subgroup membership problem in $\mathbb{G}^{k+1}$. This alternative formulation has conceptual advantages for some applications, for instance, it allowed to provide more instantiations of the original DDH-based scheme of Cramer and Shoup and it is also the most natural point of view for translating schemes originally constructed in composite order groups into prime order groups [14].

LINEAR ALGEBRA IN BILINEAR GROUPS. In its formulation as subgroup decision membership problem, the k-Lin assumption can be seen as the problem of deciding linear dependence "in the exponent." Recently, a number of works have illustrated the usefulness of a more algebraic point of view on decisional assumptions in bilinear groups, like the Dual Pairing Vector Spaces of Okamoto and Takashima [28] or the Subspace Assumption of Lewko [24]. Although these new decisional assumptions reduce to the 2-Lin Assumption, their flexibility and their algebraic description have proven to be crucial in many works to obtain complex primitives in strong security models previously unrealized in the literature, like Attribute-Based Encryption, Unbounded Inner Product Encryption and many more.

THIS WORK. Motivated by the success of this algebraic viewpoint of decisional assumptions, in this paper we explore new insights resulting from interpreting the k-Lin decisional assumption as a special case of what we call a Matrix Diffie-Hellman Assumption. The general problem states that it is hard to distinguish whether a given vector in $\mathbb{G}^\ell$ is contained in the space spanned by the columns of a certain matrix $[\mathbf{A}] \in \mathbb{G}^{\ell \times k}$, where $\mathbf{A}$ is sampled according to some distribution $\mathcal{D}_{\ell,k}$. We remark that even though all our results are stated in symmetric bilinear groups, they can be naturally extended to the asymmetric setting.

1.1 The Matrix Diffie-Hellman Assumption

A NEW FRAMEWORK FOR DDH-LIKE ASSUMPTIONS. For integers $\ell > k$ let $\mathcal{D}_{\ell,k}$ be an (efficiently samplable) distribution over $\mathbb{Z}_q^{\ell \times k}$. We define the $\mathcal{D}_{\ell,k}$-Matrix DH ($\mathcal{D}_{\ell,k}$-MDDH) Assumption as the following subgroup decision assumption:

$$\mathcal{D}_{\ell,k}\text{-MDDH}: \quad [\mathbf{A}||\mathbf{A}\mathbf{r}] \approx_c [\mathbf{A}||\mathbf{u}] \in \mathbb{G}^{\ell \times (k+1)}, \qquad (1)$$

where $\mathbf{A} \in \mathbb{Z}_q^{\ell \times k}$ is chosen from distribution $\mathcal{D}_{\ell,k}$, $\mathbf{r} \leftarrow \mathbb{Z}_q^k$, and $\mathbf{u} \leftarrow \mathbb{G}^\ell$. The $(k\text{-Lin})_{k \in \mathbb{N}}$ family corresponds to this problem when $\ell = k + 1$, and $\mathcal{D}_{\ell,k}$ is the specific distribution $\mathcal{L}_k$ (formally defined in Example 2).

GENERIC HARDNESS. Due to its linearity properties, the $\mathcal{D}_{\ell,k}$-MDDH Assumption does not hold in $k + 1$-linear groups. In Section 3.2 we give two different theorems which state sufficient conditions for the $\mathcal{D}_{\ell,k}$-MDDH Assumption to hold generically in m-linear groups. Theorem 1 is very similar to the Uber-Assumption [1,6] that characterizes hardness in bilinear groups (i.e., $m = 2$) in terms of linear independence of polynomials in the inputs. We generalize this to arbitrary m using a more algebraic language. This algebraic formulation has the advantage that one can use additional tools (e.g. Gröbner bases or resultants) to show that a distribution $\mathcal{D}_{\ell,k}$ meets the conditions of Theorem 1, which is specially important for large m. It also allows to prove a completely new result, namely Theorem 2, which states that a matrix assumption with $\ell = k + 1$ is generically hard if a certain determinant polynomial is irreducible.

NEW ASSUMPTIONS FOR BILINEAR GROUPS. We propose other families of generically hard decisional assumptions that did not previously appear in the literature, e.g., those associated to $\mathcal{C}_k, \mathcal{SC}_k, \mathcal{IL}_k$ defined below. For the most important parameters $k = 2$ and $\ell = k + 1 = 3$, we consider the following examples of distributions:

$$\mathcal{C}_2 : \mathbf{A} = \begin{pmatrix} a_1 & 0 \\ 1 & a_2 \\ 0 & 1 \end{pmatrix} \quad \mathcal{SC}_2 : \mathbf{A} = \begin{pmatrix} a & 0 \\ 1 & a \\ 0 & 1 \end{pmatrix} \quad \mathcal{L}_2 : \mathbf{A} = \begin{pmatrix} a_1 & 0 \\ 0 & a_2 \\ 1 & 1 \end{pmatrix} \quad \mathcal{IL}_2 : \mathbf{A} = \begin{pmatrix} a & 0 \\ 0 & a+1 \\ 1 & 1 \end{pmatrix},$$

for uniform $a, a_1, a_2 \in \mathbb{Z}_q$ as well as $\mathcal{U}_{3,2}$, the uniform distribution in $\mathbb{Z}_q^{3 \times 2}$ (already considered in several previous works like [15]). All assumptions are hard in generic bilinear groups. It is easy to verify that $\mathcal{L}_2$-MDDH = 2-Lin. We define 2-Casc := $\mathcal{C}_2$-MDDH (Cascade Assumption), 2-SCasc := $\mathcal{SC}_2$-MDDH (Symmetric Cascade Assumption), and 2-ILin := $\mathcal{IL}_2$-MDDH (Incremental Linear Assumption). In the full version [12], we show that 2-SCasc $\Rightarrow$ 2-Casc, 2-ILin $\Rightarrow$ 2-Lin and that $\mathcal{U}_{3,2}$-MDDH is the weakest of these assumptions (which extends the results of [15,30,14] for 2-Lin), while 2-SCasc and 2-Casc seem incomparable to 2-Lin.

EFFICIENCY IMPROVEMENTS. As a measure of efficiency, we define the *representation size* $\mathrm{RE}_{\mathbb{G}}(\mathcal{D}_{\ell,k})$ of an $\mathcal{D}_{\ell,k}$-MDDH assumption as the minimal number of group elements needed to represent $[\mathbf{A}]$ for any $\mathbf{A} \leftarrow \mathcal{D}_{\ell,k}$. This parameter is important since it affects the performance (typically the size of public/secret parameters) of schemes based on a Matrix Diffie-Hellman Assumption. 2-Lin and 2-Casc have representation size 2 (elements $([a_1], [a_2])$), while 2-ILin and 2-SCasc only 1 (element $[a]$). Hence our new assumptions directly translate into shorter parameters for a large number of applications (see the Applications in Section 4). Further, our result points out a tradeoff between efficiency and hardness which questions the role of 2-Lin as the "standard decisional assumption" over a bilinear group $\mathbb{G}$.

NEW FAMILIES OF WEAKER ASSUMPTIONS. By defining appropriate distributions $\mathcal{C}_k, \mathcal{SC}_k, \mathcal{IL}_k$ over $\mathbb{Z}_q^{(k+1) \times k}$, one can generalize all three new assumptions naturally to $(k\text{-Casc})_{k \in \mathbb{N}}$, $(k\text{-SCasc})_{k \in \mathbb{N}}$, and $(k\text{-ILin})_{k \in \mathbb{N}}$ with representation size k, 1, and 1, respectively. Using our results on generic hardness, it is easy to verify that all three assumptions are generically hard in k-linear groups. Since they are false in $k + 1$-linear groups this gives us three new families of increasingly strictly weaker assumptions. In particular, the $(k\text{-SCasc})$ and $(k\text{-ILin})$ assumption families are of great interest due to their compact representation size of only 1 element.

RELATIONS TO OTHER STANDARD ASSUMPTIONS. Surprisingly, the new assumption families can also be related to standard assumptions. The k-Casc Assumption is implied by the $(k + 1)$-Party Diffie-Hellman Assumption $((k + 1)\text{-PDDH})$ [5] which states that $([a_1], \dots, [a_{k+1}], [a_1 \cdot \dots \cdot a_{k+1}]) \approx_c ([a_1], \dots, [a_{k+1}], [z]) \in \mathbb{G}^{k+2}$. Similarly, k-SCasc is implied by the $k+1$-Exponent Diffie-Hellman Assumption $((k+1)\text{-EDDH})$ [22] which states that $([a], [a^{k+1}]) \approx_c ([a], [z]) \in \mathbb{G}^2$.

1.2   Basic Applications

We believe that all schemes based on 2-Lin can be shown to work for any Matrix Assumption. Consequently, a large class of known schemes can be instantiated more efficiently with the new more compact decisional assumptions, while offering the same generic security guarantees. To support this belief, in Section 4 we show how to construct some fundamental primitives based on any Matrix Assumption. All constructions are purely algebraic and therefore very easy to understand and prove.

- **Public-key Encryption.** We build a key-encapsulation mechanism with security against passive adversaries from any $\mathcal{D}_{\ell,k}$-MDDH Assumption. The public-key is $[\mathbf{A}]$, the ciphertext consists of the first k elements of $[z] = [\mathbf{A}r]$, the symmetric key of the last $\ell - k$ elements of $[z]$. Passive security immediately follows from $\mathcal{D}_{\ell,k}$-MDDH.
- **Hash Proof Systems.** We build a smooth projective hash proof system (HPS) from any $\mathcal{D}_{\ell,k}$-MDDH Assumption. It is well-known that HPS imply chosen-ciphertext secure encryption [9], password-authenticated key-exchange, zero-knowledge proofs, and many other things.
- **Pseudo-Random Functions.** Generalizing the Naor-Reingold PRF [26,4], we build a pseudo-random function PRF from any $\mathcal{D}_{\ell,k}$-MDDH Assumption. The secret-key consists of *transformation matrices* $\mathbf{T}_1, \ldots, \mathbf{T}_n$ (derived from independent instances $\mathbf{A}_{i,j} \leftarrow \mathcal{D}_{\ell,k}$) plus a vector h of group elements. For $x \in \{0,1\}^n$ we define $\mathsf{PRF}_K(x) = \left[\prod_{i:x_i=1} \mathbf{T}_i \cdot h\right]$. Using the random self-reducibility of the $\mathcal{D}_{\ell,k}$-MDDH Assumption, we give a tight security proof.
- **Groth-Sahai Non-Interactive Zero-Knowledge Proofs.** We show how to instantiatiate the Groth-Sahai proof system [17] based on any $\mathcal{D}_{\ell,k}$-MDDH Assumption. While the size of the proofs depends only on ℓ and k, the CRS and verification depends on the representation size of the Matrix Assumptions. Therefore our new instantiations offer improved efficiency over the 2-Lin-based construction from [17]. This application in particular highlights the usefulness of the Matrix Assumption to describe in a compact way many instantiations of a scheme: instead of having to specify the constructions for the DDH and the 2-Lin assumptions separately [17], we can recover them as a special case of a general construction.

MORE EFFICIENT PROOFS FOR CRS DEPENDENT LANGUAGES. In Section 5 we provide more efficient NIZK and NIWI proofs for concrete natural languages which are dependent on the common reference string. More specifically, the common reference string of the $\mathcal{D}_{\ell,k}$-MDDH instantiation of Groth-Sahai proofs of Section 4.4 includes as part of the commitment keys the matrix $[\mathbf{A}]$, where $\mathbf{A} \in \mathbb{Z}_q^{\ell \times k} \leftarrow \mathcal{D}_{\ell,k}$. We give more efficient proofs for several languages related to $\mathbf{A}$. Although at first glance the languages considered may seem quite restricted, they naturally appear in many applications, where typically $\mathbf{A}$ is the public key of some encryption scheme and one wants to prove statements about ciphertexts. More specifically, we obtain improvements for several kinds of statements, namely:

- **Subgroup membership proofs.** We give more efficient proofs in the language $\mathcal{L}_{\mathbf{A},\mathbb{G},\mathcal{P}} := \{[\mathbf{A}r], r \in \mathbb{Z}_q^k\} \subset \mathbb{G}^\ell$. To quantify some concrete improvement, in the 2-Lin case, our proofs of membership are half of the size of a standard Groth-Sahai proof and they require only 6 groups elements. We stress that this improvement is obtained without introducing any new computational assumption. To see which kind of statements can be proved using our result, note that a ciphertext is a rerandomization of another one only if their difference is in $\mathcal{L}_{\mathbf{A},\mathbb{G},\mathcal{P}}$. The same holds for proving that two commitments with the same key hide the same value or for showing in a publicly verifiable manner that the ciphertext of our encryption scheme opens to some known message $[m]$. This improvement has a significant impact on recent results, like [25,13], and we think many more examples can be found.

- **Ciphertext validity.** The result is extended to prove membership in the language $\mathcal{L}_{\mathbf{A},z,\mathbb{G},\mathcal{P}} = \{[c] : c = \mathbf{A}r + mz\} \subset \mathbb{G}^{\ell}$, where $z \in \mathbb{Z}_q^{\ell}$ is some public vector such that $z \notin \text{Im}(\mathbf{A})$, and the witness of the statement is $(r, [m]) \in \mathbb{Z}_q^k \times \mathbb{G}$. The natural application of this result is to prove that a ciphertext is well-formed and the prover knows the message $[m]$, like for instance in [11].

- **Plaintext equality.** We consider Groth-Sahai proofs in a setting in which the variables of the proofs are committed with different commitment keys, defined by two matrices $\mathbf{A} \leftarrow \mathcal{D}_{\ell_1,k_1}, \mathbf{B} \leftarrow \mathcal{D}'_{\ell_2,k_2}$. We give more efficient proofs of membership in the language $\mathcal{L}_{\mathbf{A},\mathbf{B},\mathbb{G},\mathcal{P}} := \{([c_A], [c_B]) : [c_A] = [\mathbf{A}r + (0, \ldots, 0, m)^T], [c_B] = [\mathbf{B}s + (0, \ldots, 0, m)^T], r \in \mathbb{Z}_q^{k_1}, s \in \mathbb{Z}_q^{k_2}\} \subset \mathbb{G}^{\ell_1} \times \mathbb{G}^{\ell_2}$. To quantify our concrete improvements, the size of the proof is reduced by 4 group elements with respect to [21]. As in the previous case, this language appears most naturally when one wants to prove equality of two committed values or plaintexts encrypted under different keys, e.g., when using Naor-Yung techniques to obtain chosen-ciphertext security [27]. Concretely, our results apply also to the encryption schemes in [18,7,10].

2   Notation

For $n \in \mathbb{N}$, we write 1^n for the string of n ones. Moreover, $|x|$ denotes the length of a bitstring x, while $|S|$ denotes the size of a set S. Further, $s \leftarrow S$ denotes the process of sampling an element s from S uniformly at random. For an algorithm A, we write $z \leftarrow \mathsf{A}(x, y, \ldots)$ to indicate that A is a (probabilistic) algorithm that outputs z on input $(x, y, \ldots)$. If $\mathbf{A}$ is a matrix we denote by a_{ij} the entries and a_i the column vectors.

Let Gen be a probabilistic polynomial time (ppt) algorithm that on input 1^{λ} returns a description $\mathcal{G} = (\mathbb{G}, q, \mathcal{P})$ of a cyclic group $\mathbb{G}$ of order q for a λ-bit prime q and a generator $\mathcal{P}$ of $\mathbb{G}$. More generally, for any fixed $k \geq 1$, let MGen_k be a ppt algorithm that on input 1^{λ} returns a description $\mathcal{MG}_k = (\mathbb{G}, \mathbb{G}_{T_k}, q, e_k, \mathcal{P})$, where $\mathbb{G}$ and $\mathbb{G}_{T_k}$ are cyclic additive groups of prime-order q, $\mathcal{P}$ a generator of $\mathbb{G}$, and $e_k : \mathbb{G}^k \to \mathbb{G}_{T_k}$ is a (non-degenerated, efficiently computable) k-linear map. For $k = 2$ we define $\mathsf{PGen} := \mathsf{MGen}_2$ to be a generator of a bilinear group $\mathcal{PG} = (\mathbb{G}, \mathbb{G}_T, q, e, \mathcal{P})$.

For an element $a \in \mathbb{Z}_q$ we define $[a] = a\mathcal{P}$ as the implicit representation of a in $\mathbb{G}$. Similarly, $[a]_{T_k} = a\mathcal{P}_{T_k}$ is its implicit representation in $\mathbb{G}_{T_k}$, where

$\mathcal{P}_{T_k} = e_k(\mathcal{P}, \ldots, \mathcal{P}) \in \mathbb{G}_{T_k}$. More generally, for a matrix $\mathbf{A} = (a_{ij}) \in \mathbb{Z}_q^{n \times m}$ we define $[\mathbf{A}]$ and $[\mathbf{A}]_{T_k}$ as the implicit representations of $\mathbf{A}$ computed elementwise.

When talking about elements in $\mathbb{G}$ and $\mathbb{G}_{T_k}$ we will always use this implicit notation, i.e., we let $[a] \in \mathbb{G}$ be an element in $\mathbb{G}$ or $[b]_{T_k}$ be an element in $\mathbb{G}_{T_k}$. Note that from $[a] \in \mathbb{G}$ it is generally hard to compute the value a (discrete logarithm problem in $\mathbb{G}$). Further, from $[b]_{T_k} \in \mathbb{G}_{T_k}$ it is hard to compute the value $b \in \mathbb{Z}_q$ (discrete logarithm problem in $\mathbb{G}_{T_k}$) or the value $[b] \in \mathbb{G}$ (pairing inversion problem). Obviously, given $[a] \in \mathbb{G}$, $[b]_{T_k} \in \mathbb{G}_{T_k}$, and a scalar $x \in \mathbb{Z}_q$, one can efficiently compute $[ax] \in \mathbb{G}$ and $[bx]_{T_k} \in \mathbb{G}_{T_k}$.

Also, all functions and operations acting on $\mathbb{G}$ and $\mathbb{G}_{T_k}$ will be defined implicitly. For example, when evaluating a bilinear pairing $e : \mathbb{G} \times \mathbb{G} \to \mathbb{G}_T$ in $[a], [b] \in \mathbb{G}$ we will use again our implicit representation and write $[z]_T := e([a], [b])$. Note that $e([a], [b]) = [ab]_T$, for all $a, b \in \mathbb{Z}_q$.

3   Matrix DH Assumptions

3.1   Definition and Basic Properties

Definition 1. *Let $\ell, k \in \mathbb{N}$ with $\ell > k$. We call $\mathcal{D}_{\ell,k}$ a matrix distribution if it outputs (in poly time, with overwhelming probability) matrices in $\mathbb{Z}_q^{\ell \times k}$ of full rank k. We define $\mathcal{D}_k := \mathcal{D}_{k+1,k}$.*

For simplicity we will also assume that, wlog, the first k rows of $\mathbf{A} \leftarrow \mathcal{D}_{\ell,k}$ form an invertible matrix.

We define the $\mathcal{D}_{\ell,k}$-matrix problem as to distinguish the two distributions $([\mathbf{A}], [\mathbf{A}\boldsymbol{w}])$ and $([\mathbf{A}], [\boldsymbol{u}])$, where $\mathbf{A} \leftarrow \mathcal{D}_{\ell,k}$, $\boldsymbol{w} \leftarrow \mathbb{Z}_q^k$, and $\boldsymbol{u} \leftarrow \mathbb{Z}_q^\ell$.

Definition 2 ($\mathcal{D}_{\ell,k}$-Matrix Diffie-Hellman Assumption $\mathcal{D}_{\ell,k}$-MDDH). *Let $\mathcal{D}_{\ell,k}$ be a matrix distribution. We say that the $\mathcal{D}_{\ell,k}$-Matrix Diffie-Hellman ($\mathcal{D}_{\ell,k}$-MDDH) Assumption holds relative to Gen if for all ppt adversaries D,*

$$\mathbf{Adv}_{\mathcal{D}_{\ell,k}, \mathsf{Gen}}(\mathsf{D}) = \Pr[\mathsf{D}(\mathcal{G}, [\mathbf{A}], [\mathbf{A}\boldsymbol{w}]) = 1] - \Pr[\mathsf{D}(\mathcal{G}, [\mathbf{A}], [\boldsymbol{u}]) = 1] = negl(\lambda),$$

where the probability is taken over $\mathcal{G} = (\mathbb{G}, q, \mathcal{P}) \leftarrow \mathsf{Gen}(1^\lambda)$, $\mathbf{A} \leftarrow \mathcal{D}_{\ell,k}, \boldsymbol{w} \leftarrow \mathbb{Z}_q^k, \boldsymbol{u} \leftarrow \mathbb{Z}_q^\ell$ and the coin tosses of adversary D.

Definition 3. *Let $\mathcal{D}_{\ell,k}$ be a matrix distribution. Let $\mathbf{A}_0$ be the first k rows of $\mathbf{A}$ and $\mathbf{A}_1$ be the last $\ell - k$ rows of $\mathbf{A}$. The matrix $\mathbf{T} \in \mathbb{Z}_q^{(\ell-k) \times k}$ defined as $\mathbf{T} = \mathbf{A}_1 \mathbf{A}_0^{-1}$ is called the transformation matrix of $\mathbf{A}$.*

We note that using the transformation matrix, one can alternatively define the advantage from Definition 2 as

$$\mathbf{Adv}_{\mathcal{D}_{\ell,k}, \mathsf{Gen}}(\mathsf{D}) = \Pr[\mathsf{D}(\mathcal{G}, \begin{bmatrix} \mathbf{A}_0 \\ \mathbf{T}\mathbf{A}_0 \end{bmatrix}, \begin{bmatrix} \boldsymbol{h} \\ \mathbf{T}\boldsymbol{h} \end{bmatrix}) = 1] - \Pr[\mathsf{D}(\mathcal{G}, \begin{bmatrix} \mathbf{A}_0 \\ \mathbf{T}\mathbf{A}_0 \end{bmatrix}, [\boldsymbol{u}]) = 1],$$

where the probability is taken over $\mathcal{G} = (\mathbb{G}, q, \mathcal{P}) \leftarrow \mathsf{Gen}(1^\lambda)$, $\mathbf{A} \leftarrow \mathcal{D}_{\ell,k}, \boldsymbol{h} \leftarrow \mathbb{Z}_q^k, \boldsymbol{u} \leftarrow \mathbb{Z}_q^{\ell-k}$ and the coin tosses of adversary D.

We can generalize Definition 2 to the m-fold $\mathcal{D}_{\ell,k}$-MDDH Assumption as follows. Given $\mathbf{W} \leftarrow \mathbb{Z}_q^{k \times m}$ for some $m \geq 1$, we consider the problem of distinguishing the distributions $([\mathbf{A}], [\mathbf{AW}])$ and $([\mathbf{A}], [\mathbf{U}])$ where $\mathbf{U} \leftarrow \mathbb{Z}_q^{\ell \times m}$ is equivalent to m independent instances of the problem (with the same $\mathbf{A}$ but different $\boldsymbol{w}_i$). This can be proved through a hybrid argument with a loss of m in the reduction, or, with a tight reduction (independent of m) via random self-reducibility.

Lemma 1 (Random self reducibility). *For any matrix distribution* $\mathcal{D}_{\ell,k}$, *$\mathcal{D}_{\ell,k}$-MDDH is random self-reducible. Concretely, for any m,*

$$\mathbf{Adv}^m_{\mathcal{D}_{\ell,k},\mathsf{Gen}}(\mathsf{D}') \leq \begin{cases} m \cdot \mathbf{Adv}_{\mathcal{D}_{\ell,k},\mathsf{Gen}}(\mathsf{D}) & 1 \leq m \leq \ell - k \\ (\ell - k) \cdot \mathbf{Adv}_{\mathcal{D}_{\ell,k},\mathsf{Gen}}(\mathsf{D}) + \dfrac{1}{q-1} & m > \ell - k \end{cases},$$

where

$$\mathbf{Adv}^m_{\mathcal{D}_{\ell,k},\mathsf{Gen}}(\mathsf{D}') = \Pr[\mathsf{D}'(\mathcal{G}, [\mathbf{A}], [\mathbf{AW}]) = 1] - \Pr[\mathsf{D}'(\mathcal{G}, [\mathbf{A}], [\mathbf{U}]) = 1],$$

and the probability is taken over $\mathcal{G} = (\mathbb{G}, q, \mathcal{P}) \leftarrow \mathsf{Gen}(1^\lambda)$, $\mathbf{A} \leftarrow \mathcal{D}_{\ell,k}, \mathbf{W} \leftarrow \mathbb{Z}_q^{k \times m}, \mathbf{U} \leftarrow \mathbb{Z}_q^{\ell \times m}$ and the coin tosses of adversary D'.

The proof is given in the full version [12].

We remark that, given $[\mathbf{A}], [\boldsymbol{z}]$ the above lemma can only be used to re-randomize the value $[\boldsymbol{z}]$. In order to re-randomize the matrix $[\mathbf{A}]$ we need that one can sample matrices $\mathbf{L}$ and $\mathbf{R}$ such that $\mathbf{A}' = \mathbf{LAR}$ looks like an independent instance $\mathbf{A}' \leftarrow \mathcal{D}_{\ell,k}$. In all of our example distributions we are able to do this.

Due to its linearity properties, the $\mathcal{D}_{\ell,k}$-MDDH assumption does not hold in $(k+1)$-linear groups.

Lemma 2. *Let $\mathcal{D}_{\ell,k}$ be any matrix distribution. Then the $\mathcal{D}_{\ell,k}$-Matrix Diffie-Hellman Assumption is false in $(k+1)$-linear groups.*

This is proven in the full version [12] by computing determinants in the target group.

3.2   Generic Hardness of Matrix DH

Let $\mathcal{D}_{\ell,k}$ be a matrix distribution as in Definition 1, which outputs matrices $\mathbf{A} \in \mathbb{Z}_q^{\ell \times k}$. We call $\mathcal{D}_{\ell,k}$ *polynomial-induced* if the distribution is defined by picking $\boldsymbol{t} \in \mathbb{Z}_q^d$ uniformly at random and setting $a_{i,j} := \mathfrak{p}_{i,j}(\boldsymbol{t})$ for some polynomials $\mathfrak{p}_{i,j} \in \mathbb{Z}_q[\boldsymbol{T}]$ whose degree does not depend on λ. E.g. for 2-Lin from Section 1.1, we have $a_{1,1} = t_1, a_{2,2} = t_2, a_{2,1} = a_{3,2} = 1$ and $a_{1,2} = a_{3,1} = 0$ with t_1, t_2 (called a_1, a_2 in Section 1.1) uniform.

We set $\mathfrak{f}_{i,j} = A_{i,j} - \mathfrak{p}_{i,j}$ and $\mathfrak{g}_i = Z_i - \sum_j \mathfrak{p}_{i,j} W_j$ in the ring $\mathcal{R} = \mathbb{Z}_q[A_{1,1}, \ldots, A_{\ell,k}, \boldsymbol{Z}, \boldsymbol{T}, \boldsymbol{W}]$. Consider the ideal $\mathcal{I}_0$ generated by all $\mathfrak{f}_{i,j}$'s and $\mathfrak{g}_i$'s and the ideal

$\mathcal{I}_1$ generated only by the $\mathfrak{f}_{i,j}$'s in $\mathcal{R}$. Let $\mathcal{J}_b := \mathcal{I}_b \cap \mathbb{Z}_q[A_{1,1}, \ldots, A_{\ell,k}, Z]$. Note that the equations $\mathfrak{f}_{i,j} = 0$ just encode the definition of the matrix entry $a_{i,j}$ by $\mathfrak{p}_{i,j}(t)$ and the equation $\mathfrak{g}_i = 0$ encodes the definition of z_i in the case $z = A\omega$. So, informally, $\mathcal{I}_0$ encodes the relations between the $a_{i,j}$'s, z_i's, t_i's and w_i's in $([A], [z] = [A\omega])$ and $\mathcal{I}_1$ encodes the relations in $([A], [z] = [u])$. For $b = 0$ ($z = A\omega$) and $b = 1$ (z uniform), $\mathcal{J}_b$ encodes the relations visible by considering only the given data (i.e. the $A_{i,j}$'s and Z_j's).

Theorem 1. *Let $\mathcal{D}_{\ell,k}$ be a polynomial-induced matrix distribution with notation as above. Then the $\mathcal{D}_{\ell,k}$-MDDH assumption holds in generic m-linear groups if and only if $(\mathcal{J}_0)_{\leq m} = (\mathcal{J}_1)_{\leq m}$, where the $\leq m$ means restriction to total degree at most m.*

Proof. Note that $\mathcal{J}_{\leq m}$ captures precisely what any adversary can generically compute with polynomially many group and m-linear pairing operations. Formally, this is proven by restating the Uber-Assumption Theorem of [1,6] and its proof more algebraically.

For a given matrix distribution, the condition $(\mathcal{J}_0)_{\leq m} = (\mathcal{J}_1)_{\leq m}$ can be verified by direct linear algebra or by elimination theory (using e.g. Gröbner bases). For the special case $\ell = k+1$, we can actually give a criterion that is simple to verify using determinants:

Theorem 2. *Let $\mathcal{D}_k$ be a polynomial-induced matrix distribution, which outputs matrices $a_{i,j} = \mathfrak{p}_{i,j}(t)$ for uniform $t \in \mathbb{Z}_q^d$. Let $\mathfrak{d}$ be the determinant of $(\mathfrak{p}_{i,j}(T)\|Z)$ as a polynomial in Z, T.*

1. *If the matrices output by $\mathcal{D}_k$ always have full rank (not just with overwhelming probability), even for t_i from the algebraic closure $\overline{\mathbb{Z}_q}$, then $\mathfrak{d}$ is irreducible over $\overline{\mathbb{Z}_q}$.*
2. *If all $\mathfrak{p}_{i,j}$ have degree at most one and $\mathfrak{d}$ is irreducible over $\overline{\mathbb{Z}_q}$ and the total degree of $\mathfrak{d}$ is $k+1$, then the $\mathcal{D}_k$-MDDH assumption holds in generic k-linear groups.*

This theorem and generalizations for non-linear $\mathfrak{p}_{i,j}$ and non-irreducible $\mathfrak{d}$ are proven in the full version [12] using tools from algebraic geometry.

3.3 Examples of $\mathcal{D}_{\ell,k}$-MDDH

Let $\mathcal{D}_{\ell,k}$ be a matrix distribution and $A \leftarrow \mathcal{D}_{\ell,k}$. Looking ahead to our applications, $[A]$ will correspond to the public-key (or common reference string) and $[A w] \in \mathbb{G}^\ell$ will correspond to a ciphertext. We define the *representation size* $\mathrm{RE}_\mathbb{G}(\mathcal{D}_{\ell,k})$ of a given polynomial-induced matrix distribution $\mathcal{D}_{\ell,k}$ with linear $\mathfrak{p}_{i,j}$'s as the minimal number of group elements it takes to represent $[A]$ for any $A \in \mathcal{D}_{\ell,k}$. We will be interested in families of distributions $\mathcal{D}_{\ell,k}$ such that that Matrix Diffie-Hellman Assumption is hard in k-linear groups. By Lemma 2 we obtain a family of strictly weaker assumptions. Our goal is to obtain such a family of assumptions with small (possibly minimal) representation.

Example 1. Let $\mathcal{U}_{\ell,k}$ be the uniform distribution over $\mathbb{Z}_q^{\ell \times k}$.

The next lemma says that $\mathcal{U}_{\ell,k}$-MDDH is the weakest possible assumption among all $\mathcal{D}_{\ell,k}$-Matrix Diffie-Hellman Assumptions. However, $\mathcal{U}_{\ell,k}$ has poor representation, i.e., $\mathsf{REG}(\mathcal{U}_{\ell,k}) = \ell k$.

Lemma 3. *Let $\mathcal{D}_{\ell,k}$ be any matrix distribution. Then $\mathcal{D}_{\ell,k}$-MDDH $\Rightarrow$ $\mathcal{U}_{\ell,k}$-MDDH.*

Proof. Given an instance $([\mathbf{A}], [\mathbf{A}w])$ of $\mathcal{D}_{\ell,k}$, if $\mathbf{L} \in \mathbb{Z}_q^{\ell \times \ell}$ and $\mathbf{R} \in \mathbb{Z}_q^{k \times k}$ are two random invertible matrices, it is possible to get a properly distributed instance of the $\mathcal{U}_{\ell,k}$-matrix DH problem as $([\mathbf{LAR}], [\mathbf{LA}w])$. Indeed, $\mathbf{LAR}$ has a distribution statistically close to the uniform distribution in $\mathbb{Z}_q^{k \times \ell}$, while $\mathbf{LA}w = \mathbf{LARv}$ for $v = \mathbf{R}^{-1}w$. Clearly, v has the uniform distribution in $\mathbb{Z}_q^k$.

Example 2 (k-Linear Assumption/k-Lin). We define the distribution $\mathcal{L}_k$ as follows

$$\mathbf{A} = \begin{pmatrix} a_1 & 0 & \dots & 0 & 0 \\ 0 & a_2 & \dots & 0 & 0 \\ 0 & 0 & \ddots & & 0 \\ \vdots & & \ddots & & \vdots \\ 0 & 0 & \dots & 0 & a_k \\ 1 & 1 & \dots & 1 & 1 \end{pmatrix} \in \mathbb{Z}_q^{(k+1) \times k},$$

where $a_i \leftarrow \mathbb{Z}_q^*$. The transformation matrix $\mathbf{T} \in \mathbb{Z}_q^{1 \times k}$ is given as $\mathbf{T} = (\frac{1}{a_1}, \dots, \frac{1}{a_k})$. Note that the distribution $(\mathbf{A}, \mathbf{A}w)$ can be compactly written as $(a_1, \dots, a_k, a_1 w_1, \dots, a_k w_k, w_1 + \dots + w_k) = (a_1, \dots, a_k, b_1, \dots, b_k, \frac{b_1}{a_1} + \dots + \frac{b_k}{a_k})$ with $a_i \leftarrow \mathbb{Z}_q^*$, $b_i, w_i \leftarrow \mathbb{Z}_q$. Hence the $\mathcal{L}_k$-Matrix Diffie-Hellman Assumption is an equivalent description of the k-linear Assumption [2,19,29] with $\mathsf{REG}(\mathcal{L}_k) = k$.

It was shown in [29] that k-Lin holds in the generic k-linear group model and hence k-Lin forms a family of increasingly strictly weaker assumptions. Furthermore, in [5] it was shown that 2-Lin $\Rightarrow$ BDDH.

Example 3 (k-Cascade Assumption/k-Casc). We define the distribution $\mathcal{C}_k$ as follows

$$\mathbf{A} = \begin{pmatrix} a_1 & 0 & \dots & 0 & 0 \\ 1 & a_2 & \dots & 0 & 0 \\ 0 & 1 & \ddots & & 0 \\ \vdots & & \ddots & & \vdots \\ 0 & 0 & \dots & 1 & a_k \\ 0 & 0 & \dots & 0 & 1 \end{pmatrix},$$

where $a_i \leftarrow \mathbb{Z}_q^*$. The transformation matrix $\mathbf{T} \in \mathbb{Z}_q^{1 \times k}$ is given as $\mathbf{T} = (\pm\frac{1}{a_1 \cdot \dots \cdot a_k}, \mp\frac{1}{a_2 \cdot \dots \cdot a_k} \dots, \frac{1}{a_k})$. Note that $(\mathbf{A}, \mathbf{A}w)$ can be compactly written as $(a_1, \dots, a_k, a_1 w_1, w_1 + a_2 w_2 \dots, w_{k-1} + a_k w_k, w_k) = (a_1, \dots, a_k, b_1, \dots, b_k, \frac{b_k}{a_k} - \frac{b_{k-1}}{a_{k-1}a_k} + \frac{b_{k-2}}{a_{k-2}a_{k-1}a_k} - \dots \pm \frac{b_1}{a_1 \cdot \dots \cdot a_k})$. We have $\mathsf{REG}(\mathcal{C}_k) = k$.

Matrix $\mathbf{A}$ bears resemblance to a cascade which explains the assumption's name. Indeed, in order to compute the right lower entry w_k of matrix $(\mathbf{A}, \mathbf{A}w)$ from the remaining entries, one has to "descent" the cascade to compute all the other entries w_i $(1 \le i \le k-1)$ one after the other.

A more compact version of $\mathcal{C}_k$ is obtained by setting all $a_i := a$.

Example 4. (Symmetric k-Cascade Assumption) We define the distribution $\mathcal{SC}_k$ as $\mathcal{C}_k$ but now $a_i = a$, where $a \leftarrow \mathbb{Z}_q^*$. Then $(\mathbf{A}, \mathbf{A}w)$ can be compactly written as
$$(a, aw_1, w_1+aw_2, \dots, w_{k-1}+aw_k, w_k) = (a, b_1, \dots, b_k, \tfrac{b_k}{a} - \tfrac{b_{k-1}}{a^2} + \tfrac{b_{k-2}}{a^3} - \dots \pm \tfrac{b_1}{a^k}).$$
We have $\mathrm{RE}_\mathbb{G}(\mathcal{C}_k) = 1$.

Observe that the same trick cannot be applied to the k-Linear assumption k-Lin, as the resulting Symmetric k-Linear assumption does not hold in k-linear groups. However, if we set $a_i := a + i - 1$, we obtain another matrix distribution with compact representation.

Example 5. (Incremental k-Linear Assumption) We define the distribution $\mathcal{IL}_k$ as $\mathcal{L}_k$ with $a_i = a + i - 1$, for $a \leftarrow \mathbb{Z}_q$. The transformation matrix $\mathbf{T} \in \mathbb{Z}_q^{1\times k}$ is given as $\mathbf{T} = (\tfrac{1}{a}, \dots, \tfrac{1}{a+k-1})$. $(\mathbf{A}, \mathbf{A}w)$ can be compactly written as $(a, aw_1, (a+1)w_2, \dots, (a+k-1)w_k, w_1 + \dots + w_k) = (a, b_1, \dots, b_k, \tfrac{b_1}{a} + \tfrac{b_2}{a+1} + \dots + \tfrac{b_k}{a+k-1})$. We also have $\mathrm{RE}_\mathbb{G}(\mathcal{IL}_k) = 1$.

The last three examples need some work to prove its generic hardness.

Theorem 3. k-Casc, k-SCasc and k-ILin *are hard in generic k-linear groups.*

Proof. We need to consider the (statistically close) variants with $a_i \in \mathbb{Z}_q$ rather that $\mathbb{Z}_q^*$. The determinant polynomial for $\mathcal{C}_k$ is $\mathfrak{d}(a_1, \dots, a_k, z_1, \dots, z_{k+1}) = a_1 \cdots a_k z_{k+1} - a_1 \cdots a_{k-1} z_k + \dots + (-1)^k z_1$, which has total degree $k + 1$. As all matrices in $\mathcal{C}_k$ have rank k, because the determinant of the last k rows in $\mathbf{A}$ is always 1, by Theorem 2 we conclude that k-Casc is hard in k-linear groups. As $\mathcal{SC}_k$ is a particular case of $\mathcal{C}_k$, the determinant polynomial for $\mathcal{SC}_k$ is $\mathfrak{d}(a, z_1, \dots, z_{k+1}) = a^k z_{k+1} - a^{k-1} z_k + \dots + (-1)^k z_1$. As before, by Theorem 2, k-SCasc is hard in k-linear groups. Finally, in the case of $\mathcal{IL}$, $\mathfrak{d}(a, z_1, \dots, z_{k+1}) = a(a+1)\cdots(a+k-1)\big(z_{k-1} - \tfrac{z_1}{a} - \tfrac{z_2}{a+1} - \dots - \tfrac{z_k}{a+k-1}\big)$, which has total degree $k + 1$. It can be shown that all matrices in $\mathcal{IL}_k$ have rank k. Indeed, matrices in $\mathcal{L}_k$ can have lower rank only if at least two parameters a_i are zero, and this cannot happen to $\mathcal{IL}_k$ matrices. Therefore, as in the previous cases, k-ILin is hard in k-linear groups.

For relations among this new security assumptions we refer the reader to the full version [12].

4   Basic Applications

Basic cryptographic definitions (key-encapsulation, hash proof systems, and pseudo-random functions) are given in the full version [12].

4.1   Public-Key Encryption

Let Gen be a group generating algorithm and $\mathcal{D}_{\ell,k}$ be a matrix distribution that outputs a matrix over $\mathbb{Z}_q^{\ell \times k}$ such that the first k-rows form an invertible matrix with overwhelming probability. We define the following key-encapsulation mechanism $\mathsf{KEM}_{\mathsf{Gen},\mathcal{D}_{\ell,k}} = (\mathsf{Gen}, \mathsf{Enc}, \mathsf{Dec})$ with key-space $\mathcal{K} = \mathbb{G}^{\ell-k}$.

- $\mathsf{Gen}(1^\lambda)$ runs $\mathcal{G} \leftarrow \mathsf{Gen}(1^\lambda)$ and $\mathbf{A} \leftarrow \mathcal{D}_{\ell,k}$. Let $\mathbf{A}_0$ be the first k rows of $\mathbf{A}$ and $\mathbf{A}_1$ be the last $\ell-k$ rows of $\mathbf{A}$. Define $\mathbf{T} \in \mathbb{Z}_q^{(\ell-k) \times k}$ as the transformation matrix $\mathbf{T} = \mathbf{A}_1 \mathbf{A}_0^{-1}$. The public/secret-key is

$$pk = (\mathcal{G}, [\mathbf{A}] \in \mathbb{G}^{\ell \times k}), \quad sk = (pk, \mathbf{T} \in \mathbb{Z}_q^{(\ell-k) \times k})$$

- Enc_{pk} picks $\boldsymbol{w} \leftarrow \mathbb{Z}_q^k$. The ciphertext/key pair is

$$[\boldsymbol{c}] = [\mathbf{A}_0 \boldsymbol{w}] \in \mathbb{G}^k, \quad [K] = [\mathbf{A}_1 \boldsymbol{w}] \in \mathbb{G}^{\ell-k}$$

- $\mathsf{Dec}_{sk}([\boldsymbol{c}] \in \mathbb{G}^k)$ recomputes the key as $[K] = [\mathbf{T}\boldsymbol{c}] \in \mathbb{G}^{\ell-k}$.

Correctness follows by the equation $\mathbf{T} \cdot \boldsymbol{c} = \mathbf{T} \cdot \mathbf{A}_0 \boldsymbol{w} = \mathbf{A}_1 \boldsymbol{w}$. The public key contains $\mathsf{RE}_{\mathbb{G}}(\mathcal{D}_{\ell,k})$ and the ciphertext k group elements.

Theorem 4. *Under the $\mathcal{D}_{\ell,k}$-MDDH Assumption $\mathsf{KEM}_{\mathsf{Gen},\mathcal{D}_{\ell,k}}$ is IND-CPA secure.*

Proof. By the $\mathcal{D}_{\ell,k}$ Matrix Diffie-Hellman Assumption, the distribution of $(pk, [\boldsymbol{c}], [K]) = ((\mathcal{G}, [\mathbf{A}]), [\mathbf{A}\boldsymbol{w}])$ is computationally indistinguishable from $((\mathcal{G}, [\mathbf{A}]), [\boldsymbol{u}])$, where $\boldsymbol{u} \leftarrow \mathbb{Z}_q^\ell$.

4.2   Hash Proof System

Let $\mathcal{D}_{\ell,k}$ be a matrix distribution. We build a universal$_1$ hash proof system $\mathsf{HPS} = (\mathsf{Param}, \mathsf{Pub}, \mathsf{Priv})$, whose hard subset membership problem is based on the $\mathcal{D}_{\ell,k}$ Matrix Diffie-Hellman Assumption.

- $\mathsf{Param}(1^\lambda)$ runs $\mathcal{G} \leftarrow \mathsf{Gen}(1^\lambda)$ and picks $\mathbf{A} \leftarrow \mathcal{D}_{\ell,k}$. Define

$$\mathcal{C} = \mathbb{G}^\ell, \quad \mathcal{V} = \{[\boldsymbol{c}] = [\mathbf{A}\boldsymbol{w}] \in \mathbb{G}^\ell \ : \ \boldsymbol{w} \in \mathbb{Z}_q^k\}.$$

The value $\boldsymbol{w} \in \mathbb{Z}_q^k$ is a witness of $[\boldsymbol{c}] \in \mathcal{V}$. Let $\mathcal{SK} = \mathbb{Z}_q^\ell$, $\mathcal{PK} = \mathbb{G}^k$, and $\mathcal{K} = \mathbb{G}$. For $sk = \boldsymbol{x} \in \mathbb{Z}_q^\ell$, define the projection $\mu(sk) = [\boldsymbol{x}^\top \mathbf{A}] \in \mathbb{G}^k$. For $[\boldsymbol{c}] \in \mathcal{C}$ and $sk \in \mathcal{SK}$ we define

$$\Lambda_{sk}([\boldsymbol{c}]) := [\boldsymbol{x}^\top \cdot \boldsymbol{c}] . \tag{2}$$

The output of Param is $params = \big(\mathcal{S} = (\mathcal{G}, [\mathbf{A}]), \mathcal{K}, \mathcal{C}, \mathcal{V}, \mathcal{PK}, \mathcal{SK}, \Lambda_{(\cdot)}(\cdot), \mu(\cdot)\big)$.
- $\mathsf{Priv}(sk, [\boldsymbol{c}])$ computes $[K] = \Lambda_{sk}([\boldsymbol{c}])$.

– Pub$(pk, [c], w)$. Given $pk = \mu(sk) = [x^\top A]$, $[c] \in \mathcal{V}$ and a witness $w \in \mathbb{Z}_q^k$ such that $[c] = [A \cdot w]$ the public evaluation algorithm Pub$(pk, [c], w)$ computes $[K] = \Lambda_{sk}([c])$ as $[K] = [(x^\top \cdot A) \cdot w]$.

Correctness follows by (2) and the definition of μ. Clearly, under the $\mathcal{D}_{\ell,k}$-Matrix Diffie-Hellman Assumption, the subset membership problem is hard in HPS.

We now show that Λ is a universal$_1$ projective hash function. Let $[c] \in \mathcal{C} \setminus \mathcal{V}$. Then the matrix $(A||c) \in \mathbb{Z}_q^{\ell \times (k+1)}$ is of full rank and consequently $(x^\top \cdot A||x^\top \cdot c) \equiv (x^\top A||u)$ for $x \leftarrow \mathbb{Z}_q^k$ and $u \leftarrow \mathbb{Z}_q$. Hence, $(pk, \Lambda_{sk}([c])) = ([x^\top A], [x^\top c]) \equiv ([x^\top A], [u]) = ([x^\top A], [K])$.

4.3 Pseudo-random Functions

Let Gen be a group generating algorithm and $\mathcal{D}_{\ell,k}$ be a matrix distribution that outputs a matrix over $\mathbb{Z}_q^{\ell \times k}$ such that the first k-rows form an invertible matrix with overwhelming probability. We define the following pseudo-random function PRF$_{\mathsf{Gen}, \mathcal{D}_{\ell,k}} = (\mathsf{Gen}, \mathsf{F})$ with message space $\{0,1\}^n$. For simplicity we assume that $\ell - k$ divides k.

– Gen(1^λ) runs $\mathcal{G} \leftarrow \mathsf{Gen}(1^\lambda)$, $h \in \mathbb{Z}_q^k$, and $\mathbf{A}_{i,j} \leftarrow \mathcal{D}_{\ell,k}$ for $i = 1, \ldots, n$ and $j = 1, \ldots, t := k/(\ell - k)$ and computes the transformation matrices $\mathbf{T}_{i,j} \in \mathbb{Z}_q^{(\ell-k) \times k}$ of $\mathbf{A}_{i,j} \in \mathbb{Z}_q^{\ell \times k}$ (cf. Definition 3). For $i = 1, \ldots, n$ define the aggregated transformation matrices

$$\mathbf{T}_i = \begin{pmatrix} \mathbf{T}_{i,1} \\ \vdots \\ \mathbf{T}_{i,t} \end{pmatrix} \in \mathbb{Z}_q^{k \times k}$$

The key is defined as $K = (\mathcal{G}, h, \mathbf{T}_1, \ldots, \mathbf{T}_n)$.
– F$_K(x)$ computes

$$\mathsf{F}_K(x) = \left[\prod_{i : x_i = 1} \mathbf{T}_i \cdot h \right] \in \mathbb{G}^k.$$

We prove the following theorem in the full version [12].

Theorem 5. *Under the $\mathcal{D}_{\ell,k}$-MDDH Assumption* PRF$_{\mathsf{Gen}, \mathcal{D}_{\ell,k}}$ *is a secure pseudo-random function.*

4.4 Groth-Sahai Non-interactive Zero-Knowledge Proofs

Groth and Sahai gave a method to construct non-interactive witness-indistinguishable (NIWI) and zero-knowledge (NIZK) proofs for satisfiability of a set of equations in a bilinear group $\mathcal{PG}$. (For formal definitions of NIWI and NIZK proofs we refer to [17].) The equations in the set can be of different types, but they can be written in a unified way as

$$\sum_{j=1}^n f(a_j, \mathsf{y}_j) + \sum_{i=1}^m f(\mathsf{x}_i, b_i) + \sum_{i=1}^m \sum_{j=1}^n f(\mathsf{x}_i, \gamma_{ij} \mathsf{y}_j) = t, \tag{3}$$

where A_1, A_2, A_T are $\mathbb{Z}_q$-modules, $\mathbf{x} \in A_1^m$, $\mathbf{y} \in A_2^n$ are the variables, $\boldsymbol{a} \in A_1^n$, $\boldsymbol{b} \in A_2^m$, $\boldsymbol{\Gamma} = (\gamma_{ij}) \in \mathbb{Z}_q^{m \times n}$, $t \in A_T$ are the constants and $f : A_1 \times A_2 \to A_T$ is a bilinear map. More specifically, equations are of either one these types i) Pairing product equations, with $A_1 = A_2 = \mathbb{G}$, $A_T = \mathbb{G}_T$, $f([\mathsf{x}], [\mathsf{y}]) = [\mathsf{xy}]_T \in \mathbb{G}_T$, ii) Multi-scalar multiplication equations, with $A_1 = \mathbb{Z}_q$, $A_2 = \mathbb{G}$, $A_T = \mathbb{G}$, $f(\mathsf{x}, [\mathsf{y}]) = [\mathsf{xy}] \in \mathbb{G}$ or iii) Quadratic equations in $\mathbb{Z}_q$, with $A_1 = A_2 = A_T = \mathbb{Z}_q$, $f(\mathsf{x}, \mathsf{y}) = \mathsf{xy} \in \mathbb{Z}_q$.

OVERVIEW. In the GS proof system the prover gives to the verifier a commitment to each element of the witness (i.e., values of the variables that satisfy the equations) and some additional information, the proof. Commitments and proof satisfy some related set of equations computable by the verifier because of their algebraic properties. To give new instantiations we need to specify the distribution of the common reference string, which includes the commitment keys and some maps whose purpose is roughly to give some algebraic structure to the commitment space. All details are postponed to the full version [12], here we only specify how to commit to scalars $\mathsf{x} \in \mathbb{Z}_q$ to give some intuition of the results in Sections 5.1, 5.2 and 5.3.

COMMITMENTS. The commitment key $[\mathbf{U}] = ([\boldsymbol{u}_1], \ldots, [\boldsymbol{u}_{k+1}]) \in \mathbb{G}^{\ell \times (k+1)}$ is either $[\mathbf{U}] = [\mathbf{A} \| \mathbf{A}\boldsymbol{w}]$ in the soundness setting (binding key) or $[\mathbf{U}] = [\mathbf{A} \| \mathbf{A}\boldsymbol{w} - \boldsymbol{z}]$ in the WI setting (hiding key), where $\mathbf{A} \leftarrow \mathcal{D}_{\ell,k}$, $\boldsymbol{w} \leftarrow \mathbb{Z}_q^k$, and $\boldsymbol{z} \in \mathbb{Z}_q^\ell$, $\boldsymbol{z} \notin \mathrm{Span}(\boldsymbol{u}_1, \ldots, \boldsymbol{u}_k)$ is a fixed, public vector. Clearly, the two types of commitment keys are computationally indistinguishable under the $\mathcal{D}_{\ell,k}$-MDDH Assumption. To commit to a scalar $\mathsf{x} \in \mathbb{Z}_q$ using randomness $\boldsymbol{s} \leftarrow \mathbb{Z}_q^k$ we define the maps $\iota' : \mathbb{Z}_q \to \mathbb{Z}_q^\ell$ and $p' : \mathbb{G}^\ell \to \mathbb{Z}_q$ as

$$\iota'(\mathsf{x}) = \mathsf{x} \cdot (\boldsymbol{u}_{k+1} + \boldsymbol{z}), \quad p'([\boldsymbol{c}]) = \boldsymbol{\xi}^\top \boldsymbol{c}, \quad \text{defining } \mathsf{com}'_{[\mathbf{U}], \boldsymbol{z}}(\mathsf{x}; \boldsymbol{s}) := [\iota'(\mathsf{x}) + \mathbf{A}\boldsymbol{s}] \in \mathbb{G}^\ell,$$

where $\boldsymbol{\xi} \in \mathbb{Z}_q^\ell$ is an arbitrary vector such that $\boldsymbol{\xi}^\top \mathbf{A} = \mathbf{0}$ and $\boldsymbol{\xi}^\top \cdot \boldsymbol{z} = 1$. On a binding key (soundness setting) we have that $p' \circ [\iota']$ is the identity map on $\mathbb{Z}_q$ and $p'([\boldsymbol{u}_i]) = 0$ for all $i = 1 \ldots k$ so the commitment is perfectly binding. On a hiding key (WI setting), $\iota'(x) \in \mathrm{Span}(\boldsymbol{u}_1, \ldots, \boldsymbol{u}_k)$ for all $x \in \mathbb{Z}_q$, which implies that the commitment is perfectly hiding. Note that, given $[\mathbf{U}]$ and x, $\iota'(\mathsf{x})$ might not be efficiently computable but $[\iota'(\mathsf{x})]$ is, which is enough to be able to compute $\mathsf{com}'(\mathsf{x}; \boldsymbol{s})$.

EFFICIENCY. We emphasize that for $\mathcal{D}_{\ell,k} = \mathcal{L}_2$ and $\boldsymbol{z} = (0, 0, 1)^\top$ and for $\mathcal{D}_{\ell,k} = \mathsf{DDH}$ and $\boldsymbol{z} = (0, 1)^\top$ (in the natural extension to asymmetric bilinear groups), we recover the 2-Lin and the SXDH instantiations of [17]. While the size of the proofs depends only on ℓ and k, both the size of the CRS and the cost of verification increase with $\mathsf{RE}_\mathbb{G}(\mathcal{D}_{\ell,k})$. In particular, in terms of efficiency, the $\mathcal{SC}_2$ Assumption is preferable to the 2-Lin Assumption.

5 More Efficient Proofs for Some CRS Dependent Languages

5.1 More Efficient Subgroup Membership Proofs

Let $[\mathbf{U}]$ be the commitment key defined in last section as part of a $\mathcal{D}_{\ell,k}$-MDDH instantiation, for some $\mathbf{A} \leftarrow \mathcal{D}_{\ell,k}$. In this section we show a new technique to obtain proofs of membership in the language $\mathcal{L}_{\mathbf{A},\mathcal{PG}} := \{[\mathbf{A}r], r \in \mathbb{Z}_q^k\} \subset \mathbb{G}^\ell$.

INTUITION. Our idea is to exploit the special algebraic structure of commitments in GS proofs, namely the observation that if $[\boldsymbol{\Phi}] = [\mathbf{A}r] \in \mathcal{L}_{\mathbf{A},\mathcal{PG}}$ then $[\boldsymbol{\Phi}] = \mathrm{com}_{[\mathbf{U}]}(0; r)$. Therefore, to prove that $[\boldsymbol{\Phi}] \in \mathcal{L}_{\mathbf{A},\mathcal{PG}}$, we proceed as if we were giving a GS proof of satisfiability of the equation $\mathrm{x} = 0$ where the randomness used for the commitment to x is r. In particular, no commitments have to be given in the proof, which results in shorter proofs. To prove zero-knowledge we rewrite the equation $\mathrm{x} = 0$ as $\mathrm{x} \cdot \delta = 0$. The real proof is just a standard GS proof with the commitment to $\delta = 1$ being $\iota'(1) = \mathrm{com}_{[\mathbf{U}]}(1; 0)$, while in the simulated proof the trapdoor allows to open $\iota'(1)$ as a commitment of 0, so we can proceed as if the equation was the trivial one $\mathrm{x} \cdot 0 = 0$, for which it is easy to give a proof of satisfiability. For the 2-Lin Assumption, our proof consists of only 6 group elements, whereas without using our technique the proof consists of 12 elements. In the full version [12] we prove the following theorem.

Theorem 6. *Let $\mathbf{A} \leftarrow \mathcal{D}_{\ell,k}$, where $\mathcal{D}_{\ell,k}$ is a matrix distribution. There exists a Non-Interactive Zero-Knowledge Proof for the language $\mathcal{L}_{\mathbf{A},\mathcal{PG}}$, with perfect completeness, perfect soundness and composable zero-knowledge of $k\ell$ group elements based on the $\mathcal{D}_{\ell,k}$-MDDH Assumption.*

APPLICATIONS. Think of $[\mathbf{A}]$ as part of the public parameters of the hash proof system of Section 4.2. Proving that a ciphertext is well-formed is proving membership in $\mathcal{L}_{\mathbf{A},\mathcal{PG}}$. For instance, in [25] Libert and Yung combine a proof of membership in 2-Lin with a one-time signature scheme to obtain publicly verifiable ciphertexts. With our result, we reduce the size of their ciphertexts from 15 to 9 group elements. We stress that in our construction the setup of the CRS can be built on top of the encryption key so that proofs can be simulated without the decryption key, which is essential in their case. Another application is to show that two ciphertexts encrypt the same message under the same public key, a common problem in electronic voting or anonymous credentials. There are many other settings in which subgroup membership problems appear, for instance when proving that a certain ciphertext is an encryption of $[m]$.

5.2 More Efficient Proofs of Validity of Ciphertexts

The techniques of the previous section can be extended to prove the validity of a ciphertext. More specifically, given $\mathbf{A} \leftarrow \mathcal{D}_{\ell,k}$, and some vector $z \in \mathbb{Z}_q^\ell$, $z \notin \mathrm{Im}(\mathbf{A})$, we show how to give a more efficient proof of membership in:

$$\mathcal{L}_{\mathbf{A},z,\mathcal{PG}} = \{[c] : c = \mathbf{A}r + mz\} \subset \mathbb{G}^\ell,$$

where $(r, [m]) \in \mathbb{Z}_q^k \times \mathbb{G}$ is the witness.

This is also a proof of membership in the subspace of $\mathbb{G}^\ell$ spanned by the columns of $[\mathbf{A}]$ and the vector $[z]$, but the techniques given in Section 5.1 do not apply. The reason is that part of the witness, $[m]$, is in the group $\mathbb{G}$ and not in $\mathbb{Z}_q$, while to compute the subgroup membership proofs as described in Section 5.1 all of the witness has to be in $\mathbb{Z}_q$. In particular, since GS are non-interactive zero-knowledge proofs of knowledge when the witnesses are group elements, the proof guarantees both that the c is well-formed and that the prover knows $[m]$.

In a typical application, $[c]$ will be the ciphertext of some encryption scheme, in which case r will be the ciphertext randomness and $[m]$ the message. Deciding membership in this space is trivial when $\mathrm{Im}(\mathbf{A})$ and z span all of $\mathbb{Z}_q^\ell$, so in particular our result is meaningful when $\ell > k + 1$. In the full version [12] we prove the following theorem:

Theorem 7. *Let $\mathcal{D}_{\ell,k}$ be a matrix distribution and let $\mathbf{A} \leftarrow \mathcal{D}_{\ell,k}$. There exists a Non-Interactive Zero-Knowledge Proof for the language $\mathcal{L}_{\mathbf{A},z,\mathcal{PG}}$ of $(k + 2)\ell$ group elements with perfect completeness, perfect soundness and composable zero-knowledge based on the $\mathcal{D}_{\ell,k}$-MDDH Assumption.*

5.3   More Efficient Proofs of Plaintext Equality

The encryption scheme derived from the KEM given in Section 4.1 corresponds to a commitment in GS proofs. That is, if $pk_A = (\mathcal{G}, [\mathbf{A}] \in \mathbb{G}^{\ell \times k})$, for some $\mathbf{A} \leftarrow \mathcal{D}_{\ell,k}$, given $r \in \mathbb{Z}_q^k$,

$$\mathsf{Enc}_{pk_A}([m]; r) = [c] = [\mathbf{A}r + (0, \ldots, 0, m)^\top] = [\mathbf{A}r + m \cdot z] = \mathsf{com}_{[\mathbf{A}||\mathbf{A}w]}([m]; s),$$

where $s^\top := (r^\top, 0)$ and $z := (0, \ldots, 0, 1)^\top$. Therefore, given two (potentially distinct) matrix distributions $\mathcal{D}_{\ell_1,k_1}, \mathcal{D}'_{\ell_2,k_2}$ and $\mathbf{A} \leftarrow \mathcal{D}_{\ell_1,k_1}, \mathbf{B} \leftarrow \mathcal{D}'_{\ell_2,k_2}$, proving equality of plaintexts of two ciphertexts encrypted under pk_A, pk_B, corresponds to proving that two commitments under different keys open to the same value. Our proof will be more efficient because we do not give any commitments as part of the proof, since the ciphertexts themselves play this role. More specifically, given $[c_A] = \mathsf{Enc}_{pk_A}([m])$ and $[c_B] = \mathsf{Enc}_{pk_B}([m])$ we will treat $[c_A]$ as a commitment to the variable $[\mathsf{x}] \in A_1 = \mathbb{G}$ and $[c_B]$ as a commitment to the variable $[\mathsf{y}] \in A_2 = \mathbb{G}$ and prove that the quadratic equation $e([\mathsf{x}], [1]) \cdot e([-1], [\mathsf{y}]) = [0]_T$ is satisfied. The zero-knowledge simulator will open $\iota_1([1]), \iota_2([-1])$ as commitments to the $[0]$ variable and simulate a proof for the equation $e([\mathsf{x}], [0]) \cdot e([0], [\mathsf{y}]) = [0]_T$, which is trivially satisfiable and can be simulated. More formally, let $r \in \mathbb{Z}_q^{k_1}, s \in \mathbb{Z}_q^{k_2}, m \in \mathbb{Z}_q, z_1 \in \mathbb{Z}_q^{\ell_1}$, and $z_1 \notin \mathrm{Im}(\mathbf{A})$ and $z_2 \in \mathbb{Z}_q^{\ell_2}, z_2 \notin \mathrm{Im}(\mathbf{B})$. Define:

$$\mathcal{L}_{\mathbf{A},\mathbf{B},z_1,z_2,\mathcal{PG}} := \{([c_A], [c_B]) : c_A = \mathbf{A}r + mz_1, c_B = \mathbf{B}s + z_2\} \subset \mathbb{G}^{\ell_1} \times \mathbb{G}^{\ell_2}.$$

In the full version [12] we prove:

Theorem 8. *Let $\mathcal{D}_{\ell_1,k_1}$ and $\mathcal{D}'_{\ell_2,k_2}$ be two matrix distributions and let $\mathbf{A} \leftarrow \mathcal{D}_{\ell_1,k_1}, \mathbf{B} \leftarrow \mathcal{D}'_{\ell_2,k_2}$. There exists a Non-Interactive Zero-Knowledge Proof for the language $\mathcal{L}_{\mathbf{A},\mathbf{B},z_1,z_2,\mathcal{PG}}$ of $\ell_1(k_2+1) + \ell_2(k_1+1)$ group elements with perfect completeness, perfect soundness and composable zero-knowledge based on the $\mathcal{D}_{\ell_1,k_1}$-MDDH and the $\mathcal{D}_{\ell_2,k_2}$-MDDH Assumption.*

APPLICATIONS. In [21], we reduce the size of the proof by 4 group elements from 18 to 22, while in [18] we save 9 elements although their proof is quite inefficient altogether. We note that even if both papers give a proof that two ciphertexts under two different 2-Lin public keys correspond to the same value, the proof in [18] is more inefficient because it must use GS proofs for pairing product equations instead of multi-scalar multiplication equations. Other examples include [7,10]. We note that our approach is easily generalizable to prove more general statements about plaintexts, for instance to prove membership in $\mathcal{L}'_{\mathbf{A},\mathbf{B},z_1,z_2,\mathcal{PG}} := \{([c_A],[c_B]) : c_A = \mathbf{A}r + (0,\ldots,0,m)^{\top}, c_B = \mathbf{B}s + (0,\ldots,0,2m)^{\top}, r \in \mathbb{Z}_q^{k_1}, s \in \mathbb{Z}_q^{k_2}\} \subset \mathbb{G}^{\ell_1} \times \mathbb{G}^{\ell_2}$ or in general to show that some linear relation between a set of plaintexts encrypted under two different public-keys holds.

References

1. Boneh, D., Boyen, X., Goh, E.-J.: Hierarchical identity based encryption with constant size ciphertext. In: Cramer, R. (ed.) EUROCRYPT 2005. LNCS, vol. 3494, pp. 440–456. Springer, Heidelberg (2005) 131, 137
2. Boneh, D., Boyen, X., Shacham, H.: Short group signatures. In: Franklin, M. (ed.) CRYPTO 2004. LNCS, vol. 3152, pp. 41–55. Springer, Heidelberg (2004) 130, 138
3. Boneh, D., Franklin, M.: Identity-based encryption from the weil pairing. In: Kilian, J. (ed.) CRYPTO 2001. LNCS, vol. 2139, pp. 213–229. Springer, Heidelberg (2001) 130
4. Boneh, D., Montgomery, H.W., Raghunathan, A.: Algebraic pseudorandom functions with improved efficiency from the augmented cascade. In: Al-Shaer, E., Keromytis, A.D., Shmatikov, V. (eds.) ACM CCS 2010, pp. 131–140. ACM Press (October 2010) 130, 133
5. Boneh, D., Sahai, A., Waters, B.: Fully collusion resistant traitor tracing with short ciphertexts and private keys. In: Vaudenay, S. (ed.) EUROCRYPT 2006. LNCS, vol. 4004, pp. 573–592. Springer, Heidelberg (2006) 132, 138
6. Boyen, X.: The uber-assumption family: A unified complexity framework for bilinear groups. In: Galbraith, S.D., Paterson, K.G. (eds.) Pairing 2008. LNCS, vol. 5209, pp. 39–56. Springer, Heidelberg (2008) 131, 137
7. Camenisch, J., Chandran, N., Shoup, V.: A public key encryption scheme secure against key dependent chosen plaintext and adaptive chosen ciphertext attacks. In: Joux, A. (ed.) EUROCRYPT 2009. LNCS, vol. 5479, pp. 351–368. Springer, Heidelberg (2009) 134, 145
8. Cramer, R., Shoup, V.: A practical public key cryptosystem provably secure against adaptive chosen ciphertext attack. In: Krawczyk, H. (ed.) CRYPTO 1998. LNCS, vol. 1462, pp. 13–25. Springer, Heidelberg (1998) 130
9. Cramer, R., Shoup, V.: Universal hash proofs and a paradigm for adaptive chosen ciphertext secure public-key encryption. In: Knudsen, L.R. (ed.) EUROCRYPT 2002. LNCS, vol. 2332, pp. 45–64. Springer, Heidelberg (2002) 130, 133

10. Dodis, Y., Haralambiev, K., López-Alt, A., Wichs, D.: Cryptography against continuous memory attacks. In: 51st FOCS, pp. 511–520. IEEE Computer Society Press (October 2010) 134, 145

11. Dodis, Y., Haralambiev, K., López-Alt, A., Wichs, D.: Efficient public-key cryptography in the presence of key leakage. In: Abe, M. (ed.) ASIACRYPT 2010. LNCS, vol. 6477, pp. 613–631. Springer, Heidelberg (2010) 134

12. Escala, A., Herold, G., Kiltz, E., Ràfols, C., Villar, J.: An algebraic framework for diffie-hellman assumptions. Cryptology ePrint Archive (2013), http://eprint.iacr.org/ 132, 136, 137, 139, 141, 142, 143, 144

13. Fischlin, M., Libert, B., Manulis, M.: Non-interactive and re-usable universally composable string commitments with adaptive security. In: Lee, D.H., Wang, X. (eds.) ASIACRYPT 2011. LNCS, vol. 7073, pp. 468–485. Springer, Heidelberg (2011) 134

14. Freeman, D.M.: Converting pairing-based cryptosystems from composite-order groups to prime-order groups. In: Gilbert, H. (ed.) EUROCRYPT 2010. LNCS, vol. 6110, pp. 44–61. Springer, Heidelberg (2010) 130, 132

15. Galindo, D., Herranz, J., Villar, J.: Identity-based encryption with master key-dependent message security and leakage-resilience. In: Foresti, S., Yung, M., Martinelli, F. (eds.) ESORICS 2012. LNCS, vol. 7459, pp. 627–642. Springer, Heidelberg (2012) 132

16. Gennaro, R., Lindell, Y.: A framework for password-based authenticated key exchange. In: Biham, E. (ed.) EUROCRYPT 2003. LNCS, vol. 2656, pp. 524–543. Springer, Heidelberg (2003), http://eprint.iacr.org/2003/032.ps.gz 130

17. Groth, J., Sahai, A.: Efficient non-interactive proof systems for bilinear groups. In: Smart, N.P. (ed.) EUROCRYPT 2008. LNCS, vol. 4965, pp. 415–432. Springer, Heidelberg (2008) 130, 133, 141, 142

18. Hofheinz, D., Jager, T.: Tightly secure signatures and public-key encryption. In: Safavi-Naini, R., Canetti, R. (eds.) CRYPTO 2012. LNCS, vol. 7417, pp. 590–607. Springer, Heidelberg (2012) 134, 145

19. Hofheinz, D., Kiltz, E.: Secure hybrid encryption from weakened key encapsulation. In: Menezes, A. (ed.) CRYPTO 2007. LNCS, vol. 4622, pp. 553–571. Springer, Heidelberg (2007) 130, 138

20. Joux, A.: A one round protocol for tripartite Diffie-Hellman. Journal of Cryptology 17(4), 263–276 (2004) 130

21. Katz, J., Vaikuntanathan, V.: Round-optimal password-based authenticated key exchange. In: Ishai, Y. (ed.) TCC 2011. LNCS, vol. 6597, pp. 293–310. Springer, Heidelberg (2011) 134, 145

22. Kiltz, E.: A tool box of cryptographic functions related to the Diffie-Hellman function. In: Pandu Rangan, C., Ding, C. (eds.) INDOCRYPT 2001. LNCS, vol. 2247, pp. 339–350. Springer, Heidelberg (2001) 132

23. Kiltz, E.: Chosen-ciphertext security from tag-based encryption. In: Halevi, S., Rabin, T. (eds.) TCC 2006. LNCS, vol. 3876, pp. 581–600. Springer, Heidelberg (2006) 130

24. Lewko, A.B.: Tools for simulating features of composite order bilinear groups in the prime order setting. In: Pointcheval, D., Johansson, T. (eds.) EUROCRYPT 2012. LNCS, vol. 7237, pp. 318–335. Springer, Heidelberg (2012) 131

25. Libert, B., Yung, M.: Non-interactive CCA-secure threshold cryptosystems with adaptive security: New framework and constructions. In: Cramer, R. (ed.) TCC 2012. LNCS, vol. 7194, pp. 75–93. Springer, Heidelberg (2012) 134, 143

26. Naor, M., Reingold, O.: Number-theoretic constructions of efficient pseudo-random functions. In: 38th FOCS, pp. 458–467. IEEE Computer Society Press (October 1997) 130, 133

27. Naor, M., Yung, M.: Public-key cryptosystems provably secure against chosen ciphertext attacks. In: 22nd ACM STOC. ACM Press (May 1990) 134

28. Okamoto, T., Takashima, K.: Fully secure functional encryption with general relations from the decisional linear assumption. In: Rabin, T. (ed.) CRYPTO 2010. LNCS, vol. 6223, pp. 191–208. Springer, Heidelberg (2010) 131

29. Shacham, H.: A cramer-shoup encryption scheme from the linear assumption and from progressively weaker linear variants. Cryptology ePrint Archive, Report 2007/074 (2007), http://eprint.iacr.org/ 130, 138

30. Villar, J.L.: Optimal reductions of some decisional problems to the rank problem. In: Wang, X., Sako, K. (eds.) ASIACRYPT 2012. LNCS, vol. 7658, pp. 80–97. Springer, Heidelberg (2012) 132

Hard-Core Predicates for a Diffie-Hellman Problem over Finite Fields

Nelly Fazio[1,2], Rosario Gennaro[1,2], Irippuge Milinda Perera[2],
and William E. Skeith III[1,2]

[1] The City College of CUNY
{fazio,rosario,wes}@cs.ccny.cuny.edu
[2] The Graduate Center of CUNY
iperera@gc.cuny.edu

Abstract. A long-standing open problem in cryptography is proving the existence of (deterministic) hard-core predicates for the Diffie-Hellman problem defined over finite fields. In this paper, we make progress on this problem by defining a very natural variation of the Diffie-Hellman problem over $\mathbb{F}_{p^2}$ and proving the unpredictability of every single bit of one of the coordinates of the secret DH value.

To achieve our result, we modify an idea presented at CRYPTO'01 by Boneh and Shparlinski [4] originally developed to prove that the LSB of the elliptic curve Diffie-Hellman problem is hard. We extend this idea in two novel ways:

1. We generalize it to the case of finite fields $\mathbb{F}_{p^2}$;
2. We prove that any bit, not just the LSB, is hard using the list decoding techniques of Akavia et al. [1] (FOCS'03) as generalized at CRYPTO'12 by Duc and Jetchev [6].

In the process, we prove several other interesting results:

- Our result also hold for a larger class of predicates, called *segment predicates* in [1];
- We extend the result of Boneh and Shparlinski to prove that every bit (and every segment predicate) of the elliptic curve Diffie-Hellman problem is hard-core;
- We define the notion of *partial one-way function* over finite fields $\mathbb{F}_{p^2}$ and prove that every bit (and every segment predicate) of one of the input coordinates for these functions is hard-core.

Keywords: Hard-Core Bits, Diffie-Hellman Problem, Finite Fields, Elliptic Curves.

1 Introduction

A long-standing open problem in cryptography is proving the existence of (deterministic) hard-core predicates for the Diffie-Hellman problem defined over finite fields. In this paper we make progress on this problem by defining a very natural extension of the Diffie-Hellman problem over $\mathbb{F}_{p^2}$ and proving that a

R. Canetti and J.A. Garay (Eds.): CRYPTO 2013, Part II, LNCS 8043, pp. 148–165, 2013.

large class of predicates (including every single bit of one of the coordinates) are unpredictable under the assumption that this problem is hard.

In their seminal paper that introduced public-key cryptography [5] Diffie and Hellman defined the following key exchange protocol, which works in arbitrary finite cyclic groups. Let G be such a group, generated by g of order n. Two parties, Alice and Bob, want to establish a secret value. Alice chooses a random value $a \in \mathbb{Z}_n$ and sends the value $A = g^a$ to Bob. Similarly Bob chooses a random value $b \in \mathbb{Z}_n$ and sends the value $B = g^b$ to Alice. At this point they share the common Diffie-Hellman secret value $K = g^{ab} = A^b = B^a$.

The *Computational Diffie-Hellman Assumption* (CDH) over the group G informally states that no efficient algorithm can compute $K = g^{ab}$ when given only $g, A = g^a, B = g^b$. The hardness of computing the *entire* value K, however does not rule out an efficient way to compute some of the bits of K, or even just predict them with a probability better than a random guess. This property is very important because without it, Alice and Bob do not have any guarantee about the "pseudorandomness" of any bit of the secret value K, and those are the properties needed by K in order to be used as a secret key in a subsequent cryptographic scheme. This problem is usually addressed by making a much stronger assumption on the hardness of the Diffie-Hellman problem: the so-called *Decisional Diffie-Hellman Assumption* (DDH) states that the value K is computationally indistinguishable from a random element of G. While the DDH guarantees that the entire value of K is pseudorandom, there are groups G where the DDH is false, even when the CDH is still conjectured to be hard.

Ideally, however, one would like to prove that certain bits (or more generally, certain predicates) of the value K are unpredictable, when given g^a and g^b, simply under the CDH assumption. Such results were established quite early for other conjectured hard-problems (e.g., Blum and Micali's result on the hardness of discrete log bits [3] and Alexi at al. work on the hardness of the RSA input bits [2]). However for the case of the Diffie-Hellman problem no such result has been proven (except for the result by Boneh and Shparlinksi [4] in a slightly different model and which we discuss below). The only hard-core predicates known for the Diffie-Hellman function are the generic "randomized" predicates which work over any computationally hard problem (e.g., the Goldreich-Levin and Näslund hard-core bits [8,10]).

HARD-CORE PREDICATES. Let $\pi : G \rightarrow \{\pm 1\}$ be a predicate[1] defined over G. To prove that π is hard-core for the CDH problem one has to construct a reduction from guessing π better than at random, to solving the CDH problem. More specifically, assume we have an oracle Ω which on input g, g^a, g^b outputs the correct $\pi(g^{ab})$ with probability (taken over the choice of a, b) substantially better than[2] $1/2$, then there is an efficient algorithm A which invokes Ω and solves the CDH problem.

---

[1] For reasons that will become clearer in the technical section of the paper, we adopt the convention that predicates map a value to ± 1 instead of $\{0, 1\}$.

[2] Let's assume for now that π is balanced. In the rest of the paper we take into account the possible bias of π.

Note that a crucial step of this reduction is to "correct" the answers of the oracle Ω which are guaranteed to be right only slightly more than half of the times. This step requires randomizing the queries to Ω while still keeping its answers useful to the solution of the underlying CDH problem. This proves somewhat difficult, due to the limited random self-reducibility of the Diffie-Hellman problem.

RANDOMIZING THE PROBLEM REPRESENTATION. Boneh and Shparlinksi in [4] achieved a breakthrough for the elliptic curve Diffie-Hellman problem, i.e., the CDH problem defined over the group G of points of an elliptic curve. They were able to prove that the least significant bit of each coordinate of the Diffie-Hellman secret value K is hard-core, when the probability space of the oracle Ω also includes a random choice for the representation of the curve.

More specifically: let p be a prime and let E be an elliptic curve defined over $\mathbb{F}_p$, the finite field with p elements. To represent E we use a short Weierstrass equation $W : y^2 = x^3 + ax + b$, with $a, b \in \mathbb{F}_p$ and $4a^3 + 27b^2 \neq 0$. Let $W(E)$ be the set of Weierstrass equations representing E. It is well known that $W(E)$ is defined by the equations W_λ of the form $y^2 = x^3 + \lambda^4 a x + \lambda^6 b$ for $\lambda \in \mathbb{F}_p^\times$. If $Q = (Q_x, Q_y)$ is a point satisfying W then the point $Q_\lambda = (Q_{\lambda,x} = \lambda^2 Q_x, Q_{\lambda,y} = \lambda^3 Q_y)$ satisfies W_λ. Furthermore, the points of E form a group under a certain operation, and the mapping $\Phi_\lambda : E \to E$ defined as $\Phi_\lambda(Q) = Q_\lambda$ is an isomorphism with respect to such group operation over E.

Let G be a cyclic subgroup of E generated by a point P. Switching to additive notation for the group operation, the elliptic curve CDH (EC-CDH) assumption says that given W, P, aP, bP it is hard to compute abP.

In [4] they prove that if there exists an oracle Ω that works on a random representation of E, i.e., such that

$$\Pr_{\lambda,a,b}\left[\Omega(\lambda, P, aP, bP) = \mathrm{LSB}([\Phi_\lambda(abP)]_x)\right] > 1/2 + \epsilon$$

for a non-negligible value ϵ, then it is possible to solve EC-CDH on any curve (a similar result holds for the y-coordinate of abP).

1.1 Our Results

Our main technical contribution is to show that the Boneh-Shparlinski idea of randomizing the representation of the underlying group for the CDH problem can be also applied to the case of finite fields $\mathbb{F}_{p^2}$.

For a given prime p, there are many different fields $\mathbb{F}_{p^2}$, but they are all isomorphic to each other. Let $h(x) = x^2 + h_1 x + h_0$ be a monic irreducible polynomial of degree 2 in $\mathbb{F}_p$. It is well known that $\mathbb{F}_{p^2}$ is isomorphic to the field $\mathbb{F}_p[x]/(h)$, and therefore elements of $\mathbb{F}_{p^2}$ can be written as linear polynomials: if $g \in \mathbb{F}_{p^2}$ then $g = g_0 + g_1 x$ and addition and multiplication are performed as polynomial operations modulo h. In the following, given $g \in \mathbb{F}_{p^2}$ we denote with $[g]_i$ the coefficient of the degree-i term.

Let $I_2(p)$ be the set of monic irreducible polynomials of degree 2 in $\mathbb{F}_p$. For $h, \hat{h} \in I_2(p)$ we know that there exists an (easily computable) isomorphism

$$\phi_{h,\hat{h}} : \mathbb{F}_p[x]/(h) \to \mathbb{F}_p[x]/(\hat{h}).$$

Finally, denote with g a generator of the multiplicative group of $\mathbb{F}_{p^2}$ which is known to be cyclic.

Our first attempt was to use the approach from [4] over $\mathbb{F}_{p^2}$. That is, we hoped to prove that given an oracle Ω which, on input random values g^a, g^b and a random description of $\mathbb{F}_{p^2}$, outputs $\mathrm{LSB}([g^{ab}]_i)$, then we can solve the CDH over $\mathbb{F}_{p^2}$. Unfortunately there are several technical complications with directly applying the approach of [4] to the finite field case, one of them being the fact that representations of an elliptic curve are in bijective correspondence with $\mathbb{F}_p$ allowing them to be represented by a single element of $\mathbb{F}_p$. Conversely the representations of $\mathbb{F}_{p^2}$ are in bijective correspondence with $I_2(p)$ which has $\approx p^2/2$ elements.

A NEW DIFFIE-HELLMAN PROBLEM. To solve these technical problems we had to define the following variant of the CDH problem over $\mathbb{F}_{p^2}$: informally we say that the *Partial-CDH* problem is hard in $\mathbb{F}_{p^2}$ if no efficient algorithm given $g, A = g^a, B = g^b \in \mathbb{F}_{p^2}$ can compute $K = [g^{ab}]_1 \in \mathbb{F}_p$ (i.e., the coefficient of the degree 1 term of g^{ab}).

We note that the Partial-CDH problem is obviously weaker than the regular CDH problem over $\mathbb{F}_{p^2}$, but that it still allows Alice and Bob to agree on a common secret value in $\mathbb{F}_p$, via the traditional Diffie-Hellman protocol.

OUR MAIN RESULT. Assuming the hardness of the Partial-CDH problem we prove that for a large class of predicates π (described below – it includes every individual bit of K), the bit $\pi(K)$ is unpredictable given g^a, g^b and a random representation of $\mathbb{F}_{p^2}$. More specifically, we prove that if there exists an oracle Ω such that for any $h \in I_2(p)$ it holds that

$$\Pr_{\hat{h},a,b}\left[\Omega\left(h,\hat{h},g,g^a,g^b\right) = \pi\left(\left[\phi_{h,\hat{h}}(g^{ab})\right]_1\right)\right] > 1/2 + \epsilon$$

for a non-negligible value ϵ, then it is possible to solve Partial-CDH on $\mathbb{F}_p[x]/(h)$.

We may define an analogous problem for the general case of $\mathbb{F}_{p^t}$ with any $t > 1$. The Partial-CDH problem is defined as outputting the coefficient of the term of degree $t - 1$. However our hard-core results hold only for the quadratic ($\mathbb{F}_{p^2}$) case. See the conclusion (Section 6) for a discussion.

OUR TECHNIQUES. To achieve our result we divert from the techniques used in [4] in another fundamental way. To prove that the predicate π is hard-core for the Partial-CDH problem in $\mathbb{F}_{p^2}$ we use the list-decoding approach pioneered by Akavia et al. [1] as extended by Duc and Jetchev in [6] to the case of prediction oracles which also take as input a random representation of the underlying group.

We describe the approach in detail in Section 3. For now we just remind the reader that as defined originally in [1] this approach allows one to prove the security of so-called *segment predicates* which include both the most and least significant bits of the input. In [9] the technique was extended to work for any input bit. So the class of predicates P described above includes every individual bit of the input and also segment predicates as defined in [1].

ADDITIONAL RESULTS. Since the list-decoding approach works for a larger class of predicates, we obtain two additional results:

1. In the elliptic curve scenario, we are able to extend the [4] result for EC-CDH to any predicate π as above, not just the LSB.
2. For the finite field case we prove that the predicates π are hard-core for a much larger class of conjectured computationally hard problems. Consider a function $f : \mathbb{F}_{p^2} \to S$ for an arbitrary set S. We say that f is a *finite field-based partial one-way function* (FFB-POWF) if the following conditions hold:
   - f is "independent" of the representation used for $\mathbb{F}_{p^2}$ (see Section 5.2 for a precise definition);
   - no efficient algorithm, given $f(x)$ can compute $[x]_1$, i.e., the coefficient of the degree 1 term of x.

   Then we can prove that if f is a FFB-POWF then it is hard to predict $\pi([x]_1)$ better than at random (over a random representation of $\mathbb{F}_{p^2}$) when given only $f(x)$.

INTERPRETATION OF OUR RESULTS. One way to interpret our results is to think of the group representation as part of the input to the computational hard problem (be it a one-way function, or the CDH problem) being used. This means that our results do not apply to the case when the Diffie-Hellman key exchange protocol is performed over a fixed representation of the finite field (or the elliptic curve). Rather it is necessary for Alice and Bob to choose a random representation (an irreducible polynomial for $\mathbb{F}_{p^2}$ or a Weierstrass equation for the curve E) over which to run the protocol.

1.2 Paper Organization

Section 2 reviews some relevant background, particularly the notion of Fourier transform for codes. In Section 3, we cover the list-decoding approach to prove hard-core predicates [1] and its generalization to the case of elliptic curves from [6]. Sections 4 and 5 present our original results. First, as a warm-up we prove that every bit of the EC-CDH problem is hard-core. Then we present our main result on the bit security of Partial-CDH over finite fields, and its extension to FFB-POWF. Finally, we conclude in Section 6 with some discussion about our results and a list of interesting problems left open by our work.

2 Background

2.1 Fourier Transforms

Let $\mathbb{Z}_n$ denote the additive group of integers modulo n. For any two functions $f, g : \mathbb{Z}_n \to \mathbb{C}$, their *inner product* is defined as $\langle f, g \rangle = \frac{1}{n} \sum_{x \in \mathbb{Z}_n} f(x)\overline{g(x)}$. Let $\mathbb{C}(\mathbb{Z}_n)$ denote the vector space formed by all functions $f : \mathbb{Z}_n \to \mathbb{C}$. The ℓ_2-norm of f on $\mathbb{C}(\mathbb{Z}_n)$ is defined as $\|f\|_2 = \sqrt{\langle f, f \rangle}$. A *character* of $\mathbb{Z}_n$ is a homomorphism $\chi : \mathbb{Z}_n \to \mathbb{C}^\times$, such that $\forall_{x,y \in \mathbb{Z}_n} \chi(x + y) = \chi(x)\chi(y)$. These characters are defined by $\chi_\alpha(x) = \omega_n^{\alpha x}$, where $\alpha \in \mathbb{Z}_n$ and $\omega_n = e^{2\pi i/n}$. The set of all characters form a group $\widehat{\mathbb{Z}}_n$. Since the members of $\widehat{\mathbb{Z}}_n$ are orthogonal and $|\widehat{\mathbb{Z}}_n| = |\mathbb{Z}_n|$, they form an orthogonal basis, termed the *Fourier basis*, for $\mathbb{C}(\mathbb{Z}_n)$. The *Fourier transform* $\widehat{f} : \widehat{\mathbb{Z}}_n \to \mathbb{C}$ of f is defined as $\widehat{f}(\chi) = \langle f, \chi \rangle$. The *Fourier expansion* of f is written as $\sum_{\chi \in \widehat{\mathbb{Z}}_n} \widehat{f}(\chi)\chi$. For $\Gamma \subset \widehat{\mathbb{Z}}_n$ the restriction of f to Γ is the function $f_{|\Gamma} : \mathbb{Z}_n \to \mathbb{C}$ defined by $f_{|\Gamma} = \sum_{\chi \in \Gamma} \widehat{f}(\chi)\chi$. The *Fourier coefficients* of f are the coefficients $\widehat{f}(\chi)$ in the Fourier basis $\widehat{\mathbb{Z}}_n$. The *weight* of a Fourier coefficient is denoted by $|\widehat{f}(\chi)|^2$. Definition 2.1 formalizes the notion of *heavy characters* with respect to f.

Definition 2.1 (τ-heavy Characters). *Let $\tau \in \mathbb{R}^+$ be a threshold and $f : \mathbb{Z}_n \to \mathbb{C}$ be an arbitrary function. We say a character $\chi \in \widehat{\mathbb{Z}}_n$ is τ-heavy if the weight of its corresponding Fourier coefficient is at least τ. The set of all such character is denoted by $\mathsf{Heavy}_\tau(f)$, i.e.,*

$$\mathsf{Heavy}_\tau(f) = \{\chi \in \widehat{\mathbb{Z}}_n : |\widehat{f}(\chi)|^2 \geq \tau\}.$$

2.2 Codes and Their Properties

In what follows, we report a few useful known definitions [6] and lemmata [1] about codes over $\mathbb{Z}_n$. As in [6], we will regard $\mathbb{Z}_n$-codes as associating an element $x \in \mathbb{Z}_n$ to a $\mathbb{Z}_n$-codeword C_x, which we will in turn see interchangeably as a function $C_x : \mathbb{Z}_n \to \{\pm 1\}$ or as a length-n sequence of $\{\pm 1\}$.

Definition 2.2 (ϵ-concentrated Function). *We say a function $f : \mathbb{Z}_n \to \{\pm 1\}$ is Fourier ϵ-concentrated if there exist a size $\mathrm{poly}(n, 1/\epsilon)$, $\epsilon > 0$, set of characters $\Gamma \subset \widehat{\mathbb{Z}}_n$ such that $\|f - f_{|\Gamma}\|_2 \leq \epsilon$. We say a function is Fourier concentrated if it is ϵ-concentrated for every $\epsilon > 0$.*

Definition 2.3 (ϵ-concentrated Code). *We say a code $C = \{C_x : \mathbb{Z}_n \to \{\pm 1\}\}$ is ϵ-concentrated if all its codewords C_x are Fourier ϵ-concentrated. We say a code is Fourier concentrated if it is ϵ-concentrated for every $\epsilon > 0$.*

Definition 2.4 (Code Recoverability). *We say a code $C = \{C_x : \mathbb{Z}_n \to \{\pm 1\}\}$ is recoverable if there exists an algorithm that, given as input a threshold τ and a character $\chi \in \widehat{\mathbb{Z}}_n$, produces a list of all elements x associated with codewords C_x for which χ is a τ-heavy coefficient, that is, $\{x \in \mathbb{Z}_n : \chi \in \mathsf{Heavy}_\tau(C_x)\}$, in time polynomial in $\log n$ and $1/\tau$.*

The following two results appear in [1]. Lemma 2.5 shows that, in a concentrated code C, any noisy version $\tilde{C}_x$ of codeword C_x share at least one heavy coefficient with C_x. Theorem 2.6 shows that one can efficiently learn all the heavy characters of any function when given query access to it. Therefore having query access to $\tilde{C}_x$ (which in our case is obtained by querying the prediction oracle Ω), one can learn at least one heavy coefficient of C_x, and that if the code is also recoverable, then one can recover x.

Lemma 2.5 (Lem. 1 of [1]). *Let $f, g : \mathbb{Z}_n \to \{\pm 1\}$ such that f is Fourier concentrated and, for some $\epsilon > 0$,*

$$\Pr_{x \in \mathbb{Z}_n} f(x) = g(x) \geq \mathsf{maj}_f + \epsilon,$$

where maj_f denotes the bias of the function f, i.e., $\mathsf{maj}_f = \max_{\{b=\pm 1\}} \Pr_{x \in \mathbb{Z}_n} f(x) = b$. Then there exist a threshold τ such that $1/\tau$ is polynomial in ϵ and $\log n$, and there exists a character $\chi \neq 0$ heavy for f and g: $\chi \in \mathsf{Heavy}_\tau(f) \wedge \mathsf{Heavy}_\tau(g)$.

Theorem 2.6 (Thm. 6 of [1]). *There exists a randomized learning algorithm over $\mathbb{Z}_n$ that, given query access to a function $w : \mathbb{Z}_n \to \{\pm 1\}$, $\tau > 0$ and $0 < \delta < 1$, returns a list of $O(1/\tau)$ characters containing $\mathsf{Heavy}_\tau(w)$ with probability at least $1 - \delta$. The probability is taken over the random coins of the algorithm, whose running time is*

$$\tilde{O}\left(\log(n)\ln^2\frac{(1/\delta)}{\tau^{5.5}}\right).$$

An overview of the above learning algorithm [1] is provided in Appendix A of the full version [7].

3   Hard-Core Predicates by List Decoding

In this section, we review the work of Akavia et al. [1] on how to prove that certain predicates are hard-core for a one-way function f using list decoding of a particular error-correcting code. We also summarize the extensions by Duc and Jetchev [6] to the case of elliptic-curve based one-way functions.

Let $f : \mathbb{Z}_n \to S$ be a one-way function and let $y = f(x)$ for $x \in \mathbb{Z}_n$. Let also $\pi : \mathbb{Z}_n \to \{\pm 1\}$ denote a predicate (with the convention that a 0 bit is encoded as $+1$). Finally we denote with β_π the bias of the predicate π, i.e., $\beta_\pi = \max_{\{b=\pm 1\}} \Pr_x[\pi(x) = b]$.

The goal is to prove that π is a hard-core predicate for the function f. The proof goes as usual by contradiction by assuming that there exists an oracle Ω which, when queried on $f(x)$, returns a bit b which is equal to $\pi(x)$ with probability $\beta_\pi + \epsilon$ for a non-negligible ϵ, and then using Ω to invert f, i.e., find x given y.

To achieve this goal, Akavia et al. in [1] define a *multiplication code*

$$\mathcal{C} = \{C_x : \mathbb{Z}_n \to \{\pm 1\}\}_{x \in \mathbb{Z}_n}, \text{ where } C_x(\lambda) = \pi(\lambda x).$$

In order for their proof to work this code needs the following properties:

Accessibility: Given $y = f(x)$, it must be possible to obtain a "noisy" version $\tilde{C}_x$ of the codeword C_x, i.e., one that agrees with the correct one with probability $\beta_\pi + \epsilon$ for a non-negligible ϵ. In [1], this is done by assuming that the one-way function has some homomorphic property, i.e given $y = f(x)$ and $\lambda \in \mathbb{Z}_n$ it is possible to compute $y_\lambda = f(\lambda x)$ (modular exponentiation has this property). Then, by querying Ω on y_λ one gets the desired accessibility property;

Concentration: Every codeword C_x must be a Fourier concentrated function. Remember that according to the definition above this means that for every ϵ there exists a polynomial (in $\log n$ and ϵ^{-1}) set Γ of Fourier characters, such that $\|C_x - C_{x,\Gamma}\| \leq \epsilon$ (where $C_{x,\Gamma}$ is the restriction of C_x to the Fourier characters in Γ);

Recoverability: There exists an algorithm that on input a Fourier character χ and a threshold τ, outputs a list L_χ containing all the values $x \in \mathbb{Z}_n$ such that χ is τ-heavy for C_x. The algorithm runs in polynomial (in $\log n$ and τ^{-1}) time, which in particular means that the size of L_χ is also "small".

Concentration and recoverability depends on the choice of the predicate π. In [1], the notion of *segment predicates* is defined and shown to be sufficient for the purpose. Later Morillo and Rafols in [9] prove that any individual input bit yields a concentrated and recoverable code (we review this in Appendix B of the full version [7]). We assume π to be one of such predicates in the following.

If the code $\mathcal{C}$ has the above properties then it is possible to prove that π is a hard-core predicate. Assume we have an oracle Ω which when queried on $f(x)$ returns a bit b which is equal to $\pi(x)$ with probability $\beta_\pi + \epsilon$ where $\epsilon = 1/poly(\ell)$ (where $\ell = |n|$). We need to show how to use Ω to invert f.

The inversion works as follows. On input $y = f(x)$, the oracle Ω allows us to access a "noisy" version $\tilde{C}_x$ of C_x, i.e., such that $\Pr_\lambda[C_x(\lambda) = \tilde{C}_x(\lambda)] > \beta_\pi + \epsilon$. By applying Lemma 2.5 we know that there exists a threshold τ which is polynomial in ϵ and at least one Fourier character χ which is τ-heavy for both C_x and $\tilde{C}_x$. Using the learning algorithm described in Theorem 2.6, we obtain a list containing all the τ-heavy Fourier characters for $\tilde{C}_x$; for each such character we use the recovery property to create a polynomial size list of possible pre-images for y which because of Lemma 2.5 must necessarily include x. The correct x can be identified by evaluating the OWF f over all the possible candidates and comparing with y. Details can be found in [1] (in any case, in Sections 4 and 5 we present the details of this algorithm as it applies to our results).

3.1 Accessibility via Elliptic Curve Isomorphisms

Taking the result of [1] as a starting point, and using techniques first developed in [4], Duc and Jetchev [6] show how to obtain the accessibility property in a different way, when the one-way function is defined over the group G of points of an elliptic curve. Their result does not require the one-way function f to have some homomorphic property; on the other hand it requires the oracle to work over a *random* description of the curve.

Let p be a prime and let E be an elliptic curve defined over $\mathbb{F}_p$. To represent E we use a short Weirstrass equation $W : y^2 = x^3 + ax + b$, with $a, b \in (\mathbb{F}_p)$ and $4a^3 + 27b^2 \neq 0$. Let $W(E)$ be the set of Weirstrass equations representing E: so $W \in W(E)$. It is well known that $W(E)$ is defined by the equations W_λ of the form $y^2 = x^3 + \lambda^4 ax + \lambda^6 b$ for $\lambda \in \mathbb{F}_p^\times$. If $Q = (Q_x, Q_y)$ is a point satisfying W then the point $Q_\lambda = (Q_{\lambda,x} = \lambda^2 Q_x, Q_{\lambda,y} = \lambda^3 Q_y)$ satisfies W_λ. It is not hard to see that the mapping $\Phi_\lambda : E \to E$ defined as $\Phi_\lambda(Q) = Q_\lambda$ is an isomorphism with respect to the group operation over E.

Boneh and Shparlinski were the first to note that this isomorphism gives raise to a natural extension of the prediction oracle Ω, by requiring that the input distribution for Ω also include λ. Following this idea, the oracle in [6] takes as input $f(Q)$ where f is a one-way function defined over the group E, and *also* a value λ (i.e., a representation W_λ of E). The oracle returns a bit b such that $b = \pi(Q_{\lambda,x})$ with probability $\beta_\pi + \epsilon$ (for a non-negligible ϵ) where the probability is not only over the choice of Q (and the internal random coins of Ω) but *also* over the choice of $\lambda \in \mathbb{F}_p^\times$.

As defined, the prediction oracle Ω gives noisy access to the *quadratic* codeword $C_Q(\lambda) = \pi(\lambda^2 Q_x)$, which would complicate matters (in particular it makes it hard to prove concentration and recovery, see [6] for a discussion). To apply the techniques of [1], we need noisy access to the multiplication code $C_Q : \mathbb{F}_p \to \{\pm 1\}$ defined as $C_Q(\lambda) = \pi(\lambda Q_x)$.

Following [4] again, Duc and Jetchev defined a modified oracle Ω' which queries Ω if λ is a square in $\mathbb{F}_p^\times$, otherwise tosses a β_π-biased coin. It is not hard to see that if Ω had advantage ϵ, then Ω' has advantage $\epsilon/2$ (see [4]).

Using Ω', the generic approach on [1] shows that π is a hard-core predicate for *any* one-way function f defined over E, provided that the output of f does not depend on the Weirstrass equation used to describe E (in other words that the function f is defined over the group of points, irrespective of its representation). Duc and Jetchev call such a function an *elliptic curve-based one-way function* (ECB-OWF) and discuss the application of their result to bilinear pairings defined over elliptic curves, which are indeed a conjectured example of ECB-OWF.

4   Hard-Core Predicates for the Diffie-Hellman Problem over Elliptic Curves

In this section, we show our first original result: if the Diffie-Hellman problem over elliptic curves is hard, then every bit (and every segment predicate) of a secret Diffie-Hellman value is unpredictable. This generalizes the result of Boneh and Shparlinski [4] which holds only for the least significant bit.

For a security parameter ℓ, consider an instance generator $\mathcal{E}$ which on input 1^ℓ outputs E_ℓ an elliptic curve defined over $\mathbb{F}_{p_\ell}$ where p_ℓ is a ℓ-bit prime, such that G_ℓ is a cyclic subgroup of E_ℓ (under the standard group operation defined over the curve points) generated by a point P_ℓ. In the following, we will drop the suffix ℓ when it is clear from the context. We also use the additive notation for the group operation over E, therefore every point $Q \in G$ can be written as $Q = aP$ for some $a \in \{1, \ldots, |G|\}$.

Assumption 4.1. *We say that the Diffie-Hellman problem over $\mathcal{E}$ is hard if for every polynomial time machine A, we have that the probability*

$$\Pr\left[A(E_\ell, P_\ell, aP_\ell, bP_\ell) = abP_\ell \mid E_\ell \leftarrow \mathcal{E}(1^\ell); \; a, b \leftarrow \{1, \ldots, |G|\}\right]$$

is negligible in ℓ.

For every point $Q \in E$ we denote with Q_x the x-coordinate of Q. As before we denote with $W(E)$ the set of short Weirstrass equations describing a curve E; recall that each $W \in W(E)$ can be uniquely associated with a $\lambda \in \mathbb{F}_p^\times$ which gives rise to the isomorphism Φ_λ defined in the previous section.

Let $B_k : \mathbb{F}_p \to \{\pm 1\}$ denote the k-th bit predicate and let β_k be the bias of B_k. We now state our first main theorem. Intuitively it says that under Assumption 4.1, every bit of the binary expansion of the x-coordinate of abP is unpredictable (e.g., pseudorandom) for a random representation of the curve E.

Theorem 4.2. *Under Assumption 4.1, for any polynomial time machine Ω,*

$$\left| \Pr\left[\Omega(\lambda, P, aP, bP) = B_k([\Phi_\lambda(abP)]_x) \mid \lambda \leftarrow \mathbb{F}_p^\times; \, a, b \leftarrow \{1, \ldots, |G|\} \right] - \beta_k \right|$$

must be negligible.

The intuition of the proof is as follows. The crucial observation is that the techniques of Duc and Jetchev [6] apply not just to ECB-OWFs but to any computation which "respects" the isomorphism Φ_λ defined by a change in the Weirstrass representation of the curve. The Diffie-Hellman problem is one such problem since applying the Diffie-Hellman transform to $\Phi_\lambda(aP), \Phi_\lambda(bP)$ yields the value $\Phi_\lambda(abP)$ – indeed this is at the basis of the result of [4]. Therefore, an oracle Ω contradicting Theorem 4.2 on input aP, bP and a curve W_λ defined by a parameter $\lambda \in \mathbb{F}_p^\times$ would output a bit equal to $B_k(\lambda^2[abP]_x)$ with non-negligible advantage. This allows us to construct a multiplication code with the required properties and apply the framework of [1] to prove that the predicate is hard-core.

Remark 4.3. The extension to segment predicates follow from using the concentration and recoverability arguments for those predicates as presented in [1].

Proof. Assume that there exists an oracle Ω such that the quantity

$$\left| \Pr\left[\Omega(\lambda, P, aP, bP) = B_k([\Phi_\lambda(abP)]_x) \mid \lambda \leftarrow \mathbb{F}_p^\times; \, a, b \leftarrow \{1, \ldots, |G|\} \right] - \beta_k \right|$$

is larger than a non-negligible quantity ϵ.

From this oracle we build a modified oracle Ω' which queries Ω if λ is a square in $\mathbb{F}_p^\times$, otherwise tosses a β_k-biased coin. It is not hard to see [4] that if Ω had advantage ϵ, then Ω' has advantage $\epsilon/2$. We now show how to use Ω' to break Assumption 4.1.

Let E be an elliptic curve defined by an equation $W \in W(E)$ over $\mathbb{F}_p$ and let G be a cyclic subgroup of $|E|$ generated by the point P. Given P, aP, bP we want to compute $Q = abP$ with non-negligible probability.

Consider the codeword:

$$C_Q : \mathbb{F}_p \to \{\pm 1\} \quad \text{defined as} \quad C_Q(\lambda) = B_k(\lambda Q_x).$$

The following properties hold for C_Q.

Accessibility: The oracle Ω' gives us access to a noisy version $\tilde{C}_Q$ of this codeword defined as $\tilde{C}_Q = \Omega'(\lambda, P, aP, bP)$. Because Ω' has advantage $\epsilon/2$ we know that $\Pr_\lambda[C_Q(\lambda) = \tilde{C}_Q(\lambda)] > \beta_k + \epsilon/2$.

Concentration: The codeword C_Q is a Fourier concentrated function. Indeed for a threshold τ the τ-heavy characters of C_Q must belong to the set

$$\Gamma_{Q,\tau} = \{\chi_\beta : \beta = \alpha Q_x \bmod p \text{ for } \alpha \in \Gamma_\tau\},$$

where Γ_τ is a set of size $O(\tau^{-2})$ containing the τ-heavy coefficients of the function B_k. We refer the reader to [6,9] for a proof of this statement and also the definition of Γ_τ which shows that the elements of Γ_τ can be easily enumerated. See also Appendix B of the full version [7].

Recoverability. Given a Fourier character χ_β we want to find a set L_β containing all the points Q such that χ_β is τ-heavy for C_Q. If χ_β is τ-heavy for C_Q then $\chi_\beta \in \Gamma_{Q,\tau}$ and therefore $Q_x = \beta\alpha^{-1} \bmod p$ for $\alpha \in \Gamma_\tau$, therefore

$$L_\beta = \{Q : Q_x = \beta\alpha^{-1} \bmod p \text{ for } \alpha \in \Gamma_\tau\}.$$

By applying Lemma 2.5 we know that there exists a threshold τ which is polynomial in ϵ and at least one Fourier character χ which is τ-heavy for both C_Q and $\tilde{C}_Q$.

We then invoke Theorem 2.6 and use the learning algorithm of [1] to learn a polynomial-size list L_Q of all the τ-heavy Fourier characters for $\tilde{C}_Q$. For each such character $\chi_\beta \in L_Q$ we use the recovery property to create a polynomial size list L_β of possible values for Q. Let $L = \cup_{\chi_\beta \in L_Q} L_\beta$; this is a polynomial-size set and because of Lemma 2.5 it must necessarily include Q.

More specifically, on input E, P, aP, bP and with access to Ω, the following algorithm produces a polynomial size list of points in E which is guaranteed to contain Q with probability $1 - \delta$:

1. Let τ be the threshold determined by Lemma 2.5 ; note that τ^{-1} is polynomial in $\ell = |p|$, since ϵ^{-1} is.
2. Learn the polynomial-size set L_Q containing all τ-heavy Fourier characters of $\tilde{C}_Q$, using the learning algorithm in [1], which is correct with probability $1 - \delta$. This algorithms uses oracle Ω' to obtain the required query access to $\tilde{C}_x$. By applying Lemma 2.5, we know that there exists at least one Fourier character χ which is τ-heavy for C_Q and $\chi \in L_Q$.
3. Use the recovery algorithm to construct a polynomial-size list of candidates values for Q. For each $\chi_\beta \in L_Q$ let

$$L_\beta = \{R \in E : \chi_\beta \text{ is } \tau\text{-heavy for } C_R\}$$
$$= \{R \in E : R_x = \beta\alpha^{-1} \bmod p \text{ for } \alpha \in \Gamma\}.$$

Let $L = \cup_{\chi_\beta \in L_Q} L_\beta$. Note that L's size is polynomial in ℓ and that $Q \in L$ with probability $1 - \delta$.

The algorithm runs in polynomial time, since the learning algorithm of [1] is efficient and all the enumerations in the algorithm are over polynomial-size lists.

To contradict Assumption 4.1 at this point, it would be sufficient to choose a random point in L. The probability to select the correct point Q is $1/|L|$ and therefore the algorithm outputs the correct Q with probability $(1-\delta)/|L|$ which is non-negligible since $|L|$ is of polynomial-size.

Another option is to use the above algorithm as a subroutine in Shoup's "self-corrector" for the Diffie-Hellman problem (Theorem 7 in [11]). Shoup shows how an algorithm A that runs in time T_A and produces a list of m points, which contains the correct Diffie-Hellman value with probability $> 7/8$ can be easily converted into an algorithm B that output only the correct Diffie-Hellman value with overwhelming probability and runs in time $T_A\ell + poly(m, \ell)$.

5   Hard-Core Predicates for the Diffie-Hellman Problem over Finite Fields

In this section, we state and prove our main result: after defining a natural (though weaker) variation of the Diffie-Hellman problem over finite fields $\mathbb{F}_{p^t}$ for $t > 1$, we prove that in the case of quadratic extensions ($t = 2$), this problem admits a large class of hard-core predicates, including every single bit of one of the coordinates of the secret value.

For a given prime p, there are many different fields $\mathbb{F}_{p^2}$, but they are all isomorphic to each other. Let $h(x) = x^2 + h_1 x + h_0$ be a monic irreducible polynomial of degree 2 in $\mathbb{F}_p$. It is well known that $\mathbb{F}_{p^2}$ is isomorphic to the field $\mathbb{F}_p[x]/(h)$, and therefore elements of $\mathbb{F}_{p^2}$ can be written as linear polynomials: if $g \in \mathbb{F}_{p^2}$ then $g = g_0 + g_1 x$ and addition and multiplication are performed as polynomial operations modulo h. In the following, given $g \in \mathbb{F}_{p^2}$ we denote with $[g]_i$ the coefficient of the degree-i term.

Let $I_2(p)$ be the set of monic irreducible polynomials of degree 2 in $\mathbb{F}_p$. For $h, \hat{h} \in I_2(p)$ we know that there exists an (easily computable) isomorphism

$$\phi_{h,\hat{h}} : \mathbb{F}_p[x]/(h) \to \mathbb{F}_p[x]/(\hat{h}).$$

Finally, denote with g a generator of the multiplicative group of $\mathbb{F}_{p^2}$ which is known to be cyclic.

A NEW DIFFIE-HELLMAN PROBLEM. Denote with g the generator of the multiplicative group of $\mathbb{F}_{p^2}$ which is known to be cyclic. We define the following variant of the CDH problem over $\mathbb{F}_{p^2}$: informally we say that the *Partial-CDH* problem is hard in $\mathbb{F}_{p^2}$ if no efficient algorithm given $g, A = g^a, B = g^b \in \mathbb{F}_{p^2}$ can compute $K = [g^{ab}]_1 \in \mathbb{F}_p$, for any representation of $\mathbb{F}_{p^2}$.

More formally, for a security parameter ℓ, consider an instance generator $\mathcal{F}$ which on input 1^ℓ outputs p_ℓ an ℓ-bit prime. Let g_ℓ be a generator of the multiplicative group of the finite field $\mathbb{F}_{p_\ell^2}$. In the following, we will drop the suffix ℓ when it is clear from the context.

Assumption 5.1. *We say that the Partial Diffie-Hellman problem over $\mathcal{F}$ is hard if for every polynomial time machine A, we have that for all $h_\ell \in I_2(p_\ell)$ the following probability is negligible in ℓ:*

$$\Pr\big[A\big(p_\ell, h_\ell, g_\ell, g_\ell^a, g_\ell^b\big) = \big[g_\ell^{ab}\big]_1 \mid p_\ell \leftarrow \mathcal{F}(1^\ell); \; a, b \leftarrow \{1, \ldots, p_\ell^2 - 1\}\big].$$

Note that A gets as input a representation h_ℓ of the field, and that A's advantage must be negligible for all representations.

We now state our second main theorem. We show that, when given an oracle Ω which predicts the kth bit of the degree-1 coefficient of the Diffie-Hellman secret with non-negligible advantage (where the probability is taken over the input pair), *as well as the representation of the field*, then one can efficiently solve the Partial Diffie-Hellman problem with non-negligible probability.

Theorem 5.2. *Under Assumption 5.1, for any polynomial time machine Ω we have that the following quantity must be negligible for all $h \in I_2(p)$:*

$$\left| \Pr\left[\Omega\left(h, \hat{h}, g, g^a, g^b\right) = B_k\left(\left[\phi_{h,\hat{h}}(g^{ab})\right]_1\right) \right.\right.$$
$$\left.\left. \mid \hat{h} \leftarrow I_2(p); \; a, b \leftarrow \{1, \ldots, p^2 - 1\}\right] - \beta_k\right|.$$

The proof of Theorem 5.2 appears in Section 5.1. Here we give an informal intuition of the proof.

Our goal is to construct a code similar to that of [6], which must be accessible by querying Ω over many different representation of the field. For an element $\alpha \in \mathbb{F}_{p^2}$, and a fixed $h \in I_2(p)$, a natural definition for a codeword is as follows:

$$C_\alpha(\hat{h}) = B_k\left(\left[\phi_{h,\hat{h}}(\alpha)\right]_1\right). \tag{1}$$

This code is accessible using Ω, however it is defined over $I_2(p)$, and it is not immediately seen to be a multiplication code like the ones used in [1,6]. Note, however, that the predicate B_k is evaluated *only on the first coordinate of $\phi_{h,\hat{h}}(\alpha)$*. In this case, it holds that $\left[\phi_{h,\hat{h}}(\alpha)\right]_1 = \lambda[\alpha]_1$ for some $\lambda \in \mathbb{F}_p^\times$ (see Lemma 5.5 below).

Consider then the following multiplication code over $\mathbb{F}_p$: for $\alpha \in \mathbb{F}_{p^2}$ and for $\lambda \in \mathbb{F}_p^\times$, set

$$C_\alpha(\lambda) = B_k(\lambda[\alpha]_1)$$

extended with $C_\alpha(0) = -1$. We stress that in light of Lemma 5.5, the above code is conceptually the same as equation (1) in that codewords are obtained by evaluating a predicate over all possible representations of elements. We've simply restricted attention to the degree-1 coordinate. Therefore the multiplication is accessible via Ω and then the proof follows similarly to the one in [1,6].

Remark 5.3 (List of Candidate Solutions). The list-decoding algorithm of [1] applied to the code above returns a polynomial size list of possible candidates for $[\alpha]_1$. In our reduction $\alpha = g^{ab}$ and therefore it will be sufficient to output a random element of the list to contradict Assumption 5.1. In contrast to Theorem 4.2, we will not be able to apply Shoup's "self-corrector" in this case to identify the correct solution with high probability, as we have only a single coordinate for g^{ab}.

Remark 5.4 (Segment Predicates). While Theorem 5.2 is stated only for the predicate B_k, it holds for any predicate π such that the corresponding code C_α can be proven to be concentrated and recoverable; in particular, it holds for the segment predicates defined in [1].

5.1   Proof of Theorem 5.2

We start with a lemma that gives a simple characterization of the isomorphisms between two different representations of the field $\mathbb{F}_{p^2}$. When describing such maps, it will be convenient for us to view them as matrices in $GL_2(\mathbb{F}_p)$.

Lemma 5.5. *For any $h \in I_2(p)$ there exists a unique function $L_h : \mathbb{F}_p \times \mathbb{F}_p^\times \to I_2(p)$ which takes a pair (a,b) to the polynomial $\hat{h} = L_h(a,b)$ such that the matrix $\left(\begin{smallmatrix} 1 & a \\ 0 & b \end{smallmatrix}\right)$ defines an isomorphism $\mathbb{F}_p[x]/(h) \to \mathbb{F}_p[x]/(\hat{h})$. Moreover, for any $\hat{h} \in I_2(p)$, $L_h^{-1}(\hat{h})$ represents the complete set of isomorphisms from $\mathbb{F}_p[x]/(h) \to \mathbb{F}_p[x]/(\hat{h})$ using the above matrix identification.*

Proof. First note that any isomorphism of fields must send the unit element to itself (and thus fix the entire base field $\mathbb{F}_p$). Thus, when viewing such an isomorphism as a linear transformation, the first basis element $\left(\begin{smallmatrix} 1 \\ 0 \end{smallmatrix}\right)$ must be fixed, which determines the first column of the matrix as $\left(\begin{smallmatrix} 1 \\ 0 \end{smallmatrix}\right)$. Since clearly we must have $b \neq 0$ if the map is to represent an isomorphism, the completeness would follow immediately, once we establish the existence and uniqueness of the map L_h. We define L_h as follows. For $a, b \in \mathbb{F}_p$ with $b \neq 0$, let $L_h(a,b)(x) = \frac{h(a+bx)}{b^2}$. To make the notation less cumbersome, we'll fix a, b in what follows, and refer to this polynomial more simply as $L_h(x)$. To see that this definition is as desired, note that to specify a homomorphism ϕ from $\mathbb{F}_p[x]/(h)$ to another field K of characteristic p it is both necessary and sufficient to choose $\phi(x) = \overline{x} \in K$ such that $h(\overline{x}) = 0$ in K. The matrix corresponding to (a,b) sends $x \mapsto a + bx$, and indeed, $a + bx$ is a root of h in the ring $\mathbb{F}_p[x]/(L_h)$ by construction. However, it remains to show that $L_h \in I_2(p)$, as well as the uniqueness of L_h. Towards the first goal: it is an elementary fact that since h was irreducible over $\mathbb{F}_p$, so is $h(a + bx)$, and hence L_h. It is easy to verify additionally that L_h is monic, and has degree 2, so that $L_h \in I_2(p)$. Thus, by the above remarks, the mapping defined by $x \mapsto a + bx$ is an isomorphism $\mathbb{F}_p[x]/(h) \to \mathbb{F}_p[x]/(L_h)$ as desired. The fact that L_h so constructed is unique (within $I_2(p)$) follows easily as well, since if $h(a + bx)$, and hence $L_h(x)$, are elements of an ideal (h') for some other $h' \in I_2(p)$, then L_h, h' are associates, and thus $L_h = h'$ since both are monic. $\square$

Remark 5.6. We actually know a little more about the distribution; in particular, we have $|L_h^{-1}(\hat{h})| = 2$ for any $\hat{h} \in I_2(\mathbb{F}_p)$. This follows at once from the fact that every isomorphism has a (unique) matrix representation as above, and that $\mathrm{Gal}(\mathbb{F}_{p^2}/\mathbb{F}_p) \cong \mathbb{Z}_2$ (so that there are precisely two isomorphisms between any two representations $\mathbb{F}_p[x]/(h), \mathbb{F}_p[x]/(\hat{h})$).

Proof Sketch (Theorem 5.2). Suppose that the theorem were false, and that an oracle Ω with an advantage that is not negligible exists. Now consider another oracle Ω' that takes as input a base representation $h \in I_2(p)$, a Diffie-Hellman triple g, g^a, g^b as well as an element of $\lambda \in \mathbb{F}_p$ (instead of $\hat{h} \in I_2(p)$), which works as follows. The oracle selects $a \leftarrow \mathbb{F}_p$ uniformly at random, and constructs an isomorphism $\hat{h}$ from the matrix $\left(\begin{smallmatrix} 1 & a \\ 0 & \lambda \end{smallmatrix}\right)$ as described in Lemma 5.5. Ω' then returns the output of $\Omega(h, \hat{h}, g, g^a, g^b)$. One can then show that

$$\left| \Pr_{\lambda,a,b}\left[\Omega'\left(h, \lambda, g, g^a, g^b\right) = B_k\left(\lambda\left[g^{ab}\right]_1\right)\right] - \beta_k \right|$$

is also not a negligible function. At this point, the proof follows closely to that of Theorem 4.2. To begin, observe that we can, for any element $\alpha \in \mathbb{F}_{p^2}$, construct the following encoding of $[\alpha]_1$ in its base polynomial representation as an element of $\mathbb{F}_p[x]/(h)$:

$$C_\alpha : \mathbb{F}_p \to \{\pm 1\} \quad \text{defined as} \quad C_\alpha(\lambda) = B_k(\lambda[\alpha]_1),$$

where $[\alpha]_1$ is taken under the representation determined by h. The fact that this code is concentrated and recoverable follows immediately from the proof of Theorem 4.2. The argument for accessibility is the same, but with the added simplification that we no longer need to restrict to squares in $\mathbb{F}_p$.

As in Theorem 4.2, we will be able to efficiently construct a list of candidates for $\left[g^{ab}\right]_1$. As mentioned, we unfortunately will not be able to apply Shoup's "self-corrector" in this case as we have only a single coordinate. Nevertheless, we still obtain a contradiction by guessing a random element of the list as the value of $\left[g^{ab}\right]_1$, since the list is of polynomial size.

5.2   Finite Field-Based One-Way Functions

The work of [6] introduces "elliptic curve-based one-way functions", and goes on to prove interesting hardness results for this *entire class* of functions. Loosely speaking, elliptic curve-based OWF's are one-way functions which are well defined on isomorphism classes of curves, and do not depend on any specific representation. Similarly, we consider *finite field-based OWF's*, which are those that do not depend on the isomorphism class. When considering only prime-order fields $\mathbb{F}_p$, this concept is somewhat trivial, since once one fix a bit representation for integers, there are no non-trivial isomorphisms. However, the situation becomes far more interesting when one considers field extensions. Even with a fixed representation for integers, there are many different representations of even a quadratic extension (see Lemma 5.5). As demonstrated in [6] for the case of elliptic curves, having a one-way function which is well defined on many different representations may give rise to a number of hardness results that apply to the entire class of functions. We demonstrate similar results, showing that for quadratic extensions, an efficient oracle that predicts the k-th bit of the input *over a random representation* of the field will imply an efficient procedure that can "partially" invert the function (i.e., if f is the one-way function, given $f(\alpha)$, it computes $[\alpha]_1$).

In order to define a function f on a finite field, we first define the function on a particular "base" representation F. Then, to define f on any other isomorphic copy F', we wish to simply compute $f \circ \psi$, where $\psi : F' \to F$ is an isomorphism. The following definition guarantees that f is well defined on isomorphism classes of finite fields.

Definition 5.7. *Let $F \cong \mathbb{F}_{p^t}$ be a concrete representation of a finite field. A function $f : F \to Y$ is said to be* finite field-based *if for any $F' \cong F$ and any two isomorphisms $\psi, \psi' : F' \to F$, we have $f \circ \psi = f \circ \psi'$.*

Remark 5.8. Note that any function f satisfying Definition 5.7 is actually defined on a quotient space, $F/\sim$, where $\alpha \sim \alpha'$ if and only if α, α' have the same minimal polynomial over $\mathbb{F}_p$. Furthermore, any function which is well defined on $F/\sim$ will satisfy the definition. Thus, an equivalent definition would be to require that $f(\alpha)$ depends only on the minimal polynomial of α. (This follows from the fact that the Galois group acts transitively on the roots of irreducible polynomials.)

We now define a natural relaxation of the notion of one-way functions over finite fields, where it is assumed to be hard to output the maximal degree coordinate of the input. While this definition makes sense for the general case p^t for $t > 1$, we only consider the case of quadratic extensions.

Consider the instance generator $\mathcal{F}$ which on input a security parameter 1^ℓ, outputs p_ℓ (an ℓ-bit prime), and a function $f_\ell : \mathbb{F}_{p_\ell^2} \to S_\ell$, where S_ℓ is an arbitrary set. We drop the suffix ℓ when it is clear from the context.

Definition 5.9. *We say that $\mathcal{F}$ is* partial one-way *if for any efficient algorithm A the following probability is negligible in ℓ for all $h_\ell \in I_2(p_\ell)$:*

$$\Pr\left[A(h_\ell, f_\ell(\alpha)) = [\alpha]_1 \mid p_\ell, f_\ell \leftarrow \mathcal{F}(1^\ell); \; \alpha \leftarrow \mathbb{F}_{p_\ell}[x]/(h_\ell)\right].$$

Again, note that A takes as input a representation of the field, but the probability must be negligible for all representations.

In the case of quadratic extensions, we can obtain results similar to what was shown in [6] for elliptic-curve based OWF. In particular, the existence of a noisy oracle which works with non-negligible probability over the point, as well as the representation of the field, will give rise to an efficient procedure which "partially" inverts f contradicting Definition 5.9. More formally, we have the following.

Theorem 5.10. *Suppose that f is a finite field-based partial one-way function, and fix a base representation $\mathbb{F}_{p^2} = \mathbb{F}_p[x]/(h)$ for some $h \in I_2(p)$. Then, for any probabilistic polynomial time machine Ω, it must be that the following quantity is negligible:*

$$\left|\Pr\left[\Omega\left(h, \hat{h}, f(\alpha)\right) = B_k\left(\left[\phi_{h,\hat{h}}(\alpha)\right]_1\right) \mid \hat{h} \leftarrow I_2(p); \; \alpha \leftarrow \mathbb{F}_p[x]/(h)\right] - \beta_k\right|.$$

The proof is a combination of the proofs of Theorems 4.2 and 5.2 and will be presented in the full version [7].

Remark 5.11. We note that the Diffie-Hellman problem does *not* satisfy the above definition: apart from the fact that the domain is actually two (or three) field elements, the value g^{ab} is not independent of the representation. However, if one modifies the usual Diffie-Hellman problem to report the minimal polynomial of g^{ab} instead, then the definition is satisfied (with the caveat regarding the input coming from a product space). We also remark that the minimal polynomial is efficiently computable; see for example the work of [12]. Finally, we note that for $\mathbb{F}_{p^t}$, each of the equivalence classes under $\sim$ has size t. Since t is usually a small constant (in our case, it is 2), the aforementioned conversion in which one "throws away" some information by only considering the minimal polynomial will not affect the problem's computational character.

6   Conclusion and Future Work

We presented a relaxed variant of the Diffie-Hellman problem over finite fields of the form $\mathbb{F}_{p^t}$ for $t > 1$ and proved that for the case of quadratic extensions $\mathbb{F}_{p^2}$, this problem admits several hard-core predicates (including every single bit of one coordinate of the secret Diffie-Hellman value) over a random representation of the field. These are the first results known for hard-core predicates for the CDH problem over finite fields. We extended this result to a larger class of computationally hard problems (which we called finite field-based partial one-way functions) over such finite fields.

We also proved that the same class of predicates is hard-core for the elliptic curve Diffie-Hellman, over a random representation of the underlying elliptic curve, thereby extending the Boneh-Shparlinski result [4] which worked only for the least significant bit.

Our results can be interpreted as "augmenting" the input to the computational hard problem (being it a one-way function, or the CDH problem) with a random description of the underlying group being used.

Our work leaves several open questions. Perhaps the most natural is to extend the results to $\mathbb{F}_{p^t}$ for $t > 2$. In the case of $t = 2$, the isomorphisms from one representation to another amounted, in some sense, to a linear change of variables: $x \mapsto a + bx$. This made the set of isomorphisms between representations easy to analyze, and enabled us to show that when restricting attention to the coefficient of x, each of these maps acts by translation for some $\lambda \in \mathbb{F}_p^\times$. For $t > 2$, this is *not* the case, and thus our original techniques must be augmented somehow. Perhaps one can find a large (enough) number of representations for which the isomorphisms have the required properties as a linear map.

Other natural questions include the study of the hardness of the Partial-CDH problem in $\mathbb{F}_{p^t}$ for $t > 1$. While it seems quite a reasonable assumption to make, the ultimate goal would be to reduce it to the "full" CDH over another platform. In particular, is it possible to reduce the Partial-CDH over $\mathbb{F}_{p^t}$ to the regular CDH problem over $\mathbb{F}_p$? A related question is if we can use the hardness of Partial-CDH over, say $\mathbb{F}_{p^2}$, to prove the unpredictability of a predicate for the traditional CDH problem over $\mathbb{F}_p$.

Finally it is our hope that the techniques presented in this paper could eventually lead to the proof that CDH over $\mathbb{F}_p$ does have a (deterministic) hard-core predicate.

Acknowledgments. The authors would like to thank Adi Akavia and Dimitar Jetchev for several useful discussions and clarifications.

Nelly Fazio's research is sponsored in part by NSF CAREER award #1253927, and by PSC-CUNY award 64578-00 42, funded by the Professional Staff Congress and CUNY. Nelly Fazio and William E. Skeith III are sponsored in part by NSF award #1117675. Any opinions, findings, and conclusions or recommendations expressed in this material are those of the authors and do not necessarily reflect the views of the National Science Foundation. Nelly Fazio and Rosario Gennaro are supported in part by the U.S. Army Research Laboratory and the U.K. Ministry of Defence under Agreement Number W911NF-06-3-0001. The views and conclusions contained in this document are those of the authors and should not be interpreted as representing the official policies, either expressed or implied, of the U.S. Army Research Laboratory, the U.S. Government, the U.K. Ministry of Defence or the U.K. Government. The U.S. and U.K. Governments are authorized to reproduce and distribute reprints for Government purposes notwithstanding any copyright notation hereon.

References

1. Akavia, A., Goldwasser, S., Safra, S.: Proving hard-core predicates using list decoding. In: IEEE Symposium on Foundations of Computer Science—FOCS, pp. 146–157 (2003)
2. Alexi, W., Chor, B., Goldreich, O., Schnorr, C.: Rsa and rabin functions: Certain parts are as hard as the whole. SIAM Journal on Computing 17(2), 194–209 (1988)
3. Blum, M., Micali, S.: How to generate cryptographically strong sequences of pseudorandom bits. SIAM Journal on Computing 13(4), 850–864 (1984)
4. Boneh, D., Shparlinski, I.E.: On the unpredictability of bits of the elliptic curve diffie–hellman scheme. In: Kilian, J. (ed.) CRYPTO 2001. LNCS, vol. 2139, pp. 201–212. Springer, Heidelberg (2001)
5. Diffie, W., Hellman, M.: New directions in cryptography. IEEE Transactions on Information Theory IT-22(6), 644–654 (1976)
6. Duc, A., Jetchev, D.: Hardness of computing individual bits for one-way functions on elliptic curves. In: Safavi-Naini, R., Canetti, R. (eds.) CRYPTO 2012. LNCS, vol. 7417, pp. 832–849. Springer, Heidelberg (2012)
7. Fazio, N., Gennaro, R., Perera, I.M., Skeith III, W.E.: Hard-core predicates for a diffie-hellman problem over finite fields. Cryptology ePrint Archive, Report 2013/134 (2013)
8. Goldreich, O., Levin, L.A.: A hard-core predicate for all one-way functions. In: ACM Symposium on Theory of Computing—STOC, pp. 25–32 (1989)
9. Morillo, P., Ràfols, C.: The security of all bits using list decoding. In: Jarecki, S., Tsudik, G. (eds.) PKC 2009. LNCS, vol. 5443, pp. 15–33. Springer, Heidelberg (2009)
10. Näslund, M.: All bits in $ax + b$ mod p are hard. In: Koblitz, N. (ed.) CRYPTO 1996. LNCS, vol. 1109, pp. 114–128. Springer, Heidelberg (1996)
11. Shoup, V.: Lower bounds for discrete logarithms and related problems. In: Fumy, W. (ed.) EUROCRYPT 1997. LNCS, vol. 1233, pp. 256–266. Springer, Heidelberg (1997)
12. Shoup, V.: Efficient computation of minimal polynomials in algebraic extensions of finite fields. In: Proceedings of the 1999 International Symposium on Symbolic and Algebraic Computation, pp. 53–58. ACM (1999)

Encoding Functions with Constant Online Rate or How to Compress Garbled Circuits Keys[*]

Benny Applebaum[1], Yuval Ishai[2], Eyal Kushilevitz[2], and Brent Waters[3]

[1] School of Electrical Engineering, Tel-Aviv University
bennyap@post.tau.ac.il
[2] Department of Computer Science, Technion
{yuvali,eyalk}@cs.technion.ac.il
[3] Department of Computer Science, University of Texas
bwaters@cs.utexas.edu

Abstract. *Randomized encodings of functions* can be used to replace a "complex" function $f(x)$ by a "simpler" randomized mapping $\hat{f}(x; r)$ whose output distribution on an input x encodes the value of $f(x)$ and hides any other information about x. One desirable feature of randomized encodings is low *online complexity*. That is, the goal is to obtain a randomized encoding $\hat{f}$ of f in which most of the output can be precomputed and published before seeing the input x. When the input x is available, it remains to publish only a short string $\hat{x}$, where the online complexity of computing $\hat{x}$ is independent of (and is typically much smaller than) the complexity of computing f. Yao's garbled circuit construction gives rise to such randomized encodings in which the online part $\hat{x}$ consists of n encryption keys of length κ each, where $n = |x|$ and κ is a security parameter. Thus, the *online rate* $|\hat{x}|/|x|$ of this encoding is proportional to the security parameter κ.

In this paper, we show that the online rate can be dramatically improved. Specifically, we show how to encode any polynomial-time computable function $f : \{0,1\}^n \to \{0,1\}^{m(n)}$ with online rate of $1 + o(1)$ and with nearly linear online computation. More concretely, the online part $\hat{x}$ consists of an n-bit string and a single encryption key. These constructions can be based on the decisional Diffie-Hellman assumption (DDH), the Learning with Errors assumption (LWE), or the RSA assumption. We also present a variant of this result which applies to *arithmetic formulas*, where the encoding only makes use of arithmetic operations, as well as several negative results which complement our positive results.

Our positive results can lead to efficiency improvements in most contexts where randomized encodings of functions are used. We demonstrate this by presenting several concrete applications. These include protocols for secure multiparty computation and for non-interactive verifiable computation in the preprocessing model which achieve, for the first time, an optimal online communication complexity, as well as non-interactive zero-knowledge proofs which simultaneously minimize the online communication and the prover's online computation.

---

[*] A preliminary full version is available at [6].

R. Canetti and J.A. Garay (Eds.): CRYPTO 2013, Part II, LNCS 8043, pp. 166–184, 2013.

1   Introduction

Suppose that we want to perform some cryptographic task which involves computation and communication on n-bit data. In many scenarios, it is beneficial to minimize the online complexity (i.e., the resources spent after seeing the data) and shift the expensive computation and communication to an offline phase. This setting has been extensively studied in many contexts including signatures [17,40], verifiable computation (delegation) [19,4,14], and secure computation [8,32,11,15,31]. The goal of the present paper is to further explore the question of minimizing the online complexity of cryptography.

Let us first consider the following concrete example from [5]. Imagine a scenario of sending a weak device U to the field in order to perform some expensive computation f on sensitive data x. The computation is too complex for U to quickly perform it on its own and, since the input x is sensitive, U cannot just send the entire input out. Ideally, we would like to have a *non-interactive* solution of the following form: In an offline phase, before sent to the field, U picks a short random secret key sk and publishes a (potentially long) related public key pk. Once it observes the input x, the device U applies some cheap computation to sk and x and sends out the result $\hat{x}$, a short "encrypted" version of x. The rest of the world should be able, at this point, to recover $f(x)$ and nothing else.

Abstracting the above, the computation of U can be described as a randomized function $\hat{f} : (x; \text{sk}) \mapsto (\text{pk}, \hat{x})$ that *encodes* the value $f(x)$ in the sense that $(\text{pk}, \hat{x})$ reveals $f(x)$ but nothing else. Using the terminology of [3], the function $\hat{f}$ is referred to as a *randomized encoding* (RE) of f. The general motivation for using REs is the hope to make $\hat{f}$ in some sense "simpler" than f, where different applications dictate different notions of simplicity. The earliest uses of REs in cryptography were in the area of secure computation [42,34,18,30]. Along the years, REs have found a diverse range of other applications to problems such as computing on encrypted data [39,13], parallel cryptography [3,2], verifiable computation [19,4], software protection [25,27,9], functional encryption [38,26], key-dependent message security [7,1,10], and others. We refer the reader to [10] for a finer-grained treatment of REs under the term "garbling schemes".

In the online/offline setting considered here, we would like to minimize the online computation and communication resources required for computing and distributing $\hat{x}$. That is, we would like the online time complexity of computing $\hat{x}$ to be much smaller than the time required for computing f, and the length of $\hat{x}$ to be not much bigger than that of x.

The best known general constructions of online-efficient REs are based on Yao's garbled circuit technique [42]. In this case, the output of $f(x)$ is encoded by an offline part pk which consists of a big "garbled circuit" and an online part $\hat{x}$ which consists of n keys $K_1, \ldots, K_n$ of size κ each, where n is the bit-length of x and κ is a security parameter. (Under a standard asymptotic security convention in which n serves both as an input length parameter and a security parameter, κ can be thought of as n^ε, for some small constant $\varepsilon > 0$.) Each key K_i is selected from a pair of keys $(K_{i,0}, K_{i,1})$ according to the i-th input bit x_i. Hence, the online computation and communication complexity are both

$O(n\kappa)$. An appealing feature is that the online computation complexity is nearly linear in the input length, independently of the complexity of f. However, an undesirable feature is that the *online rate* of the construction — i.e., the ratio between the bit-length of $\hat{x}$ and the bit length of x — grows linearly with the security parameter κ. Hence, we ask:

> Is it possible to obtain a *constant* online rate or even rate of $1 + o(1)$ (e.g., $|\hat{x}| = n + \text{poly}(\kappa)$) while keeping the online computation independent of the complexity of f?

1.1 Our Contribution

We answer the above question in the affirmative by constructing, under a variety of standard intractability assumptions, an online-efficient RE with rate $1 + o(1)$ for every polynomial-time computable function.

Theorem 1. *(Informal) Under the Decisional Diffie-Hellman Assumption (DDH), the RSA Assumption, or the Learning-with-Errors Assumption (LWE), every polynomial-time computable function $f : \{0,1\}^n \to \{0,1\}^{m(n)}$ admits an RE with online rate $1+o(1)$ and with $O(n^{1+\varepsilon})$ online computation, for any $\varepsilon > 0$.*

In more concrete terms, our constructions efficiently compile any boolean circuit C into a corresponding RE with succinct and efficiently computable online part. These constructions can be viewed as analogues of the garbled circuit construction in which the n keys determined by x are compressed into a shorter string $\hat{x}$ whose length is very close to that of x. This comes at the cost of a slight increase in the online computation complexity, which still remains nearly linear in n. An additional (related) difference is that in contrast to the standard garbled circuit construction, where each bit of $\hat{x}$ depends only on a single bit of x, in our constructions there are bits of $\hat{x}$ which depend on many bits of x. We prove that this is inherent for REs with constant or even logarithmic online rate. In particular, it is impossible to obtain a direct generalization of the garbled circuit construction in which each input bit x_i selects between a pair of keys $(K_{i,0}, K_{i,1})$ which have constant size.

The DDH and LWE based constructions are *affine* in the sense that after the private randomness is fixed in the offline phase, the remaining computation can be described as an affine function of the inputs x (over some ring R, e.g., R $= \mathbb{Z}_p$ where p is the size of a DDH group). This captures a strong form of algebraic simplicity which is useful for some of the motivating applications (e.g., secure computation).

Motivated by the concrete efficiency of encoding *arithmetic* computations, we also present an LWE-based arithmetic variant of the above result that applies to arithmetic *formulas* (i.e., circuits of fan-out 1) over large finite fields, where the encoding is restricted to applying arithmetic operations to the inputs. Specifically, we obtain an affine randomized encoding (ARE, for short) with optimal online rate (i.e., $1+o(1)$) for arithmetic mod-p formulas, assuming that elements of $\mathbb{Z}_p$ can be viewed as elements of $\mathbb{Z}_q$ for some $q \gg p$. If we insist on working in the more restricted model of [5], where the encoding should be affine over the integers, then we get a constant-rate encoding.

It should be mentioned that the online *computational* overhead of our constructions is still polynomial in the security parameter. Whether this overhead can be improved remains an interesting open question.

Lower Bounds. We further explore the complexity of REs in the online/offline setting by proving several lower bounds on the online and offline rate of REs which complement our positive results. Among other results, we study the minimal achievable online rate. The online rate is clearly lower-bounded by 1 for some functions with long outputs (this is the case, for instance, for the identity function). This leaves open the possibility of achieving a strictly better rate for boolean functions. We show that even in the case of boolean functions, the online rate of *affine* REs (satisfying the algebraic simplicity condition discussed above) cannot generally be smaller than 1. Thus, achieving rate $1 + o(1)$ is essentially optimal for affine REs. While we cannot unconditionally prove a similar result for non-affine REs with, say, quadratic online computation, such a negative result follows from the conjecture that for any $c > c'$, an input for a time-(n^c) computation cannot generally be "compressed" by a time-$(n^{c'})$ algorithm into a shorter string which contains sufficient information to recover the output. See [29,16] for related conjectures.

Adaptive Security. Informally, an offline/online RE is adaptively secure if $\hat{f}(x; r) = (\mathsf{pk}, \hat{x})$ remains private even if the online input x is adaptively chosen based on the offline part of the encoding, pk. Similarly to all other known implementations of garbled circuits with short keys, our constructions cannot be proved to satisfy this stronger notion of security unless analyzed in the (programmable) random oracle model. We prove that this is inherent to some extent: in any RE whose adaptive security holds in the plain model, the length of the online part $\hat{x}$ should grow with the output length of f. (This negative result is similar in spirit to negative results for non-committing encryption [37] or functional encryption [12].) In contrast, our constructions in the non-adaptive setting (or the adaptive setting with random oracles) have online rate of $1 + o(1)$, independently of the output length of f. Adaptive security of garbled circuits has recently been considered in the work of Bellare et al. [9]. The above negative result partially settles a question left open by [9].

On Concrete Efficiency. In concrete terms, our offline/online REs reduce the online communication of Yao's garbled circuit construction by a factor of $\kappa \approx 100$ at the expense of introducing "public-key" computations. This is not always a good tradeoff in practice. For instance, communicating 100 bits is typically less expensive than a single modular exponentiation. Luckily, our REs are also very cheap in online computation. For instance, the online encoding in the DDH-based construction involves at most one mod-p *addition* per input bit, where p is the order of the DDH group. Since a mod-p addition is typically much cheaper than the amortized cost of communicating a bit (let alone 100 bits), we improve the overall concrete online complexity by roughly a factor of 100. This is contrasted with most applications of public-key cryptography towards improving communication complexity, where the additional computational cost outweighs the savings in communication (cf. [41]). While our REs do increase the complexity of the offline encoding and online decoding, the additional overhead is insignificant when the circuit complexity of f is much bigger than its input size. Thus, our offline/online REs seem to have a true practical potential in secure computation or delegation scenarios in which a weak client (who performs the offline and online encoding) interacts with a powerful server (who performs the online decoding).

1.2   Applications

Our positive results can lead to efficiency improvements in most contexts in which randomized encodings of functions are used. We focus on three representative applications.

Secure Multiparty Computation (MPC). In the online/offline model (or preprocessing model) for MPC, there are t players who wish to securely compute some fixed public function f. In the offline phase, before the inputs "arrive", the parties are allowed to invoke some (relatively expensive) protocol; later, in the online phase, the parties get their inputs and apply an online (hopefully cheap) protocol. The close connection of REs to MPC [30] allows to translate our results into highly efficient MPC protocols in the offline/online setting. In Section 5, we further extend and optimize these reductions (exploiting the affinity property and the information-theoretic techniques from [11]). This leads to general MPC protocols in which the online phase only requires each party to broadcast a message of the same length as its input along with a message of size poly(κ), where κ is a security parameter. Again, this is information-theoretic optimal, and it beats, in terms of online communication complexity, all previously known results even in the simplest case of two semi-honest parties. We note, however, that our protocols do not offer provable security against malicious parties which adaptively choose their inputs based on the information they receive in the offline phase, except in the random oracle model or under nonstandard assumptions. See full version for further discussion.

It is instructive to compare the efficiency of our RE-based protocols to protocols which are based on fully homomorphic encryption (FHE). The following discussion is restricted to the preprocessing model, which does not seem to significantly improve the complexity of FHE-based protocols. In FHE based protocols (as well as all other general MPC protocols from the literature) the communication complexity grows at least linearly with the total input and output length $n+m$. In contrast, the online communication complexity of our protocol does *not* depend on the output length. This is particularly useful when securely computing functionalities that have a short online secret input (say, shares of a signature key) and a long output (say, signatures on many predetermined messages using the shared signature key). Furthermore, our protocols can be made completely non-interactive in certain scenarios, e.g., when part of the secret input is known offline and the online part is known in its entirety to one of the parties. This is impossible to get using FHE.[1] On the other hand, our protocols are incomparable to FHE-based protocols in terms of their online computational complexity. In the case of computing a complex function f which takes inputs from Alice and Bob and delivers an output to Alice, our approach yields two-message protocols in which Bob's online computation is very efficient (nearly linear in its input), whereas FHE provides similar protocols in which Alice's computation is very efficient (quasilinear in the input and output). From a concrete efficiency point of view, the online phase of our protocols is much "lighter" (e.g., Bob only needs to add a subset of $\mathbb{Z}_p$ elements corresponding to its input) and they can also be based on a wider variety of assumptions.

Verifiable Computation. In an online/offline protocol for *verifiable computation* (VC), a computationally weak client with an input x delegates a complex computation f to an untrusted server in a two phase manner. In the offline phase the client sends to the server a possibly long and computationally expensive message pk, and at the online phase (when the input x arrives) the client sends a message $\hat{x}$ to the server, and receives back the result of the computation y together with a certificate for correctness. This setting was studied in several works (e.g., [36,25,33,19,14,4,9]). Specifically, in [19] Yao's garbled circuit technique was used to achieve efficient VC in the online/offline model. (The security of the construction follows from standard assumptions only when the input x is picked by the client independently of pk [9].) This connection was generalized and optimized in [4]. By plugging our encodings in these protocols, we get *communication optimal* VC protocols, where the bit-length of the up-stream (online) message from the client to the server is $n + \kappa$ and the bit-length of the downstream message (from server to client) is $m + \kappa$, where n is the input length, m is the output length and κ is the security parameter. Information-theoretically, $n + m$ bits are necessary even if the server is fully trusted. To the best of our knowledge, all previous protocols, including ones which are based on fully homomorphic encryption, have a *multiplicative* overhead of κ, either with respect to n or to m.

Non-Interactive Zero-Knowledge (NIZK). The complexity of NIZK has received much attention. The length of traditional NIZK proofs for NP grows linearly with the size of a circuit $R(x, w)$ which verifies that w is a legal witness for the statement $x \in L$. Using FHE, these traditional NIZKs can be converted into ones whose length is only $|w| + \text{poly}(\kappa)$ bits [20,28]. The proof consists of an FHE encryption c of w, along with a traditional NIZK proving that the ciphertext resulting from evaluating the verification algorithm on c encrypts the result of a correct verification. Thus, the prover's computation grows linearly with the time required for verifying $R(x, w)$, which can be an arbitrary polynomial in $|w|$.

---

[1] Similarly, FHE does not yield a non-interactive solution to the motivating problem described in the beginning of the introduction.

Moreover, there seems to be no obvious way to reduce this computational cost using offline preprocessing. Our results yield offline/online NIZK proofs with online proof length of $|w| + \text{poly}(\kappa)$ bits as before, but where the prover's online computation is nearly linear in $|w| + |x|$. This is done as follows. The common reference string of the NIZK defines a function f which maps w (along with a short seed which generates the prover's secret randomness) into a NIZK proof π. Applying our offline/online REs to this f yields the desired result. We note that while the length of NIZK *arguments* can be made sublinear in $|w|$ (under nonstandard but plausible assumptions), breaking this barrier in the case of *proofs* seems highly unlikely [22].

1.3   Techniques

We briefly sketch some of the ideas used to prove Theorem 1. Our starting point is a standard garbled-circuit based encoding, such as the one from [2]. In the offline phase of this encoding, we garble the circuit f and prepare, for each input i, a pair of random secret keys (K_i^0, K_i^1). In the online phase, for each i, we use the i-th bit of x to select a key $K_i^{x_i}$ and output the selected keys. In order to reduce the online complexity of the encoding, we would like to have a compact way to reveal the selected keys. Let us consider the following "riddle" which is a slightly simpler version of this problem. In the offline phase, Alice has n vectors $M_1, \ldots, M_n \in \{0,1\}^k$. She is allowed to send Bob a long encrypted version of these vectors. Later, in the online phase, she receives a bit vector $x \in \{0,1\}^n$. Her goal is to let Bob learn only the vectors which are indexed by x, i.e., $\{M_i\}_{i:x_i=1}$ while sending only a single message of length $O(n)$ bits (or even $n + \kappa$ bits).[2]

Before solving the riddle, let us further reduce it to an algebraic version in which Alice wants to reveal a 0-1 linear combination of the vectors which are indexed by x. Observe that if we can solve the new riddle with respect to nk-bit vectors $T = (T_1, \ldots, T_n)$, then we can solve the original riddle with k-bit vectors $(M_1, \ldots, M_n)$. This is done by placing the M_i's in the diagonal of T, i.e., T_i is partitioned to k-size blocks with M_i in the i-th block and zero elsewhere. In this case, Tx simply "packs" the vectors $\{M_i\}_{i:x_i=1}$.

It turns out that the linear version of the riddle can be efficiently solved via the use of a symmetric-key encryption scheme with some (additive) homomorphic properties. Specifically, let (E, D) be a symmetric encryption scheme with both key homomorphism and message homomorphism as follows: A pair of ciphertexts $\mathsf{E}_k(x)$ and $\mathsf{E}_{k'}(x')$ can be mapped (without any knowledge of the secret keys) to a new ci-pheretxt of the form $\mathsf{E}_{k+k'}(x+x')$. Given such a primitive the answer to the riddle is easy: Alice encrypts each vector under a fresh key K_i and publishes the ciphertexts C_i. At the online phase Alice sends the sum of keys $K_x = \sum K_i x_i$ together with the indicator vector x. Now Bob can easily construct $C = \mathsf{E}_{K_x}(Mx)$ by combining the ciphertexts indexed by x and, since K_x is known, Bob can decrypt the result.

---

[2] The main difference between the riddle and the garbled-circuit problem is that in the latter case, the vector x itself should remain hidden; this gap is bridged by permuting the pairs and randomizing the vector x; see Section 4.

Intuitively, Bob learns nothing about a column M_j which is not indexed by x as the online key K_x is independent of the j-th key. Our DDH and LWE based solutions are based on (approximate) implementations of this primitive. (A somewhat different approach is used in the RSA-based construction).

The arithmetic setting is more challenging. Here, instead of computing the selection function, we should compute an affine function $Mx + v$ over the integers or over $\mathbb{Z}_p$, for some large integer p (not necessarily a prime). While it is possible to solve this via a similar encryption scheme with (stronger) additive homomorphism, there are several technical problems. Typically, all (or most) of the coordinates of x are non-zero and so we should argue that given K_x the secrecy of the key K_i was not compromised, despite the fact that K_i may participate in the linear combination K_x. This translates to some form of security under Related-Key attacks. In addition, it is harder to achieve homomorphism for integers or over $\mathbb{Z}_p$ directly, and so one should somehow embed this domain in a larger, less "friendly", message space. Still, it turns out that a variant of this gadget can be implemented based on the LWE assumption. Specifically, we use the following variant of the key-shrinking gadget of [5] (which was originally introduced as a tool for garbling arithmetic circuits). Intuitively, we create a noisy version $\hat{M}$ and $\hat{v}$ of the matrix M and the vector v, and then plant them in a random linear space W of a low dimension κ over $\mathbb{Z}_q$ (where $q \gg p$). The space W is made public. Now every linear combination of $\hat{M}$ and $\hat{v}$ lies in W, and so it can be succinctly described by its coefficients with respect to W. In particular, to reveal the output $Mx + v$, it suffices for the encoding to reveal the coefficients of its representation $\hat{M}x + \hat{v}$. The security of the construction follows from the LWE assumption.

Concurrent and Subsequent Works. The recent works [24,23] gives the first *reusable* construction of garbled circuits. This implies REs in which a single offline computation can support an arbitrary polynomial number of efficient online computations. The question of optimizing the online rate of reusable garbled circuits remains open. On a different front, improvements in the size of garbled circuits for uniform Turing Machine or RAM computations were recently given in [35,23]. These lead to REs with succinct offline outputs. Our construction can be applied on top of these constructions, yielding REs with an online output of size $n + o(n)$, nearly linear online computation, and offline outputs that are only longer by an additive term of $O(n^\varepsilon \cdot T)$ than those in [35,23], where T is the online computational complexity of the original constructions.

Organization. Section 2 gives the necessary background on randomized encodings. In Section 3, we present several constructions of succinct randomized encodings for a concrete boolean function called the subset function (SF). Later, in Section 4, we use these encodings as a building block and obtain succinct encodings for general boolean functions. In Section 5, we sketch the application of succinct randomized encodings to secure multiparty computation (MPC). Applications related to non-interactive zero-knowledge proofs (NIZK), and verifiable computation (VC) in the preprocessing model are deferred to the full version [6], which also contains the construction of succinct encoding for arithmetic formulas, some lower bounds and a detailed treatment of the issue of adaptivity.

2   Randomized Encoding of Functions

Intuitively, a randomized encoding of a function $f(x)$ is a randomized mapping $\hat{f}(x; r)$ whose output distribution depends only on the output of f. We formalize this intuition via the notion of *computationally-private perfectly-correct randomized encoding* (in short RE) from [2]. In the following, we assume that f is defined over $\mathbb{Z}_p^n$ for some integer p (by default $p = 2$), and allow the encoding $\hat{f}$ be defined over a possibly larger alphabet $\mathbb{Z}_q^n$ for $p \leq q$ under the convention that a vector $x \in \mathbb{Z}_p^n$ can be naturally identified with a vector $x \in \mathbb{Z}_q^n$.

Definition 1 (Randomized Encoding (RE)). *Let $p = p(n), q = q(n)$ where $p(n) \leq q(n) \leq 2^{\mathrm{poly}(n)}$ and $\ell = \ell(n), m = m(n), s = s(n) = \mathrm{poly}(n)$ be integer valued functions. We naturally view $\mathbb{Z}_p$ as a subset of $\mathbb{Z}_q$. Let $f : \mathbb{Z}_p^n \to \mathbb{Z}_p^\ell$ be an efficiently computable function. We say that an efficiently computable randomized function $\hat{f} : \mathbb{Z}_q^n \times \{0,1\}^m \to \mathbb{Z}_q^s$ is a perfectly-correct computationally-private randomized encoding of f (in short, RE), if there exist an efficient decoder algorithm Dec and an efficient simulator Sim that satisfy the following conditions:*

- **Perfect correctness.** *For every $x \in \mathbb{Z}_p^n$, $\Pr_r[\mathsf{Dec}(1^n, \hat{f}(x; r)) \neq f(x)] = 0$.*
- (t, ε) **privacy.** *For every sequence $\{x_n\}_n$, where $x_n \in \mathbb{Z}_p^n$, and every $t(n)$-size circuit $\mathcal{A}$*

$$\left| \Pr[\mathcal{A}(\hat{f}(x_n; r)) = 1] - \Pr[\mathcal{A}(\mathsf{Sim}(1^n, f(x_n))) = 1] \right| \leq \varepsilon(n).$$

By default, $t = n^{\omega(1)}$ and $\varepsilon = n^{-\omega(1)}$, i.e., the distributions are computationally indistinguishable (denoted by $\overset{c}{\equiv}$). The encoding is statistically secure if t is unbounded and perfectly secure if, in addition, $\varepsilon = 0$.

Remarks

- (Security parameter.) The above definition uses n both as an input length parameter and as a cryptographic "security parameter" quantifying computational privacy. When describing our constructions, it will be convenient to use a separate parameter κ for the latter, where computational privacy will be guaranteed as long as $\kappa \geq n^\varepsilon$ for some constant $\varepsilon > 0$.
- (Collections) Let $\mathcal{F}$ be a collection of functions with an associated representation (by default, a boolean or arithmetic circuit). We say that a class of randomized functions $\hat{\mathcal{F}}$ is an RE of $\mathcal{F}$ if there exists an efficient algorithm (compiler) which gets as an input a function $f \in \mathcal{F}$ and outputs (in time polynomial in the representation length $|f|$) three circuits ($\hat{f} \in \hat{\mathcal{F}}, \mathsf{Dec}, \mathsf{Sim}$) which form a $(t = n^{\omega(1)}, \varepsilon = n^{-\omega(1)})$-RE of f.

2.1 Efficiency Measures

So far the notion of RE can be trivially satisfied by taking $\hat{f} = f$ and letting the simulator and decoder be the identity functions. To make the definition non-trivial, we should impose some efficiency constraint. In this work, our main measure of efficiency is online complexity.

Online/Offline Complexity. We would like to measure separately the complexity of the outputs of $\hat{f}$ which depend solely on r (*offline* part) from the ones which depend both on x and r (*online* part). Without loss of generality, we assume that $\hat{f}$ can be written as $\hat{f}(x;r) = (\hat{f}_{off}(r), \hat{f}_{on}(x;r))$, where $\hat{f}_{off}(r)$ does not depend on x at all. The *online communication complexity* (resp., *online computational complexity*) of $\hat{f}$ is the bit-length (resp., the time complexity) of $\hat{f}_{on}(x;r)$. Similarly, the *offline communication complexity* (resp., *offline computational complexity*) of $\hat{f}$ is the bit-length (resp., the time complexity) of $\hat{f}_{off}(r)$. The *rate* of $\hat{f}$ is ρ if the online communication complexity is at most ρ-times larger than the bit-length $n \log p$ of the input of the encoded function f.

Efficient Online Encodings. Let $\hat{\mathcal{F}}$ be an encoding of the collection $\mathcal{F}$. We say that $\hat{\mathcal{F}}$ is *online-efficient* if for every function $f \in \mathcal{F}$, the online computational complexity of the encoding $\hat{f}$ is *independent* of the computational complexity (i.e., circuit size) of the encoded function f (but grows with the bit-length of the input of f). The encoding is *online-succinct* (or simply succinct) if, in addition to being online efficient, every $f \in \mathcal{F}$ is encoded by a $1 + o(1)$-rate encoding.

Remark 1 (Online Inputs). In some applications, it is natural to think of the encoded function f as having online inputs x_{on} and offline inputs x_{off}. In this case, we measure the online commuincation/computational complexity of the encoding $\hat{f}$ with respect to the outputs that depend on x_{on}. By default, we simply assume that all the input x is an online input and there is no offline part.

Some of the applications of REs further require some form of algebraic simplicity; this is captured by the notion of affinity.

Affine RE. We say that an encoding $\hat{f} : \mathbb{Z}_q^n \times \{0,1\}^m \to \mathbb{Z}_q^s$ is an *affine randomized encoding* (ARE) if, for every fixing of the randomness r, the online part of the encoding $\hat{f}_{on}(x;r)$ becomes an affine function over the ring $\mathbb{Z}_q$, i.e., $\hat{f}_{on}(x;r) = M_r \cdot x + v_r$, where M_r (resp., v_r) is a matrix (resp., vector) that depends on the randomness r. It will sometimes be the case that certain outputs of $\hat{f}$ are restricted to an interval $[0, q']$ in $\mathbb{Z}_q$. Each such entry will only contribute $\lceil \log_2 q' \rceil$ towards computing the rate.

Remark 2 (ARE vs. DARE). Previous works considered a stronger form of affinity called *decomposable affine randomized encoding* (DARE).[3] Decomposability requires that each output of $\hat{f}$ depends on a single deterministic input x_i. Hence, a decomposable affine randomized encoding can be written as

---

[3] In fact, in the conference version of [5] the term ARE was used to denote DARE.

$\hat{f}(x;r) = (\hat{f}_{\mathsf{off}}(r), \hat{f}_1(x_1;r), \ldots, \hat{f}_n(x_n;r))$ where each function $\hat{f}_i$ is affine with respect to x_i. It is known how to convert an ARE to DARE, however, the known transformation introduces a non-constant $(O(n))$ multiplicative blow-up in the online communication complexity. In the full version, we show that this is inherent and decomposability *cannot* be achieved with constant rate.

*Remark 3 (**On Adaptive Security**).* In the online/offline model, it is natural to ask if the encoding can be *adaptively* secure, namely, if security holds when the online input x is chosen based on the offline part of the encoding. In the full version, we show that, in the standard model, adaptively secure REs cannot be online-efficient, let alone have constant rate (assuming the existence of one-way functions). On the other hand, it turns out that this barrier can be bypassed via the use of a (programmable) random oracle.

3 Succinct AREs for the Subset Function

In order to succinctly encode boolean circuits, we will need a succinct encoding for the following concrete function g, called the *Subset Function*. It has length parameter n and message size κ and is defined by

$$g(M, x) = ((M_i)_{i \in x}, x),$$

where $M = (M_1, \ldots, M_n) \in (\{0,1\}^\kappa)^n$ is a vector of n "messages", and $x \in \{0,1\}^n$ is a selection vector which is viewed as the set $\{i : x_i = 1\}$. (The latter convention will be implicit through the whole section.) Our goal is to encode g by an RE of the form $\hat{g}(M, x; r) = (\hat{g}_{\mathsf{off}}(M;r), x, K(x;r))$ where $K(x;r)$ is of bit-length κ^c for some universal constant c. Security will hold as long as n is bounded by some arbitrary polynomial in κ whose degree may be *independent* of the constant c. We will construct such an encoding based on several assumptions. Specifically, we will show that such an encoding can be based on a special form of symmetric-key encryption with additive homomorphism which, in turn, can be constructed under the DDH assumption or the LWE assumption. In the full version, we also present a direct encoding (which does not go through the additive homomorphism) under the RSA assumption.

3.1 Encoding the Subset Function via Additive Homomorphism

Definition 2 (Additive Homomorphic Encryption (AHE)). *An additive homomorphic Encryption is a triple of efficient algorithms* (Setup, E, D) *for which the following hold:*

- **Syntax:** *The randomized algorithm* Setup *takes a length parameter* 1^κ *and outputs a string* param *which specifies four (additive) groups: key-space* $\mathcal{K}$, *message-space* $\mathcal{M}$, *ciphertext-space* $\mathcal{C}$ *and public randomness space* $\mathcal{W}$. *We assume that* κ-bit *strings can be efficiently embedded in* $\mathcal{M}$ *and denote the identity element of* $\mathcal{M}$ *by* $\mathbf{0}$. *The input to the encryption and decryption* *algorithms consist of a message/ciphertext, a key K, some private random-ness, and some public randomness $W \overset{R}{\leftarrow} \mathcal{W}$ which is selected during the encryption. Both algorithms also depend on the string* param. *(We make this dependency implicit to simplify notation.)*

- **Semantic Security:** *Let* param $= (\mathcal{K}, \mathcal{M}, \mathcal{C}, \mathcal{W}) \overset{R}{\leftarrow}$ Setup(1^κ). *For every $n = \text{poly}(\kappa)$ and every n-tuple of messages $M_1, \ldots, M_n \in \mathcal{M}$, we have that*

$$\left(\text{param}, (W_i, \mathsf{E}_K(M_i; W_i))_{i \in [n]}\right) \overset{c}{\equiv} \left(\text{param}, (W_i, \mathsf{E}_K(\mathbf{0}; W_i))_{i \in [n]}\right),$$

where $W_i \overset{R}{\leftarrow} \mathcal{W}$, $K \overset{R}{\leftarrow} \mathcal{K}$, and indistinguishability is parameterized by κ.

- **Additive Homomorphism:** *For every $n = \text{poly}(\kappa)$ and every n-tuple of keys $K_1, \ldots, K_n \in \mathcal{K}$, n-tuple of messages $M_1, \ldots, M_n \in \mathcal{M}$, and public randomness $W \in \mathcal{W}$, we have that*

$$\mathsf{D}_{\sum_i K_i} \left(\sum_i E_{K_i}(M_i; W); W \right) = \sum_i M_i,$$

where sums are computed over the corresponding groups. In fact, it suffices to have a relaxed form of additive homomorphism which holds in the special case where all messages, except for one, equal to $\mathbf{0} \in \mathcal{M}$.

The definition implies that the key size is *independent* of the homomorphism parameter n. This will be crucial for our applications. As a concrete example of AHE consider the following symmetric-key version of ElGamal encryption. Let $\mathcal{M} = \mathcal{C} = \mathcal{W}$ equal to a cyclic group $\mathbb{G}$ of prime order p and let $\mathcal{K} = \mathbb{Z}_p$. Using the standard multiplicative notation, encryption is defined by $\mathsf{E}_K(M; W) = W^K \cdot M$ and decryption by $\mathsf{D}_K(C; W) = C/W^K$. It is not hard to show that if the DDH assumption holds in $\mathbb{G}$ then the scheme is an AHE with relaxed homomorphism. (More details about this implementation, as well as a description of an analogous implementation under LWE appear in the full version.) We show how to encode the subset function $g(M, x)$ with length n and message size κ based on AHE.

Lemma 1. *Assume that AHE exists. Then the Subset Function $g(M, x)$, where $M \in (\{0,1\}^\kappa)^n, x \in \{0,1\}^n$, has an encoding*

$$\hat{g}(M, x; r) = (\hat{g}_{\text{off}}(M; r), x, \sum_{i \in x} K_i(r)),$$

where $\hat{g}_{\text{off}}$ outputs $O(n^2)$ ciphertexts in $\mathcal{C}$, the functions K_i output an element in $\mathcal{K}$, and the sum is computed over the key-space $\mathcal{K}$.

Proof. At the offline phase, we invoke Setup(1^κ) and obtain a specification param of $\mathcal{K}$, $\mathcal{M}$, $\mathcal{C}$ and $\mathcal{W}$. We encode each entry of the offline input $M = (M_1, \ldots, M_n)$ by an element of $\mathcal{M}$, and from now on identify M_i with its encoding. We define a diagonal $n \times n$ matrix $\{M_{i,j}\}$ whose diagonal equals to the message vector M, i.e., $M_{i,i} = M_i, \forall i \in [n]$ and $M_{i,j} = \mathbf{0}, \forall i \neq j$. Next, we select a tuple of public random elements $W = (W_1, \ldots, W_n) \overset{R}{\leftarrow} \mathcal{W}^n$, a tuple of random keys $K = (K_1, \ldots, K_n) \overset{R}{\leftarrow} \mathcal{K}^n$ and compute a matrix of "ciphertexts" $C = (C_{i,j}) \in \mathcal{C}^{n \times n}$, where $C_{i,j} = \mathsf{E}_{K_i}(M_{i,j}; W_j)$. The output of $\hat{g}_{\mathsf{off}}$ consists of the tuple (param, W, C) and the online part $\hat{g}_{\mathsf{on}}$ consists of the pair $(x, K_x = \sum_{i \in x} K_i)$.

Decoding. Given $(\mathsf{param}, W, C, x, K_x)$, we decode $(M_i)_{i \in x}$ by exploiting the homomorphism property of the above encryption. Namely, for each $j \in x$ we compute

$$Y_j = \sum_{i \in x} C_{i,j} = \sum_{i \in x} \mathsf{E}_{K_i}(M_{i,j}; W_j),$$

and output the value $\mathsf{D}_{K_x}(Y_j; W_j)$.

Simulation. For $\ell = 0, \ldots, n$ define the hybrid $H_\ell(M, x)$ exactly as in $\hat{g}$ except that

$$M_{i,i} = \begin{cases} M_i & \text{if } i < \ell \text{ or } i \in x, \\ \mathbf{0} & \text{otherwise} \end{cases}$$

The first hybrid H_0 can be sampled based on $((M_i)_{i \in x}, x)$, and so it is being used as the simulator. The last hybrid H_n corresponds to the distribution of the encoding $\hat{g}$. Hence, by a standard argument, it suffices to show that each pair of neighboring hybrids is computationally indistinguishable. Assume, towards a contradiction, that $\mathcal{A}$ distinguishes the hybrid $H_{\ell-1}$ from H_ℓ with non-negligible advantage δ. Observe that in this case $x_\ell = 0$, as otherwise the two hybrids are identically distributed. We construct a new adversary $\mathcal{B}$ that breaks the semantic security of the scheme. Given a challenge $(\mathsf{param}, \boldsymbol{w}, \boldsymbol{c})$ where $\mathsf{param} \overset{R}{\leftarrow} \mathsf{Setup}(1^\kappa)$ and $\boldsymbol{w} = (w_1, \ldots, w_n) \overset{R}{\leftarrow} \mathcal{W}^n$, the adversary $\mathcal{B}$ distinguishes between $\boldsymbol{c} \overset{R}{\leftarrow} (\mathsf{E}_K(\mathbf{0}; w_1), \ldots, \mathsf{E}_K(\mathbf{0}; w_n))$ and $\boldsymbol{c} \overset{R}{\leftarrow} (\mathsf{E}_K(\mathbf{0}; w_1), \ldots, \mathsf{E}_K(M_\ell; w_\ell), \ldots, \mathsf{E}_K(\mathbf{0}; w_n)))$ as follows. Use param to compute the hybrid $H_{\ell-1}$ where the public randomness $W_1, \ldots, W_n$ is set to $\boldsymbol{w}$, and the ℓ-th row of the ciphertext matrix C takes the value $\boldsymbol{c}$. It is not hard to verify that the resulting distribution is identical to $H_{\ell-1}$ if $\boldsymbol{c} \overset{R}{\leftarrow} (\mathsf{E}_K(\mathbf{0}; w_1), \ldots, \mathsf{E}_K(\mathbf{0}; w_n))$, and to H_ℓ if $\boldsymbol{c} \overset{R}{\leftarrow} (\mathsf{E}_K(\mathbf{0}; w_1), \ldots, \mathsf{E}_K(M_\ell; w_\ell), \ldots, \mathsf{E}_K(\mathbf{0}; w_n)))$, and the claim follows. $\qquad\square$

Complexity. To encode the online part, one has to compute n additions (over the key space) and send x together with a single key element. The cost of the offline part is n^2 encryptions/ciphertexts. One can obtain a smooth tradeoff between the offline part and the online part by partitioning the inputs to blocks (see full version). Also note that decoding costs n^2 additions over the key space (which can be reduced via the previous optimization) and n decryption operations. Finally, we mention that in our RSA-based solution the offline complexity is only linear in n but quadratic in κ. (The latter can be improved assuming subexponential hardness of RSA.)

4   Succinct AREs for Boolean Circuits

In this section, we will encode any efficiently computable function via a succinct encoding. We begin by showing that if $F : \{0,1\}^n \to \{0,1\}^\ell$ has a decomposable affine randomized encoding (DARE) then it also has a succinct encoding. In the following, let κ be a security parameter which is polynomially related to n, i.e., $\kappa = n^\delta$ for some fixed $\delta > 0$. We will employ a succinct encoding for the subset function $g(M, \hat{x})$ with length $N = 2n$ and message size κ. We will also make use of the following simple observation: if a $\kappa \times 2n$ matrix M is composed of n pairs of columns $(M_{2i-1}|M_{2i}) = (v_i^0, v_i^1)_{i \in [n]}$, then for any $x \in \{0,1\}^n$ the sub-matrix $(v_i^{x_i})_{i \in [n]}$ can be written as $(M_i)_{i \in \text{pad}(x)}$, where $\text{pad}(x)$ maps an n-bit vector x to the $2n$-bit vector $(1 - x_1, x_1, \ldots, 1 - x_n, x_n)$, and $i \in \text{pad}(x)$ if $\text{pad}(x)_i = 1$.

Lemma 2. *Let* $F : \{0,1\}^n \to \{0,1\}^\ell$ *be an efficiently computable function having a decomposable ARE* $f(x; \rho) = (f_{\text{off}}(\rho), f_1(x_1; \rho), \ldots, f_n(x_n; \rho))$, *where the output length of each* f_i *is* κ *bits. Also, assume that the subset function* $g(M, \hat{x})$ *with length* $2n$ *and message size* κ *has an RE of the form* $\hat{g}(M, \hat{x}; r) = (\hat{g}_{\text{off}}(M; r), \hat{x}, K(\hat{x}; r))$. *Then,* F *is encoded by the randomized function*

$$\hat{F}(x; \rho, s, r) = \left(f_{\text{off}}(\rho), \hat{g}_{\text{off}}(M; r), x \oplus s, K(\text{pad}(x \oplus s); r)\right),$$

where

$$M = (f_1(s_1; \rho)|f_1(s_1 \oplus 1; \rho)|\cdots|f_n(s_n; \rho)|f_n(s_n \oplus 1; \rho)) \in \{0,1\}^{\kappa \times 2n}.$$

Proof. It will be useful to start by encoding the n-wise one-out-of-two selection function H which maps an online input $x \in \{0,1\}^n$ and an offline matrix of pairs $V = (v_1^0|v_1^1|\ldots|v_n^0|v_n^1) \in \{0,1\}^{\kappa \times 2n}$ to the tuple $(v_i^{x_i})_{i \in [n]}$. Observe that the output of H is essentially the value of the subset function g applied to the matrix V and the vector $\text{pad}(x) \in \{0,1\}^{2n}$, except that H hides x whereas g reveals it. Nevertheless one can easily randomize x and then employ the subset function. Specifically, select a random mask $s \xleftarrow{R} \{0,1\}^n$, let $\hat{x} \in \{0,1\}^{2n}$ be the vector $\text{pad}(x \oplus s)$, and construct the $\kappa \times 2n$ matrix $M = (v_1^{s_1}|v_1^{s_1 \oplus 1}|\ldots|v_n^{s_n}|v_n^{s_n \oplus 1})$. It is not hard to show that the randomized mapping $h(V, x; s) \mapsto g(M, \hat{x})$ is an encoding of H. Indeed, the output distribution of $g(M, \hat{x})$ consists of the matrix $(M_i)_{i \in \hat{x}}$ and the vector $\hat{x}$ — the former simply equals to $(v_i^{x_i})_{i \in [n]}$ and the latter is just a sequence of n pairs of a random bit and its complement.

Next, let us view h as a deterministic function of V, x and s. Since h can be written as $g(M_{V,s}, \hat{x}_{x,s})$, we can encode h by the mapping $\hat{g}(M_{V,s}, \hat{x}_{x,s}; r)$. It is not hard to show that the latter encoding also encodes H. Overall, our encoding for $H(V, x)$ is defined as follows:

$$(V, x; s, r) \mapsto (\hat{g}_{\text{off}}(M_{V,s}; r), \text{pad}(x \oplus s), K(\text{pad}(x \oplus s); r)).$$

To improve the online complexity, we replace the redundant value $\text{pad}(x \oplus s)$, which is sent in the clear, with $x \oplus s$.

We can now prove the lemma. Let us view ρ as a deterministic input and encode the deterministic function $f(x, \rho)$. Since f is decomposable, we can write it as

$$(f_{\mathsf{off}}(\rho), H(V_\rho, x)), \quad \text{where } V_\rho = (f_1(0; \rho)| \ldots |f_n(0; \rho)|f_1(1; \rho)| \ldots |f_n(1; \rho))$$

and H is the n-wise one-out-of-two selection function. Using appropriate substitution and concatenation lemmas (see full version), it can be shown that f is encoded by $(f_{\mathsf{off}}(\rho), \hat{h}(V, x; s, r))$, where $\hat{h}$ encodes H. Plugging in our (improved) encoding of H, we obtain an encoding of the form

$$\hat{f}(x, \rho; s, r) = (f_{\mathsf{off}}(\rho), \hat{g}_{\mathsf{off}}(M_{s,\rho}; r), x \oplus s, K(\mathrm{pad}(x \oplus s); r)).$$

Finally, a similar (composition) argument shows that the function $\hat{f}(x; \rho, s, r)$ encodes $F(x)$ and the lemma follows.   □

It follows that F has an encoding with online complexity of $n + \mathsf{Len}(K)$, online computational complexity of $O(n + \mathsf{Comp}(K))$, and offline computational complexity of $\mathsf{Comp}(f_{\mathsf{off}}) + \mathsf{Comp}(\hat{g}_{\mathsf{off}})$, where $\mathsf{Comp}(\cdot)$ and $\mathsf{Len}(\cdot)$ measure the computational complexity (circuit size), and the output length (in bits) of a given function. Furthermore, observe that for every fixed randomness s each bit of the term $\mathrm{pad}(x \oplus s)$ can be written as x_i or as $1 - x_i$ and so if $K(\hat{x}; r)$ is affine (over some ring) then so is $\hat{F}_{\mathsf{on}}$.

In [2] it is shown that, assuming the existence of one-way functions, any efficiently computable function $F(x)$ can be encoded by a decomposable ARE $f(x; \rho) = (f_{\mathsf{off}}(\rho), f_1(x_1; \rho), \ldots, f_n(x_n; \rho))$, where the output length of the f_i's is κ bits, and the computational complexity of f_{off} is $\kappa \cdot \mathsf{Comp}(f)$. Combining this with Lemma 2 and our encodings for the Subset Function, we derive succinct encodings for general boolean functions. By using an optimized version of Lemma 1 (which encodes the subset function in blocks), we can do this while keeping the online computational complexity asymptotically "almost linear", as in the following theorem whose proof is deferred to the full version.

Theorem 2 (Theorem 1 Restated). *Assume that the DDH assumption, or LWE assumption or the RSA assumption holds. Let $\varepsilon > 0$ be an arbitrary constant. Then, every efficiently computable function $F : \{0,1\}^n \to \{0,1\}^{\ell(n)}$ has an encoding $\hat{F}$ with online communication of $n + o(n)$, online computational complexity of $O(n^{1+\varepsilon})$, and offline computational/communication complexity of $O(n^\varepsilon \mathsf{Comp}(F))$. Furthermore, in the case of LWE and DDH the encoding is affine.*

5   MPC with Optimal Online Communication

In this section, we sketch the application of succinct randomized encodings to secure multiparty computation (MPC) in the preprocessing model. We start with the two-party case, and later generalize to the multiparty case. For concreteness, we focus on distributing the DDH-based encoding obtained by combining

Lemmas 1 and 2 with the DDH-based AHE. Similar protocols can be obtained based on any succinct *Affine* RE. We do not know how to get similar results from general (non-affine) succinct REs.

Let F be a deterministic two-party functionality which takes an input $a \in \{0,1\}^{n_a}$ from Alice and an input $b \in \{0,1\}^{n_b}$ from Bob, and delivers an output c to Alice.[4] The DDH-based encoding of F can be written as

$$\hat{F}(a,b;R) = (\hat{F}_{\mathsf{off}}(R), \ a \oplus r^a, \ b \oplus r^b, \ \sum_{i=1}^{n_a} K^A_{i,a_i \oplus r^a_i} + \sum_{i=1}^{n_b} K^B_{i,b_i \oplus r^b_i} \mod p),$$

where the "masks" $r^a \in \{0,1\}^{n_a}$, $r^b \in \{0,1\}^{n_b}$, and the "keys" $K^A_{i,\sigma}, K^B_{i,\sigma} \in \mathbb{Z}_p$ are random and independent of a, b (these values are given as part of R).

In the semi-honest model, the protocol is straightforward. In the offline phase, a trusted party samples R and sends the value $\hat{F}_{\mathsf{off}}(R)$ together with the mask r^a to Alice, and the mask r^b along with the $2n_a + 2n_b$ keys $K^A_{i,\sigma}, K^B_{i,\sigma}$ to Bob. (Of course, in the real world, this step is implemented via the use of any off-the-shelf secure two-party protocol.) In the online phase, Alice sends to Bob $a \oplus r^a$ and Bob replies with $b \oplus r^b$ and $\sum_{i=1}^{n_a} K^A_{i,a_i \oplus r^a_i} + \sum_{i=1}^{n_b} K^B_{i,b_i \oplus r^b_i} \mod p$.

Alice computes the output using the decoder of $\hat{F}$. Note that the view of Bob is completely random, whereas the view of Alice contains the output of $\hat{F}$ which can be simulated given $F(a,b)$. This proves the following:

Theorem 3. *Suppose that the DDH assumption holds in a prime order group of size $p = p(\kappa)$. Let $F(a,b)$ be a polynomial-time computable functionality which delivers its output to Alice. Assume trusted preprocessing which does not depend on the inputs. Then, F can be securely realized in the semi-honest model by a protocol in which Alice sends a message of length $|a|$ and Bob sends a message of length $|b| + \lceil \log p \rceil$, independently of the length of the output or the complexity of F.*

In the full version [6], we describe an efficient extension of this protocol to the malicious model and to the multiparty model, and discuss the issue of adaptive security. Applications related to NIZKs and verifiable computation are also deferred to the full version.

Acknowledgements. The first author was supported by Alon Fellowship, ISF grant 1155/11, Israel Ministry of Science and Technology (grant 3-9094), and GIF grant 1152/2011. The second author was supported by the European Research Council as part of the ERC project CaC (grant 259426). The third author was supported by ISF grant 1361/10 and BSF grant 2008411. The fourth author was supported by NSF grants CNS-0915361 and CNS-0952692, AFOSR Grant No: FA9550-08-1-0352, DARPA through the U.S. Office of Naval Research under Contract N00014-11-1-0382, DARPA N11AP20006, Google Faculty Research award,

---

[4] The case of general two-party functionalities reduces to this case via a standard reduction, cf. [21].

the Alfred P. Sloan Fellowship, and Microsoft Faculty Fellowship, and Packard Foundation Fellowship. Any opinions, findings, and conclusions or recommendations expressed in this material are those of the author(s) and do not necessarily reflect the views of the Department of Defense or the U.S. Government.

References

1. Applebaum, B.: Key-dependent message security: Generic amplification and completeness. In: Paterson, K.G. (ed.) EUROCRYPT 2011. LNCS, vol. 6632, pp. 527–546. Springer, Heidelberg (2011)
2. Applebaum, B., Ishai, Y., Kushilevitz, E.: Computationally private randomizing polynomials and their applications. Computational Complexity 15(2), 115–162 (2006)
3. Applebaum, B., Ishai, Y., Kushilevitz, E.: Cryptography in NC^0. SIAM J. Comput. 36(4), 845–888 (2006)
4. Applebaum, B., Ishai, Y., Kushilevitz, E.: From secrecy to soundness: Efficient verification via secure computation. In: Abramsky, S., Gavoille, C., Kirchner, C., Meyer auf der Heide, F., Spirakis, P.G. (eds.) ICALP 2010, Part I. LNCS, vol. 6198, pp. 152–163. Springer, Heidelberg (2010)
5. Applebaum, B., Ishai, Y., Kushilevitz, E.: How to garble arithmetic circuits. In: FOCS, pp. 120–129 (2011)
6. Applebaum, B., Ishai, Y., Kushilevitz, E., Waters, B.: Encoding functions with constant online rate or how to compress garbled circuits keys. Cryptology ePrint Archive, Report 2012/693 (2012), http://eprint.iacr.org/
7. Barak, B., Haitner, I., Hofheinz, D., Ishai, Y.: Bounded key-dependent message security. In: Gilbert, H. (ed.) EUROCRYPT 2010. LNCS, vol. 6110, pp. 423–444. Springer, Heidelberg (2010)
8. Beaver, D.: Precomputing oblivious transfer. In: Coppersmith, D. (ed.) CRYPTO 1995. LNCS, vol. 963, pp. 97–109. Springer, Heidelberg (1995)
9. Bellare, M., Hoang, V.T., Rogaway, P.: Adaptively secure garbling with applications to one-time programs and secure outsourcing. In: Wang, X., Sako, K. (eds.) ASIACRYPT 2012. LNCS, vol. 7658, pp. 134–153. Springer, Heidelberg (2012), http://eprint.iacr.org/2012/564
10. Bellare, M., Hoang, V.T., Rogaway, P.: Foundations of garbled circuits. In: ACM Conference on Computer and Communications Security, pp. 784–796 (2012), Full version: http://eprint.iacr.org/2012/265
11. Bendlin, R., Damgård, I., Orlandi, C., Zakarias, S.: Semi-homomorphic encryption and multiparty computation. In: Paterson, K.G. (ed.) EUROCRYPT 2011. LNCS, vol. 6632, pp. 169–188. Springer, Heidelberg (2011)
12. Boneh, D., Sahai, A., Waters, B.: Functional encryption: Definitions and challenges. In: Ishai, Y. (ed.) TCC 2011. LNCS, vol. 6597, pp. 253–273. Springer, Heidelberg (2011)
13. Cachin, C., Camenisch, J., Kilian, J., Müller, J.: One-round secure computation and secure autonomous mobile agents. In: Welzl, E., Montanari, U., Rolim, J.D.P. (eds.) ICALP 2000. LNCS, vol. 1853, pp. 512–523. Springer, Heidelberg (2000)
14. Chung, K.-M., Kalai, Y., Vadhan, S.: Improved delegation of computation using fully homomorphic encryption. In: Rabin, T. (ed.) CRYPTO 2010. LNCS, vol. 6223, pp. 483–501. Springer, Heidelberg (2010)

15. Damgard, I., Pastro, V., Smart, N., Zakarias, S.: Multiparty computation from somewhat homomorphic encryption. Cryptology ePrint Archive, Report 2011/535 (2011), http://eprint.iacr.org/

16. Dubrov, B., Ishai, Y.: On the randomness complexity of efficient sampling. In: STOC, pp. 711–720 (2006)

17. Even, S., Goldreich, O., Micali, S.: On-line/off-line digital signatures. J. Cryptology 9 (1996)

18. Feige, U., Kilian, J., Naor, M.: A minimal model for secure computation. In: STOC (1994)

19. Gennaro, R., Gentry, C., Parno, B.: Non-interactive verifiable computing: Outsourcing computation to untrusted workers. In: Rabin, T. (ed.) CRYPTO 2010. LNCS, vol. 6223, pp. 465–482. Springer, Heidelberg (2010)

20. Gentry, C.: Fully homomorphic encryption using ideal lattices. In: STOC, pp. 169–178 (2009)

21. Goldreich, O.: Foundations of Cryptography. Basic Applications, vol. 2. Cambridge University Press, New York (2004)

22. Goldreich, O., Vadhan, S.P., Wigderson, A.: On interactive proofs with a laconic prover. Computational Complexity 11(1-2), 1–53 (2002)

23. Goldwasser, S., Kalai, Y.T., Popa, R.A., Vaikuntanathan, V., Zeldovich, N.: Overcoming the worst-case curse for cryptographic constructions. IACR Cryptology ePrint Archive, 2013:229 (2013)

24. Goldwasser, S., Kalai, Y.T., Popa, R.A., Vaikuntanathan, V., Zeldovich, N.: Reusable garbled circuits and succinct functional encryption. In: STOC, pp. 555–564 (2013)

25. Goldwasser, S., Kalai, Y.T., Rothblum, G.N.: Delegating computation: interactive proofs for muggles. In: STOC (2008)

26. Gorbunov, S., Vaikuntanathan, V., Wee, H.: Functional encryption with bounded collusions via multi-party computation. In: Safavi-Naini, R., Canetti, R. (eds.) CRYPTO 2012. LNCS, vol. 7417, pp. 162–179. Springer, Heidelberg (2012)

27. Goyal, V., Ishai, Y., Sahai, A., Venkatesan, R., Wadia, A.: Founding cryptography on tamper-proof hardware tokens. In: Micciancio, D. (ed.) TCC 2010. LNCS, vol. 5978, pp. 308–326. Springer, Heidelberg (2010)

28. Groth, J.: Minimizing non-interactive zero-knowledge proofs using fully homomorphic encryption. IACR Cryptology ePrint Archive, 2011:12 (2011)

29. Harnik, D., Naor, M.: On the compressibility of NP instances and cryptographic applications. SIAM J. Comput. 39(5), 1667–1713 (2010)

30. Ishai, Y., Kushilevitz, E.: Randomizing polynomials: A new representation with applications to round-efficient secure computation. In: FOCS, pp. 294–304 (2000)

31. Ishai, Y., Kushilevitz, E., Meldgaard, S., Orlandi, C., Paskin-Cherniavsky, A.: On the power of correlated randomness in secure computation. In: Sahai, A. (ed.) TCC 2013. LNCS, vol. 7785, pp. 600–620. Springer, Heidelberg (2013)

32. Ishai, Y., Prabhakaran, M., Sahai, A.: Founding cryptography on oblivious transfer – efficiently. In: Wagner, D. (ed.) CRYPTO 2008. LNCS, vol. 5157, pp. 572–591. Springer, Heidelberg (2008)

33. Kalai, Y.T., Raz, R.: Probabilistically checkable arguments. In: Halevi, S. (ed.) CRYPTO 2009. LNCS, vol. 5677, pp. 143–159. Springer, Heidelberg (2009)

34. Kilian, J.: Founding cryptography on oblivious transfer. In: STOC, pp. 20–31 (1988)

35. Lu, S., Ostrovsky, R.: How to garble RAM programs? In: Johansson, T., Nguyen, P.Q. (eds.) EUROCRYPT 2013. LNCS, vol. 7881, pp. 719–734. Springer, Heidelberg (2013)

36. Micali, S.: CS proofs (extended abstract). In: FOCS, pp. 436–453 (1994)
37. Nielsen, J.B.: Separating random oracle proofs from complexity theoretic proofs: The non-committing encryption case. In: Yung, M. (ed.) CRYPTO 2002. LNCS, vol. 2442, pp. 111–126. Springer, Heidelberg (2002)
38. Sahai, A., Seyalioglu, H.: Worry-free encryption: functional encryption with public keys. In: ACM Conference on Computer and Communications Security, pp. 463–472 (2010)
39. Sander, T., Young, A., Yung, M.: Non-interactive cryptocomputing for NC^1. In: FOCS, pp. 554–567 (1999)
40. Shamir, A., Tauman, Y.: Improved online/offline signature schemes. In: Kilian, J. (ed.) CRYPTO 2001. LNCS, vol. 2139, pp. 355–367. Springer, Heidelberg (2001)
41. Sion, R., Carbunar, B.: On the practicality of private information retrieval. In: NDSS (2007)
42. Yao, A.C.: How to generate and exchange secrets. In: FOCS, pp. 162–167 (1986)

Efficient Multiparty Protocols via Log-Depth Threshold Formulae

(Extended Abstract)

Gil Cohen[1], Ivan Bjerre Damgård[2], Yuval Ishai[3], Jonas Kölker[2],
Peter Bro Miltersen[2], Ran Raz[1], and Ron D. Rothblum[1]

[1] Weizmann Institute, Rehovot, Israel
{gil.cohen,ran.raz,ron.rothblum}@weizmann.ac.il
[2] Aarhus University, Aarhus, Denmark
{ivan,epona,bromille}@cs.au.dk
[3] Technion, Haifa, Israel
yuvali@cs.technion.ac.il

Abstract. We put forward a new approach for the design of efficient multiparty protocols:

1. Design a protocol π for a small number of parties (say, 3 or 4) which achieves security against a *single* corrupted party. Such protocols are typically easy to construct, as they may employ techniques that do not scale well with the number of corrupted parties.
2. Recursively compose π with itself to obtain an efficient n-party protocol which achieves security against a constant fraction of corrupted parties.

The second step of our approach combines the "player emulation" technique of Hirt and Maurer (J. Cryptology, 2000) with constructions of logarithmic-depth formulae which compute threshold functions using only constant fan-in threshold gates.

Using this approach, we simplify and improve on previous results in cryptography and distributed computing. In particular:

- We provide conceptually simple constructions of efficient protocols for Secure Multiparty Computation (MPC) in the presence of an honest majority, as well as broadcast protocols from point-to-point channels and a 2-cast primitive.
- We obtain new results on MPC over blackbox groups and other algebraic structures.

The above results rely on the following complexity-theoretic contributions, which may be of independent interest:

- We show that for every $j, k \in \mathbb{N}$ such that $m \triangleq \frac{k-1}{j-1}$ is an integer, there is an explicit (poly(n)-time) construction of a logarithmic-depth formula which computes a good approximation of an (n/m)-out-of-n threshold function using only j-out-of-k threshold gates and no constants.

R. Canetti and J.A. Garay (Eds.): CRYPTO 2013, Part II, LNCS 8043, pp. 185–202, 2013.

– For the special case of n-bit majority from 3-bit majority gates, a non-explicit construction follows from the work of Valiant (J. Algorithms, 1984). For this special case, we provide an explicit construction with a better approximation than for the general threshold case, and also an *exact* explicit construction based on standard complexity-theoretic or cryptographic assumptions.

1   Introduction

Secure multiparty computation (MPC) enables a set of parties to jointly accomplish some distributed computational task, while maintaining the secrecy of the inputs and the correctness of the outputs in the presence of coalitions of dishonest parties. Originating from the seminal works of [41,25,4,10], secure MPC has been the subject of an enormous body of work.

Despite this body of work, MPC protocols remain quite complicated and their security is difficult to prove. In this work we propose a new general approach to the construction of efficient[1] multiparty protocols in the presence of an honest majority. This approach enables us to obtain conceptually simple derivations of known feasibility results (or slightly weaker variants of such results), and also to obtain new results.

Our approach is inspired by and builds on the "player emulation" technique of Hirt and Maurer [28], who obtain secure MPC protocols by reducing the construction of an n-party protocol to the task of constructing a protocol π for a constant (e.g., three or four) number of parties. The motivation of [28] was to obtain n-party protocols that are secure with respect to general (non-threshold) adversary structures. A disadvantage of their n-party protocols is that their complexity grows exponentially with n. This seems inevitable when considering arbitrary adversary structures.

Our motivation is very different: We would like to use the atomic protocol π for constructing *efficient* n-party protocols in the traditional MPC setting of *threshold* adversary structures. Since π only involves a small number of parties, its design may employ simpler techniques that do not scale well with the number of corrupted parties. Thus, our goal is to simplify the design of efficient n-party protocols by reducing it to the design of a simpler atomic protocol π.

To make the approach of [28] scale with the number of parties, we introduce a new complexity-theoretic primitive: a logarithmic-depth formula[2] which is composed only of constant-size threshold gates and computes an n-input threshold function. The problem of constructing such formulae is closely related to a classical problem in complexity theory. In this work we also make a contribution to this complexity-theoretic problem, which may be of independent interest.

---

[1] Here and throughout this work, by "efficient" we mean polynomial-time in the number of parties and the input size.

[2] A formula is a circuit with fan-out 1. A logarithmic-depth formula (more precisely, infinite family of formulas) is one whose depth is $O(\log n)$, where n is the number of inputs. Throughout this paper we consider only *monotone* formulas without negations or constants.

In addition to providing conceptually simple protocols, our approach is very general and can be applied in a variety of settings and models. In contrast to most traditional MPC protocols, it is not tied to some underlying algebraic structure. We demonstrate this generality by obtaining new results on MPC over black-box groups and other algebraic structures, improving on previous results from the literature.

Before proceeding to describe the details of our approach, we note that the goal of designing MPC protocols whose complexity grows (only) polynomially with the number of parties also has relevance to *two-party* cryptography. Indeed, there are general techniques for applying MPC protocols with security in the presence of an honest majority (where the number of parties grows with the security parameter) towards two-party tasks such as zero-knowledge proofs and secure two-party computation [30,31].

1.1 Our Approach

In the following, for simplicity, we consider the case of perfect security against a *passive* adversary. In this setting, parties are honest but curious. That is, they follow the protocol but may attempt to learn secret information based on what they see. We note that, in contrast to the norm, the extension of this approach to the case of an active adversary is relatively straightforward.

We first give an overview of the player emulation technique of Hirt and Maurer [28] and then proceed to describe how we overcome the exponential blow-up incurred by [28] in the case of threshold adversary structures.

Recall that security of MPC protocols is defined by comparing a real protocol to an ideal protocol, in which, in addition to the parties involved in the computation, there is a trusted party. A protocol is deemed secure if for every adversary in the real protocol controlling a subset of the parties, there is an equivalent adversary controlling the same subset in the ideal protocol.

The technique from [28] is to reduce the design of n-party protocols to the design of protocols that support only 3 parties (the minimal number of parties for perfect security in the passive security model).

We proceed to present an informal description of the reduction. Indeed, suppose that the 3-party case has been solved. That is, for every computational task involving three parties there exists a secure protocol that securely implements this task when at most one of the parties is passively corrupted.[3] We describe how to use this protocol to securely implement computational tasks using a larger number of parties.

Consider n parties that wish to securely accomplish some joint computational task. It is best to think of this task as being specified by an ideal protocol π_0 which involves, in addition to the n parties, a trusted party τ. The ideal protocol is secure (by definition) even if the adversary controls any subset of the parties that does not contain τ.

---

[3] Since we deal with *perfect* security, the size of the secure protocol depends only on the size of the original protocol. In particular, any constant size protocol can be implemented securely in constant size.

Consider a new protocol π_1 that involves the n original parties but where we replace the trusted party τ with three new virtual parties v_1, v_2, v_3. Since in π_0, the trusted party τ is just involved in a computational task, we can use the given 3-party protocol to simulate τ using v_1, v_2, v_3. When is the new protocol π_1 secure? Since π_0 was only insecure whenever the adversary controlled τ and since the 3-party protocol is secure as long as the adversary controls at most one of the virtual parties, π_1 is secure as long as the adversary does not control two or more of the virtual parties.

We continue this process by designing a new protocol π_2 in which the virtual party v_1 is itself simulated by three new virtual parties w_1, w_2, w_3. Since π_1 is only insecure whenever the adversary controls more than one of v_1, v_2, v_3 and since the protocol for emulating v_1 is secure when at most one of w_1, w_2, w_3 is controlled by the adversary, π_2 is secure as long as the adversary does not control either v_2 and v_3 or one of v_2, v_3 and two or more of w_1, w_2, w_3.

We continue in this process simulating virtual parties by more virtual parties. The sets of corrupted parties against which the resulting protocol is secure can be described by looking at a formula composed of 3-input majority gates which we denote by Maj_3. Each wire represents a virtual party. The protocol π_1 can be represented by a simple formula F_1 consisting of a single Maj_3 gate where the three input wires correspond to the virtual parties v_1, v_2, v_3 and the output wire corresponds to τ. We assign to each input wire corresponding to an honest party a value of 0 and a value of 1 to those corresponding to dishonest parties. It can be easily verified that the protocol is secure whenever the formula F_1 evaluates to 0.

Similarly, the protocol π_2 can be represented by a formula F_2 which is constructed from F_1 by connecting the input wire corresponding to v_1 with an additional Maj_3 gate with three new input wires (corresponding to w_1, w_2, w_3). It is easy to verify that the new protocol is secure whenever the formula evaluates to 0.

Suppose that we continue on like this but instead of arbitrarily choosing which virtual party to simulate, we choose it according to some formula F, composed only of Maj_3 gates.[4] Once we reach the input layer of the formula, we associate each input variable to a real party and every remaining virtual party is simulated by the real party associated with the corresponding input wire.

As above, the protocol is secure against every set T of parties on which the formula F evaluates to 0. (Here and in the following we associate a set T with its characteristic vector χ_T.) Thus, to obtain a protocol that is secure for a particular adversary structure, it suffices to provide a formula that evaluates to 0 on all sets in the structure. Since, in contrast to [28], our goal is merely to obtain security in the presence of an honest majority, we need only to construct a formula that computes the majority function (using only Maj_3 gates and no constants).

Such a formula was implicitly constructed by Hirt and Maurer [28] for general Q_2 functions[5] and in particular for majority. Unfortunately, the formula of [28]

---

[4] Actually, [28] do not present their construction in the terminology of Maj_3 formulae; we use this presentation since it is more intuitive and is better suited for our purposes.

[5] A monotone function $f : \{0,1\}^n \to \{0,1\}$ is said to be of type Q_d if $f(x_1) = f(x_2) = \ldots = f(x_d) = 0$ implies that $x_1 \vee x_2 \vee \ldots \vee x_d \neq 1^n$.

has linear depth. This yields a protocol whose complexity grows exponentially with the number of parties, since when traversing the formula we increased the complexity of the protocol by a constant multiplicative factor (corresponding to the number of operations in the 3-party protocol) at every layer.

To overcome the exponential blowup, we replace the formula of [28] by a *logarithmic-depth* formula (which computes the majority function using only Maj_3 gates). Using the formula-based protocol described above, the logarithmic depth results in an efficient protocol, namely one whose complexity only grows polynomially with the number of parties. In Section 1.2 we describe the construction of a good "approximation" of such a formula as well as exact constructions under standard complexity-theoretic assumptions.

This approach is indeed very general and can be used in different models of secure MPC. For example, it can be used to obtain both passive security as outlined above and active security by using an underlying 4-party protocol that is secure against one active party and a log-depth threshold formula composed of two-out-of-four threshold gates (denoted by Th_2^4) which we also construct (see Section 1.2).

In fact, this reduction gives us a "cookbook" for designing secure multiparty protocols. The first step is to design a protocol for a constant number of parties that is secure against one dishonest party and the second step is to use a logarithmic-depth threshold from thresholds formula to obtain an efficient multiparty protocol that is secure against a constant fraction of corrupted parties.

We demonstrate the generality of this approach by deriving protocols in both passive and active settings and in different MPC models which differ in the type of underlying algebraic structure, including models for which no protocols were known. We also obtain conceptually simple protocols for classical problems in distributed computing such as broadcast protocols.

Simplified Feasibility Results. The classical results of Ben-Or et al. [4] and Chaum et al. [10] allow n parties to evaluate an arbitrary function, using secure point-to-point channels, with perfect security against $t < n/2$ passively corrupted parties or $t < n/3$ actively corrupted parties. We can derive conceptually simpler variants of these results by applying our approach with π being a 3-party or 4-party instance of the simple MPC protocol of Maurer [36]. On the one hand our results are slightly weaker because they either need the threshold t to be slightly sub-optimal or alternatively require (standard) complexity theoretic assumptions to construct an appropriate formula for implementing the protocol. It is instructive to note that the complexity of Maurer's protocol grows exponentially with the number of parties. Our approach makes this a non-issue, as we only use the protocol from [36] with a constant number of parties.[6]

MPC over Blackbox Algebraic Structures. There has been a considerable amount of work on implementing MPC protocols for computations over different algebraic

---

[6] While in the present work we apply our approach only to perfectly secure protocols, one could apply a similar technique to derive the result of Rabin and Ben-Or [38], namely a statistically secure protocol which tolerates $t < n/2$ actively corrupted parties.

structures such as fields, rings, and groups. Algebraic computations arise in many application scenarios. While it is possible in principle to emulate each algebraic operation by a sequence of boolean operations, this is inefficient both in theory and in practice. In particular, the communication complexity of the resulting protocols grows with the computational complexity of the algebraic operations rather than just with the bit-length of the inputs and outputs. This overhead can be avoided by designing protocols which make a *blackbox* (i.e., oracle) use of the underlying structure. The advantage of such protocols is that their communication complexity and the number of algebraic operations they employ are independent of the complexity of the structure.

MPC over Rings and k-linear Maps. The work of Cramer *et al.* [13] shows how to efficiently implement secure MPC over blackbox *rings*. We obtain a simpler derivation of such a protocol by noting that the simple protocol of Maurer [36] directly generalizes to work over a blackbox ring. As before, one could not apply this protocol directly because its complexity is exponential in the number of parties. We show how to use a similar approach for obtaining the first blackbox feasibility results for MPC over *k-linear maps*.

MPC over Groups. The problem of MPC over blackbox *groups* was introduced by Desmedt *et al.* [17] and further studied in [39,16,15]. To apply our approach in the group model, we need to specify the atomic protocol π that we use. For the case of passive security, we directly construct a simple 3-party protocol that has security against one corrupted party. This protocol is loosely based on a protocol by Feige *et al.* [19] and considerably simplifies the 3-party instance of a general result from [16].

In the active security model, we rely on the recent work of [15] who obtain the first MPC protocols with active security in the group model. The complexity of the protocol of [15] grows exponentially with the number of parties. However, we only need to employ the [15] protocol for four parties and so we do not suffer the exponential blowup. Thus, we settle the main problem left open in [15] by applying our technique to an instance of their results.

We also obtain the first *two-party* MPC protocols over blackbox groups. In the passive corruption model, we combine a group product randomization technique due to Kilian [32] with a "subset sum" based statistical secret sharing of group elements. We then get security against active corruptions by combining this two-party protocol with our efficient n-party protocol for the active model via the IPS compiler [31].

Broadcast. Broadcast is one of the most basic problems in distributed computing. Recall that in a broadcast protocol a broadcaster wants to send a message to all other parties. A broadcast protocol should end with all parties holding the same value, even if some of the parties, possibly including the broadcaster, behave adversarially. Obtaining efficient broadcast protocols is a highly nontrivial task [37,18,22]. Our generic approach for MPC protocols can be used to directly construct simple broadcast protocols for $t < n/3$ corrupted parties. We also get a simplified proof of a result of Fitzi and Maurer [21], showing that an ideal primitive allowing broadcast for 3 parties (so-called 2-cast) implies broadcast with $t < n/2$ corrupted parties. Our proof technique also yields broadcast for the more general case of Q_2 adversaries which was previously an open problem.

1.2 Threshold Formulae from Threshold Gates

Motivated by the above applications to MPC, we consider the problem of constructing a logarithmic-depth threshold formula from threshold gates. Before discussing the general problem, we first discuss the special case of constructing a logarithmic depth formula composed of Maj_3 gates that computes the majority function. Note that this is exactly the type of formula required in the setting of *passive* MPC security.

Majority from Majorities. A closely related problem was considered by Valiant [40] who proved the existence of a logarithmic-depth monotone formula that computes the majority function where the formula uses And and Or gates, both of fan-in 2. As noted independently by several authors [6,26,42,24], a slight modification of Valiant's argument shows the existence of a logarithmic-depth formula composed of Maj_3 gates that computes the majority function.

Valiant's proof is based on the probabilistic method and is non-constructive. Namely, the proof only assures us of the existence of a formula with the above properties, but does not hint on how to find it efficiently. Motivated by the applications presented in Section 1.1, we ask whether Valiant's proof can be derandomized using only Maj_3 gates and no constants.[7] We raise the following conjecture:

Conjecture 1 (Majority from Majorities). There exists an algorithm A that given an odd integer n as input, runs in poly(n)-time and generates a formula F on n inputs, with the following properties:

- F consists only of Maj_3 gates and no constants.
- $\mathsf{depth}(F) = O(\log n)$.
- F computes the majority function on n inputs.

A derandomization for Valiant's proof for formulas over And and Or gates follows from the seminal paper of Ajtai, Komlós and Szemerédi [1], though the latter does not seem to imply a derandomization in the context of Maj_3 gates, where constants are not allowed.[8]

In this paper we make a significant progress towards proving Conjecture 1. In particular, we prove that relaxed variants of the conjecture hold. In addition, we show that the conjecture follows from standard complexity assumptions, namely,

---

[7] We cannot allow the use of the constant 0, as this would correspond to assuming parties to be incorruptible. The use of the constant 1 alone is not helpful in our context.

[8] Note that And and Or gates can be implemented using Maj_3 gates and constants.

$\mathsf{E} \triangleq \mathsf{DTIME}(2^{O(n)})$ does not have $2^{\varepsilon n}$-size circuits for some constant $\varepsilon > 0$. Note that the latter follows from the existence of exponentially hard one-way functions.[9] See details in Section 2.

Threshold Formulae from Threshold Gates. Motivated by applications to the active MPC setting, and being a natural complexity-theoretic problem on its own, we initiate the study of a generalization of the majority from majorities problem, which we call *the threshold from thresholds problem*.

For integers $2 \leq j \leq k$, define the threshold function $\mathsf{Th}_j^k : \{0,1\}^k \to \{0,1\}$ as follows. $\mathsf{Th}_j^k(x) = 1$ if and only if the Hamming weight of x is at least j. Note that $\mathsf{Maj}_3 = \mathsf{Th}_2^3$.

Unlike the majority from majorities problem, it is not a priori clear what threshold function, if any, can be computed by a log-depth formula composed only of Th_j^k gates, even if no explicit construction is required.

We make significant progress also on this question. Roughly speaking, we provide an explicit construction of a logarithmic depth formula composed solely of Th_j^k gates, that well approximates $\mathsf{Th}_{n/m}^n$, where $m = \frac{k-1}{j-1}$. For further details, see Section 2.3.

Organization. In Section 2 we state our results and in Section 3 we present the proof techniques of the complexity-theoretic part. For an overview of the applications to cryptography and distributed computing, as well as formal statements and full proofs of our results, see the full version.

2   Our Results

We first describe the applications of our approach in cryptography and distributed computing, and then proceed to the complexity-theoretic results.

2.1   Cryptographic Results

We start by stating known results that we rederive using our approach, and later state our new results.

In the passive Ring-MPC model, we get the following results.

- If the majority from majorities conjecture (Conjecture 1) holds then we obtain an explicit MPC protocol that has optimal security in the passive model. That is, it is secure as long as at most a $\frac{1}{2} - \Omega(\frac{1}{n})$ fraction of the n parties (more precisely, $t < n/2$) are passively corrupted.

    As noted above and stated formally in Theorem 3, Conjecture 1 follows from widely-believed conjectures in complexity theory and cryptography.

---

[9] We find it curious that *perfectly secure* MPC results are based on the existence of (sufficiently strong) one-way functions.

– An unconditional explicit and close to optimal protocol in the passive model
  in which the fraction of dishonest parties is at most $\frac{1}{2} - 2^{-O(\sqrt{\log n})}$ out of
  the n parties (in contrast to the optimal threshold of $\frac{1}{2} - \Omega(\frac{1}{n})$).
– A randomized construction of an optimal protocol in the passive model. By
  randomized construction we mean that the protocol is constructed by a randomized algorithm which may fail with negligible (undetectable) probability,
  but otherwise outputs the description of a perfect protocol.

We obtain the following result in the *active* Ring-MPC model.

– An explicit but non-optimal protocol that is secure against any *active* ad-
  versary that controls at most a $\frac{1}{3} - \Omega(\frac{1}{\sqrt{\log n}})$ fraction of the n parties (in
  contrast to the optimal bound of $\frac{1}{3} - \Omega(\frac{1}{n})$).

Next we state our new results in the blackbox group model, introduced by
Desmedt *et al.* [17,16]. In this model the function computed by the protocol
is specified by an arithmetic circuit over a (possibly non-Abelian) group, and
the parties are restricted to making blackbox access to the group. (This includes
oracle access to the group operation, taking inverses, and sampling random group
elements.) In particular, the number of group operations performed by the protocol should not depend on the structure of the group or the complexity of
implementing a group operation using, say, a Boolean circuit.

– **Group-MPC, Passive:** The best explicit protocol of [16] offers perfect security against a $\frac{1}{n^\epsilon}$ fraction of passively corrupted parties, for any constant
  $\epsilon > 0$, where n is the total number of parties.
    We improve upon the latter by constructing an explicit protocol that has
  perfect security against an (almost optimal) $\frac{1}{2} - 2^{-O(\sqrt{\log n})}$ fraction of passively corrupted parties. Alternatively, we get an optimal bound of $\frac{1}{2} - \Omega(\frac{1}{n})$
  assuming the majority from majorities conjecture, via a non-uniform con-
  struction, or under standard derandomization or cryptographic assumptions.
– **Group-MPC, Active:** In a recent work, Desmedt *et al.* [15] constructed a
  secure MPC protocol in the group model with security against an *active*
  adversary. However, their result only gives a protocol whose complexity depends exponentially on the number of parties, regardless of the corruption
  threshold.
    We construct an *efficient* secure MPC protocol in the group model where
  an active adversary can control (an almost optimal) $\frac{1}{3} - \Omega(\frac{1}{\sqrt{\log n}})$ fraction
  of the n parties.
– **Secure Two-party Computation over Groups:** We construct the first
  secure *two-party* protocols over blackbox groups. Our protocols offer statistical security against active corruptions (assuming an oblivious transfer oracle)
  and rely on the afforementioned n-party protocols over black-box groups.

Finally, our protocols for the Ring-MPC model described above can be generalized
to yield the following new result for MPC over k-linear maps.

- **MPC over k-linear Maps:** We show that, for any constant k and any basis B of k-linear maps over finite Abelian groups, there are efficient MPC protocols for computing circuits over B which only make blackbox access to functions in B and group operations. This generalizes previous results for MPC over blackbox rings [13], which follow from the case $k = 2$, and can potentially be useful in cryptographic applications that involve complex bilinear or k-linear maps. These protocols are perfectly secure against a $\frac{1}{k} - \Omega(\frac{1}{\sqrt{\log n}})$ fraction of passively corrupted parties or a $\frac{1}{k+1} - \Omega(\frac{1}{\sqrt{\log n}})$ fraction of actively corrupted parties.

2.2 Distributed Computing Results

Broadcast. It is well known that broadcast can be implemented over point-to-point channels if and only if less than a third of the parties are actively corrupted [37,18] or, more generally, if and only if no three of the subsets the adversary may corrupt cover the entire set of parties [28,20], a so called Q_3-adversary.

In this paper we show that a trivial broadcast protocol for 4 parties where one is actively corrupted easily implies the result of [20] using existing constructions of (super-logarithmic depth) formulae. Substituting instead our own logarithmic depth formula constructions implies a simple polynomial-time broadcast protocol for less than $n(\frac{1}{3} - \Omega(\frac{1}{\sqrt{\log n}}))$ corrupted parties.

Broadcast from 2-cast. In [21], Fitzi and Maurer identify a minimal primitive that allows to improve the $\frac{n}{3}$ corruption threshold: if we are given the ability to broadcast among any subset of 3 parties for free, a so-called *2-cast* primitive, then broadcast becomes possible when less than $\frac{n}{2}$ parties are corrupted. It is natural to ask whether 2-cast also implies broadcast secure against general Q_2-adversaries (where no two corruptible subsets cover the entire set of parties). This problem was previously open.

We apply our approach to construct broadcast protocols based on a 2-cast primitive. Together with existing constructions of (super-logarithmic depth) formulae composed of Maj_3-gates, this immediately implies a construction of broadcast from 2-cast for every Q_2-adversary, resolving the above problem. Substituting instead our logarithmic-depth formula constructions, we get a simplified derivation of polynomial-time protocols for the case of an honest majority considered in [21]. We do not know if the formula based approach also implies the results in [11], which consider generalizations of the 2-cast primitive.

2.3 Complexity-Theoretic Results

In this section we describe our results on constructing threshold formulae from threshold gates. For the special case of computing majority from Maj_3 gates we obtain stronger results which we state first.

Majority from Majorities. Our first complexity-theoretic result shows that given a small promise on the bias of the input (defined as the difference between the normalized Hamming weight and $1/2$), Conjecture 1 holds.

Theorem 2. *There exists an algorithm A that given an odd integer n as input, runs in $\text{poly}(n)$-time and computes a formula F on n inputs, with the following properties:*

- F *consists only of* Maj_3 *gates and no constants.*
- $\text{depth}(F) = O(\log n)$.
- $\forall x \in \{0,1\}^n$ *such that* $\text{bias}(x) \geq 2^{-O(\sqrt{\log n})}$ *it holds that* $F(x) = \text{Maj}(x)$.

Our second result shows that under standard complexity hardness assumptions, Conjecture 1 holds.

Theorem 3. *If there exists an $\varepsilon > 0$ such that $\mathsf{E} \triangleq \text{DTIME}(2^{O(n)})$ does not have $2^{\varepsilon n}$-size circuits then Conjecture 1 holds. In particular, if there exist exponentially hard one-way functions then Conjecture 1 holds.*[10]

In fact, the proof of Theorem 3 explicitly presents an algorithm for constructing a formula as in Conjecture 1 given the truth table of any function in E, on a suitable number of inputs, that cannot be computed by $2^{\varepsilon n}$-size circuits.

Thresholds Formulae from Threshold Gates

Lemma 4. *There exists an algorithm A that given $t, j, k \in \mathbb{N}$ as input, where j, k are constants in t such that $j \geq 2$ and $k \geq 2j - 1$,[11] runs in $\exp(t)$-time and generates a formula F with the following properties:*

- F *has $mt + 1$ inputs, where $m = \lfloor \frac{k-1}{j-1} \rfloor$.*
- F *consists only of* Th_j^k *gates and no constants.*
- $\text{depth}(F) = O(t)$.
- $\forall x \in \{0,1\}^{mt+1}$ *it holds that* $F(x) = \text{Th}_{t+1}^{mt+1}(x)$.

Lemma 4 generalizes results of [2,28,3], who proved it for particular values of j and k, and uses a similar technique. We note that the depth of the formula generated in Lemma 4 is linear, which is too large for our applications. Nevertheless, the following theorem, which uses Lemma 4 as a building block, shows that a formula with *logarithmic depth* can be generated efficiently assuming a sufficient "bias" on the input.

---

[10] A one-way function f is *exponentially hard* if there exists an $\varepsilon > 0$ such that every family of $2^{\varepsilon n}$-size circuits can invert f with only $2^{-\varepsilon n}$ probability. If there exists such a function f, then the language $\mathcal{L}_f$ is in E but does not have $2^{\varepsilon n}$-size circuits, where $\mathcal{L}_f = \{(y, x', 1^n) : y \text{ has a preimage of length } n \text{ under } f \text{ which starts with } x'\}$.

[11] Throughout the paper we assume, without loss of generality, that $k \geq 2j - 1$. The complementary case can be reduced to this one by using Th_{k-j+1}^k gates and interpreting 0 as 1 and vice versa.

Theorem 5. *There exists an algorithm A that given $n, j, k \in \mathbb{N}$ as input, where j, k are constants in n such that $j \geq 2$ and $k \geq 2j - 1$, runs in $\mathrm{poly}(n)$-time and generates a formula F on n inputs, with the following properties:*

- *F consists only of Th_j^k gates and no constants.*
- *$\mathrm{depth}(F) = O(\log n)$.*
- *$\forall x \in \{0,1\}^n$ with normalized Hamming weight at least $\frac{1}{m} + \Omega(\frac{1}{\sqrt{\log n}})$, it holds that $F(x) = 1$, where $m = \lfloor \frac{k-1}{j-1} \rfloor$.*
- *$\forall x \in \{0,1\}^n$ with normalized Hamming weight at most $\frac{1}{m} - \Omega(\frac{1}{\sqrt{\log n}})$, it holds that $F(x) = 0$.*

Note that Theorem 2 is not a special case of Theorem 5 (with $j = 2, k = 3$) as the required promise on the bias in Theorem 2 is exponentially smaller than that in Theorem 5.

We do not know whether an analog of Conjecture 1 is plausible for the threshold from thresholds problem, even without the time-efficiency requirement. Theorem 5 might serve as evidence for the affirmative. However, the probabilistic argument used in the majority from majorities problem (see, e.g., [24]) breaks for this more general case. We consider this to be an interesting open problem for future research.

3   Proof Overview of Complexity-Theoretic Results

In this section we give an overview of our complexity-theoretic constructions. For simplicity, we start by giving an overview of our construction of a logarithmic-depth formula composed of Maj_3 gates, and no constants, that computes the majority function for inputs with constant bias. That is, we informally describe an efficient algorithm that given n, ε as inputs, where $\varepsilon > 0$ is constant in n, outputs a logarithmic-depth formula with n inputs which computes the majority function correctly on inputs with bias at least ε. It is not hard to see that it is enough to construct a logarithmic-depth *circuit*, since such a circuit can be efficiently converted to an equivalent logarithmic-depth formula.

To this end, we design an algorithm called ShrinkerGenerator that given n, ε as inputs, generates a *constant-depth* circuit Shrinker with n inputs and $\frac{n}{2}$ outputs, composed of Maj_3 gates and no constants, such that

$$\forall x \in \{0,1\}^n \quad \mathrm{bias}(x) \geq \varepsilon \implies \mathrm{bias}(\mathsf{Shrinker}(x)) \geq \varepsilon.$$

Thus, Shrinker shrinks the number of variables to half while maintaining the bias, assuming the input has a sufficiently large bias. By repeatedly calling ShrinkerGenerator on inputs $n, \frac{n}{2}, \frac{n}{4}, \ldots, 2$ (with the same ε) and concatenating the resulting circuits, one gets a logarithmic-depth circuit that computes the majority function assuming the input has large enough bias.

A key object we use in the design of ShrinkerGenerator is a Boolean sampler. Roughly speaking, a Boolean sampler is a randomized algorithm which on input $x \in \{0,1\}^n$ approximates the Hamming weight of x by reading only a small number of the bits of x. More precisely, a (d, ε, δ)-Boolean sampler is a randomized algorithm that on input $x \in \{0,1\}^n$ with normalized Hamming weight ω, samples at most d bits of x, and outputs $\beta \in [0,1]$ such that $\Pr[|\omega - \beta| \geq \varepsilon] \leq \delta$.

We will use a special type of samplers which take their samples in a non-adaptive fashion, and their output is simply the average of the sampled bits. For any $\varepsilon, \delta > 0$ there exist efficient (d, ε, δ)-Boolean samplers, with $d = O(\varepsilon^{-2} \cdot \delta^{-1})$, that on inputs of length n use only $\log n$ random bits.

Because such a sampler is non-adaptive and simply outputs the average of the sampled bits, it can be represented as a bipartite graph $G = (L, R, E)$, with $|L| = |R| = n$. For an input $x \in \{0,1\}^n$, the i'th vertex in L is labeled with the i'th bit of x. Each vertex in R represents one of the possible $\log n$ bit random strings used by the sampler. Each right vertex r is connected to the d left-vertices that are sampled by the algorithm when r is used as the random string.

The algorithm ShrinkerGenerator on inputs n, ε starts by constructing a graph G that represents a $(d, \frac{\varepsilon}{2}, \frac{1}{8})$-Boolean sampler, with $d = \text{poly}(\frac{1}{\varepsilon}) = O(1)$. It then arbitrarily chooses half of the right vertices in G and discards the rest. This gives a bipartite graph $G' = (L', R', E')$ with $|L'| = n$, $|R'| = \frac{n}{2}$ and constant right-degree d. The circuit Shrinker that the algorithm ShrinkerGenerator outputs is given by placing a circuit that computes the majority function on d inputs for every right vertex. The inputs of this majority circuit are the neighbors of the respective right vertex. Note that as d is constant, a constant-depth circuit that computes the majority function on d inputs can be found in constant time.

As for the correctness of the construction, assume now that $x \in \{0,1\}^n$ has some constant bias ε and, without loss of generality, assume that the bias is towards 1 (i.e., $\text{wt}(x) \geq (\frac{1}{2} + \varepsilon)n$). Then, by the guarantee of the sampler, for all but $\frac{1}{8}$ of the right vertices in the original graph G, the fraction of neighbors with label 1 of a right vertex is at least $\frac{1}{2} + \varepsilon - \frac{\varepsilon}{2} > \frac{1}{2}$. Thus, all but $\frac{1}{8}$ of the (constant-size) majority circuits located in R output 1. Hence, the fraction of majority circuits that output 0 in R' is at most $\frac{n/8}{n/2} = \frac{1}{4} \leq \frac{1}{2} - \varepsilon$, as desired.

3.1   Supporting Sub-constant Bias

For sub-constant ε, the sampler technique described above is wasteful, as it requires us to use a sequence of $O(\log n)$ layers with fan-in $O(\varepsilon^{-2})$. For sub-constant ε, this results in a circuit with a super-logarithmic depth. However, we observe that one layer of fan-in $O(\varepsilon^{-2})$ circuits is enough to amplify the bias from ε to 0.4 (rather than just keep the bias at ε). This reduces us to the constant bias case, which can be solved as above with an additional $O(\log n)$-depth.

Thus, in order to obtain an $O(\log n)$-depth circuit on n inputs, that computes majority correctly for inputs with bias at least ε, it is enough to construct an $O(\log n)$-depth circuit with $O(\varepsilon^{-2})$ inputs that computes majority correctly on all inputs.

Using a naive brute-force algorithm, one can efficiently find an optimal-depth circuit on roughly $\log n$ inputs that computes majority. By plugging this circuit into the above scheme, one immediately gets an $O(\log n)$-depth circuit that computes majority on n inputs with bias roughly $\varepsilon = \Omega(\frac{1}{\sqrt{\log n}})$.

We improve on this by using an additional derandomization idea. Specifically, we construct an $O(\log n)$-depth circuit on $2^{O(\sqrt{\log n})}$ inputs, that computes majority (under no assumption on the bias). Thus, we obtain an explicit construction of a circuit that computes majority assuming the bias is at least $\varepsilon = 2^{-O(\sqrt{\log n})}$.

We first describe a randomized construction of an $O(\log m)$-depth circuit on m inputs for majority, where m is set, in hindsight, to $2^{O(\sqrt{\log n})}$. Our construction only uses $O(\log^2 m)$ random bits (compared to $\text{poly}(m)$ random bits used in Valiant's construction). We then show how to derandomize this construction.

Our randomized construction works as follows. Consider an input $x \in \{0,1\}^m$ with bias ε. Suppose that we sample uniformly and independently at random 3 bits of x and compute their majority. It is shown in [24] that the majority's bias is at least 1.2ε (as long as ε is not too large).

Thus, by placing m majority gates of fan-in 3, and selecting their inputs from x uniformly and independently at random, the output of the m majority gates will have bias of at least 1.1ϵ with overwhelming probability. By composing $O(\log(1/\varepsilon))$ such layers, we can amplify the bias to a constant. Note that this construction uses $O(m \cdot \log m \cdot \log(1/\varepsilon))$ random bits.

To save on the number of random bits used (which is essential for the derandomization step), instead of sampling the inputs of each one of the m gates uniformly at random, we choose them in each layer using a 6-wise independent hash function. While 3-wise independence suffices for the expectation of the bias to be as before, the 6-wise independence guarantees that the outputs of the majority gates in each layer are pairwise independent. Using tail inequalities we show that, with probability $1 - o(1)$, the bias increases in each layer as before.

By composing $O(\log(1/\varepsilon))$ such layers, each of which requires $O(\log m)$ random bits, we obtain a circuit as desired. The total number of random bits used is $O(\log(m) \cdot \log(1/\varepsilon))$, which is bounded by $O(\log^2 m)$. We derandomize the construction by placing all $2^{O(\log^2 m)}$ majority circuits that can be output by the randomized construction and taking the majority vote of these circuits.

Since we have a guarantee that almost all (a $1 - o(1)$ fraction) of the circuits correctly compute majority, it is enough to compute the majority vote at the end using a circuit with $2^{O(\log^2 m)}$ inputs that works for, say, constant bias. Such a circuit, with depth $O(\log^2 m)$, can be constructed in time $2^{O(\log^2 m)}$ by the constant-bias scheme described earlier.

As we set $m = 2^{O(\sqrt{\log n})}$, we get a $\text{poly}(n)$-time uniform construction of an $O(\log n)$-depth circuit on $2^{O(\sqrt{\log n})}$ inputs that computes majority correctly on all inputs. This circuit is then used in the scheme described above.

Threshold Formulae from Thresholds Gates. The scheme described above works also in the more general setting of threshold from thresholds. Indeed, in the paper we present the scheme in the general setting. To apply the scheme in the thresholds setting, one needs to construct a small circuit that computes the required threshold formula, to be used by ShrinkerGenerator. We accomplish this by extending results of [2,28,3].

4   The Player Emulation Technique

The formulas obtained in the previous section can be used to construct efficient multiparty protocols via the "player emulation" technique from [28]. Variants of this technique, also referred to as player *virtualization* or *simulation*, were used for different purposes in several other works (e.g. [5,9,27,14,31,35,34]). While implementing player emulation in the passive security model is quite straight-forward, in the active security model it requires more care. In the following we give more details on the implementation of this technique.

Recall that in a single player emulation step, the role of a party τ participating in a protocol Π is replaced by a secure protocol π which involves a small set of parties $v_1, \ldots, v_k$, along with the parties of Π. We will typically let $k = 3$ (resp., $k = 4$) in the case of security against a passive (resp., active) adversary, and let π be a protocol which remains secure as long as at most one of the emulating parties v_i is corrupted. Furthermore, the total computational complexity of all parties in π (which is typically cast in some algebraic computation model) is only bigger by a constant factor than that of the emulated party τ in Π. As explained in the Introduction, a logarithmic-depth threshold formula defines a sequence of such player emulation steps which result in transforming an atomic protocol π for a constant number of parties into an efficient n-party protocol which tolerates an optimal or near-optimal fraction of corrupted parties.

The application of the player emulation technique in [28] is formulated in a specialized framework for secure MPC and is restricted to the protocol compiler of BGW [4].[12] However, the technique is quite insensitive to many of these details and can be applied with other protocols and notions of security from the literature.

A conceptually simple way for implementing a player emulation step is by viewing the role of τ in Π as a *reactive* ideal functionality, which interacts with the parties in Π (receiving incoming messages as inputs and delivering outgoing messages as outputs), and maintains a state information during this interaction. The protocol π emulating τ then needs to realize the corresponding functionality using the emulating parties v_i instead of τ. Note that protocol π does not only involve the players emulating τ. It also specifies how players communicating with τ should translate their messages into whatever format π uses[13].

The protocol π can satisfy any composable notion of security that applies to reactive functionalities, namely one which ensures that π can be securely used as a substitute for τ in an arbitrary execution environment if at most a single v_i

---

[12] Since the atomic protocols π we employ in this work all have a similar high-level structure to the BGW protocol, they can be used within the framework of [28] in a similar way.

[13] Alternatively, if the communication channels are modeled as an ideal functionality, one can extend the definition of this functionality so it will do the translation, and then in a final step implement the translation. This leads in some cases to a slightly simpler protocol in the end.

is corrupted. The protocols π we use in this work all satisfy the standard notion of UC-security from [8], which suffices for this purpose.[14]

Alternatively, it is possible to implement a player emulation step by only relying on protocols for secure function evaluation which satisfy the standard definitions of standalone security [7,23]. The idea is to first ensure that only a single message is sent in each round of Π, and then implement a round in which τ interacts with party P by a protocol involving P and the emulating parties v_i. The functionality realized by such a protocol is determined by the choice of a concrete (robust) secret sharing scheme which is used to distribute the state of τ between the emulating parties.

Acknowledgements. Ishai was supported by the European Research Council as part of the ERC project CaC (grant 259426). Damgaård, Kölker and Miltersen acknowledge support from the Danish National Research Foundation and The National Science Foundation of China (under the grant 61061130540) for the Sino-Danish Center for the Theory of Interactive Computation, within which part of this work was performed; and also from the CFEM research center (supported by the Danish Strategic Research Council) within which part of this work was performed. Cohen, Raz and Rothblum were supported by ISF (Israel Science Foundation) grants and by the I-CORE Program of the Planning and Budgeting Committee.

References

1. Ajtai, M., Komlós, J., Szemerédi, E.: An o(n log n) sorting network. In: STOC, pp. 1–9 (1983)
2. Akers, S., Robbins, T.: Logical design with three-input majority gates. Computer Design 45(3), 12–27 (1963)
3. Barkol, O., Ishai, Y., Weinreb, E.: On locally decodable codes, self-correctable codes, and t-private PIR. Algorithmica 58(4), 831–859 (2010)
4. Ben-Or, M., Goldwasser, S., Wigderson, A.: Completeness theorems for non-cryptographic fault-tolerant distributed computation (extended abstract). In: STOC, pp. 1–10 (1988)
5. Bracha, G.: An O(log n) expected rounds randomized byzantine generals protocol. J. ACM 34(4), 910–920 (1987)
6. Bro Miltersen, P.: Lecutre notes. Available from author (1992)
7. Canetti, R.: Security and composition of multiparty cryptographic protocols. J. Cryptology 13(1), 143–202 (2000)
8. Canetti, R.: Universally composable security: A new paradigm for cryptographic protocols. In: FOCS, pp. 136–145 (2001)

---

[14] In particular, all these protocols are perfectly secure with a straight-line black-box simulator, which was shown in [33] to imply UC-security in the case of secure function evaluation. We note that while standard UC-security is cast in an asynchronous network model and does not guarantee output delivery, it can be extended to capture synchronous protocols which guarantee output delivery (cf. [12, Chapter 4]).

9. Chaum, D.: The spymasters double-agent problem: Multiparty computations secure unconditionally from minorities and cryptograhically from majorities. In: Brassard, G. (ed.) CRYPTO 1989. LNCS, vol. 435, pp. 591–602. Springer, Heidelberg (1990)

10. Chaum, D., Crépeau, C., Damgård, I.: Multiparty unconditionally secure protocols (extended abstract). In: STOC, pp. 11–19 (1988)

11. Considine, J., Fitzi, M., Franklin, M.K., Levin, L.A., Maurer, U.M., Metcalf, D.: Byzantine agreement given partial broadcast. J. Cryptology 18(3), 191–217 (2005)

12. Cramer, R., Damgård, I., Nielsen, J.B.: Secure Multiparty Computation and Secret Sharing - An Information Theoretic Appoach (2012), Book draft, available at http://www.daimi.au.dk/~ivan/MPCbook.pdf

13. Cramer, R., Fehr, S., Ishai, Y., Kushilevitz, E.: Efficient multi-party computation over rings. In: Biham, E. (ed.) EUROCRYPT 2003. LNCS, vol. 2656, pp. 596–613. Springer, Heidelberg (2003)

14. Damgård, I., Ishai, Y., Krøigaard, M., Nielsen, J.B., Smith, A.: Scalable multiparty computation with nearly optimal work and resilience. In: Wagner, D. (ed.) CRYPTO 2008. LNCS, vol. 5157, pp. 241–261. Springer, Heidelberg (2008)

15. Desmedt, Y., Pieprzyk, J., Steinfeld, R.: Active security in multiparty computation over black-box groups. In: Visconti, I., De Prisco, R. (eds.) SCN 2012. LNCS, vol. 7485, pp. 503–521. Springer, Heidelberg (2012)

16. Desmedt, Y., Pieprzyk, J., Steinfeld, R., Sun, X., Tartary, C., Wang, H., Yao, A.C.-C.: Graph coloring applied to secure computation in non-abelian groups. J. Cryptology 25(4), 557–600 (2012)

17. Desmedt, Y., Pieprzyk, J., Steinfeld, R., Wang, H.: On secure multi-party computation in black-box groups. In: Menezes, A. (ed.) CRYPTO 2007. LNCS, vol. 4622, pp. 591–612. Springer, Heidelberg (2007)

18. Dolev, D.: The byzantine generals strike again. J. Algorithms 3(1), 14–30 (1982)

19. Feige, U., Kilian, J., Naor, M.: A minimal model for secure computation (extended abstract). In: STOC, pp. 554–563 (1994)

20. Fitzi, M., Maurer, U.M.: Efficient byzantine agreement secure against general adversaries. In: Kutten, S. (ed.) DISC 1998. LNCS, vol. 1499, pp. 134–148. Springer, Heidelberg (1998)

21. Fitzi, M., Maurer, U.M.: From partial consistency to global broadcast. In: STOC, pp. 494–503 (2000)

22. Garay, J.A., Moses, Y.: Fully polynomial byzantine agreement for $n > 3t$ processors in $t + 1$ rounds. SIAM J. Comput. 27(1), 247–290 (1998)

23. Goldreich, O.: Foundations of Cryptography. Basic Applications, vol. 2. Cambridge University Press, New York (2004)

24. Goldreich, O.: Valiant's polynomial-size monotone formula for majority (2011), http://www.wisdom.weizmann.ac.il/~oded/PDF/mono-maj.pdf

25. Goldreich, O., Micali, S., Wigderson, A.: How to play any mental game or A completeness theorem for protocols with honest majority. In: STOC, pp. 218–229. ACM (1987)

26. Gupta, A., Mahajan, S.: Using amplification to compute majority with small majority gates. Computational Complexity 6(1), 46–63 (1996)

27. Harnik, D., Ishai, Y., Kushilevitz, E., Nielsen, J.B.: OT-combiners via secure computation. In: Canetti, R. (ed.) TCC 2008. LNCS, vol. 4948, pp. 393–411. Springer, Heidelberg (2008)

28. Hirt, M., Maurer, U.M.: Player simulation and general adversary structures in perfect multiparty computation. J. Cryptology 13(1), 31–60 (2000)

29. Hoory, S., Magen, A., Pitassi, T.: Monotone circuits for the majority function. In: APPROX-RANDOM, pp. 410–425 (2006)
30. Ishai, Y., Kushilevitz, E., Ostrovsky, R., Sahai, A.: Zero-knowledge proofs from secure multiparty computation. SIAM J. Comput. 39(3), 1121–1152 (2009)
31. Ishai, Y., Prabhakaran, M., Sahai, A.: Founding cryptography on oblivious transfer – efficiently. In: Wagner, D. (ed.) CRYPTO 2008. LNCS, vol. 5157, pp. 572–591. Springer, Heidelberg (2008),
http://www.cs.illinois.edu/$\sim$mmp/research.html
32. Kilian, J.: Founding cryptography on oblivious transfer. In: STOC, pp. 20–31 (1988)
33. Kushilevitz, E., Lindell, Y., Rabin, T.: Information-theoretically secure protocols and security under composition. SIAM J. Comput. 39(5), 2090–2112 (2010)
34. Lindell, Y., Oxman, E., Pinkas, B.: The IPS compiler: Optimizations, variants and concrete efficiency. In: Rogaway, P. (ed.) CRYPTO 2011. LNCS, vol. 6841, pp. 259–276. Springer, Heidelberg (2011)
35. Lucas, C., Raub, D., Maurer, U.M.: Hybrid-secure mpc: trading information-theoretic robustness for computational privacy. In: PODC, pp. 219–228 (2010)
36. Maurer, U.M.: Secure multi-party computation made simple. Discrete Applied Mathematics 154(2), 370–381 (2006)
37. Pease, M.C., Shostak, R.E., Lamport, L.: Reaching agreement in the presence of faults. J. ACM 27(2), 228–234 (1980)
38. Rabin, T., Ben-Or, M.: Verifiable secret sharing and multiparty protocols with honest majority (extended abstract). In: Johnson, D.S. (ed.) STOC, pp. 73–85. ACM (1989)
39. Sun, X., Yao, A.C.-C., Tartary, C.: Graph design for secure multiparty computation over non-abelian groups. In: Pieprzyk, J. (ed.) ASIACRYPT 2008. LNCS, vol. 5350, pp. 37–53. Springer, Heidelberg (2008)
40. Valiant, L.G.: Short monotone formulae for the majority function. J. Algorithms 5(3), 363–366 (1984)
41. Yao, A.C.-C.: Protocols for secure computations (extended abstract). In: FOCS, pp. 160–164 (1982)
42. Zwick, U.: Lecture notes (1996),
http://www.cs.tau.ac.il/~zwick/circ-comp-new/six.ps

A Dynamic Tradeoff
between Active and Passive Corruptions
in Secure Multi-Party Computation

Martin Hirt[1], Ueli Maurer[1], and Christoph Lucas[2],[*]

[1] ETH Zurich
{hirt,maurer}@inf.ethz.ch
[2] ETH Zurich and Ergon Informatik AG
christoph.lucas@ergon.ch

Abstract. At STOC '87, Goldreich et al. presented two protocols for secure multi-party computation (MPC) among n parties: The first protocol provides *passive* security against $t < n$ corrupted parties. The second protocol provides even *active* security, but only against $t < n/2$ corrupted parties. Although these protocols provide security against the provably highest possible number of corruptions, each of them has its limitation: The first protocol is rendered completely insecure in presence of a single active corruption, and the second protocol is rendered completely insecure in presence of $\lceil n/2 \rceil$ passive corruptions.

At Crypto 2006, Ishai et al. combined these two protocols into a single protocol which provides passive security against $t < n$ corruptions and active security against $t < n/2$ corruptions. This protocol unifies the security guarantees of the passive world and the active world ("best of both worlds"). However, the corruption threshold $t < n$ can be tolerated only when *all* corruptions are passive. With a single active corruption, the threshold is reduced to $t < n/2$.

As our main result, we introduce a *dynamic tradeoff* between active and passive corruptions: We present a protocol which provides security against $t < n$ passive corruptions, against $t < n/2$ active corruptions, *and everything in between*. In particular, our protocol provides full security against k active corruptions, as long as less than $n - k$ parties are corrupted in total, for any unknown k.

The main technical contribution is a new secret sharing scheme that, in the reconstruction phase, releases secrecy *gradually*. This allows to construct non-robust MPC protocols which, in case of an abort, still provide some level of secrecy. Furthermore, using similar techniques, we also construct protocols for reactive MPC with hybrid security, i.e., different thresholds for secrecy, correctness, robustness, and fairness. Intuitively, the more corrupted parties, the less security is guaranteed.

Keywords: Multi-party computation, gradual secret sharing, computational security, mixed adversary.

---

[*] Work done while the author was at ETH Zurich.

R. Canetti and J.A. Garay (Eds.): CRYPTO 2013, Part II, LNCS 8043, pp. 203–219, 2013.

1   Introduction

1.1   Secure Multi-Party Computation

Multi-Party Computation (MPC) allows a set of n parties to securely compute a (probabilistic) function f in a distributed manner, where security means that secrecy of the inputs and correctness of the output are maintained even when some of the parties are dishonest. The dishonesty of parties is modeled with a central adversary who corrupts parties. The adversary can be *passive*, i.e., can read the internal state of the corrupted parties, or *active*, i.e., can make the corrupted parties deviate arbitrarily from the protocol. Reactive MPC considers the more general case where parties can provide inputs even after having received (intermediate) outputs.

MPC was originally proposed by Yao [Yao82]. The first general solution was provided in [GMW87], where two protocols are presented, one providing passive security against any number of corruptions, and one providing active security against a faulty minority. These protocols are computationally secure only. Information-theoretically secure MPC was considered in [BGW88, CCD88, RB89, Bea91].

1.2   Extensions of the Basic Setting

These seminal MPC results have been generalized and extended in numerous directions, among which we focus on those most relevant for us: The strict separation between active and passive adversaries was overcome in [Cha89, DDWY93, FHM98, HMZ08] by considering an adversary that corrupts some parties actively and some additional parties passively. Such a *mixed adversary* is characterized by two fixed thresholds, indicating the maximum number of actively and passively corrupted parties, respectively.

A more fine-grained analysis of the achieved security guarantees was initiated in [Cha89] and further advanced in [FHHW03, FHW04, Kat07, LRM10, HLMR11, HLMR12]: These protocols provide *hybrid security*, i.e., depending on the actual adversary, only a subset of the usual security guarantees (secrecy, correctness, robustness, fairness) are guaranteed. Intuitively, the more parties are corrupted, the less security is guaranteed.

For completeness, the considered models and achieved security levels of the mentioned protocols are summarized in Appendix A.

1.3   Prior Work

In their seminal paper [GMW87], the authors provide two different protocols, one for passive security against up to $t < n$ corruptions, and one for active security against up to $t < \frac{n}{2}$ corruptions. In [IKLP06], these two protocols are combined into a single protocol, which is secure against an adversary that either passively corrupts any number of parties or actively corrupts a minority of the parties. This combined protocol is only applicable for non-reactive functions, and it is proven that this combination is impossible for reactive MPC. For this more general setting, the authors present a protocol in the active world that provides full security up to a first threshold t, and correctness and secrecy up to a second threshold s, given that $t < n/2$ and $s + t < n$.

Note that all provided protocols are secure against an adversary that is either fully passive or fully active. In particular, the protocol for non-reactive MPC is rendered completely insecure when the adversary corrupts $\lceil n/2 \rceil$ parties, even if all but one corruptions are only passive.

1.4   Contributions

We present an MPC protocol (for non-reactive functions) with a dynamic trade-off between active and passive corruptions. As [IKLP06], the protocol provides the best possible security level in presence of a purely passive adversary (namely $t < n$) as well as in presence of a purely active adversary (namely $t < n/2$). In addition, the protocol also tolerates mixed adversaries that corrupt some parties actively and some other parties passively, as long as at most k parties are corrupted actively and at most $n - k - 1$ parties are corrupted in total. Note that k need *not* be known, as it is not a parameter of the protocol.

In order to construct such a protocol, we introduce the notion of *gradual* verifiable secret sharing (VSS). In contrast to traditional VSS, a gradual VSS reduces the number of adversaries against which secrecy is guaranteed during reconstruction in a step-wise fashion, and at the same time increases the number of adversaries against which robustness is guaranteed. By that, if the reconstruction of a secret aborts, secrecy against many adversaries is still guaranteed.

Furthermore, we generalize and extend our results in two directions: First, we work in a model with hybrid security. That means, we consider each security guarantee (correctness, secrecy, robustness, and fairness) separately, and our protocols provide each guarantee against as many corrupted parties as possible. Second, in the setting of reactive MPC, we extend the protocol from [IKLP06] with fairness and security against mixed adversaries, while at the same time removing the restriction that robustness can only be guaranteed against a corrupted minority.

1.5   Outline of the Paper

The paper is organized as follows: The model used in this work is described in Section 2. In Section 3, we briefly review the standard definition of VSS and introduce the notion of gradual reconstruction. Furthermore, we provide gradual VSS protocols for threshold adversaries. In Sections 4 and 5, we present protocols for non-reactive and reactive MPC, respectively, together with optimal bounds.

2   Model

2.1   Parties

We consider n parties $p_1, \ldots, p_n$, connected by pairwise synchronous secure channels and authenticated broadcast channels,[1] who want to implement an ideal functionality $\mathcal{F}$ computing a (probabilistic) function f over a finite field $\mathbb{F}$ with $|\mathbb{F}| > n$. Without loss of generality, we assume only public outputs (possibly several at the same time). Local outputs towards a designated party can be blinded with a random input from that party. For reactive MPC, $\mathcal{F}$ is not restricted to functions and can provide outputs before taking some other inputs. There is a central adversary with polynomially bounded computing power who corrupts some parties passively (and reads their internal state) or even actively (and makes them misbehave arbitrarily). We denote the set of actually actively (passively) corrupted parties by $\mathcal{D}^*$ ($\mathcal{E}^*$), where for ease of notation, we assume that $\mathcal{D}^* \subseteq \mathcal{E}^*$. Uncorrupted parties are called *honest*, non-actively corrupted parties are called *correct*. For ease of notation, we assume that if a party does not receive an expected message (or receives an invalid message), a default message is used instead.

2.2   Security

The security of our protocols is computational, i.e., based on some computational assumption. We say a security guarantee holds *computationally* if it holds against a computationally bounded adversary. We consider the five standard security guarantees: *Secrecy* means that the adversary learns nothing about the honest parties' inputs and outputs (except, of course, for what can be derived from the corrupted parties' inputs and outputs). *Correctness* means that all parties either output the right value or no value at all. *Robustness* means that the adversary cannot prevent the honest parties from learning their respective outputs. This last requirement turns out to be very demanding. Therefore, relaxations of full security have been proposed, where robustness is replaced by weaker output guarantees: *Fairness* means that the adversary can possibly prevent the honest parties from learning their outputs, but then also the corrupted parties do not learn their outputs. In the case of reactive MPC, fairness can only be achieved for outputs provided at the same time, i.e., for each output round, either all (honest) parties learn the outputs or also the adversary does not learn them. However, the adversary can abort the protocol after having received outputs from prior rounds. *Agreement on abort* means that the adversary can possibly prevent honest parties from learning their output, even while corrupted parties learn their outputs, but then the honest parties at least reach agreement on this

---

[1] Secure bilateral channels are usually established via standard techniques such as encryption and digital signatures. Broadcast channels are usually simulated with an appropriate protocol [DS82].

fact (and typically make no output).[2] The level of security (secrecy, correctness, fairness, robustness, agreement on abort) depends on $(\mathcal{D}^*, \mathcal{E}^*)$.

2.3 Characterization of Tolerated Adversaries

Traditionally, protocols for threshold adversaries are characterized by a single threshold t that specifies the maximal adversary that can be tolerated. This basic representation has been extended as follows: On the one hand, a mixed adversary is characterized by two thresholds (t_a, t_p), where he may corrupt up to t_p parties passively, and up to t_a of these parties even actively. To model security guarantees against incomparable maximal adversaries, we need to consider multiple pairs of thresholds. Therefore, we use multi-thresholds $T = \{(t_{a,1}, t_{p,1}), \ldots, (t_{a,k}, t_{p,k})\}$, i.e., sets of pairs of thresholds (t_a, t_p). In this model, security is guaranteed if $(|\mathcal{D}^*|, |\mathcal{E}^*|) \leq (t_a, t_p)$ for some $(t_a, t_p) \in T$, denoted by $(|\mathcal{D}^*|, |\mathcal{E}^*|) \leq T$, where $(|\mathcal{D}^*|, |\mathcal{E}^*|) \leq (t_a, t_p)$ is a shorthand for $|\mathcal{D}^*| \leq t_a$ and $|\mathcal{E}^*| \leq t_p$. On the other hand, the level of security (correctness, secrecy, robustness, and fairness) depends on the number $(|\mathcal{D}^*|, |\mathcal{E}^*|)$ of *actually* corrupted parties (hybrid security). Hence, we consider the four multi-thresholds T^c, T^s, T^r, and T^f: Correctness with agreement on abort is guaranteed for $(|\mathcal{D}^*|, |\mathcal{E}^*|) \leq T^c$, secrecy is guaranteed for $(|\mathcal{D}^*|, |\mathcal{E}^*|) \leq T^s$, robustness is guaranteed for $(|\mathcal{D}^*|, |\mathcal{E}^*|) \leq T^r$, and fairness is guaranteed for $(|\mathcal{D}^*|, |\mathcal{E}^*|) \leq T^f$. We have the assumption that $T^r \leq T^c$ and $T^f \leq T^s \leq T^c$,[3] as secrecy and robustness are not well defined without correctness, and as fairness cannot be achieved without secrecy.

2.4 Ideal Functionality

For ease of presentation, we provide our proof sketches in a property-based security model. This allows to describe our ideas in a straightforward and understandable way. All statements could be made formal in a simulation-based model. To avoid ambiguity, we sketch the ideal functionality of our non-reactive MPC protocol in Figure 1.

We stress that in a setting without secrecy (i.e., $(\mathcal{D}^*, \mathcal{E}^*) \not\leq T^s$) the adversary may learn the inputs from honest parties before he has to provide inputs from the corrupted parties. Furthermore, for probabilistic functions, the adversary can freely choose the random string.

3 Gradual Verifiable Secret Sharing

We first briefly review the standard definition of verifiable secret sharing (VSS) schemes. Then, we define a new property for VSS schemes introducing the notion

---

[2] In our constructions, all abort decisions are based on publicly known values. Hence, we have agreement on abort for free. Note that the impossibility proofs hold even when agreement on abort is not required.

[3] We write $T_1 \leq T_2$ if $\forall (t_a, t_p) \in T_1 : (t_a, t_p) \leq T_2$.

Ideal Functionality $\mathcal{F}$: Given (T^r, T^f, T^s, T^c) and $(\mathcal{D}^*, \mathcal{E}^*) \leq T^c$.

1. Receive inputs from honest parties $\mathcal{P} \setminus \mathcal{E}^*$.
2. If $(\mathcal{D}^*, \mathcal{E}^*) \nleq T^s$: send these inputs to the adversary.
3. Receive inputs from the adversary for the parties in $\mathcal{E}^*$.
4. For probabilistic functions only:
   If $(\mathcal{D}^*, \mathcal{E}^*) \leq T^s$: sample a random bit string r of appropriate length.
   Otherwise: Receive r from the adversary.
5. Evaluate the function.
6. If $(\mathcal{D}^*, \mathcal{E}^*) \nleq T^f$: send the output to the adversary.
7. If $(\mathcal{D}^*, \mathcal{E}^*) \nleq T^r$: receive a bit from the adversary, and abort if the bit is 1.
8. Send the output to the honest parties $\mathcal{P} \setminus \mathcal{E}^*$ and to the adversary.

Fig. 1. Sketch of the Ideal Functionality for non-reactive MPC

of gradual reconstruction.[4] Finally, we present schemes that achieve the new requirements.

3.1   Definitions

A *Verifiable Secret Sharing* (VSS) scheme allows a designated party (the *dealer*) to share a value s among all parties, such that the parties can jointly reconstruct the value. The following definition captures the standard, well-known properties of verifiable secret sharing:

Definition 1 (VSS). *A (T^s, T^r)-secure Verifiable Secret Sharing (VSS) is a pair of protocols* Share *and* Rec, *where* Share *takes input s from the dealer and* Rec *gives output s' to each party, if the following conditions are fulfilled:*
SECRECY: *If $(\mathcal{D}^*, \mathcal{E}^*) \leq T^s$, then in* Share *the adversary obtains no information about s.*
CORRECTNESS: *After* Share, *the dealer is bound to a value s', where $s' = s$ if the dealer is correct. Furthermore, in* Rec, *either each (correct) party outputs s' or all (correct) parties abort.*
ROBUSTNESS: *The adversary cannot abort* Share. *If $(\mathcal{D}^*, \mathcal{E}^*) \leq T^r$, then the adversary cannot abort* Rec.

For $(\mathcal{D}^*, \mathcal{E}^*) \nleq T^r$, this definition does not rule out that the reconstruction protocol aborts even in an unfair way, where the honest parties do not learn the secret but the corrupted parties do. In fact, most VSS schemes in the literature show this undesired behavior: When corrupted parties do not broadcast their shares, they still learn the shares from the honest parties and can compute the secret, but the honest parties do not obtain enough shares and abort.

Clearly, a certain level of unfairness cannot be avoided when secrecy and robustness are to be guaranteed with respect to many corruptions. In particular,

---

[4] This notion should not be confused with the notion of gradual release of secrecy as introduced by [Blu83].

whenever a sharing scheme is secret with respect to some subset $M \subseteq \mathcal{P}$ of the parties, then it cannot be robust with respect to the complement $\mathcal{P} \setminus M$ of this subset: When the parties in M have no information about the shared value after Share, and the parties in $\mathcal{P} \setminus M$ do not participate in Rec, then the value cannot be reconstructed. Hence, the collection of subsets against which a sharing is secret implicitly defines the collection of subsets that can abort reconstruction (namely, the complements). In usual reconstruction protocols, all correct parties directly broadcast their entire shares, i.e., secrecy is given up against all subsets at once, before robustness against a single subset is achieved. This means that during reconstruction, any subset of parties that can abort, can also abort in an unfair way. Our new definition below requires that the transition from secrecy to robustness is gradual, such that when a small set of parties does not broadcast their share, then only a large subset of parties jointly obtains information about the secret.

Definition 2 (Gradual VSS). *A (T^s, T^r)-secure VSS is gradual if the following conditions are fulfilled: If Rec aborts, each party outputs a non-empty set $B \subseteq \mathcal{D}^*$, and the adversary obtained no information about the secret s if $(|\mathcal{D}^*|, |\mathcal{E}^*|) \leq T^s$ and $|\mathcal{E}^*| < n - |B|$.*

3.2   A Gradual VSS Scheme

We describe a gradual VSS scheme based on the standard Shamir sharing scheme [Sha79], and extended with (homomorphic) commitments to provide verifiability (e.g. [Ped91]). To obtain the gradual property, summands $s_1, \ldots, s_d$ with $s_1 + \ldots + s_d = s$ are chosen at random and, rather than the secret itself, these summands are shared, where summand s_i is shared with degree i. Then, during reconstruction, the summands are reconstructed one by one, in decreasing order of the sharing degree. We assume that each party p_i is assigned a unique and publicly known evaluation point $\alpha_i \in \mathbb{F} \setminus \{0\}$,[5] and that the commitments are homomorphic and transferable by sending the opening information. This construction results in the scheme $\text{VSS}^d = (\text{Share}^d, \text{Rec}^d)$ for parameter d.

Definition 3 (d-sharing). *A value s is d-shared, denoted by $[s]_d$, if there are values $s_1, \ldots, s_d$, such that $s_1 + \ldots + s_d = s$ and, for all $i \in \{1, \ldots, d\}$, there is a polynomial $g_i(x)$ of degree i with $g_i(0) = s_i$, and every party p_j holds a share $s_{ij} = g_i(\alpha_j)$ and is committed to it.*

The sharing protocol from [Ped91] can be extended in a straightforward way to compute such a d-sharing. A description of the protocol can be found in Figure 2. This share protocol provides resilience even against a corrupted dealer. It turns out that in our protocols, essentially only ideal functionalities need to compute d-sharings. Trivially, given a value s, such an honest dealer can directly sample and distribute a correct sharing $[s]_d$ without running Share^d. The (probabilistic) function that samples shares of some given input s is denoted by State^d.

---

[5] This implies that the field $\mathbb{F}$ must have more than n elements.

Protocol Share^d: Given input s from the dealer, compute a d-sharing of this value.

1. The dealer chooses uniformly random summands $s_1, \ldots, s_d$ with $\sum_{i=1}^{d} s_i = s$.
2. For $i \in \{1, \ldots, d\}$:
   (a) The dealer chooses a random polynomial $g_i(x)$ of degree i with $g_i(0) = s_i$, and computes and broadcasts (homomorphic) commitments of the coefficients of $g_i(x)$.
   (b) For each share $s_{ij} = g_i(\alpha_j)$, each party locally computes a commitment c_{ij} (using the homomorphic property), and the dealer sends the corresponding opening information o_{ij} to party p_j. Then, p_j broadcasts a complaint bit, indicating whether o_{ij} opens c_{ij} to some value s'_{ij}.
   (c) For each share s_{ij} for which an inconsistency was reported, the dealer broadcasts the opening information o_{ij}, and if o_{ij} opens c_{ij}, p_j accepts o_{ij}. Otherwise, the dealer is disqualified (and a default sharing of a default value is used).
3. Each party p_j outputs its share $\big((s_{1j}, o_{1j}), \ldots, (s_{dj}, o_{dj}) \big)$ and all commitments.

Fig. 2. The share protocol for threshold adversaries

Lemma 1. *Given a parameter $d < n$ and input s, Share^d robustly computes a correct d-sharing $[s]_d$. If $|\mathcal{E}^*| \le d$, the adversary obtains no information about s.*

Proof. CORRECTNESS: Trivially, in Step 2.a, any (well-formed) commitments broadcasted by the dealer are correct. In Step 2.b, commitments to all shares are computed locally by each party directly from the commitments to the coefficients broadcasted in Step 2.a. Hence, all (correct) parties have a consistent view with correct commitments. In Steps 2.b and 2.c, due to the binding property of the commitments, the adversary cannot distribute inconsistent opening information without being detected. Hence, the sharing is correct (or the dealer is disqualified and a default sharing is used).

SECRECY: The commitments are computationally hiding. Therefore, the adversary obtains no information in Step 2.a of Share^d. Furthermore, the summand s_d is shared with degree d. Hence, in Step 2.b, if not more than d parties are passively corrupted, the adversary obtains no information about s_d, and therefore not about s. In Step 2.c, whenever a value is broadcasted, the adversary knew this value already beforehand.

ROBUSTNESS: By inspection, the share protocol does not abort.

In Figure 3, we describe the reconstruction protocol for a single sharing. Clearly, this protocol can be extended to reconstruct multiple sharings in parallel by executing the protocol on a vector of sharings, where an abort in one instance implies an immediate abort (in the same round) for all instances.

Lemma 2. *Given is a d-sharing $[s]_d$ for $d < n$. If $|\mathcal{D}^*| < n - d$, then Rec^d (robustly) outputs s to all parties. Otherwise, either it outputs s to all parties, or it aborts and outputs a non-empty set $B \subseteq \mathcal{D}^*$, and the adversary obtained no information about the secret if $|\mathcal{E}^*| < n - |B|$.*

Protocol Rec^d : Given a d-sharing of some value s, reconstruct s to all parties.
1. For $i = d$ down to 1:
   (a) Each party p_j opens the commitment to its share s_{ij} via broadcast.
   (b) If at least $i+1$ parties correctly opened the commitments to their respective shares, each party locally interpolates $g_i(x)$ and computes $s_i = g_i(0)$. Otherwise, the protocol is aborted and each party outputs the set B of parties that did not broadcast correct opening information.
2. Each party outputs $s = s_1 + \ldots + s_d$.

Fig. 3. The protocol for gradual reconstruction for threshold adversaries

Proof. CORRECTNESS: The only operation in the protocol is the opening of commitments. Hence, given a correct sharing and the binding property of the commitment scheme, incorrect parties cannot deviate without being detected.
ROBUSTNESS: To abort the reconstruction of some s_i, at least $n - i \geq n - d$ parties must refuse to correctly open their respective commitments. Hence, for $|\mathcal{D}^*| < n - d$, the protocol is robust.
GRADUAL: The reconstruction aborts (with B) only if in the ith iteration (for some i), the reconstruction of s_i failed. In that case, strictly less than $i + 1$ parties opened their commitments correctly, hence $|B| \geq n - i$. Clearly, $B \subseteq \mathcal{D}^*$, since only active parties do not open their commitments correctly. Furthermore, if $|\mathcal{E}^*| < n - |B|$, the adversary has no information about $s_{|\mathcal{E}^*|}$ since the reconstruction of $s_{|\mathcal{E}^*|}$ did not yet start (note that $|\mathcal{E}^*| < n - |B| \leq i$).

The following corollary summarizes Lemma 1 (Share^d) and Lemma 2 (Rec^d):

Corollary 1. *Given a parameter $d < n$, VSS$^d = (\text{Share}^d, \text{Rec}^d)$ is a computationally (T^s, T^r)-secure, gradual VSS where $(|\mathcal{D}^*|, |\mathcal{E}^*|) \leq T^s$ if $|\mathcal{E}^*| \leq d$, and $(|\mathcal{D}^*|, |\mathcal{E}^*|) \leq T^r$ if $|\mathcal{D}^*| < n - d$.*

4 Non-reactive Multi-Party Computation

4.1 Overview

Our protocol for non-reactive MPC is based on an idea from [IKLP06]: Given the function f and the inputs $x_1, \ldots, x_n$, the protocol first distributedly computes $y = f(x_1, \ldots, x_n)$ using a correct and secret, but non-robust MPC protocol. Yet, instead of y itself, this MPC protocol outputs a sharing of y that was computed according to some VSS scheme. Then, the parties reconstruct this sharing.

In [IKLP06], whenever the non-robust MPC protocol aborts, the computation of $y = f(x_1, \ldots, x_n)$ is repeated with a robust MPC protocol, which provides security against an actively corrupted minority. Yet, if the reconstruction of y aborts, the adversary might already have learned the output, and repeating the computation would violate security. In contrast, by using a gradual VSS scheme to share y (cf. Section 3), our protocols achieve stronger security guarantees.

Given a set of actually (actively) cheating parties, a gradual VSS allows to maintain as much secrecy as possible. Then, in case of an abort during the reconstruction, the cheaters are identified and eliminated, and if the gradual VSS still guarantees enough secrecy,[6] the computation of $y = f(x_1, \ldots, x_n)$ is repeated using again the same MPC protocol among the remaining parties. Otherwise, the execution halts.

The protocol in [GMW87] (in the following denoted by GMW) provides security with abort for $t < n$ corrupted parties (i.e., correctness and secrecy against $t < n$ corrupted parties, and, in case of an abort, each correct party outputs the same non-empty set $B \subseteq \mathcal{D}^*$). However, it can easily be seen that correctness (but not secrecy) can also be achieved for $t = n$ corrupted parties.[7] We use GMW to implement the ideal functionality computing f and then a sharing of the result y.[8]

4.2 Construction

We use the gradual VSS scheme described in Section 3.2 with degree $d = n-1$. In fact, we only require the reconstruction protocol Rec^{n-1} and the (probabilistic) function State^{n-1} that, given a value y, samples shares of y according to VSS^{n-1}. Furthermore, the protocol receives a robustness parameter e stating the number of actively corrupted parties that the protocol can eliminate (and then repeat the run) without violating security. A set of parties is eliminated by removing the parties from $\mathcal{P}$ and reducing n and e accordingly. For details see Figure 4.

Lemma 3. *Given a function f and a robustness parameter e, the protocol for non-reactive MPC computes f in presence of an adversary corrupting $(|\mathcal{D}^*|, |\mathcal{E}^*|)$. It is always correct, robust if $|\mathcal{D}^*| \leq e$, secret if $|\mathcal{D}^*| + |\mathcal{E}^*| < n$ or $|\mathcal{E}^*| < n - e$, and fair if $|\mathcal{D}^*| + |\mathcal{E}^*| < n$.*

Proof. CORRECTNESS and ROBUSTNESS follow trivially by inspection.

SECRECY: GMW is secret for any number of corrupted parties. Furthermore, since $|\mathcal{E}^*| \leq n - 1$ (otherwise there is no secrecy requirement), it follows from Corollary 1 that the output $[y]_{n-1}$ reveals no information to the adversary. Hence, he obtains no information about the inputs in Step 1. Steps 2 and 3 are independent from the inputs given the output. Therefore, if the protocol does not abort,

---

[6] The protocols are described with respect to a robustness parameter rather than a secrecy parameter as suggested here. It turns out that this simplifies the description and the proof.

[7] In particular this holds also in the setting with mixed adversaries where some parties are actively and all remaining parties are passively corrupted. This follows from the fact that each party has to prove the correctness of the messages it sends using a zero-knowledge protocol. Given instant randomness (i.e., randomness generated only when needed), even the challenges of passively corrupted parties are unpredictable to the adversary.

[8] Vanilla [GMW87] considers only Boolean circuits. However, any ideal functionality can be converted into a Boolean circuit in a straightforward way.

Non-reactive MPC: Given are a function f and a robustness parameter e.

1. Employ GMW to first compute $y = f(x_1, \ldots, x_n)$, where x_i is the input from party p_i, then evaluate $\mathtt{State}^{n-1}$ on y, and finally output to each party its corresponding share, resulting in $[y]_{n-1}$. If GMW aborts with a set B of active cheaters, repeat with $\mathcal{P} = \mathcal{P} \setminus B$, $n = n - |B|$, and $e = e - |B|$.
2. Invoke $\mathtt{Rec}^{n-1}$ on $[y]_{n-1}$. On abort with a set B of active cheaters: If $|B| \leq e$, then repeat the whole protocol with $\mathcal{P} = \mathcal{P} \setminus B$, $n = n - |B|$, and $e = e - |B|$. Otherwise, halt the execution.
3. Output y.

Fig. 4. The protocol for non-reactive MPC for threshold adversaries

secrecy is maintained. Yet, secrecy may be violated if the adversary can force a repetition of the protocol after learning the output.[9] If the protocol aborts in Step 2 with $B \subseteq \mathcal{D}^*$, then in the case $|\mathcal{D}^*| + |\mathcal{E}^*| < n$ we directly have $|\mathcal{E}^*| < n - |\mathcal{D}^*| \leq n - |B|$, hence secrecy is maintained. In the case $|\mathcal{E}^*| < n - e$, we either have that $|\mathcal{E}^*| < n - |B|$ (and secrecy is maintained), or $|\mathcal{E}^*| \geq n - |B|$, hence $|B| \geq n - |\mathcal{E}^*| > e$ and the protocol aborts, i.e. the adversary learns at most one output value.

FAIRNESS is a subcase of secrecy and therefore omitted.

Given Lemma 3, we can derive a tight bound for (non-reactive) MPC:

Theorem 1. *In the model with broadcast and multi-threshold adversaries, computationally secure (non-reactive) MPC among n parties with thresholds T^c, T^s, T^r, and T^f, where $T^f \leq T^s \leq T^c$ and $T^r \leq T^c$, is possible if either $T^s = \{(0,0)\}$, or*

$$\left(\forall (t_a^s, t_p^s) \leq T^s, (t_a^r, \cdot) \leq T^r : t_a^r + t_p^s < n \ \lor \ t_a^s + t_p^s < n \right)$$
$$\land \ \left(\forall (t_a^f, t_p^f) \leq T^f : t_a^f + t_p^f < n \right).$$

This bound is tight: If violated, there are (non-reactive) functionalities that cannot be securely computed.

Proof. The proof of necessity can be found in Section 4.3. To prove sufficiency, first consider the (trivial) case $T^s = \{(0,0)\}$. Then, every party simply broadcasts its inputs and computes the function on the broadcasted values (c.f. Section 2.4).

Otherwise, we use the protocol in Figure 4 with $e = \hat{t}_a^r$, where $\hat{t}_a^r$ is the maximal t_a^r value in T^r.

CORRECTNESS is always guaranteed, and ROBUSTNESS follows directly from the choice of e.

SECRECY: Since $(|\mathcal{D}^*|, |\mathcal{E}^*|) \leq T^s$, we immediately have that $|\mathcal{D}^*| + |\mathcal{E}^*| < n \ \lor \ \forall (t_a^r, \cdot) \leq T^r : t_a^r + |\mathcal{E}^*| < n$. Then, it follows from the choice of e that $|\mathcal{D}^*| + |\mathcal{E}^*| < n \ \lor \ e + |\mathcal{E}^*| < n$.

---

[9] In that case, the adversary may learn two evaluations of f for different inputs.

FAIRNESS: Since $(|\mathcal{D}^*|, |\mathcal{E}^*|) \leq T^f$ and $\forall (t_a^f, t_p^f) \leq T^f : t_a^f + t_p^f < n$, we immediately have $|\mathcal{D}^*| + |\mathcal{E}^*| < n$.

Theorem 2. *There exists a cryptographically secure multi-party computation protocol among n parties for non-reactive functionalities which is fully secure against all adversaries $(\mathcal{D}, \mathcal{E})$ with $|\mathcal{D}| + |\mathcal{E}| < n$.*

Proof. Apply Theorem 1 with $T^c = T^s = T^r = T^f = \{(k, n - k - 1) : k = 0, \ldots, \lfloor \frac{n-1}{2} \rfloor\}$.

4.3   Proof of Necessity for Non-reactive MPC

In this section, we prove that the bound in Theorem 1 is necessary, i.e. if violated, (non-reactive) MPC is impossible. The bound in Theorem 1 is violated if $T^s \neq \{(0,0)\}$ and $\left(\exists (t_a^s, t_p^s) \leq T^s, (t_a^r, \cdot) \leq T^r : t_a^r + t_p^s \geq n \ \wedge \ t_a^s + t_p^s \geq n \right) \ \vee \ \left(\exists (t_a^f, t_p^f) \leq T^f : t_a^f + t_p^f \geq n \right)$

Case: $\exists (t_a^s, t_p^s) \leq T^s, (t_a^r, \cdot) \leq T^r : \quad t_a^r + t_p^s \geq n \ \wedge \ t_a^s + t_p^s \geq n$.
For the sake of contradiction, assume that there is a protocol for this setting, and without loss of generality assume that $t_p^s > 0$ (there is such a t_p^s because $T^s \neq \{(0,0)\}$). For each $E \subseteq \mathcal{P}$, let ℓ_E denote the first round in the protocol in which the parties in E jointly can efficiently compute the output. Among all subsets $E \subseteq \mathcal{P}$ with $|E| = t_p^s$, let $\mathcal{E}$ denote the one with minimal ℓ_E, i.e., $\mathcal{E} \in \{E \subseteq \mathcal{P} : |E| = t_p^s \wedge \nexists E' \subseteq \mathcal{P} : |E'| = t_p^s \wedge \ell_{E'} < \ell_E\}$. Now consider an adversary actively corrupting some subset $\mathcal{D} \subseteq \mathcal{E}$ with $|\mathcal{D}| = n - t_p^s$, and let him abort all parties in $\mathcal{D}$ in round $\ell - 1$. By assumption, the remaining t_p^s parties $\mathcal{P} \setminus \mathcal{D}$ cannot compute the output (corresponding to the actual inputs). However, the protocol must not abort, as the actual adversary could be $(\mathcal{D}^*, \mathcal{E}^*) = (\mathcal{D}, \mathcal{D})$, for which robustness is guaranteed as $(|\mathcal{D}^*|, |\mathcal{E}^*|) = (n - t_p^s, n - t_p^s) \leq (t_a^r, t_a^r) \leq T^r$. Hence, the remaining parties must again take inputs and set default values for the inputs of parties in $\mathcal{D}$, but this violates secrecy if the actual adversary is $(\mathcal{D}^*, \mathcal{E}^*) = (\mathcal{D}, \mathcal{E})$ (note that $(|\mathcal{D}^*|, |\mathcal{E}^*|) = (n - t_p^s, t_p^s) \leq T^s$), who then learns the output of this run as well as the output of the next run with default input values for the parties in $\mathcal{D}$ (note that $|\mathcal{E}^*| \geq 1$).

As an example, consider the following (generalized OT-) functionality: Each party p_i inputs three bits: $a_0^{(i)}$, $a_1^{(i)}$, and $b^{(i)}$ (with default input $a_0^{(i)} = a_1^{(i)} = b^{(i)} = 0$). Let $d = b^{(1)} \oplus \ldots \oplus b^{(n)}$. The output is $y = (a_d^{(1)}, \ldots, a_d^{(n)})$. The adversary lets one actively corrupted party input $b = 1$, and all others $b = 0$. Then, with the attack described above, the adversary learns both $y_0 = (a_0^{(1)}, \ldots, a_0^{(n)})$ and $y_1 = (a_1^{(1)}, \ldots, a_1^{(n)})$, which clearly is a violation of secrecy.

Case: $\exists (t_a^f, t_p^f) \leq T^f : \quad t_a^f + t_p^f \geq n$. Again, assume that there is a protocol for this setting, and let $\mathcal{E}$ denote the subset among all subsets $E \subseteq \mathcal{P}$ with $|E| = t_p^f$ such that ℓ_E is minimal (see Section 4.3). Consider the adversary

$(\mathcal{D}^*, \mathcal{E}^*) = (\mathcal{D}, \mathcal{E})$ for $\mathcal{D} \subseteq \mathcal{E}$ with $|\mathcal{D}| = n - t_p^f$, and let him abort all parties in $\mathcal{D}^*$ in round $\ell - 1$. By assumptions, the remaining t_p^f parties in $\mathcal{P} \setminus \mathcal{D}^*$ are not able to efficiently compute the output, whereas the adversary $(\mathcal{D}^*, \mathcal{E}^*)$ does learn the output, a violation of fairness.

5 Reactive Multi-Party Computation

5.1 Overview

For our protocol for reactive MPC, we adapt an idea from [IKLP06] and modify the given functionality $\mathcal{F}$ as follows: For each output y, instead of the value itself, it outputs a sharing of y that was computed according to some VSS scheme. Then, to obtain the output, the parties reconstruct this sharing. This modified $\mathcal{F}$ is implemented using an MPC protocol that is always correct, and as robust and secret as some (second) VSS scheme.

In contrast to [IKLP06], we use a gradual VSS scheme for the modification of $\mathcal{F}$. The gradual property allows to provide fairness beyond robustness. In fact, we only require the (probabilistic) function State that, given a value y, samples shares of y according to the gradual VSS scheme. We modify $\mathcal{F}$ such that it invokes State on each output value y, and then outputs the shares of y (instead of y itself). We modify $\mathcal{F}$ to use VSS^d (Section 3.2) and denote the resulting functionality by $\mathcal{F}^d$.

The MPC protocol implementing the (modified) functionality $\mathcal{F}$ receives as parameter a (T^s, T^r)-secure VSS, and then provides correctness for any number of corrupted parties, secrecy if $(\mathcal{D}^*, \mathcal{E}^*) \leq T^s$, and robustness if $(\mathcal{D}^*, \mathcal{E}^*) \leq T^r$. Furthermore, if the protocol is aborted, then each party outputs the same non-empty set $B \subseteq \mathcal{D}^*$. Clearly, the non-robust protocol used in Section 4 can be extended accordingly with a VSS as described in [GMW87].[10] We instantiate the protocol using VSS^d (Section 3.2) and denote the resulting protocol by GMW^d. Note that for this extension of GMW, a standard, non-gradual VSS would be sufficient.

5.2 Construction

We use the gradual VSS scheme described in Section 3.2 with the same sharing degree d for both the modification of $\mathcal{F}$, resulting in $\mathcal{F}^d$, and within GMW, resulting in GMW^d.

Lemma 4. *Given a functionality $\mathcal{F}$ and a parameter $d < n$, the protocol for reactive MPC implements $\mathcal{F}$ in presence of an adversary corrupting $(|\mathcal{D}^*|, |\mathcal{E}^*|)$. It is always correct, secret if $|\mathcal{E}^*| \leq d$, robust if $|\mathcal{D}^*| < n - d$, and fair if $|\mathcal{E}^*| \leq d \wedge |\mathcal{D}^*| + |\mathcal{E}^*| < n$.*

---

[10] The original description considers only VSS with a threshold of $n/2$. However, it is easy to see that any VSS can be used. The resulting protocol inherits the robustness and secrecy properties of the corresponding VSS, while leaving the correctness properties unchanged. The same holds for the simplified protocol in [Gol04, p. 735].

Reactive MPC: Given are a functionality $\mathcal{F}$ and a sharing degree d.

1. Invoke $\texttt{GMW}^d$ implementing $\mathcal{F}^d$.
2. On each output $[y]_d$, invoke $\texttt{Rec}^d$. If it aborts, halt the execution. Otherwise, output y.

Fig. 5. The protocol for reactive MPC for threshold adversaries

Proof. CORRECTNESS follows trivially by inspection, and SECRECY and RO-BUSTNESS follow immediately from Corollary 1. FAIRNESS: Since $|\mathcal{E}^*| \leq d$, it follows from Corollary 1 that the adversary obtains no information in Step 1. Furthermore, if the reconstruction of an output value aborts with B, the gradual property guarantees that $B \subseteq \mathcal{D}^*$. Since $|\mathcal{D}^*| + |\mathcal{E}^*| < n$, we then have $|\mathcal{E}^*| < n - |\mathcal{D}^*| \leq n - |B|$, hence, the adversary did not obtain any information about y and fairness is preserved.

Given Lemma 4, we can derive a tight bound for reactive MPC:

Theorem 3. *In the model with broadcast and multi-threshold adversaries, computationally secure (reactive) MPC among n parties with thresholds T^c, T^s, T^r, and T^f, where $T^f \leq T^s \leq T^c$ and $T^r \leq T^c$, is possible if either $T^s = \{(0,0)\}$, or*

$$\forall(t_a^r, \cdot) \leq T^r, (\cdot, t_p^s) \leq T^s : t_a^r + t_p^s < n \quad \wedge \quad \forall(t_a^f, t_p^f) \leq T^f : t_a^f + t_p^f < n$$

This bound is tight: If violated, there are (reactive) functionalities that cannot be securely computed.

Proof. To prove sufficiency, first consider the (trivial) case $T^s = \{(0,0)\}$. Then, every party simply broadcasts its inputs and computes the function on the broadcasted values (c.f. Section 2.4). Otherwise, we use the protocol described in Figure 5 with $d = \hat{t}_p^s$, where $\hat{t}_p^s$ is the maximal t_p^s value in T^s. CORRECTNESS is always guaranteed, and SECRECY follows immediately from the choice of d. ROBUSTNESS: Since $(|\mathcal{D}^*|, |\mathcal{E}^*|) \leq T^r$ and $\forall(t_a^r, \cdot) \leq T^r, (\cdot, t_p^s) \leq T^s : t_a^r + t_p^s < n$, we have that $\forall(\cdot, t_p^s) \leq T^s : |\mathcal{D}^*| + t_p^s < n$. Then, it follows from the choice of d that $|\mathcal{D}^*| + d < n$.
FAIRNESS: Given is that $(|\mathcal{D}^*|, |\mathcal{E}^*|) \leq T^f$. Since $T^f \leq T^s$, we have $|\mathcal{E}^*| \leq d$. Furthermore, since $\forall(t_a^f, t_p^f) \leq T^f : t_a^f + t_p^f < n$, we immediately have $|\mathcal{D}^*| + |\mathcal{E}^*| < n$.
The proof of necessity is given in the next section.

5.3   Proof of Necessity for Reactive MPC

In this section, we prove that the bound in Theorem 3 is necessary, i.e. if violated, (reactive) MPC is impossible. The bound in Theorem 3 is violated if $T^s \neq \{(0,0)\}$ and

$$\exists(t_a^r, \cdot) \leq T^r, (\cdot, t_p^s) \leq T^s : t_a^r + t_p^s \geq n \quad \vee \quad \exists(t_a^f, t_p^f) \leq T^f : t_a^f + t_p^f \geq n$$

Case: $\exists(t_a^r, \cdot) \leq T^r, (\cdot, t_p^s) \leq T^s$: $\quad t_a^r + t_p^s \geq n.$ Assume that there is a secure protocol for this setting. Then, the adversary corrupts $(\mathcal{D}^*, \mathcal{E}^*) = (\mathcal{D}, \mathcal{D})$ with $\mathcal{D} \subseteq \mathcal{P}$ and $|\mathcal{D}| = n - t_p^s$ and has the parties in $\mathcal{D}$ stop sending messages. Since there are only t_p^s remaining parties, the state is lost and the computation cannot be continued. Hence, robustness is violated.[11]

Case: $\exists(t_a^f, t_p^f) \leq T^f$: $\quad t_a^f + t_p^f \geq n.$ Same as in the non-reactive case (Section 4.3).

6 Conclusions

In this work, we have generalized and extended known results from the literature. In particular, we improved over the work in [IKLP06] that combines optimal results from the active and the passive world. Our protocols distinguish not only whether or not active cheating occurs, but provides a dynamic tradeoff between active and passive corruptions. Hence, we achieve "the best of both worlds – and everything in between" with a single protocol.

Furthermore, we introduced the notion of *gradual* verifiable secret sharing. This notion requires that, during reconstruction, secrecy is given up gradually, one subset at a time, while immediately establishing robustness against the corresponding complement set. As a consequence, intuitively speaking, the adversary might still abort the protocol, but does not automatically learn the secret. This technique turned out to be very useful in the setting of both non-reactive and reactive MPC to provide more flexible and therefore more practical protocols.

Moreover, the use of multi-thresholds allows to unify two incomparable models for combining active and passive corruption. In the first model, used for example by [IKLP06], the adversary can corrupt parties either passively or actively, but not both at the same time. Then, for each of the two corruption options, a maximally tolerable adversary is considered. In the second model, used for example by [FHM98], the adversary can corrupt some parties actively, and additionally some parties passively, at the same time. Yet, their model only allows to consider a single maximally tolerable adversary. By using multi-thresholds, we can provide a single protocol that subsumes results for both models simultaneously.

A Comparison with Related Work

For completeness, we summarize the considered models and achieved security levels of several protocols in the literature. In case of protocols with hybrid security, we indicate in parentheses over which properties the hybridization is achieved.

---

[11] Note that the proof in [IKLP06] considers only the special case where $t_a^r \leq t_p^s$.

Paper	Prot	Adv.	Security
[Cha89]	MPC	mixed	hybrid (comp/statistical)
[DDWY93]	SMT	mixed	perfect
[FHM98]	MPC	mixed	statistical
[HMZ08]	MPC	mixed	computational or statistical
[FHHW03]	BA	active	perfect
[FHW04]	MPC	active	hybrid (comp/stat)
[Kat07]	MPC	active	hybrid (output guarantee)
[LRM10]	MPC	active	hybrid (comp/stat and robustness/fairness)
[HLMR11]	MPC	mixed	perf., hybrid (privacy/correctness/robustness/fairness)
[HLMR12]	MPC	mixed	stat., hybrid (privacy/correctness/robustness/fairness)
this work	MPC	mixed	comp., hybrid (privacy/correctness/robustness/fairness)

MPC/SMT/BA = MPC/secure message transmission/Byzantine agreement,
comp/stat/perf = computational/statistical/perfect.

References

[Bea91]   Beaver, D.: Secure multiparty protocols and zero-knowledge proof systems tolerating a faulty minority. Journal of Cryptology 4(2), 75–122 (1991)

[BGW88]   Ben-Or, M., Goldwasser, S., Wigderson, A.: Completeness theorems for non-cryptographic fault-tolerant distributed computation. In: STOC 1988, pp. 1–10. ACM (1988)

[Blu83]   Blum, M.: How to exchange (secret) keys (extended abstract). In: STOC 1983, pp. 440–447. ACM (1983)

[CCD88]   Chaum, D., Crépeau, C., Damgård, I.: Multiparty unconditionally secure protocols. In: STOC 1988, pp. 11–19. ACM (1988)

[Cha89]   Chaum, D.: The spymasters double-agent problem: Multiparty computations secure unconditionally from minorities and cryptograhically from majorities. In: Brassard, G. (ed.) CRYPTO 1989. LNCS, vol. 435, pp. 591–602. Springer, Heidelberg (1990)

[DDWY93]   Dolev, D., Dwork, C., Waarts, O., Yung, M.: Perfectly secure message transmission. Journal of the ACM 40(1), 17–47 (1993)

[DS82]   Dolev, D., Strong, H.R.: Polynomial algorithms for multiple processor agreement. In: STOC 1982, pp. 401–407. ACM (1982)

[FHHW03]   Fitzi, M., Hirt, M., Holenstein, T., Wullschleger, J.: Two-threshold broadcast and detectable multi-party computation. In: Biham, E. (ed.) EUROCRYPT 2003. LNCS, vol. 2656, pp. 51–67. Springer, Heidelberg (2003)

[FHM98]   Fitzi, M., Hirt, M., Maurer, U.M.: Trading correctness for privacy in unconditional multi-party computation (extended abstract). In: Krawczyk, H. (ed.) CRYPTO 1998. LNCS, vol. 1462, pp. 121–136. Springer, Heidelberg (1998)

[FHW04]   Fitzi, M., Holenstein, T., Wullschleger, J.: Multi-party computation with hybrid security. In: Cachin, C., Camenisch, J.L. (eds.) EUROCRYPT 2004. LNCS, vol. 3027, pp. 419–438. Springer, Heidelberg (2004)

[GMW87]   Goldreich, O., Micali, S., Wigderson, A.: How to play any mental game or a completeness theorem for protocols with honest majority. In: STOC 1987, pp. 218–229. ACM (1987)

[Gol04]   Goldreich, O.: Foundations of Cryptography. Basic Applications, vol. 2. Cambridge University Press (2004)

[HLMR11] Hirt, M., Lucas, C., Maurer, U., Raub, D.: Graceful degradation in multi-party computation (extended abstract). In: Fehr, S. (ed.) ICITS 2011. LNCS, vol. 6673, pp. 163–180. Springer, Heidelberg (2011)

[HLMR12] Hirt, M., Lucas, C., Maurer, U., Raub, D.: Passive corruption in statistical multi-party computation. In: Smith, A. (ed.) ICITS 2012. LNCS, vol. 7412, pp. 129–146. Springer, Heidelberg (2012)

[HMZ08] Hirt, M., Maurer, U., Zikas, V.: MPC vs. SFE: Unconditional and computational security. In: Pieprzyk, J. (ed.) ASIACRYPT 2008. LNCS, vol. 5350, pp. 1–18. Springer, Heidelberg (2008)

[IKLP06] Ishai, Y., Kushilevitz, E., Lindell, Y., Petrank, E.: On combining privacy with guaranteed output delivery in secure multiparty computation. In: Dwork, C. (ed.) CRYPTO 2006. LNCS, vol. 4117, pp. 483–500. Springer, Heidelberg (2006)

[Kat07] Katz, J.: On achieving the "best of both worlds" in secure multiparty computation. In: STOC 2007, pp. 11–20. ACM (2007)

[LRM10] Lucas, C., Raub, D., Maurer, U.: Hybrid-secure MPC: Trading information-theoretic robustness for computational privacy. In: PODC 2010, pp. 219–228. ACM (2010)

[Ped91] Pedersen, T.P.: Non-interactive and information-theoretic secure verifiable secret sharing. In: Feigenbaum, J. (ed.) CRYPTO 1991. LNCS, vol. 576, pp. 129–140. Springer, Heidelberg (1992)

[RB89] Rabin, T., Ben-Or, M.: Verifiable secret sharing and multiparty protocols with honest majority. In: STOC 1989, pp. 73–85. ACM (1989)

[Sha79] Shamir, A.: How to share a secret. Communications of the ACM 22(11), 612–613 (1979)

[Yao82] Yao, A.C.: Protocols for secure computations (extended abstract). In: FOCS 1982, pp. 160–164. IEEE (1982)

What Information Is Leaked under Concurrent Composition?

Vipul Goyal[1], Divya Gupta[2,*], and Abhishek Jain[3,**]

[1] Microsoft Research, India
vipul@microsoft.com
[2] UCLA
divyag@cs.ucla.edu
[3] MIT and Boston University
abhishek@csail.mit.edu

Abstract. A long series of works have established far reaching impossibility results for concurrently secure computation. On the other hand, some positive results have also been obtained according to various weaker notions of security (such as by using a super-polynomial time simulator). This suggest that somehow, "not all is lost in the concurrent setting."

In this work, we ask *what and exactly how much private information can an adversary learn by launching a concurrent attack?* Inspired by the recent works on leakage-resilient protocols, we consider a security model where the ideal world adversary (a.k.a simulator) is allowed to query the trusted party for some "leakage" on the honest party inputs. (Intuitively, the amount of leakage required by the simulator upper bounds the security loss in the real world).

We show for the first time that in the concurrent setting, it is possible to achieve *full security* for "most" of the sessions, while incurring significant loss of security in the remaining (fixed polynomial fraction of total) sessions. We also give a lower bound showing that (for general functionalities) this is essentially optimal. Our results also have interesting implications to bounded concurrent secure computation [Barak-FOCS'01], as well as to precise concurrent zero-knowledge [Pandey et al.-Eurocrypt'08] and concurrently secure computation in the multiple ideal query model [Goyal et al.-Crypto'10]

At the heart of our positive results is a new simulation strategy that is inspired by the classical set covering problem. On the other hand, interestingly, our negative results use techniques from leakage-resilient cryptography [Dziembowski-Pietrzak-FOCS'08].

1 Introduction

Concurrently Secure Computation. Traditional security notions for cryptographic protocols such as secure computation [38,16] were defined for a

---

* Work done in part while visiting Microsoft Research, India.
** Supported by NSF Contract CCF-1018064 and DARPA Contract Number: FA8750-11-2-0225. The author also thanks RISCS (Reliable Information Systems and Cyber Security) Institute. Work done in part while visiting Microsoft Research, India.

R. Canetti and J.A. Garay (Eds.): CRYPTO 2013, Part II, LNCS 8043, pp. 220–238, 2013.

stand-alone setting, where security holds only if a single protocol session is executed in isolation. Today's world, however, is driven by networks – the most important example being the Internet. In a networked environment, several protocol instances may be executed *concurrently*, and an adversary may be able to perform coordinated attacks across sessions by corrupting parties in various sessions. As such, a protocol that is secure in the classical standalone setting may become completely insecure in the network setting.

Towards that end, over the last decade, a tremendous amount of effort has been made to obtain protocols with strong composability guarantees under concurrent execution. Unfortunately, a sequence of works have demonstrated far reaching impossibility results for designing secure protocols in the concurrent setting [8,9,26,25,27,3,19,1,15]. In particular, these works have ruled out secure realization of essentially all non-trivial functionalities even in very restricted settings such as where inputs of honest parties are fixed in advance (rather than being chosen adaptively in each session), and where the adversary is restricted to corrupting parties with specific roles.

What Information is Getting Leaked to the Adversary? Many of these impossibility results work by designing an explicit "chosen protocol attack". Such an attack shows that there exists some information the concurrent adversary can learn in the real world which is impossible to obtain for the ideal adversary (a.k.a the simulator). Nevertheless, subsequent to these impossibility results, several prior works have in fact obtained positive results for concurrently secure computation according to various relaxed notions of security such as super-polynomial simulation [31,4,10,13,24], input indistinguishable computation [29,13], multiple-ideal query model [20], etc.[1] These results suggest that somehow, *not all security is lost in the concurrent setting*. Given the above, the following natural questions arise:

What and exactly how much private information can the adversary learn by launching a concurrent attack? Can we "measure" the amount of security loss that must occur in a concurrent session? Can we achieve full security in some (or even most) of the sessions fully while incurring security loss in the remaining sessions?

We believe the above questions are very natural to ask and fundamental to the understanding of concurrent composition. Indeed, despite a large body of research on the study of concurrent composition, in our opinion, the understanding of "what exactly is it that goes wrong in the concurrent setting, and, to what extent" is currently unsatisfactory. The current paper represents an attempt towards improving our understanding of this question.

A Leaky-Ideal World Approach. We adopt the "leaky-ideal world" approach of Goldreich and Petrank [17] (recently used in works on leakage-resilient protocols; see below) to quantify the information leakage to the adversary in concurrently secure computation. Specifically, generalizing the approach of [17], we

---

[1] There has also been a rich line of works on designing secure computation with some type of "setup" where, e.g., a trusted party publishes a randomly chosen string [7,2,22]. However the focus of the current work is the *plain model*.

consider a modification of the standard real/ideal paradigm where in the ideal world experiment, the simulator is allowed to query the trusted party for some "leakage" on the honest party inputs. The underlying intuition (as in [17]) is that the amount of leakage observed by the simulator in order to simulate the view of an adversary represents an upper bound on the amount of private information potentially leaked to the real adversary during the concurrent protocol executions.

We remark that the our ideal model resembles that considered in the recent works on leakage-resilient secure computation protocols [14,5,6]. However, we stress that in our setting, there is *no* physical leakage in the real world and instead there are just an (unbounded) polynomial number of concurrent sessions. Indeed, while [14,5,6] use the leaky ideal world approach to bound the security loss in the real world due to leakage attacks, we use the leaky ideal world approach to bound the security loss in the real world due to concurrent attacks. Nevertheless, we find it interesting that there is a parallel between the ideal world guarantees considered in two unrelated settings: leaky real world, and, concurrent real world.

We now describe our security model in more detail. Concretely, we consider two notions of leaky ideal world, described as follows.

Ideal World with Joint Leakage. Let there be m concurrent sessions with the honest party input in the ith session denoted by x_i. In the *joint* leakage model, the simulator is allowed to query the trusted party with efficiently computable leakage functions L_i and get $L_i(X)$ in return (where $X = (x_1, \ldots, x_m)$). The constraint is that throughout the simulation, the total number of bits leaked $\sum L_i(X)$ is at most $\epsilon|X|$. If this is the case, we say that the protocol is ϵ-secure in the joint leakage model. In this model, our main result is a positive one, as we discuss below.

Ideal World with Individual Leakage. In the *individual* leakage model, in every session i, the simulator can query with an efficiently computable leakage function L_i and get $L_i(x_i)$ in return. The constraint is that in every session i, the length of $L_i(x_i)$ is at most $\epsilon|x_i|$. If this is the case, we say that the protocol is ϵ-secure in the individual leakage model. As we discuss below, in this model, our main result is a negative one. This brings us to our next model.

1.1   Our Results

We consider the setting of unbounded concurrent composition in the plain model. We allow for static corruptions and assume that the inputs of honest parties are a priori fixed. We now describe our main results along with some applications.

I. Positive Result in the Joint Leakage Model. We obtain the following main result in the joint leakage model:

Theorem 1. *(Informally stated.) Let f be any functionality. Assuming 1-out-of-2 oblivious transfer (OT), for every polynomial $poly(n)$, there exists a protocol that $(\epsilon = \frac{1}{poly(n)})$-securely realizes f in the joint leakage model.*

The round complexity of our protocol is $\frac{\log^6 n}{\epsilon}$. We show that this is *almost optimal* w.r.t. a black-box simulator: we rule out protocols with round complexity $\frac{O(\log n)}{\epsilon}$ proven secure using a black-box simulator.

Fully Preserving the Security of most Sessions. We note that the simulator for our positive result, in fact, satisfies the following additional property: rather than leaking a small fraction of the input in each session, it leaks the entire input of a small (i.e., ϵ) fraction of sessions while *fully* preserving the security of the remaining sessions. Hence, we get the following interesting corollary:

Theorem 2. *Let f be any functionality. Assuming 1-out-of-2 OT, for every polynomial $poly(n)$, there exists a protocol that $(\epsilon = \frac{1}{poly(n)})$-securely realizes f in the joint leakage model s.t. the security of at most ϵ fraction of the sessions is compromised, while the remaining sessions are fully secure.*

In fact, our negative result in the independent leakage model (discussed below) indicates that for a general positive result, the above security guarantee is essentially optimal.

Bounded Concurrent Secure Computation with Graceful Security Degradation. Going further, observe that by choosing $\epsilon < \frac{1}{m|X|}$, we get a construction where the simulator is allowed *no leakage* at all if the number of sessions is up to m. This is because the maximum number of bits simulator is allowed to leak will be $\epsilon m|X|$ which is less than 1. Hence, positive results for bounded concurrent secure computation [25,33,32] follow as a special case of our result. However if the actual number of sessions just slightly exceed m, the simulator is allowed some small leakage on the input vector (i.e., total of only 1 bit up to $2m$ sessions, 2 bits up to $3m$ sessions, and so on). Thus, the leakage allowed grows slowly as the number of sessions grow. This phenomenon can be interpreted as *graceful degradation of security* in the concurrent setting.

Theorem 3. *(Informally Stated.) Let f be any functionality. Assuming 1-out-of-2 oblivious transfer, there exists a protocol that securely realizes f in the bounded concurrent setting. However if the actual number of sessions happen to exceed this bound, there is graceful degradation of security as the number of sessions increase.*

A Set-Cover Approach to Concurrent Extraction. In order to obtain our positive result, we take a generic "cost-centric" approach to rewinding in the concurrent setting. For example, in our context, the amount of leakage required by the simulator to simulate the protocol messages during the rewindings can be viewed as the "cost" of extraction. Thus, the goal is to perform concurrent extraction with minimal cost. With this view, we model concurrent extraction as the classical *set-covering problem* and develop, as our main technical contribution, a new **sparse rewinding strategy.** Very briefly, unlike known concurrent rewinding techniques [37,23,36,30] that are very "dense", we rewind "small intervals" of the execution transcript, while still guaranteeing extraction in all of the sessions. Very roughly, by rewinding small intervals (only a few times), we are able to minimize the cost and obtain our positive result.

Our sparse rewinding strategy also yields other interesting applications that we discuss below in (III).

II. Negative Result in the Individual Leakage Model. In the individual leakage model, our main result is negative, ruling out even *non-black-box* simulation. Specifically, we give an impossibility result for the OT functionality where the ideal leakage allowed is $(1/2 - \delta)$ fraction of the input length (for every positive constant δ). Note that this is the maximum possible leakage bound such that the ideal adversary still does not learn the entire input of the honest parties (which would otherwise result in a trivial positive result).[2]

Leakage-resilient One-Time Programs. Of independent interest, the techniques used in our negative result also yield a new construction of one-time programs [18] where the adversary can query the given hardware tokens once (as usual), and *additionally* leak once on the secrets stored in each token in any arbitrary manner (as long as the total leakage is a constant fraction of the secrets). Our key technical tool in constructing such a gadget is the intrusion-resilient secret sharing scheme of [12]. In an independent work, Jain et al. [21] also consider the problem of constructing leakage-resilient OTPs. See the full version for details.

Put together, results (I) and (II) show that in the concurrent setting, significant loss of security in some of the sessions is unavoidable if one wishes to obtain a general positive result. However on the brighter side, one can make the fraction of such sessions to be an *arbitrarily small polynomial* (while *fully* preserving the security in all other sessions).

III. Other Applications. As discussed above, along the way to developing our main positive result, we develop a new *sparse rewinding strategy* that leads to other interesting applications. We discuss them below.

Improved precise concurrent zero knowledge. In the traditional notion of zero-knowledge, the simulator may run in time which is any polynomial factor of the (worst-case) running time of the adversarial verifier. The notion of precise zero-knowledge [28] deals with studying how low this polynomial can be. In particular, can one design protocols where the running time of the simulator is only slightly higher than the actual running time of the adversary? Besides being a fundamental question on its own, the notion of precise zero-knowledge has found applications in unrelated settings such as leakage-resilient zero-knowledge [14], concurrently secure protocols [20], etc.

Pandey et al. [30] study the problem of precise *concurrent* zero-knowledge (cZK) and give a protocol with the following parameters. Let t be the actual running time of the verifier. Then, their protocol has round complexity n^{δ}

---

[2] Indeed, if the fraction of leakage allowed is 1/2, the ideal adversary can learn one of the sender inputs by making use of leakage, and, the other by making use of the "official" trusted party call.

(for any constant $\delta \leq 1$) and knowledge precision $c \cdot t$ where c is a large constant depending upon the adversary.[3]

Our sparse rewinding strategy directly leads to a new construction of precise cZK, improving upon [30] *both* in terms of round complexity as well as knowledge precision.

Theorem 4. *Assuming one way functions, there exists a cZK protocol with poly-log round-complexity and knowledge precision of $(1 + \delta)t$ (for any constant δ).*

Improved concurrently secure computation in the MIQ model. In the quest for positive results for concurrently secure computation, Goyal et al. proposed the multiple ideal query (MIQ) model, where for every session in the real world, the simulator is allowed to query the ideal functionality for the output *multiple* times (as opposed to only *once*, as in the standard definition of secure computation). They construct a protocol in this model whose security is proven w.r.t. a simulator that makes a total of $c \cdot m$ number of ideal queries in total (and c queries per session, *on an average*), where c is a large constant that depends on the adversary and m is the number of sessions.

We note that our security model is intimately connected to the MIQ model since the additional output queries in this model can simply be viewed as leakage observed by the simulator in our model. Indeed, our positive result described in (I) can be stated as an improved result in the MIQ model since leaking the function output (multiple times) is "no worse" than leaking the entire secret input of the honest party. We defer further discussion to the full version due to lack of space.

Theorem 5. *(Informally stated.) Let f be any functionality. Assuming 1-out-of-2 OT, there exists a concurrently secure protocol in the MIQ model with $(1 + \frac{1}{poly(n)})$ number of ideal queries per session (on an average).*

1.2 Our Techniques

Here we give an overview of the underlying techniques used in our positive result.

A Starting Approach. A well established approach to constructing secure computation protocols against malicious adversaries in the standalone setting is to use the GMW compiler [16]: take a semi-honest secure computation protocol and "compile" it with zero-knowledge arguments. Then, a natural starting point to construct a concurrently secure computation protocol is to follow the same principles in the concurrent setting: somehow compile a semi-honest secure computation protocol with a concurrent zero-knowledge protocol (for security in more demanding settings, compilation with concurrent non-malleable zero-knowledge [3] may be required). Does such an approach (or minor variants) already give us protocols secure according to the standard ideal/real world definition in the plain model?

---

[3] [30] also give a construction requiring only $\omega(\log n)$ rounds, however, the knowledge precision achieved in this case is super-linear.

The fundamental problem with this approach is the following. Note that known concurrent zero-knowledge simulators (in the fully concurrent setting) work by rewinding the adversarial parties. In the concurrent setting, the adversary is allowed to control the scheduling of the messages of different sessions. Then the following scenario might occur:

- Between two messages of a session s_1, there may exist entire other session s_2.
- When the simulator rewinds the session s_1, it may rewind past the beginning of session s_2. Hence throughout the simulation, the session s_2 may be executed multiple times from the beginning.
- Every time the session s_2 is executed, the adversary may choose a different input (e.g., the adversary may choose his input in session s_2 based on the entire transcript of interaction so far). In such a case, the simulator is required to leak additional information about the input of the honest party (e.g., in the form of an extra output as in [20]).

Indeed, some such problem is rather inherent as indicated by various impossibility results [27,3,19,1,15]. As stated above, our basic idea will be to use leakage on the inputs of the honest parties in order to continue in the rewindings (or look-ahead threads). Our simulator would simply request the ideal functionality for the entire input of the honest party in such a session. Subsequent to this, *such a session can appear on any number of look-ahead threads*: we can simply use the leaked input and use that to proceed honestly.

Main Technical Problem. The key technical problem we face is the following. All previous rewinding strategies are too "dense" for our purposes. These strategies do not lead to any non-trivial results in our model: the simulator will simply be required the leak the honest party input in *each session*. For example, in the oblivious rewinding strategies used in [23,36,30,20], the "main" thread of protocol execution is divided into various blocks (2 blocks in [23,36] and n blocks in [30,20]). Each given block is rewound that results in a "look-ahead thread". Each session on the main thread will also appear on these look-ahead threads (in fact, on multiple look-ahead threads). Hence, it can be shown that our strategy of leaking inputs of sessions appearing in look-ahead threads will result in leakage of inputs in all sessions. For the case of adaptive rewinding strategies [37,35,11], the problem is even more pronounced. Any given block (or an interval) of the transcript may be rewound any polynomial number of times (each time to solve a different session).

Thus, the known rewinding strategies do not yield any non-trivial results in our model (let alone allow leakage of any arbitrarily small polynomial fraction of inputs).

Main Idea: Sparse Rewinding Strategies. In order to address the above problem, we develop a new "cost-based" rewinding strategy. In particular, our main technical contribution is the development of what we call *sparse rewinding strategies* in the concurrent setting. In a sparse rewinding strategy, the main idea is to choose various *small intervals* of the transcript and rewind *only* those intervals. The main technical challenge is to show that despite rewinding only

only few locations of the transcript, extraction is still guaranteed for every session (regardless of where it lies on the transcript).

In more detail, our rewinding strategy bears similarities with the oblivious recursive rewinding strategies used in [23,36]. Our main contribution lies in showing that a "significantly stripped down" version of their strategy is still sufficient to guarantee extraction in all sessions. More specifically, recall that the recursive rewinding strategies in [23,36] have various threads of executions (also called blocks) which are at different "levels" and have different sizes. We carefully select only a small subset of these blocks and carry them out as part of our rewinding schedule (while discarding the rest). The leakage parameter ϵ and the resulting round complexity (which we show to be almost optimal w.r.t. a black-box simulator) determines what fraction of blocks (and at what levels) are picked to be carried out in the rewinding schedule. Given such a strategy, we reduce the problem of covering all sessions to a *set cover problem*: pick sufficiently many blocks (each block representing a set of sessions which are "solved" when that block is carried out as part of the rewinding schedule) such that every session is covered (i.e., extraction is guaranteed) while still keeping the overall leakage (more generally, the "cost") to be low. Indeed, this cost-centric view is what also allows us to improve upon the precision guarantees in [30].

Additional Challenges. To convert the above basic idea into an actual construction, we encounter several difficulties. The main challenge is to argue extraction in all sessions. Recall that the *swapping arguments* in prior works [23,36,34,30] crucially rely on "symmetry" between the main thread of execution and the look-ahead threads (i.e., execution threads created view rewinding). In particular, to argue extraction, [36,34] define *swap* and *undo* procedures w.r.t. execution threads that allow to transform a "bad" random tape of the simulator (that leads to extraction failure) into a "good" random tape (where extraction succeeds) and back. The idea being to show that every bad random tape, there exist super-polynomially many good random tapes; as such, with overwhelming probability, the simulator must choose a good random tape.

In our setting, using such swapping arguments becomes non-trivial. First off, note that we cannot directly employ the standard greedy strategy for set-cover problem to choose which blocks must be rewound. Very briefly, this is because once one swaps two blocks (one on the main thread, and the corresponding one on a look-ahead thread), the choice of set of blocks which should be chosen might completely change (this is because the associated "costs" of blocks may change after swapping). Indeed, any such "biased" strategy seems to be doomed for failure against adversaries that choose the schedule adaptively. Towards this end, we use a *randomized* strategy for choosing which blocks to rewind, with the goal of still keeping the extraction cost minimal. Nevertheless, despite the randomized approach, the sparse nature of our block choosing strategy still results in significant "asymmetry" across the entire rewinding schedule. This leads to difficulties in carrying out the swap and undo procedures as in [36,34]. We resolve these difficulties by using a careful "localized" swapping argument (see technical sections for details).

Our final protocol is based on compilation with concurrent non-malleable zero-knowledge [3]. We recall that there are several problems that arise with such a compilation. First, the security of the [3] construction is analyzed only for the setting where all the statements being proven by honest parties are fixed in advance. Secondly, the extractor of [3] is unsuitable for extracting inputs of the adversary since it works after the entire execution is complete on a *session-by-session* basis. Fortunately, these challenges were tackled in the work of Goyal et al. [20]. Indeed, Goyal et. al. presented an approach which can be viewed as a technique to correctly compile a semi-honest secure protocol with [3]. We adopt their approach to construct our final protocol.

2   Our Model

In this section, we present a brief overview of our security model, with details deferred to the full version. Throughout this paper, we denote the security parameter by κ.

We define our security model by extending the standard real/ideal paradigm for secure computation. Roughly speaking, we consider a relaxed notion of concurrently secure computation where the ideal world adversary (aka, the simulator) is allowed to leak on the inputs of the honest parties. Intuitively, the amount of leakage obtained by the simulator in order to simulate the view of a concurrent adversary corresponds to the "information leakage" under concurrent composition.

In this work, we consider a malicious, static adversary. The scheduling of the messages across the concurrent executions is controlled by the adversary. We allow the adversary to start arbitrarily polynomial number of concurrent session. Also, we consider the fixed input setting, i.e. the inputs of the honest party across all sessions is fixed in advance. Finally, we consider *computational* security only and therefore restrict our attention to adversaries running in probabilistic polynomial time.

We consider two security models that differ in the nature of ideal world leakage available to the simulator. In both of these security models, the real world is the same as in the standard security model for concurrently secure computation. The real concurrent execution of Π with security parameter κ, input vectors $\boldsymbol{x}$, $\boldsymbol{y}$ and auxiliary input z to $\mathcal{A}$, denoted $\text{REAL}_{\Pi,\mathcal{A}}(\kappa, \boldsymbol{x}, \boldsymbol{y}, z)$, is defined as the output pair of the honest party and $\mathcal{A}$, resulting from the above real-world process. Also, in each of the ideal world experiments described below, the ideal execution of a function $\mathcal{F}$ with security parameter κ, input vectors $\boldsymbol{x}$, $\boldsymbol{y}$ and auxiliary input z to $\mathcal{S}$, denoted $\text{IDEAL}_{\mathcal{F},\mathcal{S}}(\kappa, \boldsymbol{x}, \boldsymbol{y}, z)$, is defined as the output pair of the honest party and $\mathcal{S}$ from the ideal execution.

Concurrently Secure Computation in the Joint Leaky Ideal World Model. In this model, at any time during the ideal world experiment, adversary may send leakage queries of the form L to the trusted party. On receiving such a query, the trusted party computes $L(\boldsymbol{x})$ over honest party inputs $\boldsymbol{x}$ across all sessions and returns it to the adversary.

Definition 1 (ϵ-Joint-Ideal-Leakage Simulator). *Let S be a non-uniform probabilistic (expected)* PPT *ideal-model adversary. We say that S is a ϵ-joint-ideal-leakage simulator if it leaks at most ϵ fraction of the input vector of the honest party.*

Definition 2 (Concurrently Secure Computation in the Joint Leaky Ideal World Model). *A protocol Π evaluating a functionality $\mathcal{F}$ is said to be ϵ-secure in the joint leaky ideal world model if for every real model non-uniform* PPT *adversary $\mathcal{A}$, there exists a non-uniform (expected)* PPT *ϵ-joint-ideal-leakage simulator S such that for every polynomial $m = m(\kappa)$, every pair of input vectors $x \in X^m$, $y \in Y^m$, every $z \in \{0,1\}^* s$,*

$$\{\text{IDEAL}_{\mathcal{F},S}(\kappa, x, y, z)\}_{\kappa \in \mathbb{N}} \stackrel{c}{\equiv} \{\text{REAL}_{\Pi,\mathcal{A}}(\kappa, x, y, z)\}_{\kappa \in \mathbb{N}}$$

Concurrently Secure Computation in the Individual Leaky Ideal World Model. In this model, for every session i, the ideal adversary may send one leakage query of the form (i, L) to the trusted party and learn $L(x_i)$ (where x_i is the input of the honest party in session i).

Definition 3 (ϵ-Individual-Ideal-Leakage Simulator). *Let S be a non-uniform probabilistic (expected)* PPT *ideal-model adversary. We say that S is a ϵ-individual-ideal-leakage simulator if it leaks at most ϵ fraction of the the honest party input in each session.*

Definition 4 (Concurrently Secure Computation in the Individual Leaky Ideal World Model). *A protocol Π evaluating a functionality $\mathcal{F}$ is said to be ℓ-secure in the leaky ideal world model against joint leakage if for every real model non-uniform* PPT *adversary $\mathcal{A}$, there exists a non-uniform (expected)* PPT *ϵ-individual-ideal-leakage simulator S such that for every polynomial $m = m(\kappa)$, every pair of input vectors $x \in X^m$, $y \in Y^m$, every $z \in \{0,1\}^* s$,*

$$\{\text{IDEAL}_{\mathcal{F},S}(\kappa, x, y, z)\}_{\kappa \in \mathbb{N}} \stackrel{c}{\equiv} \{\text{REAL}_{\Pi,\mathcal{A}}(\kappa, x, y, z)\}_{\kappa \in \mathbb{N}}$$

3 Framework for Cost-Based Rewinding

Consider two players P_1 and P_2 running concurrent execution of a two party protocol Π. Π may consists of multiple executions of the extractable commitment scheme $\langle C, R \rangle$ (Section 3.1) and some other protocol messages. These other protocol messages will depend upon our underlying applications. In particular we will consider two main applications. In our application of concurrently secure computation in joint leaky ideal world model, protocol Π is simply the secure computation protocol. In precise concurrent zero-knowledge protocol, Π will be a zero-knowledge protocol.

Moreover, each message in the protocol will have an associated fixed non-zero cost based on the application. In case of concurrent execution of the secure computation protocol, any message from the adversary which causes our simulator to make an output query to the trusted functionality in the ideal world is considered a "heavy" message. All other messages are almost "free". In case of concurrent precise zero-knowledge, cost of a message is the time taken by the adversary to generate that message. All messages of the honest prover are unit cost.

We consider the scenario when exactly one of the parties is corrupted. We begin by describing the extractable commitment scheme $\langle C, R \rangle$.

3.1   Extractable Commitment Protocol $\langle C, R \rangle$

Let $\text{COM}(\cdot)$ denote the commitment function of a non-interactive perfectly binding string commitment scheme. Let κ denote the security parameter. Let $\ell = \omega(\log \kappa)$. Let $N = N(\kappa)$ which is fixed based on the application. The commitment scheme $\langle C, R \rangle$, where the committer commits to a value σ (referred to as the *preamble secret*), is described as follows.

COMMIT PHASE:
STAGE INIT: To commit to a κ-bit string σ, C chooses $(\ell \cdot N)$ independent random pairs of κ-bit strings $\{\alpha_{i,j}^0, \alpha_{i,j}^1\}_{i,j=1}^{\ell,N}$ such that $\alpha_{i,j}^0 \oplus \alpha_{i,j}^1 = \sigma$ for all $i \in [\ell], j \in [N]$. C commits to all these strings using COM, with fresh randomness each time. Let $B \leftarrow \text{COM}(\sigma)$, and $A_{i,j}^0 \leftarrow \text{COM}(\alpha_{i,j}^0)$, $A_{i,j}^1 \leftarrow \text{COM}(\alpha_{i,j}^1)$ for every $i \in [\ell], j \in [N]$.
We say that the protocol has reached Start if message in Stage Init is exchanged.

CHALLENGE-RESPONSE STAGE:
For every $j \in [N]$, do the following:

- Challenge : R sends a random ℓ-bit challenge string $v_j = v_{1,j}, \ldots, v_{\ell,j}$.
- Response : $\forall i \in [\ell]$, if $v_{i,j} = 0$, C opens $A_{i,j}^0$, else it opens $A_{i,j}^1$ by sending the decommitment information.

A **slot**$_j$ of the commitment scheme consists of the receiver's Challenge and the corresponding committer's Response message. Thus, in this protocol, there are N slots.

We say that the protocol has reached End when CHALLENGE-RESPONSE STAGE is completed and is accepted by R.

OPEN PHASE: C opens all the commitments by sending the decommitment information for each one of them. R verifies the consistency of the revealed values.

This completes the description of $\langle C, R \rangle$ which is an $\mathcal{O}(N)$ round protocol. The commit phase is said to the *valid* iff there exists an opening of commitments such that the open phase is accepted by an honest receiver.

Having defined the commitment protocol, we will describe a simulator $\mathcal{S}$ for the protocol Π that uses a rewinding schedule to "simulate" the view of the adversary. For this, we would like to prove an extraction lemma similar to [36,30] for the protocol Π, i.e., in every execution whenever a *valid* commit phase ends such that the adversary is playing the role of the committer, our simulator (using rewinding) would be able to extract the *preamble secret* with all but negligible probability. Moreover, we would like to guarantee that if the honest execution has total cost[4] C, then the cost incurred by our simulator is only $C(1 + \epsilon(N, \kappa))$, where is ϵ is a small fraction.

3.2 Description of the Simulator

We describe a new "cost-based" recursive rewinding strategy. We begin by giving some preliminary definitions that will be used in the rest of the paper.

A thread of execution (consisting of the views of all the parties) is a perfect simulation of a prefix of an actual execution. In particular, the *main thread*, is a perfect simulation of a complete execution, and this is the execution thread that is output by the simulator. In addition, our simulator will also make other threads by rewinding the adversary to a previous state and continuing the execution from that state. Such a thread shares a (possibly empty) prefix with the previous thread. We call the execution on this thread which is not shared with any of the previous threads as a *look-ahead thread*.

We now first give an overview of the main ideas underlying our simulation technique and then proceed to give a more formal description.

Overview. Consider the main thread of execution. At a high level, we divide this thread into multiple parts referred to as "sets" consisting of possibly many protocol messages. The way we define our sets is similar to previous rewinding strategies [36,30]. Essentially if the entire execution has cost c, then we divide the entire main thread into two sets of cost $c/2$ each, where cost of a set is the total cost of the messages contained in that set. Next, we divide each of these sets into two subsets, each of cost $c/4$. We continue this process recursively till we have c sets, each of unit cost[5]. Note that if each message is of unit cost, then this dividing strategy is exactly identical to [36].[6] The novel idea underlying our rewinding technique is that unlike [36,30], our simulator only rewinds a small subset of these sets while still guaranteeing extraction. In other words, unlike [36,30], ours is a "sparse" rewinding strategy.

We now describe our rewinding strategy by using an analogy to the classical set covering problem. Recall that in the set covering problem, there is a universe of elements and sets. Each set contains some elements and has a fixed cost. The goal is to choose a minimum cost collection of these sets which covers all the elements in the universe. In our setting, we think of each session as an element in the universe. If there are m concurrent sessions $\{1, 2, \ldots, m\}$, we have m elements in our universe. Now consider the sets defined above in our setting. A set is said to cover an element i if it contains a complete slot of session i. Recall

---

[4] Cost of an execution is the total cost of all the messages sent and received.

[5] Note that due to this dividing strategy, we allow a message to be "divided" across multiple sets.

[6] If cost of a message is the time taken by the adversary to generate that message, then this dividing strategy is exactly identical to [30].

that the cost of a set is the sum of the cost of messages in this set. We want to consider a minimum cost collection $\mathcal{C}$ of these sets which together covers all the elements in the universe (i.e. all the sessions). Intuitively, we wish to rewind only the sets in $\mathcal{C}$. At a high level, this is the strategy adopted by our simulator. Due to reasons as discussed in Section 1.2, we adopt a slightly modified strategy in which the collection $\mathcal{C}$ is picked via a randomized strategy. Recall that for all i there 2^i sets with cost $c/2^i$. Very briefly, for each collection of sets which have same cost, we pick a fixed small fraction of these sets. We will prove that using this strategy we will cover each session $\omega(\log \kappa)$ times in order to guarantee extraction. As we will see later on, with this strategy, we are able to guarantee that the simulator performs extraction with all but negligible probability while incurring a small overhead.

Formal Description of the Simulator. We begin by introducing some notation and terminology. Let C be the total cost of main execution[7]. Without loss of generality, let $C = 2^x$ for some $x \in \mathbb{N}$. Let $p(\kappa) = \omega(\log \kappa)$, and $q(\kappa) = \omega(1)$. Recall that N is the number of challenge-response slots in $\langle C, R \rangle$.

Thread at Recursion Level RL. We say that the main thread belongs to recursion level 0. The look-ahead threads which fork off the main thread are said to be at recursion level 1. Recursively, we say that look-ahead threads forking off a thread at recursion level RL belong to recursion level (RL + 1).

Sets and Set Levels. Let T be the main thread or a look-ahead thread with cost c at recursion level RL. We define the sets and the set levels of T as follows: The entire thread T is defined as one set at recursion level RL and set level 0 with cost c. We denote it as $\mathsf{set}_{\mathsf{RL}}^0$. Now divide $\mathsf{set}_{\mathsf{RL}}^0$ into two sets at recursion level RL and set level 1 of cost $c/2$ each. We denote the first set as $\mathsf{set}_{\mathsf{RL}}^{1,1}$ and the second set as $\mathsf{set}_{\mathsf{RL}}^{1,2}$. Let $\mathsf{set}_{\mathsf{RL}}^{i,1}, \mathsf{set}_{\mathsf{RL}}^{i,2}, \ldots, \mathsf{set}_{\mathsf{RL}}^{i,2^i}$ be 2^i sets at set level i, each of cost $c/2^i$, where $\mathsf{set}_{\mathsf{RL}}^{i,j}$ is the jth set at set level i. Divide each set $\mathsf{set}_{\mathsf{RL}}^{i,j}$ into two sets at set level $(i + 1)$ each of cost $c/2^{i+1}$. We continue this recursively till we reach set level $\log c$ where each set has cost 1. This way we have $L = \log c + 1$ set levels $(0, 1, \ldots, \log c)$ with total sets $2c - 1$.

For ease of notation, we will denote each set $\mathsf{set}_{\mathsf{RL}}^{i,j}$ as a tuple (s-point, e-point) where s-point denotes the cost of the thread T from the start of T till the start of $\mathsf{set}_{\mathsf{RL}}^{i,j}$ and e-point to denote the cost of thread from the start of T till the end of $\mathsf{set}_{\mathsf{RL}}^{i,j}$. Thus by definition cost of a set $\mathsf{set}_{\mathsf{RL}}^{i,j}$ is (e-point − s-point). This will help us in describing our simulation strategy.

Simulator $\mathcal{S}$. We now proceed to describe our simulation strategy which consists of procedures SIMULATE, PICKSET and SIMMSG. More specifically, $\mathcal{S}$ simply runs SIMULATE($C, \mathsf{st}_0, \phi, 0, r_m, r_s$) to simulate the main thread at recursion level 0 with cost C when st_0 is the initial state of $\mathcal{A}$. $\mathcal{S}$ starts with empty history of messages, i.e. hist = $\emptyset$. Also, $\mathcal{S}$ uses two separate random tapes r_m and r_s to

---

[7] This cost C is always bounded by some polynomial in κ, i.e., $C \leq \kappa^\alpha$ for some constant α.

The SIMULATE(c, st, hist, RL, r_m, r_s) Procedure.

1. Compute PSETS $= \{(\text{s-point}_j, \text{ e-point}_j)\} \leftarrow \text{PICKSET}(c, r)$, where r is randomness of appropriate size from r_s. Update $r_s = r_s \backslash r$. Let $J = |\text{PSETS}|$.

2. Create a list PSĒTS from PSETS as follows: For each entry (s-point$_j$, e-point$_j$) $\in$ PSETS, initialize st$_j = \perp$ and hist$_j = \perp$. Insert (s-point$_j$, e-point$_j$, st$_j$, hist$_j$) into PSĒTS. We order the list by increasing order of e-point.

3. If $c = 1$, (st$'$, hist$'$, r'_m, PSĒTS) $\leftarrow$ SIMMSG(0, 1, st, hist, r_m, PSĒTS). Output: (st$'$, hist$'$, r'_m, r_s).

4. Otherwise (i.e., $c > 1$),
   - Initialize ctr $= 0$.
   - While ($j < J$)
     - (st$'$, hist$'$, r'_m, PSĒTS) $\leftarrow$ SIMMSG(ctr, e-point$_j$ $-$ ctr, st, hist, r_m, PSĒTS).
     - Set ctr $=$ e-point$_j$, $r^0_m = r'_m$, $r^0_s = r_s$.
     - Let there exist ℓ entries $\{(\text{s-point}_{j_i}, \text{e-point}_{j_i}, \text{st}_{j_i}, \text{hist}_{j_i})\}_{i \in [\ell]}$ in PSĒTS such that e-point$_{j_i} =$ e-point$_j$.
     - For each $i \in [\ell]$,
       (st$'_{j_i}$, hist$'_{j_i}$, r^i_m, r^i_s) $\leftarrow$ SIMULATE((e-point$_{j_i}$ $-$ s-point$_{j_i}$), st$_{j_i}$, hist$_{j_i}$, RL $+ 1$, r^{i-1}_m, r^{i-1}_s).
     - Set hist $=$ hist$' \cup (\bigcup_i$ hist$'_{j_i})$, st $=$ st$'$, $r_m = r^\ell_m$, $r_s = r^\ell_s$ and $j = j + \ell$.
   - If (ctr $< c$)
     - (st$'$, hist$'$, r'_m, PSĒTS) $\leftarrow$ SIMMSG(ctr, $c -$ ctr, st, hist, r_m, PSĒTS).
     - Update hist $=$ hist$'$, st $=$ st$'$, $r_m = r'_m$.
   - Output: (st, hist, r_m, r_s).

Fig. 1. The cost-based content oblivious simulator SIMULATE

generate messages and choose sets respectively. Finally, $\mathcal{S}$ returns its output as the view of the adversary. We begin by describing these procedures in detail.

Procedure SIMULATE. The procedure is used to simulate any thread T at recursion level RL of cost c. It takes the following set of inputs. (a) The cost c of thread T. (b) The state st of the adversary at the beginning of T. (c) The history hist of messages seen so far in simulation. (d) Recursion level RL of T. (e) The random tape r_m which is used to generate messages of the honest party. (f) The random tape r_s used by PICKSET to choose sets.

At a high level, SIMULATE procedure when invoked on a set of inputs (c, st, hist, RL, r_m, r_s) does the following:

1. It invokes PICKSET procedure to choose a list of sets on T, say PSETS, which it will rewind. Here each set will be denotes by the corresponding tuple (s-point, e-point).

2. Next, SIMULATE augments each entry of PSETS with two additional entries to create a new list PSĒTS where each entry consists of (s-point, e-point, st, hist), where st is the state of the adversary and hist is the history of simulation at s-point. State st and history hist at s-point are populated by the procedure SIMMSG (described below) when simulation reaches s-point.

3. SIMULATE generates messages for the thread iteratively till the end of the thread is reached as follows:

1. It invokes the SIMMSG procedure to generate the messages from current point of simulation to the next e-point of some set in PSÊTS.

2. For each of the sets which end at this point, it calls SIMULATE procedure recursively to create new look-ahead threads at recursion level $RL + 1$.

3. Finally, it merges the current history of messages with messages seen on the look-ahead threads.

4. It returns $(\mathsf{st}', \mathsf{hist}', r'_m, r'_s)$, where st' is the state of the adversary at the end of the thread, hist' is the updated collection of messages, r'_m and r'_s are the unused parts of the random tapes r_m and r_s respectively.

The figure 1 gives a formal description of SIMULATE procedure.

Algorithm PICKSET. At a high level, given the main thread or a look-ahead thread T at recursion level RL with cost c, it chooses a fixed fraction of sets across all set levels of T where our simulator would rewind. More formally, on input (c, r), where c is the cost of T and r is some randomness, $\mathrm{PICKSET}(c, r)$ returns a list of sets PSETS$= \{(\mathsf{s\text{-}point}_j, \mathsf{e\text{-}point}_j)\}$ consisting of $\left\lfloor \frac{p(\kappa) \cdot q(\kappa) \cdot \log^3 \kappa}{N} \cdot 2^i \right\rfloor$ sets at random at set level i for every $i \in [\log c]$.

Note that the sets picked by PICKSET depend only on the cost c of the thread T and randomness r and not on the protocol messages of T.

Procedure SimMsg. This procedure generates the messages by running the adversary step by step[8], i.e. incurring unit cost at a time. It takes the following set of inputs. (a) The partial cost ctr of the current thread simulated so far. (b) The additional cost c for which the current thread has to be simulated. (c) The current state st of the adversary. (d) The history hist of messages seen so far in simulation. (e) The random tape r_m to be used to generate messages. (f) The list PSÊTS of the sets chosen by PICKSET for thread T.

SIMMSG generates messages on thread T one step at a time for c steps as follows:

1. If the next scheduled message is the challenge message in an instance of $\langle C, R \rangle$, it chooses a challenge uniformly at random. Also, if the next scheduled message is some other protocol message from honest party, it uses the honest party algorithm to generate the same.

2. If the next scheduled message is from $\mathcal{A}$, SIMMSG runs $\mathcal{A}$ for one step and updates st and hist. Note that it is possible that $\mathcal{A}$ may not generate a message in one step.

3. If the current point on the thread corresponds to the s-point of some sets in PSÊTS, it updates the corresponding entries with current state st of $\mathcal{A}$ and history hist of messages.

---

[8] We will assume that it is possible to run the adversary one step at a time. We elaborate on this in our applications.

Finally it outputs the final state st of the adversary, updated history hist of messages, unused part r'_m of random tape r_m and updated list PSĒTS. Procedure SimMsg is described formally in Figure 2.

The SimMsg(ctr, c, st, hist, r_m, PSĒTS) Procedure.

For $i = 1$ to c do the following:

- **Next scheduled message if from honest party to $\mathcal{A}$:** If the next scheduled message is a challenge message of $\langle C, R \rangle$, choose a random challenge message using randomness from r_m. Else, if the next message is some other message from honest party, send this message according to honest party algorithm using randomness from r_m. Feed this message to $\mathcal{A}$. **Next scheduled message is from $\mathcal{A}$:** If the next scheduled message is from $\mathcal{A}$, run $\mathcal{A}$ from its current state st for exactly 1 step. If an output, β, is received and if β is a response message in $\langle C, R \rangle$, store β in hist as a response to the corresponding challenge message. Update st to the current state of $\mathcal{A}$. If it is some other message of the protocol, store it in hist.
- If there exists k entries $\{(\text{s-point}_{j_y}, \text{e-point}_{j_y}, \text{st}_{j_y}, \text{hist}_{j_y})\}_{y \in [k]}$ in PSĒTS such that $\text{s-point}_{j_y} = \text{ctr} + i$. For each $y \in [k]$ update $\text{st}_{j_y} = \text{st}$ and $\text{hist}_{j_y} = \text{hist}$.

Let r'_m be the unused part of r_m. Output: (st, hist, r'_m, PSĒTS).

Fig. 2. The SimMsg Procedure

Lemma 1. *(Extraction lemma) Consider two parties P_1 and P_2 running polynomially many (in the security parameter) sessions of a protocol Π consisting of possibly multiple executions of the commitment scheme $\langle C, R \rangle$. Also, let one the parties, say P_2, be corrupted. Then there exists a simulator S such that except with negligible probability, in every thread of execution simulated by S, if honest P_1 accepts a commit phase of $\langle C, R \rangle$ as valid, then at the point when that commit phase is concluded, S would have already extracted the preamble secret committed by the corrupted P_2.*

Lemma 2. *Let C be the cost of the main thread. Then the cost incurred by our simulator is bounded by*
$$C \cdot \left(1 + \frac{(\log^* \kappa)^2 \log C \log^4 \kappa}{N}\right) \text{ when } \langle C, R \rangle \text{ has } N \geq \log^6 \kappa \text{ slots.}$$

4 Our Results

We now state the main results in this paper.

Positive Results. As the main result of this paper, we construct an $\mathcal{O}(N)$ round protocol Π that ϵ-securely realizes any (efficiently computable) functionality $\mathcal{F}$ in the joint leaky ideal world model for any $\epsilon > 0$. More formally, we show the following:

Theorem 6. *Assume the existence of 1-out-of-2 oblivious transfer protocol secure against honest but curious adversaries and collision resistant hash functions. Then for any $\epsilon > 0$, for any functionality $\mathcal{F}$, there exists an $\mathcal{O}(N)$ round protocol Π that ϵ-securely realizes $\mathcal{F}$ in the joint leaky ideal world model, where $N = \frac{(\log^6 \kappa)}{\epsilon}$.*

In the case when $\epsilon = 1/\mathrm{poly}(\kappa)$, we do not need to assume the existence of collision resistent hash functions. Protocol Π is essentially the protocol of [20] instantiated with N-round concurrently-extractable commitment scheme described earlier in the paper. The security analysis of the protocol is done using the simulation technique described earlier.

Negative Results. We also present strong impossibility results for achieving security in both the individual and joint leaky ideal world model. First, we prove the following result:

Theorem 7. *There exists a functionality f such that no protocol Π ϵ-securely realizes f in the individual leaky ideal world model for $\epsilon = \frac{1}{2} - \delta$, where δ is any constant fraction.*

Additionally, we prove a lower bound on the round-complexity of protocols for achieving ϵ-security in the joint leaky ideal world model, with respect to black-box simulation. Specifically, we prove the following result:

Theorem 8. *Let ϵ be any inverse polynomial. Assuming dense cryptosystems, there exists a functionality f that cannot be ϵ-securely realized with respect to black-box simulation in the joint leaky ideal world model by any $\frac{\log(\kappa)}{\epsilon}$ round protocol.*

References

1. Agrawal, S., Goyal, V., Jain, A., Prabhakaran, M., Sahai, A.: New impossibility results for concurrent composition and a non-interactive completeness theorem for secure computation. In: Safavi-Naini, R. (ed.) CRYPTO 2012. LNCS, vol. 7417, pp. 443–460. Springer, Heidelberg (2012)
2. Barak, B., Canetti, R., Nielsen, J.B., Pass, R.: Universally composable protocols with relaxed set-up assumptions. In: FOCS (2004)
3. Barak, B., Prabhakaran, M., Sahai, A.: Concurrent non-malleable zero knowledge. In: FOCS (2006)
4. Barak, B., Sahai, A.: How to play almost any mental game over the net - concurrent composition using super-polynomial simulation. In: Proc. 46th FOCS, pp. 543–552 (2005)
5. Bitansky, N., Canetti, R., Halevi, S.: Leakage-tolerant interactive protocols. In: Cramer, R. (ed.) TCC 2012. LNCS, vol. 7194, pp. 266–284. Springer, Heidelberg (2012)
6. Boyle, E., Garg, S., Jain, A., Kalai, Y.T., Sahai, A.: Secure computation against adaptive auxiliary information. In: Canetti, R., Garay, J.A. (eds.) CRYPTO 2013, Part I. LNCS, vol. 8042, pp. 316–334. Springer, Heidelberg (2013)
7. Canetti, R., Lindell, Y., Ostrovsky, R., Sahai, A.: Universally composable two-party and multi-party secure computation. In: STOC (2002)
8. Canetti, R., Fischlin, M.: Universally composable commitments. In: Kilian, J. (ed.) CRYPTO 2001. LNCS, vol. 2139, pp. 19–40. Springer, Heidelberg (2001)
9. Canetti, R., Kushilevitz, E., Lindell, Y.: On the limitations of universally composable two-party computation without set-up assumptions. In: Biham, E. (ed.) EUROCRYPT 2003. LNCS, vol. 2656, pp. 135–167. Springer, Heidelberg (2003)

10. Canetti, R., Lin, H., Pass, R.: Adaptive hardness and composable security in the plain model from standard assumptions. In: FOCS (2010)
11. Deng, Y., Goyal, V., Sahai, A.: Resolving the simultaneous resettability conjecture and a new non-black-box simulation strategy. In: FOCS (2009)
12. Dziembowski, S., Pietrzak, K.: Intrusion-resilient secret sharing. In: FOCS (2007)
13. Garg, S., Goyal, V., Jain, A., Sahai, A.: Concurrently secure computation in constant rounds. In: Pointcheval, D., Johansson, T. (eds.) EUROCRYPT 2012. LNCS, vol. 7237, pp. 99–116. Springer, Heidelberg (2012)
14. Garg, S., Jain, A., Sahai, A.: Leakage-resilient zero knowledge. In: Rogaway, P. (ed.) CRYPTO 2011. LNCS, vol. 6841, pp. 297–315. Springer, Heidelberg (2011)
15. Garg, S., Kumarasubramanian, A., Ostrovsky, R., Visconti, I.: Impossibility results for static input secure computation. In: Safavi-Naini, R. (ed.) CRYPTO 2012. LNCS, vol. 7417, pp. 424–442. Springer, Heidelberg (2012)
16. Goldreich, O., Micali, S., Wigderson, A.: How to play any mental game. In: STOC (1987)
17. Goldreich, O., Petrank, E.: Quantifying knowledge complexity. In: FOCS (1991)
18. Goldwasser, S., Kalai, Y.T., Rothblum, G.N.: One-time programs. In: Wagner, D. (ed.) CRYPTO 2008. LNCS, vol. 5157, pp. 39–56. Springer, Heidelberg (2008)
19. Goyal, V.: Positive results for concurrently secure computation in the plain model. In: FOCS (2012)
20. Goyal, V., Jain, A., Ostrovsky, R.: Password-authenticated session-key generation on the internet in the plain model. In: Rabin, T. (ed.) CRYPTO 2010. LNCS, vol. 6223, pp. 277–294. Springer, Heidelberg (2010)
21. Jain, A., Prabhakaran, M., Sahai, A., Wadia, A.: Oblivious transfer from any leaky functionality. In: Personal Communication (2013)
22. Katz, J.: Universally composable multi-party computation using tamper-proof hardware. In: Naor, M. (ed.) EUROCRYPT 2007. LNCS, vol. 4515, pp. 115–128. Springer, Heidelberg (2007)
23. Kilian, J., Petrank, E.: Concurrent and resettable zero-knowledge in poly-loalgorithm rounds. In: STOC (2001)
24. Lin, H., Pass, R.: Black-box constructions of composable protocols without set-up. In: Safavi-Naini, R. (ed.) CRYPTO 2012. LNCS, vol. 7417, pp. 461–478. Springer, Heidelberg (2012)
25. Lindell, Y.: Bounded-concurrent secure two-party computation without setup assumptions. In: STOC (2003)
26. Lindell, Y.: General composition and universal composability in secure multi-party computation. In: FOCS (2003)
27. Lindell, Y.: Lower bounds for concurrent self composition. In: Naor, M. (ed.) TCC 2004. LNCS, vol. 2951, pp. 203–222. Springer, Heidelberg (2004)
28. Micali, S., Pass, R.: Local zero knowledge. In: STOC (2006)
29. Micali, S., Pass, R., Rosen, A.: Input-indistinguishable computation. In: FOCS, pp. 367–378 (2006)
30. Pandey, O., Pass, R., Sahai, A., Tseng, W.-L.D., Venkitasubramaniam, M.: Precise concurrent zero knowledge. In: Smart, N.P. (ed.) EUROCRYPT 2008. LNCS, vol. 4965, pp. 397–414. Springer, Heidelberg (2008)
31. Pass, R.: Simulation in quasi-polynomial time, and its application to protocol composition. In: Biham, E. (ed.) EUROCRYPT 2003. LNCS, vol. 2656, Springer, Heidelberg (2003)
32. Pass, R.: Bounded-concurrent secure multi-party computation with a dishonest majority. In: STOC (2004)

33. Pass, R., Rosen, A.: Bounded-concurrent secure two-party computation in a constant number of rounds. In: FOCS (2003)
34. Pass, R., Tseng, W.-L.D., Venkitasubramaniam, M.: Concurrent zero knowledge, revisited (2012) (manuscript)
35. Pass, R., Venkitasubramaniam, M.: On constant-round concurrent zero-knowledge. In: Canetti, R. (ed.) TCC 2008. LNCS, vol. TCC, pp. 553–570. Springer, Heidelberg (2008)
36. Prabhakaran, M., Rosen, A., Sahai, A.: Concurrent zero knowledge with logarithmic round-complexity. In: FOCS, pp. 366–375 (2002)
37. Richardson, R., Kilian, J.: On the concurrent composition of zero-knowledge proofs. In: Stern, J. (ed.) EUROCRYPT 1999. LNCS, vol. 1592, pp. 415–431. Springer, Heidelberg (1999)
38. Yao, A.C.C.: How to generate and exchange secrets. In: FOCS (1986)

Non-malleable Codes from Two-Source Extractors[*]

Stefan Dziembowski[1], Tomasz Kazana[2], and Maciej Obremski[2]

[1] University of Warsaw and Sapienza University of Rome
[2] University of Warsaw

Abstract. We construct an efficient information-theoretically non-malleable code in the split-state model for one-bit messages. Non-malleable codes were introduced recently by Dziembowski, Pietrzak and Wichs (ICS 2010), as a general tool for storing messages securely on hardware that can be subject to tampering attacks. Informally, a code ($\mathsf{Enc} : \mathcal{M} \to \mathcal{L} \times \mathcal{R}, \mathsf{Dec} : \mathcal{L} \times \mathcal{R} \to \mathcal{M}$) is *non-malleable in the split-state model* if any adversary, by manipulating *independently* L and R (where (L, R) is an encoding of some message M), cannot obtain an encoding of a message M' that is not equal to M but is "related" M in some way. Until now it was unknown how to construct an information-theoretically secure code with such a property, even for $\mathcal{M} = \{0, 1\}$. Our construction solves this problem. Additionally, it is leakage-resilient, and the amount of leakage that we can tolerate can be an arbitrary fraction $\xi < 1/4$ of the length of the codeword. Our code is based on the inner-product two-source extractor, but in general it can be instantiated by any two-source extractor that has large output and has the property of being *flexible*, which is a new notion that we define.

We also show that the non-malleable codes for one-bit messages have an equivalent, perhaps simpler characterization, namely such codes can be defined as follows: if M is chosen uniformly from $\{0, 1\}$ then the probability (in the experiment described above) that the output message M' is not equal to M can be at most $1/2 + \epsilon$.

1 Introduction

Real-life attacks on cryptographic devices often do not break their mathematical foundations, but exploit vulnerabilities in their implementations. Such "physical attacks" are usually based on passive measurements such as running-time, electromagnetic radiation, power consumption (see e.g. [24]), or active tampering where the adversary maliciously modifies some part of the device (see e.g. [3]) in order to force it to reveal information about its secrets. A recent trend in theoretical cryptography, initiated by [34,31,30] is to design cryptographic schemes

---

[*] This work was partly supported by the WELCOME/2010-4/2 grant founded within the framework of the EU Innovative Economy (National Cohesion Strategy) Operational Programme. The European Research Council has provided financial support for this work under the European Community's Seventh Framework Programme (FP7/2007-2013) / ERC grant agreement no CNTM-207908.

R. Canetti and J.A. Garay (Eds.): CRYPTO 2013, Part II, LNCS 8043, pp. 239–257, 2013.

that already on the abstract level guarantee that they are secure even if implemented on devices that may be subject to such physical attacks. Contrary to the approach taken by the practitioners, security of these constructions is always analyzed formally in a well-defined mathematical model, and hence covers a broad class of attacks, including those that are not yet known, but may potentially be invented in the future. Over the last few years several models for passive and active physical attacks have been proposed and schemes secure in these models have been constructed (see e.g. [31,30,22,2,35,7,15,25]). In the passive case the proposed models seem to be very broad and correspond to large classes of real-life attacks. Moreover, several constructions secure in these models are known (including even general compliers [27] for any cryptographic functionality). The situation in the case of active attacks is much less satisfactory, usually because the proposed models include an assumption that some part of the device is tamper-proof (e.g. [26]) or because the tampering attacks that they consider are very limited (e.g. [30] or [13] consider only probing attacks, and in [37] the tampering functions are assumed to be as linear). Hence, providing realistic models for tampering attacks, and constructing schemes secure in these models is an interesting research direction.

In a recent paper [23] the authors consider a very basic question of storing messages securely on devices that may be subject to tampering. To this end they introduce a new primitive that they call the *non-malleable codes*. The motivating scenario for this concept is as follows. Imagine we have a secret message $m \in \mathcal{M}$ and we want to store it securely on some hardware $\mathcal{D}$ that may be subject to the tampering attacks. In order to increase the security, we will encode the message m by some (randomized) function Enc and store the codeword $x := \mathsf{Enc}(m)$ on $\mathcal{D}$. Since we later want to recover m from $\mathcal{D}$ we obviously also need a decoding function $\mathsf{Dec} : \mathcal{X} \to \mathcal{M} \cup \{\bot\}$ such that for every $m \in \mathcal{M}$ we have $\mathsf{Dec}(\mathsf{Enc}(m)) = m$. Now, suppose the adversary can tamper with the device in some way, which we model by allowing him to choose a function $F : \mathcal{X} \to \mathcal{X}$, from some fixed set $\mathcal{F}$ of *tampering functions* and substitute the contents of $\mathcal{D}$ by $F(x)$. Let $m' := \mathsf{Dec}(F(\mathsf{Enc}(m)))$ be the result of decoding such modified codeword.

Let us now think what kind of security properties one could expect from such an encoding scheme. Optimistically, e.g., one could hope to achieve tamper-detection by which we would mean that $m' = \bot$ if $F(x) \neq x$. Unfortunately this is usually unachievable, as, e.g., if the adversary chooses F to be a constant function equal to $\mathsf{Enc}(\tilde{m})$ then $m' = \tilde{m}$. Hence, even for very restricted classes $\mathcal{F}$ (containing only the constant functions), the adversary can force m' to be equal to some message of his choice. Therefore, if one hopes to get any meaningful security notion, one should weaken the tamper-detection requirement.

In [23] the authors propose such a weakening based on the concept of *non-malleability* introduced in the seminal paper of Dolev et al. [19]. Informally, we say that a code $(\mathsf{Enc}, \mathsf{Dec})$ is *non-malleable* if either (1) the decoded message m' is equal to m, or (2) the decoded message m' is "independent" from m. The formal definition appears in Section 3, and for an informal discussion of this concept the reader may consult [23]. As argued in [23] the non-malleable codes can have vast applications to tamper-resistant cryptography. We will not discuss them in detail here, but let us mention just one example, that looks particularly appealing to us. A common practical way of breaking cryptosystems is based on the so-called related-key attacks (see, e.g. [5,4]), where the adversary that attacks some device $\mathcal{D}(K)$ (where K is the secret key) can get access to an identical device containing a *related* key $K' = F(K)$ (by for example tampering with K). Non-malleable codes provide an attractive solution to this problem. If (Enc, Dec) is a non-malleable code secure with respect to same family $\mathcal{F}$, then we can store the key K on $\mathcal{D}$ in an encoded form, and prevent the related key attacks as long as the "relation F" is in $\mathcal{F}$. This is because, the only thing that the adversary can achieve by applying F to Enc(K) is to produce encoding of either a completely unrelated key K', or to keep $K' = K$. It is clear that both cases do not help him in attacking $\mathcal{D}(K)$.

It is relatively easy to see that if the family $\mathcal{F}$ of tampering functions is equal to the entire space of functions from $\mathcal{X}$ to $\mathcal{X}$ then it is impossible to construct such a non-malleable code secure against $\mathcal{F}$. This is because in this case the adversary can always choose $F(x) = \text{Enc}(H(\text{Dec}(x)))$ for any function $H : \mathcal{M} \to \mathcal{M}$, which yields $m' = \text{Dec}(x) = \text{Dec}(\text{Enc}(H(\text{Dec}(\text{Enc}(m))))) = H(m)$, and therefore he can relate m' to m in an arbitrary way. Therefore non-malleable codes can exist only with respect to restricted classes $\mathcal{F}$ of functions. The authors of [23] propose some classes like this and provide constructions of non-malleable codes secure with respect to them. One example is the class of bit-wise tampering functions, which tamper with every bit of x "independently", more precisely: the ith bit x_i' of x' is a function of x_i, and does not depend on any x_j for $j \neq i$. This is a very strong assumption and it would be desirable to weaken it. One natural idea for such weakening would be to allow x_i' to depend on the bits of x from positions on some larger subset $\mathcal{I}_i \subsetneq \{1, \ldots, |x|\}$. Observe that $\mathcal{I}$ always needs to be a proper subset of $\{1, \ldots, |x|\}$, as, for the reasons described above, allowing x_i to depend on entire x would render impossible any secure construction. It is of course not clear what would be the right "natural" subsets $\mathcal{S}_i$ that one could use here. The authors of [23] solve this problem in the following simple way. They assume that the codeword consists of two parts (usually of equal size), i.e.: $x = (L, R) \in \mathcal{L} \times \mathcal{R}$, and the adversary can tamper in an arbitrary way with both parts, i.e., $\mathcal{F}$ consists of *all* functions $\text{Mall}^{f,g}$ that can be defined as $\text{Mall}^{f,g}(L, R) = (f(L), g(R))$ (for some $f : \mathcal{L} \to \mathcal{L}$ and $g : \mathcal{R} \to \mathcal{R}$). In practical applications this corresponds to a scenario in which L and R are stored on two separate memory parts that can be tampered independently. A similar model has been used before in the context of leakages and is called a *split-state* model [22,14,28,16]. The authors of [23] show existence of non-malleable codes secure in this model in a non-constructive way (via the probabilistic argument). They also provide a construction of such codes in a random oracle model, and leave constructing explicit information-theoretically secure codes as an open problem. A very interesting partial solution to this problem came recently from Liu and Lysyanskaya [33] who constructed such codes with computationally-security, assuming a common reference string. Their construction comes with an additional feature of being leakage-resilient, i.e.

they allow the adversary to obtain some partial information about the codeword via memory leakage (the amount of leakage that they can tolerate is a $\frac{1}{2} - o(1)$ fraction of the length of the codeword). However, constructing the information-theoretically secure nonmalleable codes in this model remained an open problem, even if messages are of length 1 only (i.e. $\mathcal{M} = \{0, 1\}$).

Our Contribution. We show a construction of efficient information-theoretically secure non-malleable codes in the split-state model for $\mathcal{M} = \{0, 1\}$. Additionally to being non-malleable, our code is also leakage-resilient and the amount of leakage that we can tolerate is an arbitrary constant $\xi < \frac{1}{4}$ of the length of the codeword (cf. Thm. 2). Our construction is fairly simple. The codeword is divided into two parts, L and R, which are vectors from a linear space $\mathbb{F}^n$, where $\mathbb{F}$ is a field of exponential size (and hence $\log |\mathbb{F}|$ is linear). Essentially, to encode a bit $B = 0$ one chooses at a random pair $(L, R) \in \mathbb{F}^n \times \mathbb{F}^n$ of orthogonal vectors (i.e. such that $\langle L, R \rangle = 0$), and to encode $B = 1$ one chooses a random pair of non-orthogonal vectors (clearly both encoding and decoding can be done very efficiently in such a code). Perhaps surprisingly, the assumption that $\mathbb{F}$ is large is important, as our construction is *not* secure for small $\mathbb{F}$'s. An interesting consequence is that our code is "non-balanced", in the sense that a random element of the codeword space with an overwhelming probability encodes 1. We actually use this property in the proof.

Our proof also very strongly relies on the fact that the inner product over finite field is a two-source extractor (cf. Sect. 2). We actually show that in general a split-state non-malleable code for one-bit messages can be constructed from any two source-extractor with sufficiently strong parameters (we call such extractors *flexible*, cf. Sect. 2).

We also provide a simple argument that shows that our scheme is secure against affine mauling functions (that look at the entire codeword, hence *not* in the split-state model).

Typically in information-theoretic cryptography solving a certain task for one-bit messages automatically gives a solution for multi-bit messages. Unfortunately, it is not the case for the non-malleable codes. Consider for example a naive idea of encoding n bits "in parallel" using the one bit encoding function Enc, i.e. letting $\mathsf{Enc}'(m_1, \ldots, m_n) := ((L_1, \ldots, L_n), (R_1, \ldots, R_n))$, where each $(L_i, R_i) = \mathsf{Enc}(m_i)$. This encoding is obviously malleable, as the adversary can, e.g., permute the bits of m by permuting (in the same way) the blocks $L_1, \ldots, L_n$ and $R_1, \ldots, R_n$. Nevertheless we believe that our solution is an important step forward, as it may be useful as a building blocks for other, more advanced constructions, like, e.g., tamper-resilient generic compilers (in the spirit of [31,30,13,20,27]). This research direction looks especially promising since many of the leakage-resilient compliers (e.g. [20,27]) are based on the same inner-product extractor.

We also show that for one-bit messages non-malleable codes can be defined in an alternative, and perhaps simpler way. Namely we show (cf. Lemma 2) that any code (Enc, Dec) (not necessarily defined in the split-state model) in non-malleable with respect to some family $\mathcal{F}$ of functions if and only if "it is hard to negate the encoded bit B with functions from $\mathcal{F}$", by which we mean that for a bit B chosen *uniformly* from $\{0,1\}$ any $F \in \mathcal{F}$ we have that

$$P\left[\mathsf{Dec}(F(\mathsf{Enc}(B))) \neq B\right] \leq \frac{1}{2}. \tag{1}$$

(the actual lemma that we prove involves also some small error parameter ϵ both in the non-malleability definition and in (1), but for the purpose of this informal discussion let us omit them). Therefore, the problem of constructing non-malleable bit encoding in the split state model can be translated to a much simpler and perhaps more natural question: can one encode a random bit B as (L, R) in such a way that independent manipulation of L and R produces an encoding (L', R') of $\overline{B}$ with probability at most $1/2$? Observe that, of course, it is easy to negate a random bit with probability exactly $1/2$, by deterministically setting (L', R') to be an encoding of a fixed bit, 0, say. Informally speaking, $(\mathsf{Enc}, \mathsf{Dec})$ is non-malleable if this is the best that the adversary can achieve.

In the full version of this paper [21] we analyze the general relationship between the two-source extractors and the non-malleable codes in the split state model pointing out some important differences. We also compare the notion of the non-malleable codes with the *leakage-resilient storage* [14] also showing that they are fundamentally different.

Related and Subsequent Work. Some of the related work was already described in the introduction. There is no space here to mention all papers that propose theoretical countermeasures against tampering. This research was initiated by Ishai et al. [30,26]. Security against both tampering and leakage attacks were also recently considered in [32]. Unlike us, they construct concrete cryptosystems (not encoding schemes) secure against such attacks. Another difference is that their schemes are computationally secure, while in this work we are interested in the information-theoretical security.

The notion of non-malleability (introduced in [19]) is used in cryptography in several contexts. In recent years it was also analyzed in the context of randomness extractors, starting from the work of Dodis and Wichs [18] on non-malleable extractors (see also [17,12]). Informally speaking an extractor ext is non-malleable if its output $\mathsf{ext}(S, X)$ is (almost) uniform even if one knows the value $\mathsf{ext}(F(S), X)$ for some "related" seed $F(S)$ (such that $F(S) \neq S$). Unfortunately, it does not look like this primitive can be used to construct the non-malleable codes in the split-state model, as this definition does not capture the situation when X is also modified.

Constructions of non-malleable codes secure in different (not split-state) models were recently proposed in [8,9,10].

Recently, Aggarwal, Dodis and Lovett [1] solved the main open problem left in this paper, by showing a non-malleable code that works for messages of arbitrary length. This exciting result is achieved by combining the inner-product based encoding with sophisticated methods from the additive combinatorics.

Acknowledgments. We are very grateful to Divesh Aggarwal and to the anonymous CRYPTO reviewer for pointing out errors in the proof of Lemma 3 in the

244   S. Dziembowski, T. Kazana, and M. Obremski

previous versions of this paper. We also thank Yevgeniy Dodis, Konrad Durnoga and Karol Cwalina for helpful discussions.

2 Preliminaries

If $\mathcal{Z}$ is a set then $Z \leftarrow \mathcal{Z}$ will denote a random variable sampled uniformly from $\mathcal{Z}$. We start with some standard definitions and lemmas about the statistical distance. Recall that if A and B are random variables over the same set $\mathcal{A}$ then the *statistical distance between A and B* is denoted as $\Delta(A; B)$, and defined as $\Delta(A; B) = \frac{1}{2} \sum_{a \in \mathcal{A}} |P[A = a] - P[B = a]|$. If the variables A and B are such that $\Delta(A, B) \leq \epsilon$ then we say that A is ϵ-close to B, and write $A \approx_{\epsilon} B$. If $\mathcal{X}, \mathcal{Y}$ are some events then by $\Delta(A|\mathcal{X} \; ; \; B|\mathcal{Y})$ we will mean the distance between variables A' and B', distributed according to the conditional distributions $P_{A|\mathcal{X}}$ and $P_{B|\mathcal{Y}}$.

If B is a uniform distribution over $\mathcal{A}$ then $d(A|\mathcal{X}) := \Delta(A|\mathcal{X}; B)$ is called *statistical distance of A from uniform given the event $\mathcal{X}$*. If moreover C is independent from B then $d(A|C) := \Delta((A, C); (B, C))$ is called *statistical distance of A from uniform given the variable C*. More generally, if $\mathcal{X}$ is an event then $d(A|C, \mathcal{X}) := \Delta((A, C)|\mathcal{X}; (B, C)|\mathcal{X})$. It is easy to see that $d(A|C)$ is equal to $\sum_c P[C = c] \cdot d(A|C = c)$.

Extractors. As described in the introduction, the main building block of our construction is a two-source randomness extractor based on the inner product over finite fields. The two source extractors were introduced (implicitly) by Chor and Goldreich [11], who also showed that the inner product over Z_2 is a two-source extractor. The generalization to any field is shown in [36].

Our main theorem (Thm. 1) does not use any special properties of the inner product (like, e.g., the linearity), besides of the fact that it extracts randomness, and hence it will be stated in a general form, without assuming that the underlying extractor is necessarily an inner product. The properties that we need from our two-source extractor are slightly non-standard. Recall that a typical way to define a strong two-source extractor[1] (cf. e.g. [36]) is to require that $d(\mathsf{ext}(L, R)|L)$ and $d(\mathsf{ext}(L, R)|R)$ are close to uniform, provided that L and R have min-entropy at least m (for some parameter m). For the reasons that we explain below, we need a slightly stronger notion, that we call *flexible* extractors. Essentially, instead of requiring that $\mathbf{H}_{\infty}(L) \geq m$ and $\mathbf{H}_{\infty}(R) \geq m$ we will require only that $\mathbf{H}_{\infty}(L) + \mathbf{H}_{\infty}(R) \geq k$ (for some k). Note that if $k = 2m$ then this requirement is obviously weaker than the standard once, and hence the flexibility strengthens the standard definition.

Formally, let $\mathcal{L}, \mathcal{R}$ and $\mathcal{C}$ be some finite sets. A function $\mathsf{ext} : \mathcal{L} \times \mathcal{R} \to \mathcal{C}$ is a *strong flexible (k, ϵ)-two source extractor* if for every $L \in \mathcal{L}$ and $R \in \mathcal{R}$ such that $\mathbf{H}_{\infty}(L) + \mathbf{H}_{\infty}(R) \geq k$ we have that $d(\mathsf{ext}(L, R)|L) \leq \epsilon$ and $d(\mathsf{ext}(L, R)|R) \leq \epsilon$. Since we are not going to use any weaker version of this notion we will often

---

[1] Recall also that a random variable A has *min-entropy* k, denoted $\mathbf{H}_{\infty}(A) = k$ if $k = \min_a (-\log P[A = a])$.

simply call such extractors "flexible" without explicitly stating that they are strong. As it turns out the inner product over finite fields is such an extractor.

Lemma 1. *For every finite field $\mathbb{F}$ and any n we have that* $\mathsf{ext} : \mathbb{F}^n \times \mathbb{F}^n \to \mathbb{F}$ *defined as* $\mathsf{ext}_{\mathbb{F}}^n(L, R) = \langle L, R \rangle$ *is a strong flexible (k, ϵ)-extractor for any k and ϵ such that*

$$\log(1/\epsilon) = \frac{k - (n+4)\log|\mathbb{F}|}{3} - 1. \tag{2}$$

Although this lemma appears to be folklore, at least in case of the "weak" flexible extractors (i.e. when we require only that $d(\mathsf{ext}(L, R)) \leq \epsilon$), we were not able to find it in the literature for the *strong* flexible extractors. Therefore for completeness in the full version of this paper [21] we provide a proof of it (which is straightforward adaptation of the proof of Theorem 3.1 in [36]).

Note that since ϵ can be at most 1, hence (2) makes sense only if $k \geq 6+4\,|\mathbb{F}|+ n\log|\mathbb{F}|$. It is easy to see that it cannot be improved significantly, as in any flexible (k, ϵ)-extractor $\mathsf{ext} : \mathcal{L} \times \mathcal{R} \to \mathcal{C}$ we need to have $k > \max(\log|\mathcal{L}|, \log|\mathcal{R}|)$. To see why it is the case, suppose we have such a flexible (k, ϵ)-extractor ext for $k = \log|\mathcal{L}|$ (the case $k = \log|\mathcal{R}|$ is obviously symmetric). Now let L' be a random variable uniformly distributed over $\mathcal{L}$ and let $R' \in \mathcal{R}$ be constant. Then obviously $\mathbf{H}_\infty(L') + \mathbf{H}_\infty(R') = \log|\mathcal{L}| + 0 = k$, but $\mathsf{ext}(L', R')$ is a deterministic function of L', and hence $d(\mathsf{ext}(L', R')|L')$ is large. Therefore, in terms of the entropy threshold k, the inner product is optimal in the class of flexible extractors (up to a small additive constant). Note that this is in contrast with the situation with the "standard" two-source extractors where a better extractor is known [6].

The reason why we need the "flexibility" property is as follows. In the proof of Lemma 3 we will actually use in two different ways the fact that ext is an extractor. In one case (in the proof of Claim 2 within the proof of Lemma 3) we will use it in the "standard" way, i.e. we will apply it to two independent random variables with high min-entropy. In the other case (proof of Claim 1) we will use the fact that $d(\mathsf{ext}(L, R)|R) \leq \epsilon$ even if L has relatively low min-entropy ($\mathbf{H}_\infty(L) = k-|R|$) while R is completely uniform (and hence $\mathbf{H}_\infty(L) + \mathbf{H}_\infty(R) = k$).[2] Hence we will treat ext as standard seeded extractor. It should not be surprising that we can use the inner product in this way, as it is easy to see that the inner product is a universal hash function, and hence the fact that it is a seeded strong extractor follows from the leftover hash lemma [29]. Hence Lemma 1 in some sense "packs" these two properties of the inner product into one simple statement.

The observation that the inner product extractor is flexible allows us as also to talk about the sum of leakages in Section 5, instead of considering bounded leakage from L and R separately (as it is done, e.g., in [14]). We would like to stress that this is actually not the main reason for introducing the "flexibility" property, as it would be needed even if one does not incorporate leakages into the model.

---

[2] We will also use a symmetric fact for $d(\mathsf{ext}(L, R)|L)$.

3   Non-malleable Codes and the Hardness of Negation

In this section we review the definition of the non-malleable codes from [23], which has already been discussed informally in the introduction. Formally, let $(\mathsf{Enc} : \mathcal{M} \to \mathcal{X}, \mathsf{Dec} : \mathcal{X} \to \mathcal{M} \cup \{\bot\})$ be an encoding scheme. For $F : \mathcal{X} \to \mathcal{X}$ and for any $m \in \mathcal{M}$ define the experiment Tamper_m^F as:

$$\mathsf{Tamper}_m^F = \left\{ \begin{array}{c} X \leftarrow \mathsf{Enc}(m), \\ X' := F(X), \\ m' := \mathsf{Dec}(X') \\ \text{output: } m' \end{array} \right\}$$

Let $\mathcal{F}$ be a family of functions from $\mathcal{X}$ to $\mathcal{X}$. We say that an encoding scheme $(\mathsf{Enc}, \mathsf{Dec})$ is ϵ-*non-malleable with respect to* $\mathcal{F}$ if for every function $F \in \mathcal{F}$ there exists distribution D^F on $\mathcal{M} \cup \{\mathsf{same}^*, \bot\}$ such that for every $m \in \mathcal{M}$ we have

$$\mathsf{Tamper}_m^F \approx_\epsilon \left\{ \begin{array}{c} d \leftarrow D^F \\ \text{if } d = \mathsf{same}^* \text{ then output } m \\ \text{otherwise output } d. \end{array} \right\} \tag{3}$$

The idea behind the "$\bot$" symbol is that it should correspond to the situation when the decoding function detects tampering and outputs an error message. Since the codes that we construct in this paper do not need this feature, we will usually drop this symbol and have $\mathsf{Dec} : \mathcal{X} \to \mathcal{M}$. The "$\bot$" symbol is actually more useful for the *strong* non-malleable codes (another notion defined in [23]) where it is required that *any* tampering with X should be either "detected" or should produce encoding of an unrelated message. Our codes do not have this property. This is because, for example, permuting the elements of the vectors L and R in the same manner *does* change these vectors, but *does not* change their inner product. Fortunately, for all applications that we are aware of this stronger notion is not needed. The following lemma, already informally discussed in Sect. 1, states that for one-bit messages non-malleability is equivalent to the hardness of negating a random encoded bit. It turns out that such a characterization of the non-malleable codes is much simpler to deal with. We also believe that it may be of independent interest.

Lemma 2. *Suppose* $\mathcal{M} = \{0, 1\}$*. Let* $\mathcal{F}$ *be any family of functions from* $\mathcal{X}$ *to* $\mathcal{X}$*. An encoding scheme* $(\mathsf{Enc} : \mathcal{M} \to \mathcal{X}, \mathsf{Dec} : \mathcal{X} \to \mathcal{M})$ *is* ϵ-*non-malleable with respect to* $\mathcal{F}$ *if and only if for any* $F \in \mathcal{F}$ *and* $B \leftarrow \{0, 1\}$ *we have*

$$P\left[\mathsf{Dec}\left(F(\mathsf{Enc}(B))\right) \neq B\right] \leq \frac{1}{2} + \epsilon. \tag{4}$$

The proof of this lemma appears in the full version of this paper [21]. In this paper we are interested in the split-state codes. A *split-state code* is a pair $(\mathsf{Enc} : \mathcal{M} \to \mathcal{L} \times \mathcal{R}, \mathsf{Dec} : \mathcal{L} \times \mathcal{R} \to \mathcal{M})$. We say that it is ϵ-*non-malleable* if it is ϵ-non-malleable with respect to a family of *all* functions $\mathsf{Mall}^{f,g}$ defined as $\mathsf{Mall}^{f,g}(L, R) = (f(L), g(R))$.

4   The Construction

In this section we present a construction of a non-malleable code in the split-state model, together with a security proof. Before going to the technical details, let us start with some intuitions. First, it is easy to see that any such code (Enc, Dec) needs to be a 2-out-of-2 secret sharing scheme, where Enc is the sharing function, Dec is the reconstruction function, and $(L, R) = \mathsf{Enc}(M)$ are shares of a secret M. Informally speaking, this is because if one of the "shares", L, say, reveals some non-trivial information about M then by modifying L we can "negate" stored secret M with probability significantly higher than $1/2$. More precisely, suppose that $\mathcal{M} = \{0, 1\}$ and that we know that there exist some values $\ell_0, \ell_1 \in \mathcal{L}$ such that for $b = 0, 1$ if $L = \ell_b$ then M is significantly more likely to be equal to b. Then (f, g) where g is an identity and f is such that $f(\ell_0) = \ell_1$ and $f(\ell_1) = \ell_0$ would lead to $M' = \mathsf{Dec}(f(L), g(R)) = 1 - M$ with probability significantly higher than $1/2$ (this argument is obviously informal, but it can be formalized).

It is also easy to see that not every secret sharing scheme is a non-malleable code in the split-state model. As an example consider $\mathsf{Enc} : Z_a \rightarrow Z_a \times Z_a$ (for some $a \geq 2$) defined as $\mathsf{Enc}(M) := (L, L + M \pmod{a})$, where $L \leftarrow Z_a$, and $\mathsf{Dec}(L, R) := L + R \bmod m$. Obviously it is a good 2-out-of-2 secret sharing scheme. However, unsurprisingly, it is malleable, as an adversary can, e.g., easily add any constant $w \in Z_a$ to a encoded message, by choosing an identity function as f, and letting g be such that that $g(R) = R + w \bmod a$. Obviously in this case for every L and R that encode some M we have $\mathsf{Dec}(f(L), g(R)) = M + w \bmod a$.

We therefore need to use a secret sharing scheme with some extra security properties. A natural idea is to look at the two-source randomness extractors, as they may be viewed exactly as "2-out-of-2 secret sharing schemes with enhanced security", and since they have already been used in the past in the context of the leakage-resilient cryptography. The first, natural idea, is to take the inner product extractor $\mathsf{ext} : \mathbb{F}^n \times \mathbb{F}^n \rightarrow \mathbb{F}$ and use it as a code as follows: to encode a message $M \in \mathbb{F}$ take a random pair $(L, R) \in \mathbb{F}^n \times \mathbb{F}^n$ such that $\langle L, R \rangle = M$ (to decode (L, R) simply compute $\langle L, R \rangle$). This way of encoding messages is a standard method to provide leakage-resilience in the split-state model (cf. e.g. [14]). Unfortunately, it is easy to see that this scheme can easily be broken by exploiting the linearity attacks of the inner product. More precisely, if the adversary chooses $f(L) := a \cdot L$ and $g(R) := R$ (for any $a \in \mathbb{F}$) then the encoded secret gets multiplied by a. Obviously, this attack does not work for $\mathbb{F} = Z_2$, as in this case the only choices are $a = 0$ (which means that the secret is deterministically transformed to 0) and $a = 1$ (which leaves the secret unchanged). Sadly, it turns out that for $\mathbb{F} = Z_2$ another attack is possible. Consider f and g that leave their input vectors unchanged except of setting the first coordinate of the vector to 1, i.e.: $f(L_1, \ldots, L_n) := (1, L_2, \ldots, L_n)$ and $g(R_1, \ldots, R_n) := (1, R_2, \ldots, R_n)$. Then it is easy to see that $\langle f(L), g(R) \rangle \neq \langle L, R \rangle$ if and only if $L_1 \cdot R_1 = 0$, which happens with probability $3/4$ both for $M = 0$ and for $M = 1$.

Note that the last attack is specific for small $\mathbb{F}$'s, as over larger fields the probability that $L_1 \cdot R_1 = 0$ is negligible. At the first glance, this fact should not bring any hope for a solution, since, as described above, for larger fields another attack exists. Our key observation is that for one-bit messages it is possible to combine the benefits of the "large field" solution with those of the "small field" solution in such a way that the resulting scheme is secure, and in particular both attacks are impossible! Our solution works as follows. The codewords are pairs of vectors from $\mathbb{F}^n$ for a large $\mathbb{F}$. The encoding of 0 remains as before – i.e. we encode it as a pair (L, R) of orthogonal vectors. To encode 1 we choose a random pair (L, R) of non-orthogonal vectors, i.e. such that $\langle L, R \rangle$ is a random non-zero element of $\mathbb{F}$. Before going to the technical details let us first "test" this construction against the attacks described above. First, observe that multiplying L (or R) by some constant $a \neq 0$ never changes the encoded bit as $\langle a \cdot L, R \rangle = a \langle L, R \rangle$ which is equal to 0 if and only if $\langle L, R \rangle = 0$. On the other hand if $a = 0$ then $\langle a \cdot L, R \rangle = 0$, and hence the secret gets deterministically transformed to 0, which is also ok. It is also easy to see that the second attack (setting the first coordinates of both the vectors to 1) results in $\langle f(L), g(R) \rangle$ close to uniform (no matter what was the value of $\langle L, R \rangle$), and hence $\mathsf{Dec}(f(L), g(R)) = 1$ with an overwhelming probability.

Let us now define our encoding scheme formally. As already mentioned in Sect. 2 our construction uses a strong flexible two-source extractor $\mathsf{ext} : \mathcal{L} \times \mathcal{R} \rightarrow \mathcal{C}$ in a black-box way (later we show how to instantiate it with an inner product extractor, cf. Thm. 2). This in particular means that we do not use any special properties of the inner product, like the linearity. Also, since $\mathcal{C}$ does not need to be a field, hence obviously the choice to encode 0 is by a pair of vectors such that $\langle L, R \rangle = 0$ (in the informal discussion above) was arbitrary, and one can encode 0 as any pair (L, R) such that $\langle L, R \rangle = c$, for some fixed $c \in \mathbb{F}$. Let $\mathsf{ext} : \mathcal{L} \times \mathcal{R} \rightarrow \mathcal{C}$ be a strong flexible (k, ϵ)-extractor, for some parameters k and ϵ, and let $c \in \mathcal{C}$ be arbitrary. We first define the decoding function. Let $\mathsf{D}^c_{\mathsf{ext}} : \mathcal{L} \times \mathcal{R} \rightarrow \{0, 1\}$ be defined as:

$$\mathsf{D}^c_{\mathsf{ext}}(L, R) = \begin{cases} 0 \text{ if } \mathsf{ext}(X) = c \\ 1 \text{ otherwise.} \end{cases}$$

Now, let $\mathsf{E}^c_{\mathsf{ext}} : \{0, 1\} \rightarrow \mathcal{L} \times \mathcal{R}$ be an encoding function defined as $\mathsf{E}^c_{\mathsf{ext}}(b) := (L, R)$, where (L, R) is a pair chosen uniformly at random from the set $\{(L, R) : \mathsf{D}^c_{\mathsf{ext}}(L, R) = b\}$. We also make a small additional assumption about ext. Namely, we require that $\tilde{L}$ and $\tilde{R}$ are completely uniform over $\mathcal{L}$ and $\mathcal{R}$ (resp.) then $\mathsf{ext}(\tilde{L}, \tilde{R})$ is completely uniform. More formally

$$\text{for } \tilde{L} \leftarrow \mathcal{L} \text{ and } \tilde{R} \leftarrow \mathcal{R} \text{ we have } d(\mathsf{ext}(\tilde{L}, \tilde{R})) = 0. \tag{5}$$

The reason why we impose this assumption is that it significantly simplifies the proof, thanks to the following fact. It is easy to see that if ext satisfies (5), then for every $x \in \mathcal{C}$ the cardinality of each set $\{(\ell, r) : \mathsf{ext}(\ell, r) = x\}$ is exactly $1/|\mathbb{F}|$ fraction of the cardinality of $\mathcal{L} \times \mathcal{R}$. Hence, if $B \leftarrow \{0, 1\}$ and $(L, R) \leftarrow \mathsf{E}^c_{\mathsf{ext}}(B)$, then in the distribution of (L, R) every (ℓ, r) such that $\mathsf{ext}(\ell, r) = c$ is exactly $(|\mathcal{C}| - 1)$ more likely than any (ℓ', r') such that $\mathsf{ext}(\ell', r') \neq c$. Formally:

$$P[(L, R) = (\ell, r)] = (|\mathcal{C}| - 1) \cdot P[(L, R) = (\ell', r')]. \tag{6}$$

It is also straightforward to see that every extractor can be easily converted to an extractor that satisfies (5)[3]. Lemma 3 below is the main technical lemma of this paper. It states that $(\mathsf{E}^c_{\mathsf{ext}}, \mathsf{D}^c_{\mathsf{ext}})$ is non-malleable, for an appropriate choice of ext. Since later (in Sect. 5) we will re-use this lemma in the context of non-malleability with leakages, we prove it in a slightly more general form. Namely, (cf. (8)) we show that it is hard to negate an encoded bit even if one knows that the codeword (L, R) happens to be an element of some set $\mathcal{L}' \times \mathcal{R}' \subseteq \mathcal{L} \times \mathcal{R}$. Note that we do not explicitly assume any lower bound on the cardinality of $\mathcal{L}' \times \mathcal{R}'$. This is not needed, since this cardinality is bounded implicitly in (7) by the fact that in any flexible extractor the parameter k needs to be larger than $\max(\log|\mathcal{L}|, \log|\mathcal{R}|)$ (cf. Sect. 2). If one is not interested in leakages then one can read Lemma 3 and its proof assuming that $\mathcal{L}' \times \mathcal{R}' = \mathcal{L} \times \mathcal{R}$. Lemma 3 is stated abstractly, but one can, of course, obtain a concrete non-malleable code, by using as ext the two-source extractor $\mathsf{ext}^n_{\mathbb{F}}$. We postpone presenting the choice of concrete parameters $\mathbb{F}$ and n until Section 5, where it is done in a general way, also taking into account leakages.

Lemma 3. *Let $\mathcal{L}'$ and $\mathcal{R}'$ be some subsets of $\mathcal{L}$ and $\mathcal{R}$ respectively. Suppose* $\mathsf{ext} : \mathcal{L} \times \mathcal{R} \to \mathcal{C}$ *is a strong flexible (k, ϵ)-extractor that satisfies (5), where, for some parameter δ we have:*

$$k = \frac{2}{3} \cdot (\log|\mathcal{L}'| + \log|\mathcal{R}'|) - \frac{2}{3} \cdot \log(1/\delta). \tag{7}$$

Take arbitrary functions $f : \mathcal{L} \to \mathcal{L}$ and $g : \mathcal{R} \to \mathcal{R}$, let B be chosen uniformly at random from $\{0, 1\}$ and let $(L, R) \leftarrow \mathsf{E}^c_{\mathsf{ext}}(B)$. Then

$$P\left[\mathsf{D}^c_{\mathsf{ext}}(f(L), g(R)) \neq B \mid (L, R) \in (\mathcal{L}', \mathcal{R}')\right] \leq$$
$$\frac{1}{2} + \frac{3}{2}|\mathcal{C}|^{-1} + 6|\mathcal{C}|^2\epsilon + \delta/(|\mathcal{C}|^{-1} - \epsilon), \tag{8}$$

and, in particular $(\mathsf{E}^c_{\mathsf{ext}}, \mathsf{D}^c_{\mathsf{ext}})$ is $\left(\frac{3}{2}|\mathcal{C}|^{-1} + 6|\mathcal{C}|^2\epsilon + \delta/(|\mathcal{C}|^{-1} - \epsilon)\right)$-non-malleable.

Proof. Before presenting the main proof idea let us start with some simple observations. First, clearly it is enough to show (8), as then the fact that $(\mathsf{E}^c_{\mathsf{ext}}, \mathsf{D}^c_{\mathsf{ext}})$ is $\left(|\mathcal{C}|^{-1} + 2|\mathcal{C}|^2\epsilon + \delta/(|\mathcal{C}|^{-1} - \epsilon)\right)$-non-malleable can be obtained easily by assuming that $\mathcal{L}' \times \mathcal{R}' = \mathcal{L} \times \mathcal{R}$ and applying Lemma 2. Observe also that (8) implies that $\log|\mathcal{L}'| + \log|\mathcal{R}'| \geq k$, and hence, from the fact that ext is a (k, ϵ)-two source extractor we obtain that if $\tilde{L} \leftarrow \mathcal{L}'$ and $\tilde{R} \leftarrow \mathcal{R}'$ then

$$d(\mathsf{ext}(\tilde{L}, \tilde{R})) \leq \epsilon. \tag{9}$$

We will use this fact later. The basic idea behind the proof is a as follows. Denote $B' := \mathsf{Mall}^{f,g}(\mathsf{Enc}(B))$. Recall that our code is "non-balanced" in the sense that

---

[3] The inner-product extractor satisfies (5) if we assume, e.g., that the fist coordinate of $\mathcal{L}$ and the last coordinate of $\mathcal{R}$ are non-zero. In general, if $\mathsf{ext} : \mathcal{L} \times \mathcal{R} \to \mathcal{C}$ is any extractor, then $\mathsf{ext}' : (\mathcal{L} \times \mathcal{C}) \times \mathcal{R} \to \mathcal{C}$ defined as $\mathsf{ext}'((C, L), R) = ext(L, R) + C$ (assuming that $(\mathcal{C}, +)$ is a group) satisfies (5).

a random codeword $(L, R) \in \mathcal{L}' \times \mathcal{R}'$ with only negligible probability encodes 0. We will exploit this fact. Very informally speaking, we would like to prove that if $B = 1$ then the adversary cannot force B' to be equal to 0, as any independent modifications of L and R that encode 1 are unlikely to produce an encoding of 0. In other words, we would hope to show that $P[B' = 0 | B = 1]$ is small. Note that if we managed to show it, then we would obviously get that $P[B' \neq B]$ cannot be much larger than $1/2$ (recall that B is uniform), and then the proof would be finished. Unfortunately, this is too good to be true, as the adversary can choose f and g to be constant such that always $\mathsf{D}_{\mathsf{ext}}^c(f(L), g(R)) = 0$, which would result in $B' = 0$ for any value of B. Intuitively, what we will actually manage to prove is that the only way to obtain $B' = 0$ if $B = 1$ is to apply such a "constant function attack". Below we show how to make this argument formal.

Let us first observe that any attack where f and g are constant will never work against any encoding scheme, as in this case $(f(L), g(R))$ carries no information about the initial value of B. Our first key observation is that for our scheme, thanks to the fact that it is based on extractor, this last statement holds even if any of f and g is only "sufficiently close to constant". Formalizing this property is a little bit tricky, as, of course, the adversary can apply "mixed" strategies, e.g., setting f to be constant on some subset of $\mathcal{L}'$ and to be injective (and hence "very far from constant") on the rest of $\mathcal{L}'$. In order to deal with such cases we will define subsets $\mathcal{L}_{\mathsf{FFC}} \subseteq \mathcal{L}'$ and $\mathcal{R}_{\mathsf{FFC}} \subseteq \mathcal{R}'$ on which f and g (resp.) are "very far from constant". Formally, for $\tilde{L} \leftarrow \mathcal{L}'$ and $\tilde{R} \leftarrow \mathcal{R}'$ let

$$\mathcal{L}_{\mathsf{FFC}} := \left\{ \ell \in \mathcal{L}' : \mathbf{H}_\infty(\tilde{L} \mid f(\tilde{L}) = f(\ell)) < k + 1 - \log |\mathcal{R}'| \right\},$$

and

$$\mathcal{R}_{\mathsf{FFC}} := \left\{ r \in \mathcal{R}' : \mathbf{H}_\infty(\tilde{R} \mid g(\tilde{R}) = g(r)) < k + 1 - \log |\mathcal{L}'| \right\},$$

where FFC stands for "far from constant". Hence, in some sense, we define a function to be "very far from constant on some argument x" if there are only a few other arguments of this function that collide with x. We now state the following claim (whose proof appears in the full version of this paper [21]) that essentially formalizes the intuition outlined above, by showing that if either $L \notin \mathcal{L}_{\mathsf{FFC}}$ or $R \notin \mathcal{R}_{\mathsf{FFC}}$ then (f, g) cannot succeed in negating B.

Claim 1. Let $B \leftarrow \{0, 1\}$ and $(L, R) \leftarrow \mathsf{E}_{\mathsf{ext}}^c(B)$. Then:

$$P\left[\mathsf{D}_{\mathsf{ext}}^c(\mathsf{Mall}^{f,g}(L, R)) \neq B \mid L \notin \mathcal{L}_{\mathsf{FFC}} \vee R \notin \mathcal{R}_{\mathsf{FFC}} \right] \leq \frac{1}{2} + \frac{3}{4} \cdot |\mathcal{C}|^{-1} + 6 |\mathcal{C}|^2 \epsilon.$$
(10)

Hence, what remains is to analyze the case when $(L, R) \in \mathcal{L}_{\mathsf{FFC}} \times \mathcal{R}_{\mathsf{FFC}}$. We will do it only for the case $B = 1$, and when $\mathcal{L}_{\mathsf{FFC}} \times \mathcal{R}_{\mathsf{FFC}}$ is relatively large, more precisely we will assume that

$$|\mathcal{L}_{\mathsf{FFC}} \times \mathcal{R}_{\mathsf{FFC}}| \geq \delta \cdot |\mathcal{L}' \times \mathcal{R}'|.$$
(11)

This will suffice since later we will show (cf. (23)) that the probability that $\mathrm{Enc}(B) \in \mathcal{L}_{\mathsf{FFC}} \times \mathcal{R}_{\mathsf{FFC}}$ is small for small δ's (note that this is not completely trivial as (L, R) does not have a uniform distribution over $\mathcal{L}' \times \mathcal{R}'$). We now have the following claim whose proof appears in the full version of this paper [21].

Claim 2. *Let* $(L^1, R^1) \leftarrow \mathsf{E}^c_{\mathrm{ext}}(1)$ *and suppose* $\mathcal{L}_{\mathsf{FFC}}$ *and* $\mathcal{R}_{\mathsf{FFC}}$ *are such that (11) holds. Then*

$$P\left[\mathsf{D}^c_{\mathrm{ext}}\left(\mathrm{Dec}(f(L^1), g(R^1))\right) = 0 \mid (L^1, R^1) \in \mathcal{L}_{\mathsf{FFC}} \times \mathcal{R}_{\mathsf{FFC}}\right] \leq 2\,|\mathcal{C}|^{-1} + 2\epsilon. \quad (12)$$

To finish the proof we need to combine the two above claims. A small technical difficulty, that we need still to deal with, comes from the fact that Claim 2 was proven only under the assumption (11). Let us first expand the left-hand-side of (8). We have

$$P\left[\mathsf{D}^c_{\mathrm{ext}}(\mathrm{Mall}^{f,g}(L, R)) \neq B \mid (L, R) \in \mathcal{L}' \times \mathcal{R}'\right] \quad (13)$$

$$= \underbrace{P\left[\mathsf{D}^c_{\mathrm{ext}}(\mathrm{Mall}^{f,g}(L, R)) \neq B \mid L \notin \mathcal{L}_{\mathsf{FFC}} \vee R \notin \mathcal{R}_{\mathsf{FFC}}\right]}_{(*)} \quad (14)$$

$$\cdot \underbrace{P\left[L \notin \mathcal{L}_{\mathsf{FFC}} \vee R \notin \mathcal{R}_{\mathsf{FFC}}\right]}_{(**)}$$

$$+ P\left[\mathsf{D}^c_{\mathrm{ext}}(\mathrm{Mall}^{f,g}(L, R)) \neq B \mid (L, R) \in \mathcal{L}_{\mathsf{FFC}} \times \mathcal{R}_{\mathsf{FFC}}\right] \quad (15)$$

$$\cdot P\left[(L, R) \in \mathcal{L}_{\mathsf{FFC}} \times \mathcal{R}_{\mathsf{FFC}}\right]$$

From Claim 1 we get that $(*)$ is at most $\frac{1}{2} + \frac{1}{2} \cdot |\mathcal{C}|^{-1} + 2\,|\mathcal{C}|^2\,\epsilon$. Now consider two cases.

Case 1 First, suppose that (11) holds (i.e. $|\mathcal{L}_{\mathsf{FFC}} \times \mathcal{R}_{\mathsf{FFC}}| \geq \delta \cdot |\mathcal{L} \times \mathcal{R}|$). In this case we get that $(**)$ is a equal to

$$\overbrace{P\left[\mathsf{D}^c_{\mathrm{ext}}(\mathrm{Mall}^{f,g}(L, R)) \neq B \mid B = 0 \wedge (L, R) \in \mathcal{L}_{\mathsf{FFC}} \times \mathcal{R}_{\mathsf{FFC}}\right]}^{\leq\, 2|\mathcal{C}|^{-1}+2\epsilon \text{ by Claim 2}}$$

$$\cdot \overbrace{P\left[B = 0 \mid (L, R) \in \mathcal{L}_{\mathsf{FFC}} \times \mathcal{R}_{\mathsf{FFC}}\right]}^{\geq\frac{1}{2}-|\mathcal{C}|\epsilon} + \quad (16)$$

$$\underbrace{P\left[\mathsf{D}^c_{\mathrm{ext}}(\mathrm{Mall}^{f,g}(L, R)) \neq B \mid B = 1 \wedge (L, R) \in \mathcal{L}_{\mathsf{FFC}} \times \mathcal{R}_{\mathsf{FFC}}\right]}_{\leq 1}$$

$$\cdot \underbrace{P\left[B = 1 \mid (L, R) \in \mathcal{L}_{\mathsf{FFC}} \times \mathcal{R}_{\mathsf{FFC}}\right]}_{\leq\frac{1}{2}+|\mathcal{C}|\epsilon} \quad (17)$$

$$\leq \frac{1}{2} + |\mathcal{C}|^{-1} - \epsilon + |\mathcal{C}|\,(\epsilon - \epsilon^2) \leq \frac{1}{2} + |\mathcal{C}|^{-1} + |\mathcal{C}|\,\epsilon. \quad (18)$$

The inequalities in (16) and (18) follow from the fact that $\mathcal{L}_{\mathsf{FFC}} \times \mathcal{R}_{\mathsf{FFC}}$ is a large set and the fact that B depends on $\mathrm{ext}(L, R)$, where ext is a randomness extractor. The detailed proof of these inequalities appears in the full version of this paper [21]. Now, since (13) is a weighted average of (∗) and (∗∗), hence obviously

$$(13) \tag{19}$$

$$\leq \max \left(\frac{1}{2} + \frac{3}{2} \cdot |\mathcal{C}|^{-1} + 6 \, |\mathcal{C}|^2 \, \epsilon, \frac{1}{2} + |\mathcal{C}|^{-1} + |\mathcal{C}| \, \epsilon \right) \tag{20}$$

$$\leq \frac{1}{2} + \frac{3}{2} \, |\mathcal{C}|^{-1} + 6 \, |\mathcal{C}|^2 \, \epsilon. \tag{21}$$

Case 2 Now consider the case when (11) does not hold, i.e.:

$$|\mathcal{L}_{\mathsf{FFC}} \times \mathcal{R}_{\mathsf{FFC}}| < \delta \cdot |\mathcal{L} \times \mathcal{R}| \tag{22}$$

We now give a bound on the probability that (L, R) is a member of $\mathcal{L}_{\mathsf{FFC}} \times \mathcal{R}_{\mathsf{FFC}}$.

$$P\left[(L, R) \in \mathcal{L}_{\mathsf{FFC}} \times \mathcal{R}_{\mathsf{FFC}}\right]$$

$$= \frac{1}{2} \cdot P\left[\mathsf{E}^c_{\mathsf{ext}}(0) \in \mathcal{L}_{\mathsf{FFC}} \times \mathcal{R}_{\mathsf{FFC}}\right] + \frac{1}{2} \cdot P\left[\mathsf{E}^c_{\mathsf{ext}}(1) \in \mathcal{L}_{\mathsf{FFC}} \times \mathcal{R}_{\mathsf{FFC}}\right]$$

$$= \frac{1}{2} \cdot P\left[(\tilde{L}, \tilde{R}) \in \mathcal{L}_{\mathsf{FFC}} \times \mathcal{R}_{\mathsf{FFC}} \mid \mathsf{ext}(\tilde{L}, \tilde{R}) = c\right] +$$

$$\frac{1}{2} \cdot P\left[(\tilde{L}, \tilde{R}) \in \mathcal{L}_{\mathsf{FFC}} \times \mathcal{R}_{\mathsf{FFC}} \mid \mathsf{ext}(\tilde{L}, \tilde{R}) \neq c\right]$$

$$\leq \frac{1}{2} \cdot \frac{P\left[(\tilde{L}, \tilde{R}) \in \mathcal{L}_{\mathsf{FFC}} \times \mathcal{R}_{\mathsf{FFC}}\right]}{P\left[\mathsf{ext}(\tilde{L}, \tilde{R}) = c\right]} + \frac{1}{2} \cdot \frac{P\left[(\tilde{L}, \tilde{R}) \in \mathcal{L}_{\mathsf{FFC}} \times \mathcal{R}_{\mathsf{FFC}}\right]}{P\left[\mathsf{ext}(\tilde{L}, \tilde{R}) \neq c\right]}$$

$$\leq \frac{1}{2} \cdot \frac{P\left[(\tilde{L}, \tilde{R}) \in \mathcal{L}_{\mathsf{FFC}} \times \mathcal{R}_{\mathsf{FFC}}\right]}{|\mathcal{C}|^{-1} - \epsilon} + \frac{1}{2} \cdot \frac{P\left[(\tilde{L}, \tilde{R}) \in \mathcal{L}_{\mathsf{FFC}} \times \mathcal{R}_{\mathsf{FFC}}\right]}{(|\mathcal{C}| - 1) \cdot |\mathcal{C}|^{-1} - \epsilon} \tag{23}$$

$$\leq \delta / (|\mathcal{C}|^{-1} - \epsilon),$$

where in (23) we used (9). Hence, in this case, (15) is at most equal to $\delta/(|\mathcal{C}|^{-1} - \epsilon)$, and therefore, altogether, we can bound (13) by

$$(13) \leq (\ast) + \delta/(|\mathcal{C}|^{-1} - \epsilon) \tag{24}$$

$$= \frac{1}{2} + \frac{3}{2} \, |\mathcal{C}|^{-1} + 6 \, |\mathcal{C}|^2 \, \epsilon + \delta/(|\mathcal{C}|^{-1} - \epsilon) \tag{25}$$

Since analyzing both cases gave us bounds (21) and (25), hence all in all we can bound (13) by their maximum, which is at most

$$\frac{1}{2} + \frac{3}{2} \, |\mathcal{C}|^{-1} + 6 \, |\mathcal{C}|^2 \, \epsilon + \delta/(|\mathcal{C}|^{-1} - \epsilon).$$

Hence (8) is proven.

5   Adding Leakages

In this section we show how to incorporate leakages into our result. First, we need to extend the non-malleability definition. We do it in the following, straightforward way. Observe that we can restrict ourselves to the situation when the leakages happen *before* the mauling process (as it is of no help to the adversary to leak from $(f(L), g(R))$ if he can leak already from (L, R)). For any split-state encoding scheme $(\mathsf{E}^c_{\mathsf{ext}} : \mathcal{M} \to \mathcal{L} \times \mathcal{R}, \mathsf{D}^c_{\mathsf{ext}} : \mathcal{L} \times \mathcal{R} \to \mathcal{M})$, a family of functions $\mathcal{F}$, any $m \in \mathcal{M}$ and any adversary $\mathcal{A}$ define a game $\mathsf{Tamper}^{\mathcal{A}}_m$ (where λ is some parameter) as follows. First, let $(L, R) \leftarrow \mathsf{E}^c_{\mathsf{ext}}(m)$. Then the adversary $\mathcal{A}$ chooses a sequence of functions $(v^1, w^1, \ldots, v^t, w^t)$, where each v^i has a type $v^i : \mathcal{L} \to \{0, 1\}^{\lambda_i}$ and each w^i has a type $w^i : \mathcal{R} \to \{0, 1\}^{\rho_i}$ where the λ's and ρ's are some parameters such that

$$\lambda_1 + \cdots + \lambda_t + \rho_1 + \cdots \rho_t \leq \lambda. \qquad (26)$$

He learns $\mathsf{Leak}(L, R) = (v^1(L), w^1(R), \ldots, v^t(L), w^t(R))$. Moreover this process is *adaptive*, i.e. the choice of an ith function in the sequence (26) can depend on the $i - 1$ first values in the sequence $\mathsf{Leak}(L, R)$. Finally the adversary chooses functions $f : \mathcal{L} \to \mathcal{L}$ and $g : \mathcal{R} \to \mathcal{R}$. Now define the output of the game as: $\mathsf{Tamper}^{\mathcal{A}}_m := (f(L), g(R))$. We say that the encoding scheme $(\mathsf{E}^c_{\mathsf{ext}}, \mathsf{D}^c_{\mathsf{ext}})$ is ϵ-*nonmalleable with leakage* λ if for every adversary $\mathcal{A}$ there exists distribution $D^{\mathcal{A}}$ on $\mathcal{M} \cup \{\mathsf{same}^*\}$ such that for every $m \in \mathcal{M}$ we have

$$\mathsf{Tamper}^{\mathcal{A}}_m \approx_\epsilon \left\{ \begin{array}{c} d \leftarrow D^{\mathcal{A}} \\ \text{if } d = \mathsf{same}^* \text{ then output } m, \\ \text{otherwise output } d. \end{array} \right\}$$

Theorem 1. *Suppose* $\mathsf{ext} : \mathcal{L} \times \mathcal{R} \to \mathcal{C}$ *is a flexible* (k, ϵ)-*extractor that satisfies (5), where, for some parameters* δ *and* λ *we have*

$$k = \frac{2}{3} \cdot (\log |\mathcal{L}| + \log |\mathcal{R}| - \lambda) - \frac{4}{3} \cdot \log(1/\delta). \qquad (27)$$

Then the encoding scheme is $\left(\frac{3}{2} |\mathcal{C}|^{-1} + 6 |\mathcal{C}|^2 \epsilon + 2\delta/(|\mathcal{C}|^{-1} - \epsilon)\right)$-*non-malleable with leakage* λ.

The proof of this theorem appears in the full version of this paper [21]. We now show how to instantiate Theorem 1 with the inner-product extractor from Sect. 2.

Theorem 2. *Take any* $\xi \in [0, 1/4)$ *and* $\gamma > 0$ *then there exist an explicit splitstate code* $(\mathsf{Enc} : \{0, 1\} \to \{0, 1\}^{N/2} \times \{0, 1\}^{N/2}, \mathsf{Dec} : \{0, 1\}^{N/2} \times \{0, 1\}^{N/2} \to \{0, 1\})$ *that is* γ-*non-malleable with leakage* $\lambda := \xi N$ *such that* $N = \mathcal{O}(\log(1/\gamma) \cdot (1/4 - \xi)^{-1})$. *The encoding and decoding functions are computable in* $\mathcal{O}(N \cdot \log^2(\log(1/\gamma)))$ *and the constant hidden under the* $\mathcal{O}$-*notation in the formula for* N *is around* 100.

The proof of this theorem appears in the full version of this paper [21]. We would like to remark that it does not look like we could prove, with our current proof techniques, a better relative leakage bound than $\xi < \frac{1}{4}$. Very roughly speaking it is because we used the fact that the inner product is an extractor twice in the proof. On the other hand we do not know any attack on our scheme for relative leakage $\xi \in \left(\frac{1}{4}, \frac{1}{2}\right)$ (recall that for $\xi = \frac{1}{2}$ obviously any scheme is broken). Hence, it is quite possible, that with a different proof strategy (perhaps relying on some special features of the inner product function) one could show a higher leakage tolerance of our scheme.

6   Security against Affine Mauling

Interestingly, we can also show that our encoding scheme $(\mathsf{E}^c_{\mathrm{ext}}, \mathsf{D}^c_{\mathrm{ext}})$, instantiated with the inner product extractor, is secure in the model where $(L, R) \in \mathbb{F}^n \times \mathbb{F}^n$ can be mauled simultaneously (i.e. we do not use the split-model assumption), but the class of the mauling functions is restricted to the affine functions over $\mathbb{F}$, i.e. each mauling function h is of a form

$$h((L_1, \ldots, L_n), (R_1, \ldots, R_n)) = M \cdot (L_1, \ldots, L_n, R_1, \ldots, R_n)^T + V^T, \quad (28)$$

where M is an $(2n \times 2n)$-matrix over $\mathbb{F}$ and $V \in \mathbb{F}^{2n}$. We now argue informally why it is the case, by showing that every h that breaks the non-malleability of this scheme can be transformed into a pair of functions (f, g) that breaks the non malleability of the scheme

$$\left(\mathsf{E}^c_{\mathrm{ext}} : \mathcal{F}^{n+2} \times \mathcal{F}^{n+2} \to \{0, 1\}, \mathsf{D}^c_{\mathrm{ext}} : \{0, 1\} \to \mathcal{F}^{n+2} \times \mathcal{F}^{n+2}\right)$$

in the split-state model. Let $(L, R) \in \mathbb{F}^{n+2} \times \mathbb{F}^{n+2}$ denote the codeword in this scheme. Our attack works only under the assumption that it happened that $(L, R) \in \mathcal{L}' \times \mathcal{R}'$, where $\mathcal{L}' \times \mathcal{R}' := (\mathbb{F}^n \times \{0\} \times \{0\}) \times (\mathbb{F}^n \times \{0\} \times \{0\})$ (in other words: the two last coordinates of both L and R are zero). Since $\mathcal{L}' \times \mathcal{R}'$ is large, therefore this clearly suffices to obtain the contradiction with the fact that our scheme is secure even if (L, R) happen to belong to some large subdomain of the set of all codewords (cf. Lemma 3). Clearly, to finish the argument it is enough to construct the functions f and g such that

$$\langle f(L), g(R) \rangle = \langle (L'_1, \ldots, L'_{n+2}), (R'_1, \ldots, R'_{n+2}) \rangle,$$

where $(L'_1, \ldots, L'_{n+2}, R'_1, \ldots, R'_{n+2}) = h(L_1, \ldots, L_n, R_1, \ldots, R_n)$. It is easy to see that, since h is affine, hence the value of $\langle (L'_1, \ldots, L'_{n+2}), (R'_1, \ldots, R'_{n+2}) \rangle$ can be represented as a sum of monomials over variables L_i and R_j where each variable appears in power at most 1. Hence it can be rewritten as the following sum:

$$\sum_{i=1}^{n} \left(L_i \cdot \sum_{j \in J_i} R_j\right) + \sum_{j \in J_{n+1}} L_j + \sum_{i,j \in K_{n+1}} L_i L_j + y + \sum_{j \in J_{n+2}} R_j + \sum_{i,j \in K_{n+2}} R_i R_j,$$

where each J_i is a subset of the indices $\{1, \ldots, n\}$ and $y \in \mathbb{F}$ is a constant. It is also easy to see that the above sum is equal to the inner product of vectors V and W defined as:

$$V := \Big(L_1, \ldots, L_n, \sum_{j \in J_{n+1}} L_j + \sum_{i,j \in K_{n+1}} L_i L_j, 1 \Big)$$

$$W := \Big(\sum_{j \in J_1} R_j, \ldots, \sum_{j \in J_n} R_j, 1, y + \sum_{j \in J_{n+2}} R_j + \sum_{i,j \in K_{n+2}} R_i R_j \Big).$$

Now observe that V depends only on the vector L, and similarly, W depends only on R. We can therefore set $f(L) := V$ and $g(R) := W$. This finishes the argument.

References

1. Aggarwal, D., Dodis, Y., Lovett, S.: Non-malleable codes from additive combinatorics. Cryptology ePrint Archive, Report 2013/201 (2013), http://eprint.iacr.org/
2. Akavia, A., Goldwasser, S., Vaikuntanathan, V.: Simultaneous hardcore bits and cryptography against memory attacks. In: Reingold, O. (ed.) TCC 2009. LNCS, vol. 5444, pp. 474–495. Springer, Heidelberg (2009)
3. Anderson, R., Kuhn, M.: Tamper resistance - a cautionary note. In: The Second USENIX Workshop on Electronic Commerce Proceedings (November 1996)
4. Bellare, M., Kohno, T.: A theoretical treatment of related-key attacks: Rka-prps, rka-prfs, and applications. In: Biham, E. (ed.) EUROCRYPT 2003. LNCS, vol. 2656, pp. 647–647. Springer, Heidelberg (2003)
5. Biham, E.: New types of cryptanalytic attacks using related keys. Journal of Cryptology 7(4), 229–246 (1994)
6. Bourgain, J.: More on the sum-product phenomenon in prime fields and its applications. International Journal of Number Theory 1(1), 1–32 (2005)
7. Brakerski, Z., Kalai, Y.T., Katz, J., Vaikuntanathan, V.: Overcoming the hole in the bucket: Public-key cryptography resilient to continual memory leakage. In: 51st Annual IEEE Symposium on Foundations of Computer Science (FOCS), pp. 501–510. IEEE (2010)
8. Chabanne, H., Cohen, G., Flori, J., Patey, A.: Non-malleable codes from the wiretap channel. In: 2011 IEEE Information Theory Workshop (ITW), pp. 55–59. IEEE (2011)
9. Chabanne, H., Cohen, G., Patey, A.: Secure network coding and non-malleable codes: Protection against linear tampering. In: 2012 IEEE International Symposium on Information Theory Proceedings (ISIT), pp. 2546–2550 (2012)
10. Choi, S.G., Kiayias, A., Malkin, T.: BiTR: Built-in tamper resilience. In: Lee, D.H., Wang, X. (eds.) ASIACRYPT 2011. LNCS, vol. 7073, pp. 740–758. Springer, Heidelberg (2011)
11. Chor, B., Goldreich, O.: Unbiased bits from sources of weak randomness and probabilistic communication complexity. SIAM Journal on Computing 17(2), 230–261 (1988)
12. Cohen, G., Raz, R., Segev, G.: Non-malleable extractors with short seeds and applications to privacy amplification. In: Computational Complexity (CCC), pp. 298–308 (2012)

13. Dachman-Soled, D., Kalai, Y.T.: Securing circuits against constant-rate tampering. In: Safavi-Naini, R. (ed.) CRYPTO 2012. LNCS, vol. 7417, pp. 533–551. Springer, Heidelberg (2012)
14. Davì, F., Dziembowski, S., Venturi, D.: Leakage-resilient storage. In: Garay, J.A., De Prisco, R. (eds.) SCN 2010. LNCS, vol. 6280, pp. 121–137. Springer, Heidelberg (2010)
15. Dodis, Y., Haralambiev, K., Lopez-Alt, A., Wichs, D.: Cryptography against continuous memory attacks. In: 51st Annual IEEE Symposium on Foundations of Computer Science (FOCS), pp. 511–520. IEEE Computer Society (2010)
16. Dodis, Y., Lewko, A., Waters, B., Wichs, D.: Storing secrets on continually leaky devices. In: 2011 IEEE 52nd Annual Symposium on Foundations of Computer Science (FOCS), pp. 688–697. IEEE (2011)
17. Dodis, Y., Li, X., Wooley, T., Zuckerman, D.: Privacy amplification and non-malleable extractors via character sums. In: FOCS 2011, pp. 668–677 (2011)
18. Dodis, Y., Wichs, D.: Non-malleable extractors and symmetric key cryptography from weak secrets. In: STOC, pp. 601–610 (2009)
19. Dolev, D., Dwork, C., Naor, M.: Nonmalleable cryptography. SIAM Review 45(4), 727–784 (2003)
20. Dziembowski, S., Faust, S.: Leakage-resilient circuits without computational assumptions. In: Cramer, R. (ed.) TCC 2012. LNCS, vol. 7194, pp. 230–247. Springer, Heidelberg (2012)
21. Dziembowski, S., Kazana, T., Obremski, M.: Non-malleable codes from two-source extractors. Cryptology ePrint Archive (2013), Full version of this paper, http://eprint.iacr.org/
22. Dziembowski, S., Pietrzak, K.: Leakage-resilient cryptography. In: FOCS 2008, pp. 293–302. IEEE (2008)
23. Dziembowski, S., Pietrzak, K., Wichs, D.: Non-malleable codes. In: ICS, pp. 434–452 (2010)
24. ECRYPT. European Network of Excellence. Side Channel Cryptanalysis Lounge, http://www.emsec.rub.de/research/projects/sclounge
25. Faust, S., Pietrzak, K., Venturi, D.: Tamper-proof circuits: How to trade leakage for tamper-resilience. In: Aceto, L., Henzinger, M., Sgall, J. (eds.) ICALP 2011, Part I. LNCS, vol. 6755, pp. 391–402. Springer, Heidelberg (2011)
26. Gennaro, R., Lysyanskaya, A., Malkin, T., Micali, S., Rabin, T.: Algorithmic tamper-proof (ATP) security: Theoretical foundations for security against hardware tampering. In: Naor, M. (ed.) TCC 2004. LNCS, vol. 2951, pp. 258–277. Springer, Heidelberg (2004)
27. Goldwasser, S., Rothblum, G.: How to compute in the presence of leakage. In: FOCS 2012, pp. 31–40 (2012)
28. Halevi, S., Lin, H.: After-the-fact leakage in public-key encryption. In: Ishai, Y. (ed.) TCC 2011. LNCS, vol. 6597, pp. 107–124. Springer, Heidelberg (2011)
29. Håstad, J., Impagliazzo, R., Levin, L., Luby, M.: A pseudorandom generator from any one-way function. SIAM Journal on Computing 28(4), 1364–1396 (1999)
30. Ishai, Y., Prabhakaran, M., Sahai, A., Wagner, D.: Private circuits II: Keeping secrets in tamperable circuits. In: Vaudenay, S. (ed.) EUROCRYPT 2006. LNCS, vol. 4004, pp. 308–327. Springer, Heidelberg (2006)
31. Ishai, Y., Sahai, A., Wagner, D.: Private circuits: Securing hardware against probing attacks. In: Boneh, D. (ed.) CRYPTO 2003. LNCS, vol. 2729, pp. 463–481. Springer, Heidelberg (2003)
32. Kalai, Y.T., Kanukurthi, B., Sahai, A.: Cryptography with tamperable and leaky memory. In: Rogaway, P. (ed.) CRYPTO 2011. LNCS, vol. 6841, pp. 373–390. Springer, Heidelberg (2011)

33. Liu, F.-H., Lysyanskaya, A.: Tamper and leakage resilience in the split-state model. In: Safavi-Naini, R. (ed.) CRYPTO 2012. LNCS, vol. 7417, pp. 517–532. Springer, Heidelberg (2012)
34. Micali, S., Reyzin, L.: Physically observable cryptography. In: Naor, M. (ed.) TCC 2004. LNCS, vol. 2951, pp. 278–296. Springer, Heidelberg (2004)
35. Naor, M., Segev, G.: Public-key cryptosystems resilient to key leakage. In: Halevi, S. (ed.) CRYPTO 2009. LNCS, vol. 5677, pp. 18–35. Springer, Heidelberg (2009)
36. Rao, A.: An exposition of bourgain 2-source extractor. In: Electronic Colloquium on Computational Complexity (ECCC), vol. 14, p. 034 (2007)
37. Wee, H.: Public key encryption against related key attacks. In: Fischlin, M., Buchmann, J., Manulis, M. (eds.) PKC 2012. LNCS, vol. 7293, pp. 262–279. Springer, Heidelberg (2012)

Optimal Coding for Streaming Authentication and Interactive Communication

Matthew Franklin[1], Ran Gelles[2],
Rafail Ostrovsky[2,3,*], and Leonard J. Schulman[4,**]

[1] Department of Computer Science, University of California, Davis
franklin@cs.ucdavis.edu
[2] Department of Computer Science, University of California, Los Angeles
{gelles,rafail}@cs.ucla.edu
[3] Department of Mathematics, University of California, Los Angeles
[4] E&AS Division, Caltech
schulman@caltech.edu

Abstract. Error correction and message authentication are well studied in the literature, and various efficient solutions have been suggested and analyzed. This is however not the case for *data streams* in which the message is very long, possibly infinite, and not known in advance to the sender. Trivial solutions for error-correcting and authenticating data streams either suffer from a long delay at the receiver's end or cannot perform well when the communication channel is noisy.

In this work we suggest a constant-rate error-correction scheme and an efficient authentication scheme for data streams over a noisy channel (one-way communication, no feedback) in the shared-randomness model. Our first scheme does not assume shared randomness and (non-efficiently) recovers a $(1 - 2c)$-fraction prefix of the stream sent so far, assuming the noise level is at most $c < 1/2$. The length of the recovered prefix is tight.

To be able to overcome the $c = 1/2$ barrier we relax the model and assume the parties pre-share a secret key. Under this assumption we show that for any given noise rate $c < 1$, there exists a scheme that correctly decodes a $(1 - c)$-fraction of the stream sent so far with high probability, and moreover, the scheme is efficient. Furthermore, if the noise rate exceeds c, the scheme aborts with high probability. We also show that no constant-rate authentication scheme recovers more than a $(1 - c)$-fraction of the stream sent so far with non-negligible probability,

* Supported in part by NSF grants 0830803, 09165174, 1065276, 1118126 and 1136174, US-Israel BSF grant 2008411, OKAWA Foundation Research Award, IBM Faculty Research Award, Xerox Faculty Research Award, B. John Garrick Foundation Award, Teradata Research Award, and Lockheed-Martin Corporation Research Award. This material is based upon work supported by the Defense Advanced Research Projects Agency through the U.S. Office of Naval Research under Contract N00014-11-1-0392. The views expressed are those of the authors and do not reflect the official policy or position of the Department of Defense or the U.S. Government.
** Supported in part by NSF Award 1038578.

R. Canetti and J.A. Garay (Eds.): CRYPTO 2013, Part II, LNCS 8043, pp. 258–276, 2013.

thus the relation between the noise rate and recoverable fraction of the stream is tight, and our scheme is optimal.

Our techniques also apply to the task of interactive communication (two-way communication) over a noisy channel. In a recent paper, Braverman and Rao [STOC 2011] show that any function of two inputs has a constant-rate interactive protocol for two users that withstands a noise rate up to $1/4$. By assuming that the parties share a secret random string, we extend this result and construct an interactive protocol that succeeds with overwhelming probability against noise rates up to $1/2$. We also show that no constant-rate protocol exists for noise rates above $1/2$ for functions that require two-way communication. This is contrasted with our first result in which computing the "function" requires only one-way communication and the noise rate can go up to 1.

Keywords: data stream, private codes, adversarial noise, authentication, tree codes, interactive communication.

1   Introduction

The tasks of *error-correction* and of *authentication* are well studied in the literature. In both cases, a sender (Alice) wishes to send a message over a one-way, noisy channel to a receiver (Bob). To do so, Alice produces a longer, redundant message and sends it over the channel. The added redundancy helps Bob in recovering the original message if possible, or aborting otherwise. The overhead of this process is the amount of redundancy added to each message; in this work we focus on *constant-rate* schemes, i.e., schemes in which the transmitted message is at most constant-times longer.

Interestingly, in all known authentication schemes (and in many of the error-correction codes) there are two important assumptions: (1) the message to be communicated has a given length n and (2) the message is fully known to the sender in advance. These two assumptions don't hold anymore when the information to be transmitted is in the form of a *data stream*, which is a long, possibly infinite, sequence of symbols $x_1, x_2, \ldots$ over some alphabet Σ, where each x_i arrives at the sender's end at time i and is unknown beforehand.

In this paper, we investigate the question of transmitting data streams over an adversarially noisy channel. Within this framework we consider two related questions, namely, error-correction and authentication of data streams. Loosely speaking, in error-correction schemes, the receiver decodes the correct message as long as the noise level is below some threshold (but possibly outputs a wrong message if the noise exceeds that threshold). In authentication schemes, the receiver's task is to indicate whether or not the received (decoded) message is indeed the one sent to him. To see the relation between these two tasks note that if the corruption level of an adversary is guaranteed to be lower than the threshold, any error-correction guarantees that the receiver decodes the original message. However, while no constant-rate error-correction scheme can withstand a noise level higher than $1/2$, this is not the case for authentication schemes that are capable of indicating a change in the message even when the adversary has a full control of the channel. On the other hand for the task of authentication, it is generally assumed that the parties pre-share a secret key.

Standard error-correction and authentication methods do not apply directly to the model of data streams. The straightforward method to perform error-correction (or authentication) of a data stream is to cut the stream into chunks and separately encode each chunk. The problem now is that while the adversary is limited to some *global* noise rate, there is no restriction on the noise level of any local part of the stream. Specifically, the adversary can corrupt a single chunk in its entirety (while not exceeding the global amount of allowed noise), and cause Bob to decode this chunk in a wrong way. Even if this event is noticed by Bob since the chunk fails the authentication, the information carried within this chunk is lost unless Bob requests a retransmission of that chunk, i.e., unless the communication is interactive. The same problem exists (with high probability) when the noise is random rather than adversarial, given that the stream is long enough or infinite.

A possible mitigation to the above is to increase the chunks' size. This, however, has an undesirable side effect—Bob needs to wait until receiving a complete chunk in order to decode and authenticate it. This means that the information received in the very recent bits is inaccessible to Bob until the chunk is completely received. Our goal is thus, to construct a constant-rate scheme that can withstand a constant fraction of errors (globally) and still guarantee the correct decoding and authenticity of the information received so far. To the best of our knowledge, no such solution is known.

1.1   Our Results

In this work we construct optimal encoding schemes for both interactive and non-interactive (streaming) communication, and show a dramatic difference between these two cases in the following sense. For each case, we show an upper bound on the noise rates that allow a successful constant-rate communication, and construct a protocol that achieves the bound. Interestingly, the bound for one-way communication is different from the interactive one.

Specifically, our result for one-way communication is a constant-rate coding scheme for data streams that withstands noise rates of less than $1/2$. Informally, as long as the global noise rate up to some time n does not exceed some parameter $c < 1/2$, a fraction of $1 - 2c$ of the stream sent up to time n can be recovered (see Section 4). For constant-rate schemes, it is clear that $c < 1/2$ is a hard limit and no scheme can succeed when the noise is higher. In order to achieve schemes that withstand higher noise rates we must relax the model and give the users more resources. Indeed, with the use of shared randomness (i.e., a shared secret key) we can break the $c = 1/2$ barrier. To emphasize the fact that the parties are allowed to share a secret key, we refer schemes in this model as *authentication schemes* rather than error-correction schemes, based on the relation of these two tasks mentioned above (codes that assume a private shared key are also known as *private codes* [16], see Related Work).

This leads to our first main result: we construct a constant-rate authentication scheme for data streams sent over a noisy (possibly adversarial) channel. For any constant fraction of noise c less than 1, our scheme succeeds in decoding at least a $(1 - c)$-fraction of the stream so far, with high probability. The decoded part is always the *prefix* of the stream. The decoded prefix is authenticated, meaning that there is only a negligible probability that the scheme outputs a different string. Furthermore, our scheme is *efficient*. More formally (see formal theorems in Section 5), we show that for any noise rate $0 \le c < 1$ and small constant $\varepsilon > 0$:

- There exists an efficient constant-rate scheme that, at time n, decodes a prefix of length at least $(1 - c)n - \varepsilon n$ of the stream sent so far.
- Any constant-rate protocol that decodes a prefix of length $(1 - c)n + \varepsilon n$ succeeds with probability at most $2^{-\Omega(\varepsilon n)}$ in the worst case.

Our scheme is unconditionally secure and does not make any (cryptographic) assumptions, other than pre-sharing a secret random string. The amount of randomness utilized by the scheme grows with the message length, and can be unbounded if the data stream is infinite. However, if we only consider a computationally bounded adversary, the required amount of randomness is relatively small (polynomial in the security parameter). With the aid of a pseudo-random generator, the parties only need to pre-share a small seed, from which they generate randomness at will. Moreover, such a solution scales to the multiparty case by a simple public-key infrastructure construction. Each user generates a pair of a public and a secret key, and any pair of users perform Diffie-Hellman key-exchange [6] to obtain a secret shared authentication-key used as the pseudo-random generator's seed.

We apply the same techniques used in our streaming-authentication scheme onto the task of *interactive communication* to get our second main result. In the interactive communication scenario, two parties perform an arbitrary interactive protocol over a noisy channel, while keeping the amount of exchanged data only a constant factor more than an equivalent protocol for a noiseless channel (i.e., the encoding is constant-rate). This question was initially considered for both random and adversarial noise by Schulman [22,23,24] who showed a constant-rate encoding scheme that copes with a noise rate of up to 1/240, and recently revisited by Braverman and Rao [5] who showed how to deal with noise rates less than 1/4. In addition, Braverman and Rao show that 1/4 is the highest error rate any protocol can withstand, as long as the protocol defines whose turn it is to speak at every round regardless of the observed noise. The fascinating open question left by the work of Braverman and Rao is whether other methods could extend the 1/4 bound.

In this work we improve the bound obtained by [5] by allowing the parties to pre-share a secret key. Specifically, we show how to convert any interactive protocol (for noiseless channel) into a constant-rate protocol that withstands any adversarial noise level smaller than 1/2, given pre-shared randomness. We also show that for higher noise rates, no constant-rate interactive protocol exists for tasks that depend on inputs of *both* parties. Similarly to previous results for interactive communication with adversarial noise [24,5,9], our decoding scheme is inefficient. Very recently, Brakerski and Kalai [2] showed how to augment previous results of interactive communication protocols and achieved *efficient* schemes that withstand adversarial noise (the computation efficiency was further improved by Brakerski and Naor [3] to $O(N \log N)$). Note that the bounds (on adversarial noise) obtained by [2] are improved by our work as well, since we improve the bounds of the underlying schemes used by [2].

1.2   Our Methods

The Blueberry Code. The main ingredient of our construction is an error-detection code we name the *Blueberry code*[1]. The Blueberry code uses the shared randomness in order to detect corruptions made by the channel, and marks them as *erasures*. One can think about this code as a weak message authentication code (MAC) that authenticates each symbol separately with a constant probability (see [11] for a formal definition of MAC). To this end, each symbol of the input alphabet Σ is randomly and independently mapped to a larger alphabet Γ (the channel alphabet). This means that only a small subset of the channel alphabet is meaningful and the other symbols serve as "booby-traps". Since each symbol is encoded independently, any corruption is caught with constant probability $\frac{|\Sigma|-1}{|\Gamma|-1}$ and marked with a special sign $\perp$ to denote it was deleted by the channel. Most of the corruptions made by an adversary become erasures and only a small fraction (arbitrarily small, controlled by the size of $|\Gamma|$) turns into errors.

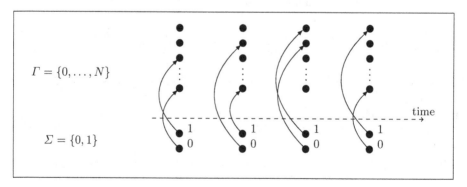

Fig. 1. A demonstration of the Blueberry code: at any given time each symbol in Σ is randomly mapped to a symbol of Γ. Symbols of Γ with no incoming arrow are "booby-traps", which serve to detect corruptions.

---

[1] The name of the Blueberry code is inspired by the children's book "The case of the hungry stranger" [1] in which a blueberry pie is gone missing, and the thief (who turns out to be the dog) is identified by his big blue grin.

The main insight that leads to our results is the different ways error correction codes deal with errors and erasures. We observe that, in terms of Hamming distance, the impact of a single error is twice as harmful as a single erasure. Indeed, assume that the Hamming distance of two strings, x and y, is m. Then if x was communicated but y is decoded it means that at least $m/2$ errors have occurred, or alternatively, at least m erasures. More generally, assuming we decode by minimizing the Hamming distance, then our decoding fails if the number of errors e and the number of erasures d satisfy $2e + d \geq m$.

Combining Blueberry Codes and Tree Codes. The second ingredient of our work is encoding via *tree codes* [24], an online encoding that has a "self-healing" property: when decoding a stream at time n, the tree will decode correctly up to a particular time t such that the stream suffix between times t and n is the longest suffix in which the error rate is high. This means, for instance, that even if all the transmissions until some time t' were corrupted (and thus the decoding failed at those times), if the noise rate up to time $n > t'$ is low enough, not only can we decode between t' and n, but we will also be able to decode the *entire* stream up to time n.

Encoding via both a tree code and a Blueberry code immediately gives a streaming authentication method: the Blueberry code prevents the adversary from corrupting too many transmissions without being noticed, and given that the noise level is low enough, the tree code correctly decodes a prefix of the stream whose length is determined by the average noise level up to that time.

Efficient Constructions. The only caveat of the above construction is that tree code decoding is not necessarily efficient and may be in the worst case exponential in the length of the received transmission. We obtain an efficient authentication scheme by splitting the stream into small segments and repeatedly sending random segments of the history. That way, even if some part of the transmission was changed by the channel, the same information will keep being retransmitted at random future times, and eventually (with high probability) will be received at the other side intact.

Roughly speaking, we use $n/\log n$ tree codes to encode chunks of the stream (each of length roughly $\log n$). Note that as n grows, so does the number of the trees in use, and the expected depth of each tree. At each time step, we randomly select one of the $n/\log n$ trees and transmit the next label of the path defined by the corresponding chunk of the stream. For most of the trees, the expected number of labels transmitted is $\Theta(\log n)$, and the decoding of the specific chunk succeeds except with polynomially small probability. Since each tree code is used to encode a word of length $O(\log n)$, the decoding can be performed efficiently by an exhaustive search.

1.3   Other Related Works

The works of Even, Goldreich and Micali [7] and Gennaro and Rohatgi [10] consider authentication of data streams, however the focus of these schemes is not only to authenticate the message but also to prevent the sender from denying having signed the information. These constructions rely on cryptographic primitives such as one-time signatures. Another related line of research [21,19,12] pursues authentication of streams over *lossy* channels, usually in the multicast setting.

Coding schemes that assume the parties pre-share some randomness (also known as *Private Codes* [16]) first appeared in [25], and were greatly analyzed since. The main advantage of such codes is that they can deal with *adversarial noise*, rather than a random noise. Langberg [16] considers private codes for adversarial channels that approach Shannon's bound and require only $O(\log n)$ randomness for block size n, as well as an $\Omega(\log n)$ lower bound for the needed randomness. The construction of Langberg also implies an efficient code with $O(n \log n)$ randomness. This result was improved to $n + o(n)$ randomness by Smith [26]. Explicit constructions with $o(n)$ randomness are yet unknown (see [26]).

Error correction codes for *computationally bounded* noise models were first addressed by Lipton [17] who constructs error-correction codes given pre-shared randomness and later considered by Micali, Peikert, Sudan and Wilson [18] who only assume sharing a short public-key, and recently by the surprising result of Guruswami and Smith [13] who assume no shared setup between the users. Locally Decodable codes with constant-rate in the public-key model were introduced by Hemenway and Ostrovsky [14] and later improved by Hemenway, Ostrovsky, Strauss and Wootters [15].

2   Preliminaries, Model and Definitions

We denote the set $\{1, 2, \ldots, n\}$ by $[n]$, and for a finite set Σ we denote by $\Sigma^{\le n}$ the set $\cup_{k=1}^{n} \Sigma^k$. The Hamming distance $\Delta(x, y)$ of two strings $x, y \in \Sigma^n$ is the number of indices i for which $x_i \ne y_i$. Throughout the paper, $\log()$ denotes the binary logarithm (base 2) and $\ln()$ denotes the natural logarithm (base e).

Shared Randomness Model. We assume the following *shared-randomness model.* The legitimate users (Alice and Bob) have access to a random string R of unbounded length, which is unknown to the adversary (Eve). Protocols in this model are thus *probabilistic*, and are required to succeed with high probability over the choice of R. We assume that all the randomness comes from R and that for a fixed R the protocols are deterministic.

Tree Codes. A d-ary *tree code* [24] over alphabet Σ is a rooted d-regular tree of arbitrary depth N whose edges are labeled with elements of Σ. For any string $x \in [d]^{\le N}$, a d-ary tree code $\mathcal{T}$ implies *an encoding* of x, $\mathsf{TCenc}_{\mathcal{T}}(x) = w_1 w_2 .. w_{|x|}$ with $w_i \in \Sigma$, defined by concatenating the labels along the path defined by x, i.e., the path that begins at the root and whose i-th node is the x_i-th child of the $(i-1)$-th node. We usually omit the subscript $\mathcal{T}$ when the tree is clear from the context. Note that tree code encoding is *online*: to communicate $\mathsf{TCenc}(x\sigma)$ where $\sigma \in [d]$ given that $\mathsf{TCenc}(x)$ was already communicated, we only need to send one symbol of Σ. Hence, if $|\Sigma| = O(1)$ the encoding scheme has a constant rate.

For any two paths (strings) $x, y \in [d]^{\leq N}$ of the same length n, let ℓ be the longest common prefix of both x and y. Denote by $anc(x,y) = n - |\ell|$ the distance from the n-th level to the least common ancestor of paths x and y. A tree code has distance α if for any $k \in [N]$ and any distinct $x, y \in [d]^k$, the Hamming distance of $\mathsf{TCenc}(x)$ and $\mathsf{TCenc}(y)$ is at least $\alpha \cdot anc(x,y)$.

For a string $w \in \Sigma^n$, decoding w using the tree code $\mathcal{T}$ means returning the string $x \in [d]^n$ whose encoding minimizes the Hamming distance to the received word, namely,

$$\mathsf{TCdec}_{\mathcal{T}}(w) = \underset{x \in [d]^n}{\operatorname{argmin}} \, \Delta(\mathsf{TCenc}_{\mathcal{T}}(x), w).$$

A theorem by Schulman [24] proves that for any d and $\alpha < 1$ there exists a d-ary tree code of unbounded depth and distance α over alphabet of size $d^{O(1/(1-\alpha))}$. However, no efficient construction of such a tree is yet known. For a given depth N, Peczarski [20] gives a randomized construction for a tree code with $\alpha = 1/2$ that succeeds with probability at least $1 - \epsilon$, and requires alphabet of size at least $d^{O(\sqrt{\log \epsilon^{-1}})}$. Braverman [4] gives a sub-exponential (in N) construction of a tree code, and Gelles, Moitra and Sahai [9] provide an efficient construction of a randomized relaxation of a tree code of depth N, namely a *potent tree code*, which is powerful enough as a substitute for a tree code in most applications.

Communication Model. Our communication model consists of a channel $ch :$ $\Sigma \to \Sigma$ subject to corruptions made by an adversary (or by the channel itself). The noise model is such that any symbol σ sent through the channel can turn into another symbol $\tilde{\sigma} \in \Sigma$. It is not allowed to insert or delete symbols. For all of our applications we assume that one symbol $\sigma_i \in \Sigma$ is sent at any time slot i.[2] We say that the adversarial *corruption rate* is c if for n transmissions, at most cn symbols were corrupted.

3   The Blueberry Code

Definition 3.1. *For $i \geq 1$ let $B_i : [L+1] \to [L+1]$ be a random and independently chosen permutation. The* Blueberry code *maps a string x of arbitrary length n to*

$$B(x) = B_1(x_1)B_2(x_2) \cdots B_n(x_n).$$

We denote such a code as $B : [L+1]^* \to [L+1]^*$.

We use the Blueberry code in the shared-randomness model where the legitimate parties share the random permutations B_i, unknown to the adversary (these kind of codes, determined by a random string unknown to the channel are referred to as *private codes* by [16]). Although B_i is a permutation on $[L+1]$, we actually

---

[2] The channel time slots need not correspond with the times in which stream symbols are received. I.e, it is possible that between the arrival of stream elements x_i and x_{i+1}, several channel-symbols are transmitted.

use it to encode strings over a smaller alphabet $[S + 1]$ with $S < L$; that is, we focus on the induced mapping $B : [S + 1]^* \to [L + 1]^*$. The adversary does not know the specific permutations B_i, and has probability of at most S/L to change a transmission into a symbol whose pre-image is in $[S + 1]$.

Definition 3.2. *Assume that at some time i, $y_i = B_i(x_i)$ is transmitted and $\tilde{y}_i \neq y_i$ is received. If $B_i^{-1}(\tilde{y}) \notin [S+1]$, we mark the transmission as an erasure (specifically, the decoding algorithm outputs $\perp$); otherwise, this event is called an error.*

Corollary 3.3. *Let $x \in [S + 1]^n$ and assume $B(x)$ is communicated over a noisy channel. Every symbol altered by the channel will cause either an error with probability S/L, or an erasure with probability $1 - S/L$.*

Assuming $S \ll L$, most of the corruptions done by the channel are marked as erasures, and only a small fraction of the corruptions percolate through the Blueberry code and cause an error.

Lemma 3.4. *Let $S, L \in \mathbb{N}$ be fixed and assume a Blueberry code $B : [S+1]^* \to [L + 1]^*$ is used to transmit a string $x \in [S + 1]^n$ over a noisy channel. For any constant $0 \leq c \leq 1$, if the channel's corruption rate c, then with probability $1 - 2^{-\Omega(n)}$ at least a $(1 - 2\frac{S}{L})$-fraction of the corruptions are marked as erasures.*

Proof. Denote by z_i the random variable which is 1 if the i-th corrupted-transmission is marked as an erasure and 0 otherwise. These are independent Bernoullis with probability $1 - \frac{S}{L}$. Let $Z = \sum_i z_i$ and note that $\mathbb{E}[Z] = cn(1 - \frac{S}{L})$. By Chernoff-Hoeffding inequality,

$$\Pr_R \left[\frac{1}{n} \sum_i z_i < c \left(1 - 2\frac{S}{L}\right) \right] < e^{-2n(cS/L)^2}.$$

Corollary 3.5. *Let $S, L \in \mathbb{N}$ be fixed. If out of n received transmissions, cn were marked as erasures by a Blueberry code $B : [S+1]^* \to [L+1]^*$, then except with probability $2^{-\Omega(n)}$ over the shared randomness, the adversarial corruption rate is at most $c/(1 - 2\frac{S}{L})$.*

We will use the Blueberry code concatenated with another (outer) code that is less sensitive to erasures than to errors. From the outer code's point of view, this effectively increases the channel's "error rate resilience" from $1 - 2c$ to $1 - c(1 + S/L)$. The construction of the code B from independent B_i's allows us to encode and decode each x_i independently, which is crucial for on-line applications in which the message x to be sent is not fully known in advance.

4   Error Correction of Data Streams

Before we reach our main result, we begin with a simple, non-efficient, constant-rate error-correction scheme for data streams that withstands noise $c < 1/2$

and decodes a prefix of length $1 - 2c$ of the stream sent so far. The scheme is obtained by simply encoding the stream via a tree code $\mathcal{T}$ with large enough distance parameter $\alpha \in (0, 1)$ and a constant-size alphabet, which depends on α.

Theorem 4.1. *For any constants $c < 1/2$ and $\varepsilon > 0$ there exists a constant-rate error-correction scheme for data stream $x_1, x_2, \ldots$ such that at any given time n the receiver outputs a string $x'_1, x'_2, \ldots, x'_n$, and if the noise rate until time n is at most c, then*

$$x'_1, x'_2, \ldots, x'_{(1-2c)n - \varepsilon n} = x_1, x_2, \ldots, x_{(1-2c)n - \varepsilon n}$$

that is, a prefix of the stream of length at least $(1 - 2c)n - \varepsilon n$ is correctly decoded.

Proof. Assume Alice encodes each stream symbol using $\mathsf{TCenc}_{\mathcal{T}}()$ using some tree code $\mathcal{T}$ whose parameters we fix shortly.

For a specific time n, consider a string $\tilde{x} \in \{0, 1\}^n$, such that $anc(x, \tilde{x}) \geq (2c + \varepsilon)n$. Due to the tree distance property, the Hamming distance between $\mathsf{TCenc}(\tilde{x})$ and $\mathsf{TCenc}(x)$ is at least $\alpha(2c + \varepsilon)n$. Assume Eve causes e errors, a maximal-likelihood decoding will prefer x over $\tilde{x}$ as long as $\lfloor \alpha(2c + \varepsilon)n \rfloor > 2e$. Since Eve's corruption rate is limited to c, we know that $e \leq cn$. By setting $\alpha > \frac{2c}{2c + \varepsilon}$ we guarantee that $\alpha(2c + \varepsilon)n > 2e$, and Bob decodes a string x' such that $anc(x, x') < (2c + \varepsilon)n$ with certainty. $\square$

5   Perpetual Authentication

Sending a data stream over a noisy channel is not a simple task, especially when the noise model is adversarial. Our goal is to design an encoding and decoding scheme such that the encoding has a constant rate and the decoding recovers the encoded transmitted stream, or else aborts. Furthermore, we wish an "authentication" guarantee, that is, if the decoding scheme did not abort, it decodes the *correct* data with high probability (note that the probability that the scheme aborts potentially differs from the probability that the decoding scheme outputs incorrect data). The amount of recoverable data depends on the noise and the goal is to output (and authenticate) the longest possible prefix of the stream, given a constant corruption rate.

Definition 5.1. *A $(c(n), \gamma(n), \kappa(n))$-Streaming Authentication Scheme with constant rate r is an encoding $e : \{0, 1\}^* \times \{0, 1\}^* \to \{0, 1\}^r$ that encodes a stream $x_1, x_2, \ldots$ into a stream $y_1 = e(x_1, R), y_2 = e(x_1 x_2, R), \ldots, y_i = e(x_1 \cdots x_i, R)$, and a decoding $d : \{0, 1\}^* \times \{0, 1\}^* \to \{0, 1\}^* \cup \{\perp\}$ such that the following holds. For any n, and for any adversary $Adv(x_1 \cdots x_n, y_1 \cdots y_n) = y'_1 \cdots y'_n$, either $d(y'_1 \cdots y'_n, R) = x'_1 x'_2 \cdots x'_n$ or $d(y'_1 \cdots y'_n, R) = \perp$, and if at most $c(n)$ transmissions were corrupted, then*

1. the scheme aborts with probability at most $\kappa(n)$,

$$\Pr_R[d(y'_1 \cdots y'_n, R) = \perp] < \kappa(n).$$

2. *if not aborted, the probability to decode an incorrect $\gamma(n)$-prefix of the stream is at most $\kappa(n)$,*

$$\Pr_R[d(y'_1 \cdots y'_n, R) \neq \perp \wedge x'_1 \cdots x'_{\gamma(n)} \neq x_1 \cdots x_{\gamma(n)}] < \kappa(n).$$

Eve is given both the raw stream and the channel transmissions, however she does not know the shared random string R used as the secret authentication key. It is desired that as long as Eve corrupts only a small fraction of the transmissions, Bob will be able to correctly decode a prefix of the stream, or otherwise be aware of the adversarial intervention and abort.

We show the following dichotomy: If the adversarial corruption rate is some constant c, then there exists a streaming authentication stream that decodes a prefix of at most $(1-c)$-fraction of the stream received so far. In addition, there does not exist a streaming authentication scheme that is capable of decoding a longer prefix with non-negligible probability.

Theorem 5.2. *In the shared-randomness model, for every constants c, ε such that $0 \leq c < 1$ and $0 < \varepsilon \leq (1-c)/2$ there exists a constant-rate $(cn, (1-c)n - \varepsilon n, 2^{-\Omega(n)})$-Streaming Authentication Scheme. Moreover, there exists an efficient constant-rate $(cn, (1-c)n - \varepsilon n, 2^{-\Omega(\log n)})$-Streaming Authentication Scheme.*

For any constant $c_{th} > c$, if the adversarial corruption rate exceeds c_{th}, the schemes abort with overwhelming probability over the shared randomness.

Theorem 5.3. *Assume that a bitstream $x_1, x_2, \ldots$ is communicated using some encoding protocol with a constant rate, and assume that at time n the receiver decodes the bitstring $x'_1, \ldots, x'_n$. If the rate of adversarial corruptions is $0 \leq c \leq 1$, then for any constant $\varepsilon > 0$,*

$$\Pr[x'_1 \cdots x'_{(1-c)n+\varepsilon n} = x_1 \cdots x_{(1-c)n+\varepsilon n}] \leq 2^{-\Omega(\varepsilon n)}$$

where the probability is over the coin tosses of the decoding algorithm, assuming $\{x_i\}$ are uniformly, independently distributed.

We now prove Theorem 5.3 and then construct the protocols guaranteed by Theorem 5.2

Proof. Consider an adversary that, starting at time $(1-c)n$, corrupts all the transmissions. It is easy to verify that the corruption rate is c. Clearly, from time $(1-c)n$ and on, the effective capacity of the channel is 0. This means that the decoder has no use of transmissions of times $\geq (1-c)n$ and he decodes only using transmissions received up to time $(1-c)n$. However, due to the streaming nature of the model, transmissions at times $< (1-c)n$ depend only on $x_1, \ldots, x_{(1-c)n}$ (the suffix of the stream is yet unknown to the sender). The receiver has no information about any bit x_i with $i > (1-c)n$ and his best strategy is to guess them. The probability to correctly guess the last εn bits is at most $2^{-\lfloor \varepsilon n \rfloor}$. □

In order to construct a streaming authentication scheme, we use two concatenated layers of online codes. The inner code is a Blueberry $B : [S+1]^* \rightarrow [L+1]^*$

code with constant S, L, and the outer code A is an online code that allows a prefix decoding in the presence of errors and erasures. The entire process can be described by

$$(x_1, \ldots) \xrightarrow{A} (y_1, \ldots) \xrightarrow{B} (z_1, \ldots) \xrightarrow{channel} (\tilde{z}_1, \ldots) \xrightarrow{B^{-1}} (\tilde{y}_1, \ldots) \xrightarrow{A^{-1}} (\tilde{x}_1, \ldots)$$

We begin with a simple and elegant construction which, although not efficient, demonstrates the power of the Blueberry code.

Proposition 5.4. *Let c, ε be constants $0 \le c < 1$, $0 < \varepsilon \le (1 - c)$ and let $A = \mathsf{TCenc}()$ be an encoding using a binary tree code and B a Blueberry code with constant parameters determined by c, ε. The concatenation of A and B is a $(cn, (1 - c)n - \varepsilon n, 2^{-\Omega(n)})$-streaming authentication scheme.*

Proof. Assume that in order to encode the bitstream $x_1, x_2, \ldots$, we use a binary tree code over alphabet $[S + 1]$ with distance α to be determined later, concatenated with a Blueberry-code $B : [S + 1]^* \to [L + 1]^*$. We show that if at time n we decode a string $\tilde{x}_1 \cdots \tilde{x}_n$ whose prefix $\tilde{x}_1 \cdots \tilde{x}_{(1-c-\varepsilon)n}$ differs from $x_1 \cdots x_{(1-c-\varepsilon)n}$, then the corruption rate was larger than c.

For a specific time n, consider a string $\tilde{x} \in \{0, 1\}^n$, such that $anc(x, \tilde{x}) \ge (c + \varepsilon)n$. Due to the tree distance property, the Hamming distance between $\mathsf{TCenc}(\tilde{x})$ and $\mathsf{TCenc}(x)$ is at least $\alpha(c + \varepsilon)n$. Assume Eve causes d erasures and e errors, a maximal-likelihood decoding will prefer x over $\tilde{x}$ as long as $\lfloor \alpha(c + \varepsilon)n \rfloor > 2e + d$.

If Eve's corruption rate is limited to c, Lemma 3.4 implies that with overwhelming probability at most $2cnS/L$ of these corruptions become errors and the rest are marked as erasures. Setting $\alpha > \frac{c}{c+\varepsilon}(1 + \frac{2S}{L})$ we guarantee that $\alpha(c + \varepsilon)n > 2 \cdot 2cnS/L + cn(1 - 2S/L),^3$ thus Bob decodes with overwhelming probability a string $\tilde{x}$ such that $anc(x, \tilde{x}) < (c + \varepsilon)n$, as claimed.

Note that the actual fraction of adversarial corruptions can be estimated out of the number of erasures marked by the Blueberry code. We abort the decoding if at a specific time n the number of erasures exceeds cn. Lemma 3.4 guarantees that if the adversary corrupts more than a $c/(1 - \frac{2S}{L})$-fraction of the transmissions, she will cause at least cn erasures, except with negligible probability. Choosing L such that $(1 - \frac{2S}{L}) \ge \frac{c}{c_{th}}$ completes the proof for the non-efficient case of Theorem 5.2. □

We note that although in the above proof we require ε to be constant, for the case of $c = 0$ (i.e., when the channel is not inherently noisy) we can let ε be smaller. For instance, if we let $\varepsilon = \kappa/n$ for a security parameter κ, the scheme is comparable to a (non-streaming) authentication scheme with the same security parameter: in order to change even a single bit in a prefix of length n, after $n + \kappa$

---

[3] It is required to have $\alpha < 1$, thus the choice of (the constant) L should depend on ε and c, specifically, $L > 2S\frac{c}{\varepsilon}$. Also note that S depends on α, however L is independent of both. For a fixed value of α (and $S = d^{O(1/(1-\alpha))}$) there is always a way to choose a constant L that satisfies the conditions.

symbols were transmitted, the adversary must change at least $\alpha\kappa/2$ transmissions, and will be caught except with probability $2^{-\Omega(\kappa)}$. Since the above holds for any time n, we get a perpetual authentication of the stream.

The case where $c > 0$ has a meaning of communicating over a noisy channel (regardless of the adversary). The users do not abort the authentication scheme although they know the message was changed by the channel. Instead, the scheme features both error-correction and authentication abilities and the parties succeed to recover (a prefix of) the original message with high probability.

5.1 Efficient Streaming Authentication

We now complete the proof of Theorem 5.2 by defining an efficient randomized code A_{eff} for prefix-decoding in the presence of errors and erasures. The protocol partitions the stream into words of logarithmic size and encodes each using a tree code. At any time n, one of the $O(n/\log n)$ words is chosen at random and its next encoded symbol is transmitted. The value n increases as the protocol progresses which means that the length of each encoded word increases as well. This however causes no problem: each word is encoded by a tree code (rather than, say, a block code), which is performed in an online manner without assuming knowledge of the word's length. Decoding can be performed efficiently by an exhaustive search since each word is of logarithmic length in the current time n. We note that the parties hold the entire stream in their memory throughout the protocol, which is different from the common practice of streaming algorithms in which there is only a single party (rather than two) which aim to compute some statistics of the stream using poly-logarithmic memory.

Proposition 5.5. *For any constants $0 \le c < 1$, $0 < \varepsilon \le (1-c)/2$ and a constant $c_1 > 0$, there exist efficient constant-rate encoding and decoding scheme such that, for any set of infinite strings $\{\mathbf{x}^1, \mathbf{x}^2, \ldots\}$ the following holds for any sufficiently large time n except with polynomially small probability in n. If the corruption rate at time n is at most c then the scheme correctly decodes a prefix of length $c_1 \log n$ of each one of the strings $\mathbf{x}^k$ with $k \in \{\lceil \frac{\varepsilon n/4}{\log \varepsilon n/4}\rceil, \ldots, \lceil \frac{(1-c-\varepsilon)n}{\log(1-c-\varepsilon)n}\rceil\}$. Moreover, up to time n the encoding scheme assumes knowledge of only strings $\mathbf{x}^k$ with $k \le n/\log n$.*

In the full version of this paper [8] we show that a protocol that satisfy Proposition 5.5 can be obtained by concatenating Protocol 1 (see below) with a Blueberry code $B : [S+1]^* \to [L+1]^*$. We show that with high probability, $\Theta(\log n)$ symbols of $\mathsf{TCenc}(\mathbf{x}^k)$ are transmitted by time n for every k in the range $K_n \triangleq \{\lceil \frac{\varepsilon n/4}{\log \varepsilon n/4}\rceil, \ldots, \lceil \frac{(1-c-\varepsilon)n}{\log(1-c-\varepsilon)n}\rceil\}$. Moreover, at least a constant fraction of these transmissions were not corrupted by the adversary. Therefore, we can use Proposition 5.4 to decode a prefix of length $O(\log n)$ of each of the codewords indexed by K_n, with high probability.

Finally, in Appendix A we show how to split the stream $x_1, x_2, \ldots$ into words $\{\mathbf{x}^1, \mathbf{x}^2, \ldots\}$, so that the prefix $x_1, \ldots, x_{(1-c-\varepsilon)n}$ completely appears in the $O(\log n)$-prefix of strings $\{\mathbf{x}^k\}$ with $k \in K_n$. This gives an efficient $(cn, (1-c)n - \varepsilon n, 2^{-\Omega(\log n)})$-authentication scheme and completes the proof of Theorem 5.2.

Let $0 \le c < 1$ and $0 < \varepsilon < (1-c)/2$ be fixed parameters of the protocol. Let c_0, c_1 be some constants which depend on c and ε. Let $\mathcal{T}$ be a tree code over alphabet $[S+1]$ with distance α to be set later.

A_{eff} **Encoding:** For every $k > 0$ set $count_k = 0$.
At any time $n > 1$, repeat the following process for $j = 1, 2, \ldots, c_0$:
  (a) randomly choose $k \in \{1, \ldots, \lfloor n/\log n \rfloor\}$.
  (b) set $count_k = count_k + 1$.
  (c) transmit $y_{n,j} \in [S+1]$, the next symbol of the encoding of $\mathbf{x}^k$ using $\mathcal{T}$, that is, the last symbol of

$$\mathsf{TCenc}(\mathbf{x}_1^k \cdots \mathbf{x}_{count_k}^k) = \mathsf{TCenc}(\mathbf{x}_1^k \cdots \mathbf{x}_{count_k-1}^k) \circ y_{n,j}.$$

A_{eff} **Decoding:** For every $(i,j) \in \mathbb{N} \times [c_0]$ we denote by $\mathsf{ID}(i,j)$ the identifier k of the string $\mathbf{x}^k$ used at iteration (i,j). For each time n, mark all the transmissions $y_{i,j}$ with $i < \varepsilon n/4$ as erasures, and decode $\mathbf{x}^k$ for $\lceil \frac{\varepsilon n/4}{\log \varepsilon n/4} \rceil \le k \le \lceil \frac{(1-c-\varepsilon)n}{\log(1-c-\varepsilon)n} \rceil$:
let $Y_k = \{(i,j) \mid \mathsf{ID}(i,j) = k\}$. Decode the received string indexed by Y_k. That is, set

$$\hat{\mathbf{x}}^k = \mathsf{TCdec}(y_{|Y_k}),$$

where $y_{|Y_k}$ is the string given by concatenating all $y_{i,j}$ with $(i,j) \in Y_k$, where $y_{i,j}$ comes before $y_{i',j'}$ if $i < i'$ or $(i = i') \wedge (j < j')$. Consider a prefix of length $c_1 \log n$ of $\hat{\mathbf{x}}^k$ and ignore the rest.

Protocol 1: An efficient protocol for communicating a logarithmic prefix of $\{\mathbf{x}^1, \mathbf{x}^2, \ldots, \}$.

5.2 Extensions for Streaming Authentication

There are several possible extensions to the above results, which we briefly discuss here. See [8] for full details and proofs.

Efficient Streaming Authentication Scheme with Exponentially Small Error. It is possible to improve the efficient scheme of Theorem 5.2 so that it aborts with polynomially small probability, however, given that it did not abort, the probability that the decoded prefix is incorrect is *exponentially small*. More accurately, the 'trust' Bob has in the decoded string *increases* with the amount of received transmissions. Thus, except for the last fraction of the stream, the decoded stream is equal to the one sent by Alice with overwhelming probability.

Theorem 5.6. *For any $0 \le c < 1$, $0 < \varepsilon \le \frac{1}{2}(1-c)$ there exists an efficient $(cn, (1-c)n - \varepsilon n, 2^{-\Omega(\log n)})$-streaming authentication protocol that, for any time n in which the decoding procedure did not abort, for any $1 \le \ell \le (1-c-\varepsilon)n$ it holds that*

$$\Pr[x'_\ell \ne x_\ell] < 2^{-\Omega(n)}.$$

Decoding a Prefix Longer than $(1 - c)n$. Although our scheme decodes a prefix of length at most $(1 - c)n$ in the worst case, the successfully decoded prefix can be in fact longer. The worst case, as demonstrated by Theorem 5.3, happens when the adversary blocks the suffix of the transmitted stream. On the other hand, if the adversary blocks the prefix of the transmissions, then the scheme of Proposition 5.4 correctly decodes the entire stream! In fact, the protocol succeeds to decode the entire prefix for any time n that satisfies the following γ-suffix condition, if the tree distance satisfies $\alpha > \gamma$.

Definition 5.7. *For any constant $0 \le \gamma < 1$, we say that time n satisfies the γ-suffix condition if any suffix $x_t \ldots x_n$ has at most $\gamma(n - t)$ corrupted transmissions.*

Definition 5.8. *Let $c < 1$ and $\gamma \in (c, 1)$ be given. For any time n let $N_\gamma(n)$ be the latest index that satisfies the γ-suffix condition. When n is clear from the context, we denote $N_\gamma(n)$ simply as N_γ.*

The following Lemma guarantees that, for any $\gamma \in (c, 1)$ it holds that $(1 - c/\gamma)n \le N_\gamma(n) \le n$.

Lemma 5.9. *For every corruption rate c and constant $1 < \xi < 1/c$ there exist a time $t > (1 - \frac{1}{\xi})n$ that satisfies the $c\xi$-suffix condition.*

For a corruption rate c and any $\varepsilon > 0$, and for any time n, if the decoding algorithm did not decode up to time n, then that time n did not satisfy the suffix condition for $\gamma = c/(c + \varepsilon)$, but then, by Lemma 5.9, there must exist a time $N_\gamma > (1 - c - \varepsilon)n$ that satisfies the γ-suffix condition, and at that time the protocol correctly decoded the entire stream (up to time N_γ). Bob does not know the value of N_γ but he can estimate it by checking the number of erasures marked by the Blueberry code.

Proposition 5.10. *Bob can efficiently compute a (lower-bound) estimation N'_γ for N_γ, such that $N'_\gamma > (1 - c - \epsilon)n$ and*

$$\Pr[N'_\gamma > N_\gamma] < 2^{-\Omega(N'_\gamma - N_\gamma)}.$$

Reducing the Amount of Shared Randomness. Our schemes rely on the fact that the parties share a secret random string whose length increases with the size of the information to be communicated. This assumption is sometimes not satisfied in practical applications, especially when considering a multiparty setting in which any two parties run a separate instance of the scheme.

We can mitigate the need for a long shared randomness if the adversary is assumed to be polynomial, assuming standard cryptographic assumptions (specifically, hardness of DDH). To this end, each user generates a pair (sk, pk) of a secret and a public key, broadcasts the public key pk and keeps sk secret. When two users initiate an authentication scheme instance, they first perform a Diffie-Hellman [6] key exchange and obtain an authentication key. They both use the authentication key as a seed to a pseudo-random-generator that generates a long random string for the authentication scheme. Under the DDH assumption, a polynomially-bounded adversary has only negligible information about the authentication key nor the generated randomness, and the authentication scheme remains secure.

6   Interactive Communication

In this section we extend our discussion to the 2-way communication model of *interactive communication*. We show that for adversarial corruption rate of $1/2$ or higher, no constant-rate protocol can compute functions that require interaction between the parties, while with the usage of the Blueberry code we show how to construct a protocol for any function assuming adversarial corruption rate below $1/2$.

Assume that Alice and Bob wish to compute some function $f : \mathcal{X} \times \mathcal{Y} \rightarrow \mathcal{Z}$, where Alice holds $x \in \mathcal{X}$ and Bob holds $y \in \mathcal{Y}$ in the shared-randomness model. The computation is performed interactively: at each round, both parties communicate a message which depends on their input and previous transmissions. At the end of the computation Alice outputs $z_A \in \mathcal{Z}$ and Bob outputs $z_B \in \mathcal{Z}$, and we say that f was correctly computed if $z_A = z_B = f(x, y)$.

In the full version [8] we prove the following separation theorems,

Theorem 6.1. *For any function f which depends on both x and y, the following holds. If the adversarial corruption rate is $\frac{1}{2}$ or higher then no constant-rate interactive protocol correctly computes f with probability higher than the probability of guessing $f(x, y)$ given only the input x (or only the input y).*

Theorem 6.2. *For any constants $\varepsilon > 0$ and for any function f and inputs x, y, there exists an interactive protocol with constant overhead such that if the adversarial corruption rate is at most $c = \frac{1}{2} - \varepsilon$, the protocol outputs $f(x, y)$ with overwhelming probability over the shared random string R.*

References

1. Bonsall, C.: The case of the hungry stranger. HarperCollins (1963)
2. Brakerski, Z., Kalai, Y.T.: Efficient interactive coding against adversarial noise. In: IEEE Annual Symposium on Foundations of Computer Science, pp. 160–166 (2012)
3. Brakerski, Z., Naor, M.: Fast algorithms for interactive coding. In: Proceedings of the 24th Annual ACM-SIAM Symposium on Discrete Algorithms, SODA 2013, pp. 443–456 (2013)
4. Braverman, M.: Towards deterministic tree code constructions. In: Proceedings of the 3rd Innovations in Theoretical Computer Science Conference, pp. 161–167. ACM (2012)
5. Braverman, M., Rao, A.: Towards coding for maximum errors in interactive communication. In: Proceedings of the 43rd Annual ACM Symposium on Theory of Computing, STOC 2011, pp. 159–166. ACM, New York (2011)

6. Diffie, W., Hellman, M.: New directions in cryptography. IEEE Transactions on Information Theory 22(6), 644–654 (1976)
7. Even, S., Goldreich, O., Micali, S.: On-line/Off-line digital signatures. In: Brassard, G. (ed.) CRYPTO 1989. LNCS, vol. 435, pp. 263–275. Springer, Heidelberg (1990)
8. Franklin, M., Gelles, R., Ostrovsky, R., Schulman, L.J.: Optimal coding for streaming authentication and interactive communication. In: Electronic Colloquium on Computational Complexity (ECCC) (2012),
   http://eccc.hpi-web.de/report/2012/104
9. Gelles, R., Moitra, A., Sahai, A.: Efficient and explicit coding for interactive communication. In: IEEE Annual Symposium on Foundations of Computer Science, pp. 768–777 (2011)
10. Gennaro, R., Rohatgi, P.: How to sign digital streams. In: Kaliski Jr., B.S. (ed.) CRYPTO 1997. LNCS, vol. 1294, pp. 180–197. Springer, Heidelberg (1997)
11. Goldreich, O.: Foundations of cryptography. Basic applications, vol. II. Cambridge University Press, New York (2004)
12. Golle, P., Modadugu, N.: Authenticating streamed data in the presence of random packet loss. In: ISOC Network and Distributed System Security Symposium, NDSS 2001 (2001)
13. Guruswami, V., Smith, A.: Codes for computationally simple channels: Explicit constructions with optimal rate. In: IEEE Annual Symposium on Foundations of Computer Science, pp. 723–732 (2010)
14. Hemenway, B., Ostrovsky, R.: Public-key locally-decodable codes. In: Wagner, D. (ed.) CRYPTO 2008. LNCS, vol. 5157, pp. 126–143. Springer, Heidelberg (2008)
15. Hemenway, B., Ostrovsky, R., Strauss, M.J., Wootters, M.: Public key locally decodable codes with short keys. In: Goldberg, L.A., Jansen, K., Ravi, R., Rolim, J.D.P. (eds.) RANDOM 2011 and APPROX 2011. LNCS, vol. 6845, pp. 605–615. Springer, Heidelberg (2011)
16. Langberg, M.: Private codes or succinct random codes that are (almost) perfect. In: Proceedings of the 45th Annual IEEE Symposium on Foundations of Computer Science, FOCS 2004, pp. 325–334. IEEE Computer Society, Washington, DC (2004)
17. Lipton, R.: A new approach to information theory. In: Enjalbert, P., Mayr, E.W., Wagner, K.W. (eds.) STACS 1994. LNCS, vol. 775, pp. 699–708. Springer, Heidelberg (1994)
18. Micali, S., Peikert, C., Sudan, M., Wilson, D.A.: Optimal error correction against computationally bounded noise. In: Kilian, J. (ed.) TCC 2005. LNCS, vol. 3378, pp. 1–16. Springer, Heidelberg (2005)
19. Miner, S., Staddon, J.: Graph-based authentication of digital streams. In: IEEE Symposium on Security and Privacy, pp. 232–246 (2001)
20. Peczarski, M.: An improvement of the tree code construction. Information Processing Letters 99(3), 92–95 (2006)
21. Perrig, A., Canetti, R., Tygar, J., Song, D.: Efficient authentication and signing of multicast streams over lossy channels. In: IEEE Symposium on Security and Privacy, pp. 56–73 (2000)
22. Schulman, L.J.: Communication on noisy channels: a coding theorem for computation. In: Annual IEEE Symposium on Foundations of Computer Science, pp. 724–733 (1992)
23. Schulman, L.J.: Deterministic coding for interactive communication. In: STOC 1993: Proceedings of the Twenty-Fifth Annual ACM Symposium on Theory of Computing, pp. 747–756. ACM, New York (1993)
24. Schulman, L.J.: Coding for interactive communication. IEEE Transactions on Information Theory 42(6), 1745–1756 (1996)

25. Shannon, C.E.: A note on a partial ordering for communication channels. Information and Control 1(4), 390–397 (1958)
26. Smith, A.: Scrambling adversarial errors using few random bits, optimal information reconciliation, and better private codes. In: Proceedings of the Eighteenth Annual ACM-SIAM Symposium on Discrete Algorithms, SODA 2007, pp. 395–404. Society for Industrial and Applied Mathematics, Philadelphia (2007)

APPENDIX

A Construction of $\{x^1, x^2, \ldots\}$

For every k, define $\mathbf{x}^k$ to be the string that contains the stream prefix $x_{t(k)}$ downto x_1 concatenated with as many zeros as needed, $\mathbf{x}^k = x_{t(k)}x_{t(k)-1}\cdots x_2 x_1 000\cdots$, where $t(k)$ is defined to be the minimal time such that $t(k)/\log t(k) > k$. We say $\mathbf{x}^k$ is *declared* at time $t(k)$, meaning that only from this time and on the algorithm may choose to send symbols of the encoding of $\mathbf{x}^k$. It is easy to verify that the string $\mathbf{x}^k$ is well defined at the time it is declared (the corresponding x_i's are known).

If some string $\mathbf{x}^k$ is declared at time $t(k)$ then $\mathbf{x}^{k+1}$ will be declared at time $t(k+1) \approx t(k) + \log t(k) + O(\log \log t(k))$. By setting $c_1 = 2$ we are guaranteed that, for every $\varepsilon n/4 \le \ell \le (1 - c - \varepsilon)n$, x_ℓ appears in a correctly decoded $c_1 \log n$-prefix of some $\mathbf{x}^k$ with $k \in K_n$.

Lemma A.1. *If $\mathbf{x}^k$ is the latest string declared at time $i > 8$, then $\mathbf{x}^{k+1}$ is declared at time sooner than $i + 2\log i$.*

Proof. Let $f(i) = \frac{i+2\log i}{\log(i+2\log i)} - \frac{i}{\log i}$. f is monotonically increasing, and $f(8) > 1$. $\blacksquare$

Corollary A.2. *For any time $n > 8$, and any ℓ, the bit x_ℓ is within the first $2\log n$ symbols of $\mathbf{x}^{\lceil \ell/\log \ell \rceil}$. Hence, every x_ℓ with $\varepsilon n/4 \le \ell \le (1 - c - \varepsilon)n$, appears in a $2\log n$-prefix of (at least) one of the strings $\{\mathbf{x}^k\}_{k \in K_n}$.*

Unfortunately, with the above choice of $\mathbf{x}^k$s, only part of the stream, namely $x_{\varepsilon n/4}, \ldots, x_{(1-c-\varepsilon)n}$, is decoded by the protocol. In order to communicate the prefix $x_1, \ldots, x_{\varepsilon n/4}$ we run another instance of the scheme guaranteed by Proposition 5.5 for the following set of infinite strings $\{\mathbf{v}^1, \mathbf{v}^2, \ldots\}$. (We explain how to combine these two instances below). Define $\mathbf{v}^k$ in the following way

$$\mathbf{v}_i^k = \begin{cases} x_1 & k = 1, \forall i \\ x_{1+(\ell \bmod \lceil t(k)/2 \rceil + 1)} & k > 1, \ i = 1 \text{ and } \mathbf{v}_{2\log t(k-1)}^{k-1} = x_\ell \\ x_{1+(\ell \bmod \lceil t(k)/2 \rceil + 1)} & k > 1, \ i > 1 \text{ and } \mathbf{v}_{i-1}^k = x_\ell \end{cases}$$

It is easy to verify that at time n, the string $\mathbf{v}^{\lfloor n/\log n \rfloor}$ is well defined and known to the encoder.

Lemma A.3. *For every time $n > 256/(1 - c - \varepsilon)$, any bit x_ℓ with $1 \le \ell \le \varepsilon n/4$ appears in a $2\log n$-prefix of (at least) one of the strings $\{\mathbf{v}^k\}_{k \in K_n}$.*

Proof. Note that the concatenation of $O(\log n)$-prefix of the $\mathbf{v}^k$s gives a string of the form $V \triangleq x_1 x_2 \ldots x_{\lceil t(k_1)/2 \rceil} x_1 x_2 \ldots x_{\lceil t(k_2)/2 \rceil} x_1 x_2 \ldots$, and V is decoded by Protocol 1 with high probability.[4] By taking $c_1 = 2$ and recalling that $\varepsilon < (1-c)/2$, (and thus, $(1-c-\varepsilon)n/4 > \varepsilon n/4$) the length of V is lower bounded by the amount of indices in prefixes of size $2 \log \frac{1}{4}(1-c-\varepsilon)n$ of $\{\mathbf{v}^{(1-c-\varepsilon)n/4}, \ldots, \mathbf{v}^{(1-c-\varepsilon)n}\}$,

$$
2 \log \frac{1}{4}(1-c-\varepsilon)n \left(\frac{(1-c-\varepsilon)n}{\log(1-c-\varepsilon)n} - \frac{\frac{1}{4}(1-c-\varepsilon)n}{\log \frac{1}{4}(1-c-\varepsilon)n} \right)
$$

$$
\geq \frac{3}{2}(1-c-\varepsilon)n - 4\frac{(1-c-\varepsilon)n}{\log(1-c-\varepsilon)n}
$$

$$
\geq (1-c-\varepsilon)n
$$

where the last inequality holds for $n > \frac{256}{1-c-\varepsilon}$. Consider the latest place in V where x_1 appears. If that place is at least $(1-c-\varepsilon)/4$ indices from the end of V, it is clear that $x_1 \ldots x_{(1-c-\varepsilon)/4}$ appears in the $(1-c-\varepsilon)/4$-suffix of the decoded V. For the other case, let the bit that precedes this x_1 be x_ℓ. By the way we defined $\mathbf{v}^k$ it follows that $\frac{3}{8}(1-c-\varepsilon) \leq \ell \leq \frac{1}{2}(1-c-\varepsilon)$ which means that $x_1 \ldots x_{(1-c-\varepsilon)/4}$ must appear in a prefix of size $3/4 \cdot (1-c-\varepsilon)n$ of V. Since $(1-c-\varepsilon)n/4 > \varepsilon n/4$, the claim holds. $\qquad\square$

One cannot run Protocol 1 twice, once for $\{\mathbf{x}\}$ and once for $\{\mathbf{v}\}$. Indeed, Eve can block all the transmissions of one of the instances, thus prevent the correct decoding of the stream with probability one, while her corruption rate does not exceed $c = 1/2$. One possible solution is to set $c_1 = 4$ and interleave the transmitted data, that is, define the set $\{\mathbf{z}^1, \mathbf{z}^2, \ldots\}$ where $\mathbf{z}^k = \mathbf{x}_1^k \mathbf{v}_1^k \mathbf{x}_2^k \mathbf{v}_2^k \ldots$, etc.

Corollary A.4. *Let c, ε be constants $0 \leq c < 1$, $0 < \varepsilon \leq (1-c)/2$, and let B be a Blueberry code with constant parameters determined by c, ε. For the strings $\{\mathbf{z}^1, \mathbf{z}^2, \ldots\}$ defined above, the concatenation of A_{eff} with B is an efficient $(cn, (1-c)n - \varepsilon n, 2^{-\Omega(\log n)})$-streaming authentication scheme.*

---

[4] To be more accurate, V is a substring of the string decoded by the scheme.

Secret Sharing, Rank Inequalities and Information Inequalities

Sebastià Martín[1], Carles Padró[2], and An Yang[2]

[1] Universitat Politècnica de Catalunya, Barcelona, Spain
[2] Nanyang Technological University, Singapore

Abstract. Beimel and Orlov proved that all information inequalities on four or five variables, together with all information inequalities on more than five variables that are known to date, provide lower bounds on the size of the shares in secret sharing schemes that are at most linear on the number of participants. We present here another negative result about the power of information inequalities in the search for lower bounds in secret sharing. Namely, we prove that all information inequalities on a bounded number of variables only can provide lower bounds that are polynomial on the number of participants.

Keywords: Secret sharing, Information inequalities, Rank inequalities, Polymatroids.

1 Introduction

Secret sharing schemes, which were independently introduced by Shamir [27] and Blakley [6], make it possible to distribute a *secret value* into *shares* among a set of *participants* in such a way that only the *qualified sets* of participants can recover the secret value, while no information at all on the secret value is provided by the shares from an unqualified set. The qualifed sets form the *access structure* of the scheme.

This work deals with the problem of the size of the shares in secret sharing schemes for general access structures. The reader is referred to [2] for an up-to-date survey on this topic. Even though there exists a secret sharing scheme for every access structure [20], all known general constructions are impractical because the size of the shares grows exponentially with the number of participants. The general opinion among the researchers in the area is that this is unavoidable. Specifically, the following conjecture, which was formalized by Beimel [2], is generally believed to be true. It poses one of the main open problems in secret sharing, and a very difficult and intriguing one.

Conjecture 1. There exists an $\epsilon > 0$ such that for every integer n there is an access structure on n participants, for which every secret sharing scheme distributes shares of length $2^{\epsilon n}$, that is, exponential in the number of participants.

Nevertheless, not many results supporting this conjecture have been proved. No proof for the existence of access structures requiring shares of superpolynomial

R. Canetti and J.A. Garay (Eds.): CRYPTO 2013, Part II, LNCS 8043, pp. 277–288, 2013.

size has been found. Moreover, the best of the known lower bounds is the one given by Csirmaz [9], who presented a family of access structures on an arbitrary number n of participants that require shares of size $\Omega(n/\log n)$ times the size of the secret.

In contrast, superpolynomial lower bounds on the size of the shares have been obtained for linear secret sharing schemes [1, 3, 17]. In a *linear secret sharing scheme*, the secret and the shares are vectors over some finite field, and both the computation of the shares and the recovering of the secret are performed by linear maps. Because of their homomorphic properties, linear schemes are needed for many applications of secret sharing. Moreover, most of the known constructions of secret sharing schemes yield linear schemes.

Similarly to the works by Csirmaz [9] and by Beimel and Orlov [5], we analyze here the limitations of the technique that has been almost exclusively used to find lower bounds on the size of the shares. This is the case of the bounds in [7–9, 21] and many other papers. Even though it was implicitly used before, the method was formalized by Csirmaz [9]. Basically, it consists of finding lower bounds on the solutions of certain linear programs. This method provides lower bounds on the *information ratio* of secret sharing schemes, that is, on the ratio between the maximum size of the shares and the size of the secret.

The constraints of those linear programs are derived from the fact that certain linear combinations of the values of the joint entropies of the random variables defining a secret sharing scheme must be nonnegative. These constraints can be divided into two classes.

1. The first class is formed by the constraints that are derived from the access structure. Namely, from the fact that the qualified subsets can recover the secret while the unqualified ones have no information about it.
2. The second class is formed by constraints derived from *information inequalities* that hold for every collection of random variables.

In the second class, the constraints derived from the so-called *Shannon inequalities* are always considered. These basic information inequalities are equivalent to the conditional mutual information being nonnegative, and equivalent also to the fact that the joint entropies of a collection of random variables define a polymatroid [15, 16].

Csirmaz [9] proved that, by taking only the Shannon inequalities in the second class, one obtains lower bounds that are at most linear on the number of participants. This was proved by showing that every such linear program admits a small solution.

One may expect that better lower bounds should be obtained by adding to the second class new constraints derived from the *non-Shannon information inequalities*, which are the ones that cannot be derived from the basic Shannon inequalities. The existence of such inequalities was unknown when Csirmaz [9] formalized that method. The first one was presented by Zhang and Yeung [30] and many others have been found subsequently [11, 13, 23, 29]. When dealing with linear secret sharing schemes, one can improve the linear program by using *rank inequalities*, which apply to configurations of vector subspaces or, equivalently, to the joint entropies of collections of random variables defined from linear maps. It is well-known that every information inequality is also a rank inequality. The first known rank inequality that cannot be derived from the Shannon inequalities was found by Ingleton [19]. Other rank inequalities have been presented afterwards [12, 22]. Indeed, better lower bounds on the information ratio have been found for some families of access structures by using non-Shannon information and rank inequalities [4, 10, 24, 25].

Nevertheless, Beimel and Orlov [5] presented a negative result about the power of non-Shannon information inequalities to provide better general lower bounds on the size of the shares. Specifically, they proved that the best lower bound that can be obtained by using all information inequalities on four and five variables, together with all inequalities on more than five variables that are known to date, is at most linear on the number of participants. Specifically, they proved that every linear program that is obtained by using these inequalities admits a small solution that is related to the solution used by Csirmaz [9] to prove his negative result. They used the fact that there exists a finite set of rank inequalities that, together with the Shannon inequalities, span all rank inequalities, and hence all information inequalities, on four or five variables [12, 18]. By executing a brute-force algorithm using a computer program, they checked that Csirmaz's solution is compatible with every rank inequality in that finite set. In addition, they manually executed their algorithm on a symbolic representation of the infinite sequence of information inequalities given by Zhang [29]. This sequence contains inequalities on arbitrarily many variables and generalizes the infinite sequences from previous works.

In particular, the results in [5] imply that all rank inequalities on four or five variables cannot provide lower bounds on the size of shares in *linear* secret sharing schemes that are better than linear on the number of participants. Unfortunately, their algorithm is not efficient enough to be applied on the known rank inequalities on six variables.

We present here another negative result about the power of information inequalities to provide general lower bounds on the size of the shares in secret sharing schemes. Namely, we prove that every lower bound that is obtained by using rank inequalities on at most r variables is $O(n^{r-2})$, and hence polynomial on the number n of participants. Since all information inequalities are rank inequalities, this negative result applies to the search of lower bounds for both linear and general secret sharing schemes. Therefore, information inequalities on arbitrarily many variables are needed to find superpolynomial lower bounds by using the method described above.

The proof is extremely simple and concise. Similarly to the proofs in [5, 9], it is based on finding small solutions to the linear programs that are obtained by using rank inequalities on a bounded number of variables. These solutions are obtained from a family of polymatroids that are uniform and Boolean. This family contains the polymatroids that were used in [5, 9].

In some sense, our result is weaker than the one in [5], because for $r = 4$ and $r = 5$, our solutions to the linear programs do not prove that the lower bounds must be linear on the number of participants, but instead quadratic and cubic, respectively. But in another sense our result is much more general because it applies to all (known or unknown) rank inequalities. In addition, our proof provides a better understanding on the limitations of the use of information inequalities in the search of lower bounds for secret sharing schemes.

2   Polymatroids, Rank Inequalities and Information Inequalities

Some basic concepts and facts about polymatroids that are used in the paper are presented here. A more detailed presentation can be found in textbooks on the topic [26, 28]. For a finite set Q, we notate $\mathcal{P}(Q)$ for the power set of Q, that is, the set of all subsets of Q.

Definition 1. *A* polymatroid *is a pair* $\mathcal{S} = (Q, f)$ *formed by a finite set* Q, *the* ground set, *and a* rank function $f \colon \mathcal{P}(Q) \to \mathbb{R}$ *satisfying the following properties.*

- $f(\emptyset) = 0$.
- f is monotone increasing: *if* $X \subseteq Y \subseteq Q$, *then* $f(X) \leq f(Y)$.
- f is submodular: $f(X \cup Y) + f(X \cap Y) \leq f(X) + f(Y)$ *for every* $X, Y \subseteq Q$.

A polymatroid is called integer *if its rank function is integer-valued.*

The following characterization of rank functions of polymatroids is a straightforward consequence of [26, Theorem 44.1].

Proposition 1. *A map* $f \colon \mathcal{P}(Q) \to \mathbb{R}$ *is the rank function of a polymatroid with ground set* Q *if and only if the following properties are satisfied.*

- $f(\emptyset) = 0$.
- *If* $X \subseteq Q$ *and* $y \in Q$, *then* $f(X) \leq f(X \cup \{y\})$.
- *If* $X \subseteq Q$ *and* $y, z \in Q$, *then* $f(X \cup \{y, z\}) + f(X) \leq f(X \cup \{y\}) + f(X \cup \{z\})$.

If $\mathcal{S} = (Q, f)$ is a polymatroid and α is a positive real number, then $\alpha \mathcal{S} = (Q, \alpha f)$ is a polymatroid too, which is called a *multiple* of $\mathcal{S}$. A polymatroid $\mathcal{S}' = (Q', g)$ is called an *extension* of a polymatroid $\mathcal{S} = (Q, f)$ if $Q \subseteq Q'$ and $g(X) = f(X)$ for every $X \subseteq Q$. In general, we will use the same symbol for the rank function of a polymatroid and the rank function of an extension.

Let V be a vector space over a field $\mathbb{K}$ and $(V_x)_{x \in Q}$ a tuple of vector subspaces of V. For $X \subseteq Q$, we notate $V_X = \sum_{x \in X} V_x$. Then the map $f \colon \mathcal{P}(Q) \to \mathbb{Z}$ defined by $f(X) = \dim V_X$ for every $X \subseteq Q$ is the rank function of an integer polymatroid $\mathcal{S}$ with ground set Q. Integer polymatroids that can be defined in this way are said to be $\mathbb{K}$-*linearly representable*, or simply $\mathbb{K}$-*linear* or $\mathbb{K}$-*representable*, and the tuple $(V_x)_{x \in Q}$ is called a $\mathbb{K}$-*linear representation* of $\mathcal{S}$. A $\mathbb{K}$-*poly-linear polymatroid* is the multiple of a $\mathbb{K}$-linear polymatroid.

For a finite set Q, consider a family of random variables $(S_x)_{x \in Q}$, where S_x is defined on a finite set E_x. For every $X \subseteq Q$, we use S_X to denote the random variable $(S_x)_{x \in X}$ on the set $\prod_{x \in X} E_x$, and $H(S_X)$ will denote its Shannon entropy. Fujishige [15, 16] found out the following connection between Shannon entropy and polymatroids.

Theorem 1. *Let $(S_x)_{x \in Q}$ be a family of random variables. Consider the mapping $h: \mathcal{P}(Q) \to \mathbb{R}$ defined by $h(\emptyset) = 0$ and $h(X) = H(S_X)$ if $\emptyset \neq X \subseteq Q$. Then h is the rank function of a polymatroid with ground set Q.*

A polymatroid $\mathcal{S} = (Q, h)$ is said to be *entropic* if there exists a family $(S_x)_{x \in Q}$ of discrete random variables such that $h(X) = H(S_X)$ for every $X \subseteq Q$. A *poly-entropic polymatroid* is a multiple of an entropic polymatroid. It is well known that, if $\mathbb{K}$ is a finite field, then every $\mathbb{K}$-poly-linear polymatroid is poly-entropic. Indeed, given a $\mathbb{K}$-vector space E, let E^* be its *dual space*, which is formed by all linear forms $\alpha : E \to \mathbb{K}$, and S the random variable given by the uniform probability distribution on E^*. For every subspace $V \subseteq E$, consider the random variable $S|_V$ on V^*, the restriction of S to V. Clearly, $H(S|_V) = \log |\mathbb{K}| \dim V$. Therefore, the $\mathbb{K}$-linear polymatroid given by a collection $(V_x)_{x \in Q}$ of subspaces of E is a multiple of the entropic polymatroid defined by $(S_x)_{x \in Q}$, where $S_x = S|_{V_x}$. The collections of random variables that can be defined in this way are said to be $\mathbb{K}$-*linear*.

Consider a finite set M and a family $(M_x)_{x \in Q}$ of subsets of M. For every $X \subseteq Q$, take $M_X = \bigcup_{x \in X} M_x$. Then the map defined by $f(X) = |M_X|$ for every $X \subseteq Q$ is the rank function of an integer polymatroid $\mathcal{S}$ with ground set Q. The family $(M_x)_{x \in Q}$ is called a *Boolean representation* of $\mathcal{S}$. *Boolean polymatroids* are those admitting a Boolean representation. Boolean polymatroids are $\mathbb{K}$-linear for every field $\mathbb{K}$. Indeed, the set $\mathbb{K}^M$ of all functions $\mathbf{v}: M \to \mathbb{K}$ is a $\mathbb{K}$-vector space. For every $w \in M$, consider the vector $\mathbf{e}^w \in \mathbb{K}^M$ given by $\mathbf{e}^w(w') = 1$ if $w' = w$ and $\mathbf{e}^w(w') = 0$ otherwise. Clearly, $(\mathbf{e}^w)_{w \in M}$ is a basis of $\mathbb{K}^M$. For every $x \in Q$, consider the vector subspace $V_x = \langle \mathbf{e}^w : w \in M_x \rangle$. Obviously, these subspaces form a $\mathbb{K}$-linear representation of $\mathcal{S}$.

We say that a polymatroid $\mathcal{S}$ with ground set Q is *uniform* if every permutation on Q is an automorphism of $\mathcal{S}$. In this situation, the rank $f(X)$ of a set $X \subseteq Q$ depends only on its cardinality, that is, there exist values $0 = f_0 \leq f_1 \leq \cdots \leq f_n$, where $n = |Q|$, such that $f(X) = f_i$ for every $X \subseteq Q$ with $|X| = i$. By Proposition 1, such a sequence $(f_i)_{1 \leq i \leq n}$ defines a uniform polymatroid if and only if $f_i - f_{i-1} \geq f_{i+1} - f_i$ for every $i = 1, \dots, n-1$. Clearly, a uniform polymatroid is univocally determined by its *increment vector* $\delta = (\delta_1, \dots, \delta_n)$, where $\delta_i = f_i - f_{i-1}$. Observe that $\delta \in \mathbb{R}^n$ is the increment vector of a uniform polymatroid if and only if $\delta_1 \geq \cdots \geq \delta_n \geq 0$. All uniform integer polymatroids are linearly representable. Specifically, a uniform integer polymatroid is $\mathbb{K}$-linear if the field $\mathbb{K}$ has at least as many elements as the ground set [14].

For a positive integer r, we notate $[r] = \{1, \dots, r\}$. Given a collection $(A_i)_{i \in [r]}$ of subsets of a set Q and $I \subseteq [r]$, we notate $A_I = \bigcup_{i \in I} A_i$. An *information inequality*, respectively *rank inequality*, on r variables consists of a collection $(\alpha_I)_{I \in \mathcal{P}([r])}$ of real numbers such that

$$\sum_{I \subseteq [r]} \alpha_I f(A_I) \geq 0$$

for every poly-entropic, respectively poly-linear, polymatroid (Q, f) and for every collection $(A_i)_{i \in [r]}$ of r subsets of Q. Observe that the number r of variables may be larger than the cardinality of the ground set Q.

Every information inequality is also a rank inequality [12]. By Theorem 1, the polymatroid axioms are information inequalities, which are called *Shannon inequalities*. The Ingleton inequality [19] was the first known example of a rank inequality that cannot be derived from Shannon-type inequalities. Zhang and Yeung [30] presented the first information inequality that cannot be derived from the Shannon inequalities. Subsequently, many other rank and information inequalities have been found in [11–13, 22, 23, 29] and other works. We need the following technical result, which is a consequence of [5, Lemma 4.3].

Lemma 1. *Let $(\alpha_I)_{I \in \mathcal{P}([r])}$ be a rank inequality. Then $\sum_{I : I \cap J \neq \emptyset} \alpha_I \geq 0$ for every $J \subseteq [r]$.*

Proof. Take $J \subseteq [r]$, a set M with $|M| = 1$, and the family $(M_i)_{i \in [r]}$ of subsets of M given by $M_i = M$ if $i \in J$ and $M_i = \emptyset$ otherwise. Let $([r], f)$ be the Boolean polymatroid defined by this family. Then $\sum_{I : I \cap J \neq \emptyset} \alpha_I = \sum_{I \subseteq [r]} \alpha_I f(I) \geq 0$ because Boolean polymatroids are linearly representable. □

3   Polymatroids and Secret Sharing

Let P be a finite set of *participants*, $p_0 \notin P$ a special participant, usually called *dealer*, and $Q = P \cup \{p_0\}$. This notation will be used from now on. An *access structure* Γ on P is a *monotone increasing* family of subsets of P, that is, if $X \subseteq Y \subseteq P$ and $X \in \Gamma$, then $Y \in \Gamma$. To avoid anomalous situations, we assume always that $\emptyset \notin \Gamma$ and $P \in \Gamma$. The members of Γ are called *qualified sets*. An access structure Γ is determined by the family $\min \Gamma$ of its minimal qualified sets. For a polymatroid $\mathcal{S} = (Q, f)$ and an element $p_0 \in Q$ with $f(\{p_0\}) > 0$, we define the access structure $\Gamma_{p_0}(\mathcal{S})$ on $P = Q \setminus \{p_0\}$ by

$$\Gamma_{p_0}(\mathcal{S}) = \{X \subseteq P : f(X \cup \{p_0\}) = f(X)\}.$$

We need also the parameter

$$\sigma_{p_0}(\mathcal{S}) = \frac{\max_{x \in P} f(\{x\})}{f(\{p_0\})}.$$

If $\Gamma = \Gamma_{p_0}(\mathcal{S})$ and, in addition, $f(X \cup \{p_0\}) = f(X) + 1$ for every unqualified set $X \subseteq P$, then $\mathcal{S}$ is said to be a Γ-*polymatroid*.

A *secret sharing scheme* Σ on P with access structure Γ is a family $(S_x)_{x \in Q}$ of random variables such that

1. $H(S_{X \cup \{p_0\}}) = H(S_X)$ if $X \in \Gamma$ and
2. $H(S_{X \cup \{p_0\}}) = H(S_X) + H(S_{p_0})$ otherwise.

The random variables S_{po} and $(S_x)_{x \in P}$ correspond, respectively, to the *secret value* and the *shares* that are distributed among the participants in P. A secret sharing scheme is $\mathbb{K}$-*linear* if it is a $\mathbb{K}$-linear collection of random variables. The *information ratio* $\sigma(\Sigma)$ of the secret sharing scheme Σ is the ratio between the maximum length of the shares and the length of the secret. Namely,

$$\sigma(\Sigma) = \frac{\max_{x \in P} H(S_x)}{H(S_{po})}.$$

The entropic polymatroid S defined by the collection $(S_x)_{x \in Q}$ is such that $\Gamma = \Gamma_{po}(S)$ and, in addition, $\sigma(\Sigma) = \sigma_{po}(S)$.

The *optimal information ratio* $\sigma(\Gamma)$ of an access structure Γ is the infimum of the information ratios of all secret sharing schemes for Γ. Clearly,

$$\sigma(\Gamma) = \inf\{\sigma_{po}(S) : S \text{ is a poly-entropic } \Gamma\text{-polymatroid}\}.$$

Therefore, the parameters

$$\kappa(\Gamma) = \inf\{\sigma_{po}(S) : S \text{ is a } \Gamma\text{-polymatroid}\}$$

and

$$\lambda(\Gamma) = \inf\{\sigma_{po}(S) : S \text{ is a poly-linear } \Gamma\text{-polymatroid}\}$$

are, respectively, a lower and an upper bound for $\sigma(\Gamma)$. Observe that $\lambda(\Gamma)$ is the infimum of the information ratios of the linear secret sharing schemes for Γ. The value $\kappa(\Gamma)$ is the solution of a linear programming problem, and hence the infimum is a minimum and $\kappa(\Gamma)$ is a rational number [25]. Most of the known lower bounds on the information ratio, as the ones from [7–9, 21], are lower bounds on $\kappa(\Gamma)$. In fact, this is the case for all lower bounds that can be obtained by using only Shannon inequalities.

Information inequalities and rank inequalities can be added to the linear program computing $\kappa(\Gamma)$ to find better lower bounds on $\sigma(\Gamma)$ and $\lambda(\Gamma)$, respectively. This has been done for several families of access structures [4, 10, 24, 25].

A polymatroid $S = (P, f)$ and an access structure Γ on a set P are said to be *compatible* if S can be extended to a Γ-polymatroid $S(\Gamma) = (Q, f)$.

Proposition 2. *An access structure Γ on P is compatible with a polymatroid $S = (P, f)$ if and only if the following conditions are satisfied.*

1. *If $X \subseteq P$ and $y \in P$ are such that $X \notin \Gamma$ and $X \cup \{y\} \in \Gamma$, then $f(X) \leq f(X \cup \{y\}) - 1$.*
2. *If $X \subseteq P$ and $y, z \in P$ are such that $X \notin \Gamma$ while both $X \cup \{y\}$ and $X \cup \{z\}$ are qualified, then $f(X \cup \{y, z\}) + f(X) \leq f(X \cup \{y\}) + f(X \cup \{z\}) - 1$.*

Proof. Suppose that S can be extended to a Γ-polymatroid $S(\Gamma) = (Q, f)$. If $X \notin \Gamma$ and $X \cup \{y\} \in \Gamma$, then

$$f(X \cup \{y\}) = f(X \cup \{y, p_0\}) \geq f(X \cup \{p_0\}) = f(X) + 1.$$

If $X \notin \Gamma$ and $X \cup \{y\}$ and $X \cup \{z\}$ are qualified, then

$$
\begin{aligned}
f(X \cup \{y\}) + f(X \cup \{z\}) &= f(X \cup \{y, p_0\}) + f(X \cup \{z, p_0\}) \\
&\geq f(X \cup \{y, z, p_0\}) + f(X \cup \{p_0\}) \\
&= f(X \cup \{y, z\}) + f(X) + 1.
\end{aligned}
$$

For the converse, assume that $\mathcal{S} = (P, f)$ satisfies the conditions in the statement and consider the extension of f to $\mathcal{P}(Q)$ determined by $f(X \cup \{p_0\}) = f(X)$ if $X \in \Gamma$ and $f(X \cup \{p_0\}) = f(X) + 1$ otherwise. We have to prove that (Q, f) is a polymatroid. Clearly, $f(X) \leq f(X \cup \{p_0\})$ and $f(X \cup \{p_0\}) \leq f(X \cup \{p_0, y\})$ for every $X \subseteq P$ and $y \in P$. Therefore, the first condition in Proposition 1 is satisfied. Moreover, it is not difficult to prove that the second condition holds by checking that $f(X \cup \{y, p_0\}) + f(X) \leq f(X \cup \{y\}) + f(X \cup \{p_0\})$ and $f(X \cup \{p_0, y, z\}) + f(X \cup \{p_0\}) \leq f(X \cup \{p_0, y\}) + f(X \cup \{p_0, z\})$ for every $X \subseteq P$ and $y, z \in P$.   □

The following result was presented by Csirmaz [9].

Proposition 3. *An access structure Γ on P is compatible with a polymatroid $\mathcal{S} = (P, f)$ if and only if the following conditions are satisfied.*

1. *If $X \subseteq Y \subseteq P$ are such that $X \notin \Gamma$ and $Y \in \Gamma$, then $f(X) \leq f(Y) - 1$.*
2. *If $X, Y \in \Gamma$ and $X \cap Y \notin \Gamma$, then $f(X \cup Y) + f(X \cap Y) \leq f(X) + f(Y) - 1$.*

Proof. Necessity can be proved in a similar way as in Proposition 2. Sufficiency is obvious from Proposition 2.   □

4   A Family of Uniform Boolean Polymatroids

We present a family of polymatroids that are uniform and Boolean. In addition, every member of this family is compatible to all access structure on its ground set. The following results are straightforward consequences of Proposition 2.

Proposition 4. *A polymatroid $\mathcal{S} = (P, f)$ is compatible with all access structures on P if and only if the following conditions are satisfied.*

1. *$f(X) \leq f(X \cup \{z\}) - 1$ for every $X \subseteq P$ and $z \in P \setminus X$.*
2. *$f(X \cup \{y, z\}) + f(X) \leq f(X \cup \{y\}) + f(X \cup \{z\}) - 1$ for every $X \subseteq P$ and $y, z \in P \setminus X$.*

Proposition 5. *Let P be a set with $|P| = n$ and let $\mathcal{S}$ be a uniform polymatroid on P. Then $\mathcal{S}$ is compatible with all access structures on P if and only if its increment vector $(\delta_1, \ldots, \delta_n)$ is such that $\delta_i \geq \delta_{i+1} + 1$ for $i = 1, \ldots, n-1$ and $\delta_n \geq 1$.*

Given a set P and an integer $r \geq 2$, let $M(P, r)$ be the set of all multisets of size r of the set P. For example, if $P = \{a, b, c\}$, then

$$
M(P, 3) = \{aaa, aab, aac, abb, abc, acc, bbb, bbc, bcc, ccc\}.
$$

Observe that $|M(P,r)| = \binom{n+r-1}{r}$ if $|P| = n$. For every $x \in P$, let $M_x(P,r)$ be the set of the multisets in $M(P,r)$ that contain x. In the previous example,

$$M_a(P,3) = \{aaa, aab, aac, abb, abc, acc\}.$$

Finally, we define $\mathcal{Z}(P,r) = (P,f)$ as the Boolean polymatroid on P defined by the family $(M_x(P,r))_{x \in P}$ of subsets of $M(P,r)$. As usual, we notate $M_X(P,r) = \bigcup_{x \in X} M_x(P,r)$ for every $X \subseteq Q$.

Clearly, every permutation on P is an automorphism of $\mathcal{Z}(P,r)$, and hence this polymatroid is uniform. For every $X \subseteq P$, the multisets in $M(P,r) \setminus M_X(P,r)$ are the ones involving only elements in $P \setminus X$. That is, $M(P,r) \setminus M_X(P,r) = M(P \setminus X, r)$, and hence

$$\begin{aligned} f(X) &= |M_X(P,r)| = |M(P,r)| - |M(P \setminus X, r)| \\ &= \binom{|P|+r-1}{r} - \binom{|P|-|X|+r-1}{r}. \end{aligned}$$

Therefore, if $|P| = n$, the increment vector $(\delta_1, \ldots, \delta_n)$ of $\mathcal{Z}(P,r)$ is given by

$$\delta_i = \binom{n-i+r}{r} - \binom{n-i+r-1}{r} = \binom{n-i+r-1}{r-1}$$

for every $i = 1, \ldots, n$. Observe that $\delta_1 > \cdots > \delta_n > 0$, and hence $\mathcal{Z}(P,r)$ is compatible with all access structures on P. In particular, $\delta_i = n - i + 1$ if $r = 2$, and hence $\kappa(\Gamma) \le n$ for every access structure Γ on n participants [9]. The *Csirmaz function* introduced in [5, Definition 3.10] coincides with the rank function of $\mathcal{Z}(P,2)$. The rank function of $\mathcal{Z}(P,2)$ is the smallest among the rank functions of all uniform polymatroids on P that are compatible with all access structures on P [5, Lemma 3.11]. Finally, observe that [5, Lemma 6.2] is a straightforward consequence of the fact that $\mathcal{Z}(P,2)$ is a Boolean polymatroid.

5 Main Result

This section is devoted to prove our main result, Theorem 2.

Proposition 6. *Let P be a set of n participants and Γ an access structure on P. For an integer $r \ge 3$, consider $\mathcal{Z}_{r-1} = \mathcal{Z}(P,r-1)$ and the Γ-polymatroid $\mathcal{Z}_{r-1}(\Gamma)$, an extension of $\mathcal{Z}_{r-1}$ to $Q = P \cup \{p_0\}$. Then $\mathcal{Z}_{r-1}(\Gamma)$ satisfies all rank inequalities on r variables.*

Proof. Let f be the rank function of $\mathcal{Z}_{r-1}(\Gamma)$ and $(\alpha_I)_{I \in \mathcal{P}([r])}$ a rank inequality on r variables. We have to prove that $\sum_{I \subseteq [r]} \alpha_I f(A_I) \ge 0$ for every r subsets $(A_i)_{i \in [r]}$ of Q. Take $B_i = A_i \setminus \{p_0\}$. If $B_i \in \Gamma$ for every $i \in [r]$, then $\sum_{I \subseteq [r]} \alpha_I f(A_I) = \sum_{I \subseteq [r]} \alpha_I f(B_I) \ge 0$ because $\mathcal{Z}_{r-1}$ is Boolean. If $B_{[r]} \notin \Gamma$, then

$$\sum_{I \subseteq [r]} \alpha_I f(A_I) = \sum_{I \subseteq [r]} \alpha_I f(B_I) + \sum_{I \,:\, p_0 \in A_I} \alpha_I \geq 0$$

by Lemma 1 with $J = \{i \in [r] \,:\, p_0 \in A_i\}$. From now on, we assume that $B_{[r]} \in \Gamma$ and that $B_i \notin \Gamma$ for some $i \in [r]$.

Consider the polymatroid $\mathcal{S} = ([r], g)$ determined by $g(I) = f(B_I)$ for every $I \subseteq [r]$. In addition, consider the access structure Λ on $[r]$ formed by the sets $I \subseteq [r]$ such that $B_I \in \Gamma$. We prove next that $\mathcal{S}$ can be extended to a linearly representable Λ-polymatroid $\mathcal{S}(\Lambda) = ([r] \cup \{0\}, g)$. This concludes the proof. Indeed, since $\mathcal{S}(\Lambda)$ is a Λ-polymatroid, $f(A_I) = g(I \cup \{0\})$ if $p_0 \in A_I$, and hence

$$\sum_{I \subseteq [r]} \alpha_I f(A_I) = \sum_{I \,:\, p_0 \notin A_I} \alpha_I f(B_I) + \sum_{I \,:\, p_0 \in A_I} \alpha_I f(A_I)$$

$$= \sum_{I \,:\, p_0 \notin A_I} \alpha_I g(I) + \sum_{I \,:\, p_0 \in A_I} \alpha_I g(I \cup \{0\}).$$

Consider the family $(C_i)_{i \in [r]}$ of subsets of $[r] \cup \{0\}$ given by $C_i = \{i, 0\}$ if $p_0 \in A_i$ and $C_i = \{i\}$ otherwise. Then

$$\sum_{I \,:\, p_0 \notin A_I} \alpha_I g(I) + \sum_{I \,:\, p_0 \in A_I} \alpha_I g(I \cup \{0\}) = \sum_{I \subseteq [r]} \alpha_I g(C_I) \geq 0$$

because $\mathcal{S}(\Lambda)$ is linearly representable.

The polymatroid $\mathcal{S}$ is Boolean. Indeed, take $M = M(P, r - 1)$ and $M_X = M_X(P, r - 1)$ for every $X \subseteq P$. Then $(M_{B_i})_{i \in [r]}$ is a Boolean representation of $\mathcal{S}$. Therefore, this polymatroid is linearly representable over every field, as proved in Section 2. For a field $\mathbb{K}$, take a basis $(\mathbf{e}^w)_{w \in M}$ of $\mathbb{K}^M$. Then the subspaces $(V_i)_{i \in [r]}$ with $V_i = \langle \mathbf{e}^w \,:\, w \in M_{B_i} \rangle$ form a $\mathbb{K}$-linear representation of $\mathcal{S}$.

Consider the dual access structure $\Lambda^* = \{J \subseteq [r] \,:\, [r] \setminus J \notin \Lambda\}$. Take $J \in \min \Lambda^*$ and $I = [r] \setminus J$. Observe that $B_I \notin \Gamma$ and $B_I \cup B_j \in \Gamma$ for every $j \in J$. In particular, this implies that $J \neq \emptyset, [r]$. Therefore, we can take an element $x_j \in B_j \setminus B_I$ for every $j \in J$. Consider a multiset $w_J \in M(P, r - 1)$ containing exactly the elements in $\{x_j \,:\, j \in J\}$, repeating some of them if necessary. Take the vector

$$\mathbf{v}_0 = \sum_{J \in \min \Lambda^*} \mathbf{e}^{w_J} \in \mathbb{K}^M$$

and the subspace $V_0 = \langle \mathbf{v}_0 \rangle$. By adding this subspace to the collection $(V_i)_{i \in [r]}$, an extension $\mathcal{S}(\Lambda) = ([r] \cup \{0\}, g)$ of $\mathcal{S}$ is obtained. Obviously, $\mathcal{S}(\Lambda)$ is $\mathbb{K}$-linearly representable.

Finally, we prove that $\mathcal{S}(\Lambda)$ is a Λ-polymatroid. Clearly, $I \in \Lambda$ if and only if $I \cap J \neq \emptyset$ for every $J \in \min \Lambda^*$. If $I \in \Lambda$, then $w_J \in M_{B_I}(P, r - 1)$ for every $J \in \min \Lambda^*$. Indeed, if $j \in I \cap J$, the element x_j in the multiset w_J is also in B_I. Therefore, $\mathbf{e}^{w_J} \in V_I$ for every $J \in \min \Lambda^*$, and hence $\mathbf{v}_0 \in V_I$ and

$g(I \cup \{0\}) = g(I)$. Suppose now that $I \not\subseteq \Lambda$ and take $J \in \min \Lambda^*$ with $I \cap J = \emptyset$. Then $w_J \notin M_{B_I}(P, r-1)$ because $x_j \notin B_I$ for every $j \in J$. Therefore, $\mathbf{v}_0 \notin V_I$ and $g(I \cup \{0\}) = g(I) + 1$. □

Theorem 2. *For an access structure Γ on n participants, the best lower bound on $\lambda(\Gamma)$ that can be obtained by using rank inequalities on r variables is at most*

$$\binom{n+r-3}{r-2}, \tag{1}$$

and hence $O(n^{r-2})$. As an immediate consequence, the same applies to the lower bounds on the optimal information ratio $\sigma(\Gamma)$ that are obtained by using information inequalities on r variables.

Proof. By Proposition 6, the polymatroid $\mathcal{Z}_{r-1}(\Gamma)$ is a feasible solution to any linear program that is obtained from rank inequalities on r variables. Therefore, every lower bound on $\lambda(\Gamma)$ derived from such a linear program is at most $\sigma_{p_0}(\mathcal{Z}_{r-1}(\Gamma)) = \delta_1$, where δ_1 is the first component of the increment vector of $\mathcal{Z}(P, r-1)$. □

Observe that we are not assuming $r \leq n$ in Theorem 2. A smaller value for the bound (1) can be proved for the case $r \leq n$ by using in the same way the uniform Boolean polymatroid defined by the set M of all subsets (instead of multisets) of P with at most $r-1$ participants and the subsets $(M_x)_{x \in P}$, where M_x consists of the subsets in M that contain x. Nevertheless, asymptotically the new bound is not better than $O(n^{r-2})$.

Acknowledgements. We thank Amos Beimel and Ilan Orlov for useful discussions about the contents of this paper. The first author's work was partially supported by the Spanish Government through the project MTM2009-07694. The second and third authors' work was supported by the Singapore National Research Foundation under Research Grant NRF-CRP2-2007-03.

References

1. Babai, L., Gál, A., Wigderson, A.: Superpolynomial lower bounds for monotone span programs. Combinatorica 19, 301–319 (1999)
2. Beimel, A.: Secret-Sharing Schemes: A Survey. In: Chee, Y.M., Guo, Z., Ling, S., Shao, F., Tang, Y., Wang, H., Xing, C. (eds.) IWCC 2011. LNCS, vol. 6639, pp. 11–46. Springer, Heidelberg (2011)
3. Beimel, A., Gál, A., Paterson, M.: Lower bounds for monotone span programs. Comput. Complexity 6, 29–45 (1997)
4. Beimel, A., Livne, N., Padró, C.: Matroids Can Be Far From Ideal Secret Sharing. In: Canetti, R. (ed.) TCC 2008. LNCS, vol. 4948, pp. 194–212. Springer, Heidelberg (2008)
5. Beimel, A., Orlov, I.: Secret Sharing and Non-Shannon Information Inequalities. IEEE Trans. Inform. Theory 57, 5634–5649 (2011)

6. Blakley, G.R.: Safeguarding cryptographic keys. In: AFIPS Conference Proceedings, vol. 48, pp. 313–317 (1979)
7. Blundo, C., De Santis, A., De Simone, R., Vaccaro, U.: Tight bounds on the information rate of secret sharing schemes. Des. Codes Cryptogr. 11, 107–122 (1997)
8. Capocelli, R.M., De Santis, A., Gargano, L., Vaccaro, U.: On the Size of Shares for Secret Sharing Schemes. J. Cryptology 6, 157–167 (1993)
9. Csirmaz, L.: The size of a share must be large. J. Cryptology 10, 223–231 (1997)
10. Csirmaz, L.: An impossibility result on graph secret sharing. Des. Codes Cryptogr. 53, 195–209 (2009)
11. Dougherty, R., Freiling, C., Zeger, K.: Six new non-Shannon information inequalities. In: 2006 IEEE International Symposium on Information Theory, pp. 233–236 (2006)
12. Dougherty, R., Freiling, C., Zeger, K.: Linear rank inequalities on five or more variables. Available at arXiv.org, arXiv:0910.0284v3 (2009)
13. Dougherty, R., Freiling, C., Zeger, K.: Non-Shannon Information Inequalities in Four Random Variables. Available at arXiv.org, arXiv:1104.3602v1 (2011)
14. Farràs, O., Metcalf-Burton, J.R., Padró, C., Vázquez, L.: On the Optimization of Bipartite Secret Sharing Schemes. Des. Codes Cryptogr. 63, 255–271 (2012)
15. Fujishige, S.: Polymatroidal Dependence Structure of a Set of Random Variables. Information and Control 39, 55–72 (1978)
16. Fujishige, S.: Entropy functions and polymatroids—combinatorial structures in information theory. Electron. Comm. Japan 61, 14–18 (1978)
17. Gál, A.: A characterization of span program size and improved lower bounds for monotone span programs. Comput. Complexity 10, 277–296 (2001)
18. Hammer, D., Romashchenko, A.E., Shen, A., Vereshchagin, N.K.: Inequalities for Shannon entropy and Kolmogorov complexity. Journal of Computer and Systems Sciences 60, 442–464 (2000)
19. Ingleton, A.W.: Representation of matroids. In: Welsh, D.J.A. (ed.) Combinatorial Mathematics and its Applications, pp. 149–167. Academic Press, London (1971)
20. Ito, M., Saito, A., Nishizeki, T.: Secret sharing scheme realizing any access structure. In: Proc. IEEE Globecom 1987, pp. 99–102 (1987)
21. Jackson, W.A., Martin, K.M.: Perfect secret sharing schemes on five participants. Des. Codes Cryptogr. 9, 267–286 (1996)
22. Kinser, R.: New inequalities for subspace arrangements. J. Combin. Theory Ser. A. 118, 152–161 (2011)
23. Matúš, F.: Infinitely many information inequalities. In: Proc. IEEE International Symposium on Information Theory, ISIT, pp. 2101–2105 (2007)
24. Metcalf-Burton, J.R.: Improved upper bounds for the information rates of the secret sharing schemes induced by the Vámos matroid. Discrete Math. 311, 651–662 (2011)
25. Padró, C., Vázquez, L., Yang, A.: Finding Lower Bounds on the Complexity of Secret Sharing Schemes by Linear Programming. Discrete Applied Mathematics 161, 1072–1084 (2013)
26. Schrijver, A.: Combinatorial optimization. Polyhedra and efficiency. Springer, Berlin (2003)
27. Shamir, A.: How to share a secret. Commun. of the ACM 22, 612–613 (1979)
28. Welsh, D.J.A.: Matroid Theory. Academic Press, London (1976)
29. Zhang, Z.: On a new non-Shannon type information inequality. Commun. Inf. Syst. 3, 47–60 (2003)
30. Zhang, Z., Yeung, R.W.: On characterization of entropy function via information inequalities. IEEE Trans. Inform. Theory 44, 1440–1452 (1998)

Linearly Homomorphic Structure-Preserving Signatures and Their Applications

Benoît Libert[1], Thomas Peters[2,*], Marc Joye[1], and Moti Yung[3]

[1] Technicolor, France
[2] Université catholique de Louvain, Crypto Group, Belgium
[3] Google Inc. and Columbia University, USA

Abstract. Structure-preserving signatures (SPS) are signature schemes where messages, signatures and public keys all consist of elements of a group over which a bilinear map is efficiently computable. This property makes them useful in cryptographic protocols as they nicely compose with other algebraic tools (like the celebrated Groth-Sahai proof systems). In this paper, we consider SPS systems with homomorphic properties and suggest applications that have not been provided before (in particular, not by employing ordinary SPS). We build linearly homomorphic structure-preserving signatures under simple assumptions and show that the primitive makes it possible to verify the calculations performed by a server on outsourced encrypted data (*i.e.*, combining secure computation and authenticated computation to allow reliable and secure cloud storage and computation, while freeing the client from retaining cleartext storage). Then, we give a generic construction of non-malleable (and actually simulation-sound) commitment from any linearly homomorphic SPS. This notably provides the first constant-size non-malleable commitment to group elements.

Keywords: Structure-preserving cryptography, signatures, homomorphism, commitment schemes, non-malleability.

1 Introduction

Composability is an important cryptographic design notion for building systems and protocols. Inside protocols, cryptographic tools need to compose well with each other in order to be used in combination. Structure-preserving cryptography [3], in turn, is a recent paradigm that takes care of composing algebraic tools, and primarily within groups supporting bilinear maps to allow smooth composition with the Groth-Sahai proof systems [41]. The notion allows for modular and simplified designs of various cryptographic protocols and primitives. In the last three years, a large body of work has analyzed the feasibility and the efficiency of structure-preserving signatures (SPS) [40,25,34,1,3,4,17,26,44,5,6], public-key encryption [18] and commitments schemes [42,2].

* This author was supported by the CAMUS Walloon Region Project.

R. Canetti and J.A. Garay (Eds.): CRYPTO 2013, Part II, LNCS 8043, pp. 289–307, 2013.
© International Association for Cryptologic Research 2013

In this paper, we consider SPS schemes with linearly homomorphic properties and argue that such primitives have many applications, even independently of Groth-Sahai proofs. Let us next review our results and then review related work.

1.1   Our Contributions

LINEARLY HOMOMORPHIC STRUCTURE-PRESERVING SIGNATURES. In this paper, we put forth the notion of linearly homomorphic structure-preserving signatures (homomorphic signatures and structure-preserving signatures have been defined before, as we review in the sequel, but the combination of the earlier notions is useful and non-trivial). These signature schemes function exactly like ordinary homomorphic signatures with the additional restriction that signatures and messages only consist of (vectors of) group elements whose discrete logarithms may not be available. We describe three constructions and prove their security under established complexity assumptions in symmetric bilinear groups.

APPLICATIONS. As in all SPS systems, the structure-preserving property makes it possible to efficiently prove knowledge of a homomorphic signature on a committed vector. However, as indicated above, we describe applications of linearly homomorphic SPS beyond their compatibility with the Groth-Sahai techniques.

First, we show that the primitive enables verifiable computation mechanisms on encrypted data.[4] Specifically, it allows a client to store encrypted files on an untrusted remote server. While the dataset is encrypted using an additively homomorphic encryption scheme, the server is able to blindly compute linear functions on the original data and provide the client with a short homomorphically derived signature vouching for the correctness of the computation. This is achieved by having the client sign each ciphertext using a homomorphic SPS scheme and handing the resulting signatures to the server at the beginning. After this initial phase, the client only needs to store a short piece of information, no matter how large the file is. Still, he remains able to authenticate linear functions on his data and the whole process is fully non-interactive. The method extends when datasets are encrypted using a CCA1-secure encryption schemes. Indeed, we will observe that linearly homomorphic SPS schemes yield simple homomorphic IND-CCA1-secure cryptosystems with publicly verifiable ciphertexts.

As a second and perhaps more surprising application, we show that linearly homomorphic SPS schemes generically yield non-malleable [31] trapdoor commitments to group elements. We actually construct a simulation-sound trapdoor commitment [35] — a primitive known (by [35,47]) to imply re-usable non-malleable commitments with respect to opening [28] — from any linearly homomorphic SPS satisfying a relatively mild condition. To our knowledge, we thus obtain the first constant-size trapdoor commitments to group elements providing re-usable non-malleability with respect to opening. Previous non-interactive commitments to group elements were either malleable [41,42] or inherently length-increasing [32]:

---

[4] Our goals are very different from those of [37], where verifiable computation on homomorphically encrypted data is also considered. We do not seek to outsource computation but rather save the client from storing large datasets.

if we disregard the trivial solution consisting of hashing the message first (which is not an option when we want to allow for efficient proofs of knowledge of an opening), no general technique has been known, to date, for committing to many group elements at once using a short commitment.

In the structure-preserving case, our transformation is purely generic as it applies to a template which any linearly homomorphic SPS necessarily satisfies in symmetric bilinear groups. We also generalize the construction so as to build simulation-sound trapdoor commitments to vectors from any pairing-based (non-structure-preserving) linearly homomorphic signature. In this case, the conversion is only semi-generic as it imposes conditions which are only met by pairing-based systems for the time being: essentially, we need the underlying signature scheme to operate over groups of finite, public order. While only partially generic, this construction of non-malleable commitments from linearly homomorphic signatures is somewhat unexpected considering that the terms "non-malleability" and "homomorphism" are antagonistic, and may be considered incompatible.

TECHNIQUES AND IDEAS. At first, the very name of our primitive may sound almost self-contradictory when it comes to formally define its security. Indeed, the security of a linearly homomorphic scheme [14] notably requires that it be infeasible to publicly compute a signature on a vector outside the linear span of originally signed vectors. The problem is that, when vector entries live in a discrete-logarithm hard group, deciding whether several vectors are independent or not is believed to be a hard problem. Yet, this will not prevent us from applying new techniques and constructing schemes with security proofs under simple assumptions and the reduction will be able to detect when the adversary has won by simply solving the problem instance it received as input.

Our first scheme's starting point is the one-time (regular) SPS scheme of Abe et al. [1]. By removing certain public key components, we obtain the desired linear homomorphism, and prove the security using information-theoretic arguments as in [1]. The key observation here is that, as long as the adversary does not output a signature on a linear combination of previously signed vectors, it will be unable to sign its target vector in the same way as the reduction would, because certain private key components will remain perfectly hidden.

Our initial scheme inherits the one-time restriction of the scheme in [1] in that only one linear subspace can be safely signed with a given public key. Nevertheless, we can extend it to build a full linearly homomorphic SPS system. To this end, we suitably combine our first scheme with Waters signatures [51]. Here, Waters signatures are used as a resting ground for fresh random exponents which are introduced in each signed vector and help us refresh the state of the system and apply each time the same argument as in the one-time scheme. We also present techniques to turn the scheme into a fully randomizable one, where a derived signature has the same distribution as a directly signed message.

In our simulation-sound commitments to group elements, the commitment generation technique appeals to the verification algorithm of the signature scheme, and proceeds by evaluating the corresponding pairing-product equations on the message, but using random group elements instead of actual signatures. The binding and simulation-binding properties, in turn, stem from the infeasibility of forging signatures while the signature homomorphism allows equivocating fake commitments when simulating the view of an adversary. It was already known how to build simulation-sound and non-malleable commitments [35,47,28,36,21] from signature schemes with efficient Σ protocols. Our method is, in fact, different and immediately yields length-reducing structure-preserving commitments to vectors without using Σ protocols.

1.2   Related Work

STRUCTURE-PRESERVING SIGNATURES. Signature schemes where messages only consist of group elements appeared for the first time — without the "structure-preserving" terminology— as ingredients of Groth's construction [40] of group signatures in the standard model. The scheme of [40] was mostly a proof of concept, with signatures consisting of thousands of group elements. More efficient realizations were given by Cathalo, Libert and Yung [25] and Fuchsbauer [34]. Abe, Haralambiev and Ohkubo [1,3] subsequently showed how to sign messages of n group elements at once using $O(1)$-size signatures. Lower bounds on the size of structure-preserving signatures were given in [4] while Abe *et al.* [7] provided evidence that optimally short SPS necessarily rely on interactive assumptions. As an ingredient for their tightly secure cryptosystems, Hofheinz and Jager [44] gave constructions based on the Decision Linear assumption [13] while similar results were independently achieved in [17,26]. Quite recently, Abe *et al.* [5,6] obtained constant-size signatures without sacrificing the security guarantees offered by security proofs under simple assumptions.

Regarding primitives beyond signature schemes, Camenisch *et al.* [18] showed a structure-preserving variant of the Cramer-Shoup cryptosystem [27] and used it to implement oblivious third parties [19]. Groth [42] described length-reducing trapdoor commitments (*i.e.*, where the commitment is shorter than the committed message) to group elements whereas [2] showed the impossibility of realizing such commitments when the commitment string lives in the same group as the message. Sakai *et al.* [49] recently suggested to use structure-preserving identity-based encryption [50] systems to restrict the power of the opening authority in group signatures.

LINEARLY HOMOMORPHIC SIGNATURES. The concept of homomorphic signatures can be traced back to Desmedt [30] while proper definitions remained lacking until the work of Johnson *et al.* [46]. Since then, constructions have appeared for various kinds of homomorphisms (see [8] and references therein).

Linearly homomorphic signatures are an important class of homomorphic signatures for arithmetic functions, whose study was initiated by Boneh, Freeman, Katz and Waters [14]. While initially motivated by applications to network coding [14], they are also useful in proofs of storage [9] or in verifiable computation mechanisms, when it comes to authenticate servers' computations on outsourced data (see, *e.g.*, [8]). The recent years, much attention was given to the notion and a variety of constructions [38,10,15,16,23,24,33,11,12] based on various assumptions have been studied.

1.3 Organization

Section 2 first gives security definitions for linearly homomorphic SPS systems, for which efficient constructions are provided in Section 3. Their applications to verifiable computation on encrypted data are explained in Section 4 while Section 5 shows how to build simulation-sound commitments to group elements.

2 Background

2.1 Definitions for Linearly Homomorphic Signatures

Let $(\mathbb{G}, \mathbb{G}_T)$ be a configuration of (multiplicatively written) groups of prime order p over which a bilinear map $e : \mathbb{G} \times \mathbb{G} \to \mathbb{G}_T$ is efficiently computable.

Following [1,3], we say that a signature scheme is *structure-preserving* if messages, signature components and public keys live in the group $\mathbb{G}$.

We consider linearly homomorphic signatures for which the message space $\mathcal{M}$ consists of pairs $\mathcal{M} := \mathcal{T} \times \mathbb{G}^n$, for some $n \in \mathbb{N}$, where $\mathcal{T}$ is a tag space. We remark that, in the applications considered in this paper, tags do not need to be group elements. We thus allow them to be arbitrary strings.

Definition 1. *A linearly homomorphic structure-preserving signature scheme over $(\mathbb{G}, \mathbb{G}_T)$ is a set of efficient algorithms $\Sigma = (\mathsf{Keygen}, \mathsf{Sign}, \mathsf{SignDerive}, \mathsf{Verify})$ for which the message space is $\mathcal{M} := \mathcal{T} \times \mathbb{G}^n$, for some $n \in \mathsf{poly}(\lambda)$ and some set $\mathcal{T}$, and with the following specifications.*

Keygen(λ, n): *is a randomized algorithm that takes in a security parameter $\lambda \in \mathbb{N}$ and an integer $n \in \mathsf{poly}(\lambda)$ denoting the dimension of vectors to be signed. It outputs a key pair $(\mathsf{pk}, \mathsf{sk})$ and the description of a tag (i.e., a file identifier) space $\mathcal{T}$.*

Sign$(\mathsf{sk}, \tau, \vec{M})$: *is a possibly probabilistic algorithm that takes as input a private key sk, a file identifier $\tau \in \mathcal{T}$ and a vector $\vec{M} \in \mathbb{G}^n$. It outputs a signature $\sigma \in \mathbb{G}^{n_s}$, for some $n_s \in \mathsf{poly}(\lambda)$.*

SignDerive$(\mathsf{pk}, \tau, \{(\omega_i, \sigma^{(i)})\}_{i=1}^{\ell})$: *is a (possibly probabilistic) signature derivation algorithm. It takes as input a public key pk, a file identifier τ as well as ℓ pairs $(\omega_i, \sigma^{(i)})$, each of which consists of a weight $\omega_i \in \mathbb{Z}_p$ and a signature $\sigma^{(i)} \in \mathbb{G}^{n_s}$. The output is a signature $\sigma \in \mathbb{G}^{n_s}$ on the vector $\vec{M} = \prod_{i=1}^{\ell} \vec{M}_i^{\,\omega_i}$, where $\sigma^{(i)}$ is a signature on $\vec{M}_i$.*

Verify$(\mathsf{pk}, \tau, \vec{M}, \sigma)$: *is a deterministic algorithm that takes in a public key pk, a file identifier $\tau \in \mathcal{T}$, a signature σ and a vector $\vec{M}$. It outputs 1 if σ is deemed valid and 0 otherwise.*

Correctness is expressed by imposing that, for all $\lambda \in \mathbb{N}$, all integers $n \in \mathsf{poly}(\lambda)$ and all triples $(\mathsf{pk}, \mathsf{sk}, \mathcal{T}) \leftarrow \mathsf{Keygen}(\lambda, n)$, the following holds:

1. For all $\tau \in \mathcal{T}$ and all n-vectors $\vec{M}$, if $\sigma = \mathsf{Sign}(\mathsf{sk}, \tau, \vec{M})$, then we have $\mathsf{Verify}(\mathsf{pk}, \tau, \vec{M}, \sigma) = 1$.

2. For all $\tau \in \mathcal{T}$, $\ell > 0$ and $\{(\omega_i, \sigma^{(i)}, \vec{M}_i)\}_{i=1}^{\ell}$, if $\mathsf{Verify}(\mathsf{pk}, \tau, \vec{M}_i, \sigma^{(i)}) = 1$ for each i, then $\mathsf{Verify}\big(\mathsf{pk}, \tau, \prod_{i=1}^{\ell} \vec{M}_i^{\omega_i}, \mathsf{SignDerive}(\mathsf{pk}, \tau, \{(\omega_i, \sigma^{(i)})\}_{i=1}^{\ell})\big) = 1$.

SECURITY. In linearly homomorphic signatures, we use the same definition of unforgeability as in [11]. This definition implies security in the stronger model used by Freeman [33] since the adversary can interleave signing queries for individual vectors belonging to distinct subspaces. Moreover, file identifiers can be chosen by the adversary (which strengthens the definition of [14]) and are not assumed to be uniformly distributed. As a result, a file identifier can be a low-entropy, easy-to-remember string such as the name of the dataset's owner.

Definition 2. *A linearly homomorphic SPS scheme* $\Sigma = (\mathsf{Keygen}, \mathsf{Sign}, \mathsf{Verify})$ *is secure if no PPT adversary has non-negligible advantage in the game below:*

1. *The adversary* $\mathcal{A}$ *chooses an integer* $n \in \mathbb{N}$ *and sends it to the challenger who runs* $\mathsf{Keygen}(\lambda, n)$ *and obtains* $(\mathsf{pk}, \mathsf{sk})$ *before sending* pk *to* $\mathcal{A}$.
2. *On polynomially-many occasions,* $\mathcal{A}$ *can interleave the following kinds of queries.*
   - *Signing queries:* $\mathcal{A}$ *chooses a tag* $\tau \in \mathcal{T}$ *and a vector* $\vec{M} \in \mathbb{G}^n$. *The challenger picks a handle* h *and computes* $\sigma \leftarrow \mathsf{Sign}(\mathsf{sk}, \tau, \vec{M})$. *It stores* $(\mathsf{h}, (\tau, \vec{M}, \sigma))$ *in a table* T *and returns* h.
   - *Derivation queries:* $\mathcal{A}$ *chooses a vector of handles* $\vec{\mathsf{h}} = (\mathsf{h}_1, \ldots, \mathsf{h}_k)$ *and a set of coefficients* $\{\omega_i\}_{i=1}^{k}$. *The challenger retrieves* $\{(\mathsf{h}_i, (\tau, \vec{M}_i), \sigma^{(i)})\}_{i=1}^{k}$ *from* T *and returns* $\bot$ *if one of these does not exist or if there exists* $i \in \{1, \ldots, k\}$ *such that* $\tau_i \neq \tau$. *Otherwise, it computes* $\vec{M} = \prod_{i=1}^{k} \vec{M}_i^{\omega_i}$ *and runs* $\sigma' \leftarrow \mathsf{SignDerive}(\mathsf{pk}, \tau, \{(\omega_i, \sigma^{(i)})\}_{i=1}^{k})$. *It also chooses a handle* h', *stores* $(\mathsf{h}', (\tau, \vec{M}), \sigma')$ *in* T *and returns* h' *to* $\mathcal{A}$.
   - *Reveal queries:* $\mathcal{A}$ *chooses a handle* h. *If no tuple of the form* $(\mathsf{h}, (\tau, \vec{M}), \sigma')$ *exists in* T, *the challenger returns* $\bot$. *Otherwise, it returns* σ' *to* $\mathcal{A}$ *and adds* $((\tau, \vec{M}), \sigma')$ *to the set* Q.
3. $\mathcal{A}$ *outputs an identifier* $\tau^\star$, *a signature* $\sigma^\star$ *and a vector* $\vec{M}^\star \in \mathbb{G}^n$. *The adversary* $\mathcal{A}$ *wins if* $\mathsf{Verify}(\mathsf{pk}, \tau^\star, \vec{M}^\star, \sigma^\star) = 1$ *and one of the conditions below is satisfied:*
   - *(Type I):* $\tau^\star \neq \tau_i$ *for any entry* $(\vec{\tau}_i, .)$ *in* Q *and* $\vec{M}^\star \neq (1_{\mathbb{G}}, \ldots, 1_{\mathbb{G}})$.
   - *(Type II):* $\tau^\star = \tau_i$ *for* $k_i > 0$ *entries* $(\tau_i, .)$ *in* Q *and* $\vec{M}^\star \notin V_i$, *where* V_i *denotes the subspace spanned by all vectors* $\vec{M}_1, \ldots, \vec{M}_{k_i}$ *for which an entry of the form* $(\tau^\star, \vec{M}_j)$, *with* $j \in \{1, \ldots, k_i\}$, *appears in* Q.

$\mathcal{A}$'s *advantage is its probability of success taken over all coin tosses.*

In our first scheme, we will consider a weaker notion of *one-time* security. In this notion, the adversary is limited to obtain signatures for only *one* linear subspace. In this case, there is no need for file identifiers and we assume that all vectors are assigned the identifier $\tau = \varepsilon$.

In the following, the adversary will be said *independent* if

- For any given tag τ, it is restricted to only query signatures on linearly independent vectors.
- Each vector is only queried at most once.

Non-independent adversaries are not subject to the above restrictions. It will be necessary to consider these adversaries in our construction of non-malleable commitments. Nevertheless, security against independent adversaries suffices for many applications — including encrypted cloud storage— since the signer can always append unit vectors to each newly signed vector.

At first, one may wonder how Definition 2 can be satisfied at all given that the challenger may not have an efficient way to check whether the adversary is successful. Indeed, in cryptographically useful discrete-logarithm-hard groups $\mathbb{G}$, deciding whether vectors $\{\vec{M}_i\}_i$ of $\mathbb{G}^n$ are linearly dependent is believed to be difficult when $n > 2$. However, it may be possible using some trapdoor information embedded in pk, especially if the adversary additionally outputs signatures on $\{\vec{M}_i\}_i$.

2.2 Hardness Assumptions

We rely on the following hardness assumptions, the first of which implies the second one.

Definition 3 ([13]). *In a group $\mathbb{G}$ of prime order p, the* Decision Linear Problem *(DLIN), consists in distinguishing the distributions $(g^a, g^b, g^{ac}, g^{bd}, g^{c+d})$ and $(g^a, g^b, g^{ac}, g^{bd}, g^z)$, with $a, b, c, d \xleftarrow{R} \mathbb{Z}_p^*$, $z \xleftarrow{R} \mathbb{Z}_p^*$. The* Decision Linear Assumption *is the intractability of DLIN for any PPT distinguisher $\mathcal{D}$.*

Definition 4. *The* Simultaneous Double Pairing *problem (SDP) in $(\mathbb{G}, \mathbb{G}_T)$ is, given a tuple of elements $(g_z, g_r, h_z, h_u) \in_R \mathbb{G}^4$, to find a non-trivial triple $(z, r, u) \in \mathbb{G}^3 \backslash \{(1_\mathbb{G}, 1_\mathbb{G}, 1_\mathbb{G})\}$ satisfying the equalities $e(g_z, z) \cdot e(g_r, r) = 1_{\mathbb{G}_T}$ and $e(h_z, z) \cdot e(h_u, u) = 1_{\mathbb{G}_T}$.*

3   Constructions of Linearly Homomorphic Structure-Preserving Signatures

As a warm-up, we begin by describing a one-time homomorphic signature, where a given public key allows signing only *one* linear subspace.

3.1   A One-Time Linearly Homomorphic Construction

In the description hereunder, since only one linear subspace can be signed for each public key, no file identifier τ is used. We thus set τ to be the empty string ε in all algorithms.

Keygen(λ, n): given a security parameter λ and the dimension $n \in \mathbb{N}$ of the subspace to be signed, choose bilinear group $(\mathbb{G}, \mathbb{G}_T)$ of prime order $p > 2^\lambda$. Then, choose generators $h, g_z, g_r, h_z \xleftarrow{R} \mathbb{G}$. Pick $\chi_i, \gamma_i, \delta_i \xleftarrow{R} \mathbb{Z}_p$, for $i = 1$ to n. Then, for each $i \in \{1, \ldots, n\}$, compute $g_i = g_z^{\chi_i} g_r^{\gamma_i}$, $h_i = h_z^{\chi_i} h^{\delta_i}$. The private key is $sk = \{\chi_i, \gamma_i, \delta_i\}_{i=1}^n$ while the public key is defined to be

$$\mathsf{pk} = \left(g_z, \ h_r, \ h_z, \ h, \ \{g_i, h_i\}_{i=1}^n \right) \in \mathbb{G}^{2n+4}.$$

Sign($sk, \tau, (M_1, \ldots, M_n)$): to sign a vector $(M_1, \ldots, M_n) \in \mathbb{G}^n$ associated with the identifier $\tau = \varepsilon$ using $sk = \{\chi_i, \gamma_i, \delta_i\}_{i=1}^n$, compute the signature consists of $\sigma = (z, r, u) \in \mathbb{G}^3$, where

$$z = \prod_{i=1}^n M_i^{-\chi_i}, \qquad r = \prod_{i=1}^n M_i^{-\gamma_i}, \qquad u = \prod_{i=1}^n M_i^{-\delta_i}.$$

SignDerive($\mathsf{pk}, \tau, \{(\omega_i, \sigma^{(i)})\}_{i=1}^\ell$): given a file identifier $\tau = \varepsilon$, the public key pk and ℓ tuples $(\omega_i, \sigma^{(i)})$, parse each $\sigma^{(i)}$ as $\sigma^{(i)} = (z_i, r_i, u_i) \in \mathbb{G}^3$ for $i = 1$ to ℓ. Compute and return $\sigma = (z, r, u) = \left(\prod_{i=1}^\ell z_i^{\omega_i}, \prod_{i=1}^\ell r_i^{\omega_i}, \prod_{i=1}^\ell u_i^{\omega_i} \right)$.

Verify($\mathsf{pk}, \sigma, \tau, (M_1, \ldots, M_n)$): given a signature $\sigma = (z, r, u) \in \mathbb{G}^3$, a vector $(M_1, \ldots, M_n)$ and a file identifier $\tau = \varepsilon$, return 1 if and only if it holds that $(M_1, \ldots, M_n) \neq (1_{\mathbb{G}}, \ldots, 1_{\mathbb{G}})$ and (z, r, u) satisfy

$$1_{\mathbb{G}_T} = e(g_z, z) \cdot e(g_r, r) \cdot \prod_{i=1}^n e(g_i, M_i), \qquad 1_{\mathbb{G}_T} = e(h_z, z) \cdot e(h, u) \cdot \prod_{i=1}^n e(h_i, M_i).$$

The security proof relies on the fact that, while the signing algorithm is deterministic, signatures are not unique. However, the reduction will be able to compute exactly one signature for each vector. At the same time, an adversary has no information about which signature the legitimate signer would compute on a vector outside the span of already signed vectors. Moreover, by obtaining two distinct signatures on a given vector, the reduction can solve a given SDP instance. The following theorem is proved in the full version of the paper.

Theorem 1. *The scheme is unforgeable if the SDP assumption holds in $(\mathbb{G}, \mathbb{G}_T)$.*

3.2 A Full-Fledged Linearly Homomorphic SPS Scheme

We upgrade our one-time construction so as to sign an arbitrary number of linear subspaces. Here, each file identifier τ is a L-bit string. The construction builds on the observation that, in the scheme of Section 3.1, signatures (z, r, u) could be re-randomized by computing $(z \cdot g_r^\theta, r \cdot g_z^{-\theta}, u \cdot h_z^{-\log_h(g_r) \cdot \theta})$, with $\theta \xleftarrow{R} \mathbb{Z}_p$, if $h_z^{-\log_h(g_r)}$ were available. Since publicizing $h_z^{-\log_h(g_r)}$ would render the scheme insecure, our idea is to use Waters signatures as a support for introducing extra randomizers in the exponent.

In the construction, the u component of each signature can be seen as an aggregation of the one-time signature of Section 3.1 with a Waters signature $(h_z^{\log_h(g_r)} \cdot H_{\mathbb{G}}(\tau)^{-\rho}, h^\rho)$ [51] on the tag τ.

Keygen(λ, n): given a security parameter λ and the dimension $n \in \mathbb{N}$ of the subspace to be signed, choose bilinear group $(\mathbb{G}, \mathbb{G}_T)$ of prime order $p > 2^\lambda$.

1. Choose $h \xleftarrow{R} \mathbb{G}$, $\alpha_z, \alpha_r, \beta_z \xleftarrow{R} \mathbb{Z}_p$. Define $g_z = h^{\alpha_z}$, $g_r = h^{\alpha_r}$, $h_z = h^{\beta_z}$.
2. For $i = 1$ to n, pick $\chi_i, \gamma_i, \delta_i \xleftarrow{R} \mathbb{Z}_p$ and compute $g_i = g_z^{\chi_i} g_r^{\gamma_i}$, $h_i = h_z^{\chi_i} h^{\delta_i}$.
3. Choose a random vector $\overline{\mathbf{w}} = (w_0, w_1, \ldots, w_L) \xleftarrow{R} \mathbb{G}^{L+1}$. The latter defines a hash function $H_{\mathbb{G}} : \{0,1\}^L \to \mathbb{G}$ which maps any L-bit string $\tau = \tau[1] \ldots \tau[L] \in \{0,1\}^L$ to $H_{\mathbb{G}}(\tau) = w_0 \cdot \prod_{k=1}^L w_k^{\tau[k]}$.

The private key is $\mathsf{sk} = \left(h_z^{\alpha_r}, \{\chi_i, \gamma_i, \delta_i\}_{i=1}^n \right)$ while the public key consists of

$$\mathsf{pk} = \left(g_z, \ g_r, \ h_z, \ h, \ \{g_i, h_i\}_{i=1}^n, \ \overline{\mathbf{w}} \right) \in \mathbb{G}^{2n+4} \times \mathbb{G}^{L+1}.$$

Sign$(\mathsf{sk}, \tau, (M_1, \ldots, M_n))$: to sign a vector $(M_1, \ldots, M_n) \in \mathbb{G}^n$ w.r.t. the file identifier τ using $sk = \left(h_z^{\alpha_r}, \{\chi_i, \gamma_i, \delta_i\}_{i=1}^n \right)$, choose $\theta, \rho \xleftarrow{R} \mathbb{Z}_p$ and output $\sigma = (z, r, u, v) \in \mathbb{G}^4$, where

$$z = g_r^\theta \cdot \prod_{i=1}^n M_i^{-\chi_i} \qquad\qquad r = g_z^{-\theta} \cdot \prod_{i=1}^n M_i^{-\gamma_i}$$

$$u = (h_z^{\alpha_r})^{-\theta} \cdot \prod_{i=1}^n M_i^{-\delta_i} \cdot H_{\mathbb{G}}(\tau)^{-\rho} \qquad\qquad v = h^\rho$$

SignDerive$(\mathsf{pk}, \tau, \{(\omega_i, \sigma^{(i)})\}_{i=1}^\ell)$: given pk, a file identifier τ and ℓ tuples $(\omega_i, \sigma^{(i)})$, parse $\sigma^{(i)}$ as $\sigma^{(i)} = (z_i, r_i, u_i, v_i) \in \mathbb{G}^4$ for $i = 1$ to ℓ. Then, choose $\rho' \xleftarrow{R} \mathbb{Z}_p$ and compute and return $\sigma = (z, r, u, v)$, where $z = \prod_{i=1}^\ell z_i^{\omega_i}$, $r = \prod_{i=1}^\ell r_i^{\omega_i}$, $u = \prod_{i=1}^\ell u_i^{\omega_i} \cdot H_{\mathbb{G}}(\tau)^{-\rho'}$ and $v = \prod_{i=1}^\ell v_i^{\omega_i} \cdot h^{\rho'}$.

Verify$(\mathsf{pk}, \sigma, \tau, (M_1, \ldots, M_n))$: given a signature $\sigma = (z, r, u, v) \in \mathbb{G}^4$, a file identifier τ and a vector $(M_1, \ldots, M_n) \in \mathbb{G}^n$, return 1 if and only if $(M_1, \ldots, M_n) \neq (1_{\mathbb{G}}, \ldots, 1_{\mathbb{G}})$ and (z, r, u, v) satisfy

$$1_{\mathbb{G}_T} = e(g_z, z) \cdot e(g_r, r) \cdot \prod_{i=1}^n e(g_i, M_i), \tag{1}$$

$$1_{\mathbb{G}_T} = e(h_z, z) \cdot e(h, u) \cdot e(H_{\mathbb{G}}(\tau), v) \cdot \prod_{i=1}^n e(h_i, M_i).$$

The security of the scheme against *non-independent* Type I adversaries is proved under the SDP assumption. In the case of Type II forgeries, we need to assume the adversary to be independent because, at some point, the simulator is only able to compute a signature for a unique value[5] of θ.

Theorem 2. *The scheme is unforgeable against independent adversaries if the SDP assumption holds in $(\mathbb{G}, \mathbb{G}_T)$. Moreover, the scheme is secure against non-independent Type I adversaries.*

---

[5] Note that this is not a problem since the signer can derive θ as a pseudorandom function of τ and $(M_1, \ldots, M_n)$ to make sure that a given vector is always signed using the same θ.

The proof of Theorem 2 is available in the full verison of the paper. It uses Waters signatures as a handle to randomize signatures. Whenever the reduction is able to compute a Waters signatures $(h_z^{\alpha r} \cdot H_{\mathbb{G}}(\tau)^{-\rho}, h^{\rho})$ on the tag τ, it can inject a fresh extra randomizer $\theta \in \mathbb{Z}_p$ in the exponent for each vector associated with τ. By doing so, with non-negligible probability, the specific vector $(\chi_1, \ldots, \chi_n)$ used by the reduction will remain undetermined from $\mathcal{A}$'s view.

Since the signature component u cannot be publicly randomized, the scheme does not have fully randomizable signatures. In the full version of the paper, we describe a fully randomizable variant. In applications like non-malleable commitments to group elements, the above scheme is sufficient however.

4 Applications

4.1 Verifiable Computation for Encrypted Cloud Storage

Linearly homomorphic schemes are known (see, e.g., [8]) to provide verifiable computation mechanisms for outsourced data. Suppose that a user has a dataset consisting of n samples $s_1, \ldots, s_n \in \mathbb{Z}_p$. The dataset can be encoded as vectors $\vec{v}_i = (\vec{e}_i | s_i) \in \mathbb{Z}_p^{n+1}$, where $\vec{e}_i \in \mathbb{Z}_p^n$ denotes the i-th unit vector for each $i \in \{1, \ldots, n\}$. The user then assigns a file identifier τ to $\{\vec{v}_i\}_{i=1}^n$, computes signatures $\sigma_i \leftarrow \mathsf{Sign}(\mathsf{sk}, \tau, \vec{v}_i)$ on the resulting vectors and stores $\{(\vec{v}_i, \sigma_i)\}_{i=1}^n$ at the server. When requested, the server can then evaluate a sum $s = \sum_{i=1}^n s_i$ and provide evidence that the latter computation is correct by deriving a signature on the vector $(1, 1, \ldots, 1, s) \in \mathbb{Z}_p^{n+1}$. Unless the server is able to forge a signature for a vector outside the span of $\{\vec{v}_i\}_{i=1}^n$, it is unable to fool the user. The above method readily extends to authenticate weighted sums or Fourier transforms.

One disadvantage of the above method is that it requires the server to retain the dataset $\{s_i\}_{i=1}^n$ in the clear. Using linearly homomorphic structure-preserving signatures, the user can apply the above technique on encrypted samples using the Boneh-Boyen-Shacham (BBS) cryptosystem [13].

The BBS cryptosystem involves a public key $(g, \tilde{g}, f = g^x, h = g^y) \in_R \mathbb{G}^4$, where $(x, y) \in \mathbb{Z}_p^2$ is the private key. The user (or anyone else knowing his public key) can first encrypt his samples $\{s_i\}_{i=1}^n$ by computing BBS encryptions $(C_{1,i}, C_{2,i}, C_{3,i}) = (f^{r_i}, h^{t_i}, \tilde{g}^{s_i} \cdot g^{r_i+t_i})$, with $r_i, t_i \xleftarrow{R} \mathbb{Z}_p$, for each $i \in \{1, \ldots, n\}$. If the user holds a linearly homomorphic structure preserving signature key pair for vectors of dimension $n + 3$, he can generate n signatures on vectors $((C_{1,i}, C_{2,i}, C_{3,i})|\vec{E}_i) \in \mathbb{G}^{n+3}$, where $\vec{E}_i = (1_{\mathbb{G}}, \ldots, 1_{\mathbb{G}}, g, 1_{\mathbb{G}}, \ldots, 1_{\mathbb{G}}) = g^{\vec{e}_i}$ for each $i \in \{1, \ldots, n\}$. The vectors $\{((C_{1,i}, C_{2,i}, C_{3,i})|\vec{E}_i)\}_{i=1}^n$ are then archived in the cloud with their signatures $\{(z_i, r_i, u_i, v_i)\}_{i=1}^n$ in such a way that the server can publicly derive a signature on $\left(f^{\sum_i r_i}, h^{\sum_i t_i}, \tilde{g}^{\sum_i s_i} \cdot g^{\sum_i (r_i+t_i)}, g, g, \ldots, g \right) \in \mathbb{G}^{n+3}$ in order to convince the client that the encrypted sum was correctly computed. Using his private key (x, y), the client can then retrieve the sum $\sum_i s_i$ as long as it remains in a sufficiently small range.

The interest of the above solution lies in that the client can dispense with the need for storing the $O(n)$-size public key of his linearly homomorphic signature.

Indeed, he can simply retain the random seed that was used to generate pk and re-compute private key elements $\{(\chi_i, \gamma_i, \delta_i)\}_{i=1}^n$ whenever he wants to verify the server's response. In this case, the verification equations (1) become

$$1_{\mathbb{G}_T} = e(g_z, z \cdot \prod_{i=1}^n M_i^{\chi_i}) \cdot e(g_r, r \cdot \prod_{i=1}^n M_i^{\gamma_i})$$

$$1_{\mathbb{G}_T} = e(h_z, z \cdot \prod_{i=1}^n M_i^{\chi_i}) \cdot e(h, u \cdot \prod_{i=1}^n M_i^{\delta_i}) \cdot e(H_{\mathbb{G}}(\tau), v),$$

so that the client only has to compute $O(1)$ pairings. Moreover, the client does not have to determine an upper bound on the size of his dataset when generating his public key. Initially, he only needs to generate $\{(g_j, h_j)\}_{j=1}^3$. When the i-th ciphertext $(C_{1,i}, C_{2,i}, C_{3,i})$ has to be stored, the client derives $(\chi_{i+3}, \gamma_{i+3}, \delta_{i+3})$ and (g_{i+3}, h_{i+3}) by applying a PRF to the index i. This will be sufficient to sign vectors of the form $((C_{1,i}, C_{2,i}, C_{3,i}) | \vec{E}_i)$.

In order to hide all partial information about the original dataset, the server may want to re-randomize the derived signature and ciphertext before returning them. This can be achieved by having the client include signatures on the vectors $(f, 1_{\mathbb{G}}, g, 1_{\mathbb{G}}, \ldots, 1_{\mathbb{G}})$, $(1_{\mathbb{G}}, h, g, 1_{\mathbb{G}}, \ldots, 1_{\mathbb{G}})$ in the outsourced dataset. Note that, in this case, the signature should be re-randomized as well. For this reason, our randomizable scheme described in the full version of the paper should be preferred.

Complete security models for "verifiable computation on encrypted data" are beyond the scope of this paper. Here, they would naturally combine the properties of secure homomorphic encryption and authenticated computing. It should be intuitively clear that a malicious server cannot trick a client into accepting an incorrect result (*i.e.*, one which differs from the actual defined linear function it is supposed to compute over the defined signed ciphertext inputs) without defeating the security of the underlying homomorphic signature.

4.2   Extension to CCA1-Encrypted Data

In the application of Section 4.1, the underlying crypotosystem has to be additively homomorphic, which prevents it from being secure against adaptive chosen-ciphertext attacks. On the other hand, the method is compatible with security against *non-adaptive* chosen ciphertext attacks. One possibility is to apply the "lite" Cramer-Shoup technique (in its variant based on DLIN) as it achieves CCA1-security while remaining homomorphic. Unfortunately, the validity of ciphertexts is not publicly verifiable, which may be annoying in applications like cloud storage or universally verifiable e-voting systems. Indeed, servers may be willing to have guarantees that they are actually storing encryptions of some message instead of random group elements.

Consider the system where $(C_1, C_2, C_3, C_4) = (f^r, h^t, g^{r+t}, \tilde{g}^m \cdot X_1^r \cdot X_2^t)$ is decrypted as $m = \log_{\tilde{g}}(C_4 \cdot C_1^{-x_1} C_2^{-x_2} C_3^{-z})$, where $X_1 = f^{x_1} g^z$ and $X_2 = $

$h^{x_2}g^z$ are part of the public key. In [45], such a system was made chosen-ciphertext secure using a *publicly* verifiable one-time simulation-sound proof that (f, h, g, C_1, C_2, C_3) forms a DLIN tuple. In the security proof, if the reduction is guaranteed not to leak $C_1^{-x_1}C_2^{-x_2}C_3^{-z}$ for an invalid triple (C_1, C_2, C_3) (*i.e.*, as long as the adversary is unable to generate a fake proof for this), the private key component z will remain perfectly hidden. Consequently, if the challenge ciphertext is computed by choosing $C_3^\star \in_R \mathbb{G}$ (so that $(f, h, g, C_1^\star, C_2^\star, C_3^\star)$ is not a DLIN tuple) and computing $C_4^\star = \tilde{g}^m \cdot C_1^{\star x_1} \cdot C_2^{\star x_2} \cdot C_3^{\star z}$, the plaintext m is independent of $\mathcal{A}$'s view. If we replace the one-time simulation-sound proofs by standard proofs of membership in the scheme of [45], we obtain a CCA1 homomorphic encryption scheme. Linearly homomorphic SPS schemes provide a simple and efficient way to do that.

The idea is to include in the public key the verification key of a one-time linearly homomorphic SPS — using the scheme of Section 3.1 — for $n = 3$ as well as signatures on the vectors $(f, 1_{\mathbb{G}}, g)$, $(1_{\mathbb{G}}, h, g) \in \mathbb{G}^3$. This will allow the sender to publicly derive a signature (z, r, u) on the vector $(C_1, C_2, C_3) = (f^r, h^t, g^{r+t})$. Each ciphertext thus consists of $(z, r, u, C_1, C_2, C_3, C_4)$. In the security proof, at each pre-challenge decryption query, the signature (z, r, u) serves as publicly verifiable evidence that (f, h, g, C_1, C_2, C_3) is a DLIN tuple. In the challenge phase, the reduction reveals another homomorphic signature $(z^\star, r^\star, u^\star)$ for a vector $(C_1^\star, C_2^\star, C_3^\star)$ that may be outside the span of $(f, 1_{\mathbb{G}}, g)$ and $(1_{\mathbb{G}}, h, g)$ but it does not matter since decryption queries are not allowed beyond this point.

We note that linearly homomorphic SPS can also be used to construct CCA1-secure homomorphic encryption schemes based on the Naor-Yung paradigm [48].

5 Non-malleable Trapdoor Commitments to Group Elements from Linearly Homomorphic Structure-Preserving Signatures

As noted in [42,43], some applications require to commit to group elements without knowing their discrete logarithms or destroying their algebraic structure by hashing them first. This section shows that, under a certain mild condition, linearly homomorphic SPS imply length-reducing non-malleable structure-preserving commitments to vectors of group elements.

As a result, we obtain the first length-reducing non-malleable structure-preserving trapdoor commitment. Our scheme is not *strictly*[6] structure-preserving (according to the terminology of [2]) because the commitment string lives in $\mathbb{G}_T$ rather than $\mathbb{G}$. Still, openings only consist of elements in $\mathbb{G}$, which makes it possible to generate efficient NIWI proofs that committed group elements satisfy certain properties. To our knowledge, the only known non-malleable commitment schemes whose openings only consist of group elements were described by

---

[6] We recall that strictly structure-preserving commitments cannot be length-reducing, as shown by Abe *et al.* [2], so that our scheme is essentially the best we can hope for if we aim at short commitment stings.

Fischlin *et al.* [32]. However, these constructions cannot be length-reducing as they achieve universal composability [20,22].

Our schemes are obtained by first constructing simulation-sound trapdoor commitments (SSTC) [35,47] to group elements. SSTC schemes were first suggested by Garay, MacKenzie and Yang [35] as a tool for constructing universally composable zero-knowledge proofs [20]. MacKenzie and Yang subsequently gave a simplified security definition which suffices to provide non-malleability with respect to opening in the sense of the definition of re-usable non-malleable commitments [28].

In a SSTC, each commitment is labeled with a tag. The definition of [47] requires that, even if the adversary can see equivocations of commitments to possibly distinct messages for several tags $tag_1, \ldots, tag_q$, it will not be able to break the binding property for a new tag $tag \notin \{tag_1, \ldots, tag_q\}$.

Definition 5 ([47]). *A simulation-sound trapdoor commitment is a tuple of algorithms* (Setup, Com, FakeCom, FakeOpen, Verify) *where* (Setup, Com, Verify) *forms a commitment scheme and* (FakeCom, FakeOpen) *are PPT algorithms with the following properties*

Trapdoor: *for any tag and any message* Msg, *the following distributions are computationally indistinguishable:*

$$D_{fake} := \{(pk, tk) \leftarrow \mathsf{Setup}(\lambda); (\widetilde{\mathsf{com}}, \mathsf{aux}) \leftarrow \mathsf{FakeCom}(pk, tk, tag);$$
$$\widetilde{\mathsf{dec}} \leftarrow \mathsf{FakeOpen}(\mathsf{aux}, tk, \widetilde{\mathsf{com}}, \mathsf{Msg}) : (pk, tag, \mathsf{Msg}, \widetilde{\mathsf{com}}, \widetilde{\mathsf{dec}})\}$$

$$D_{real} := \{(pk, tk) \leftarrow \mathsf{Setup}(\lambda); (\mathsf{com}, \mathsf{dec}) \leftarrow \mathsf{Com}(pk, tag, \mathsf{Msg}) :$$
$$(pk, tag, \mathsf{Msg}, \mathsf{com}, \mathsf{dec})\}$$

Simulation-sound binding: *for any PPT adversary* $\mathcal{A}$, *the following probability is negligible*

$$\Pr[\ (pk, tk) \leftarrow \mathsf{Setup}(\lambda); (\mathsf{com}, tag, \mathsf{Msg}_1, \mathsf{Msg}_2, \mathsf{dec}_1, \mathsf{dec}_2) \leftarrow \mathcal{A}^{\mathcal{O}_{tk,pk}}(pk) :$$
$$\mathsf{Msg}_1 \neq \mathsf{Msg}_2 \wedge \mathsf{Verify}(pk, tag, \mathsf{Msg}_1, \mathsf{com}, \mathsf{dec}_1) = 1$$
$$\wedge \mathsf{Verify}(pk, tag, \mathsf{Msg}_2, \mathsf{com}, \mathsf{dec}_2) = 1 \wedge tag \notin Q],$$

where $\mathcal{O}_{tk,pk}$ *is an oracle that maintains an initially empty set Q and operates as follows:*
- *On input* (commit, tag), *it runs* $(\widetilde{\mathsf{com}}, \mathsf{aux}) \leftarrow \mathsf{FakeCom}(pk, tk, tag)$, *stores* $(\widetilde{\mathsf{com}}, tag, \mathsf{aux})$, *returns* $\widetilde{\mathsf{com}}$ *and adds tag in Q.*
- *On input* (decommit, $\widetilde{\mathsf{com}}, \mathsf{Msg}$): *if a tuple* $(\widetilde{\mathsf{com}}, tag, \mathsf{aux})$ *was previously stored, it computes* $\widetilde{\mathsf{dec}} \leftarrow \mathsf{FakeOpen}(\mathsf{aux}, tk, tag, \widetilde{\mathsf{com}}, \mathsf{Msg})$ *and returns* $\widetilde{\mathsf{dec}}$. *Otherwise,* $\mathcal{O}_{tk,pk}$ *returns* $\perp$.

While our SSTC to group elements will be proved secure in the above sense, a *non-adaptive* flavor of simulation-sound binding security is sufficient for constructing non-malleable commitments. Indeed, Gennaro used [36] such a relaxed notion to achieve non-malleability from similar-looking multi-trapdoor commitments. In the non-adaptive notion, the adversary has to choose the set of tags $tag_1, \ldots, tag_\ell$ for which it wants to query $\mathcal{O}_{tk,pk}$ before seeing the public key pk.

5.1 Template of Linearly Homomorphic SPS Scheme

We first remark that *any* constant-size linearly homomorphic structure-preserving signature necessarily complies with the template below.

For simplicity, the template is described in terms of symmetric pairings but generalizations to asymmetric configurations are possible.

Keygen(λ, n): given λ and the dimension $n \in \mathbb{N}$ of the vectors to be signed, choose constants n_z, n_v, m. Among these, n_z and n_v will determine the signature length while m will be the number of verification equations. Then, choose $\{F_{j,\mu}\}_{j \in \{1,\ldots,m\}, \mu \in \{1,\ldots,n_z\}}$, $\{G_{j,i}\}_{i \in \{1,\ldots,n\}, j \in \{j,\ldots,m\}}$ in the group $\mathbb{G}$. The public key is $pk = (\{F_{j,\mu}\}_{j \in \{1,\ldots,m\}, \mu \in \{1,\ldots,n_z\}}, \{G_{j,i}\}_{i \in \{1,\ldots,n\}, j \in \{j,\ldots,m\}})$ while sk contains information about the representation of public elements w.r.t. specific bases.

Sign(sk, $\tau, (M_1, \ldots, M_n)$): Outputs $\sigma = (Z_1, \ldots, Z_{n_z}, V_1, \ldots, V_{n_v}) \in \mathbb{G}^{n_z + n_v}$.

SignDerive(pk, $\tau, \{(\omega_i, \sigma^{(i)})\}_{i=1}^\ell$): parses $\sigma^{(i)}$ as $(Z_1^{(i)}, \ldots, Z_{n_z}^{(i)}, V_1^{(i)}, \ldots, V_{n_v}^{(i)})$ for each $i \in \{1, \ldots, \ell\}$ and computes

$$Z_\mu = \prod_{i=1}^\ell Z_\mu^{(i)}{}^{\omega_i} \qquad V_\nu = \prod_{i=1}^\ell V_\nu^{(i)}{}^{\omega_i} \qquad \mu \in \{1, \ldots, n_z\}, \ \nu \in \{1, \ldots, n_v\}.$$

After possible extra re-randomizations, it outputs $(Z_1, \ldots, Z_{n_z}, V_1, \ldots, V_{n_v})$.

Verify(pk, $\sigma, \tau, (M_1, \ldots, M_n)$): given $\sigma = (Z_1, \ldots, Z_{n_z}, V_1, \ldots, V_{n_v}) \in \mathbb{G}^{n_z + n_v}$, a tag τ and $(M_1, \ldots, M_n)$, return 0 if $(M_1, \ldots, M_n) = (1_\mathbb{G}, \ldots, 1_\mathbb{G})$. Otherwise, do the following.

1. For each $j \in \{1, \ldots, m\}$ and $\nu \in \{1, \ldots, n_v\}$, compute one-to-one[7] encodings $T_{j,\nu} \in \mathbb{G}$ of the tag τ as a group element.
2. Return 1 if and only if $c_j = 1_{\mathbb{G}_T}$ for $j = 1$ to m, where

$$c_j = \prod_{\mu=1}^{n_z} e(F_{j,\mu}, Z_\mu) \cdot \prod_{\nu=1}^{n_v} e(T_{j,\nu}, V_\nu) \cdot \prod_{i=1}^n e(G_{j,i}, M_i) \quad j \in \{1, \ldots, m\}. \qquad (2)$$

We say that a linearly homomorphic SPS is *regular* if, for each file identifier τ, any non-trivial vector $(M_1, \ldots, M_n) \neq (1_\mathbb{G}, \ldots, 1_\mathbb{G})$ has a valid signature.

5.2 Construction of Simulation-Sound Structure-Preserving Trapdoor Commitments

Let $\Pi^{\mathsf{SPS}} = (\mathsf{Keygen}, \mathsf{Sign}, \mathsf{SignDerive}, \mathsf{Verify})$ be a linearly homomorphic SPS. We construct a simulation-sound trapdoor commitment as follows.

---

[7] This condition can be relaxed to have collision-resistant deterministic encodings. Here, we assume injectivity for simplicity.

SSTC.Setup(λ, n): given the desired dimension $n \in \mathbb{N}$ of vectors, choose public parameters pp for the linearly homomorphic SPS scheme. Then, run Π^{SPS}.Keygen(λ, n) to obtain a public key $\mathsf{pk} = \left(\{F_{j,\mu}\}_{j \in \{1,\dots,m\}, \mu \in \{1,\dots,n_z\}}, \{G_{j,i}\}_{i \in \{1,\dots,n\}, j \in \{j,\dots,m\}}\right)$, for some constants n_z, n_v, m, and a sk. The commitment key is $pk = \mathsf{pk}$ and the trapdoor tk consists of sk. Note that the public key defines a signature space $\mathbb{G}^{n_z + n_v}$, for constants n_z and n_v.

SSTC.Com($pk, tag, (M_1, \dots, M_n)$): to commit to $(M_1, \dots, M_n) \in \mathbb{G}^n$ with respect to the tag $tag = \tau$, choose $\left(Z_1, \dots, Z_{n_z}, V_1, \dots, V_{n_v}\right) \xleftarrow{R} \mathbb{G}^{n_z + n_v}$ in the signature space. Then, run step 1 of the verification algorithm and evaluate the right-hand-side member of (2). Namely, compute

$$c_j = \prod_{\mu=1}^{n_z} e(F_{j,\mu}, Z_\mu) \cdot \prod_{\nu=1}^{n_v} e(T_{j,\nu}, V_\nu) \cdot \prod_{i=1}^{n} e(G_{j,i}, M_i) \qquad j \in \{1, \dots, m\} \quad (3)$$

where $\{T_{j,\nu}\}_{j,\nu}$ form an injective encoding of $tag = \tau$ as a set of group elements. The commitment string is $\mathsf{com} = (c_1, \dots, c_m)$ whereas the decommitment is $\mathsf{dec} = \left(Z_1, \dots, Z_{n_z}, V_1, \dots, V_{n_v}\right)$.

SSTC.FakeCom(pk, tk, tag): proceeds like SSTC.Com with randomly chosen $(\hat{M}_1, \dots, \hat{M}_n) \xleftarrow{R} \mathbb{G}^n$. If $(\hat{\mathsf{com}}, \hat{\mathsf{dec}})$ denotes the resulting pair, the algorithm outputs $\widetilde{\mathsf{com}} = \hat{\mathsf{com}}$ and the auxiliary information aux, which consists of the pair $\mathsf{aux} = ((\hat{M}_1, \dots, \hat{M}_n), \hat{\mathsf{dec}})$ for $tag = \tau$.

SSTC.FakeOpen($\mathsf{aux}, tk, tag, \widetilde{\mathsf{com}}, (M_1, \dots, M_n)$): parses $\widetilde{\mathsf{com}}$ as $(\tilde{c}_1, \dots, \tilde{c}_m)$ and aux as $\left((\hat{M}_1, \dots, \hat{M}_n), (\hat{Z}_1, \dots, \hat{Z}_{n_z}, \hat{V}_1, \dots, \hat{V}_{n_v})\right)$. The algorithm first generates a linearly homomorphic signature on $(M_1/\hat{M}_1, \dots, M_n/\hat{M}_n)$ for the tag $tag = \tau$. Namely, using the trapdoor $tk = \mathsf{sk}$, compute a signature $\sigma' = (Z'_1, \dots, Z'_{n_z}, V'_1, \dots, V'_{n_v}) \leftarrow \Pi^{\mathsf{SPS}}$.Sign($\mathsf{sk}, \tau, (M_1/\hat{M}_n, \dots, M_n/\hat{M}_n)$). Since $\mathsf{aux} = \left((\hat{M}_1, \dots, \hat{M}_n), (\hat{Z}_1, \dots, \hat{Z}_{n_z}, \hat{V}_1, \dots, \hat{V}_{n_v})\right)$ satisfies

$$\tilde{c}_j = \prod_{\mu=1}^{n_z} e(F_{j,\mu}, \hat{Z}_\mu) \cdot \prod_{\nu=1}^{n_v} e(T_{j,\nu}, \hat{V}_\nu) \cdot \prod_{i=1}^{n} e(G_{j,i}, \hat{M}_i) \qquad j \in \{1, \dots, m\}, \quad (4)$$

FakeOpen runs $(\tilde{Z}_1, \dots, \tilde{Z}_{n_z}, \tilde{V}_1, \dots, \tilde{V}_{n_v}) \leftarrow$ SignDerive($\mathsf{pk}, \tau, \{(1, \sigma'), (1, \hat{\sigma})\}$), where $\hat{\sigma} = (\hat{Z}_1, \dots, \hat{Z}_{n_z}, \hat{V}_1, \dots, \hat{V}_{n_v})$. It outputs a valid de-commitment $\widetilde{\mathsf{dec}} = (\tilde{Z}_1, \dots, \tilde{Z}_{n_z}, \tilde{V}_1, \dots, \tilde{V}_{n_v})$ to $(M_1, \dots, M_n)$ with respect to $tag = \tau$.

SSTC.Verify($pk, tag, (M_1, \dots, M_n), \mathsf{com}, \mathsf{dec}$): parse com as $(c_1, \dots, c_m) \in \mathbb{G}_T^m$ and dec as $\left(Z_1, \dots, Z_{n_z}, V_1, \dots, V_{n_v}\right) \in \mathbb{G}^{n_z + n_v}$ (if these values do not parse properly, return 0). Then, compute a one-to-one encoding $\{T_{j,\nu}\}_{j,\nu}$ of $tag = \tau$. Return 1 if relations (3) hold and 0 otherwise.

In the full version of the paper, we extend this construction so as to build simulation-sound trapdoor commitment to vectors from any linearly homomorphic signature that fits a certain template. As a result, we obtain a modular construction of constant-size non-malleable commitment to vectors which preserves the feasibility of efficiently proving properties about committed values.

Theorem 3. *Assuming that the underlying linearly homomorphic SPS is regular and secure against non-independent Type I adversaries, the above construction is a simulation-sound trapdoor commitment to group elements.* (The proof is given in the full version of the paper.

A standard technique (see [35,36]) to construct a re-usable non-malleable commitment from a SSTC scheme is as follows. To commit to Msg, the sender generates a key-pair (VK, SK) for a one-time signature and generates $(com, dec) \leftarrow SSTC.Commit(pk, VK, MSg)$ using VK as a tag. The non-malleable commitment string is the pair (com, VK) and the opening is given by (dec, σ), where σ is a one-time signature on com, so that the receiver additionally checks the validity of σ. This construction is known to provide input independence [29] and thus non-malleability with respect to opening, as proved in [29,39].

In our setting, we cannot compute σ as a signature of com, as it consists of $\mathbb{G}_T$ elements. However, we can sign the pair (Msg, dec) — whose components live in $\mathbb{G}$ — as long as it uniquely determines com. To this end, we can use the one-time structure-preserving of [1, Appendix C.1] as it allows signing messages of arbitrary length using a constant-size one-time public key. Like our scheme of Section 3.2, it relies on the SDP assumption and yields a non-malleable commitment based on this sole assumption. Alternatively, we can move σ in the commitment string (which becomes (com, VK, σ)), in which case the one-time signature does not need to be structure-preserving but it has to be strongly unforgeable (as can be observed from the definition of independent commitments [29]) while the standard notion of unforgeability suffices in the former case.

References

1. Abe, M., Haralambiev, K., Ohkubo, M.: Signing on Elements in Bilinear Groups for Modular Protocol Design. Cryptology ePrint Archive: Report 2010/133 (2010)
2. Abe, M., Haralambiev, K., Ohkubo, M.: Group to Group Commitments Do Not Shrink. In: Pointcheval, D., Johansson, T. (eds.) EUROCRYPT 2012. LNCS, vol. 7237, pp. 301–317. Springer, Heidelberg (2012)
3. Abe, M., Fuchsbauer, G., Groth, J., Haralambiev, K., Ohkubo, M.: Structure-Preserving Signatures and Commitments to Group Elements. In: Rabin, T. (ed.) CRYPTO 2010. LNCS, vol. 6223, pp. 209–236. Springer, Heidelberg (2010)
4. Abe, M., Groth, J., Haralambiev, K., Ohkubo, M.: Optimal Structure-Preserving Signatures in Asymmetric Bilinear Groups. In: Rogaway, P. (ed.) CRYPTO 2011. LNCS, vol. 6841, pp. 649–666. Springer, Heidelberg (2011)
5. Abe, M., Chase, M., David, B., Kohlweiss, M., Nishimaki, R., Ohkubo, M.: Constant-Size Structure-Preserving Signatures: Generic Constructions and Simple Assumptions. In: Wang, X., Sako, K. (eds.) ASIACRYPT 2012. LNCS, vol. 7658, pp. 4–24. Springer, Heidelberg (2012)
6. Abe, M., David, B., Kohlweiss, M., Nishimaki, R., Ohkubo, M.: Tagged One-Time Signatures: Tight Security and Optimal Tag Size. In: Kurosawa, K., Hanaoka, G. (eds.) PKC 2013. LNCS, vol. 7778, pp. 312–331. Springer, Heidelberg (2013)
7. Abe, M., Groth, J., Ohkubo, M.: Separating Short Structure-Preserving Signatures from Non-interactive Assumptions. In: Lee, D.H., Wang, X. (eds.) ASIACRYPT 2011. LNCS, vol. 7073, pp. 628–646. Springer, Heidelberg (2011)

8. Ahn, J.H., Boneh, D., Camenisch, J., Hohenberger, S., Shelat, A., Waters, B.: Computing on Authenticated Data. In: Cramer, R. (ed.) TCC 2012. LNCS, vol. 7194, pp. 1–20. Springer, Heidelberg (2012)

9. Ateniese, G., Burns, R., Curtmola, R., Herring, J., Kissner, L., Peterson, Z., Song, D.: Provable data possession at untrusted stores. In: ACM-CCS 2007, pp. 598–609 (2007)

10. Attrapadung, N., Libert, B.: Homomorphic Network Coding Signatures in the Standard Model. In: Catalano, D., Fazio, N., Gennaro, R., Nicolosi, A. (eds.) PKC 2011. LNCS, vol. 6571, pp. 17–34. Springer, Heidelberg (2011)

11. Attrapadung, N., Libert, B., Peters, T.: Computing on Authenticated Data: New Privacy Definitions and Constructions. In: Wang, X., Sako, K. (eds.) ASIACRYPT 2012. LNCS, vol. 7658, pp. 367–385. Springer, Heidelberg (2012)

12. Attrapadung, N., Libert, B., Peters, T.: Efficient Completely Context-Hiding Quotable and Linearly Homomorphic Signatures. In: Kurosawa, K., Hanaoka, G. (eds.) PKC 2013. LNCS, vol. 7778, pp. 386–404. Springer, Heidelberg (2013)

13. Boneh, D., Boyen, X., Shacham, H.: Short Group Signatures. In: Franklin, M. (ed.) CRYPTO 2004. LNCS, vol. 3152, pp. 41–55. Springer, Heidelberg (2004)

14. Boneh, D., Freeman, D., Katz, J., Waters, B.: Signing a Linear Subspace: Signature Schemes for Network Coding. In: Jarecki, S., Tsudik, G. (eds.) PKC 2009. LNCS, vol. 5443, pp. 68–87. Springer, Heidelberg (2009)

15. Boneh, D., Freeman, D.: Linearly Homomorphic Signatures over Binary Fields and New Tools for Lattice-Based Signatures. In: Catalano, D., Fazio, N., Gennaro, R., Nicolosi, A. (eds.) PKC 2011. LNCS, vol. 6571, pp. 1–16. Springer, Heidelberg (2011)

16. Boneh, D., Freeman, D.: Homomorphic Signatures for Polynomial Functions. In: Paterson, K.G. (ed.) EUROCRYPT 2011. LNCS, vol. 6632, pp. 149–168. Springer, Heidelberg (2011)

17. Camenisch, J., Dubovitskaya, M., Haralambiev, K.: Efficient Structure-Preserving Signature Scheme from Standard Assumptions. In: Visconti, I., De Prisco, R. (eds.) SCN 2012. LNCS, vol. 7485, pp. 76–94. Springer, Heidelberg (2012)

18. Camenisch, J., Haralambiev, K., Kohlweiss, M., Lapon, J., Naessens, V.: Structure Preserving CCA Secure Encryption and Applications. In: Lee, D.H., Wang, X. (eds.) ASIACRYPT 2011. LNCS, vol. 7073, pp. 89–106. Springer, Heidelberg (2011)

19. Camenisch, J., Gross, T., Heydt-Benjamin, T.-S.: Rethinking accountable privacy supporting services: extended abstract. In: Digital Identity Management, DIM 2008, pp. 1–8 (2008)

20. Canetti, R.: Universally Composable Security: A New Paradigm for Cryptographic Protocols. In: FOCS 2001, pp. 136–145 (2001)

21. Canetti, R., Dodis, Y., Pass, R., Walfish, S.: Universally Composable Security with Global Setup. In: Vadhan, S.P. (ed.) TCC 2007. LNCS, vol. 4392, pp. 61–85. Springer, Heidelberg (2007)

22. Canetti, R., Fischlin, M.: Universally Composable Commitments. In: Kilian, J. (ed.) CRYPTO 2001. LNCS, vol. 2139, pp. 19–40. Springer, Heidelberg (2001)

23. Catalano, D., Fiore, D., Warinschi, B.: Adaptive Pseudo-free Groups and Applications. In: Paterson, K.G. (ed.) EUROCRYPT 2011. LNCS, vol. 6632, pp. 207–223. Springer, Heidelberg (2011)

24. Catalano, D., Fiore, D., Warinschi, B.: Efficient Network Coding Signatures in the Standard Model. In: Fischlin, M., Buchmann, J., Manulis, M. (eds.) PKC 2012. LNCS, vol. 7293, pp. 680–696. Springer, Heidelberg (2012)

25. Cathalo, J., Libert, B., Yung, M.: Group Encryption: Non-Interactive Realization in the Standard Model. In: Matsui, M. (ed.) ASIACRYPT 2009. LNCS, vol. 5912, pp. 179–196. Springer, Heidelberg (2009)

26. Chase, M., Kohlweiss, M.: A New Hash-and-Sign Approach and Structure-Preserving Signatures from DLIN. In: Visconti, I., De Prisco, R. (eds.) SCN 2012. LNCS, vol. 7485, pp. 131–148. Springer, Heidelberg (2012)

27. Cramer, R., Shoup, V.: A practical public key cryptosystem provably secure against adaptive chosen ciphertext attack. In: Krawczyk, H. (ed.) CRYPTO 1998. LNCS, vol. 1462, pp. 13–25. Springer, Heidelberg (1998)

28. Damgård, I., Groth, J.: Non-interactive and reusable non-malleable commitment schemes. In: STOC 2003, pp. 426–437 (2003)

29. Di Crescenzo, G., Ishai, Y., Ostrovsky, R.: Non-Interactive and Non-Malleable Commitment. In: STOC 1998, pp. 141–150 (1998)

30. Desmedt, Y.: Computer security by redefining what a computer is. In: New Security Paradigms Workshop, NSPW 1993, pp. 160–166 (1993)

31. Dolev, D., Dwork, C., Naor, M.: Non-malleable cryptography. In: STOC 1991, pp. 542–552. ACM Press (1991)

32. Fischlin, M., Libert, B., Manulis, M.: Non-interactive and Re-usable Universally Composable String Commitments with Adaptive Security. In: Lee, D.H., Wang, X. (eds.) ASIACRYPT 2011. LNCS, vol. 7073, pp. 468–485. Springer, Heidelberg (2011)

33. Freeman, D.: Improved security for linearly homomorphic signatures: A generic framework. In: Fischlin, M., Buchmann, J., Manulis, M. (eds.) PKC 2012. LNCS, vol. 7293, pp. 697–714. Springer, Heidelberg (2012)

34. Fuchsbauer, G.: Automorphic Signatures in Bilinear Groups and an Application to Round-Optimal Blind Signatures. Cryptology ePrint Archive: Report 2009/320 (2009)

35. Garay, J., MacKenzie, P., Yang, K.: Strengthening Zero-Knowledge Protocols Using Signatures. In: Biham, E. (ed.) EUROCRYPT 2003. LNCS, vol. 2656, pp. 177–194. Springer, Heidelberg (2003)

36. Gennaro, R.: Multi-trapdoor Commitments and Their Applications to Proofs of Knowledge Secure Under Concurrent Man-in-the-Middle Attacks. In: Franklin, M. (ed.) CRYPTO 2004. LNCS, vol. 3152, pp. 220–236. Springer, Heidelberg (2004)

37. Gennaro, R., Gentry, C., Parno, B.: Non-interactive Verifiable Computing: Outsourcing Computation to Untrusted Workers. In: Rabin, T. (ed.) CRYPTO 2010. LNCS, vol. 6223, pp. 465–482. Springer, Heidelberg (2010)

38. Gennaro, R., Katz, J., Krawczyk, H., Rabin, T.: Secure Network Coding over the Integers. In: Nguyen, P.Q., Pointcheval, D. (eds.) PKC 2010. LNCS, vol. 6056, pp. 142–160. Springer, Heidelberg (2010)

39. Gennaro, R., Micali, S.: Independent Zero-Knowledge Sets. In: Bugliesi, M., Preneel, B., Sassone, V., Wegener, I. (eds.) ICALP 2006. LNCS, vol. 4052, pp. 34–45. Springer, Heidelberg (2006)

40. Groth, J.: Simulation-Sound NIZK Proofs for a Practical Language and Constant Size Group Signatures. In: Lai, X., Chen, K. (eds.) ASIACRYPT 2006. LNCS, vol. 4284, pp. 444–459. Springer, Heidelberg (2006)

41. Groth, J., Sahai, A.: Efficient non-interactive proof systems for bilinear groups. In: Smart, N.P. (ed.) EUROCRYPT 2008. LNCS, vol. 4965, pp. 415–432. Springer, Heidelberg (2008)

42. Groth, J.: Homomorphic trapdoor commitments to group elements. Cryptology ePrint Archive: Report 2009/007 (2009)

43. Groth, J.: Efficient Zero-Knowledge Arguments from Two-Tiered Homomorphic Commitments. In: Lee, D.H., Wang, X. (eds.) ASIACRYPT 2011. LNCS, vol. 7073, pp. 431–448. Springer, Heidelberg (2011)
44. Hofheinz, D., Jager, T.: Tightly Secure Signatures and Public-Key Encryption. In: Safavi-Naini, R., Canetti, R. (eds.) CRYPTO 2012. LNCS, vol. 7417, pp. 590–607. Springer, Heidelberg (2012)
45. Libert, B., Yung, M.: Non-Interactive CCA2-Secure Threshold Cryptosystems with Adaptive Security: New Framework and Constructions. In: Cramer, R. (ed.) TCC 2012. LNCS, vol. 7194, pp. 75–93. Springer, Heidelberg (2012)
46. Johnson, R., Molnar, D., Song, D., Wagner, D.: Homomorphic Signature Schemes. In: Preneel, B. (ed.) CT-RSA 2002. LNCS, vol. 2271, pp. 244–262. Springer, Heidelberg (2002)
47. MacKenzie, P., Yang, K.: On Simulation-Sound Trapdoor Commitments. In: Cachin, C., Camenisch, J.L. (eds.) EUROCRYPT 2004. LNCS, vol. 3027, pp. 382–400. Springer, Heidelberg (2004)
48. Naor, M., Yung, M.: Public-key cryptosystems provably secure against chosen ciphertext attacks. In: STOC 1990. ACM Press (1990)
49. Sakai, Y., Emura, K., Hanaoka, G., Kawai, Y., Matsuda, T., Omote, K.: Group Signatures with Message-Dependent Opening. In: Abdalla, M., Lange, T. (eds.) Pairing 2012. LNCS, vol. 7708, pp. 270–294. Springer, Heidelberg (2013)
50. Shamir, A.: Identity-Based Cryptosystems and Signature Schemes. In: Blakely, G.R., Chaum, D. (eds.) CRYPTO 1984. LNCS, vol. 196, pp. 47–53. Springer, Heidelberg (1985)
51. Waters, B.: Efficient Identity-Based Encryption Without Random Oracles. In: Cramer, R. (ed.) EUROCRYPT 2005. LNCS, vol. 3494, pp. 114–127. Springer, Heidelberg (2005)

Man-in-the-Middle Secure Authentication Schemes from LPN and Weak PRFs

Vadim Lyubashevsky[1] and Daniel Masny[2,*]

[1] INRIA / École Normale Supérieure, Paris
lyubash@di.ens.fr
[2] Ruhr-Universitat Bochum
daniel.masny@ruhr-uni-bochum.de

Abstract. We show how to construct, from any weak pseudorandom function, a 3-round symmetric-key authentication protocol that is secure against man-in-the-middle attacks. The construction is very efficient, requiring both the secret key and communication size to be only $3n$ bits long and involving only one call to the weak-PRF. Our techniques also extend to certain classes of randomized weak-PRFs, chiefly among which are those based on the classical LPN problem and its more efficient variants such as Toeplitz-LPN and Ring-LPN. Building an efficient man-in-the-middle secure authentication scheme from any weak-PRF resolves a problem left open by Dodis et al. (Eurocrypt 2012), while building a man-in-the-middle secure scheme based on any variant of the LPN problem solves the main open question in a long line of research aimed at constructing a *practical* light-weight authentication scheme based on learning problems, which began with the work of Hopper and Blum (Asiacrypt 2001).

1 Introduction

The need for light-weight cryptography is increasing rapidly due to the growing deployment of low-cost devices, such as smart cards and RFID tags, in the real world. One of the most common cryptographic protocols required on these devices is a symmetric key authentication protocol in which the prover (usually referred to as the Tag) authenticates his identity to the verifier (usually referred to as the Reader). The most direct way in which this protocol can be constructed is by using a pseudorandom function f (e.g. AES) for which the Tag and the Reader share a common key. Then the authentication protocol simply consists of the Reader sending a challenge c to which the Tag replies with $f(c)$, and the Reader verifies that the received evaluation of c is indeed correct. The main problem with this approach is that the pseudorandom function, whether it is a "provably-secure" one based on some mathematical assumption or an "ad-hoc" block cipher like AES, is usually quite costly for light-weight devices. For this reason, researchers have worked on designing block ciphers specifically

---

* Part of this work was done while visiting École Normale Supérieure, Paris.

R. Canetti and J.A. Garay (Eds.): CRYPTO 2013, Part II, LNCS 8043, pp. 308–325, 2013.

for low-cost devices (e.g. [19,4]). A different approach for solving this problem is constructing authentication schemes from building blocks that have weaker security properties than block ciphers or pseudorandom functions. We pursue this latter avenue of research in the present work.

1.1 Authentication from LPN

The Learning Parity with Noise (LPN) problem was initially shown to have cryptographic applications by Goldreich et al. and Blum et al. in [11,3], and then used as a basis for authentication schemes by Hopper and Blum in their HB scheme [14]. In this latter paper, a simple LPN-based authentication scheme was proposed that was secure in the *passive* attack model. Later work by Juels and Weis [15], and also by Katz and Shin [16], modified this protocol (the result was called HB$^+$) to be secure against *active* adversaries. Nevertheless, even these schemes had a serious security shortcoming. If the adversary were allowed to modify the communication between the Tag and the Reader and observe the response of the reader to verification queries, then, as shown by Gilbert et al. [7], there exists a very simple attack that can recover the secret key in polynomial time. Because such a *man-in-the-middle* attack can be mounted with relatively small effort, schemes that fall to it cannot be considered secure enough for real-world applications that require some decent level of security. It was thus a major open problem to construct an efficient LPN-based authentication scheme that remains secure against man-in-the-middle attacks.

A notable advance was made by Gilbert et al. [9] who proposed a scheme (termed HB$^\#$) that was able to resist the attack from [7] and was shown to be secure against restricted man-in-the-middle adversaries. A second contribution of this work was to offer a solution to another problematic feature of previous LPN-based protocols. All protocols that are based on LPN require either the key size or the communication complexity to be square in the security parameter. Thus either the key size or the communication complexity would have to be on the order of hundreds of thousands of bits. Since the main motivation for LPN-based protocols is low-cost hardware, this is clearly unacceptable. To this end, [9] proposed a protocol based on a related assumption, called Toeplitz-LPN (see Section 2.2 for definitions), where the communication complexity was small and the secret key had some structured form which allowed for compact representations. While there has been no known weakness caused by using the Toeplitz-LPN assumption, it did turn out that the restricted man-in-the-middle model introduced in [9] was not sufficient to prevent all practical attacks, and one such attack was shown by Ouafi et al. [24].

There have been many other proposals, some without security proofs, others with claimed proofs that attempted to solve this problem, but all of these methods were ultimately shown to be flawed (see [8] for a small overview). A breakthrough finally came in a series of recent papers by Kiltz et al. [18] and Dodis et al. [6] who constructed relatively-efficient MACs based on the hardness of the LPN problem. Because MACs immediately give rise to man-in-the-middle secure authentication schemes, their work also resolved the problem of building such schemes from the LPN problem. This LPN MAC, however, suffered from the same drawback as other LPN-based schemes – the key size was prohibitively large. Thus in order to be useful in practice, the proof techniques would have to be adapted to work with more compact LPN-related assumptions, such as Toeplitz-LPN. But the constructions of [18] and [6] made use of certain algebraic structure of the LPN-problem, and the proofs turn out to be incompatible with other previously-considered versions of LPN.

1.2   Authentication from Weak-PRFs

Weak pseudo-random functions are keyed functions whose outputs on *random* inputs are indistinguishable from uniform. Weak PRFs are considered to be much "weaker" primitives than PRFs, and in particular, it is not known how to transform a weak-PRF into a PRF except by using tree techniques similar to the classical GGM construction [10]. Additionally, it also appears to be much easier to build secure weak-PRFs than PRFs. For example, the function $f_a(x) = x^a \bmod p$ is a weak-PRF based on the DDH assumption, whereas the construction of a PRF based on DDH is much less efficient [22], requiring n multiplications in addition to the exponentiation in the weak-PRF. Similarly, the recent construction of lattice-based PRFs [1] first builds a relatively efficient weak-PRF (which is just $f_A(x) = Round(Ax \bmod p)$, where $A \in \mathbb{Z}_p^{m \times n}, x \in \mathbb{Z}^n$ with $\|x\|$ small, and the $Round(\cdot)$ function drops a super-logarithmic number of least-significant bits) and then converts it to a full PRF using techniques similar to [22,23]. The resulting lattice-based PRF is both less efficient and requires a stronger computational assumption than the underlying weak-PRF.

Due to efficiency advantages and lower security requirements, there has been some research on constructions of cryptographic primitives such as symmetric encryption and stream ciphers built directly from weak-PRFs (e.g. [5,21,25]). The work along this theme that is most related to ours is the aforementioned one of Dodis et al. [6], where it is shown how to build a 3-round authentication scheme secure against *active* attacks from any weak-PRF. As we mentioned earlier, the active security model, where the adversary is not allowed to send any verification queries to the Reader, is not considered strong enough for real-world applications. And so the problem of constructing man-in-the-middle secure authentication schemes from arbitrary weak-PRFs remained open.

1.3   Our Results

Our first result is a construction, from any weak pseudorandom function, of a 3-round symmetric-key authentication protocol that is secure against man-in-the-middle attacks. Our scheme has the exact same communication complexity as the actively-secure scheme of [6], and only has one extra key element. To be more precise, the secret keys in our scheme consist of the key of the weak-PRF plus the description of a pairwise-independent hash function, which requires an additional two elements, whose size is the output length of the weak-PRF. So if we assume that both the domain and range of the weak-PRF is n bits, then the total key size is $3n$.

We then extend our construction of a weak-PRF scheme to *randomized* weak-PRFs. Randomized weak-PRFs are keyed functions that become computationally indistinguishable from uniform when their outputs are perturbed by some low-entropy noise. Noisy learning problems such as LPN and LWE [26] can be equivalently viewed as problems of distinguishing the outputs of a randomized weak-PRF from the uniform distribution. To get a man-in-the-middle secure authentication scheme from a randomized weak-PRF, we require just one more secret key element than our weak-PRF based scheme.

Our constructions, and to some extent their security proofs as well, turn out to be surprisingly simple. The main insight is that one should embed the n-bit output of the (randomized) weak-PRF into a finite field of size 2^n. Then, in addition to the secret keys associated to the function, we also create secret keys in the field which end up being masked by the presumed indistinguishability from uniform of the (randomized) weak-PRF. We then show how the interplay in the field between the weak-PRF and the additional secret keys results in protocols that have the desired man-in-the-middle security.

We prove security of our schemes in the sequential man-in-the-middle model, in which the adversary simultaneously interacts with one copy of the Tag and Reader (see Figure 1). The schemes remain secure even if the adversary has access to multiple readers (this is shown in the full version of the work), whereas concurrent access to multiple tags may result in a vulnerability.[1] In the stronger notion of concurrent man-in-the-middle security the adversary is allowed to simultaneously communicate with multiple copies of the Tag and Reader. While the concurrent model is theoretically stronger, we do not believe that it is practically relevant to the low-cost device setting considered in this paper. In particular, it is unlikely that a low-cost Tag would have the need (or ability) to simultaneously participate in more than one authentication session. Furthermore, it also seems imprudent that in an ecosystem where one wants to have relatively strong security, secret keys would be shared among the Tags. Still, constructing an efficient authentication scheme from generic weak-PRFs that is secure in the concurrent man-in-the-middle model is an interesting open problem. [2]

1.4   Comparison to Other Works

Table 1 compares the results obtained in this paper with those of previous works. Compared to the protocols that only achieve active security, our scheme achieves the much stronger man-in-the-middle security at a fairly small cost. In the case of protocols based on a generic weak-PRF, we extend the security to the man-in-the middle model at the cost of only one extra secret key element and one extra field multiplication. We get similar results when comparing our protocol with actively-secure LPN-based ones.

---

[1] In the full version of the paper, we show that for certain instantiations of a randomized weak-PRF there indeed exists a concurrent man-in-the-middle attack.

[2] Our current scheme is still secure in the concurrent model against *active* attacks.

Table 1. Authentication Protocols Based on Weak-PRFs and the LPN-related Assumptions. Listed is the amount of authentication rounds $\#r$, the security properties achieved by the protocol and its complexity (with lower order terms dropped) according to the key size and the communication. Let ϵ be the advantage in breaking the assumption, then the term depending on ϵ is proven to be the best possible advantage of breaking the protocol in the given model. Q is the amount of tag and verification queries whereas q_v is defined as the amount of verification queries, which is $q_v = 1$ in the active model. n parameterizes the hardness of the assumption and λ is the statistical security parameter. [6] gives an alternate construction of MAC_1 and MAC_2 with better computational complexity, but the rest of the properties are basically the same.

Protocol	$\#r$	Security			Complexity	
		assumption	active	MIM	key size	com.
weak-PRF [6]	3	weak-PRF	$\sqrt{\epsilon}$	?	$2n$	$3n$
weak-PRF [this work]	3	weak-PRF	$q_v \cdot \sqrt{\epsilon}$		$3n$	$3n$
HB^+ [15,16]	3	$LPN_{n,\tau}$	$\sqrt{\epsilon}$	X [7]	$2n$	$2n^2$
Random-$HB^\#$ [9]	3	$LPN_{n,\tau}$	$\sqrt{\epsilon}$	X [24]	$2n^2$	$3n$
$HB^\#$ [9]	3	Toeplitz-$LPN_{n,\tau}$	$\sqrt{\epsilon}$	X [24]	$4n$	$3n$
MAC_1 [18]	2	$LPN_{n,\tau}$	ϵ	$2^\lambda \cdot \epsilon$	$2n^2$	$4n$
MAC_2 [18]	2	$LPN_{n,\tau}$	ϵ	$Q \cdot \epsilon$	λn^2	$4n$
Lapin [13]	2	Ring-$LPN_{n,\tau}$	ϵ	?	$2n$	$3n$
MAC_1 + Lapin	2	Ring-$LPN_{n,\tau}$	ϵ	$2^\lambda \cdot \epsilon$	$6n + 2\lambda$	$4n$
LPN-based [this work]	3	$LPN_{n,\tau}$	$q_v \cdot \sqrt{\epsilon}$		n^2	$3n$
		Toeplitz-$LPN_{n,\tau}$			$5n$	
		Ring-$LPN_{n,\tau}$			$4n$	

It is also interesting to compare our LPN protocol to the MAC constructions in [18]. There are three advantages to the MAC constructions – they are only two rounds, they have slightly tighter reductions to LPN, and they are secure in the concurrent man-in-the-middle model, whereas our scheme is secure in the sequential man-in-the-middle model. The advantages of our construction are that the key sizes and the communication complexities are smaller.

The above-listed differences between our LPN scheme and the MAC schemes are, in our opinion, fairly minor with several pluses and minuses on both sides. In practice, it makes almost no difference whether the authentication scheme is 2 or 3 rounds since the Tag is the one who starts the protocol – thus a 2-round protocol essentially becomes a 3-round one. And while security tightness is certainly a desirable property, it is very unclear what effects it has in practice. Similar public key authentication schemes, such as GQ [12] and Schnorr [27], have been studied for a long time, yet do not exhibit any weaknesses due to their non-tight reductions.

The major advantage of our construction is that it is *generic* and can be instantiated with virtually any version of the LPN function or a randomized weak-PRF satisfying a few mild properties (see Section 2.1). For example, our construction allows for authentication schemes based on the fairly well-studied

Toeplitz-LPN assumption, which seems to provide a very good compromise between security and computational efficiency. The constructions of [18] and [6], on the other hand, can only construct MACs from functions with very "algebraic" properties.

The recent work of Heyse et al. [13] proposed a new LPN-type assumption, called Ring-LPN, to enable efficient constructions that are compatible with the MAC transformation in [18]. The assumption is relatively new, and its unclear at this point whether it has the same hardness as the more well-studied LPN and Toeplitz-LPN assumptions. Still, even if the Ring-LPN problem is hard, our LPN protocol can also be instantiated based on this assumption and is more efficient than the resulting MAC transformation.

2   Preliminaries and Notation

2.1   Function Families and Their Properties

In this section we define the important classes of functions that will appear in the paper. As mentioned earlier, we will be considering embeddings of function outputs into a finite field. The embedding can be arbitrary, and the simplest one is to simply think of a function output string $s \in \{0,1\}^n$ as a polynomial in a finite field $\mathbb{F} = (\mathbb{Z}_2^n, +, \times)$ for appropriately defined addition and multiplication operations. Thus, without loss of generality, we will assume that all our functions output elements to some finite field $\mathbb{F}$.

Definition 2.1. *A function family* $\mathcal{H} : \mathbb{D} \to \mathbb{F}$ *is called pairwise-independent if for all* $x_1 \neq x_2 \in \mathbb{D}, y_1, y_2 \in \mathbb{F}$,

$$\Pr_{h \xleftarrow{\$} \mathcal{H}} [h(x_1) = y_1 \wedge h(x_2) = y_2] = 1/|\mathbb{F}|^2.$$

Definition 2.2. *A function family* $\mathcal{F} : \mathbb{D} \to \mathbb{F}$ *is said to be a weak-PRF family if for any polynomial-sized* k, *randomly-chosen* $f \in \mathcal{F}$, *and randomly-chosen* $r_1, \ldots, r_k \in \mathbb{D}$, *the distribution of* $(r_1, f(r_1)), \ldots, (r_k, f(r_k))$ *is computationally indistinguishable from the uniform distribution over* $(\mathbb{D}, \mathbb{F})^k$.

Even if $(r_1, f(r_1)), \ldots, (r_k, f(r_k))$ can be distinguished from the uniform distribution over $(\mathbb{D}, \mathbb{F})^k$, it's possible that the sequence can become indistinguishable if the outputs $f(r_i)$ were perturbed by some noise. Such function families are called randomized weak-PRFs. The noise perturbation can be anything, but in this paper we will only consider noise distributions with an eye towards LPN applications. In particular, both the noise and the output of $f(r_i)$ are group elements, and the perturbation consists of adding the two together. This is still consistent with our requirement of being able to embed the output of all functions into a finite field $\mathbb{F}$ since the group needed for LPN can simply be the underlying additive group of $\mathbb{F}$ (see Section 2.2).

Definition 2.3. *For a function $f(\cdot) : \mathbb{D} \to \mathbb{F}$ and a distribution χ over $\mathbb{F}$, we will write $f^{\chi}(r)$ to mean a randomized function that generates an element $e \in \mathbb{F}$ according to the distribution χ and outputs $f(r)+e$. A function family $\mathcal{F} : \mathbb{D} \to \mathbb{F}$ is said to be a* randomized weak-PRF *family with noise χ if for any polynomial-sized k, randomly-chosen $f \in \mathcal{F}$, and randomly-chosen $r_1, \ldots, r_k \in \mathbb{D}$, the distribution of $(r_1, f^{\chi}(r_1)), \ldots, (r_k, f^{\chi}(r_k))$ is computationally indistinguishable from the uniform distribution over $(\mathbb{D}, \mathbb{F})^k$.*

In order for randomized weak-PRFs to be useful for cryptographic constructions, the range $\mathbb{F}$ and the error distribution should have certain characteristics. For example, the weak-PRFs would be of very little use if the error distribution χ was just the uniform distribution over $\mathbb{F}$. In this paper we will assume that the additive group of the field $\mathbb{F}$ and the error distribution χ satisfy the following three properties:

1. There exists a weight function $\| \cdot \| : \mathbb{F} \to \mathbb{R}^+$ such that the additive group that underlies the field $\mathbb{F}$ satisfies the triangle inequality – that is for all $a, b \in \mathbb{F}$, $\|a \pm b\| \le \|a\| + \|b\|$. Additionally, $\|a\| = 0$ if and only if $a = 0$.
2. There exists a positive real $\tau' \in \mathbb{R}$ such that $\Pr_{e \sim \chi}[\|e\| \le \tau'] = 1 - n^{-\omega(1)}$.[3]
3. For a positive real α, let $\beta(\alpha) = \{z \in \mathbb{F} : \|z\| \le \alpha\}$. We will assume that $|\beta(2\tau')|/|\mathbb{F}| = n^{-\omega(1)}$.

The first property essentially makes sure that the randomness in the randomized weak-PRF behaves "nicely" via the triangular inequality.[4] The second property determines the completeness of our protocol. Additionally, because of the way our security proof works, the completeness of the protocol also plays a role in the soundness of the protocol.[5] Thus this value should be very close to 1. The third property determines the soundness of the protocol. Intuitively, it is related to the probability that an adversary can randomly guess a response and be accepted by the verifier.

Due to their similarity, we will be presenting our authentication scheme and its proof based on weak-PRFs together with the ones based on randomized weak-PRFs. Since a weak-PRF is just a randomized weak-PRF whose error distribution χ has its support entirely on 0, it's easy to see that it can trivially be made to satisfy the above three properties. We can define the weight function as $\|x\| = 1$ for all $x \ne 0$ and set $\tau' = 0$. Thus for weak-PRFs we have $\Pr_{e \sim \chi}[\|e\| \le \tau'] = 1$ (and so the protocol will have perfect completeness) and $|\beta(2\tau')|/|\mathbb{F}| = |\{0\}|/|\mathbb{F}| = 1/|\mathbb{F}|$.

---

[3] More formally, τ' is a function of n, $\tau'(n)$, but we will omit the n throughout the paper.

[4] Even though we are using the standard notation for "norm", the weight function $\| \cdot \|$ is not quite a norm because it's not true that for all integers α, $\alpha\|a\| = \|\alpha a\|$ (since we are working over a finite field).

[5] This seems to be a common feature of protocols that have man-in-the-middle security because the simulator replies to the adversary under the assumption that properly-formed responses by the Tag are accepted by the Reader. Even though it is not stated in [18,6], the soundness of their protocols also depends on their completeness in exactly the same way as in this work.

2.2   Randomized Weak-PRFs from the LPN Problem and Its Variants

The classical decisional $LPN_{n,\tau}$ assumption states that the uniform distribution over $\mathbb{Z}_2^n \times \mathbb{Z}_2$ is computationally-indistinguishable from the following distribution: for a fixed randomly-chosen vector $s \in \mathbb{Z}_2^n$, output $(r, r \cdot s + e)$ where r is chosen uniformly random from $\mathbb{Z}_2^n$ and e is a Bernoulli random variable that is 1 with probability τ. By the hybrid argument, it is easy to see that if the fixed secret is now a matrix $S \in \mathbb{Z}_2^{m \times n}$ then the distribution $(r, Sr + e)$, where r is chosen as before and e is a vector each of whose coefficients is 1 with probability τ, is also computationally-indistinguishable from the uniform distribution over $\mathbb{Z}_2^n \times \mathbb{Z}_2^m$ (with a loss of a factor m in the reduction). We now formulate this latter statement in terms of the randomized weak-PRF notation from the previous subsection.

Let Ber_τ^m be a distribution over $\mathbb{Z}_2^m$ where every coordinate is independently chosen to be 1 with probability τ and 0 with probability $1 - \tau$.

Definition 2.4 (LPN). *Let $\mathcal{F} : \mathbb{Z}_2^n \to \mathbb{Z}_2^m$ be a function family indexed by matrices $S \in \mathbb{Z}_2^{m \times n}$. For a function $f_S \in \mathcal{F}$ and a vector $r \in \mathbb{Z}_2^n$, define $f_S(r) := Sr$. Then the $LPN_{n,\tau}$ assumption implies that $\mathcal{F}$ is a randomized weak-PRF family with noise Ber_τ^m.*

In the above definition, the domain $\mathbb{D}$ of $\mathcal{F}$ is $\mathbb{Z}_2^n$. Because we insisted in Definition 2.3 that the range of the function family $\mathcal{F}$ be a finite field (this will be used in our protocol) and the LPN problem only requires an additive group structure, we have some freedom as to how to define this field. The LPN assumption requires the range to have the group structure $(\mathbb{Z}_2^m, +)$, thus $\mathbb{F}$ can be any finite field that has $(\mathbb{Z}_2^m, +)$ as its underlying additive group. The most natural definition is $\mathbb{F} = \mathbb{Z}_2[x]/(g(x))$ where $g(x)$ is a polynomial of degree m that is irreducible over $\mathbb{Z}_2$, and addition and multiplication are just standard polynomial addition and multiplications modulo 2 and $g(x)$. Thus addition in $(\mathbb{F}, +, \times)$ exactly corresponds to addition in $(\mathbb{Z}_2^m, +)$.

The randomized weak-PRF based on LPN can also quite naturally be made to satisfy the three properties after Definition 2.3. The weight function $\| \cdot \|$ can be defined to be the Hamming weight. That is, for any element $a \in \mathbb{Z}_2^m$, $\|a\|$ is the number of 1's in a. With this definition of the weight function, one can compute, via the Chernoff bound, a τ' such that any element e chosen according to Ber_τ^m satisfies $\|e\| \leq \tau'$ with overwhelming probability. To satisfy the third property, we would need that $|\beta(2\tau')|/|\mathbb{F}| = n^{-\omega(1)}$, which is equivalent to the condition that $\left(\sum_{i=0}^{\lfloor 2\tau' \rfloor} \binom{m}{i} \right) / 2^m = n^{-\omega(1)}$. The above conditions are identical to those in other authentication protocols, such as [16,17,9], and so the LPN parameters needed to make those schemes secure, also carry over to ours.

Because the LPN problem yields rather inefficient schemes, Gilbert et al. [9] proposed protocols based on the hardness of the Toeplitz-LPN problem, which is just like the LPN problem except that the secret matrix S is a Toeplitz matrix.

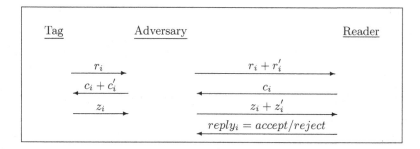

Fig. 1. Man-in-the-Middle Attack Model

Definition 2.5 (Toeplitz-LPN). *Let* $\mathcal{F} : \mathbb{Z}_2^n \to \mathbb{Z}_2^m$ *be a function family indexed by Toeplitz matrices* $S \in \mathbb{Z}_2^{m \times n}$. *For a function* $f_S \in \mathcal{F}$ *and a vector* $r \in \mathbb{Z}_2^n$, *define* $f_S(r) := Sr$. *Then the Toeplitz-LPN$_{n,\tau}$ assumption implies that* $\mathcal{F}$ *is a randomized weak-PRF family with noise* Ber_τ^m.

Heyse et al. [13] recently introduced the Ring-LPN problem, which also results in more efficient protocols. While the Ring-LPN problem has not been well-studied, it does have some resemblance to the better-studied Ring-LWE problem [20] in lattice cryptography, and so there are some reasons to believe that it might be secure.

Definition 2.6 (Ring-LPN). *Let* $g(x)$ *be a polynomial of degree* n *in* $\mathbb{Z}_2[x]$ *irreducible over* $\mathbb{Z}_2$ *and define the field* $\mathbb{F}$ *to be* $\mathbb{F} = \mathbb{Z}_2[x]/((g(x)))$. *Let* $\mathcal{F} : \mathbb{F} \to \mathbb{F}$ *be a function family indexed by polynomials* $s \in \mathbb{F}$. *For a function* $f_s \in \mathcal{F}$ *and a polynomial* $r \in \mathbb{F}$, *define* $f_s(r) := sr$. *Then the Ring-LPN$_{n,\tau}$ assumption implies that* $\mathcal{F}$ *is a randomized weak-PRF family with noise* Ber_τ^n.

2.3   Security Models

All authentication schemes are protocols in which the Tag and the Reader possess some secret key sk and then perform an interaction in which the Tag must convince the Reader of his identity. The difference in the security models depends on the strength that we give the adversary. The three most natural security models are *passive*, *active*, and *man-in-the-middle*. All three models consist of two stages. In the first stage, depending on the model, the Adversary is allowed to have some interaction with the Tag and the Reader. In the second stage, in all three models, he loses the interaction with the Tag and must interact with the Reader in hopes of getting the latter to accept the interaction.

Man-in-the-Middle Adversary. The strongest type of Adversary is one who in the first stage is able to simultaneously interact with the Tag and the Reader and make *verification queries* to the Reader. In the second stage, the Adversary loses access to the Tag, and interacts with the Reader hoping to make the latter accept.

In this paper, the protocols we will be constructing will be sigma protocols (i.e. have three rounds usually referred to as *commit, challenge,* and *response*) and will use a model that is simpler to describe and is at least as secure as the man-in-the-middle one. We now describe the security game and the Adversary's condition for winning it:

Setup: Generate a secret key and give it to the Tag $\mathcal{T}$ and the Reader $\mathcal{R}$.

Attack: Invoke the Adversary $\mathcal{A}$ who has access to $\mathcal{T}$ and $\mathcal{R}$ and let him interact with them t times. Each of the interactions is as follows (see Figure 1):
$\mathcal{A}$ receives a commitment r_i from $\mathcal{T}$ and sends a commitment $r_i + r_i'$ to $\mathcal{R}$. $\mathcal{R}$ responds with a challenge c_i and $\mathcal{A}$ sends a challenge $c_i + c_i'$ to $\mathcal{T}$. $\mathcal{T}$ answers with a valid response z_i. $\mathcal{A}$ can now send his response $z_i + z_i'$ for verification to $\mathcal{R}$. $\mathcal{R}$ answers with *accept*, if $(r_i + r_i', c_i', z_i + z_i')$ is valid according to the verification function, otherwise he answers with *reject*.

Winning Condition: We say that the Adversary $\mathcal{A}$ wins the game if at some point he makes a query to $\mathcal{R}$ such that $(r_i', c_i', z_i') \neq (0, 0, 0)$ and the Reader $\mathcal{R}$ sends *reply = accept*.

Notice that if there is an Adversary who can win the two stage Man-in-the-Middle game (i.e. where he loses access to the Tag in the second stage and must get the reader to accept), then he can also win the game described above since he can simply ignore the messages sent by the Tag in the second stage. Thus security in the model that we will be using in this paper implies security in the "more natural" two stage model.

3   Construction Based on a (Randomized) Weak-PRF

In this section we present our main construction, an authentication protocol secure against man-in-the-middle attacks from any weak-PRF or a randomized weak-PRF that satisfies the three properties stated after Definition 2.3. The protocol based on a weak-PRF is very similar to the one based on a randomized weak-PRF, and so we present them together in Figure 2. The security proofs are also very similar, and we also present them together in the next section.

The underlying building blocks of the protocol in Figure 2 are a pairwise-independent function family $\mathcal{H}$ and a family $\mathcal{F}$ of randomized weak-PRFs with noise χ. If $\mathcal{F}$ is a family of standard (non-randomized) weak-PRFs, then it's the same as a randomized weak-PRF with noise χ, where χ has all of its support on 0 – thus for all $f \in \mathcal{F}$, $f^\chi(\cdot) = f(\cdot)$. The secret keys of the authentication scheme are randomly chosen $f \in \mathcal{F}$, $h \in \mathcal{H}$, and $s \in \mathbb{F}$. In the case that $\mathcal{F}$ is a standard weak-PRF family, we do not need the extra key s, and in the protocol we can assume that $s = 1$. In the case that $\mathcal{F}$ is a randomized weak-PRF family, we assume that it satisfies the three properties after Definition 2.3. Thus there is an associated weight function $\| \cdot \|$ and a value τ' such that the error e chosen from χ satisfies $\|e\| \leq \tau'$ with overwhelming probability.

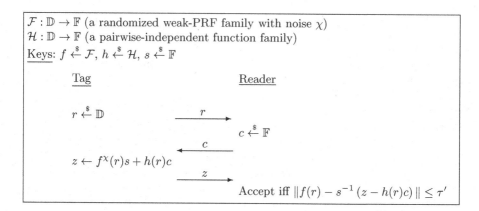

Fig. 2. Authentication Protocol Based on a (randomized) weak-PRF. If the weak-PRF is not randomized, (i.e. the support of the distribution χ is 0 and $\tau' = 0$), then we can set $s = 1$. In this case, the condition $\|f(r) - s^{-1}(z - h(r)c)\| \leq \tau'$ simplifies to $f(r) = z - h(r)c$.

In the first step of the protocol, the Tag picks a random element $r \in \mathbb{D}$ and sends it to the Reader. The reader chooses a random $c \in \mathbb{F}$ and sends it to the Tag. In its turn, the Tag evaluates $f^\chi(r)$ and $h(r)$, and sends $z = f^\chi(r)s + h(r)c$ back to the Reader, where all addition and multiplication operations take place in the field $\mathbb{F}$. In the case that $\mathcal{F}$ is a standard weak-PRF family, the response of the Tag is simply $z = f(r) + h(r)c$. The Reader accepts the Tag if $\|f(r) - s^{-1}(z - h(r)c)\| \leq \tau'$. In case of a standard weak-PRF family without noise, this condition is equivalent to $f(r) = z - h(r)c$.

Example Instantiation. We now give an example instantiation of the protocol using the $\text{LPN}_{n,\tau}$ assumption from Definition 2.4. The noise distribution χ is Ber_τ^m and to choose the secret key f, a random $S \in \mathbb{Z}_2^{n \times m}$ is picked and $f^\chi(r) := Sr + e$ where $e \sim Ber_\tau^m$. Thus f maps the domain $\mathbb{Z}_2^n$ to $\mathbb{Z}_2^m$. As in the discussion following Definition 2.4, the field $\mathbb{F}$ is defined to be $\mathbb{Z}_2[x]/(g(x))$ where $g(x)$ is any irreducible polynomial of degree m. The simplest definition of a pairwise independent function family that maps $\mathbb{Z}_2^n$ to $\mathbb{F}$ is to index the family by two polynomials in $\mathbb{F}$. To pick a random element of the family, one randomly picks $a_1, a_2 \in \mathbb{F}$ and defines $h(r) = a_1 r + a_2$, where r is treated like a polynomial in $\mathbb{F}$ and multiplication and addition is performed over $\mathbb{F}$.[6] The final secret key is a random polynomial $s \in \mathbb{F}$. Thus the secret keys are (S, a_1, a_2, s).

In the protocol, the Tag chooses an $r \in \mathbb{Z}_2^n$ and sends it to the Reader, who replies with a randomly-chosen $c \in \mathbb{F}$. The Tag receives the c computes $f^\chi(r) = Sr + e \in \mathbb{Z}_2^m$, and treats the result as a polynomial in $\mathbb{F}$. He then multiplies it by s and adds it to $h(r)c = (a_1 r + a_2)c$, and sends the resulting $z = f^\chi(r)s + h(r)c$

---

[6] To be able to treat r as an element of $\mathbb{F}$, it is important that $m \geq n$. If $m < n$, then one can define the pairwise-independent function differently (e.g. $h(r) = a_1 r_1 + \ldots a_k r_k + a_{k+1}$), where $r = r_1 | \cdots | r_k$).

to the Reader. The reader computes $f(r) = Sr$ and $s^{-1}(z - h(r)c)$, and accepts if the weight of $f(r) - s^{-1}(z - h(r)c)$ is less than or equal to τ'.

Notice that the protocol would be exactly the same for the Toeplitz-LPN$_{n,\tau}$ problem, with the only difference being how S is defined. By having S be a Toeplitz matrix, the key storage space shrinks from $mn + 3m$ to $n + 4m$, and the matrix-vector multiplication Sr can be computed more efficiently. The Ring-LPN$_{n,\tau}$ protocol would also work in essentially the same way. In this case, we set $m = n$ and have $\mathbb{D} = \mathbb{F}$. The secret key S will just be a random polynomial in $\mathbb{F}$ just like s, a_1, and a_2. Thus Sr will simply be a multiplication of two polynomials in the field $\mathbb{F}$.

Lemma 3.1. *The completeness of the authentication protocol is* $\mathrm{Pr}_{e \sim \chi}[\|e\| \leq \tau']$. *And in particular, if the weak-PRF is not randomized, the completeness is 1.*

Proof. The Tag sets $z \leftarrow f^\chi(r)s + h(r)c = (f(r) + e)s + h(r)c$, where $e \sim \chi$. Thus $f(r) - s^{-1}(z - h(r)c) = e$, and so the Reader accepts whenever $\|f(r) - s^{-1}(z - h(r)c)\| = \|e\| \leq \tau'$.

4   Security of the Authentication Scheme

Theorem 4.1 *Suppose that the authentication protocol in Figure 2 has completeness κ and there is a man-in-the-middle adversary who successfully breaks this scheme with probability ϵ while making at most q_v verification queries. Then there exists an algorithm which, in the same amount of time, has advantage $\frac{1}{2}\left(\kappa^{q_v-1}\left(\epsilon/q_v - 1/|\mathbb{F}|\right)^2 - \beta(2\tau')/|\mathbb{F}|\right)$ in breaking the (randomized) weak-PRF assumption of the family $\mathcal{F}$.*

Proof. If an adversary making q_v verification queries wins the game, then one of these q_v queries can be thought of as the "winning query". By "winning query", we mean that it is the *first* accepted query such that $(r_i', c_i', z_i') \neq (0,0,0)$ (where r_i', c_i', z_i' are as in Figure 1). Once the Adversary sends such a query, he wins the game. If the Adversary has an ϵ success probability of winning the MIM-game, then by an averaging argument there must be some integer $i^* \leq q_v$ such that the probability that the Adversary wins the game and query number i^* is the "winning query" is at least ϵ/q_v. For the rest of the proof, we will assume that we know this i^* (which can be determined a priori by running the adversary on known inputs.)

The Challenger gives us ordered pairs $(r_i, y_i) \in \mathbb{D} \times \mathbb{F}$ where the r_i are uniformly random in $\mathbb{D}$ and the y_i are either uniformly random in $\mathbb{F}$ or equal to $f^\chi(r_i)$ (where f is a randomly-chosen function from the (randomized) weak-PRF family $\mathcal{F}$ with noise χ). We will show how to use the adversary who breaks the authentication protocol with the i^*th winning query to decide which of the two distribution the Challenger is outputting.

Our security proof is most naturally divided into two cases. In the first case, the adversary does not modify the r_{i^*}, in other words, $(r_{i^*}', c_{i^*}', z_{i^*}') = (0, c_{i^*}', z_{i^*}')$. In the second case, the r_{i^*} is modified in the winning query. The manner in which

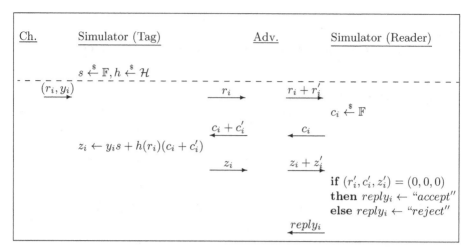

Fig. 3. Simulating the Tag and the Reader before the Adversary's i^*th verification query. If the weak-PRF is not randomized, then we set the secret key s=1 instead of choosing it at random from $\mathbb{F}$.

the simulator uses the Adversary's winning query to respond to the challenger differs based on whether r'_{i^*} is 0 or not. For the purposes of improved readability, throughout the rest of the paper, we will drop the subscript i^* from all variables in the winning query. So for example, instead of writing r'_{i^*}, we simply write r'.

Answering the Challenger when $r' = 0$. We first show how to simulate the Tag and the Reader before the Adversary's i^*th verification query (see Figure 3). We pick a random $s \in \mathbb{F}$ and $h \in \mathcal{H}$ as the secret keys, and upon receiving a pair (r_i, y_i) from the Challenger, we send r_i to the Adversary. The Adversary can then modify this and forward $r_i + r'_i$ to the Reader. The Reader picks a random $c_i \in \mathbb{F}$, sends it to the adversary, who then sends the possibly modified challenge $c_i + c'_i$ to the Tag. The Simulator playing as the Tag computes $h(r_i)(c_i + c'_i)$ using his secret key h, and then uses the y_i received from the challenger together with his other secret key s, to send $z_i = y_i s + h(r_i)(c_i + c'_i)$. After receiving z_i, the Adversary may send $z_i + z'_i$ to the verifier and make a verification query.

Notice that if the Challenger sends $(r_i, y_i) = f^\chi(r_i))$, then the responses of the Tag are exactly what they should be if the secret key were (f, s, h). Thus if $(r'_i, c'_i, z'_i) = (0, 0, 0)$, the Reader who always sends "accept" is correct with probability κ (the completeness of the protocol). And if $(r'_i, c'_i, z'_i) \neq (0, 0, 0)$, the response of "reject" is also correct since the i^*th verification query has not yet been reached. Because the simulator has faithfully simulated the valid Tag and Reader up to this point with probability $\kappa^{q_v - 1}$, the Adversary's i^*th query will be the "winning one" (i.e. $(r', c', z') \neq (0, 0, 0)$ and $\|f(r + r') - s^{-1}((z + z') - h(r + r')c)\| \leq \tau'$) with probability $\kappa^{q_v - 1}\epsilon/q_v$. Additionally, because $z = f^\chi(r)s + h(r)(c + c')$, we also have $\|f(r) - s^{-1}(z - h(r)(c + c'))\| \leq \tau'$.

Ch.	Simulator (Tag)	Adv.	Simulator (Reader)
(r,y)			
		$\xrightarrow{\quad r \quad}$ $\xrightarrow{\quad r \quad}$	$c \xleftarrow{\$} \mathbb{F}$
	$\xleftarrow{\ c+c'\ }$	$\xleftarrow{\ c\ }$	
	$z \leftarrow ys + h(r)(c+c')$		
		$\xrightarrow{\quad z \quad}$ $\xrightarrow{\ z+z'\ }$	
			if $\|s^{-1}(z'+h(r)c')\| \le 2\tau'$ **then** *reply* $\leftarrow$ *"not random"* **else** *reply* $\leftarrow$ *"random"*
	$\xleftarrow{\ reply\ }$		

Fig. 4. Answering the Challenger after the Adversary's i^*th verification query if $r'=0$

Thus by the triangular inequality, we obtain $\|s^{-1}(z'+h(r)c')\| \le 2\tau'$. If this condition is satisfied, we respond to the challenger that the ordered pairs he was sending were indeed of the form $(r_i, y_i = f^{\chi}(r_i))$.

Lemma 4.1. *If the challenger were sending valid pairs, i.e.* $(r_i, y_i = f^{\chi}(r_i))$ $\forall i$, *then* $\|s^{-1}(z'+h(r)c')\| \le 2\tau'$ *should be satisfied with probability at least* $\kappa^{q_v-1}\epsilon/q_v$.

On the other hand, if the Challenger were sending uniformly random pairs $(r_i, y_i) \in (\mathbb{D}, \mathbb{F})$ then we will show that the adversary is not be able (except with a negligible probability) to come up with $(c', z') \ne (0,0)$ that satisfy the inequality $\|s^{-1}(z'+h(r)c')\| \le 2\tau'$. Notice that if the y_i are uniform and independent of the r_i, the secret keys h, s chosen by the simulator are information-theoretically hidden throughout the interaction in Figure 3. Therefore the Adversary's behavior will be exactly the same as in the case where s and h are chosen *after* he sends his i^*th query. In Lemma 4.2, we use this to show that even an all-powerful adversary cannot produce a query $z + z'$ such that $\|s^{-1}(z'+h(r)c')\| \le 2\tau'$, except with probability $\beta(2\tau')/|\mathbb{F}|$.

Lemma 4.2. *If the ordered pairs* (r_i, y_i) *sent by the challenger are uniformly random in* $\mathbb{D} \times \mathbb{F}$, *then the probability that even an all-powerful adversary can output* $(c', z') \ne (0,0)$ *such that* $\|s^{-1}(z'+h(r)c')\| \le 2\tau'$ *is at most* $|\beta(2\tau')|/|\mathbb{F}|$.

Proof. We first handle the case where f is a weak-PRF without any noise (i.e. the support of the distribution χ is 0 and $\tau' = 0$). In this case, the extra random key s is not necessary in the protocol (i.e. $s = 1$) and so the condition $\|s^{-1}(z'+h(r)c')\| \le 2\tau'$ becomes $0 = z'+h(r)c'$. Since y is uniformly random in $\mathbb{F}$ and independent of everything else, the value z that the adversary receives is also uniformly random and independent of the pairwise independent hash function h. Thus the adversary will behave in the same way if the function

h were chosen *after* the adversary chooses c' and z'. Notice that the adversary must set $c' \neq 0$ because otherwise z' is also necessarily 0. Thus,

$$\forall r \in \mathbb{D}, z' \in \mathbb{F}, c' \in \mathbb{F} \setminus \{0\}, \Pr_h[0 = z' + h(r)c'] = \Pr_h[h(r) = -z'c'^{-1}] = 1/|\mathbb{F}|.$$

The proof for case where the support of χ is not restricted to 0 is similar, except that it also uses the unpredictability of the key s. The full proof is given in the full version. □

Answering the Challenger When $r' \neq 0$. We now deal with the case where the Adversary's winning query changes the randomness r to $r + r'$. Performing the simulation until the i^*th query is exactly the same as before (i.e. see Figure 3). Similarly, if the Challenger sends $(r_i, y_i = f^\chi(r_i))$, then the responses of the Tag are exactly what they should be if the secret key were (f, s, h). And so, as before, the Adversary's i^*th query will be the "winning one" with probability $\kappa^{q_v-1}\epsilon/q_v$. The difference from the previous part lies in how we will use the Adversary's response in the i^*th query to respond to the challenger. Unlike the previous case, we will now need to rewind the adversary and receive two responses for the same value of $r + r'$ (see Figure 5). By the Reset Lemma [2, Lemma 3.1], the adversary will respond correctly to two distinct challenges c_0 and c_1 with probability $\kappa^{q_v-1} (\epsilon/q_v - 1/|\mathbb{F}|)^2$. If the Adversary successfully replies to the two queries, then we have $\|f(r + r') - s^{-1}(z_0 + z_0' - h(r + r')c_0)\| \leq \tau'$ and $\|f(r + r') - s^{-1}(z_1 + z_1' - h(r + r')c_1)\| \leq \tau'$. Thus, by the triangle inequality, we have the condition $\|s^{-1}((z_0 + z_0') - (z_1 + z_1') - h(r + r')(c_0 - c_1))\| \leq 2\tau'$. If this is satisfied, we reply to the challenger that the ordered pairs he was sending were indeed of the form $(r_i, y_i = f^\chi(r_i))$.

Lemma 4.3. *If the Challenger were sending valid pairs, i.e. $(r_i, y_i = f^\chi(r_i))$ $\forall i$, then $\|s^{-1}((z_0 + z_0') - (z_1 + z_1') - h(r + r')(c_0 - c_1))\| \leq 2\tau'$ should be satisfied with probability at least $\kappa^{q_v-1} (\epsilon/q_v - 1/|\mathbb{F}|)^2$.*

On the other hand, if the Challenger were sending uniformly random pairs $(r_i, y_i) \in (\mathbb{D}, \mathbb{F})$ then we will show that the adversary is not be able (except with a negligible probability) to come up with (r', c', z') where $r' \neq 0$ that satisfy $\|s^{-1}((z_0 + z_0') - (z_1 + z_1') - h(r + r')(c_0 - c_1))\| \leq 2\tau'$. As before, notice that if the y_i are uniform and independent of the r_i, the secret keys h, s chosen by the simulator are information-theoretically hidden throughout the interaction in Figure 3. Therefore the Adversary's behavior will be exactly the same as in the case where s and h are chosen *after* he outputs his i^*th query *the first time*. When we rewind the Adversary, we also end up rewinding the left-hand side of the simulator, which will end up revealing some information about h. But we use the pairwise-independent property of h to show (in Lemma 4.4) that even an all-powerful adversary still cannot produce a query $z + z'$ such that $\|s^{-1}((z_0 + z_0') - (z_1 + z_1') - h(r + r')(c_0 - c_1))\| \leq 2\tau'$, except with probability $\beta(2\tau')/|\mathbb{F}|$.

Ch.	Simulator (Tag)	Adv.	Simulator (Reader)

$\underrightarrow{(r,y)}$ $\xrightarrow{r}$ $\xrightarrow{r+r'}$

$$c_0, c_1 \xleftarrow{\$} \mathbb{F}$$

- -

$\xleftarrow{c_0 + c_0'}$ $\xleftarrow{c_0}$

$z_0 \leftarrow ys + h(r)(c_0 + c_0')$

$\xrightarrow{z_0}$ $\xrightarrow{z_0 + z_0'}$

- -

$\xleftarrow{c_1 + c_1'}$ $\xleftarrow{c_1}$

$z_1 \leftarrow ys + h(r)(c_1 + c_1')$

$\xrightarrow{z_1}$ $\xrightarrow{z_1 + z_1'}$

if $\|s^{-1}((z_0 + z_0') - (z_1 + z_1') - h(r + r')(c_0 - c_1))\| \le 2\tau'$
then *reply* $\leftarrow$ *"not random"*
else *reply* $\leftarrow$ *"random"*

$\xleftarrow{reply}$

Fig. 5. Answering the Challenger after the Adversary's i^*th verification query if $r' \ne 0$

Lemma 4.4. *If the ordered pairs (r_i, y_i) sent by the challenger are uniformly random in $\mathbb{D} \times \mathbb{F}$, and $r' \ne 0$, and $c_0 \ne c_1$, then the probability that even an all-powerful adversary can output z_0' and z_1' such that $\|s^{-1}((z_0 + z_0') - (z_1 + z_1') - h(r + r')(c_0 - c_1))\| \le 2\tau'$ is at most $|\beta(2\tau')|/|\mathbb{F}|$.*

Proof. For simplicity, we will define $w = (z_0 + z_0') - (z_1 + z_1')$. The information given to the adversary (in the two rewindings) by the simulator playing as the tag is $z_0 = ys + h(r)(c_0 + c_0')$ and $z_1 = ys + h(r)(c_1 + c_1')$. This is exactly the same as receiving z_0 and $\tilde{z} = z_0 - z_1 = h(r)(c_0 + c_0' - (c_1 + c_1'))$. Notice that since z_0 contains the term ys, the value of z_0 is uniform and independent of the function h. The value of $\tilde{z}$, on the other hand, does depend on $h(r)$. So the behavior of the adversary would be unchanged if we chose z_0 uniformly at random, chose a random element u for $h(r)$ and set $\tilde{z} = h(r)(c_0 + c_0' - (c_1 + c_1'))$, and then *after* the adversary picks z_0', z_1', we finally choose h (conditioned on the already set value of $h(r)$). Thus we have that $\forall t \in \beta(2\tau'), c_0 \ne c_1 \in \mathbb{F}, r \in \mathbb{D}, r' \ne 0, w, s, u \in \mathbb{F}$

$$\Pr_h[s^{-1}(w - h(r + r')(c_0 - c_1)) = t \mid h(r) = u] = \Pr_h[h(r + r')$$
$$= (w - st)(c_0 - c_1)^{-1} \mid h(r) = u] = 1/|\mathbb{F}|$$

where $(c_0 - c_1)^{-1}$ exists since we assumed $c_0 \ne c_1$ and the last equality is true because h is a pairwise-independent function and $r' \ne 0$. □

Combining Lemmas 4.1, 4.2, 4.3, and 4.4 gives the statement of Theorem 4.1. □

5 Discussion and Open Problems

In this work we presented a very efficient 3-round authentication scheme that utilizes only one call to a (randomized) weak-PRF, and proved it secure against sequential MIM attacks. The security proof can be extended to the scenario where Adversary is allowed to concurrently interact with many copies of the Reader (this will be shown in the full version of the paper). Another simple extension is the conversion of the scheme into an interactive message authentication scheme by sending $r||\mu$ in lieu of just r in the first round and setting $z \leftarrow f^x(r)s + h(r||\mu)c$ in the third round, where μ is the message digest.

We believe that the most interesting (theoretical) question left open by our work is to construct a MIM-secure authentication scheme that is secure in the full concurrent setting (i.e. where the Adversary is also allowed to concurrently interact with multiple provers). Such a scheme can be easily constructed by first creating a PRF from a weak-PRF using $O(n)$ calls to the weak-PRF. The challenge is thus to construct such a scheme using just one, or even a constant number, of (randomized) weak-PRF invocations.

Acknowledgements. We are very grateful to Eike Kiltz and Krzysztof Pietrzak for numerous discussions pertaining to their work on the LPN problem. We also thank Daniel Wichs and the other anonymous CRYPTO 2013 reviewers for their valuable comments – in particular mentioning the application to interactive MACs.

References

1. Banerjee, A., Peikert, C., Rosen, A.: Pseudorandom functions and lattices. In: Pointcheval, D., Johansson, T. (eds.) EUROCRYPT 2012. LNCS, vol. 7237, pp. 719–737. Springer, Heidelberg (2012)
2. Bellare, M., Palacio, A.: Gq and schnorr identification schemes: Proofs of security against impersonation under active and concurrent attacks. In: Yung, M. (ed.) CRYPTO 2002. LNCS, vol. 2442, pp. 162–177. Springer, Heidelberg (2002)
3. Blum, A., Furst, M., Kearns, M., Lipton, R.J.: Cryptographic primitives based on hard learning problems. In: Stinson, D.R. (ed.) CRYPTO 1993. LNCS, vol. CRYPTO, pp. 278–291. Springer, Heidelberg (1994)
4. Bogdanov, A., Knudsen, L.R., Leander, G., Paar, C., Poschmann, A., Robshaw, M., Seurin, Y., Vikkelsoe, C.: Present: An ultra-lightweight block cipher. In: Paillier, P., Verbauwhede, I. (eds.) CHES 2007. LNCS, vol. 4727, pp. 450–466. Springer, Heidelberg (2007)
5. Damgård, I.B., Nielsen, J.B.: Expanding pseudorandom functions; or: From known-plaintext security to chosen-plaintext security. In: Yung, M. (ed.) CRYPTO 2002. LNCS, vol. 2442, pp. 449–464. Springer, Heidelberg (2002)
6. Dodis, Y., Kiltz, E., Pietrzak, K., Wichs, D.: Message authentication, revisited. In: Pointcheval, D., Johansson, T. (eds.) EUROCRYPT 2012. LNCS, vol. 7237, pp. 355–374. Springer, Heidelberg (2012)
7. Gilbert, H., Robshaw, M., Sibert, H.: An active attack against HB$^+$ - a provably secure lightweight authentication protocol. Cryptology ePrint Archive, Report 2005/237 (2005)

8. Gilbert, H., Robshaw, M.J.B., Seurin, Y.: Good variants of HB$^+$ are hard to find. In: Tsudik, G. (ed.) FC 2008. LNCS, vol. 5143, pp. 156–170. Springer, Heidelberg (2008)

9. Gilbert, H., Robshaw, M.J.B., Seurin, Y.: HB$^\#$: Increasing the security and efficiency of HB$^+$. In: Smart, N.P. (ed.) EUROCRYPT 2008. LNCS, vol. 4965, pp. 361–378. Springer, Heidelberg (2008)

10. Goldreich, O., Goldwasser, S., Micali, S.: How to construct random functions. J. ACM 33(4), 792–807 (1986)

11. Goldreich, O., Krawczyk, H., Luby, M.: On the existence of pseudorandom generators. SIAM J. Comput. 22(6), 1163–1175 (1993)

12. Guillou, L.C., Quisquater, J.-J.: A "Paradoxical" identity-based signature scheme resulting from zero-knowledge. In: Goldwasser, S. (ed.) CRYPTO 1988. LNCS, vol. 403, pp. 216–231. Springer, Heidelberg (1990)

13. Heyse, S., Kiltz, E., Lyubashevsky, V., Paar, C., Pietrzak, K.: Lapin: An efficient authentication protocol based on ring-lpn. In: Canteaut, A. (ed.) FSE 2012. LNCS, vol. 7549, pp. 346–365. Springer, Heidelberg (2012)

14. Hopper, N.J., Blum, M.: Secure human identification protocols. In: Boyd, C. (ed.) ASIACRYPT 2001. LNCS, vol. 2248, pp. 52–66. Springer, Heidelberg (2001)

15. Juels, A., Weis, S.A.: Authenticating pervasive devices with human protocols. In: Shoup, V. (ed.) CRYPTO 2005. LNCS, vol. 3621, pp. 293–308. Springer, Heidelberg (2005)

16. Katz, J., Shin, J.S.: Parallel and concurrent security of the HB and HB$^+$ protocols. In: Vaudenay, S. (ed.) EUROCRYPT 2006. LNCS, vol. 4004, pp. 73–87. Springer, Heidelberg (2006)

17. Katz, J., Shin, J.S., Smith, A.: Parallel and concurrent security of the HB and HB$^+$ protocols. J. Cryptology 23(3), 402–421 (2010)

18. Kiltz, E., Pietrzak, K., Cash, D., Jain, A., Venturi, D.: Efficient authentication from hard learning problems. In: Paterson, K.G. (ed.) EUROCRYPT 2011. LNCS, vol. 6632, pp. 7–26. Springer, Heidelberg (2011)

19. Leander, G., Paar, C., Poschmann, A., Schramm, K.: New lightweight des variants. In: Biryukov, A. (ed.) FSE 2007. LNCS, vol. 4593, pp. 196–210. Springer, Heidelberg (2007)

20. Lyubashevsky, V., Peikert, C., Regev, O.: On ideal lattices and learning with errors over rings. In: Gilbert, H. (ed.) EUROCRYPT 2010. LNCS, vol. 6110, pp. 1–23. Springer, Heidelberg (2010)

21. Maurer, U.M., Sjödin, J.: A fast and key-efficient reduction of chosen-ciphertext to known-plaintext security. In: Naor, M. (ed.) EUROCRYPT 2007. LNCS, vol. 4515, pp. 498–516. Springer, Heidelberg (2007)

22. Naor, M., Reingold, O.: Number-theoretic constructions of efficient pseudo-random functions. In: FOCS, pp. 458–467 (1997)

23. Naor, M., Reingold, O., Rosen, A.: Pseudorandom functions and factoring. SIAM J. Comput. 31(5), 1383–1404 (2002)

24. Ouafi, K., Overbeck, R., Vaudenay, S.: On the security of HB$^\#$ against a man-in-the-middle attack. In: Pieprzyk, J. (ed.) ASIACRYPT 2008. LNCS, vol. 5350, pp. 108–124. Springer, Heidelberg (2008)

25. Pietrzak, K.: A leakage-resilient mode of operation. In: Joux, A. (ed.) EUROCRYPT 2009. LNCS, vol. 5479, pp. 462–482. Springer, Heidelberg (2009)

26. Regev, O.: On lattices, learning with errors, random linear codes, and cryptography. J. ACM 56(6) (2009)

27. Schnorr, C.-P.: Efficient signature generation by smart cards. J. Cryptology 4(3), 161–174 (1991)

Achieving the Limits of the Noisy-Storage Model Using Entanglement Sampling

Frédéric Dupuis[1,2], Omar Fawzi[2], and Stephanie Wehner[3]

[1] Department of Computer Science, Aarhus University, Denmark
[2] Institute for Theoretical Physics, ETH Zürich, Switzerland
[3] Center for Quantum Technologies, National University of Singapore, Singapore

Abstract. A natural measure for the amount of quantum information that a physical system E holds about another system $A = A_1, ..., A_n$ is given by the min-entropy $H_{\min}(A|E)$. Specifically, the min-entropy measures the amount of entanglement between E and A, and is the relevant measure when analyzing a wide variety of problems ranging from randomness extraction in quantum cryptography, decoupling used in channel coding, to physical processes such as thermalization or the thermodynamic work cost (or gain) of erasing a quantum system. As such, it is a central question to determine the behaviour of the min-entropy after some process M is applied to the system A. Here we introduce a new generic tool relating the resulting min-entropy to the original one, and apply it to several settings of interest, including sampling of subsystems and measuring in a randomly chosen basis. The results on random measurements yield new high-order entropic uncertainty relations with which we prove the optimality of cryptographic schemes in the bounded quantum storage model. This is an abridged version of the paper; the full version containing all proofs and further applications can be found in [13].

1   Introduction

A central task in quantum theory is to effectively quantify the amount of information that some system E holds about some classical or quantum data A. For classical data, i.e., A is a string $X^n = X_1, \ldots, X_n$, the *min-entropy* $H_{\min}(X^n|E)$ forms a particularly relevant measure because it determines the length of a secure key that can be obtained from X^n. This is the setting typically considered in quantum key distribution where E is some information that an adversary Eve has gathered during the course of the protocol, and X^n is the so-called raw key. More precisely, the maximum number ℓ of (almost) random bits [1] that can be obtained from X^n that are both uniform and uncorrelated from E obeys $\ell \approx H_{\min}(X^n|E)$, if E is classical [15] and quantum [25]. The process by which such randomness is obtained is known as *randomness extraction* (see [30] for a survey) or privacy amplification. Classically, a (strong) randomness extractor is simply a set of functions $\mathcal{F} = \{f : \{0,1\}^n \to \{0,1\}^\ell\}$ such that for almost all functions $f \in \mathcal{F}$, its output $f(X^n)$ is close to uniform and uncorrelated from the

---

[1] We restrict ourselves to bits in the introduction, however, all our results also apply to higher dimensional alphabets.

R. Canetti and J.A. Garay (Eds.): CRYPTO 2013, Part II, LNCS 8043, pp. 326–343, 2013.

adversary, even if he learns which function was applied. That is, the output is of the form $\rho_{F(X)EF} \approx \mathrm{id}/2^n \otimes \rho_{EF}$. A well known example of such a set $\mathcal{F}$ is a set of two-universal hash functions which are used in quantum cryptography to turn a raw key X^n into a secure key $f(X^n)$. The min-entropy also has a very intuitive interpretation as it can be expressed as $\mathrm{H}_{\min}(X^n|E) = -\log P_{\mathrm{guess}}(X^n|E)$ where $P_{\mathrm{guess}}(X^n|E)$ is the probability that the adversary manages to guess X^n maximized over all measurements on E [16].

What can we say in the case of quantum data A? It turns out that the fully quantum min-entropy $\mathrm{H}_{\min}(A|E)$ provides us with a similarly useful way to quantify the amount of information that E holds about A. Its first significance is to quantum cryptography where E is again held by an adversary. More specifically, it has been shown that a quantum-to-classical extractor (QC-extractor) can produce exactly $\ell \approx \mathrm{H}_{\min}(A|E) + \log|A|$ classical bits which are uniform and uncorrelated from E [7]. Instead of applying functions to a classical string, a QC-extractor consists of a set of projective measurements on A giving a classical string as a measurement outcome. Such extractors form a useful tool in two-party quantum cryptography where one might have an estimate of $\mathrm{H}_{\min}(A|E)$, but not of the min-entropy of any classical string X^n produced from A. Thus $\mathrm{H}_{\min}(A|E)$ is directly related to the amount of cryptographic randomness that can be produced from A.

It turns out that the fully quantum min-entropy also enjoys a very appealing operational interpretation [16]. More precisely,

$$\mathrm{H}_{\min}(A|E) = -\log\left(|A| \max_{\Lambda_{E\to\bar{A}}} F(\Phi^N_{A\bar{A}}, \mathrm{id}_A \otimes \Lambda_{E\to\bar{A}}(\rho_{AE}))^2\right), \qquad (1)$$

where F is the fidelity (see below) and $\Phi^N_{A\bar{A}}$ is the normalized maximally entangled state across A and $\bar{A}$. That is, $\mathrm{H}_{\min}(A|E)$ measures how close ρ_{AE} can be brought to the maximally entangled state by performing a quantum operation on E. Intuitively, this quantifies how close the adversary E can bring himself to being quantumly maximally correlated with A — exactly analogous to maximizing his classical correlations by trying to guess X^n.

1.1 Results

Given the significance of the min-entropy in quantum information, it is a natural question to ask how the min-entropy changes if we apply a quantum operation $\mathcal{M}$ to A. More precisely, one might ask how $\mathrm{H}_{\min}(\mathcal{M}(A)|E)$ relates to $\mathrm{H}_{\min}(A|E)$, for some completely positive trace preserving map $\mathcal{M}$. At present, we know that the min-entropy satisfies $\mathrm{H}_{\min}(\mathcal{M}(A)|E) \geqslant \mathrm{H}_{\min}(A|E)$ if $\mathcal{M}$ is unital [27]. Can we make more refined statements?

Of particular interest to us is the case where the quantum system consist of n qudits $A^n = A_1, \ldots, A_n$. Our main result is to establish the following very general theorem for maps $\mathcal{M}$ with the property that we can diagonalize $((\mathcal{M}^\dagger \circ \mathcal{M}) \otimes \mathrm{id}_{\bar{A}^n})(\Phi_{A^n\bar{A}^n}) = \sum_{s\in\{0,\ldots,d^2-1\}} \lambda_s \Phi_s$ where $A^n = A_1, \ldots, A_n$, $d = |A_j|$ is the dimension of one of the individual qudits, $\Phi_{A^n\bar{A}^n}$ is again the maximally entangled state, and $\{\Phi_s\}_s$ is a basis for the space $A^n \otimes \bar{A}^n$ consisting of maximally entangled vectors, and

$\lambda_s \geq 0$ are the corresponding eigenvalues (see Sections 2 and Section 3 for precise definitions and statement of the theorem). In terms of the smooth min-entropy $\mathrm{H}_{\min}^{\varepsilon}$, which, loosely speaking, is equal to the min-entropy except with error probability ε, our first contribution can be stated as

- **Main result (Informal)** For any partition of $\{0, \ldots, d^2 - 1\}^n = \mathfrak{S}_+ \cup \mathfrak{S}_-$ into subsets $\mathfrak{S}_+, \mathfrak{S}_-$ we have $2^{-\mathrm{H}_{\min}^{\varepsilon}(\mathcal{M}(A^n)|E)} \lesssim \sum_{s \in \mathfrak{S}_+} \lambda_s 2^{-\mathrm{H}_{\min}(A^n|E)} + (\max_{s \in \mathfrak{S}_-} \lambda_s) d^n$.

At first glance, our condition on the maps $\mathcal{M}$ may seem rather unintuitive and indeed restrictive. Yet, it turns out that many interesting maps do indeed satisfy these conditions, allowing us to establish the following results.

Entanglement Sampling. In the study of classical extractors, a goal was to construct families of functions f that are *locally computable* [31]. That is, if our goal were to extract only a very small number of key bits from a long string X^n of length n, one might wonder whether this can be done efficiently in the sense that the functions f depend only on a small number of bits of X^n. Classically, a very beautiful method to answer this question is to show that the min-entropy can in fact be *sampled* [31,24]. That is, if we choose a subset S of the bits at random, then the min-entropy of the bits X_S in that subset S obeys

$$\mathrm{H}_{\min}(X_S|ES) \gtrsim |S|R(\mathrm{H}_{\min}(X^n|E)/n) \,, \tag{2}$$

for some function R. The function R can be understood as a rate function that determines the relation of the original min-entropy rate $\frac{\mathrm{H}_{\min}(X^n|E)}{n}$ to the min-entropy rate on a subset S of the bits. In other words, min-entropy sampling says that if X^n is hard to guess, then even given the choice of subset S it is tricky for the adversary to guess X_S. To see why this yields the desired functions f note that one way to construct a randomness extractor would be to first pick a random subset S, and then apply an arbitrary extractor to the much shorter bit string X_S. In the classical literature, this is known as the sample-then-extract approach [31].

Inspired by the classical results of Vadhan [31], it is a natural question whether there exists QC-extractors which are efficient in the sense that the measurements $M \in \mathcal{M}$ only act on a small number of qubits of $A^n = A_1, \ldots, A_n$. Or, even more generally, whether there exist decoupling operations which depend on only very few qubits. As before, one way to answer this question in generality is to show that even the fully quantum min-entropy can be sampled.

- **Entanglement sampling (Informal)** For any quantum state $\rho_{A^n E}$, i.e., $\mathrm{H}_{\min}^{\varepsilon}(A_S|ES) \gtrsim |S|R(\mathrm{H}_{\min}(A^n|E)/n)$ for the rate function R plotted in Figure 1. See Theorem 2 for a precise statement.

It should be noted that even the case of standard min-entropy sampling of a classical string X^n, but quantum side information E has proved challenging. The results of [4] imply that sampling of classical strings is possible when the distribution over the strings X^n is uniform (i.e., $\rho_{X^n E} = (1/2^n) \sum_{x \in \{0,1\}^n} |x\rangle\langle x| \otimes \rho_E^x$), and the size of E

is bounded, and [18] has shown that sampling of blocks (but not individual bits) is possible. This was later refined in [34] to show that bitwise sampling is also possible (see Figure 1 for a comparison of the rate function). Very roughly, the techniques used in [34] relate the adversary's ability to guess the string X^n to his ability to guess the XOR of bits in the string. Clearly, in the case of fully quantum A^n such techniques cannot be used as it is indeed unclear what the XOR of qubits even means.

As this is the first result on entanglement sampling, it required entirely novel techniques. More precisely, it inspired the even more general theorem sketched above, from which entanglement sampling follows by choosing an appropriate map $\mathcal{M}$. As a byproduct, using the same techniques, we also obtain a stronger statement of sampling a classical string X^n with respect to a quantum system E in the sense that the rate R is improved (see Figure 1 for a comparison). What's more, we are able to show an even more precise statement in terms of the entropy $H_2(A^n|E)_\rho$ - without any ε error terms. Classically, this quantity is known as the (conditional) collision entropy. In general, it is very closely related to the min-entropy, and in fact enjoys a very similar operational interpretation. More specifically, it can be expressed in the same form as (1) where the optimization over all quantum operations $\Lambda_{E \to \bar{A}^n}$ is replaced by the so-called *pretty good recovery* map $\Lambda_{E \to \bar{A}^n}^{\mathrm{pg}}$ which is close to optimal [2].

Uncertainty Relations. Another consequence of our main result is a new uncertainty relation with quantum side information for measurements of n qubits $A^n = A_1, \ldots, A_n$ in randomly chosen BB84 bases. Apart from the foundational consequences, such relations have found applications in quantum cryptography (see e.g., [7]). Our result establishes the first entropic uncertainty relation with quantum side-information that uses a high-order entropy like the min-entropy and that is nontrivial as soon as the system being measured is not maximally entangled with the observer E. In other words, this shows a quantitative bound on the probability of successfully guessing the measurement outcome that is nontrivial as soon as $H_{\min}(A^n|E) > -n$.[2]

- **High-order entropic uncertainty relation for BB84 bases** If X^n is obtained by measuring the system A^n in a random BB84 basis Θ^n, we have $H_{\min}(X^n|E\Theta^n) \geqslant n \cdot \frac{1}{2}\gamma\left(\frac{H_{\min}(A^n|E)}{n}\right)$, where the function γ is plotted in Figure 2. See Theorem 5 and Corollary 6 for precise statements.

We can also prove uncertainty relations for qudit-wise measurements in mutually unbiased bases (see full version [13]). Again, these results follow from our very general theorem sketched above, this time for a map $\mathcal{M}$ that represents randomly chosen measurements.

Applications to the Noisy-Storage Model. Our new uncertainty relations have several interesting applications to cryptography. The goal of two-party cryptography is to enable Alice and Bob to solve tasks in cooperation even if they do not trust each other. A classic example of such tasks are bit commitment and oblivious transfer. Unfortunately, it has been shown that even using quantum communication, none of these tasks can

---

[2] The fully quantum min-entropy can be negative up to $H_{\min}(A^n|E) = -n$ if $\rho_{A^n E}$ is the maximally entangled state.

be implemented securely without making assumptions [22,19]. What makes such tasks more difficult than quantum key distribution is that Alice and Bob cannot collaborate to check on any eavesdropper. Instead, each party has to fend for itself.

Nevertheless, because two-party computation is such a central part of modern cryptography, one is willing to make *assumptions* on how powerful an attacker can be in order to implement them securely. Classically, such assumptions generally take the form of computational assumptions, where we assume that a particular mathematical problem cannot be solved in polynomial time. Here, we consider *physical* assumptions that can enable us to solve such tasks. In particular, can the sole assumption of a limited storage device lead to security [21]? This is indeed the case and it was shown that security can be obtained if the attacker's *classical* storage is limited [21,9]. Yet, apart from the fact that classical storage is cheap and plentiful, assuming a limited classical storage has one rather crucial caveat: If the honest players need to store n classical bits to execute the protocol in the first place, *any* classical protocol can be broken if the attacker can store more than roughly n^2 bits [14]. Motivated by this unsatisfactory gap, it was thus suggested to assume that the attacker's *quantum* storage was bounded [5,10,11,12,8], or, more generally, noisy [32,26,17]. The central assumption of the noisy-storage model is that during waiting times Δt introduced in the protocol, the attacker can keep quantum information only in his noisy quantum storage device; otherwise he is all-powerful (see Section 4.4).

The assumption of bounded or noisy quantum storage offers significant advantages in that the proposed protocols do not require any quantum storage at all to be implemented by the honest parties. They are typically based on BB84 [17] or six-state [7] encodings, and indeed the first implementation of a bit commitment protocol has recently been performed experimentally [23]. So far it was known that there exist protocols that send n qubits encoded in either the BB84 or six-state encoding, and that are secure as long as the adversary can only store strictly less than $n/2$ or $2n/3$ noise-free qubits respectively.

Using our new techniques, we are able to show security of the primitive called *weak string erasure* [17] (see Section 4.4), which in turn can be supplemented with additional classical or quantum communication to obtain primitives such as bit commitment.

– **Application 1: Bounded storage** There exists a weak string erasure protocol transmitting n qubits that is secure as long as the adversary can store at most strictly less than $n - O(\log^2 n)$ qubits. The protocol does not require any quantum memory to be executed, and merely requires simple quantum operations and measurements. See Theorem 8 for a precise statement.

It should be noted that no such protocol can be secure as soon as the adversary can store n qubits, so our result is essentially optimal. Our result highlights the sharp contrast between the classical and the quantum bounded storage model and answers the main open question in the BQSM. The noisy-storage model offers an advantage over the case of bounded-storage not only for implementations using high-dimensional encodings such as the infinite-dimensional states sent in continuous variable experiments, but allows security even for arbitrarily large storage devices as long as the noise is large enough. Essentially, the noisy-storage model captures our intuition that security should be linked to how much information the adversary can store in his quantum memory.

The first proofs linked security to the classical capacity [17], the entanglement cost [6] and finally the quantum capacity [7]. The latter result used a protocol based on six-state encodings and required the fidelity of the device to be exponentially small in the number of qubits communicated during the protocol.

- **Application 2: Noisy storage** We prove that security in the noisy-storage model is possible basically as soon as the fidelity of the storage device is smaller than desired error parameter, which is best possible (see Section 4.4). Furthermore, we link security of a BB84-based protocol to the quantum capacity of the adversary's storage device for the first time. See Theorem 7 for a precise statement.

2 Preliminaries

2.1 Basic Concepts and Notation

In quantum mechanics, a system such as Alice's or Bob's labs are described mathematically by *Hilbert spaces*, denoted by $A, B, C, \ldots$. Here, we follow the usual convention in quantum cryptography and assume that all Hilbert spaces are finite-dimensional. We write $|A|$ for the dimension of A. A system of n qudits is also denoted as $A^n = A_1, \ldots, A_n$, where we also use $|A|$ to denote the dimension of one single qudit in A^n. The set of linear operators on A is denoted by $\mathcal{L}(A)$, and we write $\mathrm{Herm}(A)$ and $\mathrm{Pos}(A)$ for the set of hermitian and positive semidefinite operators on A respectively. We denote the adjoint of an operator M by $M^\dagger$. A *quantum state* ρ_A is an operator $\rho_A \in \mathcal{S}(A)$, where $\mathcal{S}(A) = \{\sigma_A \in \mathrm{Pos}(A) \mid \mathrm{Tr}(\sigma_A) = 1\}$. We will often make use of *operator inequalities*: whenever $X, Y \in \mathrm{Herm}(A)$, we write $X \leqslant Y$ to mean that $Y - X \in \mathrm{Pos}(A)$. A quantum operation is given by a completely positive map $\mathcal{M} : \mathcal{L}(A) \to \mathcal{L}(C)$. A map $\mathcal{M}$ is said to be completely positive if for any system B and $X \in \mathrm{Pos}(A \otimes B)$ we have $(\mathcal{M} \otimes \mathrm{id})(X) \geqslant 0$.

Throughout, we use the shorthand $[d] = \{0, 1, \ldots, d - 1\}$. We will follow the convention to use H to denote the unitary that takes the computational $\{|0\rangle, |1\rangle\}$ to the Hadamard basis: $H|0\rangle = \frac{1}{\sqrt{2}}(|0\rangle + |1\rangle), H|1\rangle = \frac{1}{\sqrt{2}}(|0\rangle - |1\rangle)$. When considering n qubits, we also use $H^{\theta^n} = H^{\theta_1} \otimes \cdots \otimes H^{\theta_n}$ for the unitary defining the basis $\theta^n \in \{0, 1\}^n$.

2.2 Entropies

Next to its operational interpretation given in (1), the *conditional min-entropy* of a state $\rho_{AB} \in \mathcal{S}(AB)$ can also be expressed as $\mathrm{H}_{\min}(A|B)_\rho = \max_{\sigma_B \in \mathcal{S}(B)} \mathrm{H}_{\min}(A|B)_{\rho|\sigma}$, with

$$\mathrm{H}_{\min}(A|B)_{\rho|\sigma} = \max\left\{\lambda \in \mathbb{R} : 2^{-\lambda} \cdot \mathrm{id}_A \otimes \sigma_B \geqslant \rho_{AB}\right\}, \tag{3}$$

where the symbol id_A refers to the identity on A. We use the subscript ρ to emphasize the state ρ_{AB} of which we evaluate the min-entropy. The smoothed version is defined by $\mathrm{H}_{\min}^\varepsilon(A|B)_\rho = \max_{\tilde{\rho}_{AB} \in \mathcal{B}^\varepsilon(\rho_{AB})} \mathrm{H}_{\min}(A|B)_{\tilde{\rho}}$, where $\mathcal{B}^\varepsilon(\rho)$ is the set of states at a distance at most ε from ρ. We use the purified distance as the distance measure [28]. We refer to [27] for a review of the properties of the min-entropy.

It is simpler to state our results in terms of the related collision entropy defined for any $\rho_{AB} \in \text{Pos}(A \otimes B)$ by

$$H_2(A|B)_\rho = -\log \text{Tr}\left[\left(\rho_B^{-1/4} \rho_{AB} \rho_B^{-1/4}\right)^2\right]. \tag{4}$$

We use relations between $H_{\min}$ and H_2 proved in the full version [13], in particular

$$H_{\min}^\varepsilon(A|B)_\rho \geq H_2(A|B)_\rho - \log(2/\varepsilon^2), \tag{5}$$

and

$$H_{\min}(X|B)_\sigma \leq H_2(X|B)_\sigma \leq 2H_{\min}(X|B)_\sigma, \tag{6}$$

for a classical-quantum state σ_{XB}. Finally, we use the binary entropy function $h(x) = -x \log x - (1-x) \log(1-x)$.

2.3   A Convenient Basis

Throughout, we make use of a very convenient basis of maximally entangled states for the space $A \otimes \bar{A}$ where $\bar{A} \simeq A$. The (unnormalized) maximally entangled state

$$|\Phi\rangle_{A\bar{A}} = \sum_a |a\rangle_A \otimes |a\rangle_{\bar{A}} \tag{7}$$

will play an important role in our analysis. Here, the vectors $|a\rangle$ label the standard basis of A. We use $|\Phi^N\rangle_{A\bar{A}}$ to denote the normalized version $|\Phi^N\rangle_{A\bar{A}} = \frac{1}{\sqrt{|A|}}|\Phi\rangle_{A\bar{A}}$. We repeatedly use the following properties. For any operators X and Y acting on A, we have

$$\text{Tr}[XY] = \text{Tr}[X \otimes \top(Y)\Phi_{A\bar{A}}] \tag{8}$$

where $\top$ denotes the transpose map in the standard basis and $\Phi_{A\bar{A}} = |\Phi\rangle\langle\Phi|_{A\bar{A}}$. Moreover, if $X : A \to C$ is a linear operator from A to C we have

$$(X \otimes \text{id}_{\bar{A}})|\Phi\rangle_{A\bar{A}} = (\text{id}_C \otimes \top(X))|\Phi\rangle_{C\bar{C}}. \tag{9}$$

Using (8) and (9) one can naturally construct an orthogonal basis of $A\bar{A}$ by applying unitary transformations to $|\Phi\rangle$ that are orthogonal with respect to the Hilbert-Schmidt inner product. Define for $s \in [|A|^2]$, $|\Phi_s\rangle = (W_s \otimes \text{id})|\Phi\rangle_{A\bar{A}}$ where W_s denote the generalized Pauli operators (see e.g., [1]), sometimes also called Weyl operators. In fact, all our results would hold for any unitary operators W_s that are orthogonal with respect to the Hilbert-Schmidt inner product. We again use $\Phi_s = |\Phi_s\rangle\langle\Phi_s|$.

In particular for $|A| = 2$, W_0, W_1, W_2, W_3 are the Pauli operators id, X, Y, Z respectively, and we obtain the well-known Bell basis.

For $n > 0$, we will denote by A^n the system $\bigotimes_{i=1}^n A_i$, where each A_i is a copy of A. Furthermore, if $S \subseteq \{1, \ldots, n\}$, we write A_S to denote $\bigotimes_{i \in S} A_i$. In other words, A^n consists of n copies of the system A, and A_S contains the copies that correspond to indices in S. In such a setting the dimension of the system A is denoted d. We can naturally define for $s \in [d^2]^n$, $|\Phi_s\rangle = \bigotimes_{i=1}^n |\Phi_{s_i}\rangle_{A_i \bar{A}_i}$. We then have that $\{\frac{1}{\sqrt{d^n}}|\Phi_s\rangle\}_s$ is an orthonormal basis of $A^n \bar{A}^n$. For such strings s, we denote $\text{supp}(s) = \{i \in \{1, \ldots, n\} : s_i \neq 0\}$ and $|s| = |\text{supp}(s)|$.

3 Evolution of H_2 under General Maps

In this section, we derive constraints on the evolution of the conditional collision entropy H_2 when the system A^n undergoes some transformation described by a completely positive map $\mathcal{M}$. Our results on entanglement sampling and uncertainty relations are obtained by evaluating this bound for particular channels $\mathcal{M}$. A statement for the smooth min-entropy follows directly by applying inequality (5).

Theorem 1. *Let $\mathcal{M}_{A^n \to C}$ be a completely positive map such that $((\mathcal{M}^\dagger \circ \mathcal{M})_{A^n} \otimes \mathrm{id}_{\bar{A}^n})(\Phi_{A^n \bar{A}^n}) = \sum_{s \in [d^2]^n} \lambda_s \Phi_s$ and let $\rho_{A^n E} \in \mathcal{S}(A^n E)$ be a state, where $A^n = A_1, \ldots, A_n$ is comprised of n qudits of dimension d. Then for any partition $[d^2]^n = \mathfrak{S}_+ \cup \mathfrak{S}_-$ into subsets $\mathfrak{S}_+$ and $\mathfrak{S}_-$, we have*

$$2^{-H_2(C|E)_{\mathcal{M}(\rho)}} \leqslant \sum_{s \in \mathfrak{S}_+} \lambda_s 2^{-H_2(A^n|E)_\rho} + (\max_{s \in \mathfrak{S}_-} \lambda_s) d^n. \tag{10}$$

The maps $\mathcal{M}$ of interest typically have some symmetry. For example, if the map $\mathcal{M}$ is invariant under permutations of the n systems $A_1, \ldots, A_n$, then the coefficients λ_s only depend on the type of s, i.e., the number of times each symbol in $[d^2]$ occurs in s. For example, for the entropy sampling result (Theorem 2), the map $\mathcal{M}$ is such that λ_s only depends on the weight $|s| = |\{i \in [n] : s_i \neq 0\}|$.

Proof. Let $\tilde{\rho}_{A^n E} = \rho_E^{-1/4} \rho_{A^n E} \rho_E^{-1/4}$, and let $\hat{\rho}_{A^n \bar{A}^n} = \mathrm{Tr}_{E\bar{E}}[(\tilde{\rho}_{A^n E} \otimes \mathsf{T}(\tilde{\rho}_{\bar{A}^n \bar{E}}))\Phi_{E\bar{E}}]$. Note that $\hat{\rho}_{A^n \bar{A}^n} \geq 0$ and $\mathrm{Tr}[\hat{\rho}_{A^n \bar{A}^n}] = \mathrm{Tr}[\tilde{\rho}_E^2] = 1$. Furthermore, define $\bar{\mathcal{M}}$ as $\bar{\mathcal{M}}(X) = \mathsf{T}(\mathcal{M}(\mathsf{T}(X)))$ for all X. Our first goal is to rewrite $H_2(C|E)_\sigma$ in terms of the basis $\{\Phi_s\}_s$. We obtain from (8)

$$
\begin{aligned}
2^{-H_2(C|E)_\sigma} &= \mathrm{Tr}[\mathcal{M}(\tilde{\rho}_{A^n E})^2] \\
&= \mathrm{Tr}[(\mathcal{M}(\tilde{\rho}_{A^n E}) \otimes \mathsf{T}(\mathcal{M}(\tilde{\rho}_{\bar{A}^n \bar{E}})))\Phi_{C\bar{C}} \otimes \Phi_{E\bar{E}}] \\
&= \mathrm{Tr}[(\mathcal{M}(\tilde{\rho}_{A^n E}) \otimes \bar{\mathcal{M}}(\mathsf{T}(\tilde{\rho}_{\bar{A}^n \bar{E}})))\Phi_{C\bar{C}} \otimes \Phi_{E\bar{E}}] \\
&= \mathrm{Tr}[(\tilde{\rho}_{A^n E} \otimes \mathsf{T}(\tilde{\rho}_{\bar{A}^n \bar{E}}))((\mathcal{M}^\dagger) \otimes (\bar{\mathcal{M}}^\dagger))(\Phi_{C\bar{C}}) \otimes \Phi_{E\bar{E}}].
\end{aligned}
$$

Now by writing a Kraus representation $\mathcal{M}(X) = \sum_i K_i X K_i^\dagger$ with operators $K_i : A \to C$ and using (9), we see that $(\mathrm{id}_C \otimes \bar{\mathcal{M}}^\dagger)(\Phi_{C\bar{C}}) = (\mathcal{M}_{A^n \to C} \otimes \mathrm{id}_{\bar{A}^n})(\Phi_{A^n \bar{A}^n})$. Thus, we obtain using the definition of $\hat{\rho}_{A^n \bar{A}^n}$ and the condition on $\mathcal{M}$

$$
\begin{aligned}
2^{-H_2(C|E)_\sigma} &= \mathrm{Tr}[(\tilde{\rho}_{A^n E} \otimes \mathsf{T}(\tilde{\rho}_{\bar{A}^n \bar{E}}))((\mathcal{M}^\dagger \circ \mathcal{M}) \otimes \mathrm{id}_{\bar{A}^n})(\Phi_{A^n \bar{A}^n}) \otimes \Phi_{E\bar{E}}] \\
&= \mathrm{Tr}[\hat{\rho}_{A^n \bar{A}^n}((\mathcal{M}^\dagger \circ \mathcal{M}) \otimes \mathrm{id}_{\bar{A}^n})(\Phi_{A^n \bar{A}^n})] \\
&= \sum_{s \in [d^2]^n} \lambda_s \mathrm{Tr}[\hat{\rho}_{A^n \bar{A}^n} \Phi_s].
\end{aligned} \tag{11}
$$

We prove the two key constraints on the terms $\mathrm{Tr}[\hat{\rho}_{A^n \bar{A}^n} \Phi_s]$ we will be using. First, we have a global constraint. Note that the set of vectors $\{\frac{1}{\sqrt{d^n}}|\Phi_s\rangle\}_{s \in [d^2]^n}$ forms an *orthonormal* basis and thus $\mathrm{id}_{A^n \bar{A}^n} = \frac{1}{d^n} \sum_{s \in [d^2]^n} \Phi_s$. This yields

$$\sum_{s \in [d^2]^n} \mathrm{Tr}[\hat{\rho}_{A^n \bar{A}^n} \Phi_s] = d^n \mathrm{Tr}[\hat{\rho}_{A^n \bar{A}^n}] = d^n. \tag{12}$$

The second observation concerns the individual terms $\mathrm{Tr}[\widehat{\rho}_{A^n \bar{A}^n} \Phi_s]$. For any s,

$$
\begin{aligned}
\mathrm{Tr}[\widehat{\rho}_{A^n \bar{A}^n} \Phi_s] &= \mathrm{Tr}[\widehat{\rho}_{A^n \bar{A}^n}(W_s \otimes \mathrm{id}_{\bar{A}^n}) \Phi_{A^n \bar{A}^n}(W_s^\dagger \otimes \mathrm{id}_{\bar{A}^n})] \\
&= \mathrm{Tr}[((W_s^\dagger \tilde{\rho}_{A^n E} W_s \otimes \top(\tilde{\rho}_{\bar{A}^n \bar{E}})) \Phi_{A^n \bar{A}^n} \otimes \Phi_{E \bar{E}}] \\
&= \mathrm{Tr}[W_s^\dagger \tilde{\rho}_{A^n E} W_s \tilde{\rho}_{A^n E})] \\
&\leqslant \mathrm{Tr}[\tilde{\rho}_{A^n E}^2] = 2^{-\mathrm{H}_2(A^n|E)_\rho},
\end{aligned}
$$

using the Cauchy-Schwarz inequality in the form $\mathrm{Tr}[XY] \leq \sqrt{\mathrm{Tr}[X^2]\,\mathrm{Tr}[Y^2]}$ with $X = W_s^\dagger \tilde{\rho}_{A^n E} W_s$ and $Y = \tilde{\rho}_{A^n E}$. Also, observe that the positivity of $\widehat{\rho}_{A^n \bar{A}^n}$ implies that $\mathrm{Tr}[\widehat{\rho}_{A^n \bar{A}^n} \Phi_s] = \langle \Phi_s | \widehat{\rho}_{A^n \bar{A}^n} | \Phi_s \rangle \geqslant 0$. Thus, we have

$$
0 \leqslant \mathrm{Tr}[\widehat{\rho}_{A^n \bar{A}^n} \Phi_s] \leqslant 2^{-\mathrm{H}_2(A^n|E)_\rho}. \tag{13}
$$

Applying inequalities (12) and (13) to (11), we obtain the desired result.

4   Applications

We now derive several interesting consequences of Theorem 1. All of these follow by making an appropriate choice for the map $\mathcal{M}$.

4.1   Quantum-Quantum Min-entropy Sampling

We now state our results on entanglement sampling. The theorem below deals with the following scenario: we have n qudits and we choose a subset of them of size k uniformly at random. We have a lower bound on the collision entropy of the whole state conditioned on some quantum side-information E; the theorem then gives a lower bound on the conditional collision entropy of the sample. The rate function obtained is plotted in Figure 1. The same figure also shows plots of classical-quantum sampling results that are discussed in Section 4.2.

Theorem 2. *Let $\rho_{A^n E} \in \mathcal{S}(A^n E)$ and $1 \leqslant k \leqslant n$, let $d = |A|$ be the dimension of a single system, and let $h_2 := \frac{\mathrm{H}_2(A^n|E)_\rho}{n}$. Then, we have for $n > d^2$*

$$
2^{-\mathrm{H}_2(A_S|ES)_\rho} = \mathbb{E}_{S \subseteq [n], |S|=k} 2^{-\mathrm{H}_2(A_S|E)_\rho} \leqslant 2^{-k R_d(h_2) + \log(n^2 + 1)}, \tag{14}
$$

where $R_d(\cdot)$ is the rate function defined as $R_d(x) := -\log(d - d f_d^{-1}(x))$, and $f_d(x) := h(x) + x \log(d^2 - 1) - \log d$. Using (5), we have for any $\varepsilon \in [0, 1)$

$$
\mathrm{H}_{\min}^\varepsilon(A_S|ES)_\rho \geqslant k R_d(h_{\min}) - \log(n^2 + 1) - \log \frac{2}{\varepsilon^2}, \tag{15}
$$

where $h_{\min} := \frac{\mathrm{H}_{\min}(A^n|E)_\rho}{n}$.

Proof. We now prove (14) by applying Theorem 1 for an appropriately chosen map $\mathcal{M}$. Naturally, $\mathcal{M}$ will (up to normalization) select a random subset S and discard all the qubits of the input except the ones in S. More formally, define $\mathcal{M}_{A^n \to A^k S}(X) =$

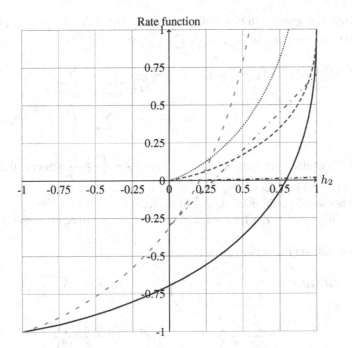

Fig. 1. Plot of our quantum-quantum rate function $R_2(h_2)$ from Theorem 2 (——), our classical-quantum rate function $C_2(h_2)$ from Theorem 4 (- - -), Wullschleger's min-entropy sampling result [34, Corollary 1] (-·-·), Vadhan's purely classical min-entropy sampling results [31, Lemma 6.2] (- · -), and the classical and quantum upper bounds we get from a state that is uniform on strings of a fixed type analyzed in the full version [13] (·······, - - -). As Vadhan's result requires a choice of parameters we chose $\tau = 0.1$, which yields a lower bound on the *smooth* min-entropy, with smoothing parameter of the order of 10^{-6} for a block size of $n = 10000$.

$$\frac{1}{\sqrt{\binom{n}{k}}} \sum_{S \subseteq [n], |S|=k} \mathrm{Tr}_{S^c}[X] \otimes |S\rangle\langle S|, \text{ for } X \in \mathcal{L}(A^n),$$ where the second register contains a classical description of the set S, and S^c denotes the complement of S in $[n]$. The reason for this normalization will be clear in the following calculation. Our first task is to relate this map to $\mathrm{H}_2(A_S|ES)_\rho$. A simple calculation reveals that

$$2^{-\mathrm{H}_2(A^k S|E)_{\mathcal{M}(\rho)}} = \mathbb{E}_{S \subseteq [n], |S|=k} \mathrm{Tr}\left[\left(\rho_E^{-1/4} \rho_{A_S E} \rho_E^{-1/4}\right)^2\right] = 2^{-\mathrm{H}_2(A_S|ES)_\rho}.$$

Our second task is to show that our choice of $\mathcal{M}$ satisfies the conditions of Theorem 1. We have

$$((\mathcal{M}^\dagger \circ \mathcal{M}) \otimes \mathrm{id}_{\bar{A}^n})(\Phi_{A^n \bar{A}^n}) = \mathcal{M}^\dagger\left(\frac{1}{\sqrt{\binom{n}{k}}} \sum_{|S|=k} |S\rangle\langle S| \otimes \Phi_{A_S \bar{A}_S} \otimes \mathrm{id}_{\bar{A}_{S^c}}\right)$$

$$= \frac{1}{\binom{n}{k}} \sum_{|S|=k} \Phi_{A_S \bar{A}_S} \otimes \mathrm{id}_{A_{S^c} \bar{A}_{S^c}}.$$

We now write this operator in terms of $\{\Phi_s\}_{s\in[d^2]^n}$. Recall that $\{\frac{1}{\sqrt{d^n}}|\Phi_s\rangle\}_s$ forms an orthonormal basis and thus $\mathrm{id}_{A^n\bar{A}^n} = \frac{1}{d^n}\sum_{s\in[d^2]^n}\Phi_s$:

$$((\mathcal{M}^\dagger\circ\mathcal{M})\otimes\mathrm{id}_{\bar{A}^n})(\Phi_{A^n\bar{A}^n}) = \frac{1}{d^{n-k}\binom{n}{k}}\sum_{|S|=k}\sum_{s:\mathrm{supp}(s)\subseteq S^c}\Phi_s$$

$$= \frac{1}{d^{n-k}\binom{n}{k}}\sum_{s:|s|\leqslant n-k}\binom{n-|s|}{k}\Phi_s.$$

As a result, the coefficients λ_s from Theorem 1 are $\lambda_s = \frac{\binom{n-|s|}{k}}{d^{n-k}\binom{n}{k}}$. Observe that λ_s only depends on $|s|$ and is a decreasing function of $|s|$. In order to apply Theorem 1, it is natural to choose the partition $\mathfrak{S}_+\cup\mathfrak{S}_-$ of the form $\mathfrak{S}_+ = \{s\in[d^2]^n : |s|\leqslant\ell_0\}$ and $\mathfrak{S}_- = \{s\in[d^2]^n : |s|>\ell_0\}$ for a value of $\ell_0\in\{0,\ldots,n\}$ to be chosen as a function of h_2.

Writing equation (10) in our case we obtain,

$$2^{-H_2(A_S|ES)_\rho} \leqslant \sum_{\ell=0}^{\ell_0}\frac{\binom{n-\ell}{k}}{d^{n-k}\binom{n}{k}}\binom{n}{\ell}(d^2-1)^\ell 2^{-h_2 n} + \frac{\binom{n-\ell_0-1}{k}}{\binom{n}{k}}d^k$$

$$= \frac{2^{-h_2 n}}{d^{n-k}}\sum_{\ell=0}^{\ell_0}\binom{n-k}{\ell}(d^2-1)^\ell + \frac{\binom{n-\ell_0-1}{k}}{\binom{n}{k}}d^k. \tag{16}$$

Now all that remains is to optimize over ℓ_0 and to find a simple expression for this quantity. Before choosing ℓ_0, we simplify the expression above. For the second term, we bound

$$\frac{\binom{n-\ell_0-1}{k}}{\binom{n}{k}}d^k \leqslant \left(\frac{n-\ell_0-1}{n}\right)^k d^k.$$

To obtain a simple bound on the first term, we use the following lemma whose proof can be found in the appendix of the full version [13].

Lemma 3. *For any $\ell_0\in\{0,\ldots,n\}$ such that $\ell_0\leqslant\frac{d^2-1}{d^2}n$ where $d^2<n$, we have*

$$\sum_{\ell=0}^{\ell_0}\binom{n-k}{\ell}(d^2-1)^\ell \leqslant n^2\binom{n}{\ell_0}(d^2-1)^{\ell_0}\max\left(\frac{n-\ell_0-1}{n},\frac{1}{d^2}\right)^k.$$

It then follows from equation (16) that

$$2^{-H_2(A_S|ES)_\rho} \leqslant \max\left(\frac{n-\ell_0-1}{n},\frac{1}{d^2}\right)^k d^k\left(\frac{2^{-h_2 n}}{d^n}n^2\binom{n}{\ell_0}(d^2-1)^{\ell_0}+1\right).$$

We now determine the value of ℓ_0 as a function of h_2. Observe that using properties of binomial coefficients, we have $\binom{n}{\ell}(d^2-1)^\ell \leqslant 2^{nh(\ell_0/n)}(d^2-1)^{\ell_0} = 2^{nf_d(\ell_0/n)}d^n$ provided $\ell_0\leqslant\frac{d^2-1}{d^2}n$. We define ℓ_0 to be the largest integer that is at most $\frac{d^2-1}{d^2}n$ such that $f_d(\ell_0/n)\leqslant h_2$. As a result, we have

$$2^{-H_2(A_S|ES)_\rho} \leqslant \max\left(\frac{n-\ell_0-1}{n},\frac{1}{d^2}\right)^k d^k\left(n^2+1\right). \tag{17}$$

Observe also that in the case where the maximum is $1/d^2$, the result follows directly as $R_d(h_2) \leq \log d$. In the case where $(n - \ell_0 - 1)/n > 1/d^2$, we observe that $(\ell_0 + 1)/n > f_d^{-1}(h_2)$ by our choice of ℓ_0. Note that if $\ell_0 + 1 \leq (d^2 - 1)/d^2 \cdot n$, this follows from the fact that f_d is nondecreasing, and otherwise it follows from the fact that by definition f_d^{-1} is always upper bounded by $(d^2 - 1)/d^2$. We now write $\left(\frac{n - \ell_0 - 1}{n} \right)^k$ in terms of the entropy rate h_2:

$$
\begin{aligned}
k \log \left(\frac{n - \ell_0 - 1}{n} \right) &= k \log \left(1 - \frac{\ell_0 + 1}{n} \right) \\
&\leq k \log(1 - f_d^{-1}(h_2)) \\
&= k \log(d - d f_d^{-1}(h_2)) - k \log d \\
&= -k R_d(h_2) - k \log d.
\end{aligned}
$$

By plugging these inequalities into (17), we obtain the desired result.

4.2 Classical-Quantum Min-entropy Sampling

Statement. Observe that in the case where the system A^n is classical, i.e., $\rho_{A^n E} = \sum_{x^n \in [d]^n} p(x^n) |x^n\rangle\langle x^n| \otimes \rho_E(x^n)$ for some distribution p and states $\rho_E(x^n)$, Theorem 2 can still be applied but in many cases it gives trivial bounds. In fact, when A^n is classical, we have $H_2(A^n|E) \geq 0$ as well as $H_2(A_S|ES) \geq 0$. In order to improve on the lower bound of Theorem 2 in the case of a classical system, we can apply Theorem 1 to a more specific map $\mathcal{M}$ that *measures* the systems A_S that are sampled. This allows us to obtain a lower bound on the collision entropy $H_2(A_S|ES)$ that is nontrivial for the entire range $H_2(A^n|E) \in [0, n \log d]$.

Theorem 4. *Let $\rho_{A^n E}$ be a classical-quantum state, and $1 \leq k \leq n$, let $d = |A|$, and let $h_2 := \frac{H_2(A^n|E)_\rho}{n}$. Then, for any $n > d$,*

$$
2^{-H_2(A_S|ES)_\rho} = \mathbb{E}_{S \subseteq [n], |S| = k} 2^{-H_2(A_S|E)_\rho} \leq 2^{-kC_d(h_2) + \log(n^2 + 1)},
$$

where $C_d(\cdot)$ is the rate function defined as $C_d(\alpha) := -\log(1 - c_d^{-1}(\alpha))$, and $c_d(\alpha) := h(\alpha) + \alpha \log(d - 1)$.

4.3 High-Order Uncertainty Relations against Quantum Side-Information

Uncertainty relations play a fundamental role in quantum information and in particular in quantum cryptography. Many of the modern security proofs for quantum key distribution are based on an uncertainty relation (see, e.g. [29]). They are also at the heart of security proofs in the bounded quantum storage model [11,10,7]. An uncertainty relation is a statement about a guaranteed uncertainty in the outcome of a measurement in a randomly chosen basis. We refer the reader to [33] for a survey on uncertainty relations.

Uncertainty Relation for BB84 Measurements. Here we consider a system A^n of n qubits. Then we measure each one of these qubits in either the standard basis (labeled 0 with vector $|0\rangle, |1\rangle$) or the Hadamard basis (labeled 1 with vectors $|+\rangle = (|0\rangle + |1\rangle)/\sqrt{2}, |-\rangle = (|0\rangle - |1\rangle)/\sqrt{2})$. More precisely, choose a random vector $\Theta^n \in \{0,1\}^n$ and measure qubit i in the basis specified by the i-th component of $\Theta^n = \Theta_1, \ldots, \Theta_n$. Call the outcome X_i. An uncertainty relation is a statement about the amount of uncertainty in the random variable $X^n = X_1, \ldots, X_n$ given the knowledge of the basis choice Θ^n. The uncertainty is often measured in terms of the Shannon entropy. However, for the applications we consider here, the measure of uncertainty needs to be stronger, i.e., we should use a higher order entropy like $\mathrm{H}_{\min}$ or H_2. Such an uncertainty relation has been established in [10]:

$$\mathrm{H}_{\min}^\varepsilon(X^n|\Theta^n) \gtrsim n/2. \tag{18}$$

The way this uncertainty relation was used in the context of the bounded storage model was to apply a chain rule to (18) to obtain $\mathrm{H}_{\min}^\varepsilon(X^n|E\Theta^n) \gtrsim n/2 - \log|E|$. There are two reasons for this inequality to be unsatisfactory: it depends on the dimension of E rather than on the correlations between A^n and E, and it becomes trivial when $\mathrm{H}_2(A^n|E) < -n/2$ as this implies $\log|E| > n/2$. An uncertainty relation for measurements in the six-state bases that depends on $\mathrm{H}_2(A^n|E)$ was established in [7], but it also becomes trivial when $\mathrm{H}_2(A^n|E) < -0.586n$.

It is simple to see that if the system A^n is maximally entangled with some system E, then the outcome X^n of this measurement can be perfectly predicted by having access to E. In other words, if the conditional entropy $\mathrm{H}_2(A^n|E) = -n$, then X^n can be correctly guessed with probability 1. The following theorem provides a converse: if $\mathrm{H}_2(A^n|E) \geqslant -(1 - \varepsilon)n$ for $\varepsilon > 0$, then X^n cannot be guessed with probability better than $2^{-n\delta(\varepsilon)}$ with $\delta(\varepsilon) > 0$ whenever $\varepsilon > 0$.

Theorem 5. Let $\rho_{A^n E} \in \mathcal{S}(A^n E)$ where A^n is an n-qubit space and define $h_2 = \frac{\mathrm{H}_2(A^n|E)_\rho}{n}$. Then we have

$$\mathrm{H}_2(X^n|E\Theta^n)_\rho \geqslant n\gamma(h_2) - 1$$

where $\rho_{X^n E\Theta^n} = \frac{1}{2^n} \sum_{x^n \in \{0,1\}^n, \theta^n \in \{0,1\}^n} |x^n\rangle\langle x^n| \langle x^n|H^{\theta^n} \rho_{A^n E} H^{\theta^n}|x^n\rangle \otimes |\theta^n\rangle\langle \theta^n|$ is the state obtained when system A^n is measured in the basis defined in the register Θ^n and the function γ (plot in Figure 2) is defined by $\gamma(h_2) = h_2$ if $h_2 \geq 1/2$ and $\gamma(h_2) = g^{-1}(h_2)$ if $h_2 < 1/2$ with $g(\alpha) = h(\alpha) + \alpha - 1$.

Proof. We apply Theorem 1 with $\mathcal{M}_{A^n \to X^n \Theta^n} = \mathcal{N}^{\otimes n}$ where $\mathcal{N}(\rho) = \frac{1}{\sqrt{2}} \sum_{x \in \{0,1\}, \theta \in \{0,1\}} |\theta\rangle\langle\theta| \otimes |x\rangle\langle x| \langle x|H^\theta \rho H^\theta|x\rangle$. We have

$$2^{-\mathrm{H}_2(X^n\Theta^n|E)_{\mathcal{M}(\rho)}} = \mathrm{Tr}\left[\left(\rho_E^{-1/4}(\mathcal{N}^{\otimes n} \otimes \mathrm{id})(\rho_{A^n E})\rho_E^{-1/4}\right)^2\right]$$

$$= \frac{1}{2^n} \sum_{\theta^n \in \{0,1\}^n} \mathrm{Tr}\left[\left(\rho_E^{-1/4} \sum_{x^n \in \{0,1\}^n} |\theta^n\rangle\langle\theta^n| \otimes |x^n\rangle\langle x^n| \langle x^n|H^{\theta^n} \rho H^{\theta^n}|x^n\rangle \rho_E^{-1/4}\right)^2\right]$$

$$= 2^{-\mathrm{H}_2(X^n|E\Theta^n)_\rho}.$$

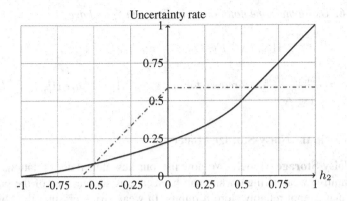

Fig. 2. Plot of the function $\gamma(h_2)$ (——) from Theorem 5 giving a lower bound on the uncertainty of the outcome of BB84 measurement as a function of the entropy rate h_2 of the state being measured. For comparison, we also plot the uncertainty rate function proved in [7] for measurements in the six-state bases (-·-).

We then evaluate the state $(\mathcal{N}^\dagger \circ \mathcal{N} \otimes \mathrm{id})(\Phi) = \frac{1}{2}\left(\Phi_0 + \frac{1}{2}\Phi_1 + \frac{1}{2}\Phi_3\right)$, where Φ_i are defined Section 2.3. In the notation of Theorem 1, we have for the map $\mathcal{M}$ and for $s \in \{0,1,3\}^n$, $\lambda_s = \frac{1}{2^n} \cdot \frac{1}{2^{|s|}}$. For $s \notin \{0,1,3\}^n$, $\lambda_s = 0$. As a result, when applying Theorem 1, it is natural to choose the partition $\mathfrak{S}_+ \cup \mathfrak{S}_-$ of the form $\mathfrak{S}_+ = \{s \in [d^2]^n : |s| \leq \ell_0\}$ and $\mathfrak{S}_- = \{s \in [d^2]^n : |s| > \ell_0\}$ for a value of $\ell_0 \in \{0, \ldots, n\}$ to be chosen as a function of h_2. We obtain for any ℓ_0

$$2^{-\mathrm{H}_2(X^n|E\Theta^n)_\rho} \leq \sum_{\ell=0}^{\ell_0} \binom{n}{\ell} 2^{-h_2 n - n} + 2^{-\ell_0 - 1}\delta_{\ell_0 \leq n-1}, \tag{19}$$

where $\delta_{\ell_0 \leq n-1} = 1$ if $\ell_0 \leq n-1$ and 0 if $\ell_0 = n$. If $h_2 \geq 1/2$, let $\ell_0 = n$, in which case we obtain a bound of $2^{-\mathrm{H}_2(X^n|E\Theta^n)_\rho} \leq 2^{-h_2 n}$.

If $h_2 < 1/2$, then we are going to choose $\ell_0 \leq n/2$. Define the function $g(\alpha) = h(\alpha) + \alpha - 1$ and let $\alpha_0 \leq 1/2$ be such that $g(\alpha_0) = h_2$. We then choose $\ell_0 = \lfloor \alpha_0 n \rfloor$. As a result,

$$\sum_{\ell=0}^{\ell_0} \binom{n}{\ell} 2^{-h_2 n - n} \leq 2^{n(h(\ell_0/n) - h_2 - 1)}$$

$$\leq 2^{n(h(\alpha_0) - h_2 - 1)} = 2^{n(-\alpha_0 + 1 + h_2 - h_2 - 1)} = 2^{-\alpha_0 n}.$$

In addition, we have $2^{-\ell_0 - 1} \leq 2^{-\alpha_0 n}$. Using these bounds in (19), we obtain in this case $2^{-\mathrm{H}_2(X^n|E\Theta^n)_\rho} \leq 2^{-\alpha_0 n + 1}$. Taking the logarithm leads to the desired result.

The following corollary expresses the uncertainty relation described in Theorem 5 in terms of min-entropies, which will be more convenient for the cryptographic applications.

Corollary 6. *Using the same notation as in Theorem 5, we have*

$$\mathrm{H}_{\min}(X^n|E\Theta^n)_\rho \geqslant \frac{1}{2}(n\gamma(h_{\min}) - 1), \tag{20}$$

where $h_{\min} = \frac{\mathrm{H}_{\min}(A^n|E)_\rho}{n}$. *Moreover, for any* $\varepsilon \in (0,1]$, *we have* $\mathrm{H}^\varepsilon_{\min}(X^n|E\Theta^n)_\rho \geqslant n\gamma(h_{\min}) - 1 - \log\frac{2}{\varepsilon^2}$.

4.4 Security in the Noisy-Storage Model

General Noisy Storage Model. We now use our new uncertainty relations to prove that the primitive weak string erasure can be secure as soon as one of the parties has a memory that cannot reliably store n qubits. In weak string erasure, the objective is to generate a string X^n such that Alice holds X^n and Bob holds a random subset $I \subseteq [n]$ and the bits X_I of X^n corresponding to the indices in I. Randomly chosen here means that each index $i \in [n]$ has probability $1/2$ of being in I. The security criterion is that at the end of the protocol, a cheating Bob should have a state satisfying $\mathrm{H}_{\min}(X^n|B) \geqslant \lambda n$ where B represents Bob's system, and a cheating Alice should not learn anything about I. To summarize all relevant parameters, we speak of an (n, λ)-WSE scheme and refer to [17] for a definition.[3] It is proved in [17] that bit commitment can be implemented using weak string erasure and classical communication.

Protocol. The protocol we use here is the same as the one of [17]. Alice prepares a random string $X^n \in \{0,1\}^n$ and encodes each bit X_i in either the standard basis $\Theta_i = 0$ or the Hadamard basis $\Theta_i = 1$, each with probability $1/2$. Then Bob measures these qubits in randomly chosen bases Θ'_i. After the waiting time, Alice reveals both X^n and Θ^n. The set I is defined by $I = \{i : \Theta_i = \Theta'_i\}$. For a more detailed description of the protocol, we refer the reader to [17].

To state the result, we first define the notion of *channel fidelity* introduced by [3] which is perhaps the most widely used quantity to measure how good a channel is at sending quantum information. For a channel $\mathcal{N} : \mathcal{S}(Q) \to \mathcal{S}(Q')$, the channel fidelity F_c quantifies how well $\mathcal{N}$ preserves entanglement with a reference:

$$F_c(\mathcal{N}) = F(\Phi^N_{Q'A}, [\mathcal{N} \otimes \mathrm{id}_A](\Phi^N_{QA})), \tag{21}$$

where Φ^N_{QA} is a normalized maximally entangled state. For example, one way of defining the (one-shot) quantum capacity with free classical forward communication of a channel $\mathcal{F}_{B \to C}$ is by the maximum of $\log|Q|$ over all encodings $\mathcal{E} : \mathcal{S}(Q) \to \mathcal{S}(B \otimes M)$ and decodings $\mathcal{D} : \mathcal{S}(C \otimes M) \to \mathcal{S}(Q')$ such that $F_c(\mathcal{D} \circ (\mathcal{F} \otimes \overline{\mathrm{id}}_M) \circ \mathcal{E}) \geqslant 1 - \varepsilon$ for small enough ε. Here $\overline{\mathrm{id}}_M$ refers to a noiseless classical channel.

The following theorem states that as soon as the storage device of Bob cannot send quantum information with reliability better than η, then we can perform two-party computation securely provided $\eta \leqslant 2^{-c(\log^2 n + \log n \log(1/\varepsilon))}$ for some large enough constant c. Previously, this was only known when $\eta < 2^{-(2-\log 3)n}$ [7]. Before that, security was analyzed in terms of other more specific quantities like the ability of the storage device to transmit *classical* information [17], or to simulate noiseless quantum

---

[3] Note that the original definition includes a security error ε, which in our case is $\varepsilon = 0$.

channels [6]. As the ability to transmit quantum information is a stronger requirement, the results we prove here apply to more general settings and give better bounds.

Theorem 7. *Let Bob's storage device be given by* $\mathcal{F} : \mathcal{S}(\mathcal{H}_{in}) \to \mathcal{S}(B)$, *and let* $\eta \in (0, 1)$. *Assume that we have*

$$\max_{\mathcal{D}, \mathcal{E}} F_c(\mathcal{D} \circ (\mathcal{F} \otimes \overline{id}_M) \circ \mathcal{E})^2 \leqslant \eta \qquad (22)$$

where the maximum is over all quantum channels $\mathcal{E} : \mathcal{S}\left((\mathbb{C}^2)^{\otimes n}\right) \to \mathcal{S}(\mathcal{H}_{in} \otimes M)$ *and* $\mathcal{D} : \mathcal{S}(B \otimes M) \to \mathcal{S}((\mathbb{C}^2)^{\otimes n})$.

Then, the protocol described above implements a (n, λ)-*WSE for*

$$\lambda = \frac{1}{2} \left(\gamma \left(-1 + \log(1/\eta)/n \right) - \frac{1}{n} \right).$$

Proof. The proof of correctness of the protocol, and security against dishonest Alice is identical to [17] and does not lead to any error terms. For the security against dishonest Bob, it is convenient to imagine a purification of the protocol, in which Alice prepares n EPR pairs $\Phi_{A^n Q}^N$, where she sends Q to Bob and later measures her n qubits A^n in randomly chosen BB84 bases. Bob's general attack can be modeled as performing some encoding on Q and obtaining some classical output M together with a quantum output that has to be stored in the device described by $\mathcal{F}$. The output of this device is denoted B. We use the uncertainty relation in Equation (20), with $E = BM\Theta^n$ on $\rho_{A^n BM\Theta^n}$. In order to do that, we first derive a lower bound on $h_{min} = \frac{H_{min}(A^n|BM\Theta^n)_\rho}{n}$. Note that because Θ^n is independent of $A^n BM$, we have $H_{min}(A^n|BM\Theta^n)_\rho = H_{min}(A^n|BM)_\rho$. We now use Condition (22) to obtain a lower bound on $H_{min}(A^n|BM)$. In fact, we use an operational interpretation of the conditional min-entropy due to [16]:

$$H_{min}(A^n|BM)_\rho = -\log|A^n| \max_{\Lambda_{BM \to \bar{A}^n}} F(\Phi_{A^n \bar{A}^n}^N, id_{A^n} \otimes \Lambda(\rho_{A^n BM}))^2, \qquad (23)$$

where $\Phi_{A^n \bar{A}^n}^N$ is the normalized maximally entangled state across $A^n \bar{A}^n$. That is, the min-entropy is directly related to the "amount" of entanglement between A^n and BM. The map Λ in (23) can be understood as a decoding attack $\mathcal{D}$ aiming to restore entanglement with Alice.

Further, note that the expression in (23) is the same as

$$\max_{\mathcal{D}, \mathcal{E}} F\left(\Phi_{A^n B}^N, id_{A^n} \otimes \left[\mathcal{D} \circ (\mathcal{F} \otimes \overline{id}_M) \circ \mathcal{E} \right](\Phi_{A^n Q}^N) \right) = \max_{\mathcal{D}, \mathcal{E}} F_c(\mathcal{D} \circ (\mathcal{F} \otimes \overline{id}_M) \circ \mathcal{E}).$$

By the assumption on the storage device $\mathcal{F}$, we obtain that for any encoding $\mathcal{E}$ and decoding $\mathcal{D}$ attack of Bob

$$H_{min}(A^n|BM)_\rho \geqslant -\log 2^n F_c(\mathcal{D} \circ (\mathcal{F} \otimes \overline{id}_M) \circ \mathcal{E})^2 \geqslant -(n - \log(1/\eta)).$$

Then, using the uncertainty relation (20), we obtain $H_{min}(X^n|BM\Theta^n)_\rho \geqslant \frac{1}{2}(n\gamma(-1 + \log(1/\eta)/n) - 1)$, which proves the desired result.

Special Case: Bounded Storage Model. The next theorem simply states the result in the important special case of the bounded storage model.

Theorem 8 (WSE in the bounded storage model). *If Alice has q qubits of quantum memory then the protocol described in the previous section implements (n, λ)-WSE with $\lambda = \frac{1}{2}\left(\gamma(-q/n) - \frac{1}{n}\right)$.*

Previously, in this case, security was only proven when $q < \frac{2n}{3}$ [20] with a variant of this protocol that uses a six-state encoding. Using simple estimates for the function γ, the previous theorem shows that $q < n - c\log^2 n$ for some large enough c would be sufficient to perform WSE securely. Using the construction of [17], this leads to a secure bit commitment provided $q < n - c\log^2 n - c\log n \log(1/\varepsilon)$ for some large enough constant c and where ε is the failure probability.

Acknowledgments. We thank Oleg Szehr, Marco Tomamichel and Thomas Vidick for useful discussions. OF is supported by the European Research Council grant No. 258932. SW thanks ETH Zürich for their hospitality. SW is supported by the National Research Foundation and the Ministry of Education, Singapore. FD acknowledges support from the Danish National Research Foundation and The National Science Foundation of China (under the grant 61061130540) for the Sino-Danish Center for the Theory of Interactive Computation, within which part of this work was performed; and also from the CFEM research center (supported by the Danish Strategic Research Council) within which part of this work was performed.

References

1. Bandyopadhyay, S., Boykin, P., Roychowdhury, V., Vatan, F.: A new proof for the existence of mutually unbiased bases. Algorithmica 34(4), 512–528 (2002) arXiv:quant-ph/0103162
2. Barnum, H., Knill, E.: Reversing quantum dynamics with near-optimal quantum and classical fidelity. J. Math. Phys. 43, 2097 (2002)
3. Barnum, H., Knill, E., Nielsen, M.A.: On quantum fidelities and channel capacities. IEEE Trans. Inform. Theory 46, 1317–1329 (2000) arXiv:quant-ph/9809010
4. Ben-Aroya, A., Regev, O., de Wolf, R.: A hypercontractive inequality for matrix-valued functions with applications to quantum computing and LDCs. In: Proc. IEEE FOCS (2008) arXiv:0705.3806
5. Bennett, C.H., Brassard, G.: Quantum cryptography: Public key distribution and coin tossing. In: Proc. International Conference on Computers, Systems and Signal Processing (1984)
6. Berta, M., Brandao, F., Christandl, M., Wehner, S.: Entanglement cost of quantum channels (2011) arXiv:1108.5357
7. Berta, M., Fawzi, O., Wehner, S.: Quantum to classical randomness extractors. In: Safavi-Naini, R., Canetti, R. (eds.) CRYPTO 2012. LNCS, vol. 7417, pp. 776–793. Springer, Heidelberg (2012)
8. Bouman, N.J., Fehr, S., González-Guillén, C., Schaffner, C.: An all-but-one entropic uncertainty relation, and application to password-based identification. In: Iwama, K., Kawano, Y., Murao, M. (eds.) TQC 2012. LNCS, vol. 7582, pp. 29–44. Springer, Heidelberg (2012)
9. Cachin, C., Maurer, U.M.: Unconditional security against memory-bounded adversaries. In: Kaliski Jr., B.S. (ed.) CRYPTO 1997. LNCS, vol. 1294, pp. 292–306. Springer, Heidelberg (1997)

10. Damgård, I., Fehr, S., Renner, R., Salvail, L., Schaffner, C.: A tight high-order entropic quantum uncertainty relation with applications. In: Menezes, A. (ed.) CRYPTO 2007. LNCS, vol. 4622, pp. 360–378. Springer, Heidelberg (2007)
11. Damgård, I., Fehr, S., Salvail, L., Schaffner, C.: Cryptography in the bounded quantum-storage model. In: Proc. IEEE FOCS, pp. 449–458 (2005) arXiv:quant-ph/0508222
12. Damgård, I.B., Fehr, S., Salvail, L., Schaffner, C.: Secure identification and QKD in the bounded-quantum-storage model. In: Menezes, A. (ed.) CRYPTO 2007. LNCS, vol. 4622, pp. 342–359. Springer, Heidelberg (2007)
13. Dupuis, F., Fawzi, O., Wehner, S.: Entanglement sampling and applications (2013) arXiv:1305.1316
14. Dziembowski, S., Maurer, U.: On generating the initial key in the bounded-storage model. In: Cachin, C., Camenisch, J.L. (eds.) EUROCRYPT 2004. LNCS, vol. 3027, pp. 126–137. Springer, Heidelberg (2004)
15. Impagliazzo, R., Levin, L., Luby, M.: Pseudo-random generation from one-way functions. In: Proc. ACM STOC, pp. 12–24. ACM (1989)
16. König, R., Renner, R., Schaffner, C.: The operational meaning of min- and max-entropy. IEEE Trans. Inform. Theory 55, 4674–4681 (2009) arXiv:0807.1338
17. König, R., Wehner, S., Wullschleger, J.: Unconditional security from noisy quantum storage. IEEE Trans. Inform. Theory 58(3), 1962–1984 (2012) arXiv:0906.1030
18. König, R., Renner, R.: Sampling of min-entropy relative to quantum knowledge. IEEE Trans. Inform. Theory 57(7), 4760–4787 (2011) arXiv:0712.4291
19. Lo, H.-K., Chau, H.F.: Is quantum bit commitment really possible? Phys. Rev. Lett. 78, 3410 (1997)
20. Mandayam, P., Wehner, S.: Achieving the physical limits of the bounded-storage model. Phys. Rev. A. 83, 022329 (2011) arXiv:1009.1596
21. Maurer, U.: Conditionally-perfect secrecy and a provably-secure randomized cipher. J. Cryptol. 5, 53–66 (1992)
22. Mayers, D.: Unconditionally secure quantum bit commitment is impossible. Phys. Rev. Lett. 78, 3414–3417 (1997)
23. Ng, N., Joshi, S., Chia, C., Kurtsiefer, C., Wehner, S.: Experimental implementation of bit commitment in the noisy-storage model. Nat. Comm. 3, 1326 (2012)
24. Nisan, N., Zuckerman, D.: Randomness is linear in space. J. Comput. Syst. Sci. 52(1), 43–52 (1996)
25. Renner, R.: Security of quantum key distribution. Int. J. Quantum Inf. 6, 1 (2008) arXiv:quant-ph/0512258
26. Schaffner, C., Terhal, B., Wehner, S.: Robust cryptography in the noisy-quantum-storage model. Quantum Inf. Comput. 9, 11 (2008) arXiv:0807.1333
27. Tomamichel, M.: A Framework for Non-Asymptotic Quantum Information Theory. PhD thesis, ETH Zürich (2012) arXiv:1203
28. Tomamichel, M., Colbeck, R., Renner, R.: A fully quantum asymptotic equipartition property. IEEE Trans. Inform. Theory 55, 5840–5847 (2009) arXiv:0811.1221
29. Tomamichel, M., Lim, C.C.W., Gisin, N., Renner, R.: Tight finite-key analysis for quantum cryptography. Nat. Comm. 3, 634 (2012)
30. Vadhan, S.: Pseudorandomness
31. Vadhan, S.: Constructing locally computable extractors and cryptosystems in the bounded-storage model. J. Cryptol. 17, 43–77 (2004)
32. Wehner, S., Schaffner, C., Terhal, B.: Cryptography from noisy storage. Phys. Rev. Lett 100, 220502 (2008) arXiv:0711.2895
33. Wehner, S., Winter, A.: Entropic uncertainty relations—a survey. New J. Phys. 12, 025009 (2010) arXiv:0907.3704
34. Wullschleger, J.: Bitwise quantum min-entropy sampling and new lower bounds for random access codes (2010) arXiv:1012.2291

Quantum One-Time Programs

(Extended Abstract)

Anne Broadbent[1], Gus Gutoski[2], and Douglas Stebila[3]

[1] Institute for Quantum Computing and
Department of Combinatorics and Optimization
University of Waterloo, Waterloo, Ontario, Canada
albroadb@iqc.ca
[2] Institute for Quantum Computing and School of Computer Science
University of Waterloo, Waterloo, Ontario, Canada
gus.gutoski@uwaterloo.ca
[3] School of Electrical Engineering and Computer Science and
School of Mathematical Sciences, Science and Engineering Faculty
Queensland University of Technology, Brisbane, Queensland, Australia
stebila@qut.edu.au

Abstract. A *one-time program* is a hypothetical device by which a user may evaluate a circuit on exactly one input of his choice, before the device self-destructs. One-time programs cannot be achieved by software alone, as any software can be copied and re-run. However, it is known that every circuit can be compiled into a one-time program using a very basic hypothetical hardware device called a *one-time memory*. At first glance it may seem that quantum information, which cannot be copied, might also allow for one-time programs. But it is not hard to see that this intuition is false: one-time programs for classical or quantum circuits based solely on quantum information do not exist, even with computational assumptions.

This observation raises the question, "what assumptions are required to achieve one-time programs for *quantum* circuits?" Our main result is that any quantum circuit can be compiled into a one-time program assuming only the *same basic one-time memory devices* used for classical circuits. Moreover, these quantum one-time programs achieve statistical universal composability (UC-security) against any malicious user. Our construction employs methods for computation on authenticated quantum data, and we present a new quantum authentication scheme called the *trap scheme* for this purpose. As a corollary, we establish UC-security of a recent protocol for delegated quantum computation.

1 Introduction

A *one-time program (OTP)* for a function f, as introduced by Goldwasser, Kalai, and Rothblum [1], is a cryptographic primitive by which a user may evaluate f on only one input chosen by the user at run time. (See also Refs. [2,3] for subsequent improvements.) No adversary, after evaluating the one-time program on x, should be able to learn anything about $f(x')$ for any $x' \neq x$ beyond what

R. Canetti and J.A. Garay (Eds.): CRYPTO 2013, Part II, LNCS 8043, pp. 344–360, 2013.

can be inferred from $f(x)$. One-time programs cannot be achieved by software alone, as any classical software can be re-run. Thus, any hope of achieving any one-time property must necessarily rely on an additional assumptions such as secure hardware or quantum mechanics: computational assumptions alone do not suffice.

Classically, it has been shown [1,2,3] how to construct a one-time program for any function f using a hypothetical hardware device called a *one-time memory (OTM)*. An OTM is non-interactive idealization of oblivious transfer: it stores two secret strings (or bits) s_0, s_1; a receiver can specify a bit c, obtain s_c, and then the OTM self-destructs so that $s_{\bar{c}}$ is lost forever. OTMs are an attractive minimal hardware assumption: their specification is independent of any specific function f, so they could theoretically be mass-produced.

OTPs are a special form of *non-interactive secure two-party computation* [3], in which two parties evaluate a publicly known function $f(x, y)$ as follows: the *sender* uses her input string x to prepare a *program* $p(x)$ for the *receiver*, who uses this program and his input y to compute $f(x, y)$. A malicious receiver should not be able to learn anything about $f(x, y')$ beyond what can be inferred from $f(x, y)$, for any y'. We use the term "OTP" interchangeably with "non-interactive secure two-party computation".

In this extended abstract we study *quantum one-time programs (QOTPs)*, in which the sender and receiver evaluate a publicly known channel $\Phi : (A, B) \to C$ specified by a quantum circuit acting on registers A (the sender's input), B (the receiver's input), and C (the receiver's output). The security goal is similar in spirit to that for classical functions: for each joint state ρ of the input registers (A, B), a malicious receiver should not be able to learn anything about $\Phi(\rho')$ beyond what can be inferred from $\Phi(\rho)$, for any ρ'.

Can quantum one-time programs be constructed? If so, how? If not, why not, and under what additional assumptions can they be achieved? QOTPs, if they do exist, would be useful for a variety of secure quantum computation tasks, such as providing *copy protection* of software [4] and implementing verification for quantum coin schemes [5]. (Note that QOTPs are different from the task of *program obfuscation*, which is known to be impossible classically [6] but remains an open question quantumly.)

Our main contributions are as follows: (i) We present a universally composable QOTP protocol for *any quantum channel*, assuming only the *same single-bit one-time memories* used in classical OTPs. Our protocol employs *quantum computation on authenticated data (QCAD)*, a technique of independent interest in quantum cryptography. (ii) We present a new quantum authentication scheme called the *trap scheme* and show that it allows for QCAD. (iii) We identify pathological classes of "unlockable" classical functions and quantum channels that admit trivial OTPs without any hardware assumptions. The remainder of this section elaborates upon these contributions.

1.1   Quantum One-Time Programs from Classical One-Time Memories

Unlike ordinary classical information, quantum information cannot in general be copied. This no-cloning property prompts one to ask: does quantum information allow for one-time programs without hardware assumptions? (When there are no hardware assumptions, we refer to this as the *plain quantum model*.)

For both classical functions and quantum channels, a moment's thought reveals a negative answer to this question: for any function f or channel Φ, a quantum "program state" for f or Φ can always be re-constructed by a reversible receiver after each use to obtain the evaluation of f or Φ on multiple distinct inputs. Computational assumptions do not help.

Given that one-time programs do *not* exist for arbitrary quantum channels in the plain quantum model, and that one-time programs *do* exist for arbitrary classical functions assuming secure OTMs, we ask: what additional assumptions are required to achieve one-time programs for quantum channels? Our main result answers this question.

Theorem 1 (Main result, informal). *For each channel $\Phi : (A, B) \to C$ specified by a quantum circuit there is a non-interactive two-party protocol for the evaluation of Φ, assuming classical one-time memory devices. The run time of this protocol is polynomial in the size of the circuit specifying Φ and the protocol achieves statistical quantum universal composability (UC-security) against a malicious receiver.*

Since all communication is one-way from sender to receiver, a malicious sender cannot learn anything about the receiver's portion of the input state ρ. The question of security against a malicious sender who tries to convince the receiver to accept an output state other than $\Phi(\rho)$ is left for future work. We restrict our attention to the case of *non-reactive* quantum one-time programs. The more general scenario of *bounded reactive* programs which can be queried a bounded number of times (including the case of an n-use program) may be implemented using standard techniques as is done in the classical case. Most of the components of our QOTP for Φ are independent of the sender's input register A and so can be compiled by the sender before she receives her input. As a corollary of our main result we obtain the UC-security of the protocol for *delegated quantum computations* (DQC) from Ref. [7]. Composable security for other variants of DQC was independently studied in Ref. [8].

1.2   A New Authentication Scheme That Admits Universal Computation

Our protocol employs a method for quantum computation on authenticated data (QCAD), which refers to the application of quantum gates to authenticated quantum data without knowing the authentication key. We propose a new authentication scheme, called the *trap scheme*, and show that it allows for QCAD. Our trap scheme also seems to provide a concrete and efficient realization of the "hidden subspaces" used in the public-key quantum money scheme of Ref. [9].

Prior to our work, the only authentication scheme known to admit QCAD was the *signed polynomial scheme* [10,7]. Recently, and independently of our work, it was shown in Ref. [11] that the *Clifford authentication scheme* can be used to authenticate two-party quantum computations. However, that protocol requires two parties to process quantum information and so cannot be used for QCAD or QOTPs.

Our QOTP protocol calls for the receiver to use QCAD to apply the gates of Φ to the authenticated input registers (A, B). In general, QCAD can only be performed if the receiver (who holds the authenticated data) is allowed to exchange classical messages with the sender (who knows the authentication key). To keep our protocol non-interactive, all the classical interaction is encapsulated by a *bounded, reactive classical one-time program (BR-OTP)* prepared by the sender, the existence of which follows straightforwardly from the work of Ref. [3] and is described in detail in the full version of this extended abstract [12]. This program for the BR-OTP depends upon the authentication key chosen for the sender's input register, but *not* on the contents of that register. By selecting this key in advance, the BR-OTP can be prepared before the sender gets his input register.

To implement QCAD, the receiver's input must be authenticated prior to computation. This is accomplished non-interactively by having the sender prepare a pair of registers in a special "teleport-through-encode" state. The authentication key is determined by the (classical) result of the Bell measurement used for teleportation. The receiver non-interactively de-authenticates the output at the end of the computation by means of a special "teleport-through-decode" state, also prepared by the sender. In order to successfully de-authenticate, the receiver's messages to the BR-OTP must be consistent with the secret authentication key held by the BR-OTP. Otherwise, the BR-OTP simply declines to reveal the final decryption key for the receiver's output.

1.3   Unlockable Functions and Channels

Curiously, our study has uncovered a pathological class of functions and channels that can *never* be made into a one-time program. For example, the function $f : (x, y) \mapsto x + y$ cannot have a one-time program because a receiver can use his knowledge of y to deduce x from $f(x, y)$. Once he has deduced x, the receiver is free to evaluate $f(x, y')$ for any y' of his choosing. This function is an example of what we call an *unlockable* function. Technically, it is incorrect to say that such a function can never be made into a one-time program. Rather, such functions admit *trivial* one-time programs in the *plain model*—a technicality arising from the standard simulation-based definition of security. This phenomenon is somewhat akin to trivially obfuscatable functions [6].

We propose a definition of unlockability and prove that a (classical) function f admits a one-time program in the plain quantum model if and only if it is unlockable. For quantum channels the situation is quite interesting. We define two classes of channels called *weakly* and *strongly* unlockable. We prove that every strongly unlockable channel admits a trivial one-time program in the plain quantum model. Conversely, we prove that any channel admitting a one-time program in the plain quantum model must be weakly unlockable. It is easy to see that every strongly unlockable channel is also weakly unlockable; we conjecture that the two classes are equal. To summarize, we prove that no "useful" function or channel admits a one-time program without any hardware assumptions.

2   Security of Quantum One-Time Programs

Intuitively, a QOTP for a channel Φ is secure if anything that any (possibly cheating) receiver could learn by processing the program state prepared by the sender could also be learned by interacting with a simulator that uses only one-time access to an idealized black box for Φ. Thus, no receiver can learn anything beyond what can be inferred from this ideal functionality for Φ.

Formally, we define security in the quantum UC framework as defined by Unruh [13]. Our main ideal functionality, $\mathcal{F}_\Phi^{\text{OTP}}$, is specified in Functionality 1 and involves two parties, the *sender* and the *receiver*. The functionality may exist in multiple instances and involve various parties.[1]

Functionality 1. Ideal functionality $\mathcal{F}_\Phi^{\text{OTP}}$ for a quantum channel $\Phi : (\mathsf{A}, \mathsf{B}) \to \mathsf{C}$

---

1. **Create:** Upon input register A from the sender, send `create` to the receiver and store the contents of register A.
2. **Execute:** Upon input register B from the receiver, evaluate Φ on registers A, B and send the contents of the output register C to the receiver. Delete any trace of this instance.

The map Φ that is computed is a public parameter of the functionality and it takes an input from the sender and an input from the receiver, so $\mathcal{F}_\Phi^{\text{OTP}}$ hides the sender's *input* only. If the intention is to hide the map Φ itself—as in the intuitive notion of one-time programs—then we can consider a universal map U that takes as part of the sender's input a representation of Φ (see [14,15,16]). Sometimes we emphasize the fact that the ideal functionality may be called only a single time by saying "one-shot access to an ideal functionality for Φ". The functionality $\mathcal{F}_\Phi^{\text{OTP}}$ is *sender-oblivious* since it delivers the result of the functionality to the receiver but not the sender.

We now give some intuition on how the notions of UC translate to the context of QOTPs.

Functionality. A *non-interactive protocol for evaluation of a channel* $\Phi : (\mathsf{A}, \mathsf{B}) \to \mathsf{C}$ consists of (i) an *encoding channel* enc $: \mathsf{A} \to \mathsf{P}$ applied by the sender on its input A that prepares a program state P, and (ii) a *decoding channel*

---

[1] Formally, instances are denoted by *session identifiers* and each instance involves labelled parties. For simplicity, we have omitted these identifiers as they are implicit from the context.

dec : (P, B) → C applied by the receiver on the program state P and its input B such that dec ∘ enc and Φ are indistinguishable.

When P consists solely of a quantum register, we call this the *plain quantum model*. In the *bounded reactive OTP-quantum-hybrid model*, the program state is a quantum register P augmented with one or more BR-OTPs. (For our construction, it suffices to consider a single BR-OTP.) In this setting, the actions of any receiver (honest or otherwise) can be viewed as the serialization of a multi-round "interaction" in which the first message consists of the quantum registers from the sender and subsequent messages consist of purely classical data exchanged with the BR-OTP.

Security. By the completeness of the dummy-adversary [13], in order to show security, it suffices to consider only the adversary that relays messages between the environment and the honest parties (we can see the environment as performing the attack). Thus, security of a non-interactive protocol for the evaluation of Φ in the BR-OTP-quantum-hybrid model corresponds to the existence of a *simulator* that can mimic the sender's message, combined with the interactive behaviour of the BR-OTP, using only one-shot, black-box access to Φ with register A fixed.

A key result of Unruh [13] is the *quantum lifting theorem* which establishes that, in the statistical case, classical-UC-secure protocols are quantum-UC-secure. We apply this result to the protocol of Goyal, Ishai, Sahai, Venkatesan, and Wadia [3], which establishes statistically classical-UC-secure one-time programs in the OTM-hybrid model (i.e., assuming one-time memories); by quantum lifting, this protocol is also statistically quantum-UC-secure and hence we can use it our construction. Ideal functionalities for OTMs, OTPs, and BR-OTPs, as well as a proof extending Goyal et al.'s result for OTPs to BR-OTPs, appear in the full version [12].

3   The Trap Authentication Scheme

In this section we present a new quantum authentication scheme called the *trap scheme* and argue that it admits quantum computation on authenticated data (QCAD). A *quantum authentication scheme* consists of procedures for encoding and decoding quantum information with a secret classical key k such that an adversary with no knowledge of k who tampers with encoded data will be detected with high probability. Quantum authentication codes were first introduced by Barnum, Crépeau, Gottesman, Smith and Tapp [17].

3.1   Trap Codes Yield a Secure Authentication Scheme

Our trap scheme is based on any fixed quantum error-detecting code C that encodes one logical qubit into n physical qubits with distance d (an $[[n, 1, d]]$-*code*). Each such code induces a different trap scheme. Authentication and de-authentication operations for the trap scheme based on a code C are specified in Protocol 1.

Protocol 1. Authentication and de-authentication for the trap scheme based on an $[[n, 1, d]]$-code C

Classical key. A pair (π, P) consisting of a permutation π on $3n$ elements and a (description of a) $3n$-qubit Pauli operator P.

Authentication. *Input:* one qubit. *Output:* $3n$ qubits.
1. Encode the data qubit under C, producing an n-qubit register.
2. Introduce two new n-qubit *trap registers* in states $|0\rangle^{\otimes n}$, $|+\rangle^{\otimes n}$, respectively.
3. Permute all $3n$ qubits according to π.
4. Encrypt all $3n$ qubits by applying P.

De-authentication. *Input:* $3n$-qubits. *Output:* one qubit and "accept"; or "reject".
1. Decrypt all $3n$ qubits by applying P.
2. Permute all $3n$ qubits according to π^{-1}.
3. Decode the data qubit under C.
4. Measure the trap registers to ensure they are in their proper states. If these measurements succeed and if C indicated no error syndrome then "accept" and output the data qubit, otherwise "reject".

The trap scheme is an example of a class of authentication schemes that we call *encode-encrypt schemes*, owing to a two-step authentication process of encoding followed by encryption. Encode-encrypt schemes have many desirable properties, chief among them the fact that an arbitrary attack on such a scheme is equivalent to a probabilistic mixture of Pauli attacks on the underlying family $\mathscr{E}$ of codes. Thus, by the encode-encrypt mechanism, in order to construct a secure quantum authentication scheme it suffices to exhibit a family $\mathscr{E}$ of codes that is secure against Pauli attacks.

In the trap scheme, the family $\mathscr{E}$ consists of all codes obtained by permuting data encoded under C together with registers in states $|0\rangle^{\otimes n}$, $|+\rangle^{\otimes n}$. We call $\mathscr{E}$ a *family of trap codes*. The first use of these codes was implicit in the Shor–Preskill security proof for quantum key distribution [18]. (See also Ref. [19].) We establish security of this family against Pauli attacks, from which the security of the trap scheme follows.

Proposition 1 (Security of trap codes against Pauli attacks). *The family $\mathscr{E}$ of trap codes based on a code of distance d is $(2/3)^{d/2}$-secure against Pauli attacks.*

That is, for each fixed choice of $3n$-qubit Pauli operation Q it holds that the probability—taken over a uniformly random choice of code $E \in \mathscr{E}$—that Q acts nontrivially on logical data and yet has no error syndrome is at most $(2/3)^{d/2}$.

See the full version [12] for proofs of Proposition 1 and several other properties of encode-encrypt schemes.

3.2   The Trap Scheme Admits Quantum Computing on Authenticated Data

Authentication schemes that also allow for QCAD—the implementation of a universal set of quantum gates on authenticated data without knowing the key—hold great promise for a host of cryptographic applications. In this section we argue that the trap scheme allows for QCAD for appropriate choices of the underlying code C.

It helps to think of two parties: a trusted *verifier* who prepares authenticated data with secret classical key k and a malicious *attacker* who is to act upon the authenticated data without knowledge of k. The goal is to construct a scheme with the property that for each gate G belonging to some universal set of gates there exists a *gadget* circuit $\tilde{G}$ that the attacker can apply to authenticated data so as to implement a logical G. Furthermore, we require that the gadget $\tilde{G}$ be *independent of the choice of classical key k* so that it may be implemented by an attacker without knowledge of k.

Normally, any non-identity gadget $\tilde{G}$ would invalidate the authenticated state. We therefore require a scheme which allows the verifier to validate the state again simply by updating the classical key $k \mapsto k'$. Moreover, by updating the key in this way the verifier effectively *forces* the attacker to apply the desired gadget $\tilde{G}$ as otherwise the state would fail verification under the updated key k'.

Following the example of the polynomial scheme of Ben-Or et al. [10], gadget design for our trap scheme is inspired by methods for fault-tolerant quantum computation. In the full version [12] we present gadgets for the universal gate set consisting of Pauli gates, controlled-NOT, Hadamard, i-shift phase $K : |a\rangle \mapsto i^a|a\rangle$, and $\pi/8$-phase $T : |a\rangle \mapsto e^{ai\pi/4}|a\rangle$.

Some gates, such as the controlled-NOT, admit straightforward bitwise gadgets. Others, such as the $\pi/8$ gate, require authenticated "magic states" and the ability to measure authenticated data in the computational basis. For these gadgets the verifier must interpret the classical measurement result for the attacker so that he may complete the gadget. Thus, these gadgets require classical interaction between verifier and attacker.

Our gadgets require that the underlying code C allow bitwise implementation of logical controlled-NOT and Hadamard gates—that is, that C be a self-dual CSS code. For a concrete example, it suffices that C be the seven-qubit Steane code nested a sufficient number of levels so as to achieve distance d.

4 Protocol for Quantum One-Time Programs

In this section we present our protocol for quantum one-time programs in the quantum BR-OTP hybrid model. In particular, we specify how an honest sender prepares her quantum registers and BR-OTP for the receiver and how an honest receiver should use these objects to recover the action of Φ. The protocol requires an encode-encrypt scheme that admits QCAD such as the trap scheme presented in Section 3, but is completely independent of the specific choice of scheme.

We assume without loss of generality that the channel Φ has the form $\Phi : (\mathsf{A}, \mathsf{B}) \to \mathsf{B}$ so that the receiver's output register $\mathsf{C} \cong \mathsf{B}$ has the same size as the input register and that Φ is specified by a unitary circuit U acting on registers $(\mathsf{A}, \mathsf{B}, \mathsf{E})$. The extra register E is an auxiliary register initialized to the $|0_\mathsf{E}\rangle$ state. The action of Φ is recovered from U by discarding registers (A, E) so that $\Phi : \rho \mapsto \mathrm{Tr}_{\mathsf{AE}}(U\,(\rho \otimes |0_\mathsf{E}\rangle\langle 0_\mathsf{E}|)\,U^*)$.

Given a circuit U one can efficiently find a circuit for the controlled-U operation, which we denote $c\text{-}U$. This circuit acts on registers $(\mathsf{A}, \mathsf{B}, \mathsf{E})$ plus an extra control qubit, which we bundle into the auxiliary register E for convenience. Our protocol calls for the receiver to apply $c\text{-}U$ to authenticated data with the control qubit always initialized to the $|\text{on}\rangle$ state. The purpose of this technicality is to better facilitate the proof of security. We also have an alternate protocol in which logical U is implemented directly with no need for $c\text{-}U$. However, the security proof for this alternate protocol is more technically cumbersome than our protocol for $c\text{-}U$, so we have elected to present only the protocol for $c\text{-}U$ in this extended abstract.

4.1   Protocol for an Honest Sender

Let r be the number of gates in $c\text{-}U$ that require magic states. After the parties have received their input registers A, B, a non-interactive protocol for $c\text{-}U$ consists of a single message from the sender to the receiver containing the following objects:

1. Quantum registers $\tilde{\mathsf{A}}, \mathsf{B}_{\mathsf{in}}, \tilde{\mathsf{B}}_{\mathsf{in}}, \mathsf{B}_{\mathsf{out}}, \tilde{\mathsf{B}}_{\mathsf{out}}, \tilde{\mathsf{E}}, \tilde{\mathsf{M}} = (\tilde{\mathsf{M}}_1, \dots, \tilde{\mathsf{M}}_r)$.
2. An $(r+1)$-round BR-OTP.

The sender prepares these objects as specified in Protocol 2 and Figure 1.

(a) Teleport-through-authentication

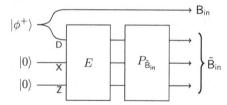

(b) Teleport-through-de-authentication

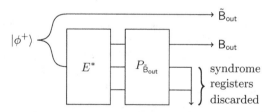

Fig. 1. Circuits for teleporting through authentication and de-authentication. Here the Pauli operations $P_{\tilde{\mathsf{B}}_{\mathsf{in}}}, P_{\tilde{\mathsf{B}}_{\mathsf{out}}}$ refer to the portions of P acting on registers $\tilde{\mathsf{B}}_{\mathsf{in}}, \tilde{\mathsf{B}}_{\mathsf{out}}$, respectively.

Protocol 2. Message preparation for an honest sender

Secret classical key. Authentication key for registers $(\tilde{A}, \tilde{B}_{in}, \tilde{B}_{out}, \tilde{E}, \tilde{M})$. In particular, a random pair (E, P) consisting of a code $E \in \mathscr{E}$ and Pauli P acting on these registers.

Registers prepared by the sender. Given the input register A the sender prepares the following registers:

$(\tilde{B}_{in}, \tilde{B}_{in})$:   Teleport-through-authentication state of Figure 1(a).

$(\tilde{B}_{out}, B_{out})$:   Teleport-through-de-authentication state of Figure 1(b).

$\tilde{A}$:   Authenticated input register A.

$\tilde{E}$:   Authenticated ancilla in logical state $|0\rangle|on\rangle$.

$\tilde{M}$:   Authenticated ancilla in logical state $|\mu\rangle = |\mu_1\rangle \cdots |\mu_r\rangle$ where $|\mu_1\rangle, \ldots, |\mu_r\rangle$ are the r magic states required for $c\text{-}U$.

BR-OTP prepared by the sender.

1. Receive (a classical description of) a purported teleport-through-authentication correction Pauli T^{in}.
2. For $i = 1, \ldots, r$:
   (a) Receive a classical bit string c_i—a purported measurement result of the ith authenticated magic state register $\tilde{M}_i$.
   (b) Decode c_i into a classical bit a_i as dictated by T^{in} and the authentication key (E, P). If the decoding process indicates a non-zero error syndrome then cheating has been detected. Return the decoded bit a_i to the user.
3. Receive a purported teleport-through-de-authentication correction Pauli T^{out}. If cheating was never detected in step 2b then return a decryption Pauli $\hat{S}$. Otherwise return random bits.

4.2 Protocol for an Honest Receiver

An honest receiver can recover $\Phi(\rho)$ from an honest sender's message as specified in Protocol 3.

5 Simulator and Proof of UC Security

The simulator must not pre-process the sender's input register A. Instead, the simulator is permitted only one-shot, black-box access to the "ideal functionality" for Φ. We represent this ideal functionality by a single call to an oracle for U acting on registers (A, B, E) prepared by the simulator. The rules for permissible preparation and disposal of these registers are as follows:

1. The simulator must pass the input register A directly to U without any pre-processing.
2. The simulator must prepare the ancillary register E in pure state $|0\rangle$.
3. Upon receiving the output registers (A, B, E) from the oracle for U, the simulator must discard registers A, E without any post-processing.

The simulator is specified in Protocol 4. The main idea is that our simulator will use the control qubit contained in register $\tilde{E}$ to "switch off" the application of U

Protocol 3. Protocol for an honest receiver

1. Perform a Bell measurement on $(\mathsf{B}, \mathsf{B_{in}})$ so as to teleport-through-authentication. Let T^{in} be the correction Pauli indicated by this measurement. Send T^{in} as the first message to the BR-OTP. *[At this time the contents of* B *have been authenticated and placed in register* $\tilde{\mathsf{B}}_{\mathsf{in}}$*.]*
2. Apply a logical $c\text{-}U$ to the authenticated registers $(\tilde{\mathsf{A}}, \tilde{\mathsf{B}}_{\mathsf{in}}, \tilde{\mathsf{E}}, \tilde{\mathsf{M}})$. Explicitly:
   (a) Apply the gates of $c\text{-}U$ occurring before the first magic state measurement.
   (b) For $i = 1, \ldots, r$:
      i. Measure the ith magic state register in the computational basis and send the result to the BR-OTP.
      ii. The BR-OTP provides a single bit indicating the proper correction.
      iii. Apply the gates of $c\text{-}U$ occurring after the ith magic state measurement but before the $(i+1)$th magic state measurement.
   [The implementation of $c\text{-}U$ *is now complete. At this time the register* $(\tilde{\mathsf{A}}, \tilde{\mathsf{B}}_{\mathsf{in}}, \tilde{\mathsf{E}})$ *holds the authenticated version of* $(\mathsf{A}, \mathsf{B}, \mathsf{E})$ *with* $c\text{-}U$ *applied.]*
3. Perform a Bell measurement on $(\tilde{\mathsf{B}}_{\mathsf{in}}, \tilde{\mathsf{B}}_{\mathsf{out}})$ so as to teleport-through-deauthentication. Let T^{out} be the correction Pauli indicated by this measurement. Send T^{out} as the final message to the BR-OTP. *[At this time the register* $\mathsf{B_{out}}$ *holds the receiver's output. This register is encrypted but not authenticated.]*
4. For its final output, the BR-OTP provides the Pauli decryption key $\hat{S}$. Apply this Pauli to $\mathsf{B_{out}}$ to recover the output of Φ.

that would have been implemented by an honest receiver. Instead, the blackbox call to the ideal functionality will be embedded at the proper time so as to recover the required action of U. An additional teleportation step is required so that our simulator can embed U at the proper time.

We now sketch a proof that the simulator of Protocol 4 certifies the security of the QOTP protocol presented in Section 4. We begin with a formal restatement of Theorem 1 in the language of UC-security. Details appear in the full version [12].

Theorem 2 (Main theorem, formal). *For each channel* $\Phi : (\mathsf{A}, \mathsf{B}) \to \mathsf{C}$ *specified by a quantum circuit, there is an efficient, non-interactive, quantum protocol in the OTM-hybrid model that statistically quantum-UC-emulates* $\mathcal{F}_\Phi^{\mathrm{OTP}}$ *against a malicious receiver.*

Proof (sketch). As discussed in Section 2, UC-security of our protocol is established by proving that that no entity (or *environment*) could possibly distinguish an interaction with our simulator from an interaction with a real sender. We employ a highly technical, "brute-force" approach to this end. In particular, we begin by writing down a general form that every environment must have. From such a description we derive an expression for the final state of all the registers in the environment's possession at the end of an interaction with a real sender. We perform a similar analysis for the environment's final state at the end of an interaction with our simulator. Finally, we argue that these two final states are statistically indistinguishable—that is, the trace distance between them is proportional to the security parameter of the underlying encode-encrypt

Protocol 4. Simulator

Secret classical key. Authentication key (E, P) for registers $(\tilde{A}, \tilde{B}_{in}, \tilde{B}_{out}, \tilde{E}, \tilde{M})$ as in Protocol 2.

Registers prepared by the simulator. Given the input register A, the simulator constructs the following registers:

(B_{in}, S_{in}): Simple EPR pairs $|\phi^+\rangle$ for teleportation.

$(S_{out}, \tilde{B}_{in})$: Teleport-through-authentication state of Figure 1(a).

$(\tilde{B}_{out}, B_{out})$: Teleport-through-de-authentication state of Figure 1(b).

$\tilde{A}$: Authenticated dummy input register in logical state $|0\rangle$.

$\tilde{E}$: Authenticated dummy ancillary register in logical state $|0\rangle|\text{off}\rangle$.

$\tilde{M}$: Authenticated magic states as in Protocol 2.

E : To be used in the call to the ideal functionality. Ancillary register in state $|0\rangle$.

Execution of the simulator.

1. Send the registers $B_{in}, \tilde{B}_{in}, \tilde{B}_{out}, B_{out}, \tilde{A}, \tilde{E}, \tilde{M}$ to the environment.
2. The environment responds with a Pauli T^{in}. Apply T^{in} to register S_{in}. Then use the ideal black-box to apply U to (A, S_{in}, E).
3. Perform a Bell measurement on (S_{in}, S_{out}) so as to teleport the contents of S_{in} through authentication and place the result in $\tilde{B}_{in}$. Let T^{sim} denote the teleportation Pauli indicated by this measurement.
4. Execute the BR-OTP of Protocol 2 under the assumption that T^{sim} was received in the first round.

scheme upon which our protocol is based. (For example, if the trap scheme built on a code of distance d is used with our protocol then this trace distance is exponentially small in d.) $\qquad\qquad\square$

6 Impossibility of Non-trivial OTPs in the Plain Model

In this section we propose a definition of unlockability for quantum channels, from which a definition for classical functions arises as a special case. We then prove two complementary results on channels that admit one-time programs in the plain model. Our *possibility* result (Theorem 3) is that every strongly unlockable channel admits a trivial one-time program in the plain quantum model, and in fact that this protocol is UC-secure. Our *impossibility* result (Theorem 4) is that every channel that is not weakly unlockable does not admit a one-time program in the plain quantum model. The latter result holds even if we relax to an approximate case or allow computational assumptions.[2]

As we will see, it is easy to establish that the weak and strong unlockability notions are equivalent for classical functions. Whether these notions are equivalent

---

[2] Although our impossibility result is stated in the UC framework, the impossibility is not an artifact of the high security required by UC, but seems inherent in the notion of OTPs, and the impossibility argument applied for any relaxation we attempted.

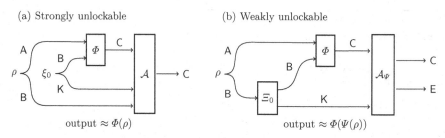

(a) Strongly unlockable   (b) Weakly unlockable

output $\approx \Phi(\rho)$ output $\approx \Phi(\Psi(\rho))$

Fig. 2. (a) For a strongly unlockable channel Φ, there exists a key state ξ_0 and a recovery algorithm $\mathcal{A}$ that allows computation of $\Phi(\rho)$ for any ρ. (b) For a weakly unlockable channel Φ, there exists a key channel Ξ_0 such that for any channel Ψ there exists a recovery algorithm $\mathcal{A}_\Psi$ that allows computation of $\Phi(\Psi(\rho))$ for any ρ.

for quantum channels is an open question, which appears to be an interesting and deep question related to invertible subspaces of a channel.

6.1   Definitions of Unlockability

Informally, a function or channel is unlockable if there is a key^3 input for the receiver that unlocks enough information to fully simulate the map. For quantum channels we present two variants of unlockability, the difference being whether the key that unlocks the channel is a state (*strongly unlockable*) or a channel that transforms a given input (*weakly unlockable*).

Definition 1 (Strongly/weakly unlockable channels). *A channel $\Phi : (A, B) \to C$ is* strongly unlockable *if there exists a register K, a key state ξ_0 of (B, K) and a recovery algorithm (i.e., channel) $\mathcal{A} : (C, K, B) \to C$ with the property that $\mathcal{A} \circ \Phi_0 \approx \Phi$, where the channel Φ_0 is specified by $\Phi_0 : A \to (C, K) : A \mapsto (\Phi \otimes \mathbb{1}_K)(A \otimes \xi_0)$.*

A channel $\Phi : (A, B) \to C$ is weakly unlockable *if there exists a register K and a key channel $\Xi_0 : B \to (B, K)$ such that the channel $\Phi \circ \Xi_0$ has the following property: for every choice of registers E and channels $\Psi : B \to (B, E)$ for the receiver there exists a recovery algorithm (i.e., channel) $\mathcal{A}_\Psi : (C, K) \to (C, E)$ such that $\mathcal{A}_\Psi \circ \Phi \circ \Xi_0 \approx \Phi \circ \Psi$.*

See Figure 2 for graphical depictions of these definitions. Here, $\approx$ can denote perfect, statistical, or (for polynomial-time uniform families of channels $\{\Phi_n\}$) computational indistinguishability; in all cases, channels Φ_0, $\mathcal{A}$, Ξ_0, and $\mathcal{A}_\Psi$ must have circuits of size polynomial in the size of the circuit for Φ.

It is easy to see that every strongly unlockable channel is also weakly unlockable: if ξ_0 is the key state for Φ, then the key channel Ξ_0 generates ξ_0, sends the B register of ξ_0 to ideal functionality Φ and the K register of ξ_0 and the B register of ρ to $\mathcal{A}_\Psi = \mathcal{A} \circ \Psi$.

---

[3] Note we use "key" not in the cryptographic sense of a secret key, but in the metaphorical sense of something that unlocks a lock.

When the channel Φ is entirely classical, the definitions of strongly unlockable and weakly unlockable are equivalent. A simplification for the classical case is as follows (we restrict to the perfect case for clarity). A classical function $f : A \times B \to C$ is *unlockable* if there exists a *key input* $b_0 \in B$ and a *recovery algorithm* $\mathcal{A} : C \times B \to C$ such that, for all $a \in A$ and $b \in B$, we have that $f(a, b) = \mathcal{A}(f(a, b_0), b)$. Intuitively, for an unlockable classical function there exists an algorithm that can compute all values of $f(a, \cdot)$ given a one-time program for $f(a, \cdot)$, but this is okay, because a simulator, given one-shot oracle access to $f(a, \cdot)$, can also compute $f(a, b)$ for all b: this function is "learnable" in one shot, and so a simulator can do everything any algorithm can.

Simple examples of strongly unlockable channels include all unitary channels of the form $\Phi : X \mapsto UXU^*$ for some unitary U and all constant channels of the form $\Phi : X \mapsto \mathrm{Tr}(X)\sigma$ for some fixed state σ. Simple examples of unlockable functions include permutations.

6.2   Trivial One-Time Programs for Unlockable Channels

We can now see that strongly unlockable channels have OTPs; but again, trivially so.

Theorem 3. *Let $\Phi : (\mathsf{A}, \mathsf{B}) \to \mathsf{C}$ be a channel specified by a circuit. If Φ is strongly unlockable then there exists an efficient, quantum non-interactive protocol which quantum-UC-emulates $\mathcal{F}_\Phi^{\mathrm{OTP}}$ in the plain quantum model. This holds in the perfect, statistical and computational cases.*

Theorem 3 is in the quantum setting; it follows from the proof that if Φ is in fact a classical channel, then the resulting protocol is a purely classical protocol.

6.3   Impossibility of One-Time Programs for Arbitrary Channels

Having seen that, in the plain model, strongly unlockable channels admit one-time programs, we now see that every channel which admits a one-time program must be weakly unlockable.

Theorem 4. *Let $\Phi : (\mathsf{A}, \mathsf{B}) \to \mathsf{C}$ be a channel specified by a circuit and suppose that Φ admits an efficient, non-interactive quantum protocol which quantum-UC-emulates $\mathcal{F}_\Phi^{\mathrm{OTP}}$ in the plain model. Then Φ is weakly unlockable. This holds in the perfect, statistical and computational cases.*

The intuition of the proof is as follows. If a channel has a one-time program, then for any adversary there exists a simulator that can match the behaviour of the adversary. In particular, there must be a simulator that matches the behaviour of the dummy-adversary that just outputs the program state: thus, there must be an algorithm that can reconstruct the program state given the output of the channel, thus allowing computation for any output, meeting the definition of a weakly unlockable channel.

An alternate intuition for the impossibility result for classical functions can be found by considering rewinding. Any correct one-time program state ρ_x for a classical function $f(x, \cdot)$ must result in the receiver obtaining an output state $\rho_{x,y}$ that is (almost) diagonal in the basis in which the receiver measures it, because the measurement of $\rho_{x,y}$ results in $f(x,y)$ with (almost) certainty. As a result, measurement does not disturb the state (much), so the receiver can reverse the computation to obtain (almost) the program state again, and then rerun the computation to obtain (close to) $f(x,y')$ for a different y'. It is possible to give a proof for impossibility of OTPs for classical functions in the plain quantum model using this rewinding argument. Impossibility for classical functions also follows as a special case of the impossibility shown in Ref. [20].

6.4 A Conjecture on Unlockable Channels

As noted earlier, every strongly unlockable channel is also weakly unlockable. We conjecture that the converse also holds. Though we do not yet have a formal proof of this conjecture for arbitrary Φ, we can nonetheless provide a high-level outline of a direction that might lead to a proof. See the full version for details [12].

Conjecture 1. Every channel $\Phi : (\mathsf{A}, \mathsf{B}) \to \mathsf{C}$ that is weakly unlockable is also strongly unlockable.

7 UC-Security of Delegated Quantum Computations

Several protocols have been designed to allow a computationally weak client to interface with a quantum computer in order to remotely accomplish a quantum computation while maintaining privacy of the user's input [21,19,7]. These works, however, do not consider composability. (Recently, Dunjko, Fitzsimons, Portmann and Renner [8] showed the composability of the blind quantum computing protocol of Ref. [19].)

In this section we show that our main proof technique can be used to establish the statistical quantum-UC security of a family of protocols for delegated quantum computations, closely related to the protocol of Aharonov et al. [7]. Originally studied in the context of *quantum interactive proof systems*, the protocol of Aharonov et al., which provides a mechanism to ensure both privacy of the user's input *and* verifiability of the computation, was not originally shown to be secure according to any rigorous cryptographic security definition.

We generalize the protocol of Aharonov et al. to support delegated quantum computation (in contrast to only deciding membership in a language) by making two minor modifications. First we instantiate the protocol using any encode-encrypt quantum authentication scheme that admits computing on authenticated data (such as the trap scheme or the signed polynomial scheme as used by Aharonov et al.). Analogously to our main protocol, we also introduce as an aid in the proof a control bit so that the circuit being implemented is a controlled-unitary.

The ideal functionality we achieve is described in Functionality 2. Following [7], we describe the functionality in terms of a *prover* and *verifier*.

Functionality 2. Ideal functionality $\mathcal{F}_{\Phi}^{\text{delegated}}$ for a quantum channel $\Phi : \mathsf{A} \to \mathsf{C}$

1. **Create:** Upon input register A from the verifier, send **create** to the prover and store the contents of register A.
2. **Execute:** The prover provides an input in $\{\texttt{execute}, \texttt{abort}\}$. Upon input **execute**, evaluate Φ on register A, and send the contents of the output register C to the verifier; upon input **abort**, output $\bot$ to the verifier.

Theorem 5. *Let Φ be a channel specified by a circuit. There exists an efficient quantum interactive protocol in the plain model that statistically quantum-UC-emulates $\mathcal{F}_{\Phi}^{\text{delegated}}$ against a malicious prover. Furthermore, the only quantum power required of the verifier is to encode the input and auxiliary quantum registers and to decode the output. In particular, all the interaction is classical except for the first and last messages.*

The proof of Theorem 5 follows as a special case of our main result about QOTPs (Theorem 2). In the case of a general Φ, the registers that the verifier prepares in Theorem 5 are polynomial-size in the security parameter. In the interactive proof scenario of Aharonov et al., the input to Φ is the all-$|0\rangle$ product state, the output is a single classical bit, and it suffices to implement $\mathcal{F}_{\Phi}^{\text{delegated}}$ with only constant security. Given these assumptions, the only quantum power required of the verifier is the ability to prepare constant-sized quantum registers in the first round.

Acknowledgements. We gratefully acknowledge helpful discussions with Harry Buhrman, Daniel Gottesman, Christopher Portmann, Bruce Richmond, Christian Schaffner and Dominique Unruh. A.B. acknowledges support from the Canadian Institute for Advanced Research (CIFAR), Canada's NSERC and Industry Canada. G.G. acknowledges support from Industry Canada, Ontario's Ministry of Research and Innovation, NSERC, DTO-ARO, CIFAR, and QuantumWorks. D.S. acknowledges support from the Australian Research Council. Part of this research conducted while D.S. was a visitor at the University of Waterloo's IQC.

References

1. Goldwasser, S., Kalai, Y., Rothblum, G.: One-time programs. In: Wagner, D. (ed.) CRYPTO 2008. LNCS, vol. 5157, pp. 39–56. Springer, Heidelberg (2008)
2. Bellare, M., Hoang, V.T., Rogaway, P.: Adaptively secure garbling with applications to one-time programs and secure outsourcing. In: Wang, X., Sako, K. (eds.) ASIACRYPT 2012. LNCS, vol. 7658, pp. 134–153. Springer, Heidelberg (2012), Full version available at http://eprint.iacr.org/2012/564
3. Goyal, V., Ishai, Y., Sahai, A., Venkatesan, R., Wadia, A.: Founding cryptography on tamper-proof hardware tokens. In: Micciancio, D. (ed.) TCC 2010. LNCS, vol. 5978, pp. 308–326. Springer, Heidelberg (2010), Full version available at http://eprint.iacr.org/2010/153

4. Aaronson, S.: Quantum copy-protection and quantum money. In: Proc. 24th IEEE Conference on Computational Complexity, CCC 2009, pp. 229–242 (2009)
5. Mosca, M., Stebila, D.: Quantum coins. In: Error-Correcting Codes, Finite Geometries and Cryptography. Contemporary Mathematics, vol. 523, pp. 35–47. American Mathematical Society (2010)
6. Barak, B., Goldreich, O., Impagliazzo, R., Rudich, S., Sahai, A., Vadhan, S., Yang, K.: On the (im)possibility of obfuscating programs. In: Kilian, J. (ed.) CRYPTO 2001. LNCS, vol. 2139, pp. 1–18. Springer, Heidelberg (2001), Full version available at http://www.wisdom.weizmann.ac.il/~oded/p_obfuscate.html
7. Aharonov, D., Ben-Or, M., Eban, E.: Interactive proofs for quantum computations. In: Proc. Innovations in Computer Science (ICS) 2010, pp. 453–469 (2010)
8. Dunjko, V., Fitzsimons, J.F., Portmann, C., Renner, R.: Composable security of delegated quantum computation (2013), arXiv.org/abs/1301.3662 (quant-ph)
9. Aaronson, S., Christiano, P.: Quantum money from hidden subspaces. In: Proc. 44th Symposium on Theory of Computing (STOC) 2012, pp. 41–60 (2012), Full version available as arXiv:1203.4740 (quant-ph)
10. Ben-Or, M., Crépeau, C., Gottesman, D., Hassidim, A., Smith, A.: Secure multi-party quantum computation with (only) a strict honest majority. In: FOCS 2006, pp. 249–260 (2006)
11. Dupuis, F., Nielsen, J.B., Salvail, L.: Actively secure two-party evaluation of any quantum operation. In: Safavi-Naini, R., Canetti, R. (eds.) CRYPTO 2012. LNCS, vol. 7417, pp. 794–811. Springer, Heidelberg (2012)
12. Broadbent, A., Gutoski, G., Stebila, D.: Quantum one-time programs (2013), (full version) arXiv:1211.1080 (quant-ph)
13. Unruh, D.: Universally composable quantum multi-party computation. In: Gilbert, H. (ed.) EUROCRYPT 2010. LNCS, vol. 6110, pp. 486–505. Springer, Heidelberg (2010), Full version available as arXiv:0910.2912 (quant-ph)
14. Bera, D., Fenner, S., Green, F., Homer, S.: Efficient universal quantum circuits. Quantum Information and Computation 10(1), 16–28 (2010)
15. Nielsen, M.A., Chuang, I.L.: Programmable quantum gate arrays. Physical Review Letters 79, 321–324 (1997)
16. de Sousa, P.B., Ramos, R.V.: Universal quantum circuit for N-qubit quantum gate: a programmable quantum gate. Quantum Information and Computation 7(3), 228–242 (2007)
17. Barnum, H., Crépeau, C., Gottesman, D., Smith, A., Tapp, A.: Authentication of quantum messages. In: FOCS 2002, pp. 449–458 (2002), Full version available as arXiv:quant-ph/0205128
18. Shor, P., Preskill, J.: Simple proof of security of the BB84 quantum key distribution protocol. Physical Review Letters 85, 441–444 (2000)
19. Broadbent, A., Fitzsimons, J., Kashefi, E.: Universal blind quantum computation. In: FOCS 2009, pp. 517–526. IEEE (2009)
20. Buhrman, H., Christandl, M., Schaffner, C.: Complete insecurity of quantum protocols for classical two-party computation. Physical Review Letters 109, 160501 (2012)
21. Childs, A.: Secure assisted quantum computation. Quantum Information and Computation 5, 456–466 (2005)

Secure Signatures and Chosen Ciphertext Security in a Quantum Computing World

Dan Boneh and Mark Zhandry

Stanford University
{dabo,zhandry}@cs.stanford.edu

Abstract. We initiate the study of *quantum*-secure digital signatures and *quantum* chosen ciphertext security. In the case of signatures, we enhance the standard chosen message query model by allowing the adversary to issue *quantum* chosen message queries: given a superposition of messages, the adversary receives a superposition of signatures on those messages. Similarly, for encryption, we allow the adversary to issue *quantum* chosen ciphertext queries: given a superposition of ciphertexts, the adversary receives a superposition of their decryptions. These adversaries model a natural ubiquitous quantum computing environment where end-users sign messages and decrypt ciphertexts on a personal quantum computer.

We construct classical systems that remain secure when exposed to such quantum queries. For signatures, we construct two compilers that convert classically secure signatures into signatures secure in the quantum setting and apply these compilers to existing post-quantum signatures. We also show that standard constructions such as Lamport one-time signatures and Merkle signatures remain secure under quantum chosen message attacks, thus giving signatures whose quantum security is based on generic assumptions. For encryption, we define security under quantum chosen ciphertext attacks and present both public-key and symmetric-key constructions.

Keywords: Quantum computing, signatures, encryption, quantum security.

1 Introduction

Recent progress in building quantum computers [IBM12] gives hope for their eventual feasibility. Consequently, there is a growing need for quantum-secure cryptosystems, namely classical systems that remain secure against quantum computers. Post-quantum cryptography generally studies the settings where the adversary is armed with a quantum computer, but users only have classical machines. In this paper, we go a step further and study the eventuality where end-user machines are quantum. In these settings, an attacker may interact with honest parties using quantum queries, as discussed below, potentially giving the attacker more power. The challenge is to construct cryptosystems that remain secure when exposed to such quantum queries. We emphasize that all the systems

R. Canetti and J.A. Garay (Eds.): CRYPTO 2013, Part II, LNCS 8043, pp. 361–379, 2013.

we consider are classical and can be easily implemented on a classical computer. Our goal is to construct classical systems that remain secure even when implemented on a quantum computer, thereby potentially giving the attacker the ability to issue quantum queries.

Along these lines, Zhandry [Zha12b] showed how to construct pseudorandom functions (PRFs) that remain secure even when the adversary is allowed to issue *quantum* queries to the PRF. A quantum query is a superposition of inputs $\sum_x \psi_x |x\rangle$ of the attacker's choice. The response is a superposition $\sum_x \psi_x |x, F(k, x)\rangle$ where $F(k, x)$ is the value of the PRF at a point x under key k. Zhandry showed that certain PRFs are secure even under such a powerful query model. More recently, Boneh and Zhandry [BZ13a] showed how to construct message authentication codes (MACs) that remain secure even when the attacker is allowed to issue *quantum* chosen message queries. That is, for a superposition of messages $\sum_m \psi_m |m\rangle$ of the attacker's choice, the attacker is given $\sum_m \psi_m |m, S(k, m)\rangle$ where $S(k, m)$ is the tag on message m using key k. They showed that some classically secure MACs become insecure under quantum chosen message queries and they constructed several quantum-secure MAC families.

Our Contributions. In this paper, we construct the first quantum-secure signatures and quantum-secure chosen ciphertext encryption systems.

We begin by defining security for digital signatures under a *quantum* chosen message attack. A quantum chosen message query [BZ13a] gives the attacker the signatures on all messages in a quantum superposition. In more detail, a quantum chosen message query is the transformation

$$\sum_m \psi_m |m\rangle \quad \longrightarrow \quad \sum_m \psi_m |m, \; S(\mathsf{sk}, m)\rangle$$

where $S(\mathsf{sk}, x)$ is the signature on x using signing key sk. The attacker can sample the response to such a query and obtain one valid message-signature pair. After q such queries, it can obtain q valid message-signature pairs. We say that a signature scheme is existentially unforgeable under a *quantum* chosen message attack if, after q quantum chosen message queries, the attacker cannot produce $q + 1$ valid message-signature pairs.

Next, we present several compilers that convert a signature scheme that is secure under *classical* queries into one secure under *quantum* queries. In particular, we give the following constructions:

- Using a chameleon hash [KR00], we show how to transform any signature that is existentially unforgeable under a *classical random* message into a signature scheme that is existentially unforgeable under a *quantum chosen* message attack. We apply this conversion to several existing signature schemes, giving constructions whose quantum security is based on the quantum hardness of lattice problems.
- We show that any *universally* unforgeable signature under a *classical random* message attack can be made *existentially* unforgeable under a *quantum*

chosen message attack in the random oracle model. For example, this conversion applies to a randomized variant of GPV signatures [GPV08], proving security of the scheme even under a *quantum* chosen message attack. We also separately show that the basic deterministic GPV scheme is secure in this setting.

- Finally, we prove that classical constructions such as Lamport one-time signatures and Merkle signatures are existentially unforgeable under a *quantum* chosen message attack. These results show how to build quantum-secure signatures from any collision resistant hash function. We leave open the problem of basing security on one-way functions. We also note that the version of Lamport signatures that we prove secure is non-optimized, and can potentially be made more efficient using standard combinatorial techniques. Unfortunately, we cannot prove quantum-security of an optimized Lamport signature and leave that as an interesting open problem.

Turning to encryption, we first explain how to adapt the chosen ciphertext security game to the quantum setting. In the classical game, the attacker is given classical access to a decryption oracle used to answer chosen ciphertext queries and to an encryption oracle used to create challenge ciphertexts. In the quantum setting, the decryption oracle accepts a superposition of ciphertexts and returns a superposition of their decryptions:

$$\sum_m \psi_c \, |c\rangle \quad \longrightarrow \quad \sum_c \psi_c \, |c, \, D(\mathsf{sk}, c)\rangle \; .$$

One might also try to allow quantum access to the encryption oracle; however, we show that the resulting concept is unsatisfiable. We therefore restrict the encryption oracle to be classical.

Armed with this definition of security, we construct quantum-secure chosen ciphertext systems in both the public-key and symmetric-key settings:

- Our symmetric-key construction is built from any secure PRF, and follows the encrypt-then-MAC paradigm. The classical proof that encrypt-then-MAC is secure for generic encryption and generic MAC schemes does not carry over to the quantum setting, but we are able to prove security for our specific construction.
- We show that public-key quantum chosen ciphertext security can be obtained from any identity-based encryption scheme that is selectively secure under a quantum chosen identity attack. Such an identity-based encryption scheme can, in turn, be built from lattice assumptions. This construction is the quantum analogue of the CHK transformation from identity-based encryption to public-key chosen ciphertext security [BCHK04].

Motivation. Allowing the adversary to issue quantum queries is a natural and conservative security model and is therefore an interesting one to study. Constructing signature and encryption schemes that remain secure in these models gives confidence in the event that end-user computing devices eventually become quantum. Nevertheless, one might imagine that in a future where all computers are quantum, the last step in a signature or decryption procedure is to sample the final quantum state. This ensures that the results are always classical, thereby preventing quantum superposition attacks. Security in this case relies on a physical hardware assumption, namely that the final "classicalization" step is implemented correctly and cannot be circumvented by a quantum adversary. In contrast, using systems that are inherently secure against superposition attacks frees the hardware designer from worrying about the security of the classicalization step.

As further motivation, we note that our results are the tip of a large emerging area with many open questions. For any cryptographic primitive modeled as an interactive game, one can ask how to design primitives that remain secure when the interaction between the adversary and its given oracles is quantum. For example, can we design quantum-secure threshold signatures and group signatures? Can we construct a quantum-secure PRF for a large domain from a quantum-secure PRF for a small domain? In particular, do the CBC-MAC or NMAC constructions give quantum-secure PRFs?

Other Related Work. Several recent works study the security of cryptographic primitives when the adversary can issue quantum queries. Boneh et al. [BDF+11] and Zhandry [Zha12a] prove the classical security of signatures, encryption, and identity-based encryption schemes in the *quantum* random oracle model, where the adversary can query the random oracle on superpositions of inputs. In these papers, the interaction with the challenger is classical. These results show that many, but not all, random oracle constructions remain secure in the quantum random oracle model. The quantum random oracle model has also been used to prove security of Merkle's Puzzles in the quantum setting [BS08, BHK+11]. Damgård et al. [DFNS11] examine secret sharing and multiparty computation in a model where an adversary may corrupt a superposition of subsets of players, and build zero knowledge protocols that are secure, even when a dishonest verifier can issue challenges on superpositions.

Some progress toward identifying sufficient conditions under which classical protocols are also quantum immune has been made by Unruh [Unr10] and Hallgren et al. [HSS11]. Unruh shows that any scheme that is statistically secure in Cannetti's universal composability (UC) framework [Can01] against classical adversaries is also statistically secure against quantum adversaries. Hallgren et al. show that for many schemes, this is also true in the computational setting. These results, however, do not apply to cryptographic primitives such as signatures and encryption and do not consider quantum superposition attacks.

2 Preliminaries: Background and Techniques

We will let $[n]$ denote the set $\{1, ..., n\}$. Functions will be denoted by capital letters (such as F), and sets by capital script letters (such as $\mathcal{X}$). We will let $x \xleftarrow{R} D$ for some distribution D denote drawing x according to D, and $x \xleftarrow{R} \mathcal{X}$

for some set $\mathcal{X}$ denote drawing a random element from $\mathcal{X}$. Given a function $F : \mathcal{X} \to \mathcal{Y}$ and a subset $\mathcal{S} \subseteq \mathcal{X}$, the restriction of F to $\mathcal{S}$ is the function $F_{\mathcal{S}} : \mathcal{S} \to \mathcal{Y}$ where $F_{\mathcal{S}}(x) = F(x)$ for all $x \in \mathcal{S}$. A distribution D on F induces a distribution $D_{\mathcal{S}}$ on $F_{\mathcal{S}}$. We say that D is k-wise independent if each of the distributions $D_{\mathcal{S}}$ are truly random distributions on functions from $\mathcal{S}$ to $\mathcal{Y}$, for all sets $\mathcal{S}$ of size at most k. A set $\mathcal{F}$ of functions from $\mathcal{X}$ to $\mathcal{Y}$ is k-wise independent if the uniform distribution on $\mathcal{F}$ is k-wise independent. A non-negative function $f(n)$ is negligible if, for any c, $f(n) < 1/n^c$ for all sufficiently large n. If a function $g(n)$ can be written as $h(n) \pm f(n)$ where $f(n)$ is negligible, we write $g(n) = h(n) \pm \mathsf{negl}$.

2.1   Quantum Computation

We give a short introduction to quantum computation. A quantum system A is a complex Hilbert space $\mathcal{H}$ together with and inner product $\langle \cdot | \cdot \rangle$. The state of a quantum system is given by a vector $|\psi\rangle$ of unit norm ($\langle \psi | \psi \rangle = 1$). Given quantum systems $\mathcal{H}_1$ and $\mathcal{H}_2$, the joint quantum system is given by the tensor product $\mathcal{H}_1 \otimes \mathcal{H}_2$. Given $|\psi_1\rangle \in \mathcal{H}_1$ and $|\psi_2\rangle \in \mathcal{H}_2$, the product state is given by $|\psi_1\rangle|\psi_2\rangle \in \mathcal{H}_1 \otimes \mathcal{H}_2$. Given a quantum state $|\psi\rangle$ and an orthonormal basis $B = \{|b_0\rangle, ..., |b_{d-1}\rangle\}$ for $\mathcal{H}$, a measurement of $|\psi\rangle$ in the basis B results in the value i with probability $|\langle b_i | \psi \rangle|^2$, and the quantum state collapses to the basis vector $|b_i\rangle$. If $|\psi\rangle$ is actually a state in a joint system $\mathcal{H} \otimes \mathcal{H}'$, then $|\psi\rangle$ can be written as

$$|\psi\rangle = \sum_{i=0}^{d-1} \alpha_i \, |b_i\rangle|\psi_i'\rangle$$

for some complex values α_i and states $|\psi_i'\rangle$ over $\mathcal{H}'$. Then, the measurement over $\mathcal{H}$ obtains the value i with probability $|\alpha_i|^2$ and in this case the resulting quantum state is $|b_i\rangle|\psi_i'\rangle$.

A unitary transformation over a d-dimensional Hilbert space $\mathcal{H}$ is a $d \times d$ matrix $\mathbf{U}$ such that $\mathbf{U}\mathbf{U}^\dagger = \mathbf{I}_d$, where $\mathbf{U}^\dagger$ represents the conjugate transpose. A quantum algorithm operates on a product space $\mathcal{H}_{in} \otimes \mathcal{H}_{out} \otimes \mathcal{H}_{work}$ and consists of n unitary transformations $\mathbf{U}_1, ..., \mathbf{U}_n$ in this space. $\mathcal{H}_{in}$ represents the input to the algorithm, $\mathcal{H}_{out}$ the output, and $\mathcal{H}_{work}$ the work space. A classical input x to the quantum algorithm is converted to the quantum state $|x, 0, 0\rangle$. Then, the unitary transformations are applied one-by-one, resulting in the final state

$$|\psi_x\rangle = \mathbf{U}_n...\mathbf{U}_1|x, 0, 0\rangle \ .$$

The final state is then measured, obtaining the tuple (a, b, c) with probability $|\langle a, b, c | \psi_x \rangle|^2$. The output of the algorithm is b. We say that a quantum algorithm is efficient if each of the unitary matrices $\mathbf{U}_i$ come from some fixed basis set, and n, the number of unitary matrices, is polynomial in the size of the input.

Quantum-accessible Oracles. We will implement an oracle $O : \mathcal{X} \to \mathcal{Y}$ by a unitary transformation $\mathbf{O}$ where

$$\mathbf{O}|x, y, z\rangle = |x, y + O(x), z\rangle$$

where $+ : \mathcal{X} \times \mathcal{X} \to \mathcal{X}$ is some group operation on $\mathcal{X}$. Suppose we have a quantum algorithm that makes quantum queries to oracles $O_1, ..., O_q$. Let $|\psi_0\rangle$ be the input state of the algorithm, and let $\mathbf{U}_0, ..., \mathbf{U}_q$ be the unitary transformations applied between queries. Note that the transformations $\mathbf{U}_i$ are themselves possibly the products of many simpler unitary transformations. The final state of the algorithm will be

$$\mathbf{U}_q \mathbf{O}_q ... \mathbf{U}_1 \mathbf{O}_1 \mathbf{U}_0 |\psi_0\rangle$$

We can also have an algorithm make classical queries to O_i. In this case, the input to the oracle is measured before applying the transformation $\mathbf{O}_i$. We call a quantum oracle algorithm efficient if the number of queries q is a polyomial, and each of the transformations $\mathbf{U}_i$ between queries can be written as the product polynomially many unitary transformations from some fixed basis set.

Tools. Next we state several lemmas and definitions that we will use throughout the paper. Some have been proved in other works, and the rest are proved in the full version [BZ13b]. The first concerns partial measurements, and will be used extensively throughout the paper:

Lemma 1. *Let A be a quantum algorithm, and let $\Pr[x]$ be the probability that A outputs x. Let A' be another quantum algorithm obtained from A by pausing A at an arbitrary stage of execution, performing a partial measurement on the state of A that obtains one of k outcomes, and then resuming A. Let $\Pr'[x]$ be the probability A' outputs x. Then $\Pr'[x] \geq \Pr[x]/k$.*

This lemma means, for example, that if you measure just one qubit, the probability of a particular output drops by at most a factor of two. We also make use of the following lemma, proved by Zhandry [Zha12a], which allows us to simulate random oracle efficiently using k-wise independent functions:

Lemma 2 ([Zha12a]). *Let H be an oracle drawn from a $2q$-wise independent distribution. Then the advantage any quantum algorithm making at most q queries to H has in distinguishing H from a truly random function is identically 0.*

The next definition and lemma are given by Zhandry [Zha12b] and allow for the efficient simulation of an exponentially-large list of samples, given only a polynomial number of samples:

Definition 1 (Small-range distributions [Zha12b]). *Fix sets $\mathcal{X}$ and $\mathcal{Y}$ and a distribution D on $\mathcal{Y}$. Fix an integer r. Let $\mathbf{y} = (y_1, ..., y_r)$ be a list of r samples from D and let P be a random function from $\mathcal{X}$ to $[r]$. The distributions on $\mathbf{y}$ and P induce a distribution on functions $H : \mathcal{X} \to \mathcal{Y}$ defined by $H(x) = y_{P(x)}$. This distribution is called a small-range distribution with r samples of D.*

Lemma 3 ([Zha12b]). *There is a universal constant C_0 such that, for any sets $\mathcal{X}$ and $\mathcal{Y}$, distribution D on $\mathcal{Y}$, any integer ℓ, and any quantum algorithm A making q queries to an oracle $H : \mathcal{X} \to \mathcal{Y}$, the following two cases are indistinguishable, except with probability less than $C_0 q^3 / \ell$:*

- *$H(x) = y_x$ where $\mathbf{y}$ is a list of samples of D of size $|\mathcal{X}|$.*
- *H is drawn from the small-range distribution with ℓ samples of D.*

3   Quantum-Secure Signatures

Our goal is to construct signatures that are resistant to a *quantum* chosen mes-sage attack, where the adversary submits quantum superpositions of messages and receives the corresponding superpositions of signatures in return. First, we need a suitable definition of what a signature scheme is in our setting, and what it means for such a scheme to be secure. Correctness for a stateless signature scheme is identical to the classical setting: any signature produced by the sign-ing algorithm must verify. There is some subtlety, however, for stateful signature schemes. If the state of the signing algorithm depends on the messages signed, and if the adversary mounts a quantum chosen message attack, the signing al-gorithm and adversary will become entangled. To keep the state of the signing algorithm classical and unentangled with the adversary, we therefore restrict the state to be independent of the messages signed so far. We note that many stateful signature schemes, such as stateful Merkle signatures, satisfy this requirement. We arrive at the following definition:

Definition 2. *A signature scheme S is a tuple of efficient classical algorithms* $(\mathsf{G}, \mathsf{Sign}, \mathsf{Ver})$ *where*

- $\mathsf{G}(\lambda)$ *generates a private/public key pair* $(\mathsf{sk}, \mathsf{pk})$.
- $\mathsf{Sign}(\mathsf{sk}, m, \textit{state})$ *outputs a signature σ and new state* $\textit{state}'$. *If the output* *state is ever non-empty, we say that algorithm* Sign *is stateful and we require* *that the state does not depend in any way on the messages that have been* *signed so far. If the output* *state is always empty, we say that* Sign *is stateless* *and we drop the* *state variables altogether.*
- $\mathsf{Ver}(\mathsf{pk}, m, \sigma)$ *either accepts or rejects. We require that valid signatures are al-ways accepted, that is if σ is the output of* $\mathsf{Sign}(\mathsf{sk}, m, \textit{state})$ *then* $\mathsf{Ver}(\mathsf{pk}, m, \sigma)$ *accepts.*

For security, we use a notion similar to that for message authentication codes defined by Boneh and Zhandry [BZ13a]. There are two issues in defining security under a quantum chosen message attack:

- **Randomness.** When using a randomized signature scheme, there are several choices for how the randomness is used. One option is to choose a single randomness value for each chosen message query, and sign every message in the superposition with that randomness. Another approach is to choose fresh randomness for each message in the superposition. Using a single randomness value for each query is much simpler for implementers, and we therefore design signature schemes secure in this setting.

  Fortunately, there is a simple transformation that converts a scheme requir-ing independent randomness for every message into a scheme that is secure when a single randomness value is used for an entire query: when signing, choose a fresh random key k for a quantum pseudorandom function (QPRF). This will be the single per-query randomness value. To sign a superposition of messages, sign each message m in the superposition using randomness obtained by applying the QPRF to m using the key k. From the adversary's point of view, this is indistinguishable from choosing independent randomness for each message. Using Lemma 2, we can replace the QPRF with a function drawn from a pairwise independent function family, which is far more efficient than using a QPRF. Hence, requiring global randomness per query does not complicate the signature scheme much, but greatly simplifies its implementation.

- **Forgeries.** Each quantum chosen message query can be a superposition of every message in the message space. Sampling the returned superposition will result in a single message/signature pair for a random message. Therefore, the classical notion of existential forgery being a signature on a *new* message is ill-defined when we allow quantum access. Instead, for security we require that the adversary cannot produce $q + 1$ valid message/signature pairs with q quantum chosen message queries. Security definitions in this style were previously used in the context of blind signatures [PS96].

We arrive at the following definition of security:

Definition 3 (Quantum Security). *A signature scheme $\mathcal{S} = (\mathsf{G}, \mathsf{Sign}, \mathsf{Ver})$ is strongly existentially unforgeable under a quantum chosen-message attack (EUF-qCMA secure) if, for any efficient quantum algorithm A and any polynomial q, A's probability of success in the following game is negligible in λ:*

Key Gen. *The challenger runs $(\mathsf{sk}, \mathsf{pk}) \leftarrow \mathsf{G}(\lambda)$, and gives pk to A.*

Signing Queries. *The adversary makes a polynomial q chosen message queries. For each query, the challenger chooses randomness r, and responds by signing each message in the query using r as randomness:*

$$\sum_{m,t} \psi_{m,t} |m,\ t\rangle \qquad \longrightarrow \qquad \sum_{m,t} \psi_{m,t} |m,\ t \oplus \mathsf{Sign}(\mathsf{sk}, m; r)\rangle$$

Forgeries. *The adversary is required to produce $q + 1$ message/signature pairs. The challenger then checks that all the signatures are valid, and that all message/signature pairs are distinct. If so, the challenger reports that the adversary wins.*   □

In this paper, we will also be using several weaker notions of security. The first is for a classical chosen message attack:

Definition 4. *$\mathcal{S}$ is existentially unforgeable under a classical random message attack (EUF-CMA secure) if every signing query is measured before signing, so that only a single classical message is signed per query.*

Next, we define random message security:

Definition 5. *$\mathcal{S}$ is existentially unforgeable under a random message attack (EUF-RMA secure) if the adversary is not allowed any signing queries, but instead receives q message/signature pairs for uniform random messages at the beginning of the game.*

We can weaken the security definition even further, to get universal unforgeability:

Definition 6. *S is universally unforgeable under a random message attack (UUF-RMA secure) if, along with receiving q message/signature pairs for random messages, the adversary receives n additional random messages, and all of the $q + 1$ messages for which a signature is forged must be among the $q+n$ messages received.*

All of the above security definitions also have weak variants, where in addition to requiring that message/signature forgery pairs be distinct, we also require that the messages themselves be distinct. Finally, all of the above security definitions also have k-time variants for any constant k, where the value of q is bounded to at most k. When the distinction is required, we refer to the standard unbounded q notion as many-time security.

Separation from Classical Security. In the full version [BZ13b], we present a signature scheme that is secure under classical queries, but completely insecure once an adversary can make quantum queries.

The idea is to augment a classically secure scheme by choosing a random secret prime p and storing p in the secret signing key. We modify the signature scheme so that the signature on the message $m = p$ includes the entire secret key. As long as the adversary does not learn p, she should not be able to learn the secret key. Following ideas from Zhandry [Zha12b], we also add some auxiliary information to the signatures such that, under classical queries, p is hidden, but a single quantum query suffices to recover p. Since classically, signatures can be built from one-way functions, we immediately get the following theorem:

Theorem 1. *Assuming the existence of one-way functions, there exists a signature scheme S that is existentially unforgeable under a classical chosen message attack, but is totally broken under a quantum chosen message attack.*

3.1 Quantum-Secure Signatures from Classically-Secure Signatures

Now we move to actually building signature schemes that are secure against quantum chosen message attacks. In this section, we show a general transformation from classically secure signatures to quantum secure signatures. The building blocks for our construction are chameleon hash functions and signatures that are secure against a classical random message attack. First, we will define a chameleon hash function. The definition we use is slightly different from the original definition from Krawczyk and Rabin [KR00], but is satisfied by the known lattice constructions:

Definition 7. *A chameleon hash function $\mathcal{H}$ is a tuple of efficient algorithms $(\mathsf{G}, \mathsf{H}, \mathsf{Inv}, \mathsf{Sample})$ where:*

- $\mathsf{G}(\lambda)$ *generates a secret/public key pair* $(\mathsf{sk}, \mathsf{pk})$.
- $\mathsf{H}(\mathsf{pk}, m, r)$ *maps messages to some space* $\mathcal{Y}$

- Sample(λ) *samples* r *from some distribution such that, for every* pk *and* m, H(pk, m, r) *is uniformly distributed.*
- Inv(sk, h, m) *produces an* r *such that* H(pk, m, r) = h, *and* r *is distributed negligibly-close to* Sample(λ) *conditioned on* H(pk, m, r) = h

We say that a chameleon hash function is collision resistant if no efficient quantum algorithm, given only pk, can find collisions in H(pk, ·, ·). Cash et al. [CHKP10] build a simple lattice-based chameleon hash function, and prove that it is collision resistant, provided that the *Shortest Integer Solution* problem (SIS) is hard for an appropriate choice of parameters. The idea behind our construction is to first hash the message with the chameleon hash function and then sign the hash. In order to be secure against quantum queries, care has to be taken in how the randomness for the hash and the signature scheme is generated. In what follows, for any randomized algorithm A, we let $A(x; r)$ denote running A on input x with randomness r.

Construction 2. *Let* $\mathcal{H} = (\mathsf{G}_H, \mathsf{H}, \mathsf{Inv}, \mathsf{Sample})$ *be a chameleon hash function, and* $\mathcal{S}_c = (\mathsf{G}_c, \mathsf{Sign}_c, \mathsf{Ver}_c)$ *a signature scheme. Let* $\mathcal{Q}$ *and* $\mathcal{R}$ *be families of pairwise independent functions mapping messages to randomness used by* Inv *and* Sign_c, *respectively. We define a new signature scheme* $\mathcal{S} = (\mathsf{G}, \mathsf{Sign}, \mathsf{Ver})$ *where:*

$$\mathsf{G}(\lambda) : (\mathsf{sk}_H, \mathsf{pk}_H) \xleftarrow{R} \mathsf{G}_H(\lambda), \ (\mathsf{sk}_c, \mathsf{pk}_c) \xleftarrow{R} \mathsf{G}_c(\lambda)$$
$$\text{output } \mathsf{sk} = (\mathsf{pk}_H, \mathsf{sk}_c), \ \mathsf{pk} = (\mathsf{pk}_H, \mathsf{pk}_c)$$

$$\mathsf{Sign}((\mathsf{pk}_H, \mathsf{sk}_c), m) : Q \xleftarrow{R} \mathcal{Q}, R \xleftarrow{R} \mathcal{R}$$
$$r \leftarrow \mathsf{Sample}(\lambda; R(m)), \ s \leftarrow Q(m), \ h \leftarrow \mathsf{H}(\mathsf{pk}_H, m, r)$$
$$\sigma \leftarrow \mathsf{Sign}(\mathsf{pk}_c, h; s), \ \text{output } (r, \sigma)$$

$$\mathsf{Ver}((\mathsf{pk}_H, \mathsf{pk}_c), m, (r, \sigma)) : h \leftarrow \mathsf{H}(\mathsf{pk}_H, m, r), \ \text{output } \mathsf{Ver}(\mathsf{pk}_c, h, \sigma)$$

We note that the chameleon secret key is not used in Construction 2, though it will be used in the security proof. Classically, this method of hashing with a chameleon hash and then signing converts any non-adaptively secure scheme into an adaptive one. We show that the resulting scheme is actually secure against an adaptive *quantum* chosen message attack.

Theorem 3. *If* $\mathcal{S}_c$ *is weakly (resp. strongly) EUF-RMA secure and* $\mathcal{H}$ *is a secure chameleon hash function, then* $\mathcal{S}$ *in Construction 2 is weakly (resp. strongly) EUF-qCMA secure. Moreover, if* $\mathcal{S}_c$ *is only one-time secure, then* $\mathcal{S}$ *is also one-time secure.*

Theorem 3 shows that we can take a classically EUF-RMA secure signature scheme, combine it with a a chameleon hash, and obtain a quantum-secure signature scheme. In particular, the following constructions will be quantum secure, assuming SIS is hard:

- A slight modification to the signature scheme of Cash et al. [CHKP10], which combines their chameleon hash function with an EUF-RMA secure signature scheme. The only difference in their scheme is that the values r and s are sampled directly, rather than setting them to be the outputs of pairwise independent functions.

- A modification of the scheme of Agrawal, Boneh, and Boyen [ABB10], where we hash the message using a chameleon hash before applying the signature.

We now prove Theorem 3:

Proof. We first sketch the proof idea. Given an $\mathcal{S}_c$ signature σ on a random hash h, we can construct an $\mathcal{S}$ signature on any given message m: use the chameleon secret key sk_H to compute a randomness r such that $\mathsf{H}(\mathsf{pk}_H, m, r) = h$, and output the signature (r, σ). Thus, we can respond to a classical chosen message attack, given only signatures on random messages.

If the adversary issues a *quantum* chosen message query, we need to sign each of the exponentially many messages in the query superposition. Therefore, using the above technique directly would require signing an exponential number of random hashes. Instead, we use small-range distributions and Lemma 3 to reduce the number of signed hashes required to a polynomial. The problem is that the number of hashes signed is still a very large polynomial, whereas the number of signatures produced by our adversary is only $q + 1$, so we cannot rely on the pigeon-hole principle to argue that one of the $\mathcal{S}$ forgeries is in fact a $\mathcal{S}_c$ forgery. We can, however, argue that two of the forgeries must, in some sense, correspond to the same query. If we knew which query, we could perform a measurement, observing which of the (polynomially many) random hashes were signed. Lemma 1 shows that the adversary's advantage is reduced by only a polynomial factor. For this query, we now only sign a single random hash, but the adversary produces two forgeries. Therefore, one of these forgeries must be a forgery for $\mathcal{S}_c$. Of course, we cannot tell ahead of time which query to measure, so we just pick the query at random, and succeed with probability $1/q$.

We now give the complete proof. There are four variants to the theorem (one-time vs many time, strong vs weak). We will prove the many-time strong security variant, the other proofs being similar. Let A be an adversary breaking the EUF-qCMA security of $\mathcal{S}$ in Construction 2 with non-negligible probability ϵ. We prove security through a sequence of games.

Game 0. This is the standard attack experiment, where A receives pk_c and pk_H, and is allowed to make a polynomial number of quantum chosen message queries. For query i, the challenger produces pairwise independent functions $R^{(i)}$ and $Q^{(i)}$, and responds to each message in the query superposition as follows:

- Let $r_m^{(i)} = \mathsf{Sample}(\lambda; R^{(i)}(m))$ and $s_m^{(i)} = Q^{(i)}(m)$.
- Compute $h_m^{(i)} = \mathsf{H}(\mathsf{pk}_H, m, r_m^{(i)})$
- Compute $\sigma_m^{(i)} = \mathsf{Sign}_c(\mathsf{sk}_c, h_m^{(i)}; s_m^{(i)})$
- Respond with the signature $(r_m^{(i)}, \sigma_m^{(i)})$.

In the end, A must produce $q + 1$ distinct triples $(m_k^*, r_k^*, \sigma_k^*)$ such that $\mathsf{Ver}(\mathsf{pk}_c, \mathsf{H}(\mathsf{pk}_H, m_k^*, r_k^*), \sigma_k^*)$ accepts. By definition, A wins with probability ϵ, which is non-negligible. Therefore, there is some polynomial $p = p(\lambda)$ such that $p(\lambda) > 1/\epsilon(\lambda)$ for infinitely-many λ.

Game 1. We make two modifications: first, we choose $R^{(i)}$ and $Q^{(i)}$ as truly random functions, which amounts to generating $r_m^{(i)} \leftarrow \mathsf{Sample}(\lambda)$ and picking $s_m^{(i)}$ at random for each i, m. According to Lemma 2, the view of the adversary is unchanged. Second, we modify the conditions in which A wins by requiring that no two (m_k^*, r_k^*) pairs form a collision for H. The security of $\mathcal{H}$ implies that A succeeds in Game 1 with probability at least $\epsilon - \mathsf{negl}$.

Game 2. Generate $s_m^{(i)}$ as before, but now draw $h_m^{(i)}$ uniformly at random. Additionally, draw uniform randomness $t_m^{(i)}$. We will sample $r_m^{(i)}$ from the set of randomness making $\mathsf{H}(\mathsf{pk}, m, r_m^{(i)}) = h_m^{(i)}$. That is, let $r_m^{(i)} = \mathsf{Inv}(\mathsf{sk}, h_m^{(i)}, m; t_m^{(i)})$. The only difference from A's perspective is the distribution of the $r_m^{(i)}$ values. For each m, the distribution of $r_m^{(i)}$ is negligibly-close to that of Game 1, and we show in the full version [BZ13b] that this implies Games 1 and 2 are indistinguishable. Therefore, the success probability is at least $\epsilon - \mathsf{negl}$.

Game 3. Let $\ell = 2C_0 qp$ where C_0 is the constant from Lemma 3. At the beginning of the game, for $i = 1, ..., q$ and $j = 1, ..., \ell$, sample values $\hat{h}_j^{(i)}$ and let $\hat{\sigma}_j^{(i)} = \mathsf{Sign}_c(\mathsf{sk}_c, \hat{h}_j^{(i)})$. Also pick q random functions O_i mapping m to $[\ell]$. Then let $h_m^{(i)} = \hat{h}_{O_i(m)}^{(i)}$ and $\sigma_m^{(i)} = \hat{\sigma}_{O_i(m)}^{(i)}$. Let T_i be random functions, and let $t_m^{(i)} = T_i(m)$. The only difference between Game 2 and Game 3 is that the $h_m^{(i)}$ and $\sigma_m^{(i)}$ values were generated by q small-range distributions on ℓ samples. Each of the small-range distributions is only queried once, so Lemma 3 implies that the success probability is still at least $\epsilon - \mathsf{negl} - 1/2p$.

Game 4. Let the O_i and T_i be pairwise independent functions. The adversary cannot tell the difference.

Notice that Game 4 can now be simulated efficiently, and A wins in this game with probability $\epsilon - \mathsf{negl} - 1/2p$. Let $h_k^* = \mathsf{H}(\mathsf{pk}, m_k^*, r_k^*)$ be the hashes of the forgeries. Since we have no collisions in H, the pairs (h_k^*, σ_k^*) are distinct. Let $\mathcal{H}^{(i)} = \{\hat{h}_j^{(i)}\}$ be the set of $\hat{h}$ values used to answer query i, and $\mathcal{H}$ be the union of the $\mathcal{H}^{(i)}$. There are two possibilities:

- At least one of the h_k^* is not in $\mathcal{H}$, or two of them are equal. This means that one of the h_k^* was never signed, or one of them was signed once, but two signatures were produced for it. In either case, it is straightforward to construct a forger B_0 for $\mathcal{S}_c$ that wins in this case. Since $\mathcal{S}_c$ is secure, this event only happens with negligible probability.
- All of the h_k^* values are distinct and lie in $\mathcal{H}$. In this case, there is some i such that two h_k^* values are in $\mathcal{H}^{(i)}$ for the same i. Notice that this event happens, and all the forgeries are valid, with probability $\epsilon - \mathsf{negl} - 1/2p$.

Game 5. Now we guess a random query i^* and add a check that all the h_k^* values lie in $\mathcal{H}$, and that two of them are distinct and lie in $\mathcal{H}^{(i^*)}$. Without loss of generality, assume these two h^* values are h_0^* and h_1^*. A then wins in this game with probability $\epsilon/q - \mathsf{negl} - 1/2pq$. Let j_b^* be the j such that $h_b^* = \hat{h}_{j_b^*}^{(i^*)}$ for $b = 0, 1$.

Game 6. On query i^*, measure the value of $O_i(m)$, to get a value j^*. O_i takes values in $[\ell]$, so Lemma 1 says the adversary's success probability is still at least $\epsilon/q\ell - \mathsf{negl} - 1/2pq\ell$. Notice now that for query i^*, the challenger only needs to sign $\hat{h}_{j^*}^{(i^*)}$, and therefore, one of the $h_b^* = \hat{h}_j^{(i^*)}$ values was never signed.

Game 7. Now guess at the beginning of the game the value of j^*, and at the end, check that the guess was correct. The adversary still wins with probability $\epsilon/q\ell^2 - \mathsf{negl} - 1/2pq\ell^2$.

If the adversary wins in Game 7, it produced two signatures on $\hat{h}^{(i^*)}$ values, while only one of them was signed. It is straightforward to construct a forger B_1 for $\mathcal{S}_c$ that wins in this case. B_1 has success probability $\epsilon/q\ell^2 - \mathsf{negl} - 1/2pq\ell^2$, and the security of $\mathcal{S}_c$ implies that this quantity is negligible. Thus $\epsilon - 1/2p$ is negligible. Since $\epsilon > 1/p$ infinitely often, we then have $1/2p < \mathsf{negl}$ infinitely often, a contradiction. Therefore, ϵ is negligible.   □

We note that for one-time security, this security reduction signs only a single message, so we only need to rely on the one-time security of $\mathcal{S}_c$.

3.2   Signatures in the Quantum Random Oracle Model

In this section we present a simple generic conversion from any classical signature scheme to a scheme secure against quantum chosen message attacks in the quantum random oracle model.

Recall that when a random oracle scheme is implemented in the real-world, the random oracle is replaced by a concrete hash function H, thereby enabling a quantum adversary to evaluate H on a superposition of inputs. Therefore, security proofs in the random oracle model must allow all parties, including the adversary, to issue *quantum* queries to H. This model is called the *quantum random oracle model* [BDF+11] and is the one we use here.

Our construction is quite simple: use the random oracle to hash the message along with a random salt, and send the signature on the hash, together with the salt. This construction is very appealing since messages are often hashed anyway before signing. The results in this section then show that only minor modifications to existing schemes are necessary to make them quantum immune.

Construction 4. *Let $\mathcal{S}_c = (\mathsf{G}_c, \mathsf{Sign}_c, \mathsf{Ver}_c)$ be a signature scheme, H be a hash function, and $\mathcal{Q}$ be a family of pairwise independent functions mapping messages to the randomness used by Sign_c, and k some polynomial in λ. Define $\mathcal{S} =$*

$(\mathsf{G}, \mathsf{Sign}, \mathsf{Ver})$ *where:*

$$\mathsf{G}(\lambda) = \mathsf{G}_c(\lambda)$$

$$\mathsf{Sign}(\mathsf{sk}, m) : Q \xleftarrow{R} \mathcal{Q},\ r \xleftarrow{R} \{0,1\}^k$$
$$s \leftarrow Q(m),\ h \leftarrow H(m,r),\ \sigma \leftarrow \mathsf{Sign}_c(\mathsf{sk}, h; s),\ output\ (r, \sigma)$$

$$\mathsf{Ver}(\mathsf{pk}, m, (r, \sigma)) : h \leftarrow H(m,r),\ output\ \mathsf{Ver}_c(\mathsf{pk}, h, \sigma)$$

We note that Construction 4 is similar to Construction 2: instead of the chameleon hash $\mathsf{H}(\mathsf{pk}, \cdot, \cdot)$ we have a random oracle $H(\cdot, \cdot)$, and instead of generating a different r for each message in the superposition, we just generate a single r for the entire superposition. We can achieve security for Construction 4, assuming only a very weak form of security for $\mathcal{S}_c$, namely, universal unforgeability under a random message attack (UUF-RMA security):

Theorem 5. *If $\mathcal{S}_c$ is strongly (resp. weakly) UUF-RMA secure, then $\mathcal{S}$ in Construction 4 is strongly (resp. weakly) EUF-qCMA secure in the quantum random oracle model. Moreover, if $\mathcal{S}_c$ is only one-time secure, then $\mathcal{S}$ is also one-time secure.*

We prove Theorem 5 in the full version [BZ13b]. Given that Construction 4 is similar to Construction 2, the security proofs are similar. Now, we explain how to realize the strong UUF-RMA notion of security. We note that any strongly EUF-RMA or EUF-CMA secure signature scheme satisfies this security notion. We also note that some weaker primitives do as well, such as pre-image sampleable functions (PSFs) defined by Gentry et al. [GPV08]. Roughly, PSFs are many-to-one functions F such that, with the secret key, a random pre-image can be sampled. For security, we require that without the secret key, the function is one-way and collision resistant. If we sign a message m by sampling a random pre-image of m, and verify a signature σ by checking that $F(\sigma) = m$, then one-wayness plus collision resistance implies strong UUF-RMA security.

Corollary 1. *If PSF is a collision resistant and one-way PSF, then Construction 4 instantiated with PSF is strongly EUF-qCMA secure in the quantum random oracle model.*

Gentry et al. [GPV08] show how to construct a PSF that is collision-resistant and one-way under the assumption that SIS is hard. Therefore, we can construct efficient signatures in the quantum random oracle model based on SIS. In the full version [BZ13b], we also show that the basic GPV signature scheme is secure in the quantum random oracle model, though the proof is very different.

Next, it is straightforward to show that any adversary A breaking the universal unforgeability of $\mathcal{S}_c$ by mounting a random message attack can easily be transformed into an adversary B breaking Construction 4 under a *classical* chosen message attack in the *classical* random oracle model. Together with Theorem 5, we get the following:

Corollary 2. *If S in Construction 4 is weakly (resp. strongly) existentially unforgeable under a classical chosen message attack performed by a quantum adversary, then it is also weakly (resp. strongly) exististentially unforgeable under a quantum chosen message attack.*

Therefore, if a scheme matches the form of Construction 4, it is only necessary to prove classical security.

3.3 Signatures from Generic Assumptions

We briefly explain how to construct signatures from generic assumptions. We first construct one-time signatures from one-way functions using the basic Lamport construction [Lam79]. In the classical setting, the next step would be to use target collision resistance to expand the message space. Unfortunately, target collision resistance ceases to make sense in the quantum setting, so we resort to collision resistance to expand the message space. Finally, we plug these one-time signatures into the Merkle signature scheme [Mer87]. The end result is a signature scheme whose quantum security relies only on the existence of collision-resistant functions. The following is proved in the full version [BZ13b]:

Theorem 6. *If there exists a collision-resistant hash function, then there exists a strongly EUF-qCMA secure signature scheme.*

4 Quantum-Secure Encryption Schemes

We now turn to encryption schemes where we first discuss an adequate notion of security under quantum queries. In what follows, we will discuss symmetric key schemes; the discussion for public key schemes is similar. At a high level, our notion of security allows quantum encryption and decryption queries, but requires challenge queries to be *classical*:

Definition 8. *A symmetric key encryption scheme $\mathcal{E} = (\mathsf{Enc}, \mathsf{Dec})$ is indistinguishable under a quantum chosen message attack (IND-qCCA secure) if no efficient adversary A can win in the following game, except with probability at most $1/2 + \mathsf{negl}$:*

Key Gen. *The challenger picks a random key k and a random bit b. It also creates a list $\mathcal{C}$ which will store challenger ciphertexts.*

Queries. *A is allowed to make three types of queries:*

    Challenge queries. *A sends two messages m_0, m_1, to which the challenger responds with $c^* = \mathsf{Enc}(k, m_b)$. The challenger also adds c^* to $\mathcal{C}$.*

    Encryption queries. *For each such query, the challenger chooses randomness r, and encrypts each message in the superposition using r as randomness:*

$$\sum_{m,c} \psi_{m,c} |m,\ c\rangle \quad \longrightarrow \quad \sum_{m,c} \psi_{m,\ c} |m, c \oplus \mathsf{Enc}(k, m; r)\rangle$$

Decryption queries. *For each such query, the challenger decrypts all ciphertexts in the superposition, except those that were the result of a challenge query:*

$$\sum_{c,m} \psi_{c,m} |c,\ m\rangle \quad \longrightarrow \quad \sum_{c,m} \psi_{c,m} |c,\ m \oplus f(c)\rangle$$

where

$$f(c) = \begin{cases} \perp & \textit{if } c \in \mathcal{C} \\ \mathsf{Dec}(k,c) & \textit{otherwise} \end{cases}$$

Guess. *A produces a bit b', and wins if $b = b'$.*

In the above definition, we need to define the operation $m \oplus \perp$. Since the query responses will XOR $\perp$ with different messages, we need a convention that makes this operation reversible. Taking $\perp$ to be some bit string that lies outside of the message space and $\perp \oplus m$ to be the bitwise XOR will suffice.

Note that we implicitly assume that the decryption algorithm is deterministic. This will be true of our encryption schemes. We note that this is not a limiting assumption since we can make the decryption algorithm deterministic by deriving the randomness for decryption from a PRF applied to the ciphertext. Also, as in the classical case, a simple hybrid argument shows that the above definition is equivalent to the case where the number of encryption queries is limited to one. Lastly, it is straightforward to modify the above definition for public key encryption schemes.

Quantum Challenge Queries. One might hope to enhance Definition 8 by making the security game entirely quantum, where challenge queries are quantum as well. This leads to several difficulties. First, with quantum challenge queries, it is no longer possible to record the challenge ciphertext. This makes it difficult to check that the adversary only asks decryption queries on ciphertexts other than the challenge ciphertexts. The second difficulty is more serious: allowing quantum challenge queries results in definitions of security that are unachievable, even if we disallow decryption queries. In the full version [BZ13b], we show several attempts at defining security with quantum challenges, and show that each of these definitions is insecure.

Separation from Classical Security. Similar to the case for signatures, quantum chosen ciphertext queries give the adversary more power than classical queries. The following is proved in the full version [BZ13b]:

Theorem 7. *If there exists a symmetric (resp. public) key encryption scheme $\mathcal{E}$ that is secure against a* classical *chosen ciphertext attack, then there is a symmetric (resp. public) key encryption scheme $\mathcal{E}'$ that is secure under a classical chosen ciphertext attack, but totally insecure under a* quantum *chosen ciphertext attack.*

4.1   Symmetric CCA Security

In this section, we construct symmetric-key CCA secure encryption. We will follow the encrypt-then-MAC paradigm. Ideally, we would like to show that encrypt-then-MAC, when instantiated with any quantum chosen *plaintext* secure encryption scheme and any EUF-qCMA secure MAC, would be quantum chosen *ciphertext* secure. However, it is not obvious how to prove security, as the reduction algorithm has no way to tell which ciphertexts the adversary received as the result of an encryption query, and no way to decrypt the ciphertexts if it has received them. To remedy these problems, we choose a specific encryption scheme and MAC and leave the general security proof as an open question. The encryption scheme allows us to efficiently check if the adversary has seen a particular ciphertext as a result of an encryption query, and to decrypt in this case. The construction is as follows:

Construction 8. *Let F and G be pseudorandom functions. We construct the following encryption scheme $\mathcal{E} = (\mathsf{Enc}, \mathsf{Dec})$ where:*

$$\mathsf{Enc}((k_1, k_2), m) : r \xleftarrow{R} \{0, 1\}^\lambda$$
$$c_1 \leftarrow F(k_1, r) \oplus m, \ c_2 \leftarrow G(k_2, (r, m))$$
$$output \ (r, c_1, c_2)$$
$$\mathsf{Dec}((k_1, k_2), (r, c_1, c_2)) : m \leftarrow c_1 \oplus F(k_1, r), \ c_2' \leftarrow G(k_2, (r, m))$$
$$if \ c_2 \neq c_2', \ output \perp$$
$$otherwise, \ output \ m$$

For security, we require F to be a classically secure PRF, and G to be quantum secure — secure against queries on a superposition of inputs. Zhandry [Zha12b] shows how to construct PRFs meeting this strong notion of security.

Theorem 9. *If F and G are quantum-secure pseudrandom functions, then $\mathcal{E}$ in Construction 8 is qCCA-secure.*

Theorem 9 is proved in the full version [BZ13b]. As demonstrated by Zhandry [Zha12b], quantum-secure pseudorandom functions can be built from any one-way function. Therefore, Theorem 9 shows that quantum chosen ciphertext security can be obtained from the minimal assumption that one-way functions exist.

4.2   Public-key CCA Security

In the full version [BZ13b], we construct CCA-secure signatures in the public-key setting. We follow the generic transformation from identity-based encryption (IBE) to CCA security due to Boneh et al.[BCHK04], which uses a selectively secure IBE scheme and a strong one-time signature scheme. The one-time signature scheme only needs to be classically secure, and can hence be built from any one-way function. In contrast, we need the IBE scheme to be secure against *quantum* chosen identity queries. We observe that the IBE scheme of Agrawal, Boneh, and Boyen [ABB10] meets this notion of security, assuming the hardness of the Learning With Errors (LWE) problem. We obtain the following:

Theorem 10. *If the* LWE *problem is hard for quantum computers, then there exists a public-key encryption scheme that is IND-qCCA secure.*

5   Conclusion and Open Problems

We defined the notions of a quantum chosen message attack for signatures and quantum chosen ciphertext attack for encryption. We gave the first constructions of signatures and encryption schemes meeting these strong notions of security. For signatures, we presented two simpler compilers that transform classically secure schemes into quantum-secure schemes. We also showed that signatures can be built from any collision resistant hash function. For encryption, we presented both a symmetric-key and a public-key construction. There are many directions for future work. First, can we base quantum security for signatures on the minimal assumption of one-way functions? Also, it may be possible to mount quantum superposition attacks against many cryptographic primitives. For example, can we build identification protocols or functional encryption that remain secure in the presence of such attacks?

Acknowledgments. We thank Luca Trevisan and Amit Sahai for helpful conversations about this work. This work was supported by NSF, DARPA, the Air Force Office of Scientific Research (AFO SR) under a MURI award, Samsung, and a Google Faculty Research Award. The views and conclusions contained herein are those of the authors and should not be interpreted as necessarily representing the official policies or endorsements, either expressed or implied, of DARPA or the U.S. Government.

References

[ABB10]   Agrawal, S., Boneh, D., Boyen, X.: Efficient lattice (H)IBE in the standard model. In: Gilbert, H. (ed.) EUROCRYPT 2010. LNCS, vol. 6110, pp. 553–572. Springer, Heidelberg (2010)

[BCHK04]   Canetti, R., Halevi, S., Katz, J.: Chosen-ciphertext security from identity-based encryption. In: Cachin, C., Camenisch, J.L. (eds.) EUROCRYPT 2004. LNCS, vol. 3027, pp. 207–222. Springer, Heidelberg (2004)

[BDF+11]   Boneh, D., Dagdelen, Ö., Fischlin, M., Lehmann, A., Schaffner, C., Zhandry, M.: Random Oracles in a Quantum World. In: Lee, D.H., Wang, X. (eds.) ASIACRYPT 2011. LNCS, vol. 7073, pp. 41–69. Springer, Heidelberg (2011)

[BHK+11]   Brassard, G., Høyer, P., Kalach, K., Kaplan, M., Laplante, S., Salvail, L.: Merkle Puzzles in a Quantum World. In: Rogaway, P. (ed.) CRYPTO 2011. LNCS, vol. 6841, pp. 391–410. Springer, Heidelberg (2011)

[BS08]   Brassard, G., Salvail, L.: Quantum Merkle Puzzles. In: Second International Conference on Quantum, Nano and Micro Technologies (ICQNM 2008), pp. 76–79 (February 2008)

[BZ13a]   Boneh, D., Zhandry, M.: Quantum-secure message authentication codes. In: Johansson, T., Nguyen, P.Q. (eds.) EUROCRYPT 2013. LNCS, vol. 7881, pp. 592–608. Springer, Heidelberg (2013), Full version available at the Electronic Colloquium on Computational Complexity: http://eccc.hpi-web.de/report/2012/136

[BZ13b]   Boneh, D., Zhandry, M.: Secure signatures and chosen ciphertext security in a quantum computing world. In: Canetti, R., Garay, J.A. (eds.) CRYPTO 2013, Part II. LNCS, vol. 8043, pp. 361–379. Springer, Heidelberg (2013), Full version available at the Cryptology ePrint Archives (2013), http://eprint.iacr.org/2013/088

[Can01]   Canetti, R.: Universally composable security: A new paradigm for cryptographic protocols. In: Proceedings of FOCS. IEEE (2001)

[CHKP10]   Cash, D., Hofheinz, D., Kiltz, E., Peikert, C.: Bonsai Trees, or How to Delegate a Lattice Basis. In: Gilbert, H. (ed.) EUROCRYPT 2010. LNCS, vol. 6110, pp. 523–552. Springer, Heidelberg (2010)

[DFNS11]   Damgård, I., Funder, J., Nielsen, J.B., Salvail, L.: Superposition attacks on cryptographic protocols. CoRR, abs/1108.6313 (2011)

[GPV08]   Gentry, C., Peikert, C., Vaikuntanathan, V.: Trapdoors for Hard Lattices and New Cryptographic Constructions. In: Proceedings of the 40th Annual ACM symposium on Theory of computing (STOC), p. 197 (2008)

[HSS11]   Hallgren, S., Smith, A., Song, F.: Classical cryptographic protocols in a quantum world. In: Rogaway, P. (ed.) CRYPTO 2011. LNCS, vol. 6841, pp. 411–428. Springer, Heidelberg (2011)

[IBM12]   IBM Research. IBM research advances device performance for quantum computing (February 2012), http://www-03.ibm.com/press/us/en/pressrelease/36901.wss

[KR00]   Krawczyk, H., Rabin, T.: Chameleon hashing and signatures. In: Proc. of NDSS, pp. 1–22 (2000)

[Lam79]   Lamport, L.: Constructing digital signatures from a one-way function. Technical Report SRI-CSL-98 (1979)

[Mer87]   Merkle, R.C.: A Digital Signature Based on a Conventional Encryption Function. In: Pomerance, C. (ed.) CRYPTO 1987. LNCS, vol. 293, pp. 369–378. Springer, Heidelberg (1988)

[PS96]   Pointcheval, D., Stern, J.: Provably secure blind signature schemes. In: Kim, K.-C., Matsumoto, T. (eds.) ASIACRYPT 1996. LNCS, vol. 1163, pp. 1–12. Springer, Heidelberg (1996)

[Unr10]   Unruh, D.: Universally Composable Quantum Multi-Party Computation. In: Gilbert, H. (ed.) EUROCRYPT 2010. LNCS, vol. 6110, pp. 486–505. Springer, Heidelberg (2010)

[Zha12a]   Zhandry, M.: Secure identity-based encryption in the quantum random oracle model. In: Safavi-Naini, R., Canetti, R. (eds.) CRYPTO 2012. LNCS, vol. 7417, pp. 758–775. Springer, Heidelberg (2012), Full version available at the Cryptology ePrint Archives: http://eprint.iacr.org/2012/076/

[Zha12b]   Zhandry, M.: How to construct quantum random functions. In: Proceedings of FOCS (2012), Full version available at the Cryptology ePrint Archives: http://eprint.iacr.org/2012/182/

Everlasting Multi-party Computation

Dominique Unruh

University of Tartu

Abstract. A protocol has everlasting security if it is secure against adversaries that are computationally unlimited *after* the protocol execution. This models the fact that we cannot predict which cryptographic schemes will be broken, say, several decades after the protocol execution. In classical cryptography, everlasting security is difficult to achieve: even using trusted setup like common reference strings or signature cards, many tasks such as secure communication and oblivious transfer cannot be achieved with everlasting security. An analogous result in the quantum setting excludes protocols based on common reference strings, but not protocols using a signature card. We define a variant of the Universal Composability framework, everlasting quantum-UC, and show that in this model, we can implement secure communication and general multiparty computation using signature cards as trusted setup.

1 Introduction

Everlasting Security. Computers and algorithms improve over time and so does the ability of an adversary to break cryptographic complexity assumptions and protocols. It may be feasible to make a good estimate as to which computational problems are hard *today*, and which encryption schemes unbroken. But it is very difficult to make more than an educated guess as to which cryptographic schemes will be secure, say, ten years from now. Key length recommendations (e.g., [1,2,3]) can only be made based on the assumption that progress continues at a similar rate as today; unexpected algorithmic progress and future technologies like quantum computers can render even the most paranoid choices for the key length obsolete.

This situation is very problematic if we wish to run cryptographic protocols on highly sensitive data such as medical or financial data or government secrets. Such data often has to stay confidential for many decades. But an adversary might intercept messages from a protocol that is secure today, store them, and some decades later, when the underlying cryptosystems have been broken, decrypt them. For highly sensitive data, this would not be an acceptable risk.

One way out is to use protocols with unconditional (information-theoretical) security that are not based on any computational hardness assumptions. For many tasks, however, unconditionally secure protocols simply do not exist (in particular if we cannot assume an majority of honest participants). A compromise is the concept of *everlasting security*. In a nutshell, a protocol is everlastingly

R. Canetti and J.A. Garay (Eds.): CRYPTO 2013, Part II, LNCS 8043, pp. 380–397, 2013.

secure if it cannot be broken by an adversary that becomes computationally unlimited *after* the protocol execution. This guarantees that all assumptions need only to hold *during* the protocol execution, sensitive data is not threatened by possible future attacks on today's schemes. We only need to reliably judge the *current* state of the art, not future technologies.

Unfortunately, also for everlasting security, we have strong impossibility results. It is straightforward to see that everlastingly secure public key encryption is not possible, symmetric encryption needs keys as long as the transmitted messages, and most secure multi-party computations (MPC) are impossible (e.g., oblivious transfer, see Section 3).

Quantum Cryptography. Since the inception of quantum key distribution (QKD) by Bennett and Brassard [4], it has been known that quantum cryptography can achieve tasks that are impossible in a classical setting: a shared key can be agreed upon between two parties such that even a computationally unlimited eavesdropper does not learn that key. Classically, this is easily seen to be impossible. Crépeau and Kilian [5] showed how, given only a commitment scheme, we can securely realize an oblivious transfer (OT), which in turn, using ideas from Kilian [6] can be used to implement arbitrary unconditionally secure MPC. Classically, given only a commitment, it is impossible to construct arbitrary unconditionally secure MPC (or even everlastingly secure ones, see Section 3). Initial enthusiasm was, however, dampened by strong impossibility results. Mayers [7] showed that it is impossible to construct an unconditionally secure commitment from scratch. Similar impossibilities hold for OT and many other function evaluations (Lo [8]). So the goal to get unconditionally secure MPC is not achievable, even with quantum cryptography.

Also, the usefulness of QKD has been challenged (e.g., by Bernstein [9], who also raises other concerns than the following). To run a QKD protocol, an authenticated channel is needed. But how to implement such a channel? If we use a public key infrastructure for signing messages, we lose unconditional security and thus the main advantage of QKD. If we use shared key authentication, a key needs to be exchanged beforehand. (And, if we exchange an authentication key in a personal meeting, why not just exchange enough key material for one-time pad encryption – storage is cheap.)

Everlasting Quantum Security. A simple change of focus resolves the problems described in the previous paragraph. Instead of seeing the goal of quantum cryptography in achieving unconditional security, we can see it as achieving *everlasting security*. For example, if we run a QKD protocol and authenticate all messages using signatures and a public key infrastructure, then we do not get an unconditionally secure protocol, but we do get everlasting security: only the signatures are vulnerable to unlimited adversaries, but breaking the security of the signatures after the protocol execution does not help the adversary to recover the key. (Experience and the discussion on composition below show that one has to be careful: we need to check that signatures and QKD indeed play together well and compose securely. We answer this positively in Section 4: we achieve everlastingly secure universally composable security.)

What about secure MPC? Recall that for constructing unconditionally secure MPC in the quantum setting, the only missing ingredient was a commitment. Once we have a commitment, unconditionally secure MPC protocols exist [10]. Unconditionally secure commitments do not exist, but everlastingly secure ones do! Consider a statistically hiding commitment. That is, the binding property may be subject to computational assumptions, but the hiding property holds with respect to unlimited adversaries. Such a scheme is in fact everlastingly secure. Being able to break the binding property of a commitment after the protocol end is of no use – the recipient of the commitment is not listening any more. And the hiding property, i.e., the secrecy of the committed data, holds forever. So a statistically hiding commitment is in fact everlastingly secure. It seems that we have all ingredients for everlastingly secure quantum MPC. The next paragraph, however, shows that the situation is considerably more subtle.

We stress that the neither the concept of everlasting security nor the idea of combining it with quantum cryptography is original to this paper. For example, [11] already suggested to combine QKD with computational authenticated, albeit without proof or analysis of composition problems.

Everlasting Security and Composition – A Cautionary Tale. As discussed above, statistically hiding commitments are in fact everlastingly secure, and there are quantum protocols that construct unconditionally secure OT (among other things). Thus, composing a statistically hiding commitment with such a protocol will give us an everlastingly secure OT in the bare model (i.e., not using any trusted setup). But it turns out that this reasoning is wrong! Lo's impossibility of OT [8] can be easily modified to show that unconditional OT is impossible, even if we consider only passive (semi-honest) adversaries. But everlasting security implies unconditional security against passive adversaries: A passive adversary is one that during the protocol follows the protocol (and thus in particular is computationally bounded) but after the protocol may perform unlimited computations. Thus Lo's impossibility excludes the existence of everlastingly secure OTs.

What happened? The problem is that although statistically hiding commitments are everlastingly secure on their own, they lose their security when composed. Composition problems are common in cryptography, but we find this case particularly instructive: The commitment does not lose its security only when composed with some contrived protocol, but instead in a natural construction. And not only does a particular construction break down, we are faced with a general impossibility. And the resulting protocol is insecure in a strong sense: an unlimited adversary can guess either Alice's or Bob's input. (As opposed to a situation where the "break" consists solely of the non-existence of a required simulator.)

One may be tempted to suggest that the failure is not related to the everlasting security, but to the non-composability of the commitments. Damgård and Nielsen [12] present commitment schemes that are universally composable (we elaborate on this notion below, it is a security notion that essentially guarantees "worry-free" composition), that only need a predistributed common reference strings

(CRS), and that are statistically hiding.[1] Yet, when using these commitments to get everlastingly secure OT, we run into the same problem again: We would get an everlastingly secure OT using a CRS, but a generalization of Lo's impossibility shows that no everlastingly secure OT protocols exist even given a CRS (see Section 3).[2] (See also page 391 for another view on the problem in the quantum case.)

Quantum Everlasting Universal Composability. The preceding paragraph shows that, in the setting of everlasting security, it is vital to find definitions that guarantee composability. One salient approach is the Universal Composability (UC) framework by Canetti [14]. In the UC framework, we compare a protocol π against a so-called ideal functionality $\mathcal{F}$ which describes what π should ideally do. (E.g., $\mathcal{F}$ could be a commitment functionality that registers the value Alice commits to, but forwards it to Bob only when Alice requests an open.) We say π UC-emulates $\mathcal{F}$ if for any adversary Adv (that attacks π) there is a simulator Sim (that "attacks" $\mathcal{F}$) we have that no machine $\mathcal{Z}$ (the environment) can distinguish π running with Adv (real model) from $\mathcal{F}$ running with Sim (ideal model). The intuition behind this is that Adv can perform only attacks that can be mimicked by Sim. Since $\mathcal{F}$ is secure by definition, Adv can perform no "harmful" attacks. A salient property of the UC framework is that UC secure protocols can be composed in arbitrary ways (universal composition). By tweaking the details of the definition, we get various variants of UC: If $\mathcal{Z}$, Sim, Adv are polynomial-time, we have computational UC. If they are unlimited, statistical UC (modeling unconditional security). Unlimited quantum machines lead to the definition of statistical quantum-UC [10].

Müller-Quade and Unruh [13] showed that the UC framework can also be adapted to the setting of everlasting security: We quantify over $\mathcal{Z}$, Sim, Adv that are polynomial-time, but we say that $\mathcal{Z}$ distinguishes the real and ideal model if the distribution of $\mathcal{Z}$'s output is not *statistically* indistinguishable. That is, a protocol is considered insecure if one can distinguish real and ideal model when being polynomial-time during the protocol, but unlimited afterwards (statistical indistinguishability means that no *unlimited* machine can distinguish).

The ideas from [13] can be easily adapted to the quantum case. In Section 2, we introduce everlasting quantum UC (eqUC). Here $\mathcal{Z}$, Sim, Adv are quantum-polynomial-time machines (representing the fact that adversaries are limited during the protocol run), but we require that the quantum state output by $\mathcal{Z}$ in the real and ideal model is trace-indistinguishable (two quantum states are trace-indistinguishable if no unlimited quantum machine can distinguish them). The eqUC security notion inherits all composability properties from the UC notion. Also, protocols that are secure with respect to statistical classical or statistical quantum UC are also eqUC-secure. In particular, known quantum protocols for

---

[1] The schemes given in [12] were only shown secure classically. But we think it likely that similar protocols can be constructed in the quantum setting, too.

[2] That Damgård and Nielsen's commitment does not compose well in an everlasting security setting was already observed in [13]. Their example, however, only shows insecurity when composing with contrived protocols.

constructing MPC from commitments [10] are also eqUC secure. Thus, if we find an eqUC-secure commitment protocol, we immediately get eqUC-secure MPC protocols by composition.

Everlasting Quantum-UC Commitments. The problem of everlasting UC commitments in the classical setting was already studied in [13]. Their protocol uses a signature card as trusted setup.[3] Here a signature card is a trusted device (modeled as a functionality) such that the owner of the card can sign messages, everyone can access the public key, and no-one (not even the owner) can get the secret key.[4] Their protocol is, however, only known to be secure in the classical setting. In fact, when we try to prove the protocol secure in a quantum setting, we stumble upon an interesting difficulty in the interplay of zero-knowledge proofs of knowledge and signature schemes.

A core step in the protocol is that Alice performs a proof of knowledge P showing that she knows a certain signature σ. In the security proof, we then show that Alice must have obtained σ from the signature card: Assume Alice successfully performs P without requesting σ first. Since P is a proof of knowledge, there is an extractor E (using Alice and indirectly the signing oracle as a black box) that returns a valid witness, i.e., the signature σ. Since E returns the signature without requesting it from the signing oracle, we have a contradiction to the unforgeability of the signature scheme.

It seems that the same reasoning applies against quantum adversaries if we use quantum proofs of knowledge instead. Unfortunately, this is not the case. In a quantum proof of knowledge (as defined by Unruh [17]), an extractor with black box access to the prover executes both the prover (modeled as a unitary operation) as well as its inverse (i.e., the inverse of that unitary). This is the quantum analogue of classical rewinding. So the extractor E will invoke not only the signing oracle, but also its inverse! But unforgeability will not guarantee that there are no forgeries when the adversary accesses the inverse of the signing oracle. Hence the security proof fails.

To avoid this problem, we need a new protocol which does not require rewinding in the same places of the security proof where we use the unforgeability of the signature scheme. We present such a protocol; it is considerably more involved than the one from [13]. We believe that our approach is of independent interest because it shows one way around the limitations of quantum proofs of knowledge.

Bounded Quantum Storage Model. We quickly compare the concept of everlasting security in this paper with the bounded quantum storage model (BQSM; [18]). The BQSM achieves very similar goals. Security in the BQSM guarantees that the protocol cannot be broken by an adversary that has limited quantum memory during the protocol execution and unlimited quantum memory after the execution. The BQSM is thus analogous to everlasting security as discussed

---

[3] It is impossible to construct UC commitments without using some trusted setup such as a CRS [15]. [13] shows that for everlasting UC, even a CRS is not sufficient.

[4] The last property is mandated, e.g., by the German signature card law [16].

here, except that it considers quantum memory where we consider computational power. The advantage of the BQSM over our model is that when using a BQSM protocol, we only need to make assumptions about the power of the adversary (its quantum memory). In contrast, in our model we need to assume that the computational power is limited *and* that certain mathematical problems are hard. In our view, the main disadvantage of the BQSM is that it might be useful only for a limited time: currently, we may assume a small limit on the adversary's quantum memory. Should quantum technology advance, though, quantum memory might become cheap, and at that point BQSM protocols must not be used any more. In contrast, with everlasting security as in this paper, if an assumption we use in a protocol is broken, it is likely that there still are other assumptions that can be used – we can then fix the protocol by switching the underlying problem. Also, BQSM protocol tend to have a high communication complexity, and composition is more involved (in particular when we wish for universal composability [19]). Then again, our approach requires trusted setup (signature cards). An interesting goal would be protocols that are simultaneously secure in our model and the BQSM.

In the classical setting, the bounded storage model can also be used [20] but has very high communication complexity (quadratic in the memory bound). [21] shows that if we combine bounded storage with temporary computational assumptions, then in the random oracle model we achieve lower communication complexity (but they also show impossibilities when not using the random oracle model). In contrast, our work uses quantum communication and temporary computational assumptions, but no bounded storage.

Further Related Work. [22] also considers the problem of using an unconditionally hiding computationally binding commitment to construct a quantum OT (as opposed to using directly a functionality). They show that with such a commitment, OT can be realized (no impossibility results are given). However, their OT protocol only computationally hides the sender's inputs (although one may be tempted to assume otherwise as the commitments that are used are unconditionally hiding). In fact, our impossibility results imply that their OT cannot be everlastingly secure.

Organization and Contribution. In Section 2 we present the everlasting quantum UC model and the corresponding composition theorem. In Section 3 we show the impossibility of everlastingly secure OT in the classical and the quantum setting using various functionalities. In Section 4 we show that using signature cards or a public key infrastructure, an everlastingly quantum-UC-secure secure channel can be implemented. In Section 5 we implement arbitrary everlastingly quantum-UC-secure multi-party computation using signature cards. Many details and proofs are omitted for space reasons, these are given in the full version [23].

2   Everlasting Quantum UC

We now give a terse overview of the definition of everlasting quantum UC (eqUC). Our definition is based on the modeling of UC in the quantum case from [10]. For a full definition, see [23]. The only difference between the definition from [10] and ours is that we allow the environment to output a quantum state and that we require that state to be trace-indistinguishable between real and ideal model. See also [13] for additional discussion on how to model everlasting security in the UC framework.

The basic concept is that of a network. A *network* $\mathbf{N}$ is a set of quantum machines. Each machine maintains a quantum state and can send and receive messages from other machines in the network. A message can be a quantum state. In a network, there is a distinguished machine $\mathcal{Z}$. This machine is initially activated with some input z. When a machine is activated by an incoming message, it can apply an arbitrary quantum operation to the message and its state, producing a new state and an outgoing message. Then the recipient of that message is activated. If a machine sends no outgoing message, $\mathcal{Z}$ is activated. At any point, $\mathcal{Z}$ may terminate with some output quantum state. We denote by $\mathrm{QExec}_{\mathbf{N}}(\eta, z)$ the state output by $\mathcal{Z}$ after an execution of the network $\mathbf{N}$ when $\mathcal{Z}$ gets initial input z and the security parameter is η. We call two networks $\mathbf{N}, \mathbf{N}'$ *trace-indistinguishable* $\mathrm{TD}(\mathrm{QExec}_{\mathbf{N}}(\eta, z), \mathrm{QExec}_{\mathbf{N}'}(\eta, z)) \leq \mu(k)$ is negligible for all $z \in \{0,1\}^*$ and $k \in \mathbb{N}$ where $\mathrm{TD}(\rho, \rho')$ denotes the so-called trace-distance between two quantum states.

A protocol π is a network without $\mathcal{Z}$ or adversary. We cannot execute π itself, but given machines Adv and $\mathcal{Z}$, we can run $\pi \cup \{\mathrm{Adv}, \mathcal{Z}\}$. Given a set C of party identities, let π^C denote the result of replacing, for each $id \in C$, the party with id id by the *corruption party* P_{id}^C. This corruption party just forwards all its communication to the adversary and is controlled by it.

We can now specify everlasting quantum-UC-security. The fact that in this definition, we require the networks to be trace-indistinguishable (i.e., even an unlimited machine cannot distinguish the output states of $\mathcal{Z}$ in real and ideal model), models the fact that in everlasting security, we allow unlimited computations *after* the protocol execution. During the protocol execution, environment, adversary, and simulator are quantum-polynomial-time.

Definition 1 (Everlasting quantum-UC-security). *Let protocols π and ρ be given. We say π everlastingly quantum-UC-emulates (short eqUC-emulates) ρ iff for every set C of party ids and for every quantum-polynomial-time adversary Adv there is a quantum-polynomial-time simulator Sim such that for every quantum-polynomial-time environment $\mathcal{Z}$, the networks $\pi^C \cup \{\mathrm{Adv}, \mathcal{Z}\}$ and $\rho^C \cup \{\mathrm{Sim}, \mathcal{Z}\}$ are trace-indistinguishable.*

We can now define security by comparing a protocol π with some ideal functionality $\mathcal{F}$. If we say that π eqUC-emulates a functionality $\mathcal{F}$, we mean that π eqUC-emulates $\rho_{\mathcal{F}}$ where the ideal protocol $\rho_{\mathcal{F}}$ is the protocol consisting of the functionality $\mathcal{F}$ plus the so-called *dummy-parties*. For each party in π, there is a dummy-party $\tilde{P}$ that just forwards messages between the environment $\mathcal{Z}$ and the functionality $\mathcal{F}$. The reason for introducing dummy-parties is that dummy-parties can be corrupted. By corrupting Alice in the ideal protocol, the simulator controls the dummy-party and thus effectively Alice's inputs to $\mathcal{F}$ and also gets the outputs from $\mathcal{F}$ to Alice.

If, e.g., we wish to express the fact that π is a eqUC-secure commitment, we say that π eqUC-emulates $\mathcal{F}_{\mathrm{COM}}$ where $\mathcal{F}_{\mathrm{COM}}$ is the commitment functionality defined below. We specify two functionalities that will be used in this paper.

Definition 2 (Commitment). *Let A and B be two parties. The functionality $\mathcal{F}_{\mathrm{COM}}^{A\to B,\ell}$ behaves as follows: Upon (the first) input (`commit`, x) with $x \in \{0,1\}^{\ell(k)}$ from A, send* `committed` *to B. Upon input* `open` *from A send (`open`, x) to B. All communication/input/output is classical.*

We call A the sender and B the recipient.

Definition 3 (Signature card). *Let $\mathfrak{S} = (\mathrm{KG}, \mathrm{Sign}, \mathrm{Verify})$ be a signature scheme. Let A be a party. Then the functionality $\mathcal{F}_{\mathrm{SC}}^{\mathfrak{S},A}$ (signature card for scheme $\mathfrak{S}$ with owner A) behaves as follows: Upon the first activation, $\mathcal{F}_{\mathrm{SC}}^{\mathfrak{S},A}$ chooses a verification/signing key pair (pk, sk) using the key generation algorithm $\mathrm{KG}(1^\lambda)$. Upon a message (`getpk`) from a party P or the adversary, it sends pk to P or the adversary, respectively. Upon a message (`sign`, m) from A $\mathcal{F}_{\mathrm{SC}}^{\mathfrak{S},A}$ computes $\sigma \leftarrow \mathrm{Sign}(sk, m)$ and sends (pk, σ) to A.*

All communication/input/output is classical.

One of the salient features of the UC model is the universal composition theorem. It says that if π eqUC-emulates $\mathcal{F}$, then we can replace $\mathcal{F}$ by π in any context. (Thus allowing for modular protocol design.) The proof of the following theorem follows the lines of that for quantum UC [10].

Theorem 1 (Universal composition theorem). *Let $\mathcal{F}$ and $\mathcal{G}$ be quantum-polynomial-time functionalities. Let π and $\sigma^{\mathcal{F}}$ be quantum-polynomial-time protocols. Here the notation $\sigma^{\mathcal{F}}$ means that σ invokes (possibly many) instances of $\mathcal{F}$. Assume π eqUC-emulates $\mathcal{F}$. Assume further that $\sigma^{\mathcal{F}}$ eqUC-emulates $\mathcal{G}$. Then σ^π eqUC-emulates $\mathcal{G}$. (Here σ^π is the result of replacing $\mathcal{F}$ by the protocol π in $\sigma^{\mathcal{F}}$.)*

3 Impossibilities

In Section 5, we show that by using signature cards and a quantum channel, we can construct general everlastingly secure MPC protocols. The question arises whether both signature cards and quantum channels are needed. We answer this question positively by showing that (a) in the classical setting, most typical trusted setup (including signature cards) is not sufficient to implement everlasting OT and that (b) in the quantum setting, typical trusted setup such as a CRS is not sufficient to implement everlasting OT. The impossibilities even apply if we do not try to achieve UC security but only to implement a stand-alone OT.

For space reasons, we only give a short overview here. For precise statements and proofs see [23].

Classical Impossibilities. The basic observation underlying our impossibility result is that a protocol that is everlastingly secure is also secure against unlimited passive adversaries. This is due to the fact that a passive adversary follows the protocol during the protocol execution (and is thus polynomial-time) and only after the protocol execution performs an unlimited computation. Thus if an unlimited passive adversary could break the protocol, the protocol would not be everlastingly secure either.

We call a functionality $\mathcal{F}$ passively-realizable if there is a protocol that realizes $\mathcal{F}$ with respect to unlimited passive adversaries. We show that the following functionalities are passively-realizable: the coin-toss $\mathcal{F}_{CT}$, the common reference string $\mathcal{F}_{CRS}$, the public key infrastructure $\mathcal{F}_{PKI}$, the commitment $\mathcal{F}_{COM}$, and the signature card $\mathcal{F}_{SC}$.

Assume now an everlastingly secure OT protocol π that uses a passively-realizable functionality $\mathcal{F}$. Then π is also secure against passive unlimited adversaries. Let ρ be the protocol that realizes $\mathcal{F}$ (passively). Then π', resulting from replacing $\mathcal{F}$ by ρ, will still be an OT secure against passive unlimited adversaries. (Here, of course, we have to be careful with our definition of passively realizing a functionality – the notion needs to compose such that π' is still secure.) But π' does not use any functionality, and we know that no OT protocol in the bare model can be secure against unlimited passive adversaries.

Concluding, we get:

Theorem 2 (Simplified). *There is no everlastingly secure OT protocol which only uses arbitrarily many instances of $\mathcal{F}_{CT}$ (coin-toss), $\mathcal{F}_{CRS}$ (common reference string), $\mathcal{F}_{COM}$ (commitment), $\mathcal{F}_{PKI}$ (public key infrastructure), and $\mathcal{F}_{SC}$ (signature cards).*

Quantum Impossibilities. The impossibility in the quantum case follows similar lines. However, the classical notion of passive adversaries does not make sense in the quantum case. (A passive adversary copies all data, this is not possible in the quantum case.) To solve this issue, we consider only protocols that perform no measurements (unitary protocols). Any protocol can be transformed into such a protocol at the expense of additional quantum memory. We call a functionality $\mathcal{F}$ quantum-passively-realizable if there is a unitary protocol π that realizes $\mathcal{F}$ with respect to passive unlimited adversaries (that follow the protocol exactly and do not even copy information). Notice that the requirement that π has to be unitary has the effect that the protocol cannot just throw away information. Thus an adversary that is passive will still have some information left over after the protocol execution. The following functionalities turn out to be quantum-passively-realizable: coin toss $\mathcal{F}_{CT}$, predistributed EPR pairs $\mathcal{F}_{EPR}$, public key infrastructure $\mathcal{F}_{PKI}$ (assuming the secret key is uniquely determined by the public key). However, signature cards and commitments are not! (The reason being that signature cards and commitments do not allow to commit/sign superpositions of messages and thus enforce measurements. This cannot be realized with a unitary protocol.)

Then we can proceed as in the classical case: Assume an everlasting quantum OT protocol π using a quantum-passively-realizable functionality $\mathcal{F}$. This protocol is also secure against unlimited passive adversaries (in the above sense). By replacing $\mathcal{F}$ by the protocol ρ that realizes $\mathcal{F}$, we get a quantum OT protocol π' not using any functionality that is secure against unlimited passive adversaries. But Lo [8] shows that such protocols do not exist. Thus we get:

Theorem 3 (Simplified). *There is no quantum-polynomial-time everlastingly secure OT protocol which only uses arbitrarily many instances of $\mathcal{F}_{CT}$ (cointoss), $\mathcal{F}_{CRS}$ (common reference string), $\mathcal{F}_{EPR}$ (predistributed EPR pair), $\mathcal{F}_{PKI}$ (public key infrastructure; assuming that the secret key is uniquely determined by the public key).*

4   Everlasting Quantum Key Distribution

The first application of quantum everlasting security we present in this paper is a new view on quantum key distribution (QKD). Instead of thinking of QKD as a method for getting unconditionally secure message transmission (but then being stuck with the problem of how to realize authenticated channels), we can combine QKD with a computationally secure authenticated channel to get everlastingly secure message transmission. This was already suggested in [11, Section 3.1], but no formal statement or proof was given. We only give a short overview here, for details see [23]. The first step is to implement an authenticated channel from, say, a signature card. (All results in this section also hold with a normal public key infrastructure instead of a signature card.)

Lemma 1 (Authenticated channels from signature cards). *Let $\mathfrak{S}$ be a quantum existentially unforgeable signature-scheme. Then there is a polynomial-time classical protocol π using one instance of $\mathcal{F}_{SC}^{\mathfrak{S},A}$ such that π eqUC-emulates $\mathcal{F}_{auth}^{A \to B}$. Here $\mathcal{F}_{auth}^{A \to B}$ denotes an authenticated channel from A to B.*

The proof presents no surprises. Using $\mathcal{F}_{auth}^{A \to B}$ and $\mathcal{F}_{auth}^{B \to A}$, we can implement (with statistical quantum-UC-security) a bidirectional secure channel between Alice and Bob using existing protocols from the literature [24,25,26].

Corollary 1 (Secure channels from signature cards). *Let $\mathfrak{S}$ be a quantum existentially unforgeable signature-scheme. Then there is a polynomial-time protocol π using one instance of $\mathcal{F}_{SC}^{A,\mathfrak{S}}$ and $\mathcal{F}_{SC}^{B,\mathfrak{S}}$ each such that π eqUC-emulates $\mathcal{F}_{secchan}$. (Here $\mathcal{F}_{secchan}$ denotes a bidirectional secure channel between A and B.)*

5   Everlasting Quantum Multi-party Computation

Classical Everlasting UC Commitments. In the classical setting, Müller-Quade and Unruh [13] presented a protocol that everlastingly *classical*-UC-emulates (called "long-term UC-emulates" there, ecUC-emulates in the following) the commitment functionality $\mathcal{F}_{\text{COM}}$ and that uses a signature card $\mathcal{F}_{\text{SC}}$. There protocol cannot be proven secure in the quantum setting (at least we do not know how), but it is instructive to understand their protocol before we present ours.[5]

In order for a commitment protocol to be everlastingly UC secure, we need to achieve the following: Obviously, it needs to be statistically hiding and computationally binding. Furthermore we need that the protocol is extractable: a simulator who controls the signature card can find out what value Alice committed to. And the protocol needs to be equivocal: a simulator who controls the signature card can cheat the binding property and open to a different value. The simulators need to behave in a way that is statistically indistinguishable from the honest behavior of the parties.

The difficulty lies in the extractability. If the committed value can be extracted by the simulator from the interaction, then it must be somehow contained in that interaction, and an unlimited entity can extract it. But that would contradict the statistical hiding property. The approach is to use the signature card $\mathcal{F}_{\text{SC}}^A$. When Alice wishes to commit to a value m, we force her to obtain a signature on m. Since the simulator controls $\mathcal{F}_{\text{SC}}$, and since Alice can only sign using $\mathcal{F}_{\text{SC}}$ (even the owner of the signature card does not know the secret key), the simulator will learn m. How do we force Alice to sign m? First, Alice commits to m using a commitment COM. Then Alice obtains a signature σ on (m, u) from $\mathcal{F}_{\text{SC}}$ where u is the opening information for $\text{COM}(m)$. And then Alice proves that she knows a signature σ on (m, u) for some u that opens $\text{COM}(m)$ as m. (Here COM is statistically hiding, and the proof is a statistically witness-indistinguishable argument of knowledge.) **Commit phase:**

$$A \xrightarrow{\begin{array}{c} c := \text{COM}(m) \\ \hline \text{Proof: I know signature } \sigma \text{ on } (m, u) \text{ s.t. } u \text{ opens } c \text{ as } m \\ \text{or I know the secret key of } \mathcal{F}_{\text{SC}} \end{array}} B$$

We now have extractability: Alice can only succeed in the proof if she gets a signature on (m, u). But then all the simulator has to do is to check which query (m, u) to $\mathcal{F}_{\text{SC}}$ opens the commitment c, and then he knows m. (We explain the "or I know the secret key"-part in a moment.) In the open phase, we cannot just send u, then we would not have equivocality. Instead, Alice proves that she *could* open c as m. **Open phase:**

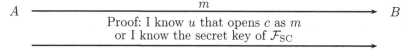

$$A \xrightarrow{\begin{array}{c} m \\ \hline \text{Proof: I know } u \text{ that opens } c \text{ as } m \\ \text{or I know the secret key of } \mathcal{F}_{\text{SC}} \end{array}} B$$

Now, if the simulator wishes to equivocate, he simply commits to 0, and later he produces a fake proof that he can open c as m. To produce this fake proof, we

---

[5] [13] actually first construct a ecUC zero-knowledge proof and use that one to construct an ecUC commitment. For clarity, we present and discuss a direct construction instead. An analogous discussion applies to their original zero-knowledge protocol.

have added the "or I know the secret key sk"-part. Since the simulator knows sk (he controls $\mathcal{F}_{SC}$), he can always perform the proof using sk as witness. (While Alice, not knowing sk, is forced to prove the part of the statement before the "or".)

Another (Quantum) View on the Problem. It has been pointed out (by an anonymous reviewer) that in the quantum case, the problem is actually the following: Using a standard unconditionally hiding commitment scheme fails to achieve everlasting security when using it to construct an OT. But this is not due to composability issues, but to the fact that commitment schemes do not force the committer to commit to a classical value, allowing commitments to superpositions instead. In contrast, an ideal commitment functionality would not allow the commit to occur in superposition. This also matches what we do in our quantum-secure protocol below: The signature card forces the committed message to be classical.

We believe this view to be correct, too. Indeed, our protocol would not work if the signature card would allow the adversary to sign superpositions of messages. Yet, this view only partially explains the situation: Even in the purely classical case described above, standard commitments are not sufficient. But in the classical case, the possibility of committing to superpositions obviously cannot be the reason for the problem, indicating that composition is at least part of the problem. In fact, we believe that non-composition and the possibility to commit to superpositions might actually be two sides of the same coin. For example, composition usually requires extractability, i.e., the fact that the adversary can only commit to values he knows. But if the adversary can commit to superpositions, he cannot know what he commits to. It would be interesting (but beyond the scope of this work) to explore this connection further.

Difficulties in the Quantum Case. Now assume we wish to prove the above protocol secure in the quantum case. Then instead of an argument of knowledge, we need to use a quantum argument of knowledge. But then we run into problems when showing extractability. To show extractability, we need to show that Alice cannot perform the first proof without first sending (m, u) to $\mathcal{F}_{SC}$. To do so, consider an execution where Alice performs the proof without sending (m, u) to $\mathcal{F}_{SC}$. We can then consider Alice as a prover $A^{\mathcal{O}}$ with access to a signing oracle $\mathcal{O}$. Applying the extractor E from the argument of knowledge to Alice, we get that $E^{A^{\mathcal{O}}}$ outputs a witness to the statement that is proven. I.e., either a signature on (m, u) or the secret key sk of $\mathcal{O}$. Since $E^{A^{\mathcal{O}}}$ has only black-box access to $\mathcal{O}$, and since $A^{\mathcal{O}}$ and thus also $E^{A^{\mathcal{O}}}$ never signs (m, u), both possibilities contradict the existential unforgeability of the signature scheme. This reasoning works in the classical case. In the quantum case (following [17]), however, the extractor $E^{A^{\mathcal{O}}}$, while rewinding, does the following: It applies both U and U^{-1} where U is the unitary transformation describing the operation of $A^{\mathcal{O}}$. Thus, indirectly $E^{A^{\mathcal{O}}}$ invokes not only $\mathcal{O}$, but also its inverse. Existential unforgeability makes no statement in this case. It could well be that given access to the inverse of $\mathcal{O}$, we can efficiently construct forgeries or even extract the secret key.

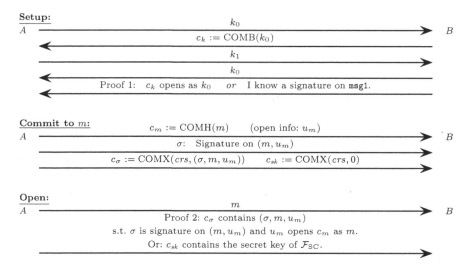

Fig. 1. The commitment protocol based on signature cards – overview. Proof 1 is a witness indistinguishable argument of knowledge, proof 2 is a statistically witness indistinguishable argument. COMH is a statistically hiding quantum-computationally binding commitment. COMB is a quantum-computationally hiding perfectly binding commitment. COMX is a dual-mode commitment.

Note: At a first glance, it might seem that invoking the inverse of $\mathcal{O}$ is not a problem due to the following reasoning. An oracle $\mathcal{O}$ implementing a function $f(x)$ is usually modeled as a unitary mapping $|x\rangle|y\rangle$ to $|x\rangle|y \oplus f(x)\rangle$. That unitary is self-inverse, so applying $\mathcal{O}^{-1}$ is equivalent to applying $\mathcal{O}$.

However, if the signing oracle $\mathcal{O}$ is modeled in this way, then it can be queried on superposition. Instead, $\mathcal{O}$ should measure the message to be signed first. This could be realised by copying the message (using CNOTs) into fresh ancillae bits. But then $\mathcal{O}$ is not self-inverse any more. Furthermore, to formulate the existential unforgeability, $\mathcal{O}$ additionally needs to keep track of all messages that were signed (otherwise it is not possible to define a "fresh" forgery). Applying the inverse of $\mathcal{O}$ will remove messages from this list, making the notion of a fresh message meaningless.

Our Approach. To solve this problem, we need to construct a new protocol in whose security proof we do not need to rewind the signing oracle. A protocol overview is given in Figure 1. We now explain the intuition behind the protocol. As explained above, the main challenge is the extractability of the protocol: Alice commits to m using a commitment scheme COMB, the unveil information is u_m. We need to make sure that Alice is forced to sign (m, u_m) in order to complete the protocol. We cannot just perform a proof of knowledge that Alice knows such a signature σ on (m, u_m) – it might be that Alice proves that she knows a signatures without actually knowing it. To force Alice to actually know the signature, we use the following approach: During the commit phase, Alice commits to (σ, m, u_m) using a commitment scheme COMX. $(c_\sigma := \text{COMX}((\sigma, m, u_m)).)$

And additionally, we let Alice prove ("proof 2" in Figure 1) that the resulting commitment c_σ indeed contains a valid signature σ on (m, u_m). However, we seem to have the same problem as before: How do we guarantee that Alice knows the content of the resulting commitment c_σ? We cannot use rewinding for the same reason as before. Instead, we use a so-called dual-mode commitment for c_σ. A dual-mode commitment COMX depends on a public parameter crs: If crs is honestly chosen, then COMX is statistically hiding (we need this as otherwise the overall protocol would not be statistically hiding and thus not everlastingly secure). But crs can also be chosen in a special way together with a trapdoor td such that using td, we can efficiently compute (σ, m, u_m) given $c_\sigma = \text{COMX}(crs, (\sigma, m, u_m))$.

Then we can prove extractability of the eqUC commitment protocol roughly as follows:

1. For extracting, the simulator looks at the list of signing queries to $\mathcal{F}_{SC}$ and finds a suitable pair (m, u_m). We need to show that if Alice opens successfully, there must have been such a signing query for (m, u_m) during the commit phase.
2. To show that, consider a game consisting of an execution with corrupted Alice and that simulator. We change the game such that instead of picking crs honestly, we pick it together with a trapdoor td. (We discuss below how to do that.)
   Note: the new game will only be computationally indistinguishable from the preceding one. But this does not contradict everlasting security: we are in a side-arm of the proof in order to bound the probability of a certain event ("Alice opens without signing (m, u_m)"). The extracting simulator will still be statistically indistinguishable from an honest recipient of the commitment since the extracting simulator just passively looks at the signing queries.
3. We use the soundness of "proof 2" to show that c_σ contains with overwhelming probability a valid signature σ on (m, u_m). (In the full proof, we need to additionally exclude that Alice proves the alternative option that c_{sk} contains the secret key.)
   Note: we do not claim at this point that Alice knows σ, we only show that whatever is extracted from c_σ using td is a valid signature on (m, u_m). In particular, we do not use the unforgeability of the signature scheme in this step.
4. Now we use the unforgeability: We have derived that extracting c_σ using td produces a signature on (m, u_m). If this would be the case without having sent (m, u_m) to $\mathcal{F}_{SC}$, we would have produced a forgery, contradicting unforgeability.
5. So Alice always signs (m, u_m), hence the simulator from Step 1 succeeds with overwhelming probability in extracting.

One thing is missing in this description: How to pick crs in a way that we can choose it together with a trapdoor in Step 2? For this, we have the setup phase in Figure 1. Here crs is chosen using a coin toss that is designed such that Bob, if he knows a signature on a special message msg1, can cheat and choose crs arbitrarily. In Step 2, this allows us to pick crs together with a trapdoor by requesting a signature msg1 from $\mathcal{F}_{SC}$. (Here msg1 is an arbitrary fixed bitstring, but syntactically different from all other messages occurring in the protocol.)

Notice that "proof 1" in the coin toss protocol needs to be "of knowledge" (more precisely, a witness-indistinguishable argument of knowledge). However, we do not run into problems with the combination of rewinding and unforgeability this time, because during the execution of "proof 1", the signature card is not accessed by the honest verifier Alice. (And thus the signing oracle is not accessed by the extractor at all.)

Thus, the protocol from Figure 1 is extractable.

Finally, we need to see how to achieve equivocality. Fortunately, this is easy: The equivocating simulator commits to the secret key sk of $\mathcal{F}_{SC}$ in the commitment c_{sk} (he knows it since he controls $\mathcal{F}_{SC}$) and commits to 0 in c_σ. Then, in the open phase, to open as an arbitrary m, the simulator just performs "proof 2" using the fact that c_{sk} indeed contains sk. Thus the protocol is equivocal, too. (No fake CRS is needed in this case.)

5.1   The Full Protocol Description

We fix the following notation for interactive commitment schemes: If COM is a commitment scheme, we denote by $(c, u) \leftarrow \text{COM}_{C,R}(1^\eta, m)$ an execution of the commit phase with sender C and recipient R where C commits to the message m. After the protocol execution, both C and R know the value c (e.g., c could be the protocol transcript), intuitively c represents the commitment itself. Furthermore, C gets the value u, the opening information. We assume that the opening phase consists of C sending (m, u), and R verifying the open phase via a deterministic function $\text{COMVerify}(c, m, u)$. For commitments that take a public parameter crs, we add this parameter as an additional argument to $\text{COM}_{C,R}$ and COMVerify.

We now give a definition of dual-mode commitments. The definition is close to that of dual-mode commitments in [27]. The main difference is that we additionally require that the honestly chosen CRS is uniformly chosen from a set CRS. As discussed in [27], dual-mode commitments (also according to our definition) can be constructed from Regev's cryptosystem [28].

Definition 4. *A dual-mode commitment* COM *is an interactive commitment with a public common reference string crs and which has the following properties:*

- *The common reference string crs is chosen from a set CRS such that one can efficiently sample elements of CRS that are statistically indistinguishable from uniform, and such that CRS is endowed with an arbitrary group operation $*$ (e.g., CRS could be $\{0,1\}^n$ or $\mathbb{Z}_n$ for some n). The operation $*$ is efficiently computable, and inverses with respect to $*$ are efficiently computable.*
- Statistical hiding: *For crs chosen uniformly from CRS,* COM *is statistically hiding.*

- Fake-CRS: *There is an algorithm* $(crs, td) \leftarrow \text{COMFakeCRS}(1^\eta)$ *such that crs is quantum-computationally indistinguishable from being uniformly distributed on CRS.*
- Extractability: *There is an efficient algorithm* COMExtract *such that for any quantum-polynomial-time A, we have that the following probability is negligible:*

$$\Pr[\exists u, m. \ (m \neq m' \wedge \text{COMVerify}(crs, c, m, u) = 1) :$$
$$(crs, td) \leftarrow \text{COMFakeCRS}(1^\eta),$$
$$c \leftarrow \text{COM}_{A,R}(crs), m' \leftarrow \text{COMExtract}(td, c)]$$

Here $c \leftarrow \text{COM}_{A,R}(crs)$ *stands, in abuse of notation, for a commit phase between the adversary A and an honest recipient R. The value c is the value R gets at the end of the commit phase.*

Furthermore, we will need a signature scheme $\mathfrak{S}$ that has some (very natural) additional properties besides quantum existential unforgeability. First, we will need deterministic verification. This just means that the verification algorithm is not randomized. Second, we will need that $\mathfrak{S}$ has a matchingKeys-predicate. This means that there is a predicate matchingKeys that can be decided in deterministic polynomial time, and such that for pk, sk chosen according to the key generation algorithm, we have matchingKeys$(pk, sk) = 1$ with overwhelming probability. And given pk as chosen by the key generation, a quantum polynomial-time algorithm outputs sk with matchingKeys$(pk, sk) = 1$ only with negligible probability. (Intuitively, this just means that there is a well-defined concept of whether a given secret key matches a given public key.)

Theorem 4 (Commitments from signature cards). *Let A and B be parties. Let ℓ be an integer. Assume the existence of: quantum-computationally witness-indistinguishable quantum arguments of knowledge, statistically witness-indistinguishable arguments,*[6] *statistically hiding quantum-computationally binding commitments, quantum-computationally hiding perfectly binding commitments, dual-mode commitments. Assume that $\mathfrak{S}$ is a quantum existentially unforgeable signature scheme with deterministic verification and with matchingKeys-predicate.*

Then there is a protocol π using secure channels and one instance of $\mathcal{F}_{\text{SC}}^{A,\mathfrak{S}}$ such that π eqUC-emulates $(\mathcal{F}_{\text{COM}}^{A \to B, \ell})^*$. (Here $(\mathcal{F}_{\text{COM}}^{A \to B, \ell})^*$ is the functionality consisting of many instances of $\mathcal{F}_{\text{COM}}^{A \to B, \ell}$. I.e., we can perform many commitments using a single signature card.)

The protocol π is shown in Figure 1. A more precise description and a security proof are given in [23].

---

[6] Quantum-computational witness-indistinguishability is defined analogously to the computational witness-indistinguishability (as in, e.g., [29]). Quantum arguments and quantum arguments of knowledge are defined like quantum proofs [30] and quantum proofs of knowledge [17], except that we consider only quantum-polynomial-time provers instead of unlimited provers.

Corollary 2 (Everlasting two-party computation). *Let A and B be parties. Let $\mathcal{G}$ be a well-formed[7] classical probabilistic-polynomial-time functionality involving A and B. Under the conditions from Theorem 4, there is a protocol $\pi_{\mathcal{G}}$ using one instance of $\mathcal{F}_{SC}^{A,\mathfrak{S}}$ such that $\pi_{\mathcal{G}}$ eqUC-emulates $\mathcal{G}^*$.*

This corollary follows from combining Theorem 4 with known statistically secure protocols from [31,32,10]. Analogously, we get everlasting multi-party computation at the price of using more instances of $\mathcal{F}_{SC}$.

Acknowledgments. This work was funded by institutional research grant IUT2-1 from the Estonian Research Council, and by European Regional Development Fund and the Estonian ICT program 2011-2015 (3.2.1202.12-0001), by the European Social Fund's Doctoral Studies and Internationalisation Programme DoRa, by the European Regional Development Fund through the Estonian Center of Excellence in Computer Science, EXCS.

References

1. ECRYPT II: Yearly report on algorithms and keysizes. D.SPA.17 Rev. 1.0, ICT-2007-216676 (June 2011)
2. NIST: Recommendation for key management. Special Publication 800-57 Part 1 Rev. 3 (May 2011)
3. Bundesnetzagentur, B.S.I.: Algorithms for qualified electronic signatures (May 2011)
4. Bennett, C.H., Brassard, G.: Quantum cryptography: Public-key distribution and coin tossing. In: IEEE International Conference on Computers, Systems and Signal Processing 1984, pp. 175–179. IEEE (1984)
5. Crépeau, C., Kilian, J.: Achieving oblivious transfer using weakened security assumptions (extended abstract). In: FOCS 1988, pp. 42–52. IEEE (1988)
6. Kilian, J.: Founding cryptography on oblivious transfer. In: STOC 1988, pp. 20–31. ACM (1988)
7. Mayers, D.: Unconditionally secure quantum bit commitment is impossible. Physical Review Letters 78(17), 3414–3417 (1997)
8. Lo, H.K.: Insecurity of quantum secure computations. Phys. Rev. A. 56, 1154–1162 (August 1997) Eprint on arXiv:quant-ph/9611031v2
9. Bernstein, D.: Cost-benefit analysis of quantum cryptography. Classical and Quantum Information Assurance Foundations and Practice, Dagstuhl Seminar 09311 (2009), Abstract at `http://drops.dagstuhl.de/opus/volltexte/2010/2365`, slides at `http://cr.yp.to/talks/2009.07.28/slides.pdf`
10. Unruh, D.: Universally composable quantum multi-party computation. In: Gilbert, H. (ed.) EUROCRYPT 2010. LNCS, vol. 6110, pp. 486–505. Springer, Heidelberg (2010)
11. Alleaume, R., et al.: Secoqc white paper on quantum key distribution and cryptography. arXiv:quant-ph/0701168v1 (2007)

---

[7] Well-formedness describes certain technical restrictions stemming from the proof by Ishai et al. [31]: Whenever the functionality gets an input, the adversary is informed about the length of that input. Whenever the functionality makes an output, the adversary is informed about the length of that output and may decide when this output is to be scheduled.

12. Damgård, I., Nielsen, J.B.: Perfect hiding and perfect binding universally composable commitment schemes with constant expansion factor. In: Yung, M. (ed.) CRYPTO 2002. LNCS, vol. 2442, pp. 581–596. Springer, Heidelberg (2002)
13. Müller-Quade, J., Unruh, D.: Long-term security and universal composability. Journal of Cryptology 23(4), 594–671 (2010)
14. Canetti, R.: Universally composable security: A new paradigm for cryptographic protocols. In: FOCS 2001, pp. 136–145. IEEE (2001)
15. Canetti, R., Fischlin, M.: Universally composable commitments. In: Kilian, J. (ed.) CRYPTO 2001. LNCS, vol. 2139, pp. 19–40. Springer, Heidelberg (2001)
16. Gesetz über Rahmenbedingungen für elektronische Signaturen. Bundesgesetzblatt I 2001, 876 (May 2001), http://bundesrecht.juris.de/sigg_2001/index.html
17. Unruh, D.: Quantum proofs of knowledge. In: Pointcheval, D., Johansson, T. (eds.) EUROCRYPT 2012. LNCS, vol. 7237, pp. 135–152. Springer, Heidelberg (2012)
18. Damgård, I., Fehr, S., Salvail, L., Schaffner, C.: Cryptography in the bounded quantum-storage model. In: FOCS 2005, pp. 449–458 (2005)
19. Unruh, D.: Concurrent composition in the bounded quantum storage model. In: Paterson, K.G. (ed.) EUROCRYPT 2011. LNCS, vol. 6632, pp. 467–486. Springer, Heidelberg (2011)
20. Maurer, U.: Conditionally-perfect secrecy and a provably-secure randomized cipher. Journal of Cryptology 5(1), 53–66 (1992)
21. Harnik, D., Naor, M.: On everlasting security in the hybrid bounded storage model. In: Bugliesi, M., Preneel, B., Sassone, V., Wegener, I. (eds.) ICALP 2006. LNCS, vol. 4052, pp. 192–203. Springer, Heidelberg (2006)
22. Crépeau, C., Dumais, P., Mayers, D., Salvail, L.: Computational collapse of quantum state with application to oblivious transfer. In: Naor, M. (ed.) TCC 2004. LNCS, vol. 2951, pp. 374–393. Springer, Heidelberg (2004)
23. Unruh, D.: Everlasting multi-party computation. IACR ePrint 2012/177, Full version of this paper (2013)
24. Renner, R., König, R.: Universally composable privacy amplification against quantum adversaries. In: Kilian, J. (ed.) TCC 2005. LNCS, vol. 3378, pp. 407–425. Springer, Heidelberg (2005)
25. Ben-Or, M., Horodecki, M., Leung, D.W., Mayers, D., Oppenheim, J.: The universal composable security of quantum key distribution. In: Kilian, J. (ed.) TCC 2005. LNCS, vol. 3378, pp. 386–406. Springer, Heidelberg (2005)
26. Raub, D., Steinwandt, R., Müller-Quade, J.: On the security and composability of the one time pad. In: Vojtáš, P., Bieliková, M., Charron-Bost, B., Sýkora, O. (eds.) SOFSEM 2005. LNCS, vol. 3381, pp. 288–297. Springer, Heidelberg (2005)
27. Damgård, I., Fehr, S., Lunemann, C., Salvail, L., Schaffner, C.: Improving the security of quantum protocols via commit-and-open. In: Halevi, S. (ed.) CRYPTO 2009. LNCS, vol. 5677, pp. 408–427. Springer, Heidelberg (2009)
28. Regev, O.: On lattices, learning with errors, random linear codes, and cryptography. J. ACM 56(6) (2009)
29. Goldreich, O.: Foundations of Cryptography – (Basic Tools), vol. 1. Cambridge University Press (August 2001)
30. Watrous, J.: Zero-knowledge against quantum attacks. SIAM J. Comput. 39(1), 25–58 (2009)
31. Ishai, Y., Prabhakaran, M., Sahai, A.: Founding cryptography on oblivious transfer – efficiently. In: Wagner, D. (ed.) CRYPTO 2008. LNCS, vol. 5157, pp. 572–591. Springer, Heidelberg (2008)
32. Wullschleger, J.: Oblivious-Transfer Amplification. PhD thesis, ETH Zurich, arXiv:cs/0608076v3 [cs.CR] (March 2007)

Instantiating Random Oracles via UCEs

Mihir Bellare[1], Viet Tung Hoang[2], and Sriram Keelveedhi[1]

[1] Dept. of Computer Science & Engineering, University of California San Diego
[2] Dept. of Computer Science, University of California Davis

Abstract. This paper provides a (standard-model) notion of security for (keyed) hash functions, called UCE, that we show enables instantiation of random oracles (ROs) in a fairly broad and systematic way. Goals and schemes we consider include deterministic PKE; message-locked encryption; hardcore functions; point-function obfuscation; OAEP; encryption secure for key-dependent messages; encryption secure under related-key attack; proofs of storage; and adaptively-secure garbled circuits with short tokens. We can take existing, natural and efficient ROM schemes and show that the instantiated scheme resulting from replacing the RO with a UCE function is secure in the standard model. In several cases this results in the first standard-model schemes for these goals. The definition of UCE-security itself is quite simple, asking that outputs of the function look random given some "leakage," even if the adversary knows the key, as long as the leakage does not permit the adversary to compute the inputs.

1 Introduction

The core contribution of this paper is a new notion of security for (keyed) hash functions called UCE (Universal Computational Extractor). UCE-security is the first well-defined, standard-model security attribute of a hash function shown to permit the latter to securely instantiate ROs across a fairly broad spectrum of schemes and goals.

Under the random-oracle paradigm of Bellare and Rogaway (BR93) [14], a "real-world" or instantiated scheme is obtained by implementing the RO of the overlying ROM scheme via a cryptographic hash function. The central (and justified) critique of the paradigm [36] is that the instantiated scheme has only heuristic security. This paper offers *proven* security for the (standard model) instantiated schemes. The proof is based on the (standard-model) assumption that the instantiating function is UCE-secure.

UCE of course does not *always* work. But we show that it works across a fairly large, diverse and interesting spectrum of schemes and goals including deterministic PKE; message-locked encryption; hardcore predicates; point-function obfuscation; encryption of key-dependent messages; encryption secure under related-key attack; OAEP; correlated-input secure hashing; adaptively-secure garbled circuits; and proofs of safe storage. In all these cases we can use UCE to obtain standard-model solutions, in most cases instantiating known, natural and

R. Canetti and J.A. Garay (Eds.): CRYPTO 2013, Part II, LNCS 8043, pp. 398–415, 2013.

efficient schemes, and in several cases getting the first standard-model schemes for the goals in question.

UCE is quite simple and natural, yet powerful. The basic intuition is that the output of a UCE-secure function looks random even given the key and some "leakage," as long as the inputs are not computable from the leakage. Let us now step back to provide some background and then return to our contributions.

BACKGROUND. The random-oracle paradigm of BR93 [14] has two steps: (1) Design your scheme, and prove it secure, in the ROM, where the scheme algorithms and adversary have access to a RO denoted RO (2) Instantiate the RO to get the standard model scheme that is actually implemented and used. We will consider instantiation via a family of functions H, which means that the instantiated scheme is obtained by replacing RO calls of the ROM-scheme algorithms by evaluations of the deterministic function $H.Ev(hk, \cdot)$ specified by a key $hk \leftarrow_{\$} H.Kg(1^\lambda)$, where λ is the security parameter. The key hk is put in the public key of the instantiated scheme if the latter is public key, else enters in some scheme-dependent way. The suggestion of BR93 was that if H "behaved like a RO," the instantiated scheme would be secure in the standard model. They suggested to obtain such instantiations, heuristically, via cryptographic hash functions. The fundamental subsequent concern has been the lack of a proof of security for the instantiated scheme. Canetti, Goldreich and Halevi (CGH98) [36] show that this lack in some cases cannot be overcome because there exist schemes secure in the ROM but which no family of functions can securely instantiate. Advocates for the defense counter by pointing out that the counter-example schemes are artificial, and in-use instantiations of "natural" ROM schemes are unbroken. This has led to examples that are in one way or another less artificial [7, 37, 42, 51, 60, 64].

It is not the purpose of this paper to take sides in this debate. We want instead to make a scientific contribution towards better grounding the security of instantiated ROM schemes.

THE CORE PROBLEM AND PREVIOUS WORK. The lack of a proof of security for the instantiated scheme is, we submit, a consequence of an even more fundamental lack, namely that of a *definition*, of what it means for a family of functions to "behave like a RO," that could function as an assumption on which to base the proof. The PRF definition [50], which has worked so well in the symmetric setting, is inadequate here because PRF-security relies on the adversary not knowing the key. And collision-resistance (CR) is far from sufficient in any non-trivial usage of a RO.

Canetti [34] was the first to articulate this position and seek a standard-model primitive sufficient to capture some usages of a RO. Notions such as Perfectly One-Way Probabilistic Hash Functions (POWHFs) [34,35,39] and non-malleable hash functions [19] have however proven of limited applicability [21]. Another direction has been to try to instantiate the RO in particular schemes like OAEP [15], again with limited success [21,22] or under strong assumptions on RSA [59].

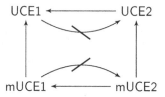

Fig. 1. Relations between UCE security notions. Letting S denote the set of all families H that are S-secure, an arrow A → B represents A ⊆ B, meaning any H that is A-secure is also B-secure. A barred arrow A ↛ B represents A ⊄ B, meaning there is an H that is A-secure but not B-secure. (Assuming of course that some A-secure H exists.)

Our position is philosophically different from that of [34, 39]. These works aimed for security notions that they could achieve under standard assumptions. Expectedly, applicability was limited. We aim to maximize applicability and are willing to see our notion (UCE) as an assumption rather than something to achieve under other assumptions.

UCE. Our definition considers an adversary S, called the source, who is given an oracle HASH, the latter being H.Ev(hk, ·) for key $hk \leftarrow_{\$} $ H.Kg(1^λ) if the challenge bit b is 1, and a RO otherwise. If security now asks that S not figure out b, then, if we deny it hk, we would be back to PRFs, and if we give it hk, security would be unachievable. So we don't ask S to figure out b. Instead, it must pass to an accomplice adversary D, called the distinguisher, some information L called the leakage. The distinguisher *is given the key* hk and must figure out b.

Clearly, security is not achievable for arbitrary leakage. (The source could include in L a point x and the result $y = $ HASH(x) of its oracle on x, and D, having hk, can test whether or not $y = $ H.Ev(hk, x).) We put an extra condition on the source that we call unpredictability. It requires that it be computationally infeasible for a predictor adversary P, given the leakage produced by the source in the *random* ($b = 0$) game, to find any of the inputs queried by the source to its oracle. Note that unpredictability is a property of the source, not of the family of functions H, the latter not figuring in the definition at all.

Security, finally, requires that for any PT *unpredictable* source S, and any PT distinguisher D, the advantage of S, D in figuring out b is negligible. See Section 4 for a formal definition of this notion that we call UCE1. A variant called UCE2, introduced in [11], preserves the source-distinguisher framework of UCE1 but replaces the unpredictability condition with a weaker condition we call reset-security. ("Weaker" because any unpredictable source is reset-secure. This makes UCE stronger: any UCE2-secure family is UCE1-secure.) Both UCE1 and UCE2 involve a single hashing key. We define natural multi-key extensions mUCE1 and mUCE2 as well.

In [11] we examine the relation between UCE and standard security notions for families of functions such as PRF-security and collision-resistance (CR). We show that UCE (of whatever form) neither implies, nor is implied by, any

Goal	Result	UCE
D-PKE	Instantiation of the ROM EwH scheme of [6] to obtain the first standard model deterministic PKE scheme providing full IND [9] and PRIV [6] security.	UCE1
MLE	Instantiation of the ROM convergent encryption scheme of [12, 43], showing this in-use message-locked encryption scheme meets the IND$-CDA goal of [12].	UCE1
HC	Any UCE1-secure family is hardcore for any one-way function and allows for extraction of any number of hardcore bits.	UCE1
BR93 PKE	Instantiation of a natural ROM PKE scheme from BR93 [14] showing it is IND-CPA-secure.	UCE1
PFOB	Instantiation of a ROM point-function obfuscation scheme of [38] to obtain a secure standard-model scheme.	mUCE1
KDM	Instantiation of the ROM BRS scheme [18] to get an efficient and natural standard-model symmetric scheme for encryption of key-dependent messages.	mUCE1
RKA	An efficient standard-model symmetric encryption scheme providing best-possible security against related-key attacks.	mUCE1
CIH	Construction from UCE1 of correlation-intractable hash functions meeting the strongest notion of [54].	UCE1
STORE	Instantiation of a natural ROM proof of storage scheme from [67].	UCE1
OAEP	IND-CPA-KI security of OAEP [15] assuming partial one-wayness (with UCE1) or one-wayness (with UCE2) of the underlying trapdoor function.	UCE1/2
GB	Standard-model adaptively secure garbling with short tokens.	UCE2

Fig. 2. Applications of UCE: We summarize results for different goals, the last column indicating the form of UCE used

of these. We also investigate the relations between the different forms of UCE we have introduced. Our findings are summarized in Fig. 1. As indicated there, UCE2 implies UCE1 but not vice versa, and analogously mUCE2 implies mUCE1 but not vice versa. Of course mUCE1 implies UCE1 and mUCE2 implies UCE2. We do not know whether UCE1 implies mUCE1, and analogously for UCE2 and mUCE2.

APPLICATIONS. Fig. 2 summarizes the applications we now discuss.

1. Deterministic PKE. The EwH deterministic PKE (D-PKE) ROM scheme of BBO07 [6] encrypts message m under public key ek by applying the RO to $ek\|m$ to get coins r and then encrypting m with an IND-CPA PKE scheme under ek and coins r. They showed that this achieved their PRIV notion of security in the ROM. Our instantiation adds $hk \leftarrow_\$ \mathsf{H.Kg}(1^\lambda)$ to

the public key and then replaces the RO with $\mathsf{H.Ev}(hk, \cdot)$. We show that if H is UCE1-secure then this instantiated D-PKE scheme is PRIV-secure in the standard model. This is not only the first standard-model PRIV-secure scheme (previous standard-model D-PKE schemes achieve only restricted notions of blocksource-PRIV-security [9,20,32,47]) but also the most practical. Our proof makes crucial use of the equivalence between PRIV and an indistinguishability-style notion IND of D-PKE security [9].

2. <u>Message-locked encryption</u>. In convergent encryption (CE) [12,43], message m is encrypted using a deterministic symmetric encryption scheme with the key derived, via a RO, from the message itself. CE is the most natural and prominent embodiment of message-locked encryption (MLE) and is in current use by commercial cloud-storage providers to provide secure deduplicated storage. The scheme is shown in [12] to meet, in the ROM, a formal notion of MLE-security called PRV\$-CDA. We instantiate with a UCE1-family, putting the key in public parameters, and show that the resulting MLE scheme is PRV\$-CDA in the standard model.

3. <u>Hardcore functions</u>. A RO is an ideal hardcore function, with $\mathsf{RO}(x)$ returning any number of bits that remain pseudorandom given $f(x)$ where f is one-way. UCE1 families can securely instantiate the RO here, meaning are secure hardcore functions for any one-way function, able to extract as many bits as desired.

4. <u>BR93 PKE</u>. A simple and natural ROM IND-CPA PKE scheme from [14] encrypts m by picking random x and returning $(f(x), \mathsf{RO}(x) \oplus m)$ where f is a trapdoor function in the public key. We show that instantiating the RO with a UCE1-secure family preserves the IND-CPA security.

5. <u>Point-function obfuscation</u>. A *point function* has non-$\perp$ output on just one point. Canetti, Kalai, Varia, and Wichs [38] give a ROM point-function obfuscation scheme. We mUCE1-instantiate their construction to obtain a standard-model point-function obfuscation scheme.

6. <u>KDM-secure SE</u>. Black, Rogaway and Shrimpton (BRS) [18] showed that the following simple and efficient symmetric encryption (SE) scheme is KDM-secure in the ROM: to encrypt message m under key K, pick a random r and return $(r, \mathsf{RO}(r\|K) \oplus m)$. We instantiate by letting the random value r in the BRS scheme take on the role of a fresh hash key, so that, to encrypt m, we pick $hk \leftarrow_\$ \mathsf{H.Kg}(1^\lambda)$ and return $(hk, \mathsf{H.Ev}(hk, K) \oplus m)$. We prove that if H is mUCE1-secure then this instantiated scheme is KDM secure in the standard model. (We achieve non-adaptive KDM security, but this includes popular cases such as key-cycles.) This scheme is more practical than other standard-model KDM-secure encryption schemes such as [1,2,4,31,62].

7. <u>RKA-secure SE</u>. Symmetric encryption schemes secure against related-key attack (RKA) must preserve security even when encryption is performed under keys derived from the original key by application of a key-deriving function. Previous schemes [3,13] provided security for algebraic key-deriving functions such as linear or polynomial functions over a keyspace that is a particular group depending on the scheme. We provide a scheme that has

"best possible" security, in that key-deriving functions are arbitrary subject only to a condition necessary for security, namely to have unpredictable outputs. Furthermore, in our scheme, keys are binary strings rather than group elements, so we cover the most common practical attacks, such as XORing a constant to the key. We assume only a mUCE1-secure family of functions.

8. **Correlation-intractable secure hashing.** Goyal, O'Neill and Rao (GOR) introduced the notion of correlated-input hash (CIH) function families [54] and proposed several notions of security for them. GOR provided constructions achieving limited CIH security from the q-DHI assumption of [25] and from RKA-secure blockciphers, but achieving full CIH security in the standard model has remained open. We solve this problem, showing that UCE1-secure function families are selective (pseudorandomness) CIH secure in the terminology of GOR.

9. **Secure storage.** Ristenpart, Shacham and Shrimpton [67] give a ROM protocol allowing a client to check that a server is storing its file in its entirety, its interest being that constructions indifferentiable from a RO [63] may fail to securely replace the RO. In contrast, we show that UCE1 instantiation succeeds.

10. **OAEP.** OAEP [15] has been a benchmark for RO instantiation [21, 22, 59]. We instantiate OAEP by adding $hk \leftarrow\!\!\text{\$}\ \mathsf{H.Kg}(1^\lambda)$ to the public key and then implementing both the ROs via $\mathsf{H.Ev}(hk, \cdot)$. Under UCE1, we get IND-CPA-KI security under the partial-domain one-wayness, and hence by [46] under standard one-wayness, of RSA; under UCE2 we get it directly under standard one-wayness. IND-CPA-KI is IND-CPA when challenge messages are not allowed to depend on the public key. (This limitation arises because in UCE the strings being hashed by the source cannot depend on the hashing key. We note that this UCE feature does not *always* prevent us from achieving full IND-CPA. Indeed, we do achieve it for the BR93 PKE scheme, because there the inputs to the RO do not depend on the messages.) Kiltz, O'Neill and Smith (KOS) [59] show that RSA-OAEP is IND-CPA-secure if its two ROs are replaced with t-wise independent hash functions and RSA is Φ-hiding [33]. In comparison our results for RSA are under the standard one-wayness assumption.

11. **Adaptively-secure garbling.** Verifiable outsourcing [48], as well as one-time programs [52], call for garbling schemes that are adaptively secure [10]. Standard-model adaptively-secure garbling has however so far been at the cost of large tokens, meaning ones as large as the circuit being garbled [10, 53]. This is not only inefficient but makes the resulting verifiable outsourcing "trivial" in that the client does as much work as the server. We provide a UCE2-based garbling scheme that is adaptively secure and has short tokens. This is the first standard-model garbling scheme with these properties and it results in the first non-trivial instantiation of the outsourcing scheme of [48]. Our garbling scheme is obtained by instantiating a ROM garbled circuit construction of [66].

CONSTRUCTING UCE-SECURE FAMILIES. We provide a ROM construction of a family of functions shown to achieve both mUCE1 and mUCE2. (And thereby UCE1 and UCE2.)

This at first may seem like a step backwards; wasn't the purpose of UCE to avoid the ROM? As explained in more depth in Section 2, it is a step forward because the security we require from families of functions in implementations has moved from something heuristic and vague, namely to "behave like a RO," to something well defined, namely to be UCE-secure.

In practice we would aim to instantiate UCE-secure families via blockciphers or cryptographic hash functions. We explain that direct instantiation with a blockcipher (e.g. AES) is not secure due to the invertibility of the blockcipher. Cryptographic hash functions, being unkeyed, do not directly provide instantiations either. We suggest instead to use HMAC [5,8].

THIS EXTENDED ABSTRACT. Due to space limitations, this extended abstract will provide only the UCE1 definition and detail only one application from Fig. 2. We refer the reader to our full paper [11] for definitions of mUCE1, UCE2 and mUCE2 and for the 10 omitted applications.

2   Perspective and Discussion

We explain why UCE is step forward even if we can (currently) only achieve it in the ROM, and how UCE relates to other assumptions.

LAYERED CRYPTOGRAPHY. Currently, RO-based design *directly* proves schemes (for end goals) secure in the ROM. We are instead advocating and using what we call a *layered* approach. In this approach, *base primitives* with standard-model security definitions are validated in the ROM. End goals are then reached from the base primitives purely in the standard model, the ROM being entirely dispensed with in the second step. This is illustrated in Fig. 3. We are showing that UCE can function as such a base primitive, and a powerful one at that, since many goals may be reached from it.

In implementations, we would continue to instantiate families assumed UCE-secure via appropriately-keyed cryptographic hash functions, but we claim this layered approach is still an important advance on direct ROM-based design. This is because the property we desire from the object (family of functions) actually being used in the implementation has moved from something heuristic and vague ("behave like a random oracle") to something precise and meaningful (be UCE-secure). Cryptanalytic validation of UCE security, even if difficult, is at least meaningful, while cryptanalytic evaluation of "behaving like a RO" is not even meaningful because the phrase in quotes is not well defined.

We make an analogy with pairing-based cryptography. Here we have seen the proposal of a large number of standard-model assumptions, including BDH [28], DLIN [27], SDH [27], BDHE [26] and SD (Subgroup Decision) [30] to name just a small fraction. These assumptions are (ubiquitously) validated in the generic-group model, end goals then reached from the assumptions in the standard

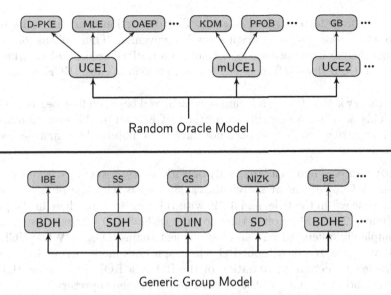

Fig. 3. The layered-cryptography paradigm for the ROM (left) and for pairing-based cryptography (right). Assumptions are validated in the idealized model and then used to attain end goals entirely in the standard model. SS refers to the short signatures of [24]; BE refers to the broadcast encryption scheme of [29]; NIZK refers to the NIZK arguments of [55]. See text for other abbreviations.

model. But the generic-group model is subject to issues, critiques and counter-examples analogous to those for the ROM, if not worse [41, 44]. We believe that the (deserved) success and acceptance of pairing-based cryptography, and that it has not come under as much fire as ROM-based cryptography, are due in part to what, in our terminology, is its layered approach (again illustrated in Fig. 3). Namely, schemes for end goals, rather than being directly validated in the generic model (the un-layered or direct approach), are based on standard-model assumptions that are themselves validated in the generic-group model and amenable to cryptanalysis.

It is perhaps curious that the layered approach has not been explicitly articulated and widely used for ROM-based cryptography, while it has been widely used (even if not explicitly articulated) in pairing-based cryptography. The benefits are identical in the two cases. We view our work as making layered cryptography an explicit approach for ROM-based design.

ASSUMPTION DEGREE AND ACHIEVING UCE. In the UCE definition, the adversary consists of stages (source and distinguisher) that (due to the unpredictability condition) cannot completely share state. We refer to this as a second-degree assumption, as opposed to a first-degree assumption, where the adversary is a single algorithm. Put another way, a first-degree assumption can be specified via an interaction (game) between an adversary and a challenger. (In some places [57, 65] this is called a "standard" assumption, but we think this is less clear than "first degree.") UCE cannot. This distinction is crucial to its power and to why various negative results are circumvented. Thus, Wichs [68] shows that first-degree assumptions do not suffice for PRIV-secure D-PKE, but our proof that UCE does suffice is not a contradiction because UCE is not first-degree.

A corollary is that UCE itself cannot be achieved based on first-degree assumptions. This does not necessarily mean that UCE is an implausible assumption. (A second-degree assumption does not have to be implied by a first-degree one to be true.)

WITHOUT ROS. There is a large body of work on cryptography without random oracles. (A Google Scholar search shows 286 papers with the phrase "without random oracles" in the title, and 3,640 with this phrase somewhere in the paper, as of June 6, 2013.) More often than not, the without-RO schemes of such works are completely different from, and less efficient than, RO ones. While UCE also serves, of course, to get without-RO schemes, it does more, permitting these to be obtained by actual instantiation of the RO in a ROM scheme, so that the efficiency and practicality of the starting ROM scheme is preserved.

DIRECTIONS. We believe that achieving UCE under other assumptions is an interesting and important direction for future work. We suggest to begin by targeting restricted versions of UCE, for example UCE1 for block sources. This we may hope to achieve under first-degree assumptions. Hope is lent to the enterprise by the fact that D-PKE that is PRIV-secure for block sources has been achieved under standard assumptions [20,32,47]. Full UCE security would, of course, require second-degree assumptions.

UCE is a framework permitting definitional variants beyond the four we have formalized. One could define variants with extractability, which may be useful for further applications. A tempting variant is to allow some communication back from the distinguisher to the source. This opens the door to many interesting applications, but is a dangerous path to tread, for any version we, at least, have formalized, we have also broken, even for forms of communication that seemed highly restricted.

DISCUSSION, LIMITATIONS AND RELATED WORK. That the source adversary in UCE does not get the key is important in avoiding impossibility results like those in [36,63]. (For example, UCE does not imply correlation intractability as defined, and shown to be unachievable in the standard model, by [36].)

UCE is not a panacea in the sense that it can replace ROs everywhere. UCE helps in cases where the RO is applied to inputs hidden (at least in part) from the adversary. As far as we know, UCE will not help for tasks like instantiating the RO in FDH signatures [16]. This is consistent with impossibility results [42].

Curiously, UCE-based proofs for instantiated schemes are sometimes simpler than the proofs for the starting ROM schemes. This is the case for D-PKE. The intuition for the ROM security of the EwH scheme of [6] is simple enough, but a rigorous ROM proof is in our view less straightforward than our proofs for the UCE1-based instantiation of EwH.

The term "computational extractor" has been used for primitives that extract pseudorandomness from distributions that have computational min-entropy [40, 45, 61]. A UCE-secure family instead extracts pseudorandomness from unpredictable distributions. These may or may not have computational min-entropy in the formal sense the latter is defined [56] but we view unpredictability as we defined it as another computational relaxation of min-entropy so preserved the "extractor" name. "Universal" refers to the ability to do this from *any* starting (unpredictable) distribution.

Programmable hash functions [58] are an information-theoretic tool that in some way mimic the "programmability" of ROs and were used by [58] to build signature schemes with short signatures in the standard model. They do not serve to instantiate ROs in the kinds of applications we consider. Several works [23, 49] define new security properties of hash functions tailored for their own particular applications.

3   Preliminaries

By $\lambda \in \mathbb{N}$ we denote the security parameter and by 1^λ its unary representation. We denote the number of coordinates of a vector $\mathbf{x}$ by $|\mathbf{x}|$, and the length of a string $x \in \{0,1\}^*$ by $|x|$. Algorithms are randomized unless otherwise indicated. Running time is worst case. "PT" stands for "polynomial-time," whether for randomized algorithms or deterministic ones. If A is an algorithm, we let $y \leftarrow A(x_1, \ldots; r)$ denote running A with random coins r on inputs $x_1, \ldots$ and assigning the output to y. We let $y \leftarrow_{\$} A(x_1, \ldots)$ be the resulting of picking r at random and letting $y \leftarrow A(x_1, \ldots; r)$. We let $[A(x_1, \ldots,)]$ denote the set of all possible outputs of A when invoked with inputs $x_1, \ldots$.

We use the code based game playing framework of [17] augmented with explicit MAIN procedures as in [67]. (See Fig. 4 for an example.) By $G^A(\lambda)$ we denote the event that the execution of game G with adversary A and security parameter λ results in output true, the game output being what is returned by MAIN.

4   UCE1

We define UCE1 security of a family of functions and provide a simplified but equivalent form of unpredictability. In [11] we provide further basic results and also define mUCE1.

SYNTAX. A family of functions H specifies the following. On input the unary representation 1^λ of the security parameter $\lambda \in \mathbb{N}$, key generation algorithm H.Kg returns a key $hk \in \{0,1\}^{\text{H.KI}(\lambda)}$, where H.KI: $\mathbb{N} \to \mathbb{N}$ is the keylength function associated to H. The deterministic, PT evaluation algorithm H.Ev takes 1^λ, a key $hk \in [\text{H.Kg}(1^\lambda)]$, an input $x \in \{0,1\}^*$ with $|x| \in \text{H.IL}(\lambda)$, and a unary encoding 1^ℓ of an output length $\ell \in \text{H.OL}(\lambda)$ to return an output $\text{H.Ev}(1^\lambda, hk, x, 1^\ell) \in \{0,1\}^\ell$. (The syntax in the Introduction had simplified by dropping the first and last inputs.) Here H.IL is the input-length function associated to H, so that $\text{H.IL}(\lambda) \subseteq \mathbb{N}$

MAIN $\mathrm{UCE}_H^{S,D}(\lambda)$	MAIN $\mathrm{Pred}_S^P(\lambda)$	MAIN $\mathrm{SPred}_S^{P'}(\lambda)$
$b \leftarrow\!\!\$ \{0,1\}$; $hk \leftarrow\!\!\$ \text{H.Kg}(1^\lambda)$	$done \leftarrow false$; $Q \leftarrow \emptyset$	$Q \leftarrow \emptyset$
$L \leftarrow\!\!\$ S^{\mathrm{HASH}}(1^\lambda)$	$L \leftarrow\!\!\$ S^{\mathrm{HASH}}(1^\lambda)$	$L \leftarrow\!\!\$ S^{\mathrm{HASH}}(1^\lambda)$
$b' \leftarrow\!\!\$ D(1^\lambda, hk, L)$	$done \leftarrow true$	$x \leftarrow\!\!\$ P'(1^\lambda, L)$
Return $(b' = b)$	$Q' \leftarrow\!\!\$ P^{\mathrm{HASH}}(1^\lambda, L)$	Return $(x \in Q)$
	Return $(Q \cap Q' \neq \emptyset)$	
$\underline{\mathrm{HASH}(x, 1^\ell)}$		$\underline{\mathrm{HASH}(x, 1^\ell)}$
If $T[x, \ell] = \perp$ then	$\underline{\mathrm{HASH}(x, 1^\ell)}$	$Q \leftarrow Q \cup \{x\}$
If $b = 1$ then	If $done = false$ then	If $T[x, \ell] = \perp$ then
$T[x, \ell] \leftarrow \text{H.Ev}(1^\lambda, hk, x, 1^\ell)$	$Q \leftarrow Q \cup \{x\}$	$T[x, \ell] \leftarrow\!\!\$ \{0,1\}^\ell$
Else $T[x, \ell] \leftarrow\!\!\$ \{0,1\}^\ell$	If $T[x, \ell] = \perp$ then	Return $T[x, \ell]$
Return $T[x, \ell]$	$T[x, \ell] \leftarrow\!\!\$ \{0,1\}^\ell$	
	Return $T[x, \ell]$	

Fig. 4. Games UCE, Pred used to define UCE1 security of family of functions H, and game SPred defining the simplified but equivalent form of unpredictability. Here S is the source, D is the distinguisher, P is the predictor and P' is the simple predictor.

is the (non-empty) set of allowed input lengths, and similarly H.OL is the output-length function associated to H, so that $\text{H.OL}(\lambda) \subseteq \mathbb{N}$ is the (non-empty) set of allowed output lengths. The latter allows us to cover fixed output length (FOL) functions, captured by $\text{H.OL}(\lambda)$ being a set of size one, or variable output length (VOL) functions, where $\text{H.OL}(\lambda)$ could be larger and even be $\mathbb{N}$. We say that H has input-length $\ell\colon \mathbb{N} \to \mathbb{N}$ if $\text{H.IL}(\lambda) = \{\ell(\lambda)\}$ for all $\lambda \in \mathbb{N}$, and if such an ℓ exists we denote it by H.il. We say H has output-length $\ell\colon \mathbb{N} \to \mathbb{N}$ if $\text{H.OL}(\lambda) = \{\ell(\lambda)\}$ for all $\lambda \in \mathbb{N}$, and if such an ℓ exists we denote it by H.ol.

UCE1 SECURITY. We define what it means for a family of functions H to be UCE1-secure. Let S be an adversary called the *source* and D an adversary called the *distinguisher*. We associate to them and H the game $\mathrm{UCE}_H^{S,D}(\lambda)$ of Fig. 4. The source has access to an oracle HASH and we require that any query $x, 1^\ell$ made to this oracle satisfy $|x| \in \text{H.IL}(\lambda)$ and $\ell \in \text{H.OL}(\lambda)$. When the challenge bit b is 1 (the "real" case) the oracle responds via H.Ev under a key hk that is chosen by the game and *not* given to the source. When $b = 0$ (the "random" case) it responds as a RO. The source communicates to its accomplice distinguisher a string $L \in \{0,1\}^*$ we call the *leakage*. The distinguisher *does* get the key hk as input and must now return its guess $b' \in \{0,1\}$ for b. The game returns true iff $b' = b$, and the UCE1 advantage of (S, D) is defined for $\lambda \in \mathbb{N}$ via $\text{Adv}_{H,S,D}^{uce}(\lambda) = 2\Pr[\mathrm{UCE}_H^{S,D}(\lambda)] - 1$. One's first thought may now be to say that H is UCE1-secure if $\text{Adv}_{H,S,D}^{uce}(\cdot)$ is negligible for all PT S and all PT D. But an obvious attack shows that no H can meet this definition. Indeed, S can pick some x and ℓ, let $h \leftarrow \mathrm{HASH}(x, 1^\ell)$ and return leakage $L = (x, h, 1^\ell)$ to D. The latter, knowing hk, can return 1 if $h = \text{H.Ev}(1^\lambda, hk, x, 1^\ell)$ and 0 otherwise. We obtain a meaningful and useful definition of UCE1-security for H by restricting attention to sources that are what we call "unpredictable." The formalization considers game $\mathrm{Pred}_S^P(\lambda)$ of Fig. 4 associated to source S and an adversary P called a *predictor*. Given the leakage, the latter outputs a set Q'. It wins if this set contains any HASH-query of the source. For $\lambda \in \mathbb{N}$ we let $\mathsf{Adv}_{P,S}^{\mathrm{pred}}(\lambda) = \Pr[\mathrm{Pred}_S^P(\lambda)]$. We say that source S is *unpredictable* if $\mathsf{Adv}_{P,S}^{\mathrm{pred}}(\cdot)$ is negligible for all PT predictors P. We stress that in the prediction game, the HASH oracle of the source is a RO like in the random game, and the predictor gets the same oracle. The family H is not involved in this definition; unpredictability is a property of the source. Finally, we say that H is UCE1-secure if $\mathsf{Adv}_{\mathsf{H},S,D}^{\mathrm{uce}}(\cdot)$ is negligible for all unpredictable, PT sources S and all PT distinguishers D. It is convenient to let UCE1 denote the set of all function families H that are UCE1-secure.

SIMPLE UNPREDICTABILITY. Applications of UCE1 will involve proving the unpredictability of sources we construct. This task is simplified by using a simpler formulation of unpredictability, called simple unpredictability, that is equivalent to the original. The formalization considers game $\mathrm{SPred}_S^{P'}(\lambda)$ of Fig. 4 associated to source S and an adversary P' called a *simple predictor*. There are two simplifications: the simple predictor does not have access to the RO HASH, and its output is a single string x rather than a set of strings. It wins if x is a HASH-query of the source. For $\lambda \in \mathbb{N}$ we let $\mathsf{Adv}_{P',S}^{\mathrm{spred}}(\lambda) = \Pr[\mathrm{SPred}_S^{P'}(\lambda)]$. We say that source S is *simple unpredictable* if $\mathsf{Adv}_{P',S}^{\mathrm{spred}}(\cdot)$ is negligible for all PT simple predictors P'. The following, whose proof is in [11], says that simple unpredictability is equivalent to unpredictability.

Lemma 1. Let S be a source. Then S is unpredictable if and only if it is simple unpredictable.

FROM FOL TO VOL. In [11] we show how to build a UCE1-secure family with variable output length (VOL) from a UCE1-secure family with fixed output length (FOL) in a simple way using a PRF.

5   Applications of UCE1

We detail one of the 11 applications of Fig. 2. For the rest, see [11].

DETERMINISTIC ENCRYPTION. EwH is a simple and natural D-PKE scheme from [6] that deterministically encrypts m by encrypting m with a randomized IND-CPA scheme with the coins derived by applying a RO to m. In the ROM the scheme is PRIV-secure [6] and equivalently IND-secure [9]. We show that instantiating the RO with a UCE1 hash family results in a scheme meeting the same notion of security in the standard model. Previous standard model schemes have met notions providing security only when one assumes messages are drawn from a blocksource, meaning each message has high min-entropy even given previous ones [20,32]. Instantiated EwH however meets the original and full notions of [6,9] which only make the necessary assumption that each individual

MAIN $\mathrm{IND}_{\mathsf{PKE}}^A(\lambda)$	$\mathsf{DE.Kg}(1^\lambda)$
$b \leftarrow_\$ \{0,1\}$	$(ek, dk) \leftarrow_\$ \mathsf{RE.Kg}(1^\lambda)$; $hk \leftarrow_\$ \mathsf{H.Kg}(1^\lambda)$
$(ek, dk) \leftarrow_\$ \mathsf{DE.Kg}(1^\lambda)$	Return $((ek, hk), dk)$
$(\mathbf{m}_0, \mathbf{m}_1) \leftarrow_\$ A_1(1^\lambda)$	
For $i = 1$ to $|\mathbf{m}_b|$ do	$\mathsf{DE.Enc}(1^\lambda, (ek, hk), m)$
$\quad \mathbf{c}[i] \leftarrow_\$ \mathsf{DE.Enc}(1^\lambda, ek, \mathbf{m}_b[i])$	$r \leftarrow \mathsf{H.Ev}(1^\lambda, hk, ek \,\|\, m, 1^{\mathsf{RE.rl}(\lambda)})$
$b' \leftarrow_\$ A_2(1^\lambda, ek, \mathbf{c})$	$c \leftarrow \mathsf{RE.Enc}(1^\lambda, ek, m; r)$; Return c
Return $(b = b')$	$\mathsf{DE.Dec}(1^\lambda, dk, c)$
	$m \leftarrow \mathsf{RE.Dec}(1^\lambda, dk, c)$; Return m

Fig. 5. Left: The IND game. **Right:** D-PKE scheme $\mathsf{DE} = \mathsf{EwH}[\mathsf{H}, \mathsf{RE}]$.

message has high min-entropy, but allow messages to be arbitrarily correlated. This is the first standard-model scheme meeting the PRIV and IND notions.

A PKE scheme PKE specifies a triple of PT algorithms. Via $(ek, dk) \leftarrow_\$ \mathsf{PKE.Kg}(1^\lambda)$ we generate keys. Via $c \leftarrow_\$ \mathsf{PKE.Enc}(1^\lambda, ek, m)$ we can encrypt a message $m \in \{0,1\}^{\mathsf{PKE.il}(\lambda)}$ where $\mathsf{PKE.il} \colon \mathbb{N} \to \mathbb{N}$ is the message-length function of the scheme. Via $m \leftarrow \mathsf{PKE.Dec}(1^\lambda, dk, c)$ we deterministically decrypt. We say PKE is a D-PKE scheme if the encryption algorithm $\mathsf{PKE.Enc}$ is deterministic. The game defining the IND notion of security for D-PKE scheme DE, following [9], is in Fig. 5. An IND adversary $A = (A_1, A_2)$ is a pair of PT algorithms, where A_1 on input 1^λ returns a pair $(\mathbf{m}_0, \mathbf{m}_1)$ of vectors of messages. It is required that there are functions v, ℓ, depending on the adversary, such that $|\mathbf{m}_0| = |\mathbf{m}_1| = v(\lambda)$ and $|\mathbf{m}_b[i]| = \ell(\lambda)$ for all $b \in \{0,1\}$ and $i \in [v(\lambda)]$. It is also required that the strings (messages) $\mathbf{m}_0[1], \ldots, \mathbf{m}_0[|\mathbf{m}_0|]$ are distinct and the strings (messages) $\mathbf{m}_1[1], \ldots, \mathbf{m}_1[|\mathbf{m}_1|]$ are distinct. The guessing probability $\mathrm{Guess}_A(\cdot)$ of A is the function that on input $\lambda \in \mathbb{N}$ returns the maximum, over all b, i, m, of $\Pr[\mathbf{m}_b[i] = m]$, the probability over $(\mathbf{m}_0, \mathbf{m}_1) \leftarrow_\$ A_1(1^\lambda)$. We say that A has *high min-entropy* if $\mathrm{Guess}_A(\cdot)$ is negligible. We let $\mathsf{Adv}_{\mathsf{DE}, A}^{\mathsf{ind}}(\lambda) = 2\Pr[\mathrm{IND}_{\mathsf{DE}}^A(\lambda)] - 1$ and say that DE is IND-secure if $\mathsf{Adv}_{\mathsf{DE}, A}^{\mathsf{ind}}(\cdot)$ is negligible for all PT A of high min-entropy. Let IND be the set of all IND-secure D-PKE schemes.

Let RE be a PKE scheme. Let $\mathsf{RE.rl} \colon \mathbb{N} \to \mathbb{N}$ denote its randomness-length function, meaning $\mathsf{RE.Enc}(1^\lambda, \cdot, \cdot)$ draws its coins at random from $\{0,1\}^{\mathsf{RE.rl}(\lambda)}$. Let H be a family of functions with $\mathsf{H.IL} = \mathbb{N}$ and $\mathsf{RE.rl}(\lambda) \in \mathsf{H.OL}(\lambda)$ for all $\lambda \in \mathbb{N}$. Our standard-model instantiation of the ROM encrypt-with-hash transform of BBO07 [6] associates to RE and H the (standard-model) D-PKE scheme $\mathsf{DE} = \mathsf{EwH}[\mathsf{H}, \mathsf{RE}]$ described in Fig. 5. The message length of DE is that of RE. The following theorem says that the transform yields an IND-secure D-PKE scheme if H is UCE1-secure and RE is IND-CPA-secure. Here IND-CPA denotes the set of all IND-CPA-secure PKE schemes.

Theorem 2. *If* $\mathsf{H} \in \mathsf{UCE1}$ *and* $\mathsf{RE} \in \mathsf{IND\text{-}CPA}$ *then* $\mathsf{EwH}[\mathsf{H}, \mathsf{RE}] \in \mathsf{IND}$.

The proof of Theorem 2 is in [11]. Here we give a sketch. Let $\mathsf{PKE} = \mathsf{EwH}[\mathsf{H}, \mathsf{RE}]$. Given a high min-entropy adversary $A = (A_1, A_2)$ for game $\mathrm{IND}^A_{\mathsf{PKE}}(\lambda)$, we build a source S and distinguisher D as follows. The source $S^{\mathrm{HASH}}(1^\lambda)$ picks $(ek, dk) \leftarrow_\$ \mathsf{RE.Kg}(1^\lambda)$ and $d \leftarrow_\$ \{0, 1\}$. It runs $A_1(1^\lambda)$ to get $(\mathbf{m}_0, \mathbf{m}_1)$ and lets $n \leftarrow |\mathbf{m}_d|$. For $i = 1, \ldots, n$ it obtains coins $\mathbf{r}[i]$ by calling its HASH oracle with $ek \| \mathbf{m}_d[i], 1^{\mathsf{RE.rl}(\lambda)}$. It then creates ciphertexts $\mathbf{c}[i] \leftarrow \mathcal{E}(1^\lambda, ek, \mathbf{m}_d[i]; \mathbf{r}[i])$ for $i = 1, \ldots, n$. It would like now to run A_2 on $\mathbf{c}$ but cannot since A_2 needs the public key, which includes hk. Accordingly, S returns as leakage $L \leftarrow (ek, d, \mathbf{c})$. Distinguisher $D(1^\lambda, hk, L)$ can create public key (ek, hk). It now lets $d' \leftarrow_\$ A_2(1^\lambda, (ek, hk), \mathbf{c})$. If $d = d'$ it sets $b' \leftarrow 1$, else $b' \leftarrow 0$. It returns b'. When the challenge bit in game $\mathrm{UCE}^{S,D}_{\mathsf{H}}(\lambda)$ is $b = 1$, adversaries S, D are simulating game $\mathrm{IND}^A_{\mathsf{PKE}}(\lambda)$, so that $2\Pr[d' = d \mid b = 1] - 1 = \mathsf{Adv}^{\mathrm{ind}}_{\mathsf{DE}, A}(\lambda)$. If $b = 0$ then A_2 is seeing ciphertexts under the randomized RE scheme, and the assumed IND-CPA security of RE can be used to show that $2\Pr[d' = d \mid b = 0] - 1$ is negligible. This will allow us to upper bound $\mathsf{Adv}^{\mathrm{ind}}_{\mathsf{DE}, A}(\cdot)$ by $2\mathsf{Adv}^{\mathrm{uce}}_{\mathsf{H}, S, D}(\cdot)$ plus a negligible amount. To conclude it suffices to show that $\mathsf{Adv}^{\mathrm{uce}}_{\mathsf{H}, S, D}(\cdot)$ is negligible. This follows if we show that S is unpredictable. By Lemma 1 it suffices to show that S is simple-unpredictable. Since oracle queries of S include messages created by A_1, (simple) unpredictability may seem at first to follow from the high min-entropy assumption on A. However we will additionally exploit (once again) the assumed IND-CPA security of the randomized RE scheme. This is because the leakage contains the ciphertexts. Overall, we exploit the IND-CPA security of RE in two places, building two corresponding adversaries.

Acknowledgments. We thank the Crypto 2013 PC for their many valuable comments and suggestions. We thank Dan Boneh and Adam O'Neill for their comments.

References

1. Applebaum, B.: Key-dependent message security: Generic amplification and completeness. In: Paterson, K.G. (ed.) EUROCRYPT 2011. LNCS, vol. 6632, pp. 527–546. Springer, Heidelberg (2011)
2. Applebaum, B., Cash, D., Peikert, C., Sahai, A.: Fast cryptographic primitives and circular-secure encryption based on hard learning problems. In: Halevi, S. (ed.) CRYPTO 2009. LNCS, vol. 5677, pp. 595–618. Springer, Heidelberg (2009)
3. Applebaum, B., Harnik, D., Ishai, Y.: Semantic security under related-key attacks and applications. In: Yao, A.C.-C., Yao, A.C.-C. (eds.) ICS 2011. Tsinghua University Press (2011)
4. Barak, B., Haitner, I., Hofheinz, D., Ishai, Y.: Bounded key-dependent message security. In: Gilbert, H. (ed.) EUROCRYPT 2010. LNCS, vol. 6110, pp. 423–444. Springer, Heidelberg (2010)
5. Bellare, M.: New proofs for NMAC and HMAC: Security without collision-resistance. In: Dwork, C. (ed.) CRYPTO 2006. LNCS, vol. 4117, pp. 602–619. Springer, Heidelberg (2006)

6. Bellare, M., Boldyreva, A., O'Neill, A.: Deterministic and efficiently searchable encryption. In: Menezes, A. (ed.) CRYPTO 2007. LNCS, vol. 4622, pp. 535–552. Springer, Heidelberg (2007)

7. Bellare, M., Boldyreva, A., Palacio, A.: An uninstantiable random-oracle-model scheme for a hybrid-encryption problem. In: Cachin, C., Camenisch, J.L. (eds.) EUROCRYPT 2004. LNCS, vol. 3027, pp. 171–188. Springer, Heidelberg (2004)

8. Bellare, M., Canetti, R., Krawczyk, H.: Keying hash functions for message authentication. In: Koblitz, N. (ed.) CRYPTO 1996. LNCS, vol. 1109, pp. 1–15. Springer, Heidelberg (1996)

9. Bellare, M., Fischlin, M., O'Neill, A., Ristenpart, T.: Deterministic encryption: Definitional equivalences and constructions without random oracles. In: Wagner, D. (ed.) CRYPTO 2008. LNCS, vol. 5157, pp. 360–378. Springer, Heidelberg (2008)

10. Bellare, M., Hoang, V.T., Rogaway, P.: Adaptively secure garbling with applications to one-time programs and secure outsourcing. In: Wang, X., Sako, K. (eds.) ASIACRYPT 2012. LNCS, vol. 7658, pp. 134–153. Springer, Heidelberg (2012)

11. Bellare, M., Hoang, V.T., Keelveedhi, S.: Instantiating random oracles via UCEs. Cryptology ePrint Archive (2013)

12. Bellare, M., Keelveedhi, S., Ristenpart, T.: Message-locked encryption and secure deduplication. In: Johansson, T., Nguyen, P.Q. (eds.) EUROCRYPT 2013. LNCS, vol. 7881, pp. 296–312. Springer, Heidelberg (2013)

13. Bellare, M., Paterson, K., Thomson, S.: RKA Security beyond the Linear Barrier: IBE, Encryption and Signatures. In: Wang, X., Sako, K. (eds.) ASIACRYPT 2012. LNCS, vol. 7658, pp. 331–348. Springer, Heidelberg (2012)

14. Bellare, M., Rogaway, P.: Random oracles are practical: A paradigm for designing efficient protocols. In: Ashby, V. (ed.) ACM CCS 1993, pp. 62–73. ACM Press (November 1993)

15. Bellare, M., Rogaway, P.: Optimal asymmetric encryption. In: De Santis, A. (ed.) EUROCRYPT 1994. LNCS, vol. 950, pp. 92–111. Springer, Heidelberg (1995)

16. Bellare, M., Rogaway, P.: The exact security of digital signatures: How to sign with RSA and Rabin. In: Maurer, U.M. (ed.) EUROCRYPT 1996. LNCS, vol. 1070, pp. 399–416. Springer, Heidelberg (1996)

17. Bellare, M., Rogaway, P.: The security of triple encryption and a framework for code-based game-playing proofs. In: Vaudenay, S. (ed.) EUROCRYPT 2006. LNCS, vol. 4004, pp. 409–426. Springer, Heidelberg (2006)

18. Black, J., Rogaway, P., Shrimpton, T.: Encryption-scheme security in the presence of key-dependent messages. In: Nyberg, K., Heys, H.M. (eds.) SAC 2002. LNCS, vol. 2595, pp. 62–75. Springer, Heidelberg (2003)

19. Boldyreva, A., Cash, D., Fischlin, M., Warinschi, B.: Foundations of non-malleable hash and one-way functions. In: Matsui, M. (ed.) ASIACRYPT 2009. LNCS, vol. 5912, pp. 524–541. Springer, Heidelberg (2009)

20. Boldyreva, A., Fehr, S., O'Neill, A.: On notions of security for deterministic encryption, and efficient constructions without random oracles. In: Wagner, D. (ed.) CRYPTO 2008. LNCS, vol. 5157, pp. 335–359. Springer, Heidelberg (2008)

21. Boldyreva, A., Fischlin, M.: Analysis of random oracle instantiation scenarios for OAEP and other practical schemes. In: Shoup, V. (ed.) CRYPTO 2005. LNCS, vol. 3621, pp. 412–429. Springer, Heidelberg (2005)

22. Boldyreva, A., Fischlin, M.: On the security of OAEP. In: Lai, X., Chen, K. (eds.) ASIACRYPT 2006. LNCS, vol. 4284, pp. 210–225. Springer, Heidelberg (2006)

23. Boneh, D., Boyen, X.: Secure identity based encryption without random oracles. In: Franklin, M. (ed.) CRYPTO 2004. LNCS, vol. 3152, pp. 443–459. Springer, Heidelberg (2004)

24. Boneh, D., Boyen, X.: Short signatures without random oracles and the SDH assumption in bilinear groups. Journal of Cryptology 21(2), 149–177 (2008)
25. Boneh, D., Boyen, X.: Efficient selective identity-based encryption without random oracles. Journal of Cryptology 24(4), 659–693 (2011)
26. Boneh, D., Boyen, X., Goh, E.-J.: Hierarchical identity based encryption with constant size ciphertext. In: Cramer, R. (ed.) EUROCRYPT 2005. LNCS, vol. 3494, pp. 440–456. Springer, Heidelberg (2005)
27. Boneh, D., Boyen, X., Shacham, H.: Short group signatures. In: Franklin, M. (ed.) CRYPTO 2004. LNCS, vol. 3152, pp. 41–55. Springer, Heidelberg (2004)
28. Boneh, D., Franklin, M.K.: Identity based encryption from the Weil pairing. SIAM Journal on Computing 32(3), 586–615 (2003)
29. Boneh, D., Gentry, C., Waters, B.: Collusion resistant broadcast encryption with short ciphertexts and private keys. In: Shoup, V. (ed.) CRYPTO 2005. LNCS, vol. 3621, pp. 258–275. Springer, Heidelberg (2005)
30. Boneh, D., Goh, E.-J., Nissim, K.: Evaluating 2-DNF formulas on ciphertexts. In: Kilian, J. (ed.) TCC 2005. LNCS, vol. 3378, pp. 325–341. Springer, Heidelberg (2005)
31. Boneh, D., Halevi, S., Hamburg, M., Ostrovsky, R.: Circular-secure encryption from decision diffie-hellman. In: Wagner, D. (ed.) CRYPTO 2008. LNCS, vol. 5157, pp. 108–125. Springer, Heidelberg (2008)
32. Brakerski, Z., Segev, G.: Better security for deterministic public-key encryption: The auxiliary-input setting. In: Rogaway, P. (ed.) CRYPTO 2011. LNCS, vol. 6841, pp. 543–560. Springer, Heidelberg (2011)
33. Cachin, C., Micali, S., Stadler, M.A.: Computationally private information retrieval with polylogarithmic communication. In: Stern, J. (ed.) EUROCRYPT 1999. LNCS, vol. 1592, pp. 402–414. Springer, Heidelberg (1999)
34. Canetti, R.: Towards realizing random oracles: Hash functions that hide all partial information. In: Kaliski Jr., B.S. (ed.) CRYPTO 1997. LNCS, vol. 1294, pp. 455–469. Springer, Heidelberg (1997)
35. Canetti, R., Dakdouk, R.R.: Extractable perfectly one-way functions. In: Aceto, L., Damgård, I., Goldberg, L.A., Halldórsson, M.M., Ingólfsdóttir, A., Walukiewicz, I. (eds.) ICALP 2008, Part II. LNCS, vol. 5126, pp. 449–460. Springer, Heidelberg (2008)
36. Canetti, R., Goldreich, O., Halevi, S.: The random oracle methodology, revisited (preliminary version). In: 30th ACM STOC, pp. 209–218. ACM Press (May 1998)
37. Canetti, R., Goldreich, O., Halevi, S.: On the random-oracle methodology as applied to length-restricted signature schemes. In: Naor, M. (ed.) TCC 2004. LNCS, vol. 2951, pp. 40–57. Springer, Heidelberg (2004)
38. Canetti, R., Tauman Kalai, Y., Varia, M., Wichs, D.: On symmetric encryption and point obfuscation. In: Micciancio, D. (ed.) TCC 2010. LNCS, vol. 5978, pp. 52–71. Springer, Heidelberg (2010)
39. Canetti, R., Micciancio, D., Reingold, O.: Perfectly one-way probabilistic hash functions (preliminary version). In: 30th ACM STOC, pp. 131–140. ACM Press (May 1998)
40. Dachman-Soled, D., Gennaro, R., Krawczyk, H., Malkin, T.: Computational extractors and pseudorandomness. In: Cramer, R. (ed.) TCC 2012. LNCS, vol. 7194, pp. 383–403. Springer, Heidelberg (2012)
41. Dent, A.W.: Adapting the weaknesses of the random oracle model to the generic group model. In: Zheng, Y. (ed.) ASIACRYPT 2002. LNCS, vol. 2501, pp. 100–109. Springer, Heidelberg (2002)

42. Dodis, Y., Oliveira, R., Pietrzak, K.: On the generic insecurity of the full domain hash. In: Shoup, V. (ed.) CRYPTO 2005. LNCS, vol. 3621, pp. 449–466. Springer, Heidelberg (2005)

43. Douceur, J., Adya, A., Bolosky, W., Simon, P., Theimer, M.: Reclaiming space from duplicate files in a serverless distributed file system. In: Proceedings of the 22nd International Conference on Distributed Computing Systems, 2002, pp. 617–624. IEEE (2002)

44. Fischlin, M.: A note on security proofs in the generic model. In: Okamoto, T. (ed.) ASIACRYPT 2000. LNCS, vol. 1976, pp. 458–469. Springer, Heidelberg (2000)

45. Fouque, P.-A., Pointcheval, D., Zimmer, S.: HMAC is a randomness extractor and applications to TLS. In: Abe, M., Gligor, V. (eds.) ASIACCS 2008, pp. 21–32. ACM Press (March 2008)

46. Fujisaki, E., Okamoto, T., Pointcheval, D., Stern, J.: RSA-OAEP is secure under the RSA assumption. Journal of Cryptology 17(2), 81–104 (2004)

47. Fuller, B., O'Neill, A., Reyzin, L.: A unified approach to deterministic encryption: New constructions and a connection to computational entropy. In: Cramer, R. (ed.) TCC 2012. LNCS, vol. 7194, pp. 582–599. Springer, Heidelberg (2012)

48. Gennaro, R., Gentry, C., Parno, B.: Non-interactive verifiable computing: Outsourcing computation to untrusted workers. In: Rabin, T. (ed.) CRYPTO 2010. LNCS, vol. 6223, pp. 465–482. Springer, Heidelberg (2010)

49. Gennaro, R., Halevi, S., Rabin, T.: Secure hash-and-sign signatures without the random oracle. In: Stern, J. (ed.) EUROCRYPT 1999. LNCS, vol. 1592, pp. 123–139. Springer, Heidelberg (1999)

50. Goldreich, O., Goldwasser, S., Micali, S.: How to construct random functions. Journal of the ACM 33, 792–807 (1986)

51. Goldwasser, S., Kalai, Y.T.: On the (in)security of the Fiat-Shamir paradigm. In: 44th FOCS, pp. 102–115. IEEE Computer Society Press (October 2003)

52. Goldwasser, S., Kalai, Y.T., Rothblum, G.N.: One-time programs. In: Wagner, D. (ed.) CRYPTO 2008. LNCS, vol. 5157, pp. 39–56. Springer, Heidelberg (2008)

53. Goyal, V., Ishai, Y., Sahai, A., Venkatesan, R., Wadia, A.: Founding cryptography on tamper-proof hardware tokens. In: Micciancio, D. (ed.) TCC 2010. LNCS, vol. 5978, pp. 308–326. Springer, Heidelberg (2010)

54. Goyal, V., O'Neill, A., Rao, V.: Correlated-input secure hash functions. In: Ishai, Y. (ed.) TCC 2011. LNCS, vol. 6597, pp. 182–200. Springer, Heidelberg (2011)

55. Groth, J., Ostrovsky, R., Sahai, A.: Perfect non-interactive zero knowledge for NP. In: Vaudenay, S. (ed.) EUROCRYPT 2006. LNCS, vol. 4004, pp. 339–358. Springer, Heidelberg (2006)

56. Håstad, J., Impagliazzo, R., Levin, L.A., Luby, M.: A pseudorandom generator from any one-way function. SIAM Journal on Computing 28(4), 1364–1396 (1999)

57. Hofheinz, D.: Possibility and impossibility results for selective decommitments. Journal of Cryptology 24(3), 470–516 (2011)

58. Hofheinz, D., Kiltz, E.: Programmable hash functions and their applications. In: Wagner, D. (ed.) CRYPTO 2008. LNCS, vol. 5157, pp. 21–38. Springer, Heidelberg (2008)

59. Kiltz, E., O'Neill, A., Smith, A.: Instantiability of RSA-OAEP under chosen-plaintext attack. In: Rabin, T. (ed.) CRYPTO 2010. LNCS, vol. 6223, pp. 295–313. Springer, Heidelberg (2010)

60. Kiltz, E., Pietrzak, K.: On the security of padding-based encryption schemes – or – why we cannot prove OAEP secure in the standard model. In: Joux, A. (ed.) EUROCRYPT 2009. LNCS, vol. 5479, pp. 389–406. Springer, Heidelberg (2009)

61. Krawczyk, H.: Cryptographic extraction and key derivation: The HKDF scheme. In: Rabin, T. (ed.) CRYPTO 2010. LNCS, vol. 6223, pp. 631–648. Springer, Heidelberg (2010)

62. Malkin, T., Teranishi, I., Yung, M.: Efficient circuit-size independent public key encryption with KDM security. In: Paterson, K.G. (ed.) EUROCRYPT 2011. LNCS, vol. 6632, pp. 507–526. Springer, Heidelberg (2011)

63. Maurer, U.M., Renner, R.S., Holenstein, C.: Indifferentiability, impossibility results on reductions, and applications to the random oracle methodology. In: Naor, M. (ed.) TCC 2004. LNCS, vol. 2951, pp. 21–39. Springer, Heidelberg (2004)

64. Nielsen, J.B.: Separating random oracle proofs from complexity theoretic proofs: The non-committing encryption case. In: Yung, M. (ed.) CRYPTO 2002. LNCS, vol. 2442, pp. 111–126. Springer, Heidelberg (2002)

65. Pass, R.: Limits of provable security from standard assumptions. In: Fortnow, L., Vadhan, S.P. (eds.) 43rd ACM STOC, pp. 109–118. ACM Press (June 2011)

66. Pinkas, B., Schneider, T., Smart, N.P., Williams, S.C.: Secure two-party computation is practical. In: Matsui, M. (ed.) ASIACRYPT 2009. LNCS, vol. 5912, pp. 250–267. Springer, Heidelberg (2009)

67. Ristenpart, T., Shacham, H., Shrimpton, T.: Careful with composition: Limitations of the indifferentiability framework. In: Paterson, K.G. (ed.) EUROCRYPT 2011. LNCS, vol. 6632, pp. 487–506. Springer, Heidelberg (2011)

68. Wichs, D.: Barriers in cryptography with weak, correlated and leaky sources. In: ITCS, 2013. Cryptology ePrint Archive, Report 2012/459 (2013)

Obfuscating Conjunctions

Zvika Brakerski[1] and Guy N. Rothblum[2]

[1] Stanford University
[2] Microsoft Research

Abstract. We show how to securely obfuscate the class of *conjunction functions* (functions like $f(x_1, \ldots, x_n) = x_1 \wedge \neg x_4 \wedge \neg x_6 \wedge \cdots \wedge x_{n-2}$). Given any function in the class, we produce an obfuscated program which preserves the input-output functionality of the given function, but reveals nothing else.

Our construction is based on multilinear maps, and can be instantiated using the recent candidates proposed by Garg, Gentry and Halevi (EUROCRYPT 2013) and by Coron, Lepoint and Tibouchi (CRYPTO 2013). We show that the construction is secure when the conjunction is drawn from a distribution, under mild assumptions on the distribution. Security follows from multilinear entropic variants of the Diffie-Hellman assumption. We conjecture that our construction is secure for *any* conjunction, regardless of the distribution from which it is drawn. We offer supporting evidence for this conjecture, proving that our obfuscator is secure for any conjunction against *generic adversaries*.

1 Introduction

Code obfuscation is the problem of compiling a computer program so as to make it unintelligible to an adversary, or impossible to reverse-engineer, while preserving its input-output functionality. Obfuscation has been of long-standing interest to both the cryptography and security communities. However, despite the importance of the problem, and its many exciting applications, very few techniques or effective heuristics are known. In particular, the theoretical study of the problem (in the "virtual black-box model" [2]) led to a handful of known constructions, which apply to very limited classes of functions. These include the class of point functions, and extensions such as multi-point functions, "lockers" and constant-dimension hyperplanes.

In this work, we present an obfuscator for a new and different class: *conjunction functions*. These are functions that take n-bit strings as input and only accept if a subset of these bits are set to predefined values. Our construction relies on *(asymmetric) multilinear maps*, and is instantiated using the new candidate construction due to Garg, Gentry and Halevi [14].

Previous Results. The goal of an obfuscator is generating a program that preserves the functionality of the original program, but reveals nothing else. One commonly used formalization of this objective is "virtual black box" obfuscation,

R. Canetti and J.A. Garay (Eds.): CRYPTO 2013, Part II, LNCS 8043, pp. 416–434, 2013.

due to Barak et al. [2]. Our work uses this formalization , as well as alternative formalizations from subsequent works (see below).

In their work, [2] also proved the impossibility of general-purpose obfuscators (i.e. ones that work for any functionality) in the virtual black box model. This impossibility result was extended in [15]. While these negative results show serious limitations on the possibility of general-purpose obfuscation, they focus on specific (often cryptographic or contrived) functionalities. Thus, they do not rule out that obfuscation may be possible for many programs of interest.

Positive results on obfuscation focus on specific, simple programs. One program family, which has received extensive attention, is that of "point functions": password checking programs that only accept a single input string, and reject all others. Starting with the work of Canetti [6], several works have shown obfuscators for this family under various assumptions [8,20,25], as well as extensions [7,3]. Canetti, Rothblum and Varia [9] showed how to obfuscate a function that checks membership in a hyperplane of constant dimension (over a large finite field). Other works showed how to obfuscate cryptographic function classes under different definitions and formalizations. These function classes include checking proximity to a hidden point [12], vote mixing [1], and re-encryption [18]. Several works [6,8,17,18] relaxed the security requirement so that obfuscation only holds for a random choice of a program from the family, we will also use this relaxation for one of our results. A different relaxation, known as "best-possible obfuscation", which allows the obfuscation to leak non black-box information was presented in [16].

This Work: Obfuscating Conjunctions. Our main contribution is a new obfuscator for *conjunctions*. A conjunction $C = (W, V)$ is a function on n bit inputs, specified by a set $W \subseteq [n]$ of "wildcard" entries, and a vector $V \in \{0,1\}^n$ of target values for non-wildcard entries. The conjunction accepts an input $\vec{x} \in \{0,1\}^n$ if for all $i \in ([n] \setminus W)$, $\vec{x}[i] = V[i]$, i.e. if for all non-wildcard entries in $\vec{x}$, their values equal those specified in V. We use the convention that if $W[i] = 1$ then $V[i] = 0$ (wildcard entries are ignored, so this does not effect the conjunction's functionality).

The class of conjunctions, while obviously quite limited, has a rich combinatorial and computational expressive power. They are studied in a multitude of settings throughout computer science (e.g. in learning theory [19]). One significant distinction from previous function classes for which obfuscators were known, is that a conjunction may ignore some of its input bits (the wildcard entries). An obfuscator for conjunctions needs to produce a program that hides which bits are ignored, and which ones are influential.

As an example of the applications of a conjunction obfuscator, consider the following setting. There are k passwords, each controlling access to a particular type of resource. Each individual knows some subset of the k passwords, which corresponds to the resources it is allowed to access. A gatekeeper wishes to check whether an individual has access to some combination of resources, i.e. whether the individual knows a particular subset $S \subset [k]$ of the passwords, without revealing to an observer which combination it is checking. A conjunction, which takes as input k concatenated passwords, can check whether the passwords for resources in S are correct, while ignoring passwords for resources not in S. An obfuscation of this conjunction can be made public, and used to check whether an individual has access to that combination of resources, without revealing which resources are being checked (nor, of course, what any of the passwords are).

1.1   Our Construction and Its Security

The main tool in our construction is multilinear maps. In particular, we utilize a recent candidate for graded encoding (a generalization of multilinear maps) due to [14].[1] We prove the security of our obfuscator when the conjunction is chosen from a distribution with sufficient entropy: namely, when sampling $C = (W, V)$ from the distribution, even given the wildcard locations W, there is sufficient (superlogarithmic) entropy in V. We stress that this *does not* imply that the attacker is allowed to learn W; on the contrary, we prove that if C is drawn from a distribution with the aforementioned property, the adversary cannot learn anything, wildcard locations included.[2] As noted above, here we follow several works [6,8,17,18] which relax the security requirement to hold only when the circuit to be obfuscated is drawn from a distribution from a certain class (usually one with sufficient entropy).

We prove the above under two security assumptions on graded encodings schemes: The first is a translation of the SXDH assumption on bilinear groups to the setting of graded encoding schemes.[3] The second assumption is reminiscent of "Canetti's Assumption" [6] on Diffie-Hellman groups, which was introduced for the purpose of obfuscating point functions.

We conjecture that the construction is secure for *every* conjunction, but we were unable to produce a proof based on a well-established assumption (naturally, one can always take the security of the obfuscator as an assumption). As supportive evidence for the conjectured security, we prove that the obfuscator is secure against *generic adversaries*: Ones that only use the group structure and not the representation of the group elements. This is similar to the generic group model of [24,21]. The proof of security against generic adversaries is non-trivial, and we view this as one of our main technical contributions. We note that previous works on obfuscation [20,9] have also used the random oracle and generic group models to provide evidence for the security of constructions.

---

[1] We use the *asymmetric* variant of the encoding scheme, where there are several distinct "source groups".

[2] We remark that in this case nothing at all can be learned from black-box access to the function since it is infeasible to find an accepting input. We also remark that, for example, the conjunctions used for the k-resource application above naturally satisfy this condition, because of the entropy in each password.

[3] This assumption is actually known to be *false* for the construction and formulation of [14]. However, we show a more careful definition of the scheme and the assumption for which no attack is known. Also, no attack is known for the recent construction of Coron, Lepoint and Tibouchi [10].

We proceed with an overview of our construction and results. As we explained, the obfuscator uses the recent construction of multi-linear maps via graded encoding schemes [14]. We begin with a high-level overview on the properties of multilinear maps that will be used. We then proceed with an overview of our construction, and state our two main results.

Multilinear Maps and Graded Encoding Schemes: Background. We begin by recalling the notion of multilinear maps, due to Boneh and Silverberg [5]. Rothblum [23] considered the asymmetric case, where the groups may be different (this is crucial for our construction).

Definition 1.1 (Asymmetric Multilinear Map [5,23]).

For $\tau + 1$ cyclic groups $G_1, \ldots, G_\tau, G_T$ of the same order p, a τ-multilinear map $e : G_1 \times \ldots \times G_\tau \to G_T$ has the following properties:

1. *For elements $\{g_i \in G_i\}_{i=1,\ldots,\tau}$, index $i \in [\tau]$ and integer $\alpha \in \mathbb{Z}_p$, it holds that:*
$$e(g_1, \ldots, \alpha \cdot g_i, \ldots, g_\tau) = \alpha \cdot g(g_1, \ldots, g_\tau)$$

2. *The map e is non-degenerate: when its inputs are all generators of their respective groups $\{G_i\}$, then its output is a generator of the target group G_T.*

Recently, [14] suggested a candidate for graded encoding, a generalization of (symmetric or asymmetric) multilinear maps. See Section 2.2 for a more complete overview of these objects. For this introduction, we treat them as a generalization of asymmetric multilinear maps in the following way. For a τ-multilinear map e, for the group G_i of prime order p, we consider the ring $\mathbb{Z}_p$. For an element $\sigma \in \mathbb{Z}_p$, we can think of g_i^σ as an "encoding" of σ in G_i. We denote this by $\mathsf{enc}_i(\sigma)$. We note that this encoding is easy to compute, but (presumably) hard to invert. The multilinear map e lets us take τ encodings $\{\mathsf{enc}_i(\sigma_i)\}_{i \in [\tau], \sigma_i \in \mathbb{Z}_p}$, and compute the target group encoding $\mathsf{enc}_T(\prod_i \sigma_i)$. Graded encoding schemes afford a similar functionality, albeit with randomized and noisy encodings, and with a procedure for testing equality of encoded elements in the target group.

Our Construction. For a conjunction $C = (W, V)$ on n-bits inputs, the obfuscator uses the graded encoding scheme to obtain the above generalization to an $(n+1)$-multilinear map. For each input entry $i \in [n]$, the obfuscator picks ring elements $(\rho_{i,0}, \rho_{i,1}, \alpha_{i,0}, \alpha_{i,1})$ distributed as follows: if $i \notin W$, namely the entry isn't a wildcard, then the ring elements are independent and uniformly random. If $i \in W$, namely the entry is a wildcard, then the ring elements are uniformly random under the constraint that $\alpha_{i,0} = \alpha_{i,1}$. After picking the ring elements, the obfuscator outputs two pairs of encodings for each $i \in [n]$:

$$\{(w_{i,b} = \mathsf{enc}_i(\rho_{i,b}), u_{i,b} = \mathsf{enc}_i(\rho_{i,b} \cdot \alpha_{i,b}))\}_{i \in [n], b \in \{0,1\}}$$

Note that if $i \in W$, then the ratio between the ring elements encoded in $u_{i,0}$ and $w_{i,0}$, is equal to the ratio between the ring elements encoded in $u_{i,1}$ and $w_{i,1}$

(these ratios are, respectively, $\alpha_{i,0}$ and $\alpha_{i,1}$, which are equal when $i \in W$). We remark that this part of the obfuscation depends only on the wildcards W, but not on the values V.

To complete the obfuscation, the obfuscator picks independent and uniformly random ring element ρ_{n+1}, and outputs a pair of encodings:

$$(w_{n+1} = \mathsf{enc}_{n+1}(\rho_{n+1}), u_{n+1} = \mathsf{enc}_{n+1}(\rho_{n+1} \cdot \underbrace{\prod_{i \in [n]} \alpha_{i,V[i]}}_{=\alpha_{n+1}}))$$

To evaluate the obfuscated program on an input $\vec{x} \in \{0,1\}^n$, we test equality between two multilinear products:[4]

$$e(\ldots, u_{i,\vec{x}[i]}, \ldots, w_{n+1}) \stackrel{?}{=} e(\ldots, w_{i,\vec{x}[i]}, \ldots, u_{n+1}) \tag{1}$$

The full construction is in Section 3.

Correctness. Examining the two multilinear products in Eq. (1), the element encoded in the left-hand side is $(\prod_{i \in [n]} \rho_{i,\vec{x}[i]} \cdot \alpha_{i,\vec{x}[i]}) \cdot \rho_{n+1}$. The element encoded in the right-hand side is $(\prod_{i \in [n]} \rho_{i,\vec{x}[i]} \cdot \alpha_{i,V[i]}) \cdot \rho_{n+1}$. Thus, Eq. (1) holds if and only if:

$$\prod_{i \in [n]} \alpha_{i,\vec{x}[i]} = \prod_{i \in [n]} \alpha_{i,V[i]} \tag{2}$$

For $i \in W$ we have $\alpha_{i,0} = \alpha_{i,1}$, the contributions from the i-th group to both products in Eq. (2) are identical. For $i \notin W$, the contribution from the i-th group in the left-hand side of Eq. (2) is $\alpha_{i,\vec{x}[i]}$. In the right-hand side, the contribution is $\alpha_{i,V[i]}$. Except for a negligible probability of error, Eq. (2) holds if and only if all these contributions are identical, i.e. if and only if $\forall i \notin W : \vec{x}[i] = V[i]$.

Security. Security is not as straightforward. A slightly misleading intuition for security, is that if a DDH-like assumption holds within each group G_i separately, then no observer can distinguish from that group's encodings whether $\alpha_{i,0} = \alpha_{i,1}$. This is true for each group in isolation, but it is insufficient because the obfuscation also includes encodings, in group G_{n+1}, of items that are correlated with the items encoded in group i. The multilinear map e might allow an adversary to distinguish whether the i-th entry is a wildcard.

For example, if in C all the entries are wildcards, the adversary can pick a random input, run the obfuscation, see that it accepts, and then by flipping the input bits one-by-one it can determine that all of the entries are wildcards. This attack clearly demonstrates that (for some conjunctions) an adversary can determine which entries are wildcards and which aren't. Note, however, that (for the specific example of a conjunction that is all-wildcards) this could also be accomplished using black-box access to the conjunction.

---

[4] For the candidate of [14], the encodings are randomized, but there is a procedure for testing equality between encoded elements in the target group.

Indeed, we prove the security of the obfuscator when the conjunction is drawn from a distribution, under mild assumptions on the distribution's entropy. We conjecture that the obfuscator is actually secure for *any* conjunction, and as supporting evidence we show that it is secure against generic adversaries. An overview on both results follows.

Security for High Entropy. We prove the security of our scheme in the case where $C = (W, V)$ is drawn from a distribution where the entropy of V given W is superlogarithmic. We do so by resurrecting the flawed argument described above: We use the entropy to remove the dependence between the elements in G_{n+1} and those in the other groups, and then apply DDH in each group.

We start by noting that this dependence is due to the relation

$$\alpha_{n+1} = \prod_{i \in [n]} \alpha_{i, V[i]} \,, \tag{3}$$

and if we could replace α_{n+1} with a completely uniform variable, independent of the other α's, we'd be done. To this end, we notice that Eq. (3) describes an (almost) pairwise independent hash function, whose seed are the values $\alpha_{i,b}$ and whose input is V. We show that such a hash function is a good entropy condenser, so that almost all of the entropy in V is preserved in α_{n+1}. (It is important to notice that the distinguisher has side information which depends on W, and therefore we must require that the *conditional* entropy is high.)

Once we establish that α_{n+1} has superlogarithmic entropy, we use a "Canetti-like Assumption" [6]: we assume a high-entropy element in the exponent of a random group generator is indistinguishable from uniform.[5] We thus isolate α_{n+1} from the dependence on the other α's, which allows us to apply DDH in groups $G_1, \ldots, G_n$, and obtain the final result: that the obfuscated program comes from a distribution that can be efficiently simulated. The security proof is in Section 4.

Security in The Generic Model. We prove security against *generic adversaries*. A generic adversary is one that succeeds regardless of the representation of the encoding scheme. This is modeled by allowing it to only manipulate encodings in the graded encoding scheme via oracle access to an oracle for the operations that are available using the *evparams* parameters. We show that for any generic adversary $\mathcal{A}$, which takes as input an obfuscation and outputs a single bit, there exists a generic simulator $\mathcal{S}$ s.t. for any conjunction C, the adversary's output on an obfuscation of C is statistically close to the simulator's output given only black-box access to C. The distribution of the adversary is taken over the choice of a *random* graded encoding scheme oracle: an oracle that represents each encoding in each group using a (long enough) uniformly random string.

In this model, since each element's encoding is uniformly random, and the obfuscation contains the encodings of distinct ring elements, the obfuscation of any conjunction is simply a collection of uniformly random strings. Thus

---

[5] Wee [25] showed that these types of assumptions (hardness given only super-logarithmic entropy) are essential even for obfuscating point functions.

simulating the obfuscator's output is easy. The main challenge is that *the outputs to oracle calls* on the string in the obfuscation are highly dependant on the conjunction C. It is thus not clear how the simulator can simulate the oracle's outputs. For example, each accepting input $\vec{x}$ for C specifies two possible inputs to the oracle implementing the multilinear map, which should both yield the same encoding in the target group. Indeed, simulating the oracle call outputs proves challenging. Moreover, the more generalized notion of graded encoding schemes permits more general generic operations.

The simulator $\mathcal{S}$ operates as follows. It feeds the adversary $\mathcal{A}$ with a "dummy obfuscation" containing uniformly random strings. It then follows $\mathcal{A}$'s calls to the graded encoding scheme (GES) oracle, and tries to simulate the output. For each call made by $\mathcal{A}$, we show how $\mathcal{S}$ can (efficiently) identify a polynomial size set X of inputs, such that if $\forall \vec{x} \in X, C(\vec{x}) = 0$, then the oracle's output is essentially independent of C and can be simulated. On the other hand, if there exists $\vec{x} \in X$ s.t. $C(\vec{x}) = 1$, then the simulator can use its black-box access to C to identify this input,. Once an accepting input is identified, the simulator can further use its block-box access to C to retrieve the conjunction's explicit description (W, V) (see Claim 3 below). Once the simulator knows (W, V) it can (perfectly) simulate the adversary's behavior. We view this proof of security for generic adversaries as one of our main technical contributions.

The full specification and treatment of the generic GES model, as well as the proof of security for generic adversaries, are deferred to the full version due to space constraints.

2   Preliminaries

Notation. We use $\Delta(\cdot, \cdot)$ to indicate total variation distance (statistical distance). We use $\vec{1}$ (respectively $\vec{0}$) to denote the all-1 (all-0) vector (the dimension will be clear from the context).

2.1   Min-entropy and Extraction

The following are information theoretic tools that will be required in our proof. The main notion of entropy used in this work is that of average min-entropy from [11], as well as its smooth version (see Definitions 2.1 and 2.2 below). We then show that applying a pairwise independent hash function with a large enough image on an average min-entropy source, roughly preserves the average min-entropy (that is, it is an entropy condenser). This is derived from the generalized "crooked" leftover hash lemma [13,4].

We start by defining average min-entropy.

Definition 2.1 (average min-entropy [11]). *Let X, Z be (possibly dependent) random variables, the average min entropy of X conditioned on Z is:*

$$\widetilde{\mathbf{H}}_\infty(X|Z) = -\log\left(\operatorname*{\mathbb{E}}_{z \leftarrow Z}\left[2^{-\mathbf{H}_\infty(X|Z=z)}\right]\right)$$

It follows from the definition that for every deterministic function f (that may depend on Z):

$$\widetilde{\mathbf{H}}_\infty(f(X)|Z) \leq \widetilde{\mathbf{H}}_\infty(X|Z) . \tag{4}$$

We also use a smooth variant introduced in [11, Appendix A] following [22].

Definition 2.2 (smooth average min-entropy [11]). *Let X, Z be as above and let $\epsilon > 0$, then*

$$\widetilde{\mathbf{H}}_\infty^\epsilon(X|Z) = \max_{(X',Z'):\Delta((X,Z),(X',Z'))\leq\epsilon} \widetilde{\mathbf{H}}_\infty(X'|Z') .$$

We will next show that pairwise independent functions condense average min-entropy in the following way.

Lemma 2.3. *Let X, Z be random variables, let $\mathcal{H}$ be a pairwise independent hash family with output length $\geq \left\lfloor \widetilde{\mathbf{H}}_\infty(X|Z) - 2\log(1/\epsilon) + 2 \right\rfloor$ (represented as binary string), for some $\epsilon > 0$. Then letting $h \leftarrow \mathcal{H}$ be a properly sampled function from this family, it holds that*

$$\widetilde{\mathbf{H}}_\infty^\epsilon(h(X)|Z, h) \geq \widetilde{\mathbf{H}}_\infty(X|Z) - 2\log(1/\epsilon) + 1 .$$

Proof. Let $X, Z, \mathcal{H}, h, \epsilon$ be as in the lemma statement. Our goal is to show that there exists a random variable Y such that

$$\Delta((h(X), Z, h), (Y, Z, h)) \leq \epsilon ,$$

and

$$\widetilde{\mathbf{H}}_\infty(Y|Z, h) \geq \widetilde{\mathbf{H}}_\infty(X|Z) - 2\log(1/\epsilon) + 1 .$$

Let f be the function that outputs the first k bits of its input, for

$$k = \left\lfloor \widetilde{\mathbf{H}}_\infty(X|Z) - 2\log(1/\epsilon) + 2 \right\rfloor \geq \widetilde{\mathbf{H}}_\infty(X|Z) - 2\log(1/\epsilon) + 1 ,$$

and note that $f(U)$ is uniform over $\{0,1\}^k$ (in fact, we can use any function that has this property).

We recall that the generalized "crooked" leftover hash lemma [4, Lemma 7.1] implies that

$$\Delta((f(h(X)), Z, h), (f(U), Z, h)) \leq \epsilon .$$

Now, consider a 2-step process for sampling the joint distribution $(h(X), Z, h)$: first, sample $(f(h(X)), Z, h)$ from the appropriate marginal distribution; and then sample $h(X)$ conditioned on the previously sampled values.

We define Y using using the following process: First, sample a tuple according to the distribution $(f(U), Z, h)$, and then apply the second stage of the sampling process from above. The result will be the distribution (Y, Z, h). Clearly,

$$\Delta((h(X), Z, h), (Y, Z, h)) = \Delta((f(h(X)), Z, h), (\underbrace{f(Y)}_{=f(U)}, Z, h)) \leq \epsilon ,$$

where the first equality is since there is a deterministic mapping (f) from the left hand side to the right hand side, and a randomized mapping (the second step sampler) from the right hand side to the left hand side.

To conclude, we notice that

$$\widetilde{\mathbf{H}}_\infty(Y|Z,h) \geq \widetilde{\mathbf{H}}_\infty(f(Y)|Z,h) = \underbrace{\widetilde{\mathbf{H}}_\infty(\{0,1\}^k|Z,h)}_{=k} \geq \widetilde{\mathbf{H}}_\infty(X|Z)-2\log(1/\epsilon)+1 \ .$$

2.2 Graded Encoding Schemes and Assumptions

We begin with the definition of a graded encoding scheme, due to Garg, Gentry and Halevi [14]. While their construction is very general, for our purposes a more restricted setting is sufficient as defined below.

Definition 2.4 (τ-Graded Encoding Scheme [14]). *A τ-encoding scheme for a ring R is a collection of sets $\mathcal{S} = \{S_\mathbf{v}^{(\alpha)} \subset \{0,1\}^* : \mathbf{v} \in \{0,1\}^\tau, \alpha \in R\}$, with the following properties:*

1. *For every index $\mathbf{v} \in \{0,1\}^\tau$, the sets $\{S_\mathbf{v}^{(\alpha)} : \alpha \in R\}$ are disjoint, and so they are a partition of the indexed set $S_\mathbf{v} = \bigcup_{\alpha \in R} S_\mathbf{v}^{(\alpha)}$.*
2. *There are binary operations "$+$" and "$-$" such that for all $\mathbf{v} \in \{0,1\}^\tau$, $\alpha_1, \alpha_2 \in R$ and for all $u_1 \in S_\mathbf{v}^{(\alpha_1)}$, $u_2 \in S_\mathbf{v}^{(\alpha_2)}$:*

$$u_1 + u_2 \in S_\mathbf{v}^{(\alpha_1+\alpha_2)} \quad and \quad u_1 - u_2 \in S_\mathbf{v}^{(\alpha_1-\alpha_2)} \ ,$$

   where $\alpha_1 + \alpha_2$ and $\alpha_1 - \alpha_2$ are addition and subtraction in R.
3. *There is an associative binary operation "$\times$" such that for all $\mathbf{v}_1, \mathbf{v}_2 \in \{0,1\}^\tau$ such that $\mathbf{v}_1 + \mathbf{v}_2 \in \{0,1\}^\tau$, for all $\alpha_1, \alpha_2 \in R$ and for all $u_1 \in S_{\mathbf{v}_1}^{(\alpha_1)}$, $u_2 \in S_{\mathbf{v}_2}^{(\alpha_2)}$, it holds that*

$$u_1 \times u_2 \in S_{\mathbf{v}_1+\mathbf{v}_2}^{(\alpha_1 \cdot \alpha_2)},$$

   where $\alpha_1 \cdot \alpha_2$ is multiplication in R.

In this work, the ring R will always be $\mathbb{Z}_p$ for a prime p.

To the reader who is familiar with the [14] work, we note that the above is the special case of the [14] construction in which we consider only binary index vectors (in the [14] notation, this corresponds to setting $\kappa = 1$), and we construct our encoding schemes to be *asymmetric* (as will become apparent below when we define our zero-text index $\mathbf{vzt} = \vec{1}$).

Definition 2.5 (Efficient Procedures for a τ-Graded Encoding Scheme [14]). *We consider τ-graded encoding schemes (see above) where the following procedures are efficiently computable.*

- *Instance Generation:* InstGen($1^\lambda, 1^\tau$) *outputs the set of parameters params, a description of a τ-Graded Encoding Scheme. (Recall that we only consider Graded Encoding Schemes over the set indices $\{0,1\}^\tau$, with zero testing in the set $S_{\bar{1}}$). In addition, the procedure outputs a subset evparams $\subset$ params that is sufficient for computing addition, multiplication and zero testing[6] (but possibly insufficient for encoding or for randomization).*

- *Ring Sampler:* samp($params$) *outputs a "level zero encoding" $a \in S_0^{(\alpha)}$ for a nearly uniform $\alpha \in_R R$.*

- *Encode and Re-Randomize:[7]* encRand($params, i, a$) *takes as input an index $i \in [\tau]$ and a $a \in S_0^{(\alpha)}$, and outputs an encoding $u \in S_{e_i}^{(\alpha)}$, where the distribution of u is (statistically close to being) only dependent on α and not otherwise dependent of a.*

- *Addition and Negation:* add($evparams, u_1, u_2$) *takes $u_1 \in S_v^{(\alpha_1)}, u_2 \in S_v^{(\alpha_2)}$, and outputs $w \in S_v^{(\alpha_1 + \alpha_2)}$. (If the two operands are not in the same indexed set, then* add *returns $\bot$). We often use the notation $u_1 + u_2$ to denote this operation when evparams is clear from the context. Similarly,* negate($evparams, u_1$) $\in S_v^{(-\alpha_1)}$.

- *Multiplication:* mult($evparams, u_1, u_2$) *takes $u_1 \in S_{\mathbf{v}_1}^{(\alpha_1)}, u_2 \in S_{\mathbf{v}_2}^{(\alpha_2)}$. If $\mathbf{v}_1 + \mathbf{v}_2 \in \{0,1\}^\tau$ (i.e. every coordinate in $\mathbf{v}_1 + \mathbf{v}_2$ is at most 1), then* mult *outputs $w \in S_{\mathbf{v}_1 + \mathbf{v}_2}^{(\alpha_1 \cdot \alpha_2)}$. Otherwise,* mult *outputs $\bot$. We often use the notation $u_1 \times u_2$ to denote this operation when evparams is clear from the context.*

- *Zero Test:* isZero($evparams, u$) *outputs 1 if $u \in S_{\bar{1}}^{(0)}$, and 0 otherwise.*

In the [14,10] constructions, encodings are noisy and the noise level increases with addition and multiplication operations, so one has to be careful not to go over a specified noise bound. However, the parameters can be set so as to support $O(\tau)$ operations, which are sufficient for our purposes. We therefore ignore noise management throughout this manuscript. An additional subtle issue is that with negligible probability the initial noise may be too big. However this can be avoided by adding rejection sampling to samp *and therefore ignored throughout the manuscript as well.*

It is important to notice that our definition deviates from that of [14] as we define two sets of parameters *params* and *evparams*. While the former will be used by the obfuscator in our construction (and therefore will not be revealed to an external adversary), the latter will be used when evaluating an obfuscated program (and thus will be known to an adversary). When instantiating our definition, the guideline is to make *evparams* minimal so as to give the least amount of information to the adversary. In particular, in the known candidates [14,10], *evparams* only needs to contain the zero-test parameter **pzt** (as well as the global modulus).

---

[6] The "zero testing" parameter **pzt** defined in [14] is a part of *evparams*.

[7] This functionality is not explicitly provided by [14], however it can be obtained by combining their encoding and re-randomization procedures.

Hardness Assumptions. In this work, we will use two hardness assumptions over graded encoding schemes. The first, which we call "graded external DDH" (or GXDH, Assumption 2.6 below) is an analog of the symmetric external DH assumption (SXDH), instantiated for the multilinear case. The second assumption (GCAN Assumption 2.7) is an analog of Canetti's assumption [6], that taking a random generator to a high-entropy power results in a random-looking element. We note that we make these assumptions against non-uniform adversaries.

Assumption 2.6 (Graded External DH). *Letting* $(params, evparams) \leftarrow$ $\mathsf{InstGen}(1^\lambda, 1^\tau)$, *for all* $i = 1, \ldots, \tau$, *sample* $r_{i,0}, r_{i,1}, a_{i,0}, a_{i,1} \leftarrow \mathsf{samp}(params)$ *and consider the following values:*

$$w_{i,0} \leftarrow \mathsf{encRand}(params, i, r_{i,0}) \qquad w_{i,1} \leftarrow \mathsf{encRand}(params, i, r_{i,1})$$
$$u_{i,0} \leftarrow \mathsf{encRand}(params, i, r_{i,0} \times a_{i,0}) \quad u_{i,1} \leftarrow \mathsf{encRand}(params, i, r_{i,1} \times a_{i,1})$$
$$u'_{i,1} \leftarrow \mathsf{encRand}(params, i, r_{i,1} \times a_{i,0})$$

The GXDH assumption is that for every choice of $\tau \in \mathbb{N}$ *and* $i^* \in [\tau]$, *no ensemble of polynomial time adversaries can have have non-negligible advantage in distinguishing the distributions:*

$$\left(evparams, \{(w_{i,0}, u_{i,0}, w_{i,1}, u_{i,1}, u'_{i,1})\}_{i \neq i^*}, (w_{i^*,0}, u_{i^*,0}, w_{i^*,1}, u_{i^*,1})\right)$$

$$and$$

$$\left(evparams, \{(w_{i,0}, u_{i,0}, w_{i,1}, u_{i,1}, u'_{i,1})\}_{i \neq i^*}, (w_{i^*,0}, u_{i^*,0}, w_{i^*,1}, u'_{i^*,1})\right)$$

We note that a stronger version of this assumption, where the distinguisher is given access to *params* rather than *evparams*, was presented in the the early versions of [14]. It was later shown that this stronger assumption is false, see later versions of [14] for the attack. We emphasize that *no attacks are known if only evparams is given as above.* Furthermore, the new candidate of [10] is not known to be sensitive to such attacks even if *params* is given.

Since we only provide our distinguisher with *evparams*, it may not be able to generate DDH tuples by itself. We therefore provide it with correctly labeled DDH samples for all groups except i^*. This is the minimal assumption that is required for our construction, however we conjecture that a stronger variant where the adversary is allowed to receive an unbounded number of labeled samples at any group (including i^*) is also true.

For our next assumption, we introduce the following notation. Consider a distribution D over $S_{\mathbf{v}} = \cup_{\alpha \in R} S_{\mathbf{v}}^{(\alpha)}$. The distribution $\mathsf{enc}^{-1}(D)$ is defined by the following process: Sample $x \leftarrow D$, let α be such that $x \in S_{\mathbf{v}}^{(\alpha)}$, output α. We also recall the definition of smooth average min-entropy (see Definition 2.2 above).

Assumption 2.7 ("Graded Canetti"). *Let* $(params, \mathbf{pzt}) \leftarrow \mathsf{InstGen}(1^\lambda, 1^\tau)$ *and let* $\{(D_\lambda, Z_\lambda)\}_{\lambda \in \mathbb{N}}$ *be a distribution ensemble over* $S_0 \times \{0,1\}^*$, *such that*

$$\widetilde{\mathbf{H}}_\infty^\epsilon(\mathsf{enc}^{-1}(D_\lambda)|Z_\lambda) \geq h(\lambda) \ ,$$

for some $\epsilon = negl(\lambda)$ *and function* $h(\lambda) = \omega(\log \lambda)$.

The GCAN assumption is that no ensemble of polynomial time adversaries and indices i can have non-negligible advantage in distinguishing the distributions

$$(params, evparams, w, u, z) \text{ and } (params, evparams, w, u', z) \text{ ,}$$

where we let: $(params, evparams) \leftarrow \mathsf{InstGen}(1^\lambda, 1^\tau)$, $r \leftarrow \mathsf{samp}(params)$, $w \leftarrow \mathsf{encRand}(params, i, r)$, $(x, z) \leftarrow (D_\lambda, Z_\lambda)$, $u \leftarrow \mathsf{encRand}(params, i, r \times x)$, $u' \leftarrow \mathsf{encRand}(params, i, \mathsf{samp}(params))$. *(In this definition, the distinguisher is given both params and evparams.)*

This assumption is consistent with out knowledge on candidate graded encoding schemes. However, if we want to make an even weaker assumption, we can set the minimal entropy requirement to be higher than just $\omega(\log \lambda)$. The constructions in this paper can trivially be adapted to such weaker variants (with the expected degradation in security).

2.3   Obfuscation

Definition 2.8 (Virtual Black-Box Obfuscator [2]).

Let $\mathcal{C} = \{\mathcal{C}_n\}_{n \in \mathbb{N}}$ *be a family of polynomial-size circuits, where* $\mathcal{C}_n$ *is a set of boolean circuits operating on inputs of length n. And let $\mathcal{O}$ be a PPTM algorithm, which takes as input an input length $n \in \mathbb{N}$, a circuit $C \in \mathcal{C}_n$, a security parameter $\lambda \in \mathbb{N}$, and outputs a boolean circuit $\mathcal{O}(C)$ (not necessarily in $\mathcal{C}$).*

$\mathcal{O}$ *is an* obfuscator *for the circuit family $\mathcal{C}$ if it satisfies:*

1. Preserving Functionality: *For every $n \in \mathbb{N}$, and every $C \in \mathcal{C}_n$, and every $\vec{x} \in \{0, 1\}^n$, with all but $negl(\lambda)$ probability over the coins of $\mathcal{O}$:*

$$(\mathcal{O}(C, 1^n, \lambda))(\vec{x}) = C(\vec{x})$$

2. Polynomial Slowdown: *For every $n, \lambda \in \mathbb{N}$ and $C \in \mathcal{C}$, the circuit $\mathcal{O}(C, 1^n, 1^\lambda)$ is of size at most $poly(|C|, n, \lambda)$.*

3. Virtual Black-Box: *For every (non-uniform) polynomial size adversary $\mathcal{A}$, there exists a (non-uniform) polynomial size simulator $\mathcal{S}$, such that for every $n \in \mathbb{N}$ and for every $C \in \mathcal{C}_n$:*

$$\left| \Pr_{\mathcal{O}, \mathcal{A}}[\mathcal{A}(\mathcal{O}(C, 1^n, 1^\lambda)) = 1] - \Pr_{\mathcal{S}}[\mathcal{S}^C(1^{|C|}, 1^n, 1^\lambda) = 1] \right| = negl(\lambda)$$

Remark 2.9. A stronger notion of functionality, which also appears in the literature, requires that with overwhelming probability the obfuscated circuit is correct on *every input simultaneously*. We use the relaxed requirement that for every input (individually) the obfuscated circuit is correct with overwhelming probability (in both cases the probability is only over the obfuscator's coins). We note that our construction can be modified to achieve the stronger functionality property (by using a ring of sufficiently large size and the union bound).

Definition 2.10 (*Average-Case* Secure Virtual Black-Box).
Let $\mathcal{C} = \{\mathcal{C}_n\}_{n \in \mathbb{N}}$ be a family of circuits and $\mathcal{O}$ a PPTM as in Definition 2.8. Let $\mathcal{D} = \{\mathcal{D}_n\}_{n \in \mathbb{N}}$ be an ensemble of distribution families $\mathcal{D}_n$, where each $D \in \mathcal{D}_n$ is a distribution over $\mathcal{C}_n$.

$\mathcal{O}$ is an obfuscator for the distribution class $\mathcal{D}$ over the circuit family $\mathcal{C}$, if it satisfies the functionality and polynomial slowdown properties of Definition 2.8 with respect to $\mathcal{C}$, but the virtual black-box property is replaced with:

3. *Distributional Virtual Black-Box: For every (non-uniform) polynomial size adversary $\mathcal{A}$, there exists a (non-uniform) polynomial size simulator $\mathcal{S}$, such that for every $n \in \mathbb{N}$, every distribution $D \in \mathcal{D}_n$ (a distribution over $\mathcal{C}_n$), and every predicate $P : C_n \rightarrow \{0,1\}$:*

$$\left| \Pr_{\substack{C \sim D_n, \\ \mathcal{O}, \mathcal{A}}} [\mathcal{A}(\mathcal{O}(C, 1^n, 1^\lambda)) = P(C)] - \Pr_{\substack{C \sim D_n, \\ \mathcal{S}}} [\mathcal{S}^C(1^{|C|}, 1^n, 1^\lambda) = P(C)] \right| = negl(\lambda)$$

Remark 2.11. Our proof of average-case security for the conjunction obfuscator (Theorem 4.2) is in fact stronger. We show a simulator $\mathcal{S}$ that does not even require black-box access to the circuit C. Rather, for a circuit C drawn from a distribution in $\mathcal{D}$, the probability of predicting $P(C)$ from an obfuscation of C, is the same as the probability of predicting $P(C)$ from a "dummy obfuscation" that is independent of C. See the proof for further details.

3   Obfuscating Conjunctions

In this section we present our obfuscator for conjunctions ConjObf (Figure 1). We provide a proof of security for functions that are not determined by the locations of the wildcards in Section 4. In the full version, we provide evidence of the security of our construction for *any* conjunction, by proving that it is secure against generic adversaries that do not use the representation of the specific graded encoding scheme.

We start by defining the class of conjunctions, and a useful property thereof.

Definition 3.1 (*n*-bit Conjunction).
For an input length n, a conjunction $C = (W, V) : \{0,1\}^n \rightarrow \{0,1\}$ is a predicate on n-bit inputs, which is defined by two vectors $W, V \in \{0,1\}^n$. For every input $\vec{x} \in \{0,1\}^n$, $C(\vec{x}) = 1$ iff for all $i \in [n]$, $W[i] = 1$ or $V[i] = \vec{x}[i]$. For the sake of unity of representation, we require that whenever $W[i] = 1$, it holds that $V[i] = 0$.

We often alternate between treating W as an index vector and treating it as a subset of $[n]$. If $i \in W$ then we say that i is a wildcard location.

Definition 3.2 (Conjunction Ensemble).
A conjunction ensemble $\mathcal{C} = \{C_n\}_{n \in \mathbb{N}}$ is a collection of conjunctions $C_n : \{0,1\}^n \rightarrow \{0,1\}$, one for each input length.

Claim. There exists an efficient algorithm $\mathcal{B}$, such that for any conjunction $C = (W, V)$, and any accepting input $\vec{x}$ of C, $\mathcal{B}$ can recover (W, V):

$$\forall C = (W, V), \forall \vec{x} : C(\vec{x}) = 1, \quad \mathcal{B}^C(\vec{x}) = (W, V)$$

Proof. Take $n = |\vec{x}|$, the algorithm $\mathcal{B}$ enumerates over the bits of $\vec{x}$. For each bit i, it flips the i-th bit of $\vec{x}$: $\vec{x}^{(i)} = \vec{x} \oplus \mathbf{e}_i$, and checks whether $C(\vec{x}^{(i)}) = 1$. If so, then i must be a wildcard: $W[i] = 1$ and $V[i] = 0$. Otherwise, i is not a wildcard: $W[i] = 0$ and $V[i] = \vec{x}[i]$.

Obfuscator ConjObf, on input $(1^\lambda, 1^n, C = (W, V))$

1. generate $(params, evparams) \leftarrow \mathsf{InstGen}(1^\lambda, 1^{n+1})$
2. for $i \in [n]$:
   if $i \in W$, then: $a_{i,0} = a_{i,1} \leftarrow \mathsf{samp}(params) \in S_0^{(\alpha_{i,0})} = S_0^{(\alpha_{i,1})}$
   if $i \notin W$, then: $a_{i,0} \leftarrow \mathsf{samp}(params) \in S_0^{(\alpha_{i,0})}$, $a_{i,1} \leftarrow \mathsf{samp}(params) \in S_0^{(\alpha_{i,1})}$
3. for $i \in [n]$:
   $r_{i,0} \leftarrow \mathsf{samp}(params) \in S_0^{(\rho_{i,0})}, r_{i,1} \leftarrow \mathsf{samp}(params) \in S_0^{(\rho_{i,0})}$
   $s_{i,0} \leftarrow r_{i,0} \times a_{i,0} \in S_0^{(\rho_{i,0} \cdot \alpha_{i,0})}, s_{i,1} \leftarrow r_{i,1} \times a_{i,1} \in S_0^{(\rho_{i,1} \cdot \alpha_{i,1})}$
   $w_{i,0} \leftarrow \mathsf{encRand}(params, i, r_{i,0}) \in S_{\mathbf{e}_i}^{(\rho_{i,0})}$
   $w_{i,1} \leftarrow \mathsf{encRand}(params, i, r_{i,1}) \in S_{\mathbf{e}_i}^{(\rho_{i,1})}$
   $u_{i,0} \leftarrow \mathsf{encRand}(params, i, s_{i,0}) \in S_{\mathbf{e}_i}^{(\rho_{i,0} \cdot \alpha_{i,0})}$
   $u_{i,1} \leftarrow \mathsf{encRand}(params, i, s_{i,1}) \in S_{\mathbf{e}_i}^{(\rho_{i,1} \cdot \alpha_{i,1})}$
4. $a_{n+1} \leftarrow \left(\Pi_{i \in [n]} a_{i, V[i]} \right) \in S_0^{(\Pi_{i \in [n]} \alpha_{i, V[i]})}$
   $r_{n+1} \leftarrow \mathsf{samp}(params) \in S_0^{(\rho_{n+1})}$
   $s_{n+1} \leftarrow r_{n+1} \times a_{n+1} \in S_0^{(\rho_{n+1} \cdot \Pi_{i \in [n]} \alpha_{i, V[i]})}$
   $w_{n+1} \leftarrow \mathsf{encRand}(params, n+1, r_{n+1}) \in S_{\mathbf{e}_{n+1}}^{(\rho_{n+1})}$
   $u_{n+1} \leftarrow \mathsf{encRand}(params, n+1, s_{n+1}) \in S_{\mathbf{e}_{n+1}}^{(\rho_{n+1} \cdot \Pi_{i \in [n]} \alpha_{i, V[i]})}$
5. output the obfuscation:

$$(evparams, \{(w_{i,0}, u_{i,0}), (w_{i,1}, u_{i,1})\}_{i \in [n]}, (w_{n+1}, u_{n+1}))$$

Evaluation, on input $\vec{x} \in \{0, 1\}^n$

1. $\mathbf{t} \leftarrow \left(w_{n+1} \times \Pi_{i \in [n]} u_{i, \vec{x}[i]} \right) \in S_{(1,1,\ldots,1)}^{\left(\rho_{n+1} \cdot (\Pi_{i \in [n]} \rho_{i, \vec{x}[i]} \cdot \alpha_{i, \vec{x}[i]}) \right)}$
2. $\mathbf{t}' \leftarrow \left(u_{n+1} \times \Pi_{i \in [n]} w_i \right) \in S_{(1,1,\ldots,1)}^{\left(\rho_{n+1} \cdot (\Pi_{i \in [n]} \alpha_{i, V[i]}) \cdot (\Pi_{i \in [n]} \rho_{i, \vec{x}[i]}) \right)}$
3. output the bit: $\mathsf{isZero}(evparams, (\mathbf{t} - \mathbf{t}'))$.

Fig. 1. Obfuscator for Conjunctions

Our obfuscator for the class of conjunctions is presented in Figure 1. Correctness follows in a straightforward manner as described in the following lemma (the proof is omitted). We note that the error is one sided, it is always the case that if $C(\vec{x}) = 1$ then for the obfuscated program $\mathcal{O}_C(\vec{x}) = 1$ as well.

Lemma 3.3 (Obfuscator Functionality). *Let C be an n-variable conjunction and consider its obfuscation $\mathcal{O}_C = \mathsf{ConjObf}(C, 1^n, 1^\lambda)$. Then for all $\vec{x}$,*

$$\Pr[\mathcal{O}_C(\vec{x}) \neq C(\vec{x})] \leq poly(n)/p \ ,$$

where $p = 2^{\Omega(\lambda)}$ is the order of the group in the graded encoding scheme, and the probability is taken over the randomness of $\mathsf{ConjObf}$.

As a concluding remark, we note that if our graded encoding scheme has the property that $p \gg 2^n$ (which is indeed achievable in the candidate of [14]), then a stronger correctness guarantee, as mentioned in Remark 2.9, can be achieved by using the union bound. In this parameter range, the proof of security also becomes somewhat simpler (see Section 4). However, we want to present our scheme in the most generic way so as to be compatible with possible choices of the security parameter and with future graded encoding schemes.

4   Security from GXDH and GCAN

In this section we prove that $\mathsf{ConjObf}$ is a secure distributional black box obfuscator for any distribution over the conjunctions family for which the function is hard to determine (i.e. has super-logarithmic entropy) even if the locations of all the wildcards are known. Namely, there is sufficient min-entropy in V even given W (recall that $V[i] = 0$ wherever $W[i] = 1$).

Definition 4.1 (equivocality given wildcards). *Let $\mathcal{C}$ be the class of conjunctions, and let $\mathcal{D} = \{\mathcal{D}_\lambda\}$ be an ensemble of families of distributions. We say that $\mathcal{D}$ is equivocal given the wildcards if there exists $h(\lambda) = \omega(\log \lambda)$ such that for all $D \in \mathcal{D}_\lambda$, if $(V, W) \leftarrow D$ then*

$$\widetilde{\mathbf{H}}_\infty(V|W) \geq h(\lambda) \ .$$

We will prove the security of $\mathsf{ConjObf}$ for such functions under the GXDH and GCAN assumptions (see Section 2.2).

Theorem 4.2. *Based on the GXDH and GCAN assumptions, the algorithm $\mathsf{ConjObf}$ is an average-case black-box obfuscator for ensembles of distribution families that are equivocal given the wildcards.*

Proof. We start by stating a claim that will be used later on in the proof.

Claim. Let p be prime and let $k \in \mathbb{N}$ be integer. Consider the hash family $\mathcal{H} \subseteq \{0,1\}^k \to \mathbb{Z}_p^*$, where each function in $\mathcal{H}$ is defined by a sequence $a_0, a_1, \ldots, a_k \in \mathbb{Z}_p^*$ and

$$\mathcal{H}_{a_0, a_1, \ldots, a_k}(x_1, \ldots, x_k) = a_0 \cdot \prod_{i \in [k]} a_i^{x_i} \ ,$$

then $\mathcal{H}$ is pairwise independent.

This claim follows in a straightforward manner since $\mathcal{H}$ (defined therein) is a random linear function "in the exponent".

Consider a function $C = (W, V)$ drawn from a distribution $\mathcal{D}_\lambda$, and consider the distribution of a properly obfuscated program $\mathsf{ConjObf}(C)$. We will show, using a sequence of hybrids, that this distribution is computationally indistinguishable from one that does not depend on C, even for a distinguisher who knows the value of the predicate $P(C)$. This will immediately imply a simulator. We note that our proof works even for $P(C)$ with multiple-bit output, so long as $h(\lambda) - |P(C)| = \omega(\log \lambda)$.

1. In this hybrid, we use $\mathsf{ConjObf}$ as prescribed:

$$\mathcal{O}_C = \mathsf{ConjObf}(C) = \left(params, \mathbf{pzt}, \{(w_{i,b}, u_{i,b})\}_{i \in [n], b \in \{0,1\}}, (w_{n+1}, u_{n+1}) \right)$$

2. We change the algorithm so that $a_{i,b} \notin S_0^{(0)}$. This is implemented efficiently by rejection sampling, using the zero-test procedure. In this hybrid, therefore, $\alpha_{i,b}$ is uniform in $\mathbb{Z}_p^*$.

   This hybrid only incurs a negligible $poly(\lambda)/p$ statistical distance in the distribution of $\mathcal{O}_C$ compared to the previous hybrid.

3. We change step 4 of the obfuscator. In particular, we will now sample $a_{n+1} \leftarrow samp(params)$ (conditioned on it not being zero, as above). This means that $a_{n+1} \in S_0^{(\alpha_{n+1})}$ for a random $\alpha_{n+1} \in \mathbb{Z}_p^*$.

   We will now show that the resulting $\mathcal{O}_C$ distribution is computationally indistinguishable from the previous hybrid under the GCAN assumption (Assumption 2.7), even when the distinguisher knows $P(C)$. Namely, we will show that for some negligible ϵ, the distributions in the previous hybrid are such that

$$\widetilde{\mathbf{H}}_\infty^\epsilon (\alpha_{n+1} | \{(w_{i,b}, u_{i,b})\}_{i \in [n], b \in \{0,1\}}, P(C)) = \omega(\log \lambda) , \tag{5}$$

which will allow us to apply GCAN and conclude that α_{n+1} can be replaced by a uniform variable.

To show that Eq. (5) holds, we present a slightly different way to generate the variables $\alpha_{i,b}$ (note that from this point and on, we are a completely information-theoretic setting, so we will not worry about computational aspects). We will first sample $\{\hat{\alpha}_{i,b}\}_{i \in [n], b \in \{0,1\}}$ completely uniformly in $\mathbb{Z}_p^*$, and then set $\alpha_{i,b}$ as follows. If $W[i] = 0$ then $\alpha_{i,0} = \hat{\alpha}_{i,0}$, $\alpha_{i,1} = \hat{\alpha}_{i,1}$; and if $W[i] = 1$ then $\alpha_{i,0} = \alpha_{i,1} = \hat{\alpha}_{i,0}$. Note that the resulting distribution of the α's is exactly as prescribed. Further notice that

$$\alpha_{n+1} = \prod_{i \in [n]} \alpha_{i,0}^{1-V[i]} \alpha_{i,1}^{V[i]} = \prod_{i \in [n]} \hat{\alpha}_{i,0}^{1-V[i]} \hat{\alpha}_{i,1}^{V[i]} = \prod_{i \in [n]} \hat{\alpha}_{i,0} \cdot \prod_{i \in [n]} (\hat{\alpha}_{i,1}/\hat{\alpha}_{i,0})^{V[i]}$$

$$\tag{6}$$

where the second equality is since α and $\hat{\alpha}$ only differ where $W[i] = V[i] = 0$. By Claim 4 it follows, therefore, that α_{n+1} is the output of a pairwise-independent hash function applied to V.

We proceed to apply Lemma 2.3. Note that $\widetilde{\mathbf{H}}_\infty(V|W, P(C)) \geq \widetilde{\mathbf{H}}_\infty(V|W) - |P(C)| \geq h(\lambda) - 1$. Therefore there must exist $h'(\lambda) = \omega(\log \lambda)$ such that $h'(\lambda) \leq h(\lambda) - 1$, and in addition the length of α_{n+1} is at least $h'(\lambda)/3 + 2$. We can thus apply Lemma 2.3 with $\epsilon = 2^{-h'(\lambda)/3} = negl(\lambda)$ to argue that

$$\widetilde{\mathbf{H}}_\infty^\epsilon(\alpha_{n+1}|W, P(C), \{\hat{\alpha}_{i,b}\}) \geq h'(\lambda)/3 = \omega(\log \lambda) \ . \tag{7}$$

Finally, Eq. (5) follows by noticing that there is an invertible mapping between $W, \{\hat{\alpha}_{i,b}\}$ and $\{(w_{i,b}, u_{i,b})\}_{i\in[n], b\in\{0,1\}}$.

It is interesting to note that this hybrid (and therefore our entire argument) works not only for predicates. In fact, ℓ-bit functions of the circuit C can be used, so long as $h(\lambda) - \ell = \omega(\log \lambda)$.

At this point, $\mathcal{O}_C$ does not depend on V anymore, however it still depends on W via step 2 of ConjObf.

4. We again allow $a_{i,b}$ to be zero. The statistical difference is $poly(\lambda)/p = negl(\lambda)$, as above.

5. We change step 2 of the obfuscator to always act as if $i \notin W$, namely $\alpha_{i,0}$ and $\alpha_{i,1}$ are uniform and independent.

   A sequence of n hybrids will show that any adversary distinguishing this distribution from the previous one, can be used to break GXDH with only a factor n loss in the advantage. This implies that the hybrids are computationally indistinguishable assuming GXDH. Note that knowledge of $P(C)$ (or even of C in its entirety) is useless for the distinguisher at this point.

After the last hybrid, we are at a case where all $a_{i,b}, r_{i,b}, a_{n+1}, r_{n+1}$ are completely independent of each other, and are sampled in the same way regardless of (V, W). It follows that our final distribution is independent of C, but produces $\mathcal{O}_C$ indistinguishable from ConjObf(C) (even given $P(C)$). Since this distribution is efficiently sampleable (via the process we describe in the proof), the theorem follows.

Acknowledgments. We thank the reviewers of CRYPTO 2013 for their comments. We thank Masayuki Abe for pointing us to the zeroing attack on [14], and the authors of [10] for discussing its (in)applicability to their scheme. The first author wishes to thank the Simons Foundation and DARPA for their support.

References

1. Adida, B., Wikström, D.: How to shuffle in public. In: Vadhan, S.P. (ed.) TCC 2007. LNCS, vol. 4392, pp. 555–574. Springer, Heidelberg (2007)
2. Barak, B., Goldreich, O., Impagliazzo, R., Rudich, S., Sahai, A., Vadhan, S.P., Yang, K.: On the (im)possibility of obfuscating programs. J. ACM 59(2), 6 (2012); Preliminary version in: Kilian, J. (ed.) CRYPTO 2001. LNCS, vol. 2139, pp. 1–18. Springer, Heidelberg (2001)

3. Bitansky, N., Canetti, R.: On strong simulation and composable point obfuscation. In: Rabin, T. (ed.) CRYPTO 2010. LNCS, vol. 6223, pp. 520–537. Springer, Heidelberg (2010)

4. Boldyreva, A., Fehr, S., O'Neill, A.: On notions of security for deterministic encryption, and efficient constructions without random oracles. In: Wagner, D. (ed.) CRYPTO 2008. LNCS, vol. 5157, pp. 335–359. Springer, Heidelberg (2008)

5. Boneh, D., Silverberg, A.: Applications of multilinear forms to cryptography. IACR Cryptology ePrint Archive 2002, 80 (2002)

6. Canetti, R.: Towards realizing random oracles: Hash functions that hide all partial information. In: Kaliski Jr., B.S. (ed.) CRYPTO 1997. LNCS, vol. 1294, pp. 455–469. Springer, Heidelberg (1997)

7. Canetti, R., Dakdouk, R.R.: Obfuscating point functions with multibit output. In: Smart, N.P. (ed.) EUROCRYPT 2008. LNCS, vol. 4965, pp. 489–508. Springer, Heidelberg (2008)

8. Canetti, R., Micciancio, D., Reingold, O.: Perfectly one-way probabilistic hash functions (preliminary version). In: STOC, pp. 131–140 (1998)

9. Canetti, R., Rothblum, G.N., Varia, M.: Obfuscation of hyperplane membership. In: Micciancio, D. (ed.) TCC 2010. LNCS, vol. 5978, pp. 72–89. Springer, Heidelberg (2010)

10. Coron, J.-S., Lepoint, T., Tibouchi, M.: Practical multilinear maps over the integers. In: Canetti, R., Garay, J.A. (eds.) CRYPTO 2013, Part I. LNCS, vol. 8042, pp. 476–493. Springer, Heidelberg (2013)

11. Dodis, Y., Ostrovsky, R., Reyzin, L., Smith, A.: Fuzzy extractors: How to generate strong keys from biometrics and other noisy data. SIAM J. Comput. 38(1), 97–139 (2008); Preliminary version in: Cachin, C., Camenisch, J.L. (eds.) EUROCRYPT 2004. LNCS, vol. 3027, pp. 523–540. Springer, Heidelberg (2004)

12. Dodis, Y., Smith, A.: Correcting errors without leaking partial information. In: Gabow, H.N., Fagin, R. (eds.) STOC, pp. 654–663. ACM (2005)

13. Dodis, Y., Smith, A.: Correcting errors without leaking partial information. In: Gabow, H.N., Fagin, R. (eds.) STOC, pp. 654–663. ACM (2005)

14. Garg, S., Gentry, C., Halevi, S.: Candidate multilinear maps from ideal lattices. In: Johansson, T., Nguyen, P.Q. (eds.) EUROCRYPT 2013. LNCS, vol. 7881, pp. 1–17. Springer, Heidelberg (2013)

15. Goldwasser, S., Kalai, Y.T.: On the impossibility of obfuscation with auxiliary input. In: FOCS, pp. 553–562 (2005)

16. Goldwasser, S., Rothblum, G.N.: On best-possible obfuscation. In: Vadhan, S.P. (ed.) TCC 2007. LNCS, vol. 4392, pp. 194–213. Springer, Heidelberg (2007)

17. Hofheinz, D., Malone-Lee, J., Stam, M.: Obfuscation for cryptographic purposes. J. Cryptology 23(1), 121–168 (2010)

18. Hohenberger, S., Rothblum, G.N., Shelat, A., Vaikuntanathan, V.: Securely obfuscating re-encryption. J. Cryptology 24(4), 694–719 (2011)

19. Kearns, M.J., Vazirani, U.V.: An introduction to computational learning theory. MIT Press, Cambridge (1994)

20. Lynn, B., Prabhakaran, M., Sahai, A.: Positive results and techniques for obfuscation. In: Cachin, C., Camenisch, J.L. (eds.) EUROCRYPT 2004. LNCS, vol. 3027, pp. 20–39. Springer, Heidelberg (2004)

21. Maurer, U.M.: Abstract models of computation in cryptography. In: Smart, N.P. (ed.) Cryptography and Coding 2005. LNCS, vol. 3796, pp. 1–12. Springer, Heidelberg (2005)

22. Renner, R., Wolf, S.: Smooth renyi entropy and applications. In: IEEE International Symposium on Information Theory, p. 233 (2004)
23. Rothblum, R.D.: On the circular security of bit-encryption. In: Sahai, A. (ed.) TCC 2013. LNCS, vol. 7785, pp. 579–598. Springer, Heidelberg (2013)
24. Shoup, V.: Lower bounds for discrete logarithms and related problems. In: Fumy, W. (ed.) EUROCRYPT 1997. LNCS, vol. 1233, pp. 256–266. Springer, Heidelberg (1997)
25. Wee, H.: On obfuscating point functions. In: Gabow, H.N., Fagin, R. (eds.) STOC, pp. 523–532. ACM (2005)

Fully, (Almost) Tightly Secure IBE
and Dual System Groups*

Jie Chen[1,**] and Hoeteck Wee[2,***]

[1] Nanyang Technological University, Singapore
s080001@e.ntu.edu.sg
[2] George Washington University
hoeteck@alum.mit.edu

Abstract. We present the first fully secure Identity-Based Encryption scheme (IBE) from the standard assumptions where the security loss depends only on the security parameter and is independent of the number of secret key queries. This partially answers an open problem posed by Waters (Eurocrypt 2005). Our construction combines the Waters' dual system encryption methodology (Crypto 2009) with the Naor-Reingold pseudo-random function (J. ACM, 2004) in a novel way. The security of our scheme relies on the DLIN assumption in prime-order groups. Along the way, we introduce a novel notion of *dual system groups* and a new randomization and parameter-hiding technique for prime-order bilinear groups.

1 Introduction

In an Identity-Based Encryption (IBE) scheme [27], encryption requires only the identity of the recipient (e.g. an email address or an IP address) and a set of global public parameters, thus eliminating the need to distribute a separate public key for each user in the system. The first realizations of IBE were given in 2001; the security of these schemes were based on either Bilinear Diffie-Hellman or QR in the random oracle model [7, 13]. Since then, tremendous progress has been made towards obtaining IBE and HIBE schemes that are secure in the standard model based on pairings [8, 5, 6, 28, 15, 29] as well as lattices [16, 9, 2, 3]. Specifically, starting with [29], we now have very efficient constructions of IBE based on standard assumptions which achieve the strongest security notion of full (adaptive) security, where the adversary may choose the challenge identity after seeing both the public parameters and making key queries.

In this work, we focus on the issue of security reduction and security loss in the construction of fully secure IBE. Consider an IBE scheme with a security reduction showing that attacking the scheme in time t with success probability ϵ implies breaking

---

* This is a merge of two papers [10] and [11].
** Supported in part by the National Research Foundation of Singapore under Research Grant NRF-CRP2-2007-03.
*** Supported by NSF CAREER Award CNS-1237429. Part of this work was done while visiting NTU.

R. Canetti and J.A. Garay (Eds.): CRYPTO 2013, Part II, LNCS 8043, pp. 435–460, 2013.

some conjectured hard problem in time roughly t with success probability ϵ/L; we refer to L as the security loss, and a tight reduction is one where L is a constant. All known constructions of fully secure IBE schemes from standard assumptions incur a security loss that is at least linear in the number of key queries q; the only exceptions are constructions in the random oracle model [7] and those based on q-type assumptions [15]. Motivated by this phenomenon, Waters [28] posed the following problem in 2005 (reiterated in [15, 4]):

"Design an IBE with a tight security reduction to a standard assumption."

That is, we are interested in constructions based on "static" assumptions like the Decisional Linear (DLIN) assumption or the subgroup decisional assumption and which do not rely on random oracles. Note that an IBE with a tight security reduction would also imply signatures with a tight security reduction via the Naor's transformation [7]; indeed, the latter were the focus in a series of very recent works [1, 19, 17].

We stress that tight reductions are not just theoretical issues for IBE, rather they are of utmost practical importance: as L increases, we need to increase the size of the underlying groups in order to compensate for the security loss, which in turn increases the running time of the implementation. Note that the impact on performance is quite substantial, as exponentiation in a r-bit group takes time roughly $\mathcal{O}(r^3)$.

While the ultimate goal is to achieve constant security loss (i.e. $L = \mathcal{O}(1)$), even achieving $L = \mathrm{poly}(\lambda)$ and independent of q is already of both practical and theoretical interest. For typical settings of parameters (e.g. $\lambda = 128$ and $q = 2^{20}$), λ is much smaller than q. From the theoretical stand-point, we currently have two main techniques for obtaining fully secure IBE from standard assumptions: random partitioning [28] and dual system encryption framework [29]. For the former, we now know that an $\Omega(q)$ security loss is in fact inherent [18]. For the latter, all known instantiations also incur an $\Omega(q)$ security loss; an interesting theoretical question is whether this is in fact inherent to the dual system encryption framework.

1.1 Our Results

Our main result is an IBE scheme based on the d-LIN assumption with security loss $\mathcal{O}(\lambda)$ for λ-bit identities:

Theorem 1. *There exists an IBE scheme for identity space $\{0,1\}^n$ based on the d-LIN assumption with the following property: for any adversary $\mathcal{A}$ that makes at most q key queries against the IBE scheme, there exist an adversary $\mathcal{B}$ such that:*

$$\mathsf{Adv}_{\mathcal{A}}^{\mathrm{IBE}}(\lambda) \le (2n+1) \cdot \mathsf{Adv}_{\mathcal{B}}^{d\text{-}\mathrm{LIN}}(\lambda) + 2^{-\Omega(\lambda)}$$

and

$$\mathsf{Time}(\mathcal{B}) \approx \mathsf{Time}(\mathcal{A}) + q \cdot \mathrm{poly}(\lambda, n),$$

where $\mathrm{poly}(\lambda, n)$ is independent of $\mathsf{Time}(\mathcal{A})$.

We compare our scheme with prior constructions in Figure 1. Applying the Naor transform, we also obtain a d-LIN-based signature scheme with constant-size signatures and security loss independent of the number of signature queries. This yields an alternative construction for an analogous result in [17].

Reference	|MPK|	security loss	additive overhead	assumption
BB1 [5]	$\mathcal{O}(1)$	$\mathcal{O}(2^n)$	$q \cdot \text{poly}(\lambda, n)$	DBDH
Waters [28]	$\mathcal{O}(n)$	$\mathcal{O}(qn)$	$q^2 \epsilon^{-2} \cdot \text{poly}(\lambda, n)$	DBDH
Gentry [15]	$\mathcal{O}(1)$	$\mathcal{O}(1)$	$q^2 \cdot \text{poly}(\lambda, n)$	q-ABDHE
BR [4]	$\mathcal{O}(n)$	$\mathcal{O}(qn/\epsilon)$	$q \cdot \text{poly}(\lambda, n)$	DBDH
LW[29, 22, 20]	$\mathcal{O}(1)$	$\mathcal{O}(q)$	$q \cdot \text{poly}(\lambda, n)$	DLIN or composite
Ours	$\mathcal{O}(n)$	$\mathcal{O}(n)$	$q \cdot \text{poly}(\lambda, n)$	DLIN or composite
	$\mathcal{O}(d^2 n)$	$\mathcal{O}(n)$	$d^2 q \cdot \text{poly}(\lambda, n)$	d-LIN

Fig. 1. Comparison amongst IBE schemes, where $\{0, 1\}^n$ is the identity space, q is the number of adversary's key queries, and ϵ is the adversary's advantage. In all of these constructions, $|\text{SK}| = |\text{CT}| = \mathcal{O}(1)$.

Our Approach. The inspiration for our construction comes from a recent connection between predicate encryption and one-time symmetric-key primitives [30] — namely one-time MACs in the case of IBE — via dual system encryption [29]. Our key observation is to extend this connection to "reusable MACs", namely that if we start with an appropriate pseudorandom function (PRF) with security loss L, we may derive an IBE with the security loss $\mathcal{O}(L)$. More concretely, we begin with the Naor-Reingold DDH-based PRF [24] which has security loss n for input domain $\{0, 1\}^n$, and obtain a fully secure IBE with security loss $\mathcal{O}(n)$ via a novel variant of the dual system encryption methodology. Our IBE scheme is essentially that obtained by embedding Waters' fully secure IBE based on DBDH [28] into composite-order groups, and then converting this to a prime-order scheme following [10, 25, 20, 14] (along with some new technical ideas). Here, we exploit the fact that the Waters' IBE and the Naor-Reingold PRF share a similar algebraic structure based on bit-by-bit encoding of the identity and PRF input respectively.

1.2   Technical Overview

We provide a more technical overview of our main results, starting with the proof idea and then the construction. Here, we assume some familiarity with prior works.

Proof Idea. Our security proof combines Waters' dual system encryption methodology [29] with ideas from the analysis of the Naor-Reingold PRF. In a dual system encryption scheme [29], there are two types of keys and ciphertexts: normal and semi-functional. A key will decrypt a ciphertext properly unless both the key and the ciphertext are semi-functional, in which case decryption will fail with overwhelming probability. The normal keys and ciphertexts are used in the real system, and keys are gradually introduced in the hybrid security proof, one at a time. Ultimately, we arrive at a security game in which the simulator only has to produce semi-functional objects and security can be proved directly. In all prior instantiations of this methodology, the semi-functional keys are introduced one at a time. As a result, we require q hybrid games to switch all of the keys from normal to semi-functional, leading to an $\Omega(q)$ security loss, since each step requires a computational assumption.

We deviate from the prior paradigm by using only n hybrid games, iterating over the bits in the bit-by-bit encoding of the identity, as was done in the Naor-Reingold PRF. That is, we introduce n types of semi-functional ciphertexts and keys, where type i objects appear in game i, while gradually increasing the entropy in the semi-functional components in each game. This strategy introduces new challenges specific to the IBE setting, namely that the adversary could potentially use the challenge ciphertext to test whether we have switched from type $i-1$ keys to type i keys. Prior works exploit the fact that we only switch a *single* key in each step, whereas we could be switching up to q keys in each step.

We overcome this difficulty as follows. At step i of the hybrid game, we guess the i'th bit b_i of the challenge identity ID*, and abort if our guess is incorrect. This results in a security loss of 2, which we can afford. If our guess b_i is correct,

- for all identities whose i'th bit equals b_i, the corresponding type $i-1$ and type i object are the same;
- for all other identities, we increase the entropy of the keys going from type $i-1$ to type i (via a tight reduction to a computational assumption).

The first property implies that the adversary cannot use the challenge ciphertext to distinguish between type $i-1$ and type i keys; in the proof, the simulator will not be able to generate type $i-1$ or type i ciphertexts for identities whose i'th bit is different from b_i (c.f. Remark 3 and Section 4.4). Interestingly, decryption capabilities remain unchanged throughout the hybrid games: a type i key for ID* can decrypt a type i ciphertext for ID* (c.f. Remark 5). This is again different from prior instantiations of the dual system encryption methodology where decryption fails for semi-functional objects.

In the final transition, a semi-functional type n object for identity ID has semi-functional component $R_n(\text{ID})$ where R_n is a truly random function. In particular, the semi-functional ciphertext has semi-functional component $R_n(\text{ID}^*)$. Moreover, $R_n(\text{ID}^*)$ is truly random from the adversary's view-point because it only learns SK_{ID} and thus $R_n(\text{ID})$ for ID $\neq$ ID*. We can then argue that the message which is masked by $R_n(\text{ID}^*)$ is information-theoretically hidden.

Construction. To achieve a modular analysis, we introduce a novel notion of nested dual system groups (see Section 3.1 for an overview). Our construction proceeds into two steps: the first builds an (almost) tight IBE from nested dual system groups where we rely on the Naor-Reingold PRF argument and the dual system encryption methodology; the second builds nested dual system groups from d-LIN where we handle all of the intricate linear algebra associated with simulating composite-order groups in prime-order groups from [10, 20] and with achieving a tight reduction via random self-reducibility.

Property	Where it is used	
	nested dual system groups	dual system groups
projective	correctness	correctness
	normal to type 0 (Lemma 1)	normal to semi-functional CT
associative	correctness	correctness
orthogonality	normal to type 0 (Lemma 2)	final transition
non-degeneracy	final transition (Lemma 4)	pseudo-normal to pseudo-SF keys
		final transition
$\mathbb{H}$-subgroup	type $i-1$ to type i (Lemma 3)	key delegation
left subgroup	normal to type 0 (Lemma 1)	normal to semi-functional CT
nested-hiding	type $i-1$ to type i (Lemma 3)	*unavailable*
right subgroup	*unavailable*	normal to pseudo-normal keys
		pseudo-SF to semi-functional keys
parameter-hiding	*unavailable*	pseudo-normal to pseudo-SF keys

Fig. 2. Summary of dual system groups (c.f. Section 3 and Appendix B)

Perspective. In spite of the practical motivation for tight security reductions, we clarify that our contributions are largely of theoretical and conceptual interest. This is because any gain in efficiency from using smaller groups is overwhelmed by the loss from the bit-by-bit encoding of identities. Our work raises the following open problems:

- Can we reduce the size of the public parameters to a constant?
- Can we achieve tight security, namely $L = \mathcal{O}(1)$?

We note that progress on either problem would likely require improving on the Naor-Reingold PRF: namely, reducing respectively the seed length and the security loss to a constant, both of which are long-standing open problems. We also note that the present blow-up in public parameters and security loss arise only in using the Naor-Reingold approach to build an IBE from nested dual system groups; our instantiation of nested dual system groups do achieve tight security.

1.3 Additional Results

As a pre-cursor to nested dual system groups, we introduce a basic notion of *dual system groups*. We present

- a generic construction of compact HIBE from dual system groups similar to the Lewko-Waters scheme over composite-order groups [22]; and
- instantiations of dual system groups under the d-LIN assumption in prime-order bilinear groups and the subgroup decisional assumption in composite-order

bilinear groups respectively. Along the way, we provide a new randomization and parameter-hiding technique for prime-order groups.

Putting the two together, we obtain a new construction of compact HIBE in prime-order groups, as well as new insights into the structural properties needed for Waters' dual system encryption methodology [29]. We proceed to present an overview of dual system groups, our new techniques for prime-order groups and then an overview of nested dual system groups.

Dual System Groups. Informally, dual system groups contain a triple of groups $(\mathbb{G}, \mathbb{H}, \mathbb{G}_T)$ and a non-generate bilinear map $e : \mathbb{G} \times \mathbb{H} \to \mathbb{G}_T$. For concreteness, we may think of $(\mathbb{G}, \mathbb{H}, \mathbb{G}_T)$ as composite-order bilinear groups. Dual system groups take as input a parameter 1^n (think of n as the depth of the HIBE) and satisfy the following properties:

(subgroup indistinguishability.) There are two computationally indistinguishable ways to sample correlated $(n+1)$-tuples from $\mathbb{G}^{n+1}$: the "normal" distribution, and a higher-entropy distribution with "semi-functional components". An analogous statement holds for $\mathbb{H}^{n+1}$.

(associativity.) For all $(g_0, g_1, \ldots, g_n) \in \mathbb{G}^{n+1}$ and all $(h_0, h_1, \ldots, h_n) \in \mathbb{H}^{n+1}$ drawn from the respective normal distributions, we have that for all $i = 1, \ldots, n$,

$$e(g_0, h_i) = e(g_i, h_0).$$

(parameter-hiding.) Both normal distributions can be efficiently sampled given the public parameters; on the other hand, given only the public parameters, the higher-entropy distributions contain n "units" of information-theoretic entropy (in the semi-functional component), one unit for each of the n elements in the $(n + 1)$-tuple apart from the first.

The key novelty in the framework lies in identifying the role of associativity in the prior instantiations of the dual system encryption methodology in composite-order groups [22].

Instantiation in Prime-Order Groups. We present a new randomization and parameter-hiding technique for prime-order bilinear groups, which we use to instantiate dual system groups. This technique allows us to hide arbitrarily large amounts of entropy while working with a vector space of constant dimensions, whereas prior works require a linear blow-up in dimensions.

To motivate the new technique, we begin with a review of composite-order bilinear groups. Let (G_N, G_T) denote a composite-order bilinear group of order $N = p_1 p_2$ which is the product of two primes, endowed with an efficient bilinear map $e : G_N \times G_N \to G_T$. Let g denote an element of G_N of order p_1. A useful property of composite-order groups, especially in the context of dual system encryption [22, 23], is that we can perform randomization by raising a group element to the power of a random exponent $a \leftarrow_{\text{R}} \mathbb{Z}_N$. This operation satisfy the following useful properties:

(parameter-hiding.) given g, g^a, the quantity $a \pmod{p_2}$ is completely hidden;

(associativity.) for all $u \in G_N$, we have $e(g^a, u) = e(g, u^a)$.

We show how to achieve randomization in the prime-order setting under the d-LIN assumption. Fix a prime-order bilinear group (G, G_T) of order p, endowed with an efficient bilinear map $e : G \times G \to G_T$. Let g denote an element of G of order p. Elements in G_N correspond to elements in G^{d+1} and we consider the bilinear map $e : G^{d+1} \times G^{d+1} \to G_T$ given by $e(g^{\mathbf{x}}, g^{\mathbf{y}}) := e(g, g)^{\mathbf{x}^\top \mathbf{y}}$. Following [25, 14], we pick a random pair of orthogonal basis $(\mathbf{B}, \mathbf{B}^*) \leftarrow_{\mathrm{R}} \mathsf{GL}_{d+1}(\mathbb{Z}_p) \times \mathsf{GL}_{d+1}(\mathbb{Z}_p)$ so that $\mathbf{B}^\top \mathbf{B}^*$ is the identity matrix. We consider the projection maps π_L, π_R that map a $(d+1) \times (d+1)$ matrix to the left d columns and right-most column; they correspond to projecting $a \in \mathbb{Z}_N$ to $a \pmod{p_1}$ and $a \pmod{p_2}$ respectively.

We randomize a basis $(\mathbf{B}, \mathbf{B}^*)$ as follows: pick a random $\mathbf{A} \leftarrow_{\mathrm{R}} \mathbb{Z}_p^{(d+1) \times (d+1)}$ and replace $(\mathbf{B}, \mathbf{B}^*)$ with $(\mathbf{BA}, \mathbf{B}^* \mathbf{A}^\top)$. Observe that this transformation satisfy the following properties similar to those in the composite-order setting:

(parameter-hiding.) given $g^{\pi_L(\mathbf{B})}, g^{\pi_L(\mathbf{BA})}, g^{\pi_L(\mathbf{B}^*)}, g^{\pi_L(\mathbf{B}^* \mathbf{A}^\top)}$, the bottom-right entry of $\mathbf{A}$ is completely hidden;

(associativity.) for all $(\mathbf{B}, \mathbf{B}^*)$ and all $\mathbf{A} \in \mathbb{Z}_p^{(d+1) \times (d+1)}$, we have

$$e(g^{\mathbf{BA}}, g^{\mathbf{B}^*}) = e(g^{\mathbf{B}}, g^{\mathbf{B}^* \mathbf{A}^\top}) \left(= e(g, g)^{\mathbf{A}^\top} \right)$$

where $e(g^{\mathbf{X}}, g^{\mathbf{Y}}) := e(g, g)^{\mathbf{X}^\top \mathbf{Y}}$.

We also establish a subspace indistinguishability assumption similar to those in prior works [26, 20, 12].

Nested Dual System Groups. In nested dual system groups, we require a so-called *nested-hiding* property. Roughly speaking, this property says that it is computationally infeasible to distinguish q samples from some distribution with another; specifically, it allows us to boost the entropy of the semi-functional components. In the instantiation, we will need to establish this property with a tight reduction to some standard assumption. The nested-hiding property allows us to "embed" the Naor-Reingold analysis into the semi-functional space of a dual system encryption scheme. We stress that the nested-hiding property even for $q = 1$ is *qualitatively* different from right subgroup indistinguishability in dual system groups.

We outline the instantiations of dual system groups in the composite-order and prime-order settings:

- The composite-order instantiation is very similar to that as before. We rely on composite-order group whose order is the product of three primes p_1, p_2, p_3. The subgroup G_{p_1} of order p_1 serves as the "normal space" and G_{p_2} of order p_2 serves as the "semi-functional space". We also require a new static, generically secure assumption, which roughly speaking, states that DDH is hard in the G_{p_2} subgroup. Here, we extend the techniques from [24] to establish nested-hiding indistinguishability without losing a factor of q in the security reduction. Our IBE analysis may also be viewed as instantiating the Naor-Reingold PRF in the G_{p_2} subgroup.

- For the prime-order instantiation based on d-LIN, we extend the prior instantiation in several ways. First, we work with $2d \times 2d$ matrices instead of $(d+1) \times (d+1)$ matrices. In both constructions, the first d dimensions serve as the "normal space"; in our construction, we require a d-dimensional semi-functional space instead of a 1-dimensional one so that we may embed the d-LIN assumption into the semi-functional space. Next, we extend the techniques from [24, 21] to establish nested-hiding indistinguishability without losing a factor of q in the security reduction.

Perspective. In developing the framework for dual system groups, we opted to identify the minimal properties needed for the application to dual system encryption in the most basic setting of (H)IBE; we adopted an analogous approach also for nested dual system groups. An alternative approach would have been to maximize the properties satisfied by both the composite-order and prime-order instantiations, with the hope of capturing a larger range of applications. In choosing the minimalist approach, we believe we can gain better insights into how and why dual system encryption works, as well as guide potential lattice-based instantiations. In addition, we wanted the framework to be as concise as possible and the instantiations to be as simple as possible. Nonetheless, the framework remains fairly involved and we hope to see further simplifications in future work.

Organization. We present nested dual system groups in Section 3, our IBE scheme in Section 4 and a self-contained description of our d-LIN-based scheme in Appendix A. For completeness, we included a formal description of dual system group in Appendix B. We defer all other details to the full versions of this paper [11, 10].

2 Preliminaries

Notation. We denote by $s \leftarrow_R S$ the fact that s is picked uniformly at random from a finite set S and by $x, y, z \leftarrow_R S$ that all x, y, z are picked independently and uniformly at random from S. By PPT, we denote a probabilistic polynomial-time algorithm. Throughout, we use 1^λ as the security parameter. We use $\cdot$ to denote multiplication (or group operation) as well as component-wise multiplication. We use lower case boldface to denote (column) vectors over scalars or group elements and upper case boldface to denote vectors of group elements as well as matrices.

Identity-Based Encryption. An IBE scheme consists of four algorithms (Setup, Enc, KeyGen, Dec):

Setup$(1^\lambda, 1^n) \to$ (MPK, MSK). The setup algorithm takes in the security parameter 1^λ and the length parameter 1^n. It outputs public parameters MPK and a master secret key MSK.

Enc(MPK, $\mathbf{x}$, m) $\rightarrow$ CT$_\mathbf{x}$. The encryption algorithm takes in the public parameters MPK, an identity $\mathbf{x}$, and a message m. It outputs a ciphertext CT$_\mathbf{x}$.

KeyGen(MPK, MSK, $\mathbf{y}$) $\rightarrow$ SK$_\mathbf{y}$. The key generation algorithm takes in the public parameters MPK, the master secret key MSK, and an identity $\mathbf{y}$. It outputs a secret key SK$_\mathbf{y}$.

Dec(MPK, SK$_\mathbf{y}$, CT$_\mathbf{x}$) $\rightarrow$ m. The decryption algorithm takes in the public parameters MPK, a secret key SK$_\mathbf{y}$ for an identity $\mathbf{y}$, and a ciphertext CT$_\mathbf{x}$ encrypted under an identity $\mathbf{x}$. It outputs a message m if $\mathbf{x} = \mathbf{y}$.

Correctness. For all (MPK, MSK) $\leftarrow$ Setup(1^λ, 1^n), all identities $\mathbf{x}$, all messages m, all decryption keys SK$_\mathbf{y}$, all $\mathbf{x}$ such that $\mathbf{x} = \mathbf{y}$, we have

$$\Pr[\text{Dec}(\text{MPK}, \text{SK}_\mathbf{y}, \text{Enc}(\text{MPK}, \mathbf{x}, m)) = m] = 1.$$

Security Model. The security game is defined by the following experiment, played by a challenger and an adversary $\mathcal{A}$.

Setup. The challenger runs the setup algorithm to generate (MPK, MSK). It gives MPK to the adversary $\mathcal{A}$.

Phase 1. The adversary $\mathcal{A}$ adaptively requests keys for any identity $\mathbf{y}$ of its choice. The challenger responds with the corresponding secret key SK$_\mathbf{y}$, which it generates by running KeyGen(MPK, MSK, $\mathbf{y}$).

Challenge. The adversary $\mathcal{A}$ submits two messages m_0 and m_1 of equal length and a challenge identity $\mathbf{x}^*$ with the restriction that $\mathbf{x}^*$ is not equal to any identity requested in the previous phase. The challenger picks $\beta \leftarrow_R \{0, 1\}$, and encrypts m_β under $\mathbf{x}^*$ by running the encryption algorithm. It sends the ciphertext to the adversary $\mathcal{A}$.

Phase 2. $\mathcal{A}$ continues to issue key queries for any identity $\mathbf{y}$ as in Phase 1 with the restriction that $\mathbf{y} \neq \mathbf{x}^*$.

Guess. The adversary $\mathcal{A}$ must output a guess β' for β.

The advantage Adv$_\mathcal{A}^{\text{IBE}}(\lambda)$ of an adversary $\mathcal{A}$ is defined to be $| \Pr[\beta' = \beta] - 1/2 |$.

Definition 1. *An IBE scheme is* fully secure *if all PPT adversaries $\mathcal{A}$, Adv$_\mathcal{A}^{\text{IBE}}(\lambda)$ is a negligible function in λ.*

3 Nested Dual System Groups

In this section, we present nested dual system groups, a variant of dual system groups with a notable difference: we require (computational) nested-hiding indistinguishability, in place of (computational) right subgroup indistinguishability and (information-theoretic) parameter-hiding. As noted in the introduction, the nested-hiding property even for $q = 1$ is *qualitatively* different from right subgroup indistinguishability in dual system groups.

3.1 Overview

Informally, nested dual system groups contain a triple of groups $(\mathbb{G}, \mathbb{H}, \mathbb{G}_T)$ and a non-generate bilinear map $e : \mathbb{G} \times \mathbb{H} \to \mathbb{G}_T$. For concreteness, we may think of $(\mathbb{G}, \mathbb{H}, \mathbb{G}_T)$ as composite-order bilinear groups. Nested dual system groups take as input a parameter 1^n and satisfy the following properties:

(left subgroup $\mathbb{G}$.) There are two computationally indistinguishable ways to sample correlated $(n + 1)$-tuples from $\mathbb{G}^{n+1}$: the "normal" distribution, and a higher-entropy distribution with "semi-functional components". We sample the normal distribution using SampG and the semi-functional components using $\widehat{\mathsf{SampG}}$.

(right subgroup $\mathbb{H}$.) There is a single algorithm SampH to sample correlated $(n+1)$-tuples from $\mathbb{H}^{n+1}$. We should think of these tuples as already having semi-functional components, generated by some distinguished element $h^* \in \mathbb{H}$. It is convenient to think of h^* as being orthogonal to each component in the normal distribution over $\mathbb{G}$ (c.f. orthogonality and Remark 1). On the other hand, we require that h^* is *not* orthogonal to the semi-functional components in $\mathbb{G}$ (c.f. non-degeneracy) in order to information-theoretically hide the message in the final transition.

(nested-hiding.) We require a computational assumption over $\mathbb{H}$ which we refer to as *nested-hiding*, namely that for each $i = 1, \dots, n$,

$$(h_0, h_i) \quad \text{and} \quad (h_0, h_i \cdot (h^*)^\gamma)$$

are computationally indistinguishable, where $(h_0, h_1, \dots, h_n)$ is sampled using SampH and γ is a random exponent. In the formal definition, we provide the adversary with q samples from these distributions, and in the instantiations, we provide a tight reduction (independent of q) to a static assumption such as DLIN.

(associativity.) For all $(g_0, g_1, \dots, g_n) \in \mathbb{G}^{n+1}$ and all $(h_0, h_1, \dots, h_n) \in \mathbb{H}^{n+1}$ sampled using SampG and SampH respectively, we have that for all $i = 1, \dots, n$,

$$e(g_0, h_i) = e(g_i, h_0).$$

We require this property for correctness.

3.2 Definitions

Syntax. Nested dual system groups consist of five randomized algorithms given by (SampP, SampGT, SampG, SampH) along with $\widehat{\mathsf{SampG}}$:

SampP($1^\lambda, 1^n$): On input $(1^\lambda, 1^n)$, output public and secret parameters (PP, SP), where:

- PP contains a triple of groups $(\mathbb{G}, \mathbb{H}, \mathbb{G}_T)$ and a non-generate bilinear map $e : \mathbb{G} \times \mathbb{H} \to \mathbb{G}_T$, a linear map μ defined on $\mathbb{H}$, along with some additional parameters used by SampG, SampH;
- given PP, we know $\mathrm{ord}(\mathbb{H})$ (i.e. the order of the group, which is independent of n) and can uniformly sample from $\mathbb{H}$;
- SP contains $h^* \in \mathbb{H}$ (where $h^* \neq 1$), along with some additional parameters used by $\widehat{\mathsf{SampG}}$;

SampGT : $\mathrm{Im}(\mu) \to \mathbb{G}_T$. (As a concrete example, suppose $\mu : \mathbb{H} \to \mathbb{G}_T$ and $\mathrm{Im}(\mu) = \mathbb{G}_T$.)

SampG(PP): Output $\mathbf{g} \in \mathbb{G}^{n+1}$.

SampH(PP): Output $\mathbf{h} \in \mathbb{H}^{n+1}$.

$\widehat{\mathsf{SampG}}$(PP, SP): Output $\hat{\mathbf{g}} \in \mathbb{G}^{n+1}$.

The first four algorithms are used in the actual scheme, whereas the last algorithm is used only in the proof of security. We define SampG_0 to denote the first group element in the output of SampG, and we define $\widehat{\mathsf{SampG}}_0$ analogously.

Correctness. The requirements for correctness are as follows:

(projective.) For all $h \in \mathbb{H}$ and all coin tosses s, we have $\mathsf{SampGT}(\mu(h); s) = e(\mathsf{SampG}_0(\mathrm{PP}; s), h)$.

(associative.) For all

$$(g_0, g_1, \ldots, g_n) \leftarrow \mathsf{SampG}(\mathrm{PP}), \quad (h_0, h_1, \ldots, h_n) \leftarrow \mathsf{SampH}(\mathrm{PP}),$$

and for all $i = 1, \ldots, n$, we have $e(g_0, h_i) = e(g_i, h_0)$.

Security. The requirements for security are as follows (we defer a discussion to the end of this section):

(orthogonality.) $\mu(h^*) = 1$.

(non-degeneracy.) With probability $1 - 2^{-\Omega(\lambda)}$ over $\hat{g}_0 \leftarrow \widehat{\mathsf{SampG}}_0(\mathrm{PP}, \mathrm{SP})$, we have that $e(\hat{g}_0, h^*)^\alpha$ is identically distributed to the uniform distribution over $\mathbb{G}_T$, where $\alpha \leftarrow_{\mathrm{R}} \mathbb{Z}_{\mathrm{ord}(\mathbb{H})}$.

($\mathbb{H}$-subgroup.) The output distribution of SampH(PP) is the uniform distribution over a subgroup of $\mathbb{H}^{n+1}$.

(left subgroup indistinguishability.) For any adversary $\mathcal{A}$, we define the advantage function:

$$\mathsf{Adv}^{\mathrm{LS}}_{\mathcal{A}}(\lambda) := \left| \Pr[\,\mathcal{A}(\mathrm{PP}, \boxed{\mathbf{g}}) = 1\,] - \Pr[\,\mathcal{A}(\mathrm{PP}, \boxed{\mathbf{g} \cdot \hat{\mathbf{g}}}) = 1\,] \right|$$

where

$$(\text{PP}, \text{SP}) \leftarrow \mathsf{SampP}(1^\lambda, 1^n);$$

$$\mathbf{g} \leftarrow \mathsf{SampG}(\text{PP}); \quad \hat{\mathbf{g}} \leftarrow \widehat{\mathsf{SampG}}(\text{PP}, \text{SP}).$$

For any $\mathbf{g} = (g_0, \ldots, g_n) \in \mathbb{G}^{n+1}$, and any $i \in [n]$, we use $\mathbf{g}_{-i}$ to denote $(g_0, \ldots, g_{i-1}, g_{i+1}, \ldots, g_n) \in \mathbb{G}^n$.

(nested-hiding indistinguishability.) For any adversary $\mathcal{A}$, we define the advantage function:

$$\mathsf{Adv}_{\mathcal{A}}^{\mathsf{NS}}(\lambda, q) := \max_{i \in [n]} \big| \Pr[\, \mathcal{A}(\text{PP}, h^*, \hat{\mathbf{g}}_{-i}, \boxed{\mathbf{h}^1, \ldots, \mathbf{h}^q}\,) = 1\,]$$

$$- \Pr[\, \mathcal{A}(\text{PP}, h^*, \hat{\mathbf{g}}_{-i}, \boxed{\mathbf{h}'^1, \ldots, \mathbf{h}'^q}\,) = 1\,] \big|$$

where

$$(\text{PP}, \text{SP}) \leftarrow \mathsf{SampP}(1^\lambda, 1^n);$$

$$\hat{\mathbf{g}} \leftarrow \widehat{\mathsf{SampG}}(\text{PP}, \text{SP});$$

$$\mathbf{h}^j := (h_{0,j}, h_{1,j}, \ldots, \boxed{h_{i,j}}, \ldots, h_{n,j}) \leftarrow \mathsf{SampH}(\text{PP}), \; j = 1, \ldots, q;$$

$$\mathbf{h}'^j := (h_{0,j}, h_{1,j}, \ldots, \boxed{h_{i,j} \cdot (h^*)^{\gamma_j}}, \ldots, h_{n,j}), \; \gamma_j \leftarrow_{\mathsf{R}} \mathbb{Z}_{\mathrm{ord}(\mathbb{H})}, \; j = 1, \ldots, q.$$

Discussion. We provide additional justification and discussion on the preceding security properties.

Remark 1 (orthogonality). We may deduce from $\mu(h^*) = 1$ that $e(g_0, h^*) = 1$ for all $g_0 = \mathsf{SampG}_0(\text{PP}; s)$: for all $\gamma \in \{0, 1\}$,

$$\begin{aligned} e(g_0, (h^*)^\gamma) &= \mathsf{SampGT}(\mu((h^*)^\gamma); s) && \text{(by } \textit{projective}) \\ &= \mathsf{SampGT}(\mu(h^*)^\gamma; s) && \text{(by linearity of } \mu) \\ &= \mathsf{SampGT}(1; s) && \text{(by } \textit{orthogonality}) \end{aligned}$$

Thus, we have $e(g_0, h^*) = e(g_0, 1) = 1$. For the instantiation from composite-order groups, h^* is orthogonal to each element in the output of SampG, that is,

$$e(g_0, h^*) = e(g_1, h^*) = \cdots = e(g_n, h^*) = 1$$

for all $(g_0, g_1, \ldots, g_n) \leftarrow \mathsf{SampG}(\text{PP})$. On the other hand, for the instantiation from prime-order groups, h^* is in general not orthogonal to $g_1, \ldots, g_n$.

Remark 2 ($\mathbb{H}$-subgroup). We rely on $\mathbb{H}$-subgroup to re-randomize the secret keys in the proof of security for queries that share the same i-bit prefix; see Section 4.4 case 3.

Remark 3 (indistinguishability). Observe that in left subgroup indistinguishability, the distinguisher does not get h^*; otherwise, it is possible to distinguish between the two distributions using orthogonality. It is also crucial that for nested-hiding, the distinguisher gets $\hat{\mathbf{g}}_{-i}$ and not $\hat{\mathbf{g}} := (\hat{g}_0, \hat{g}_1, \ldots, \hat{g}_n)$. (Looking ahead to the proof in Section 4.4, not having $\hat{\mathbf{g}}$ means that the simulator cannot generate ciphertexts to distinguish between Type $i-1$ and Type i secret keys.) Otherwise, given $\hat{g}_i$, it is possible to distinguish between $\mathbf{h}^j$ and $\mathbf{h}'^j$ by using the relation:

$$e(g_0 \cdot \hat{g}_0, h_{i,j}) = e(g_i \cdot \hat{g}_i, h_{0,j}).$$

This relation follows from associative and left subgroup indistinguishability.

4   (Almost) Tight IBE from Nested Dual System Groups

We provide a construction of an IBE scheme from nested dual system groups where the ciphertext comprises two group elements in $\mathbb{G}$ and one in $\mathbb{G}_T$.

Overview. We begin with an informal overview of the scheme. Fix a bilinear group with a pairing $e : G \times G \to G_T$. The starting point of our scheme is the following variant of Waters' IBE [28] with identity space $\{0,1\}^n$:

$$\text{MPK} := (g, u_1, \ldots, u_{2n}, e(g,g)^\alpha)$$

$$\text{CT}_{\mathbf{x}} := (g^s, (\prod_{k=1}^{n} u_{2k-x_k})^s, e(g,g)^{\alpha s} \cdot m)$$

$$\text{SK}_{\mathbf{y}} := (g^r, \text{MSK} \cdot (\prod_{k=1}^{n} u_{2k-y_k})^r)$$

Note that MPK contains $2n + 1$ group elements in G, which we will generate using $\mathsf{SampP}(1^\lambda, \boxed{1^{2n}})$. We will use $\mathsf{SampG}(\text{PP})$ to generate the terms $(g^s, u_1^s, \ldots, u_{2n}^s)$ in the ciphertext, and $\mathsf{SampH}(\text{PP})$ to generate the terms $(g^r, u_1^r, \ldots, u_{2n}^r)$ in the secret key.

4.1   Construction

Let $\{0,1\}^n$ be the identity space.

- Setup($1^\lambda, 1^n$): On input length parameter 1^n, first sample

$$(\text{PP}, \text{SP}) \leftarrow \mathsf{SampP}(1^\lambda, 1^{2n}).$$

Pick MSK $\leftarrow_{\text{R}} \mathbb{H}$ and output the master public and secret key pair

$$\text{MPK} := (\ \text{PP},\ \mu(\text{MSK})\ ) \quad \text{and} \quad \text{MSK}.$$

- Enc(MPK, $\mathbf{x}$, m): On input an identity $\mathbf{x} := (x_1, \ldots, x_n) \in \{0,1\}^n$ and $m \in \mathbb{G}_T$, sample

$$(g_0, g_1, \ldots, g_{2n}) \leftarrow \mathsf{SampG}(\text{PP}; s), \quad g_T' \leftarrow \mathsf{SampGT}(\mu(\text{MSK}); s)$$

and output

$$\text{CT}_{\mathbf{x}} := (\ C_0 := g_0,\ C_1 := g_{2-x_1} \cdots g_{2n-x_n},\ C_2 := g_T' \cdot m\ ) \in (\mathbb{G})^2 \times \mathbb{G}_T.$$

– KeyGen($\text{MPK}, \text{MSK}, \mathbf{y}$): On input an identity $\mathbf{y} \in \{0,1\}^n$, sample

$$(h_0, h_1, \ldots, h_{2n}) \leftarrow \mathsf{SampH}(\text{PP})$$

and output

$$\text{SK}_{\mathbf{y}} := (\ K_0 := h_0, \ K_1 := \text{MSK} \cdot h_{2-y_1} \cdots h_{2n-y_n}\ ) \in (\mathbb{H})^2.$$

– Dec($\text{MPK}, \text{SK}_{\mathbf{y}}, \text{CT}_{\mathbf{x}}$): If $\mathbf{x} = \mathbf{y}$, compute

$$e(g_0, \text{MSK}) \leftarrow e(C_0, K_1)/e(C_1, K_0)$$

and recover the message as

$$m \leftarrow C_2 \cdot e(g_0, \text{MSK})^{-1} \in \mathbb{G}_T.$$

Correctness. Fix $\mathbf{x} := (x_1, \ldots, x_n) \in \{0,1\}^n$, observe that

$$e(C_0, K_1)/e(C_1, K_0)$$
$$= e(g_0, \text{MSK} \cdot h_{2-x_1} \cdots h_{2n-x_n}) \cdot e(g_{2-x_1} \cdots g_{2n-x_n}, h_0)^{-1}$$
$$= e(g_0, \text{MSK}) \cdot \Big(e(g_0, h_{2-x_1}) \cdots e(g_0, h_{2n-x_n}) \Big) \cdot \Big(e(g_{2-x_1}, h_0) \cdots e(g_{2n-x_n}, h_0) \Big)^{-1}$$
$$= e(g_0, \text{MSK})$$

where the last equality relies on *associative*, namely, $e(g_0, h_{2i-x_i}) = e(g_{2i-x_i}, h_0)$. In addition, by *projective*, we have $g_T' = e(g_0, \text{MSK})$. Correctness follows readily.

4.2 Proof of Security

We prove the following theorem:

Theorem 2. *Under the left subgroup and nested-hiding indistinguishability (described in Section 3), our IBE scheme in Section 4.1 is fully secure (in the sense of Definition 1). More precisely, for any adversary $\mathcal{A}$ that makes at most q key queries against the IBE scheme, there exist adversaries $\mathcal{B}_1, \mathcal{B}_2$ such that:*

$$\mathsf{Adv}^{\text{IBE}}_{\mathcal{A}}(\lambda) \le \mathsf{Adv}^{\text{LS}}_{\mathcal{B}_1}(\lambda) + 2n \cdot \mathsf{Adv}^{\text{NS}}_{\mathcal{B}_2}(\lambda, q) + 2^{-\Omega(\lambda)}$$

and

$$\max\{\mathsf{Time}(\mathcal{B}_1), \mathsf{Time}(\mathcal{B}_2)\} \approx \mathsf{Time}(\mathcal{A}) + q \cdot \mathrm{poly}(\lambda, n),$$

where $\mathrm{poly}(\lambda, n)$ is independent of $\mathsf{Time}(\mathcal{A})$.

Remark 4. In our instantiations of nested dual system groups, the quantity $\mathsf{Adv}^{\text{NS}}_{\mathcal{B}_2}(\lambda, q)$ will be related to the advantage function corresponding to some static assumption, with a constant overhead independent of q. Putting the two together, this means that $\mathsf{Adv}^{\text{IBE}}_{\mathcal{A}}(\lambda)$ is independent of q, as stated in Theorem 1.

The proof follows via a series of games, summarized in Figure 3. To describe the games, we must first define semi-functional keys and ciphertexts. Following [10, 30], we first define two auxiliary algorithms, and define the semi-functional distributions via these auxiliary algorithms.

Auxiliary Algorithms. We consider the following algorithms:

$\widehat{\mathsf{Enc}}(\mathrm{PP}, \mathbf{x}, m; \mathrm{MSK}', \mathbf{t})$: On input $\mathbf{x} := (x_1, \ldots, x_n) \in \{0,1\}^n$, $m \in \mathbb{G}_T$, $\mathrm{MSK}' \in \mathbb{H}$, and $\mathbf{t} := (T_0, T_1, \ldots, T_{2n}) \in \mathbb{G}^{2n+1}$, output

$$\mathrm{CT}_{\mathbf{x}} := \left(T_0, \ \prod_{k=1}^{n} T_{2k-x_k}, \ e(T_0, \mathrm{MSK}') \cdot m \right).$$

$\widehat{\mathsf{KeyGen}}(\mathrm{PP}, \mathrm{MSK}', \mathbf{y}; \mathbf{t})$: On input $\mathrm{MSK}' \in \mathbb{H}$, $\mathbf{y} := (y_1, \ldots, y_n) \in \{0,1\}^n$, and $\mathbf{t} := (T_0, T_1, \ldots, T_{2n}) \in \mathbb{H}^{2n+1}$, output

$$\mathrm{SK}_{\mathbf{y}} := \left(T_0, \ \mathrm{MSK}' \cdot \prod_{k=1}^{n} T_{2k-y_k} \right).$$

Auxiliary Distributions. For $i = 0, 1, \ldots, n$, we pick a random function $R_i : \{0,1\}^i \rightarrow \langle h^* \rangle$ (we use $\{0,1\}^0$ to denote the singleton set containing just the empty string ε). More concretely, given (PP, h^*), we sample the function R_i by first choosing a random function $R_i' : \{0,1\}^i \rightarrow \mathbb{Z}_{\mathrm{ord}(\mathbb{H})}$ (via lazy sampling), and define $R_i(x) := (h^*)^{R_i'(x)}$ for all $x \in \{0,1\}^i$.

Pseudo-normal ciphertext.

$$\widehat{\mathsf{Enc}}(\mathrm{PP}, \mathbf{x}, m; \mathrm{MSK}, \boxed{\mathbf{g} \cdot \hat{\mathbf{g}}}),$$

where $\mathbf{g} \leftarrow \mathsf{SampG}(\mathrm{PP})$ and $\boxed{\hat{\mathbf{g}} \leftarrow \widehat{\mathsf{SampG}}(\mathrm{PP}, \mathrm{SP})}$; we can also write this distribution more explicitly as

$$\left(g_0 \cdot \hat{g}_0, \ \prod_{k=1}^{n} (g_{2k-x_k} \cdot \hat{g}_{2k-x_k}), \ e(g_0 \cdot \hat{g}_0, \mathrm{MSK}) \cdot m \right),$$

where $(g_0, g_1, \ldots, g_{2n}) \leftarrow \mathsf{SampG}(\mathrm{PP})$ and $(\hat{g}_0, \hat{g}_1, \ldots, \hat{g}_{2n}) \leftarrow \widehat{\mathsf{SampG}}(\mathrm{PP}, \mathrm{SP})$.

Semi-functional ciphertext type i (for $i = 0, 1, \ldots, n$).

$$\widehat{\mathsf{Enc}}(\mathrm{PP}, \mathbf{x}, m; \boxed{\mathrm{MSK} \cdot R_i(\mathbf{x}|_i)}, \mathbf{g} \cdot \hat{\mathbf{g}}),$$

where $\mathbf{g} \leftarrow \mathsf{SampG}(\mathrm{PP})$ and $\hat{\mathbf{g}} \leftarrow \widehat{\mathsf{SampG}}(\mathrm{PP}, \mathrm{SP})$ and $\mathbf{x}|_i$ denotes the i-bit prefix of $\mathbf{x}$; we can also write this distribution more explicitly as

$$\left(g_0 \cdot \hat{g}_0, \ \prod_{k=1}^{n} (g_{2k-x_k} \cdot \hat{g}_{2k-x_k}), \ e(g_0 \cdot \hat{g}_0, \mathrm{MSK} \cdot R_i(\mathbf{x}|_i)) \cdot m \right),$$

where $(g_0, g_1, \ldots, g_{2n}) \leftarrow \mathsf{SampG}(\mathrm{PP})$ and $(\hat{g}_0, \hat{g}_1, \ldots, \hat{g}_{2n}) \leftarrow \widehat{\mathsf{SampG}}(\mathrm{PP}, \mathrm{SP})$.

Game	Ciphertext $\mathrm{CT}_{\mathbf{x}^*}$	Secret Key $\mathrm{SK}_{\mathbf{y}}$
0	$\mathsf{Enc}(\mathrm{MPK}, \mathbf{x}^*, m_\beta)$	$\mathsf{KeyGen}(\mathrm{MPK}, \mathrm{MSK}, \mathbf{y})$
	$(g_0, \prod g_{2k-x_k}, e(g_0, \mathrm{MSK}) \cdot m_\beta)$	$(h_0, \mathrm{MSK} \cdot \prod h_{2k-y_k})$
1	$\widehat{\mathsf{Enc}}(\mathrm{PP}, \mathbf{x}^*, m_\beta; \mathrm{MSK}, \boxed{\mathbf{g} \cdot \hat{\mathbf{g}}})$	$\widehat{\mathsf{KeyGen}}(\mathrm{PP}, \mathrm{MSK}, \mathbf{y}; \mathbf{h})$
	$(g_0\hat{g}_0, \prod(g_{2k-x_k}\hat{g}_{2k-x_k}), e(g_0\hat{g}_0, \mathrm{MSK}) \cdot m_\beta)$	$(—, —)$
2,i	$\widehat{\mathsf{Enc}}(\mathrm{PP}, \mathbf{x}^*, m_\beta; \boxed{\mathrm{MSK} \cdot R_i(\mathbf{x}^*\vert_i)}, \mathbf{g} \cdot \hat{\mathbf{g}})$	$\widehat{\mathsf{KeyGen}}(\mathrm{PP}, \boxed{\mathrm{MSK} \cdot R_i(\mathbf{y}\vert_i)}, \mathbf{y}; \mathbf{h})$
	$(—, —, e(g_0\hat{g}_0, \mathrm{MSK} \cdot R_i(\mathbf{x}^*\vert_i)) \cdot m_\beta)$	$(—, \mathrm{MSK} \cdot R_i(\mathbf{y}\vert_i) \cdot \prod h_{2k-y_k})$
3	$\widehat{\mathsf{Enc}}(\mathrm{PP}, \mathbf{x}^*, \boxed{\mathrm{random}}; \mathrm{MSK} \cdot R_n(\mathbf{x}^*), \mathbf{g} \cdot \hat{\mathbf{g}})$	$\widehat{\mathsf{KeyGen}}(\mathrm{PP}, \mathrm{MSK} \cdot R_n(\mathbf{y}), \mathbf{y}; \mathbf{h})$
	$(—, —, e(g_0\hat{g}_0, \mathrm{MSK} \cdot R_n(\mathbf{x}^*)) \cdot \mathrm{random})$	$(—, \mathrm{MSK} \cdot R_n(\mathbf{y}) \cdot \prod h_{2k-y_k})$

Fig. 3. Sequence of games, where we drew a box to highlight the differences between each game and the preceding one, a dash (—) means the same as in the previous game. Recall that $R_i : \{0,1\}^i \to \langle h^* \rangle$ is a random function. Here, the product Π denotes $\Pi_{k=1}^n$. We transition from Game_0 to Game_1 and from $\mathsf{Game}_{2,i-1}$ to $\mathsf{Game}_{2,i}$ using a computational argument via left subgroup and nested-hiding respectively; for the remaining transitions, we use a statistical argument via orthogonality and non-degeneracy.

Semi-functional secret key type i (for $i = 0, 1, \dots, n$).

$$\widehat{\mathsf{KeyGen}}(\mathrm{PP}, \boxed{\mathrm{MSK} \cdot R_i(\mathbf{y}\vert_i)}, \mathbf{y}; \mathbf{h}),$$

where a fresh $\mathbf{h} \leftarrow \mathsf{SampH}(\mathrm{PP})$ is chosen for each secret key; we can also write this distribution more explicitly as

$$\left(h_0, \ \mathrm{MSK} \cdot R_i(\mathbf{x}\vert_i) \cdot \prod_{k=1}^{n} h_{2k-y_k} \right)$$

where $(h_0, h_1, \dots, h_{2n}) \leftarrow \mathsf{SampH}(\mathrm{PP})$.

Remark 5 (decryption capabilities). As noted in the introduction, decryption capabilities remain the same through the hybrid games. Observe that a type i secret key for $\mathbf{x}^*$ can decrypt a type i ciphertext for $\mathbf{x}^*$ since they share $R_i(\mathbf{x}^*\vert_i)$. In addition, a type i secret key for $\mathbf{x}^*$ can decrypt a normal ciphertext for $\mathbf{x}^*$ because $e(g_0, R_i(\mathbf{x}^*\vert_i)) = 1$, which follows readily from $R_i(\mathbf{x}^*\vert_i) \in \langle h^* \rangle$ and $e(g_0, h^*) = 1$ (see Remark 1).

Game Sequence. We present a series of games. We write $\mathsf{Adv}_{\mathrm{xx}}(\lambda)$ to denote the advantage of $\mathcal{A}$ in $\mathsf{Game}_{\mathrm{xx}}$.

- Game_0: is the real security game (c.f. Section 2).
- Game_1: is the same as Game_0 except that the challenge ciphertext is pseudo-normal.
- $\mathsf{Game}_{2,i}$ for i from 0 to n, $\mathsf{Game}_{2,i}$ is the same as Game_1 except that the challenge ciphertext and all secret keys are of type i.

- Game$_3$: is the same as Game$_{2,n}$, except that the challenge ciphertext is a semi-functional encryption of a random message in $\mathbb{G}_T$.

In Game$_3$, the view of the adversary is statistically independent of the challenge bit β. Hence, $\mathsf{Adv}_3(\lambda) = 0$. We complete the proof by establishing the following sequence of lemmas.

4.3 Normal to Pseudo-normal to Type 0

Lemma 1 (Game$_0$ to Game$_1$). *For any adversary $\mathcal{A}$ that makes at most q key queries, there exists an adversary $\mathcal{B}_1$ such that:*

$$|\mathsf{Adv}_0(\lambda) - \mathsf{Adv}_1(\lambda)| \leq \mathsf{Adv}_{\mathcal{B}_1}^{\mathsf{LS}}(\lambda),$$

and $\mathsf{Time}(\mathcal{B}_1) \approx \mathsf{Time}(\mathcal{A}) + q \cdot \mathrm{poly}(\lambda, n)$ *where* $\mathrm{poly}(\lambda, n)$ *is independent of* $\mathsf{Time}(\mathcal{A})$.

Proof. The adversary $\mathcal{B}_1$ gets as input

$$(\mathrm{PP}, \mathbf{t}),$$

where $\mathbf{t}$ is either $\mathbf{g}$ or $\mathbf{g} \cdot \hat{\mathbf{g}}$ and

$$\mathbf{g} \leftarrow \mathsf{SampG}(\mathrm{PP}), \quad \hat{\mathbf{g}} \leftarrow \widehat{\mathsf{SampG}}(\mathrm{PP}, \mathrm{SP}),$$

and proceeds as follows:

Setup. Pick $\mathrm{MSK} \leftarrow_{\mathrm{R}} \mathbb{H}$ and output

$$\mathrm{MPK} := (\ \mathrm{PP}, \ \mu(\mathrm{MSK})\ ).$$

Key Queries. On input the j'th secret key query $\mathbf{y}$, output

$$\mathrm{SK}_{\mathbf{y}} \leftarrow \widehat{\mathsf{KeyGen}}(\mathrm{PP}, \mathrm{MSK}, \mathbf{y}; \mathsf{SampH}(\mathrm{PP})).$$

Ciphertext. Upon receiving a challenge identity $\mathbf{x}^*$ and two equal length messages m_0, m_1, pick $\beta \leftarrow_{\mathrm{R}} \{0, 1\}$ and output

$$\mathrm{CT}_{\mathbf{x}^*} \leftarrow \widehat{\mathsf{Enc}}(\mathrm{PP}, \mathbf{x}^*, m_\beta; \mathrm{MSK}, \mathbf{t}).$$

Guess. When $\mathcal{A}$ halts with output β', $\mathcal{B}_1$ outputs 1 if $\beta' = \beta$ and 0 otherwise.

Observe that when $\mathbf{t} = \mathbf{g}$, $\mathrm{CT}_{\mathbf{x}^*}$ is properly distributed as $\mathsf{Enc}(\mathrm{MPK}, \mathbf{x}^*, m_\beta)$ from *projective*, the output is identical to that in Game$_0$; and when $\mathbf{t} = \mathbf{g} \cdot \hat{\mathbf{g}}$, the output is identical to that in Game$_1$. We may therefore conclude that: $|\mathsf{Adv}_0(\lambda) - \mathsf{Adv}_1(\lambda)| \leq \mathsf{Adv}_{\mathcal{B}_1}^{\mathsf{LS}}(\lambda)$. $\square$

Lemma 2 (Game$_1$ to Game$_{2,0}$). *For any adversary $\mathcal{A}$,*

$$\mathsf{Adv}_1(\lambda) = \mathsf{Adv}_{2,0}(\lambda)$$

Proof. Observe that MSK and $\mathrm{MSK} \cdot R_0(\varepsilon)$ (where $\mathrm{MSK} \leftarrow_{\mathrm{R}} \mathbb{H}$) are identically distributed, so we may replace MSK in Game$_1$ by $\mathrm{MSK} \cdot R_0(\varepsilon)$. The resulting distribution is identically distributed to that in Game$_{2,0}$ except we use $\mu(\mathrm{MSK} \cdot R_0(\varepsilon))$ instead of $\mu(\mathrm{MSK})$ in MPK. Now, by *orthogonality*, these two quantities are in fact equal. $\square$

4.4 Type $i-1$ to Type i

We begin with an informal overview of our proof strategy. For simplicity, suppose the adversary only requests secret keys for two identities $\mathbf{y}_0$ and $\mathbf{y}_1$ that differ only in the i'th bit, that is,

$$\mathbf{y}_0 = (y_1, \ldots, y_{i-1}, \boxed{0}, y_{i+1}, \ldots, y_n) \quad \text{and} \quad \mathbf{y}_1 = (y_1, \ldots, y_{i-1}, \boxed{1}, y_{i+1}, \ldots, y_n)$$

Recall that Type $i-1$ secret keys for $\mathbf{y}_0$ and $\mathbf{y}_1$ are of the form:

$$\text{SK}_{\mathbf{y}_0} = \left(h_0, \text{MSK} \cdot \boxed{R_{i-1}(y_1, \ldots, y_{i-1})} \cdot h_{2-y_1} \cdots \boxed{h_{2i}} \cdots h_{2n-y_n} \right) \quad \text{and}$$

$$\text{SK}_{\mathbf{y}_1} = \left(h_0, \text{MSK} \cdot \boxed{R_{i-1}(y_1, \ldots, y_{i-1})} \cdot h_{2-y_1} \cdots \boxed{h_{2i-1}} \cdots h_{2n-y_n} \right)$$

whereas Type i secret keys for $\mathbf{y}_0$ and $\mathbf{y}_1$ are of the form:

$$\text{SK}_{\mathbf{y}_0} = \left(h_0, \text{MSK} \cdot \boxed{R_i(y_1, \ldots, y_{i-1}, 0)} \cdot h_{2-y_1} \cdots \boxed{h_{2i}} \cdots h_{2n-y_n} \right) \quad \text{and}$$

$$\text{SK}_{\mathbf{y}_1} = \left(h_0, \text{MSK} \cdot \boxed{R_i(y_1, \ldots, y_{i-1}, 1)} \cdot h_{2-y_1} \cdots \boxed{h_{2i-1}} \cdots h_{2n-y_n} \right)$$

In order to show that Type $i-1$ and Type i secret keys for $\mathbf{y}_0$ and $\mathbf{y}_1$ are indistinguishable, it suffices to show that

$$(R_{i-1}(y_1, \ldots, y_{i-1}) \cdot h_{2i}, R_{i-1}(y_1, \ldots, y_{i-1}) \cdot h_{2i-1}) \quad \text{and}$$
$$(R_i(y_1, \ldots, y_{i-1}, 0) \cdot h_{2i}, R_i(y_1, \ldots, y_{i-1}, 1) \cdot h_{2i-1})$$

are computationally indistinguishable (*).

Now, suppose for simplicity that the i'th bit of the identity $\mathbf{x}^*$ for challenge ciphertext is 1. Then, *nested-hiding indistinguishability* with index $2i$ tells us that

$$h_{2i} \quad \text{and} \quad h_{2i} \cdot (h^*)^{\gamma}$$

are computationally indistinguishable, where $\gamma \leftarrow_{\text{R}} \mathbb{Z}_{|\mathbb{H}|}$. Moreover, this holds even if the distinguisher is given $\hat{g}_{-2i}$, which we will need to simulate the semi-functional ciphertext for $\mathbf{x}^*$. (On the other hand, given only $\hat{g}_{-2i}$, we cannot simulate semi-functional ciphertext for identities whose i'th bit is 0.) This means that

$$(R_{i-1}(y_1, \ldots, y_{i-1}) \cdot h_{2i}, R_{i-1}(y_1, \ldots, y_{i-1}) \cdot h_{2i-1}) \quad \text{and}$$
$$(R_{i-1}(y_1, \ldots, y_{i-1}) \cdot h_{2i} \cdot (h^*)^{\gamma}, R_{i-1}(y_1, \ldots, y_{i-1}) \cdot h_{2i-1})$$

are computationally indistinguishable, even given the semi-functional ciphertext for $\mathbf{x}^*$.

To achieve (*), we can then implicitly set:

$$R_i(y_1, \ldots, y_{i-1}, 0) := R_{i-1}(y_1, \ldots, y_{i-1}) \cdot (h^*)^{\gamma} \quad \text{and}$$
$$R_i(y_1, \ldots, y_{i-1}, 1) := R_{i-1}(y_1, \ldots, y_{i-1})$$

This corresponds to Case 2 and Case 1 below respectively.

More generally, we guess at random the i'th bit of $\mathbf{x}^*$ to be b_i and use nested-hiding indistinguishability with index $2i - \overline{b_i}$. In addition, we need to handle q keys and not just two keys, along with an additional complication arising from the fact that multiple queries may share the same i-bit prefix (see Case 3 below).

Lemma 3 ($\mathsf{Game}_{2,i-1}$ **to** $\mathsf{Game}_{2,i}$). *For $i = 1, \ldots, n$, for any adversary $\mathcal{A}$ that makes at most q key queries, there exists an adversary $\mathcal{B}_2$ such that:*

$$|\mathsf{Adv}_{2,i-1}(\lambda) - \mathsf{Adv}_{2,i}(\lambda)| \leq 2\mathsf{Adv}_{\mathcal{B}_2}^{\mathsf{NS}}(\lambda, q),$$

and $\mathsf{Time}(\mathcal{B}_2) \approx \mathsf{Time}(\mathcal{A}) + q \cdot \mathrm{poly}(\lambda, n)$ *where* $\mathrm{poly}(\lambda, n)$ *is independent of* $\mathsf{Time}(\mathcal{A})$.

Proof. On input $i \in [n]$, $\mathcal{B}_2$ picks a random bit $b_i \leftarrow_{\mathsf{R}} \{0, 1\}$ (that is, it guesses the i'th bit of the challenge identity $\mathbf{x}^*$) and requests nested-hiding instantiation for index $2i - \overline{b_i}$. The adversary $\mathcal{B}_2$ gets as input

$$\left(\mathsf{PP}, h^*, \hat{\mathbf{g}}_{-(2i-\overline{b_i})}, \mathbf{t}_1, \ldots, \mathbf{t}_q\right),$$

where $(\mathbf{t}^1, \ldots, \mathbf{t}^q)$ is either $(\mathbf{h}^1, \ldots, \mathbf{h}^q)$ or $(\mathbf{h}'^1, \ldots, \mathbf{h}'^q)$ and

$$\mathbf{h}^j := (h_{0,j}, h_{1,j}, \ldots, h_{2n,j}) \leftarrow \mathsf{SampH}(\mathsf{PP}),$$
$$\mathbf{h}'^j := (h_{0,j}, h_{1,j}, \ldots, h_{2i-\overline{b_i},j} \cdot (h^*)^{\gamma_j}, \ldots, h_{2n,j}),$$

and proceeds as follows:

Setup. Pick $\mathsf{MSK} \leftarrow_{\mathsf{R}} \mathbb{H}$, and output

$$\mathsf{MPK} := (\ \mathsf{PP}, \ \mu(\mathsf{MSK})\ ).$$

Programming R_{i-1}, R_i. Pick a random function $\tilde{R}_{i-1} : \{0, 1\}^{i-1} \to \langle h^* \rangle$ (which we use to program R_{i-1}, R_i). Recall that we can sample a uniformly random element in $\langle h^* \rangle$ by raising h^* to a uniformly random exponent in $\mathbb{Z}_{\mathrm{ord}(\mathbb{H})}$. For all prefixes $\mathbf{x}' \in \{0, 1\}^{i-1}$, we implicitly set

$$R_i(\mathbf{x}'\|b_i) := \tilde{R}_{i-1}(\mathbf{x}') \quad \text{and} \quad R_{i-1}(\mathbf{x}') := \tilde{R}_{i-1}(\mathbf{x}').$$

(We set $R_i(\mathbf{x}'\|\overline{b_i})$ later.) This means that for any $\mathbf{x} = (x_1, \ldots, x_n)$ such that $x_i = b_i$, we have:

$$R_i(\mathbf{x}|_i) = R_{i-1}(\mathbf{x}|_{i-1}) = \tilde{R}_{i-1}(\mathbf{x}|_{i-1}).$$

Key Queries. On input the j'th secret key query $\mathbf{y} = (\mathbf{y}|_{i-1}, y_i, \ldots, y_n)$, we consider three cases:

- Case 1: $y_i = b_i$. Here, $\mathcal{B}_2$ can compute

$$R_i(\mathbf{y}|_i) = R_{i-1}(\mathbf{y}|_{i-1}) = \tilde{R}_{i-1}(\mathbf{y}|_{i-1})$$

and simply outputs

$$\widehat{\mathsf{KeyGen}}(\mathsf{PP}, \mathsf{MSK} \cdot \tilde{R}_{i-1}(\mathbf{y}|_{i-1}), \mathbf{y}; \tilde{\mathbf{h}}^j),$$

where $\tilde{\mathbf{h}}^j \leftarrow \mathsf{SampH}(\mathsf{PP})$.

- Case 2: $y_i = \overline{b_i}$ and $R_i(\mathbf{y}|_i)$ has not been previously set. Here, we implicitly set

$$R_i(\mathbf{y}|_{i-1}\|\overline{b_i}) := \tilde{R}_{i-1}(\mathbf{y}|_{i-1}) \cdot (h^*)^{\gamma_j},$$

  where γ_j is as defined in the nested-hiding instantiation. Observe that this is the correct distribution since $R_i(\mathbf{y}|_{i-1}\|b_i)$ and $R_i(\mathbf{y}|_{i-1}\|\overline{b_i})$ are two independently random values. Then $\mathcal{B}_2$ outputs:

$$\widehat{\mathsf{KeyGen}}(\mathrm{PP}, \mathrm{MSK} \cdot \tilde{R}_{i-1}(\mathbf{y}|_{i-1}), \mathbf{y}; \mathbf{t}^j).$$

- Case 3: $y_i = \overline{b_i}$ and $R_i(\mathbf{y}|_i)$ has been previously set. Let j' be the index of key query in which we set $R_i(\mathbf{y}|_i)$, recall that

$$R_i(\mathbf{y}|_{i-1}\|\overline{b_i}) := \tilde{R}_{i-1}(\mathbf{y}|_{i-1}) \cdot (h^*)^{\gamma_{j'}}.$$

Then $\mathcal{B}_2$ outputs:

$$\widehat{\mathsf{KeyGen}}(\mathrm{PP}, \mathrm{MSK} \cdot \tilde{R}_{i-1}(\mathbf{y}|_{i-1}), \mathbf{y}; \mathbf{t}^{j'} \cdot \hat{\mathbf{h}}^j).$$

  where $\hat{\mathbf{h}}^j \leftarrow \mathsf{SampH}(\mathrm{PP})$. Here, we rely on the $\mathbb{H}$-*subgroup* property to re-randomize $\mathbf{t}^{j'}$.

Ciphertext. Upon receiving a challenge identity $\mathbf{x}^* := (x_1^*, \ldots, x_n^*)$ and two equal length messages m_0, m_1 from $\mathcal{A}$, output a random bit and halt if $x_i^* \neq b_i$. Observe that up to the point when $\mathcal{A}$ submits $\mathbf{x}^*$, its view is statistically independent of b_i. Therefore, the probability that we halt is exactly $1/2$. Suppose that we do not halt, which means we have $x_i^* = b_i$. Hence, $\mathcal{B}_2$ knows

$$R_i(\mathbf{x}^*|_i) = R_{i-1}(\mathbf{x}^*|_{i-1}) = \tilde{R}_{i-1}(\mathbf{x}^*|_{i-1}).$$

Then, $\mathcal{B}_2$ picks $\beta \leftarrow_{\mathrm{R}} \{0,1\}$ and outputs the semi-functional challenge ciphertext as:

$$\widehat{\mathsf{Enc}}(\mathrm{PP}, \mathbf{x}^*, m_\beta; \mathrm{MSK} \cdot \tilde{R}_{i-1}(\mathbf{x}^*|_{i-1}), \mathbf{g} \cdot \hat{\mathbf{g}}),$$

Here, $\mathcal{B}_2$ picks $\mathbf{g} \leftarrow \mathsf{SampG}(\mathrm{PP})$, whereas $\mathbf{g}$ is as defined in the nested-hiding instantiation. Observe that $\mathcal{B}_2$ can compute the output of $\widehat{\mathsf{Enc}}$ using just $\hat{\mathbf{g}}_{-(2i-\overline{b_i})}$ since since $x_i^* = b_i$.

Guess. When $\mathcal{A}$ halts with output β', $\mathcal{B}_2$ outputs 1 if $\beta' = \beta$ and 0 otherwise.

Suppose $x_i^* = b_i$. Then, when $(\mathbf{t}^1, \ldots, \mathbf{t}^q) = (\mathbf{h}^1, \ldots, \mathbf{h}^q)$, the output is identical to that in $\mathsf{Game}_{2,i-1}$; and when $(\mathbf{t}^1, \ldots, \mathbf{t}^q) = (\mathbf{h}'^1, \ldots, \mathbf{h}'^q)$, the output is identical to that in $\mathsf{Game}_{2,i}$. Hence,

$$\mathsf{Adv}_{\mathcal{B}_2}^{\mathrm{NS}}(\lambda, q)$$

$$= \Big| \Pr[x_i^* \neq b_i] \cdot 0 + \Pr[x_i^* = b_i]$$

$$\cdot (\Pr[\mathcal{A} \text{ outputs } \beta' = \beta \text{ in } \mathsf{Game}_{2,i-1}] - \Pr[\mathcal{A} \text{ outputs } \beta' = \beta \text{ in } \mathsf{Game}_{2,i}]) \Big|$$

$$= 1/2 \cdot \Big| \Pr[\mathcal{A} \text{ outputs } \beta' = \beta \text{ in } \mathsf{Game}_{2,i-1}] - \Pr[\mathcal{A} \text{ outputs } \beta' = \beta \text{ in } \mathsf{Game}_{2,i}] \Big|$$

$$\geq 1/2 \cdot |\mathsf{Adv}_{2,i-1}(\lambda) - \mathsf{Adv}_{2,i}(\lambda)|.$$

We may therefore conclude that $|\mathsf{Adv}_{2,i-1}(\lambda) - \mathsf{Adv}_{2,i}(\lambda)| \leq 2\mathsf{Adv}_{\mathcal{B}_2}^{\mathrm{NS}}(\lambda, q)$. $\square$

4.5 Final Transition

Lemma 4 (Game$_{2,n}$ to Game$_3$). *For any adversary $\mathcal{A}$:*

$$|\mathsf{Adv}_{2,n}(\lambda) - \mathsf{Adv}_3(\lambda)| \le 2^{-\Omega(\lambda)}.$$

Proof. Observe that the challenge ciphertext in Game$_{2,n}$ is given by:

$$\widehat{\mathsf{Enc}}(\mathsf{PP}, \mathbf{x}^*, m_\beta; \mathsf{MSK} \cdot R_n(\mathbf{x}^*), \mathbf{g} \cdot \hat{\mathbf{g}}) = (C_0, C_1, C_2' \cdot m_\beta),$$

where (C_0, C_1) depend only on $\mathbf{g} \cdot \hat{\mathbf{g}} = (g_0 \cdot \hat{g}_0, \dots)$, and C_2' is given by:

$$C_2' = e(g_0 \cdot \hat{g}_0, \mathsf{MSK} \cdot R_n(\mathbf{x}^*)) = e(g_0 \cdot \hat{g}_0, \mathsf{MSK}) \cdot \boxed{e(\hat{g}_0, R_n(\mathbf{x}^*))},$$

where in the last equality, we use the fact that $e(g_0, R_n(\mathbf{x}^*)) = 1$ (see Remarks 1 and 5). In addition, MPK and all of the secret key queries reveal no information about $R_n(\mathbf{x}^*)$. Then, by *non-degeneracy*, with probability $1 - 2^{-\Omega(\lambda)}$ over $\hat{g}_0$, we have $e(\hat{g}_0, R_n(\mathbf{x}^*))$ is uniformly distributed over $\mathbb{G}_T$. This implies that the challenge ciphertext is identically distributed to a semi-functional encryption of a random message in $\mathbb{G}_T$, as in Game$_3$. We may then conclude that: $|\mathsf{Adv}_{2,n}(\lambda) - \mathsf{Adv}_3(\lambda)| \le 2^{-\Omega(\lambda)}$. □

Remark 6. In our composite-order instantiation, we only have the weaker guarantee that $e(\hat{g}_0, R_n(\mathbf{x}^*))$ has at least 2λ bits of min-entropy, instead of being uniform over $\mathbb{G}_T$. We will modify the IBE scheme as follows: the message space is now $\{0,1\}^\lambda$, and we replace the term $g_T' \cdot m$ in the ciphertext with:

$$\mathsf{H}(g_T') \oplus m,$$

where $\mathsf{H} : \mathbb{G}_T \to \{0,1\}^\lambda$ is a pairwise independent hash function. By the left-over hash lemma, we still have $|\mathsf{Adv}_{2,n}(\lambda) - \mathsf{Adv}_3(\lambda)| \le 2^{-\Omega(\lambda)}$.

Acknowledgments. We thank Dennis Hofheinz and the anonymous reviewers for helpful feedback on the write-up.

References

[1] Abdalla, M., Fouque, P.-A., Lyubashevsky, V., Tibouchi, M.: Tightly-secure signatures from lossy identification schemes. In: Pointcheval, D., Johansson, T. (eds.) EUROCRYPT 2012. LNCS, vol. 7237, pp. 572–590. Springer, Heidelberg (2012)

[2] Agrawal, S., Boneh, D., Boyen, X.: Efficient lattice (H)IBE in the standard model. In: Gilbert, H. (ed.) EUROCRYPT 2010. LNCS, vol. 6110, pp. 553–572. Springer, Heidelberg (2010)

[3] Agrawal, S., Boneh, D., Boyen, X.: Lattice basis delegation in fixed dimension and shorter-ciphertext hierarchical IBE. In: Rabin, T. (ed.) CRYPTO 2010. LNCS, vol. 6223, pp. 98–115. Springer, Heidelberg (2010)

[4] Bellare, M., Ristenpart, T.: Simulation without the artificial abort: Simplified proof and improved concrete security for Waters' IBE scheme. In: Joux, A. (ed.) EUROCRYPT 2009. LNCS, vol. 5479, pp. 407–424. Springer, Heidelberg (2009)

[5] Boneh, D., Boyen, X.: Efficient selective-ID secure identity-based encryption without random oracles. In: Cachin, C., Camenisch, J.L. (eds.) EUROCRYPT 2004. LNCS, vol. 3027, pp. 223–238. Springer, Heidelberg (2004)

[6] Boneh, D., Boyen, X.: Secure identity based encryption without random oracles. In: Franklin, M. (ed.) CRYPTO 2004. LNCS, vol. 3152, pp. 443–459. Springer, Heidelberg (2004)

[7] Boneh, D., Franklin, M.K.: Identity-based encryption from the Weil pairing. SIAM J. Comput. 32(3), 586–615 (2003)

[8] Canetti, R., Halevi, S., Katz, J.: A forward-secure public-key encryption scheme. In: Biham, E. (ed.) EUROCRYPT 2003. LNCS, vol. 2656, pp. 255–271. Springer, Heidelberg (2003)

[9] Cash, D., Hofheinz, D., Kiltz, E., Peikert, C.: Bonsai trees, or how to delegate a lattice basis. In: Gilbert, H. (ed.) EUROCRYPT 2010. LNCS, vol. 6110, pp. 523–552. Springer, Heidelberg (2010)

[10] Chen, J., Wee, H.: Dual system groups and its applications — compact HIBE and more. IACR Cryptology ePrint Archive (2013)

[11] Chen, J., Wee, H.: Fully (almost) tightly secure IBE from standard assumptions. IACR Cryptology ePrint Archive (2013)

[12] Chen, J., Lim, H.W., Ling, S., Wang, H., Wee, H.: Shorter IBE and signatures via asymmetric pairings. In: Abdalla, M., Lange, T. (eds.) Pairing 2012. LNCS, vol. 7708, pp. 122–140. Springer, Heidelberg (2013)

[13] Cocks, C.: An identity based encryption scheme based on quadratic residues. In: Honary, B. (ed.) Cryptography and Coding 2001. LNCS, vol. 2260, pp. 360–363. Springer, Heidelberg (2001)

[14] Freeman, D.M.: Converting pairing-based cryptosystems from composite-order groups to prime-order groups. In: Gilbert, H. (ed.) EUROCRYPT 2010. LNCS, vol. 6110, pp. 44–61. Springer, Heidelberg (2010)

[15] Gentry, C.: Practical identity-based encryption without random oracles. In: Vaudenay, S. (ed.) EUROCRYPT 2006. LNCS, vol. 4004, pp. 445–464. Springer, Heidelberg (2006)

[16] Gentry, C., Peikert, C., Vaikuntanathan, V.: Trapdoors for hard lattices and new cryptographic constructions. In: STOC, pp. 197–206 (2008)

[17] Hofheinz, D., Jager, T.: Tightly secure signatures and public-key encryption. In: Safavi-Naini, R., Canetti, R. (eds.) CRYPTO 2012. LNCS, vol. 7417, pp. 590–607. Springer, Heidelberg (2012)

[18] Hofheinz, D., Jager, T., Knapp, E.: Waters signatures with optimal security reduction. In: Fischlin, M., Buchmann, J., Manulis, M. (eds.) PKC 2012. LNCS, vol. 7293, pp. 66–83. Springer, Heidelberg (2012)

[19] Kakvi, S.A., Kiltz, E.: Optimal security proofs for full domain hash, revisited. In: Pointcheval, D., Johansson, T. (eds.) EUROCRYPT 2012. LNCS, vol. 7237, pp. 537–553. Springer, Heidelberg (2012)

[20] Lewko, A.: Tools for simulating features of composite order bilinear groups in the prime order setting. In: Pointcheval, D., Johansson, T. (eds.) EUROCRYPT 2012. LNCS, vol. 7237, pp. 318–335. Springer, Heidelberg (2012)

[21] Lewko, A.B., Waters, B.: Efficient pseudorandom functions from the decisional linear assumption and weaker variants. In: ACM Conference on Computer and Communications Security, pp. 112–120 (2009)

[22] Lewko, A., Waters, B.: New techniques for dual system encryption and fully secure HIBE with short ciphertexts. In: Micciancio, D. (ed.) TCC 2010. LNCS, vol. 5978, pp. 455–479. Springer, Heidelberg (2010)

[23] Lewko, A., Okamoto, T., Sahai, A., Takashima, K., Waters, B.: Fully secure functional encryption: Attribute-based encryption and (hierarchical) inner product encryption. In: Gilbert, H. (ed.) EUROCRYPT 2010. LNCS, vol. 6110, pp. 62–91. Springer, Heidelberg (2010)

[24] Naor, M., Reingold, O.: Number-theoretic constructions of efficient pseudo-random functions. J. ACM 51(2), 231–262 (2004)

[25] Okamoto, T., Takashima, K.: Hierarchical predicate encryption for inner-products. In: Matsui, M. (ed.) ASIACRYPT 2009. LNCS, vol. 5912, pp. 214–231. Springer, Heidelberg (2009)

[26] Okamoto, T., Takashima, K.: Fully secure functional encryption with general relations from the decisional linear assumption. In: Rabin, T. (ed.) CRYPTO 2010. LNCS, vol. 6223, pp. 191–208. Springer, Heidelberg (2010)

[27] Shamir, A.: Identity-based cryptosystems and signature schemes. In: Blakely, G.R., Chaum, D. (eds.) CRYPTO 1984. LNCS, vol. 196, pp. 47–53. Springer, Heidelberg (1985)

[28] Waters, B.: Efficient identity-based encryption without random oracles. In: Cramer, R. (ed.) EUROCRYPT 2005. LNCS, vol. 3494, pp. 114–127. Springer, Heidelberg (2005)

[29] Waters, B.: Dual system encryption: Realizing fully secure IBE and HIBE under simple assumptions. In: Halevi, S. (ed.) CRYPTO 2009. LNCS, vol. 5677, pp. 619–636. Springer, Heidelberg (2009)

[30] Wee, H.: Dual system encryption via predicate encodings. Manuscript (2013)

A   Concrete IBE Scheme from d-LIN in Prime-Order Groups

In this section, we show how the concrete IBE scheme from d-LIN works in prime-order bilinear groups (G_1, G_2, G_T, e). Recall that $\pi_L : \mathbb{Z}_p^{2d \times 2d} \to \mathbb{Z}_p^{2d \times d}$ is the projection map that maps a $2d \times 2d$ matrix to the left d columns.

Setup($1^\lambda, 1^n$): On input $(1^\lambda, 1^n)$, sample

$$\mathbf{B}, \mathbf{B}^*, \mathbf{R} \leftarrow_{\mathrm{R}} \mathsf{GL}_{2d}(\mathbb{Z}_p), \ \mathbf{A}_1, \dots, \mathbf{A}_{2n} \leftarrow_{\mathrm{R}} \mathbb{Z}_p^{(2d) \times (2d)}, \ \mathbf{k} \leftarrow_{\mathrm{R}} \mathbb{Z}_p^{2d}$$

such that $\mathbf{B}^\top \mathbf{B}^* = \mathbf{I}$, and output the master public and secret key pair

$$\mathrm{MPK} := \left(g_1^{\pi_L(\mathbf{B})}, g_1^{\pi_L(\mathbf{B}\mathbf{A}_1)}, \dots, g_1^{\pi_L(\mathbf{B}\mathbf{A}_{2n})}; e(g_1, g_2)^{\mathbf{k}^\top \pi_L(\mathbf{B})} \right)$$
$$\in (G_1^{2d \times d})^{2n+1} \times G_T^d,$$
$$\mathrm{MSK} := \left(g_2^{\mathbf{k}}, g_2^{\mathbf{B}^* \mathbf{R}}, g_2^{\mathbf{B}^* \mathbf{A}_1^\top \mathbf{R}}, \dots, g_2^{\mathbf{B}^* \mathbf{A}_{2n}^\top \mathbf{R}} \right) \in G_2^{2d} \times (G_2^{2d \times 2d})^{2n+1}.$$

Enc(MPK, $\mathbf{x}$, m): On input an identity vector $\mathbf{x} := (x_1, \dots, x_n) \in \mathbb{Z}_p^n$ and $m \in G_T$, pick $\mathbf{s} \leftarrow_{\mathrm{R}} \mathbb{Z}_p^d$ and output

$$\mathrm{CT}_{\mathbf{x}} := \left(\begin{array}{l} C_0 := g_1^{\pi_L(\mathbf{B})\mathbf{s}}, \ C_1 := g_1^{\pi_L(\mathbf{B}(\mathbf{A}_{2-x_1} + \dots + \mathbf{A}_{2n-x_n}))\mathbf{s}} \\ C_2 := e(g_1, g_2)^{\mathbf{k}^\top \pi_L(\mathbf{B})\mathbf{s}} \cdot m \end{array} \right) \in (G_1^{2d})^2 \times G_T.$$

KeyGen($\text{MPK}, \text{MSK}, \mathbf{y}$): On input an identity vector $\mathbf{y} := (y_1, \ldots, y_n) \in \mathbb{Z}_p^n$, pick $\mathbf{r} \leftarrow_R \mathbb{Z}_p^{2d}$ and output

$$\text{SK}_{\mathbf{y}} := \left(K_0 := g_2^{\mathbf{B}^*\mathbf{Rr}}, \quad K_1 := g_2^{\mathbf{k}+\mathbf{B}^*(\mathbf{A}_{2-y_1}+\cdots+\mathbf{A}_{2n-y_n})^\top \mathbf{Rr}} \right) \in (G_2^{2d})^2.$$

Dec($\text{MPK}, \text{SK}_{\mathbf{y}}, \text{CT}_{\mathbf{x}}$): If $\mathbf{x} = \mathbf{y}$, compute

$$e(g_1, g_2)^{\mathbf{k}^\top \pi_L(\mathbf{B})\mathbf{s}} \leftarrow e(C_0, K_1)/e(C_1, K_0),$$

and recover the message as

$$m \leftarrow C_2 \cdot e(g_1, g_2)^{-\mathbf{k}^\top \pi_L(\mathbf{B})\mathbf{s}} \in G_T.$$

B Dual System Groups

Syntax. Dual system groups consist of six randomized algorithms given by (SampP, SampGT, SampG, SampH) along with ($\widehat{\text{SampG}}, \widehat{\text{SampH}}$):

SampP($1^\lambda, 1^n$): On input $(1^\lambda, 1^n)$, output public and secret parameters (PP, SP), where:

- PP contains a triple of groups $(\mathbb{G}, \mathbb{H}, \mathbb{G}_T)$ and a non-generate bilinear map $e : \mathbb{G} \times \mathbb{H} \to \mathbb{G}_T$, a linear map μ defined on $\mathbb{H}$, along with some additional parameters used by SampG, SampH;
- given PP, we know $\text{ord}(\mathbb{H})$ (i.e. the order of the group, which is independent of n) and can uniformly sample from $\mathbb{H}$;
- SP contains $h^* \in \mathbb{H}$ (where $h^* \neq 1$), along with some additional parameters used by $\widehat{\text{SampG}}$;

SampGT : $\text{Im}(\mu) \to \mathbb{G}_T$.

SampG(PP): Output $\mathbf{g} \in \mathbb{G}^{n+1}$.

SampH(PP): Output $\mathbf{h} \in \mathbb{H}^{n+1}$.

$\widehat{\text{SampG}}$(PP, SP): Output $\hat{\mathbf{g}} \in \mathbb{G}^{n+1}$.

$\widehat{\text{SampH}}$(PP, SP): Output $\hat{\mathbf{h}} \in \mathbb{H}^{n+1}$.

The first four algorithms are used in the actual scheme, whereas the last two algorithms are used only in the proof of security. We define SampG_0 to denote the first group element in the output of SampG, and we define $\widehat{\text{SampG}}_0, \widehat{\text{SampH}}_0$ analogously.

Correctness. The requirements for correctness are as follows:

(projective.) For all $h \in \mathbb{H}$ and all coin tosses s, we have $\mathsf{SampGT}(\mu(h); s) = e(\mathsf{SampG}_0(\mathrm{PP}; s), h)$.

(associative.) For all $(g_0, g_1, \ldots, g_n) \leftarrow \mathsf{SampG}(\mathrm{PP})$ and $(h_0, h_1, \ldots, h_n) \leftarrow \mathsf{SampH}(\mathrm{PP})$ and for all $i = 1, \ldots, n$, we have $e(g_0, h_i) = e(g_i, h_0)$.

($\mathbb{H}$-subgroup.) The output distribution of $\mathsf{SampH}(\mathrm{PP})$ is the uniform distribution over a subgroup of $\mathbb{H}^{n+1}$.

Security. The requirements for security are as follows:

(orthogonality.) $\mu(h^*) = 1$.

(non-degeneracy.) For all $\hat{h}_0 \leftarrow \widehat{\mathsf{SampH}}_0(\mathrm{PP}, \mathrm{SP})$, h^* lies in the group generated by $\hat{h}_0$. For all $\hat{g}_0 \leftarrow \widehat{\mathsf{SampG}}_0(\mathrm{PP}, \mathrm{SP})$, we have $e(\hat{g}_0, h^*)^\alpha$ is identically distributed to the uniform distribution over $\mathbb{G}_T$, where $\alpha \leftarrow_\mathrm{R} \mathbb{Z}_{\mathrm{ord}(\mathbb{H})}$.

(left subgroup indistinguishability.) For any adversary $\mathcal{A}$, we define the advantage function:

$$\mathsf{Adv}^{\mathsf{LS}}_{\mathcal{A}}(\lambda) := \left| \Pr[\, \mathcal{A}(\mathrm{PP}, \boxed{\mathbf{g}}) = 1\,] - \Pr[\, \mathcal{A}(\mathrm{PP}, \boxed{\mathbf{g} \cdot \hat{\mathbf{g}}}) = 1\,] \right|$$

where

$$(\mathrm{PP}, \mathrm{SP}) \leftarrow \mathsf{SampP}(1^\lambda, 1^n);$$
$$\mathbf{g} \leftarrow \mathsf{SampG}(\mathrm{PP}); \quad \hat{\mathbf{g}} \leftarrow \widehat{\mathsf{SampG}}(\mathrm{PP}, \mathrm{SP}).$$

(right subgroup indistinguishability.) For any adversary $\mathcal{A}$, we define the advantage function:

$$\mathsf{Adv}^{\mathsf{RS}}_{\mathcal{A}}(\lambda) := \left| \Pr[\, \mathcal{A}(\mathrm{PP}, h^*, \mathbf{g} \cdot \hat{\mathbf{g}}, \boxed{\mathbf{h}}) = 1\,] - \Pr[\, \mathcal{A}(\mathrm{PP}, h^*, \mathbf{g} \cdot \hat{\mathbf{g}}, \boxed{\mathbf{h} \cdot \hat{\mathbf{h}}}) = 1\,] \right|$$

where

$$(\mathrm{PP}, \mathrm{SP}) \leftarrow \mathsf{SampP}(1^\lambda, 1^n);$$
$$\mathbf{g} \leftarrow \mathsf{SampG}(\mathrm{PP}); \quad \hat{\mathbf{g}} \leftarrow \widehat{\mathsf{SampG}}(\mathrm{PP}, \mathrm{SP});$$
$$\mathbf{h} \leftarrow \mathsf{SampH}(\mathrm{PP}); \quad \hat{\mathbf{h}} \leftarrow \widehat{\mathsf{SampH}}(\mathrm{PP}, \mathrm{SP}).$$

(parameter-hiding.) The following distributions are identically distributed

$$\{\mathrm{PP}, h^*, \boxed{\hat{\mathbf{g}}, \hat{\mathbf{h}}}\} \quad \text{and} \quad \{\mathrm{PP}, h^*, \boxed{\hat{\mathbf{g}} \cdot \hat{\mathbf{g}}', \hat{\mathbf{h}} \cdot \hat{\mathbf{h}}'}\}$$

where

$$(\text{PP}, \text{SP}) \leftarrow \mathsf{SampP}(1^\lambda, 1^n);$$
$$\hat{\mathbf{g}} = (\hat{g}_0, \ldots) \leftarrow \widehat{\mathsf{SampG}}(\text{PP}, \text{SP});$$
$$\hat{\mathbf{h}} = (\hat{h}_0, \ldots) \leftarrow \widehat{\mathsf{SampH}}(\text{PP}, \text{SP});$$
$$\gamma_1, \ldots, \gamma_n \leftarrow_{\text{R}} \mathbb{Z}_{\text{ord}(\mathbb{H})};$$
$$\hat{\mathbf{g}}' := (1, \hat{g}_0^{\gamma_1}, \ldots, \hat{g}_0^{\gamma_n}) \in \mathbb{G}^{n+1};$$
$$\hat{\mathbf{h}}' := (1, \hat{h}_0^{\gamma_1}, \ldots, \hat{h}_0^{\gamma_n}) \in \mathbb{H}^{n+1}.$$

Function-Private Identity-Based Encryption: Hiding the Function in Functional Encryption*

Dan Boneh, Ananth Raghunathan, and Gil Segev

Computer Science Department
Stanford University, Stanford, CA 94305

Abstract. We put forward a new notion, *function privacy*, in identity-based encryption and, more generally, in functional encryption. Intuitively, our notion asks that decryption keys reveal essentially no information on their corresponding identities, beyond the absolute minimum necessary. This is motivated by the need for providing *predicate privacy* in public-key searchable encryption. Formalizing such a notion, however, is not straightforward as given a decryption key it is always possible to learn some information on its corresponding identity by testing whether it correctly decrypts ciphertexts that are encrypted for specific identities.

In light of such an inherent difficulty, any meaningful notion of function privacy must be based on the *minimal* assumption that, from the adversary's point of view, identities that correspond to its given decryption keys are sampled from somewhat unpredictable distributions. We show that this assumption is in fact *sufficient* for obtaining a strong and realistic notion of function privacy. Loosely speaking, our framework requires that a decryption key corresponding to an identity sampled from any sufficiently unpredictable distribution is indistinguishable from a decryption key corresponding to an independently and uniformly sampled identity.

Within our framework we develop an approach for designing function-private identity-based encryption schemes, leading to constructions that are based on standard assumptions in bilinear groups (DBDH, DLIN) and lattices (LWE). In addition to function privacy, our schemes are also anonymous, and thus yield the first public-key searchable encryption schemes that are provably *keyword private*: A search key sk_w enables to identify encryptions of an underlying keyword w, while not revealing any additional information about w beyond the minimum necessary, as long as the keyword w is sufficiently unpredictable.

1 Introduction

Public-key searchable encryption is needed when a proxy is asked to route encrypted messages based on their content. For example, consider a payment gateway that needs to route transactions based on the transaction type. Transactions for benign items are routed for quick processing while transactions for sensitive

---

* Due to space limitations the reader is referred to the full version [19].

R. Canetti and J.A. Garay (Eds.): CRYPTO 2013, Part II, LNCS 8043, pp. 461–478, 2013.

items are routed for special processing. Similarly, consider an email gateway that routes emails based on the contents of the subject line. Urgent emails are routed to the user's mobile device, while less urgent mails are routed to the user's desktop. When the data is encrypted a simple design is to give such gateways full power to decrypt all ciphertexts, but this clearly exposes more information than necessary.

A better solution, called public-key searchable encryption (introduced by Boneh, Di Crescenzo, Ostrovsky and Persiano [17]), is to give the gateway a trapdoor that enables it to learn the information it needs and nothing else. In recent years many elegant public-key searchable encryption systems have been developed [17,36,1,21,47,39,6,24,2,4] supporting a wide variety of search predicates.

Private Searching. Beyond the standard notions of data privacy, it is often also necessary to guarantee *predicate privacy*, i.e., to keep the specific search predicate hidden from the gateway. For example, in the payment scenario it may be desirable to keep the list of sensitive items secret, and in the email scenario users may not want to reveal the exact criteria they use to classify an email as urgent. Consequently, we want the trapdoor given to the gateway to reveal as little as possible about the search predicate.

While this question has been considered before [48,44,14,46], it is often noted that such a notion of privacy cannot be achieved in the public-key setting. For example, to test if an email from "spouse" is considered urgent the gateway could simply use the public key to create an email from the spouse and test if the trapdoor classifies it as urgent. More generally, the gateway can encrypt messages of its choice and apply the trapdoor to the resulting ciphertexts, thereby learning how the search functionality behaves on these messages. Hence, leaking some information about the search predicate is unavoidable.

As a concrete example, consider the case of keyword search [17]: A search key sk_w corresponds to a particular keyword w, and the search matches a ciphertext $\mathsf{Enc}(pk, m)$ if and only if $m = w$. In this case, it may be possible to formalize and realize a notion of "private keyword search" asking that a search key reveals no more information than what can be learned by invoking the search algorithm.

Function-private IBE: A New Notion of Security. Motivated by the challenge of hiding the search predicates in public-key searchable encryption, in this paper we introduce a new notion of security, *function privacy*, for identity-based encryption.[1] The standard notion of security for anonymous IBE schemes (e.g., [18,22,31,32,3,11]), asks that a ciphertext $c = \mathsf{Enc}(pp, id, m)$ reveals essentially

---

[1] As observed by Abdalla et al. [1], any anonymous IBE scheme can be used as a public-key searchable encryption scheme by defining the search key sk_w for a keyword w as the IBE secret key for the identity $id = w$. A keyword w' is encoded as $c = \mathsf{Enc}(pp, w', 0)$ and one tests if c matches the keyword w by invoking the IBE decryption algorithm on c with the secret key sk_w. The IBE anonymity property ensures that c reveals nothing else about the payload w'. For this reason we focus on *anonymous* IBE schemes, although we note that our notion of function privacy does not require anonymity.

no information on the pair (id, m) as long as a secret key sk_{id} corresponding to the identity id is not explicitly provided (but secret keys corresponding to other identities may be provided). Our notion of function privacy takes a step forward by asking that it should not be possible to learn any information, beyond the absolute minimum necessary, on the identity id corresponding to a given secret key sk_{id}.

Formalizing a realistic notion of function privacy, however, is not straightforward due to the actual functionality of identity-based encryption. Specifically, assuming that an adversary who is given a secret key sk_{id} has some *a priori* information that the corresponding identity id belongs to a small set S of identities (e.g., $S = \{\text{id}_0, \text{id}_1\}$), then the adversary can fully recover id: The adversary simply needs to encrypt a (possibly random) message m for each $\text{id}' \in S$, and then run the decryption algorithm on the given secret key sk_{id} and each of the resulting ciphertexts $c' = \text{Enc}(\text{pp}, \text{id}', m)$ to identify the one that decrypts correctly. In fact, as long as the adversary has some *a-priori* information according to which the identity id is sampled from a distribution whose min-entropy is at most logarithmic in the security parameter, there is a non-negligible probability for a full recovery.

Our Contributions. In light of the above inherent difficulty, any notion of function privacy for IBE schemes would have to be based on the *minimal* assumption that, from the adversary's point of view, identities that correspond to its given secret keys are sampled from distributions with a certain amount of min-entropy (which has to be at least super-logarithmic in the security parameter). Our work shows that this necessary assumption is in fact *sufficient* for obtaining a strong and meaningful indistinguishability-based notion of function privacy.

Our work formalizes this new notion of security (we call it *function privacy* to emphasize the fact that sk_{id} hides the functionality that it provides). Loosely speaking, our basic notion of function privacy requires that a secret key sk_{id}, where id is sampled from any sufficiently unpredictable (adversarially-chosen) distribution,[2] is indistinguishable from a secret key corresponding to an independently and uniformly sampled identity. In addition, we also consider a stronger notion of function privacy, to which we refer as *enhanced* function privacy. This enhanced notion addresses the fact that in various applications (such as searching on encrypted data), an adversary may obtain not only a secret key sk_{id}, but also encryptions $\text{Enc}(\text{pp}, \text{id}, m)$ of messages m. Our notion of enhanced function privacy asks that even in such a scenario, it should not be possible to learn any unnecessary information on the identity id.

---

[2] We emphasize that the distribution is allowed to depend on the public parameters of the scheme. This is in contrast to the setting of deterministic public-key encryption (DPKE) [8], where similar inherent difficulties arise when formalizing notions of security. Nevertheless, our notion is inspired by that of [8], and we refer the reader to Section 2 for an elaborate discussion (in particular, we discuss a somewhat natural DPKE-based approach for designing function-private IBE schemes which fails to satisfy our notion of security and only satisfies a weaker, less realistic, one).

We refer the reader to Section 2 for the formal definitions, and for descriptions of simple attacks exemplifying that the anonymous IBE schemes presented in [18,32,3,40] do not satisfy even our basic notion of function privacy.[3]

Within our framework we develop an approach for designing identity-based encryption schemes that satisfy our notions of function private. Our approach leads to constructions that are based on standard assumptions in bilinear groups (DBDH, DLIN) and lattices (LWE). In particular, our schemes yield keyword searchable public-key encryption schemes that *do not reveal the keywords*: A search key sk_w reveals nothing about its corresponding keyword w beyond the minimum necessary, as long as the keyword w is chosen from a sufficiently unpredictable distribution.

The Bigger Picture: Functional Encryption and Obfuscation. Our notion of function privacy for IBE naturally generalizes to functional encryption systems [20,43,12,37,5,34], where we obtain an additional security requirement on such systems. Here, a functional secret key sk_f corresponding to a function f enables to compute $f(m)$ given an encryption $c = \mathsf{Enc}_{pk}(m)$. Functional encryption systems, however, need not be predicate private and sk_f may leak unnecessary information about f. Intuitively, we say that a functional encryption system is *function private* if such a functional secret key sk_f does not reveal information about f beyond what is already known and what can be obtained by running the decryption algorithm on test ciphertexts. This can be formalized within a suitable framework of program obfuscation (e.g., [25,7,41,35,50,26] and the references therein) by asking, for example, that any adversary that receives a functional secret key sk_f learns no more information than a simulator that has oracle access to the function f.

In this setting, our identity-based encryption schemes provide function privacy for the class of functions defined as

$$f_{\mathrm{id}^*}(\mathrm{id}, m) = \begin{cases} m & \text{if id} = \mathrm{id}^* \\ \bot & \text{otherwise} \end{cases}$$

where id^* is sampled from an unpredictable distribution. A fascinating direction for future work is to extend our results to more general classes of functions.

Non-Adaptive Function Privacy and Deterministic Encryption. The inherent difficulty discussed above in formalizing function privacy is somewhat similar to the one that arises in the context of deterministic public-key encryption (DPKE), introduced by Bellare, Boldyreva, and O'Neill [8] (see also [10,15,9,23,30,42,51,45]). In that setting one would like to capture as-strong-as-possible notions of security that can be satisfied by public-key encryption

---

[3] We note that other anonymous IBE schemes, such as [31,22,11] for which we were not able to find such simple attacks, can always be *assumed* to be function private based on somewhat non-standard entropy-based assumptions (such assumptions would essentially state that the schemes satisfy our definition). In this paper we are interested in schemes whose function privacy can be based on standard assumptions (e.g., DBDH, DLIN, LWE).

schemes whose encryption algorithms are deterministic. Similarly to our setting, if an adversary has some *a priori* information that a ciphertext $c = \mathsf{Enc}_{pk}(m)$ corresponds to a plaintext m that is sampled from a low-entropy source (e.g., $m \in \{m_0, m_1\}$), then the plaintext can be fully recovered: the adversary simply needs to encrypt all "likely" plaintexts and to compare each of the resulting ciphertexts to c. Therefore, any notion of security for DPKE has to be based on the assumption that plaintexts are sampled from distributions with a certain amount of min-entropy (which has to be at least super-logarithmic in the security parameter).

However, unlike in our setting, in the setting of DPKE it is also necessary to limit the dependency of plaintexts on the public-key of the scheme.[4] In our setting, as the key-generation algorithm is allowed to be randomized, such limitations are not inherent: we allow adversaries to specify identity distributions in an adaptive manner after seeing the public parameters of the scheme.

This crucial difference between our setting and the setting of DPKE rules out, in particular, the following natural approach for designing anonymous IBE schemes providing function privacy: encapsulate all identities with a DPKE scheme, and then use any existing anonymous IBE scheme treating the ciphertexts of the DPKE scheme as its identities. That is, for encrypting to identity id, first encrypt id using a DPKE scheme and then treat the resulting ciphertext as an identity for an anonymous IBE system. This approach clearly preserves the standard security of the underlying IBE scheme. Moreover, as secret keys are now generated as sk_c, where $c = \mathsf{Enc}_{pk}(\mathsf{id})$ is a deterministic encryption of id, instead of as $\mathsf{sk}_{\mathsf{id}}$, one could hope that $\mathsf{sk}_{\mathsf{id}}$ does not reveal any unnecessary information on id as long as id is sufficiently unpredictable.

This approach, however, fails to satisfy our notion of function privacy and only satisfies a weaker, "non-adaptive", one.[5] Specifically, the notion of function privacy that is satisfied by such a two-tier construction is that secret keys do not reveal any unnecessary information on their corresponding identities as long as the identities are essentially independent of the public parameters of the scheme. In the full version [19] we formalize this non-adaptive notion and present a generic transformation satisfying it based on any IBE scheme. In fact, observing that the DPKE-based construction described above never actually uses the decryption algorithm of the DPKE scheme, in our generic transformation we show that above idea can be realized without using a DPKE scheme. Instead, we only need to assume the existence of collision-resistant hash functions (and also use any pairwise independent family of permutations).

---

[4] Intuitively, the reason is that plaintexts distributions that can depend on the public key can use any deterministic encryption algorithm as a subliminal channel for leaking information on the plaintexts (consider, for example, sampling a uniform plaintext m for which the most significant bit of $c = \mathsf{Enc}_{pk}(m)$ agrees with that of m). We refer the reader to [8,45] for an in-depth discussion.

[5] As discussed above, any DPKE becomes insecure once plaintext distributions (which here correspond to identity distributions) are allowed to depend on the public key of the scheme.

1.1   Our Approach: "Extract-Augment-Combine"

Our approach consists of three main steps: "extract," "augment," and "combine." We begin with a description of the main ideas underlying each step, and then provide an example using a concrete IBE scheme.

Given any anonymous IBE scheme Π = (Setup, KeyGen, Enc, Dec), we use the exact same setup algorithm Setup, and our first step is to modify its key-generation algorithm KeyGen as follows: Instead of generating a secret key for an identity id, first apply a strong randomness extractor Ext to id using a randomly chosen seed s, then generate a secret key $\mathsf{sk}_{\mathsf{id}_s}$ for the identity $\mathsf{id}_s := \mathsf{Ext}(\mathsf{id}, s)$, and output the pair $(s, \mathsf{sk}_{\mathsf{id}_s})$ as a secret for id in the new scheme. This steps clearly guarantees function privacy: as long as the identity id is sampled from a sufficiently unpredictable distribution,[6] the distribution (s, id_s) is statistically close to uniform, and therefore the pair $(s, \mathsf{sk}_{\mathsf{id}_s})$ reveals no information on the identity id.

This extraction step, however, may hurt the data privacy of the underlying scheme. For example, since randomness extractors are highly non-injective by definition, an adversary that is given a secret key $(s, \mathsf{sk}_{\mathsf{id}_s})$ may be able to find an identity id$'$ such that $\mathsf{Ext}(\mathsf{id}, s) = \mathsf{Ext}(\mathsf{id}', s)$. In this case, the same secret key is valid for both id and id$'$, contradicting the data privacy of the resulting scheme. Therefore, for overcoming this problem we make sure that the extractor is *at least* collision resistant: although many collisions exist, a computationally-bounded adversary will not be able to find one. This is somewhat natural to achieve in the random-oracle model [13], but significantly more challenging in the standard model.

An even more challenging problem is that the extraction step hurts the decryption of the underlying scheme. Specifically, when encrypting a message m for an identity id, the encryption algorithm does not know which seed s will be chosen (or was already chosen) when generating a secret key for id. In other words, the correctness of the decryption algorithm Dec should hold for any choice of seed s by the key-generation algorithm KeyGen, although s is not known to the encryption algorithm Enc. One possibility, is to modify the encryption algorithm such that it outputs an encryption of m for id_s for all possible seeds s. This clearly fails, as the number of seeds is inherently super-polynomial in the security parameter. We overcome this problem by augmenting ciphertexts of the underlying scheme with various additional pieces of information. These will enable the new decryption algorithm to combine the pieces in a particular way for generating an encryption of m for the identity id_s for any given s, and then simply apply the underlying decryption algorithm using the specific seed s chosen by the key-generation algorithm.[7]

---

[6] Note that the new scheme assumes a slightly larger identity space compared to the underlying scheme.

[7] In fact, in some of our schemes the decryption algorithm combines the pieces to generate an encryption of a related message m' from which m can be easily recovered (e.g., $m' = 2m$).

Our approach introduces the following two main challenges that we overcome in each of our constructions:

- Augmenting the ciphertexts of the underlying scheme with additional pieces of information may hurt the data privacy of the underlying scheme.
- Combining the additional pieces of information for generating an encryption for id_s for any given s requires using an extractor Ext that exhibits a particular interplay with the underlying encryption and decryption algorithms.

Our constructions in this paper are obtained by applying our approach to various known anonymous IBE schemes [18,32,3,40]. To do so, we overcome the two main challenges mentioned above in ways that are "tailored" specifically to each scheme. Using our approach we provide the following constructions (see also Table 1):

- In the random-oracle model we give fully-secure constructions from pairings and lattices by building upon the systems of Boneh and Franklin [18] (based on the DBDH assumption) and of Gentry, Peikert and Vaikuntanathan [32] (based on the LWE assumption).
- In the standard model we give selectively-secure constructions from pairings and lattices based on the constructions of Agrawal, Boneh and Boyen [3] (based on the LWE assumption) and of Kurosawa and Phong [40] (based on the DLIN assumption), which we then generalize to a fully-secure construction (based on the DLIN assumption[8]).

In all instances our constructions are based on the same complexity assumptions as the underlying systems.

Table 1. Our IBE schemes

Scheme	Model	Data Privacy	Function Privacy
DBDH	Random Oracle	Full	Statistical
LWE1	Random Oracle	Full	Statistical
DLIN1	Standard	Selective	Statistical + Non-adaptive enhanced
LWE2	Standard	Selective	Statistical
DLIN2	Standard	Full	Statistical + Enhanced
CRH	Standard	Full	Non-adaptive statistical enhanced

A Concrete Example. We conclude this section by exemplifying our approach using our DBDH-based construction in the random-oracle model. (We refer the reader to the full version [19] for a more formal description of the scheme and its proofs of data privacy and function privacy.) The scheme is obtained by applying our approach to the anonymous IBE scheme of Boneh and Franklin [18].

---

[8] We note that a similar generalization can also be applied to our selectively-secure LWE-based scheme in the standard model.

- The setup algorithm in the scheme of Boneh and Franklin samples $\alpha \leftarrow \mathbb{Z}_p^*$, and lets $h = g^\alpha$, where g is a generator of a group $\mathbb{G}$ of prime order p. The public parameters are g and h, and the master secret key is α. Our scheme has exactly the same setup algorithm.
- The key-generation algorithm in the scheme of Boneh and Franklin computes a secret key for an identity id as $\mathsf{sk}_{\mathsf{id}} = H(\mathsf{id})^\alpha$, where H is a random oracle mapping identities into the group $\mathbb{G}$. As discussed above our first step is to extract from id. First, we use a random oracle mapping identities into $\mathbb{G}^\ell$ for some $\ell > 1$. Then, for $H(\mathsf{id}) = (h_1, \ldots, h_\ell) \in \mathbb{G}^\ell$, we sample an extractor seed $s = (s_1, \ldots, s_\ell) \leftarrow \mathbb{Z}_p^\ell$, and output the secret key $(s, (\mathsf{Ext}(H(\mathsf{id}), s)^\alpha)$ where we use the specific extractor $\mathsf{Ext}((h_1, \ldots, h_\ell), (s_1, \ldots, s_\ell)) = \prod_{j=1}^\ell h_j^{s_j}$. Note that Ext is, in particular, collision resistant based on the discrete logarithm assumption in the group $\mathbb{G}$.
- An encryption of a message m for an identity id in the scheme of Boneh and Franklin is a pair (c_0, c_1), defined as $c_0 = g^r$ and $c_1 = \hat{e}(h, H(\mathsf{id}))^r \cdot m$. In our scheme, an encryption of a message m for an identity id consists of $\ell + 1$ components $(c_0, \ldots, c_\ell)$ defined as $c_0 = g^r$, and $c_i = \hat{e}(h, h_i)^r \cdot m$ for every $i \in [\ell]$, where $H(\mathsf{id}) = (h_1, \ldots, h_\ell)$. This is exactly using the encryption algorithm of Boneh and Franklin for separately encrypting m for each of the h_i's while re-using the same randomness r. The main technical challenge that is left is showing that such augmented ciphertexts still provide data privacy.
- Our decryption algorithm on input a ciphertext $c = (c_0, \ldots, c_\ell)$, and a secret key $\mathsf{sk}_{\mathsf{id}} = (s_1, \ldots, s_\ell, z)$, combines $c_1, \ldots, c_\ell$ by computing

$$\prod_{i=1}^\ell c_i^{s_i} = \hat{e}(h, \prod_{i=1}^\ell h_i^{s_i})^r \cdot m^{s_1 + \cdots + s_\ell} = \hat{e}(h, \mathsf{id}_s)^r \cdot m^{s_1 + \cdots + s_\ell},$$

where $\mathsf{id}_s = \mathsf{Ext}(H(\mathsf{id}), s)$, as before. Note that the pair $(c_0, \prod_{i=1}^\ell c_i^{s_i})$ is exactly an encryption of the message $m' = m^{s_1 + \cdots + s_\ell}$ for the identity id_s in the scheme of Boneh and Franklin. This allows to invoke the decryption algorithm of Boneh and Franklin for recovering m', and then to easily recover m (as the s_i's are given in the clear).

1.2   Related Work

Searchable encryption has been studied in both the symmetric settings [48,29,46] and public-key settings [17,36,1,21,47,39,6,24,4]. Public-key searching on encrypted data now supports equality testing, disjunctions and conjunctions, range queries, CNF/DNF formulas, and polynomial evaluation. These schemes, however, are not function private in that their secret searching keys reveal information about their corresponding predicates. Indeed, until this work, predicate privacy seemed impossible in the public-key settings.

The impossibility argument does not apply in the symmetric key settings where the encryptor and decryptor have a shared secret key. In this setting the entity searching over ciphertexts does not have the secret key and cannot (passively) test the searching key on ciphertexts of its choice. Indeed, in the symmetric-key setting predicate privacy is possible and a general solution to private searching on encrypted data was provided by Goldreich and Ostrovsky [33] in their construction of an oblivious RAM. More efficient constructions are known for equality testing [48,27,29,28,49,38] and inner product testing [46]. The latter enables CNF/DNF formulas, polynomial evaluation, and exact thresholds.

A closely related problem called *private stream searching* asks for the complementary privacy requirements: the data is available in the clear, but the search predicate must remain hidden. Constructions in these settings support efficient equality testing [44,14] and can be viewed as a more expressive variant of private information retrieval.

1.3 Notation

Throughout the paper we use the following standard notation. For an integer $n \in \mathbb{N}$ we denote by $[n]$ the set $\{1, \ldots, n\}$, and by U_n the uniform distribution over the set $\{0,1\}^n$. For a random variable X we denote by $x \leftarrow X$ the process of sampling a value x according to the distribution of X. Similarly, for a finite set S we denote by $x \leftarrow S$ the process of sampling a value x according to the uniform distribution over S. We denote by $\mathbf{x}$ (and sometimes $\boldsymbol{x}$) a vector $(x_1, \ldots, x_{|\mathbf{x}|})$. We denote by $\mathbf{X} = (X_1, \ldots, X_T)$ a joint distribution of T random variables, and by $\mathbf{x} = (x_1, \ldots, x_T)$ a sample drawn from $\mathbf{X}$. For two bit-strings x and y we denote by $x \| y$ their concatenation. A non-negative function $f : \mathbb{N} \to \mathbb{R}$ is negligible if it vanishes faster than any inverse polynomial. For a real number $x \in \mathbb{R}$ we define $\lfloor x \rceil = \lfloor x + 1/2 \rfloor$ (i.e., the nearest integer to x). For a group $\mathbb{G}$ of order p with generator g and any $\mathbf{X} \in \mathbb{Z}_p^{n \times m}$, we denote the matrix whose (i,j)-th entry is $(g^{x_{i,j}})$ by $g^{\mathbf{X}}$.

The *min-entropy* of a random variable X is $\mathbf{H}_\infty(X) = -\log(\max_x \Pr[X = x])$. A *k-source* is a random variable X with $\mathbf{H}_\infty(X) \geq k$. A $(k_1, \ldots, k_T)$-*source* is a random variable $\mathbf{X} = (X_1, \ldots, X_T)$ where each X_i is a k_i-source. A (T, k)-*block-source* is a random variable $\mathbf{X} = (X_1, \ldots, X_T)$ where for every $i \in [T]$ and $x_1, \ldots, x_{i-1}$ it holds that $X_i|_{X_1 = x_1, \ldots, X_{i-1} = x_{i-1}}$ is a k-source.

1.4 Paper Organization

The remainder of this paper is organized as follows. In Section 2 we formally define our notion of function privacy for identity-based encryption. In Section 3 we present a selectively-secure DLIN-based scheme in the standard model, and in Section 4 we discuss several extensions and open problems. Due to space limitations we refer the reader to the full version [19].

2   Modeling Function Privacy for IBE

In this section we introduce our notions of function privacy for anonymous IBE schemes.[9] Recall that the standard notion of security for anonymous IBE schemes, anon-IND-ID-CPA, asks that a ciphertext $c = \mathsf{Enc}(\mathsf{pp}, \mathsf{id}, m)$ reveals essentially no information on the pair (id, m) as long as a secret key $\mathsf{sk}_{\mathsf{id}}$ corresponding to the identity id is not explicitly provided (but secret keys corresponding to other identities may be provided). We refer to this notion of security as *data privacy*. As discussed in Section 1, we put forward three notions of function privacy: a basic notion, an "enhanced" notion, and a non-adaptive notion. Due to space limitations, in this section we focus on our basic notion, and refer the reader to the full version [19] for our enhanced and non-adaptive notions.

Throughout this section we let T, k, and $k_1, \dots, k_T$ be functions of the security parameter $\lambda \in \mathbb{N}$. In addition, we note that in the random-oracle model, all algorithms, adversaries, oracles, and distributions are given access to the random oracle.

Our basic notion of function privacy asks that it should not be possible to learn any information, beyond the absolute minimum necessary, on the identity id corresponding to a given secret key $\mathsf{sk}_{\mathsf{id}}$. Specifically, our notion considers adversaries that are given the public parameters of the scheme, and can interact with a "real-or-random" function-privacy oracle $\mathsf{RoR}^{\mathsf{FP}}$. This oracle takes as input any adversarially-chosen distribution over vectors of identities, and outputs secret keys either for identities sampled from the given distribution or for independently and uniformly distributed identities.[10] We allow adversaries to adaptively interact with the real-or-random oracle, for any polynomial number of queries, as long as the distributions have a certain amount of min-entropy. At the end of the interaction, we ask that adversaries have only a negligible probability of distinguishing between the "real" and "random" modes of the oracle. The following definitions formally capture our basic notion of function privacy.

Definition 2.1 (Real-or-random function-privacy oracle). *The real-or-random function-privacy oracle* $\mathsf{RoR}^{\mathsf{FP}}$ *takes as input triplets of the form* $(\mathsf{mode}, \mathsf{msk}, \boldsymbol{ID})$, *where* $\mathsf{mode} \in \{\mathsf{real}, \mathsf{rand}\}$, msk *is a master secret key, and* $\boldsymbol{ID} = (ID_1, \dots, ID_T) \in \mathcal{ID}^T$ *is a circuit representing a joint distribution over* $\mathcal{ID}^T$. *If* $\mathsf{mode} = \mathsf{real}$ *then the oracle samples* $(\mathsf{id}_1, \dots, \mathsf{id}_T) \leftarrow \boldsymbol{ID}$ *and if* $\mathsf{mode} = \mathsf{rand}$ *then the oracle samples* $(\mathsf{id}_1, \dots, \mathsf{id}_T) \leftarrow \mathcal{ID}^T$ *uniformly. It then invokes the algorithm* $\mathsf{KeyGen}(\mathsf{msk}, \cdot)$ *on each of* $\mathsf{id}_1, \dots, \mathsf{id}_T$ *and outputs a vector of secret keys* $(\mathsf{sk}_{\mathsf{id}_1}, \dots, \mathsf{sk}_{\mathsf{id}_T})$.

Definition 2.2 (Function-privacy adversary). *Let* $X \in \{(T, k)\text{-}block, (k_1, \dots, k_T)\}$. *An* X-source function-privacy adversary $\mathcal{A}$ *is an algorithm that is*

---

[9] We focus on *anonymous* IBE schemes as our motivating application is public-key *searchable* encryption, to which anonymity is crucial [1].

[10] We note that the resulting notion of security is polynomially equivalent to the one obtained by using a "left-or-right" oracle instead of a "real-or-random" oracle, as for example, in the case of semantic security for public-key encryption schemes.

given as input a pair $(1^\lambda, \text{pp})$ and oracle access to $\text{RoR}^{\text{FP}}(\text{mode}, \text{msk}, \cdot)$ for some mode $\in \{\text{real}, \text{rand}\}$, and to $\text{KeyGen}(\text{msk}, \cdot)$, and each of its queries to RoR^{FP} is an X-source.

Definition 2.3 (Function privacy). Let $X \in \{(T, k)\text{-block}, (k_1, \ldots, k_T)\}$. An identity-based encryption scheme $\Pi = (\text{Setup}, \text{KeyGen}, \text{Enc}, \text{Dec})$ is X-source function private if for any probabilistic polynomial-time X-source function-privacy adversary $\mathcal{A}$, there exists a negligible function $\nu(\lambda)$ such that

$$\mathbf{Adv}_{\Pi, \mathcal{A}}^{\text{FP}}(\lambda) \stackrel{\text{def}}{=} \left| \Pr\left[\text{Expt}_{\text{FP}, \Pi, \mathcal{A}}^{\text{real}}(\lambda) = 1 \right] - \Pr\left[\text{Expt}_{\text{FP}, \Pi, \mathcal{A}}^{\text{rand}}(\lambda) = 1 \right] \right| \leq \nu(\lambda),$$

where for each mode $\in \{\text{real}, \text{rand}\}$ and $\lambda \in \mathbb{N}$ the experiment $\text{Expt}_{\text{FP}, \Pi, \mathcal{A}}^{\text{mode}}(\lambda)$ is defined as follows:

1. $(\text{pp}, \text{msk}) \leftarrow \text{Setup}(1^\lambda)$.
2. $b \leftarrow \mathcal{A}^{\text{RoR}^{\text{FP}}(\text{mode}, \text{msk}, \cdot), \text{KeyGen}(\text{msk}, \cdot)}(1^\lambda, \text{pp})$.
3. Output b.

In addition, such a scheme is statistically X-source function private if the above holds for any computationally-unbounded X-source enhanced function-privacy adversary making a polynomial number of queries to the RoR^{FP} oracle.

Multi-shot vs. Single-Shot Adversaries. Note that Definition 2.3 considers adversaries that query the function-privacy oracle for any polynomial number of times. In fact, as adversaries are also given access to the key-generation oracle, this "multi-shot" definition is polynomially equivalent to its "single-shot" variant in which adversaries query the real-or-random function-privacy oracle RoR^{FP} at most once. This is proved via a straightforward hybrid argument, where the hybrids are constructed such that only one query is forwarded to the function-privacy oracle, and all other queries are answered using the key-generation oracle.

Known Schemes That Are Not Function Private. To exercise our notion of function privacy we demonstrate that the anonymous IBE schemes of Boneh and Frankin [18], Gentry, Peikert and Vaikuntanathan [32], Agrawal, Boneh and Boyen [3], and Kurosawa and Phong [40] are not function private. We present simple and efficient attacks showing that the schemes [18,32] do not satisfy Definition 2.3, and note that almost identical attacks can be carried on [3,40]. As discussed in Section 1, other anonymous IBE schemes such as [31,22] for which we were not able to find such simple attacks, can always be *assumed* to be function private based on somewhat non-standard entropy-based assumptions (such assumptions would essentially state that the schemes satisfy our definition). In this paper we are interested in schemes whose function privacy can be based on standard assumptions.

The Boneh-Franklin scheme uses a random oracle $H : \mathcal{ID} \rightarrow \mathbb{G}$ and the secret key for id is $\text{sk}_{\text{id}} = H(\text{id})^\alpha$ where $\alpha \leftarrow \mathbb{Z}_p$ is the master secret. The public parameters are g and $h = g^\alpha$ for some generator g of $\mathbb{G}$. Consider an adversary that queries the real-or-random oracle with the circuit of the distribution that samples a uniformly distributed id for which the most significant bit of $\hat{e}(g^\alpha, H(\text{id}))$ is 0. Clearly, this distribution has almost full entropy, and can be described by a circuit of polynomial size given the public parameters.[11] Then, given $\text{sk}_{\text{id}} = H(\text{id})^\alpha$ the adversary outputs 0 if the most significant bit of $\hat{e}(g, \text{sk}_{\text{id}})$ is 0 and outputs 1 otherwise. Since $\hat{e}(g, \text{sk}_{\text{id}}) = \hat{e}(g^\alpha, H(\text{id}))$ it is easy to see that the adversary has advantage $1/2$ in distinguishing the real mode from the rand mode, thereby breaking function privacy. In Section 1.1 we presented a modification of this scheme which is function private, and the reader is referred to the full version [19] for its proof of security.

In the scheme of Gentry, Peikert and Vaikuntanathan, the public parameters consist of a matrix $\mathbf{A} \leftarrow \mathbb{Z}_q^{n \times m}$ and the master secret key is a short basis for the lattice $\Lambda_q^\perp(\mathbf{A})$. A secret key corresponding to an identity id is a short vector $\mathbf{e} \in \mathbb{Z}^m$ such that $\mathbf{Ae} = H(\text{id}) \in \mathbb{Z}_q^n$, where $H : \mathcal{ID} \rightarrow \mathbb{Z}_q^n$ is a random oracle. Consider an adversary that queries the real-or-random oracle with the circuit of the distribution that samples a uniformly distributed id for which the most significant bit of $H(\text{id})$ is 0. Then, given $\text{sk}_{\text{id}} = \mathbf{e}$ the adversary outputs 0 if the most significant bit of $\mathbf{Ae}$ is 0 and outputs 1 otherwise. Since $\mathbf{Ae} = H(\text{id})$ it is easy to see that the adversary has advantage $1/2$ in distinguishing the real mode from the rand mode, thereby breaking function privacy. In the full version [19] we present a modification of this scheme which is function private.

3   A Selectively-Secure DLIN-Based Scheme

In this section we present an IBE scheme based on the DLIN assumption in the standard model. For emphasizing the main ideas underlying our approach, we present here a *selectively* data private scheme, and refer the reader to for full version [19] for its extension to *full* data privacy. The scheme is based on the DLIN-based IBE of Kurosawa and Phong [40], which is an adaptation of the LWE-based IBE of Agrawal, Boneh and Boyen [3] to bilinear groups. The scheme is obtained by applying our "extract-augment-combine" approach, as discussed in Section 1.1.

The Scheme. Let GroupGen be a probabilistic polynomial-time algorithm that takes as input a security parameter 1^λ, and outputs $(\mathbb{G}, \mathbb{G}_T, p, g, \hat{e})$ where $\mathbb{G}$ and $\mathbb{G}_T$ are groups of prime order p, $\mathbb{G}$ is generated by g, p is a λ-bit prime number, and $\hat{e} : \mathbb{G} \times \mathbb{G} \rightarrow \mathbb{G}_T$ is a non-degenerate efficiently computable bilinear map. The scheme $\mathcal{IBE}_{\text{DLIN1}} = (\text{Setup}, \text{KeyGen}, \text{Enc}, \text{Dec})$ is parameterized by the security parameter $\lambda \in \mathbb{N}$. For any such $\lambda \in \mathbb{N}$, the scheme has parameters $m \geq 3$ and $\ell \geq 2$, identity space $\mathcal{ID}_\lambda = \mathbb{Z}_p^\ell$, and message space $\mathcal{M}_\lambda = \mathbb{G}_T$.

– **Setup:** On input 1^λ sample $(\mathbb{G}, \mathbb{G}_T, p, g, \hat{e}) \leftarrow \text{GroupGen}(1^\lambda)$, $\mathbf{A}_0, \mathbf{A}_1, \ldots,$ $\mathbf{A}_\ell, \mathbf{B} \leftarrow \mathbb{Z}_p^{2 \times m}$, and $\mathbf{u} \leftarrow \mathbb{Z}_p^2$. Output $\text{pp} = \left(g, g^{\mathbf{A}_0}, g^{\mathbf{A}_1}, \ldots, g^{\mathbf{A}_\ell}, \mathbf{B}, g^{\mathbf{u}} \right)$ and $\text{msk} = (\mathbf{A}_0, \mathbf{A}_1, \ldots, \mathbf{A}_\ell, \mathbf{u})$.

---

[11] More specifically, rejection sampling can be used to obtain a sufficiently good approximation.

- **Key generation:** On input the master secret key msk and an identity $\mathbf{id} = (\mathrm{id}_1, \ldots, \mathrm{id}_\ell) \in \mathbb{Z}_p^\ell$, sample $s_1, \ldots, s_\ell \leftarrow \mathbb{Z}_p$ and computes

$$\mathbf{F}_{\mathbf{id},(s_1,\ldots,s_\ell)} = \left[\mathbf{A}_0 \,\middle|\, \left(\sum_{i\in[\ell]} s_i \mathbf{A}_i\right) + \left(\sum_{i\in[\ell]} s_i \cdot \mathrm{id}_i\right)\mathbf{B}\right] \in \mathbb{Z}_p^{2\times 2m}.$$

Then, sample $\mathbf{v} \leftarrow \mathbb{Z}_p^{2m}$ such that $\mathbf{F}_{\mathbf{id},(s_1,\ldots,s_\ell)} \cdot \mathbf{v} = \mathbf{u} \pmod p$ and set $\mathbf{z} = g^{\mathbf{v}} \in \mathbb{G}^{2m}$. Outputs $\mathrm{sk}_{\mathbf{id}} = (s_1, \ldots, s_\ell, \mathbf{z})$.
- **Encryption:** On input the public parameters pp, an identity $\mathbf{id} = (\mathrm{id}_1, \ldots, \mathrm{id}_\ell) \in \mathbb{Z}_p^\ell$, and a message $\mathfrak{m} \in \mathbb{G}_T$, sample $\mathbf{r} \leftarrow \mathbb{Z}_p^2$. Set $\mathbf{c}_0^\mathsf{T} = g^{\mathbf{r}^\mathsf{T}\mathbf{A}_0} \in \mathbb{G}^{1\times m}$, $\mathbf{c}_i^\mathsf{T} = g^{\mathbf{r}^\mathsf{T}[\mathbf{A}_i + \mathrm{id}_i\mathbf{B}]} \in \mathbb{G}^{1\times m}$ for all $i \in [\ell]$, $c_{\ell+1} = \hat{e}(g,g)^{\mathbf{r}^\mathsf{T}\mathbf{u}} \cdot \mathfrak{m} \in \mathbb{G}_T$, and output $(\mathbf{c}_0, \mathbf{c}_1, \ldots, \mathbf{c}_\ell, c_{\ell+1}) \in \mathbb{G}^{(\ell+1)m} \times \mathbb{G}_T$.
- **Decryption:** On input a ciphertext $c = (\mathbf{c}_0, \mathbf{c}_1, \ldots, \mathbf{c}_\ell, c_{\ell+1})$ and a secret key $\mathrm{sk} = (s_1, \ldots, s_\ell, \mathbf{z})$, output

$$\mathfrak{m} = c_{\ell+1} \cdot \hat{e}\left(\left[\mathbf{c}_0 \,\middle|\, \prod_{i\in[\ell]} \mathbf{c}_i^{s_i}\right], \mathbf{z}\right)^{-1}.$$

Correctness. Note that

$$\mathbf{d}^\mathsf{T} = \left[\mathbf{c}_0^\mathsf{T} \,\middle|\, \prod_{i\in[\ell]}(\mathbf{c}_i^\mathsf{T})^{s_i}\right] = g^{\mathbf{r}^\mathsf{T}[\mathbf{A}_0 | \sum_{i\in[\ell]} s_i\mathbf{A}_i + (\sum_{i\in[\ell]} s_i\cdot\mathrm{id}_i)\mathbf{B}]} = g^{\mathbf{r}^\mathsf{T}\mathbf{F}_{\mathbf{id},(s_1,\ldots,s_\ell)}}.$$

We have $\hat{e}(\mathbf{d}, \mathbf{z}) = \hat{e}(g,g)^{\mathbf{r}^\mathsf{T}\mathbf{F}_{\mathbf{id},(s_1,\ldots,s_\ell)}\cdot\mathbf{v}} = \hat{e}(g,g)^{\mathbf{r}^\mathsf{T}\mathbf{u}}$. Therefore, dividing $c_{\ell+1}$ by $\hat{e}(\mathbf{d},\mathbf{z})$ eliminates the term $\hat{e}(g,g)^{\mathbf{r}^\mathsf{T}\mathbf{u}}$ which recovers $\mathfrak{m}$ correctly.

Security. Due to space limitations we refer the reader to the full version [19] for the proof of the following theorem. Below we briefly highlight the main ideas underlying its proof.

Theorem 3.1. *The scheme $\mathcal{IBE}_{\mathsf{DLIN1}}$ is selectively data private based on the DLIN assumption, and is function private for:*

1. *(T,k)-block-sources for any $T = \mathrm{poly}(\lambda)$ and $k \geq \lambda + \omega(\log\lambda)$.*
2. *$(k_1, \ldots, k_T)$-sources for any $T = \mathrm{poly}(\lambda)$ and $(k_1, \ldots, k_T)$ such that $k_i \geq i \cdot \lambda + \omega(\log\lambda)$ for every $i \in [T]$.*

Proof Overview. The function privacy of the scheme follows quite naturally from our "extract" step, as discussed in Section 1.1. To prove selective data privacy under the DLIN assumption, given the challenge identity $\mathbf{id}^*$, we set up the public parameters $\{g^{\mathbf{A}_i}\}_{i\in[\ell]}$, $\mathbf{B}$, and $g^{\mathbf{u}}$ such that the matrix $\mathbf{G}_{\mathbf{id},\mathbf{s}} \stackrel{\mathrm{def}}{=} \left[\left(\sum_{i\in[\ell]} s_i\mathbf{A}_i\right) + \left(\sum_{i\in[\ell]} s_i\cdot\mathrm{id}_i\right)\mathbf{B}\right]$ is equipped with a 'punctured' trapdoor. This trapdoor allows us to sample a vector such that $\mathbf{F}_{\mathbf{id},\mathbf{s}}\cdot\mathbf{v} = \mathbf{u}$ whenever $\mathbf{G}_{\mathbf{id},\mathbf{s}}$

contains a non-zero scalar multiple of $\mathbf{B}$. This occurs whenever $\sum_{i \in [\ell]} s_i(\mathrm{id}_i - \mathrm{id}_i^*) \neq 0$. Thus, with all but a negligible probability, we can simulate the adversary's key-generation queries with specially chosen matrices as above.

To embed the DLIN challenge, the first two rows of the DLIN challenge are used to construct the public parameter $g^{\mathbf{A}_0}$. The third row of the challenge is either linearly dependent on the first two rows or chosen uniformly at random and independently. The third row of the challenge is embedded into the augmented challenge ciphertext that is either well-formed or uniform and independent of the adversary's view depending on the DLIN challenge. This is done by choosing secret matrices $\mathbf{R}_i^*$ and having $\mathbf{A}_i = \mathbf{A}_0 \mathbf{R}_i^* - \mathrm{id}_i^* \mathbf{B}$. This generalizes the ideas of [3,40] to fit our "extract-augment-combine" approach and therefore provide function privacy.

4   Extensions and Open Problems

Our framework for function privacy yields a variety of extensions and open problems, both conceptual ones regarding our new notions, and technical ones regarding our specific approach and its resulting constructions. We now discuss several such extensions and open problems.

Chosen-Ciphertext Security. In terms of data privacy, in this paper we considered the standard notion of anonymity and message indistinguishability under an adaptive chosen-identity chosen-plaintext attack (known as anon-IND-ID-CPA). A natural extension of our results is to guarantee data privacy even against chosen-ciphertext attacks (known as anon-IND-ID-CCA). We note that our IBE schemes can be extended, using standard techniques, into two-level hierarchical IBE schemes that are anon-IND-ID-CPA-secure and their first level is function private. Then, by applying the generic transformation of Boneh, Canetti, Halevi and Katz [16], any such scheme can be used to construct an IBE scheme that is anon-IND-ID-CCA-secure and function private.

Applying Our Approach to other IBE Schemes. In Section 2 we presented simple attacks exemplifying that the anonymous IBE schemes presented in [18,32,3,40] are not function private. Nevertheless, we were able to rely on these schemes for designing new ones that are function private using our "extract-augment-combine" approach. For other anonymous IBE schemes, such as [31,22,11], we were not able to find attacks against their function privacy. An interesting open problem is to explore whether these schemes can be modified (possibly by applying our "extract-augment-combine" approach) to be function private based on standard assumptions. More generally, a natural open problem is to identify a specific property of identity-based encryption schemes that make them amenable to our "extract-augment-combine" approach.

Extension to Other Classes of Functions. As discussed in Section 1, in the general setting of functional encryption our schemes provide function privacy for the class of functions f_{id^*} defined as $f_{id^*}(id, m) = m$ if $id = id^*$, and $f_{id^*}(id, m) = \bot$ otherwise. A fascinating open problem is to construct schemes that are function private for other classes of functions. A possible starting point is to consider function privacy for other, rather simple, functionalities, such as inner-product testing [39].

Robustness of Our Schemes. As pointed out by Abdalla, Bellare, and Neven [2], when using an anonymous IBE scheme as a public-key searchable encryption scheme [17,1], it is often desirable to use a "robust" IBE scheme: It should be difficult to produce a ciphertext that is valid for more than one identity. We note that our schemes do not satisfy such a notion of robustness. However, Abdalla et al. showed two generic transformations that transform any given IBE scheme into a robust one. In particular, these transformations can be applied to each of our schemes to make them robust (these transformations do not change the decryption keys, and thus function privacy is preserved). We leave it as an open problem to directly design function-private IBE schemes that are robust.

Acknowledgements. We thank the anonymous CRYPTO '13 reviewers for many useful comments. This work was supported by NSF, the DARPA PROCEED program, an AFOSR MURI award, a grant from ONR, an IARPA project provided via DoI/NBC, and by Samsung. Opinions, findings and conclusions or recommendations expressed in this material are those of the author(s) and do not necessarily reflect the views of DARPA or IARPA. Distrib. Statement "A:" Approved for Public Release, Distribution Unlimited.

References

1. Abdalla, M., Bellare, M., Catalano, D., Kiltz, E., Kohno, T., Lange, T., Malone-Lee, J., Neven, G., Paillier, P., Shi, H.: Searchable encryption revisited: Consistency properties, relation to anonymous IBE, and extensions. Journal of Cryptology 21(3), 350–391 (2008)
2. Abdalla, M., Bellare, M., Neven, G.: Robust encryption. In: Micciancio, D. (ed.) TCC 2010. LNCS, vol. 5978, pp. 480–497. Springer, Heidelberg (2010)
3. Agrawal, S., Boneh, D., Boyen, X.: Efficient lattice (H)IBE in the standard model. In: Gilbert, H. (ed.) EUROCRYPT 2010. LNCS, vol. 6110, pp. 553–572. Springer, Heidelberg (2010)
4. Agrawal, S., Freeman, D.M., Vaikuntanathan, V.: Functional encryption for inner product predicates from learning with errors. In: Lee, D.H., Wang, X. (eds.) ASIACRYPT 2011. LNCS, vol. 7073, pp. 21–40. Springer, Heidelberg (2011)
5. Agrawal, S., Gorbunov, S., Vaikuntanathan, V., Wee, H.: Functional encryption: New perspectives and lower bounds. In: Canetti, R., Garay, J.A. (eds.) CRYPTO 2013, Part II. LNCS, vol. 8043, pp. 500–518. Springer, Heidelberg (2013)
6. Baek, J., Safavi-Naini, R., Susilo, W.: Public key encryption with keyword search revisited. In: Gervasi, O., Murgante, B., Laganà, A., Taniar, D., Mun, Y., Gavrilova, M.L. (eds.) ICCSA 2008, Part I. LNCS, vol. 5072, pp. 1249–1259. Springer, Heidelberg (2008)
7. Barak, B., Goldreich, O., Impagliazzo, R., Rudich, S., Sahai, A., Vadhan, S.P., Yang, K.: On the (im)possibility of obfuscating programs. Journal of the ACM 59(2), 6 (2012)

8. Bellare, M., Boldyreva, A., O'Neill, A.: Deterministic and efficiently searchable encryption. In: Menezes, A. (ed.) CRYPTO 2007. LNCS, vol. 4622, pp. 535–552. Springer, Heidelberg (2007)

9. Bellare, M., Brakerski, Z., Naor, M., Ristenpart, T., Segev, G., Shacham, H., Yilek, S.: Hedged public-key encryption: How to protect against bad randomness. In: Matsui, M. (ed.) ASIACRYPT 2009. LNCS, vol. 5912, pp. 232–249. Springer, Heidelberg (2009)

10. Bellare, M., Fischlin, M., O'Neill, A., Ristenpart, T.: Deterministic encryption: Definitional equivalences and constructions without random oracles. In: Wagner, D. (ed.) CRYPTO 2008. LNCS, vol. 5157, pp. 360–378. Springer, Heidelberg (2008)

11. Bellare, M., Kiltz, E., Peikert, C., Waters, B.: Identity-based (Lossy) trapdoor functions and applications. In: Pointcheval, D., Johansson, T. (eds.) EUROCRYPT 2012. LNCS, vol. 7237, pp. 228–245. Springer, Heidelberg (2012)

12. Bellare, M., O'Neill, A.: Semantically-secure functional encryption: Possibility results, impossibility results and the quest for a general definition. Cryptology ePrint Archive, Report 2012/515 (2012)

13. Bellare, M., Rogaway, P.: Random oracles are practical: A paradigm for designing efficient protocols. In: Proceedings of the 1st ACM Conference on Computer and Communications Security, pp. 62–73 (1993)

14. Bethencourt, J., Song, D., Waters, B.: New techniques for private stream searching. ACM Transactions on Information and System Security 12(3) (2009)

15. Boldyreva, A., Fehr, S., O'Neill, A.: On notions of security for deterministic encryption, and efficient constructions without random oracles. In: Wagner, D. (ed.) CRYPTO 2008. LNCS, vol. 5157, pp. 335–359. Springer, Heidelberg (2008)

16. Boneh, D., Canetti, R., Halevi, S., Katz, J.: Chosen-ciphertext security from identity-based encryption. SIAM Journal on Computing 36(5), 1301–1328 (2007)

17. Boneh, D., Di Crescenzo, G., Ostrovsky, R., Persiano, G.: Public key encryption with keyword search. In: Cachin, C., Camenisch, J.L. (eds.) EUROCRYPT 2004. LNCS, vol. 3027, pp. 506–522. Springer, Heidelberg (2004)

18. Boneh, D., Franklin, M.K.: Identity-based encryption from the Weil pairing. SIAM Journal on Computing 32(3), 586–615 (2003); Preliminary version in Kilian, J. (ed.): CRYPTO 2001. LNCS, vol. 2139, pp. 213–229. Springer, Heidelberg (2001)

19. Boneh, D., Raghunathan, A., Segev, G.: Function-private identity-based encryption: Hiding the function in functional encryption. Cryptology ePrint Archive, Report 2013/283 (2013)

20. Boneh, D., Sahai, A., Waters, B.: Functional encryption: Definitions and challenges. In: Ishai, Y. (ed.) TCC 2011. LNCS, vol. 6597, pp. 253–273. Springer, Heidelberg (2011)

21. Boneh, D., Waters, B.: Conjunctive, subset, and range queries on encrypted data. In: Vadhan, S.P. (ed.) TCC 2007. LNCS, vol. 4392, pp. 535–554. Springer, Heidelberg (2007)

22. Boyen, X., Waters, B.: Anonymous hierarchical identity-based encryption (Without random oracles). In: Dwork, C. (ed.) CRYPTO 2006. LNCS, vol. 4117, pp. 290–307. Springer, Heidelberg (2006)

23. Brakerski, Z., Segev, G.: Better security for deterministic public-key encryption: The auxiliary-input setting. In: Rogaway, P. (ed.) CRYPTO 2011. LNCS, vol. 6841, pp. 543–560. Springer, Heidelberg (2011)

24. Camenisch, J., Kohlweiss, M., Rial, A., Sheedy, C.: Blind and anonymous identity-based encryption and authorised private searches on public key encrypted data. In: Jarecki, S., Tsudik, G. (eds.) PKC 2009. LNCS, vol. 5443, pp. 196–214. Springer, Heidelberg (2009)

25. Canetti, R.: Towards realizing random oracles: Hash functions that hide all partial information. In: Kaliski Jr., B.S. (ed.) CRYPTO 1997. LNCS, vol. 1294, pp. 455–469. Springer, Heidelberg (1997)

26. Canetti, R., Kalai, Y.T., Varia, M., Wichs, D.: On symmetric encryption and point obfuscation. In: Micciancio, D. (ed.) TCC 2010. LNCS, vol. 5978, pp. 52–71. Springer, Heidelberg (2010)

27. Chang, Y.-C., Mitzenmacher, M.: Privacy preserving keyword searches on remote encrypted data. In: Ioannidis, J., Keromytis, A.D., Yung, M. (eds.) ACNS 2005. LNCS, vol. 3531, pp. 442–455. Springer, Heidelberg (2005)

28. Chase, M., Kamara, S.: Structured encryption and controlled disclosure. In: Abe, M. (ed.) ASIACRYPT 2010. LNCS, vol. 6477, pp. 577–594. Springer, Heidelberg (2010)

29. Curtmola, R., Garay, J.A., Kamara, S., Ostrovsky, R.: Searchable symmetric encryption: Improved definitions and efficient constructions. Journal of Computer Security 19(5), 895–934 (2011)

30. Fuller, B., O'Neill, A., Reyzin, L.: A unified approach to deterministic encryption: New constructions and a connection to computational entropy. In: Cramer, R. (ed.) TCC 2012. LNCS, vol. 7194, pp. 582–599. Springer, Heidelberg (2012)

31. Gentry, C.: Practical identity-based encryption without random oracles. In: Vaudenay, S. (ed.) EUROCRYPT 2006. LNCS, vol. 4004, pp. 445–464. Springer, Heidelberg (2006)

32. Gentry, C., Peikert, C., Vaikuntanathan, V.: Trapdoors for hard lattices and new cryptographic constructions. In: Proceedings of the 40th Annual ACM Symposium on Theory of Computing, pp. 197–206 (2008)

33. Goldreich, O., Ostrovsky, R.: Software protection and simulation on oblivious rams. Journal of the ACM 43(3), 431–473 (1996)

34. Goldwasser, S., Kalai, Y.T., Popa, R.A., Vaikuntanathan, V., Zeldovich, N.: Reusable garbled circuits and succinct functional encryption. In: Proceedings of the 45th Annual ACM Symposium on Theory of Computing (to appear, 2013)

35. Goldwasser, S., Kalai, Y.T.: On the impossibility of obfuscation with auxiliary input. In: Proceedings of the 46th Annual IEEE Symposium on Foundations of Computer Science, pp. 553–562 (2005)

36. Golle, P., Staddon, J., Waters, B.: Secure conjunctive keyword search over encrypted data. In: Jakobsson, M., Yung, M., Zhou, J. (eds.) ACNS 2004. LNCS, vol. 3089, pp. 31–45. Springer, Heidelberg (2004)

37. Gorbunov, S., Vaikuntanathan, V., Wee, H.: Functional encryption with bounded collusions via multi-party computation. In: Safavi-Naini, R., Canett, R. (eds.) CRYPTO 2012. LNCS, vol. 7417, pp. 162–179. Springer, Heidelberg (2012)

38. Kamara, S., Papamanthou, C., Roeder, T.: Dynamic searchable symmetric encryption. In: ACM Conference on Computer and Communications Security, pp. 965–976 (2012)

39. Katz, J., Sahai, A., Waters, B.: Predicate encryption supporting disjunctions, polynomial equations, and inner products. In: Smart, N.P. (ed.) EUROCRYPT 2008. LNCS, vol. 4965, pp. 146–162. Springer, Heidelberg (2008)

40. Kurosawa, K., Phong, L.T.: Maximum leakage resilient IBE and IPE. Cryptology ePrint Archive, Report 2011/628 (2011)

41. Lynn, B., Prabhakaran, M., Sahai, A.: Positive results and techniques for obfuscation. In: Cachin, C., Camenisch, J.L. (eds.) EUROCRYPT 2004. LNCS, vol. 3027, pp. 20–39. Springer, Heidelberg (2004)

42. Mironov, I., Pandey, O., Reingold, O., Segev, G.: Incremental deterministic public-key encryption. In: Pointcheval, D., Johansson, T. (eds.) EUROCRYPT 2012. LNCS, vol. 7237, pp. 628–644. Springer, Heidelberg (2012)

43. O'Neill, A.: Definitional issues in functional encryption. IACR Cryptology ePrint Archive, Report 2010/556 (2010)

44. Ostrovsky, R., Skeith III., W.E.: Private searching on streaming data. Journal of Cryptology 20(4), 397–430 (2007)

45. Raghunathan, A., Segev, G., Vadhan, S.: Deterministic public-key encryption for adaptively chosen plaintext distributions. In: Johansson, T., Nguyen, P.Q. (eds.) EUROCRYPT 2013. LNCS, vol. 7881, pp. 93–110. Springer, Heidelberg (2013)

46. Shen, E., Shi, E., Waters, B.: Predicate privacy in encryption systems. In: Reingold, O. (ed.) TCC 2009. LNCS, vol. 5444, pp. 457–473. Springer, Heidelberg (2009)

47. Shi, E., Bethencourt, J., Chan, H.T.-H., Song, D., Perrig, A.: Multi-dimensional range query over encrypted data. In: IEEE Symposium on Security and Privacy, pp. 350–364 (2007)

48. Song, D.X., Wagner, D., Perrig, A.: Practical techniques for searches on encrypted data. In: IEEE Symposium on Security and Privacy, pp. 44–55 (2000)

49. van Liesdonk, P., Sedghi, S., Doumen, J., Hartel, P., Jonker, W.: Computationally efficient searchable symmetric encryption. In: Jonker, W., Petković, M. (eds.) SDM 2010. LNCS, vol. 6358, pp. 87–100. Springer, Heidelberg (2010)

50. Wee, H.: On obfuscating point functions. In: Proceedings of the 37th Annual ACM Symposium on Theory of Computing, pp. 523–532 (2005)

51. Wee, H.: Dual projective hashing and its applications — lossy trapdoor functions and more. In: Pointcheval, D., Johansson, T. (eds.) EUROCRYPT 2012. LNCS, vol. 7237, pp. 246–262. Springer, Heidelberg (2012)

Attribute-Based Encryption
for Circuits from Multilinear Maps

Sanjam Garg[1,*], Craig Gentry[2,**], Shai Halevi[2,**], Amit Sahai[1,*],
and Brent Waters[3,***]

[1] UCLA
{sanjamg,sahai}@cs.ucla.edu
[2] IBM Research
{cbgentry,shaih}@us.ibm.com
[3] UT Austin
bwaters@cs.utexas.edu

Abstract. In this work, we provide the first construction of Attribute-Based Encryption (ABE) for general circuits. Our construction is based on the existence of multilinear maps. We prove selective security of our scheme in the standard model under the natural multilinear generalization of the BDDH assumption. Our scheme achieves both Key-Policy and Ciphertext-Policy variants of ABE. Our scheme and its proof of security directly translate to the recent multilinear map framework of Garg, Gentry, and Halevi.

* Research supported in part from a DARPA/ONR PROCEED award, NSF grants 1228984, 1136174, 1118096, 1065276, 0916574 and 0830803, a Xerox Faculty Research Award, a Google Faculty Research Award, an equipment grant from Intel, and an Okawa Foundation Research Grant. This material is based upon work supported by the Defense Advanced Research Projects Agency through the U.S. Office of Naval Research under Contract N00014-11-1-0389. The views expressed are those of the author and do not reflect the official policy or position of the Department of Defense, the National Science Foundation, or the U.S. Government.

** This work was supported by the Intelligence Advanced Research Projects Activity (IARPA) via Department of Interior National Business Center (DoI/NBC) contract number D11PC20202. The U.S. Government is authorized to reproduce and distribute reprints for Governmental purposes notwithstanding any copyright annotation thereon. Disclaimer: The views and conclusions contained herein are those of the authors and should not be interpreted as necessarily representing the official policies or endorsements, either expressed or implied, of IARPA, DoI/NBC, or the U.S. Government.

*** Supported by NSF CNS-0915361 and CNS-0952692, CNS-1228599 DARPA via Office of Naval Research under Contract N00014-11-1-0382, DARPA N11AP20006, the Alfred P. Sloan Fellowship, and Microsoft Faculty Fellowship, and Packard Foundation Fellowship. Any opinions, findings, and conclusions or recommendations expressed in this material are those of the author(s) and do not necessarily reflect the views of the Department of Defense or the U.S. Government.

R. Canetti and J.A. Garay (Eds.): CRYPTO 2013, Part II, LNCS 8043, pp. 479–499, 2013.

1   Introduction

In traditional public key encryption a sender will encrypt a message to a targeted individual recipient using the recipient's public key. However, in many applications one may want to have a more general way of expressing who should be able to view encrypted data. Sahai and Waters [SW05] introduced the notion of Attribute-Based Encryption (ABE). There are two variants of ABE: Key-Policy ABE and Ciphertext-Policy ABE [GPSW06]. (We will consider both these variants in this work.) In a Key-Policy ABE system, a ciphertext encrypting a message M is associated with an assignment x of boolean variables. A secret key SK is issued by an authority and is associated with a boolean function f chosen from some class of allowable functions $\mathcal{F}$. A user with a secret key for f can decrypt a ciphertext associated with x, if and only if $f(x) = 1$.

Since the introduction of ABE there have been advances in multiple directions. These include: new proof techniques to achieve adaptive security [LOS$^+$10, OT10, LW12], decentralizing trust among multiple authorities [Cha07, CC09, LW11], and applications to outsourcing computation [PRV12].

However, the central challenge of expanding the *class* of allowable boolean functions $\mathcal{F}$ has been very resistant to attack. Viewed in terms of circuit classes, the work of Goyal *et al* [GPSW06] achieved the best result until now; their construction achieved security essentially for circuits in the complexity class $\mathbf{NC^1}$. This is the class of circuits with depth $\log n$, or equivalently, the class of functions representable by polynomial-size boolean formulas. Achieving ABE for general circuits is arguably the central open direction in this area[1].

Difficulties in Achieving Circuit ABE and the Backtracking Attack. To understand why achieving ABE for general circuits has remained a difficult problem, it is instructive to examine the mechanisms of existing constructions based on bilinear maps. Intuitively, a bilinear map allows one to decrypt using group elements as keys (or key components) as opposed to exponents. By handing out a secret key that consists of group elements, an authority is able to computationally hide some secrets embedded in that key from the key holder herself. In contrast, if a secret key consists of exponents in $\mathbb{Z}_p$ for a prime order group p, as in say an ElGamal type system, then the key holder or collusion of key holders can solve for these secrets using algebra. This computational hiding in bilinear map based systems allows an authority to personalize keys to a user and prevent collusion attacks, which are the central threat.

Using GPSW [GPSW06] as a canonical example we illustrate some of the main principles of decryption. In their system, private keys consist of bilinear group elements for a group of prime order p and are associated with random values $r_y \in \mathbb{Z}_p$ for each leaf node y in the boolean formula f. A ciphertext

---

[1] We note that if collusions between secret key holders are bounded by a publicly known polynomially-bounded number in advance, then even stronger results are known [SS10, GVW12]. However, throughout this paper we will deal only with the original setting of ABE where unbounded collusions are allowed between adversarial users.

encrypted to descriptor x has randomness $s \in \mathbb{Z}_p$. The decryption algorithm begins by applying a pairing operation to each "satisfied" leaf node and obtains $e(g,g)^{r_y s}$ for each satisfied node y. From this point onward decryption consists solely of finding if there is a linear combination (in the exponent) of the r_y values that can lead to computing $e(g,g)^{\alpha s}$, which will be the "blinding factor" hiding the message M. (The variable $e(g,g)^{\alpha}$ is defined in the public parameters.) The decryption algorithm should be able to find such a linear combination only if $f(x) = 1$. Of particular note is that once the $e(g,g)^{r_y s}$ values are computed the pairing operation plays no further role in decryption. Indeed, it cannot since it is intuitively "used up" on the initial step.

Let's now take a closer look at how GPSW structures a private key for a given boolean formula. Suppose inside a particular boolean formula there exists an OR gate T that received inputs from gates A and B. Then the authority will associate gate T with a value r_T and gates A, B with values $r_A = r_B = r_T$ to match the OR functionality. Now suppose that on a certain input assignment x that gate A evaluates to 1, but gate B evaluates to 0. The decryptor will then learn the "decryption value" $e(g,g)^{s r_A}$ for gate A and can interpolate up by simply by noting that $e(g,g)^{s r_T} = e(g,g)^{s r_A}$. While this structure reflects an OR gate, it also has a critical side effect. The decryption algorithm also learns the decryption value $e(g,g)^{s r_B}$ for gate B *even though gate B evaluates to 0* on input x. We call such a discovery a *backtracking attack*.

Boolean formulas are circuits with fanout one. If the fanout is one, then the backtracking attack produces no ill effect since an attacker has nowhere else to go with this information that he has learned. However, suppose we wanted to extend this structure with circuits of fanout of two or more, and that gate B also fed into an AND gate R. In this case the backtracking attack would allow an attacker to act like B was satisfied in the formula even though it was not. This misrepresentation can then be propagated up a different path in the circuit due to the larger fanout. (Interestingly, this form of attack does not involve collusion with a second user.)

We believe that such backtracking attacks are the principle reason that the functionality of existing ABE systems has been limited to circuits of fanout one. Furthermore, we conjecture that since the pairing operation is used up in the initial step, that there is no black-box way of realizing general ABE for circuits from bilinear maps.

Our Results. We present a new methodology for constructing Attribute-Based Encryption systems for circuits of arbitrary fanout. Our method is described using multilinear maps. Cryptography with multilinear maps was first postulated by Boneh and Silverberg [BS02] where they discussed potential applications such as one round, n-way Diffie-Hellman key exchange. However, they also gave evidence that it might be difficult or not possible to find useful multilinear forms within the realm of algebraic geometry. For this reason there has existed a general reluctance among cryptographers to explore multilinear map constructions even though in some constructions such as the Boneh-Goh-Nissim [BGN05] slightly homomorphic encryption system, or the Boneh-Sahai-Waters [BSW06] Traitor

Tracing scheme, there appears to exist direct generalizations of bilinear map solutions.

Very recently, Garg, Gentry, and Halvei [GGH13a] (see [GGH12b] for full version) announced a surprising result. Using ideal lattices they produced a candidate mechanism that would approximate or be the moral equivalent of multilinear maps for many applications. Speculative applications include translations of existing bilinear map constructions and direct generalizations as well as future applications. While the development and cryptanalysis of their tools is at a nascent stage, we believe that their result opens an exciting opportunity to study new constructions using a multilinear map abstraction. The promise of these results is that such constructions can be brought over to their framework or a related future one. We believe that building ABE for circuits is one of the most exciting of these problems due to the challenges discussed above and that existing bilinear map constructions do not have a direct generalization.

Our circuit ABE construction and its proof of security directly translate to the framework of [GGH12b].

We construct an ABE system of the Key-Policy variety where ciphertext descriptors are an n-tuple x of boolean variables and keys are associated with boolean circuits of a max depth ℓ, where both ℓ and n are polynomially bounded and determined at the time of system setup. Our main construction exposition is for circuits that are layered (where gates at depth j get inputs from gates at depth $j-1$) and monotonic (consisting only of AND plus OR gates). Neither one of these impacts our general result as a generic circuit can be transformed into a layered one for the same function with a small amount of overhead. In addition, using De Morgan's law one can build a general circuit from a monotone circuit with negation only appearing at the input wires. We sketch this in Section 2. We finally note that using universal circuits we can realize "Ciphertext-Policy" style ABE systems for circuits.

We use a framework of leveled multilinear maps is that a party can call a group generator $\mathcal{G}(1^\lambda, k)$ to obtain a sequence of groups $\boldsymbol{G} = (\mathbb{G}_1, \ldots, \mathbb{G}_k)$ each of large prime[2] order $p > 2^\lambda$ where each comes with a canonical generator $g = g_1, \ldots, g_k$. Slightly abusing notation, if $i + j \leq k$ we can compute a bilinear map operation on $g_i^a \in \mathbb{G}_i, g_j^b \in \mathbb{G}_j$ as $e(g_i^a, g_j^b) = g_{i+j}^{ab}$. These maps can be seen as implementing multilinear maps[3]. It is the need to commit to a certain k value which will require the setup algorithm of our construction to commit to a maximum depth $\ell = k-1$. We will prove security under a generalization of the decision BDH assumption that we call the decision k-multilinear assumption. Roughly, it states that given $g, g^s, g^{c_1}, \ldots, g^{c_k}$ it is hard to distinguish $T = g_k^{s \prod_{j \in [1,k]} c_k}$ from a random element of $\mathbb{G}_k$.

---

[2] We stress that our techniques do not rely on the groups being of prime order; we only need that certain randomization properties hold in a statistical sense (which hold perfectly over groups of prime order). Therefore, our techniques generalize to other algebraic settings.

[3] We technically consider the existence of a set of bilinear maps $\{e_{i,j} : \mathbb{G}_i \times \mathbb{G}_j \to \mathbb{G}_{i+j} \mid i, j \geq 1; \ i + j \leq k\}$, but will often abuse notation for ease of exposition.

Our Techniques. As discussed there is no apparent generalization of the GPSW methods for achieving ABE for general circuits. We develop new techniques with a focus on preventing the backtracking attacks we described above. Intuitively, we describe our techniques as "move forward and shift"; this *replaces and subsumes* the linear interpolation method of GPSW decryption. In particular, our schemes do not rely on any sophisticated linear secret sharing schemes, as was done by GPSW.

Consider a private key for a given monotonic[4] circuit f with max depth ℓ that works over a group sequence $(\mathbb{G}_1, \ldots, \mathbb{G}_k)$. Each wire w in f is associated by the authority with a random value $r_w \in \mathbb{Z}_p$. A ciphertext for descriptor x will be associated with randomness $s \in \mathbb{Z}_p$. A user should with secret key for f should be able to decrypt if and only if $f(x) = 1$.

The decryption algorithm works by computing $g_{j+1}^{s r_w}$ for each wire w in the circuit that evaluates to 1 on input x. If the wire is 0, the decryptor should not be able to obtain this value. Decryption works from the bottom up. For each input wire w at depth 1, we compute $g_2^{s r_w}$ using a very similar mechanism to GPSW.

We now turn our attention to OR gates to illustrate how we prevent backtracking attacks. Suppose wire w is the output of an OR gate with input wires $A(w), B(w)$ at depth j. Furthermore, suppose on a given input x the wire $A(w)$ evaluates to true and $B(w)$ to false so that the decryptor has $g_j^{s r A(w)}$, but not $g_j^{s r B(w)}$. The private key components associated with wire w are:

$$g^{a_w}, \quad g^{b_w}, \quad g_j^{r_w - a_w \cdot r_{A(w)}}, \quad g_j^{r_w - b_w \cdot r_{B(w)}}$$

for random a_w, b_w. To move decryption onward the algorithm first computes

$$e\left(g^{a_w}, g_j^{s r_{A(w)}}\right) = g_{j+1}^{s a_w r_{A(w)}}.$$

This is the move forward step. Then it computes

$$e\left(g^s, g_j^{r_w - a_w \cdot r_{A(w)}}\right) = g_{j+1}^{s(r_w - a_w r_{A(w)})}.$$

This is the shift step. Multiplying these together gives the desired term $g_{j+1}^{s r_w}$.

Let's examine backtracking attacks in this context. Recall that the attacker's goal is to compute $g_j^{s r B(w)}$ even though wire $B(w)$ is 0, and propagate this forward. From the output term and the fourth key component the attacker can actually inverse the shift process on the B side and obtain $g_{j+1}^{s b_w r_{B(w)}}$. However, since the map e works only in the "forward" direction, it is not possible to invert the move forward step and complete the attack. The crux of our security lies in this idea.

The AND gate mechanism has a similar shift and move forward structure, but requires both inputs for decryption. If this process is applied iteratively to

---

[4] Recall that assuming that the circuit is monotonic is without loss of generality. Our method also applies to general circuits that involve negations. See Section 2.

an output gate $\tilde{w}$, then one obtains $g_k^{sr_{\tilde{w}}}$. A final header portion of the key and decryption mechanism is used to obtain the message. This portion is similar to prior work.

1.1 Other Related Work

Other recent functionality in a similar vain to ABE includes spatial encryption [Ham11] and regular language functionality [Wat12]. Neither of these seem to point to a path for achieving the general case of circuits. Indeed, [Wat12] argues that backtracking attacks are the reason that the constructions can only support Deterministic Finitie Automata and not Nondeterministic Finite Automata.

An interesting challenge going forward is whether new techniques can be applied to the general case of functional encryption [SW08, BSW11]. In this setting we would like to hide the input x as well as the message. So far the strongest functionality in this setting has been the inner product functionality of Katz, Sahai, and Waters [KSW08] and different variants of this [OT12].

There have been different lattice based constructions of IBE, HIBE, Fuzzy IBE, and ABE [CHKP10, ABB10, ABV$^+$12, Boy13]. While the high level proof structures of these systems follow the earlier bilinear map counterparts closely, the analogies seem to break down at lower level mechanisms. For example, there is more asymmetry in the construction of keys and ciphertexts — in bilinear maps they were both bilinear group elements. Rothblum [Rot12] considers the problem of circular security from bit encryption systems from ℓ-multilinear maps. He considers a different form than us where ℓ group elements of different types are input at once to a multilinear map function. The assumption used is a variant of XDH.

Parno, Raykova and Vaikuntanathan [PRV12] note that delegation from ABE can be achieved from a system that is not collusion resistant, however, they were not able to leverage this to go beyond the boolean formulas of [GPSW06]. The fact that the backtracking attacks described above do not use collusion attacks, but are attacks within a key might help explain this. In our construction the size of group elements and computational cost of group operations grows with the sequence number k and thus the depth of the circuit. Using our system combined with the PRV techniques one can achieve delegated computation where the delegator's work grows only with the depth of the circuit and not the size of the circuit. Since the number of multilinear levels must be bounded at setup, it is not clear if our techniques can be used to improve ABE-type applications in the uniform setting [Wat12].

Concurrent Work. Concurrent to and independent of our work Gorbunov, Vaikuntanathan, and Wee [GVW13] achieve ABE for circuits[5]. One nice feature

---

[5] Historical note: The present paper which merges [GGH12a] and [SW12] contains only a technical scheme and analysis already present in these works, with some additional elaboration. Thus the scheme and analysis presented here remains independent of [GVW13], and was developed concurrently to it.

of their result is that they reduce security to the Learning with Errors (LWE) problem [Reg05]. Both our result and theirs has "succinct" ciphertexts in that the ciphertext size grows with the maximum depth of the circuits and not the size. Goldwasser, Kalai, Popa, Vaikuntanathan, and Zeldovich [GKP+13] show how to combine such an ABE with fully homomorphic encryption into a succinct single use functional encryption scheme. This in turn implies results for reusable Yao garbled circuits and other applications.

Subsequent Work. Subsequent to our work Garg, Gentry, Sahai, and Waters [GGSW13] showed that a general primitive they termed witness encryption implies circuit ABE if we have witness indistinguishable proofs. Their techniques of moving from witness encryption to ABE are quite different from our direct construction. A drawback of using witness encryption is that current GGSW constructions rely on a different assumption for each NP instance.

1.2 Roadmap

We start by providing preliminary definition in Section 2. We give our construction based on (ideal) multilinear maps in Section 3 which is then translated to the GGH framework [GGH12b] in Section 4. We refer the reader to the full version [GGH+13b] for the proofs of security.

2 Preliminaries

In this section we provide some preliminaries. These include definition of ABE for circuits, discussion of monotone versus general circuits, our multilinear map convention and assumptions, and our circuit notation.

2.1 Definitions for ABE for Circuits

We now give a formal definition of our Attribute-Based Encryption for circuits. Our security definition essentially follows [GPSW06] with the exception that access structures are circuits. Our definition is fit for bounded circuits.

Setup$(1^\lambda, n, \ell)$**.** The setup algorithm takes as input the security parameter, the length n of input descriptors from the ciphertext and a bound ℓ on the circuit depth. It outputs the public parameters PP and a master key MSK.

Encrypt$(\text{PP}, x \in \{0,1\}^n, M)$**.** The encryption algorithm takes as input the public parameters PP, a bit string $x \in \{0,1\}^n$ representing the assignment of boolean variables, and a message m. It outputs a ciphertext CT.

Key Generation$(\text{MSK}, f = (n, q, A, B, \texttt{GateType}))$**.** The key generation algorithm takes as input the master key MSK and a description of a circuit f, where the depth of f is at most ℓ. The algorithm outputs a private key SK.

Decrypt(SK, CT)**.** The decryption algorithm takes as input a secret key SK and ciphertext CT. The algorithm attempts to decrypt and outputs a message M if successful; otherwise, it outputs a special symbol $\perp$.

Correctness. Consider all messages M, strings $x \in \{0,1\}^n$, and depth ℓ circuits f where $f(x) = 1$. If $Encrypt(\text{PP}, x, M) \to \text{CT}$ and $KeyGen(\text{MSK}, f) \to$ SK where PP, MSK were generated from a call to the setup algorithm, then $Decrypt(\text{SK}, \text{CT}) = M$.

Security Model for ABE for Circuits. We now briefly describe our security model of *selective security* for ABE for general circuits. We refer the reader to [GGH+13b] for a formal treatment. The selective security definition requires that the attacker first specifies the string x^* and later queries on multiple secret keys, but not ones that can trivially be used to decrypt a ciphertext encrypted under x^*. In particular the adversary can ask secret keys corresponding to any circuit f of his choice, such that $f(x^*) = 0$. The goal of the adversary is then to break semantic security of a challenge ciphertext encrypted under the string x^*.

2.2   General Circuits vs. Monotone Circuits

We begin by observing that there is a folklore transformation that uses De Morgan's rule to transform any general Boolean circuit into an equivalent monotone Boolean circuit, with negation gates only allowed at the inputs. For completeness, we sketch the construction here.

Given a Boolean circuit C, consider the Boolean circuit $\tilde{C}$ that computes the negation of C. Note that such a circuit can be generated by simply recursively applying De Morgan's rule to each gate of C starting at the output gate. The crucial property of this transformation is that in this circuit $\tilde{C}$ each wire computes the negation of the corresponding original wire in C.

Now, we can construct a monotone circuit M by combining C and $\tilde{C}$ as follows: take each negation gate inside C, eliminate it, and replace the output of the negation gate by the corresponding wire in $\tilde{C}$. Do the same for negation gates in $\tilde{C}$, using the wires from C. In the end, this will yield a monotone circuit M with negation gates remaining only at the input level, as desired. The size of M will be no more than twice the original size of C, and the depth of M will be identical to the depth of C, where depth is computed ignoring negation gates. The correctness of this transformation follows trivially from De Morgan's rule.

As a result, we can focus our attention on monotone circuits. Note that inputs to the circuit correspond to boolean variables x_i, and we can simply introduce explicit separate attributes corresponding to $x_i = 0$ and $x_i = 1$. Honest encryptors are instructed to only set one of these two attributes for each variable x_i.

Because of this simple transformation, in the sequel we will only consider ABE for monotone circuits.

2.3   Multilinear Maps

We assume the existence of a group generator $\mathcal{G}$, which takes as input a security parameter n and a positive integer k to indicate the number of allowed pairing operations. $\mathcal{G}(1^\lambda, k)$ outputs a sequence of groups $\mathbb{G} = (\mathbb{G}_1, \ldots, \mathbb{G}_k)$ each of large prime order $p > 2^\lambda$. In addition, we let g_i be a canonical generator of $\mathbb{G}_i$ (and is known from the group's description). We let $g = g_1$.

We assume the existence of a set of bilinear maps $\{e_{i,j} : \mathbb{G}_i \times \mathbb{G}_j \to \mathbb{G}_{i+j} \mid i,j \geq 1;\ i+j \leq k\}$. The map $e_{i,j}$ satisfies the following relation:

$$e_{i,j}\left(g_i^a, g_j^b\right) = g_{i+j}^{ab} \ :\ \forall a,b \in \mathbb{Z}_p.$$

We observe that one consequence of this is that $e_{i,j}(g_i, g_j) = g_{i+j}$ for each valid i,j.

When the context is obvious, we will sometimes abuse notation drop the subscripts i,j, For example, we may simply write:

$$e\left(g_i^a, g_j^b\right) = g_{i+j}^{ab}.$$

We define the k-Multilinear Decisional Diffie-Hellman (k-MDDH) assumption as follows:

Assumption 1 (k-Multilinear Decisional Diffie-Hellman: k-MDDH).
The k-Multilinear Decisional Diffie-Hellman (k-MDDH) problem states the following: A challenger runs $\mathcal{G}(1^\lambda, k)$ to generate groups and generators of order p. Then it picks random $s, c_1, \ldots, c_k \in \mathbb{Z}_p$.

The assumption then states that given $g = g_1, g^s, g^{c_1}, \ldots, g^{c_k}$ it is hard to distinguish $T = g_k^{s \prod_{j \in [1,k]} c_j}$ from a random group element in $\mathbb{G}_k$, with better than negligible advantage (in security parameter λ).

2.4   Circuit Notation

We now define our notation for circuits that adapts the model and notation of Bellare, Hoang, and Rogaway [BHR12] (Section 2.3). For our application we restrict our consideration to certain classes of boolean circuits. First, our circuits will have a single output gate. Next, we will consider layered circuits. In a layered circuit a gate at depth j will receive both of its inputs from wires at depth $j-1$. Finally, we will restrict ourselves to monotonic circuits where gates are either AND or OR gates of two inputs. [6]

Our circuits will be a five-tuple $f = (n, q, A, B, \texttt{GateType})$. We let n be the number of inputs and q be the number of gates. We define inputs $= \{1, \ldots, n\}$, Wires $= \{1, \ldots, n+q\}$, and Gates $= \{n+1, \ldots, n+q\}$. The wire $n+q$ is the designated output wire. $A : \text{Gates} \to \text{Wires/outputwire}$ is a function where $A(w)$ identifies w's first incoming wire and $B : \text{Gates} \to \text{Wires/outputwire}$ is a function where $B(w)$ identifies w's second incoming wire. Finally, $\texttt{GateType} : \text{Gates} \to \{\text{AND}, \text{OR}\}$ is a function that identifies a gate as either an AND or OR gate.

We require that $w > B(w) > A(w)$. We also define a function $\texttt{depth}(w)$ where if $w \in$ inputs $\texttt{depth}(w) = 1$ and in general $\texttt{depth}(w)$ of wire w is equal to the shortest path to an input wire plus 1. Since our circuit is layered we require that for all $w \in$ Gates that if $\texttt{depth}(w) = j$ then $\texttt{depth}(A(w)) = \texttt{depth}(B(w)) = j-1$.

---

[6] These restrictions are mostly useful for exposition and do not impact functionality. General circuits can be built from non-monotonic circuits. In addition, given a circuit an equivalent layered exists that is larger by at most a polynomial factor.

We will abuse notation and let $f(x)$ be the evaluation of the circuit f on input $x \in \{0,1\}^n$. In addition, we let $f_w(x)$ be the value of wire w of the circuit on input x.

3   Our Construction: Multilinear maps

We now describe our construction. Our main construction is of the Key-Policy form where a key generation algorithm takes in the description of a circuit f and encryption takes in an input x and message M. A user with secret key for f can decrypt if and only if $f(x) = 1$. The system is of the "public index" variety in that only the message M is hidden, while x can be efficiently discovered from the ciphertext, as is standard for ABE. We will also discuss how our KP-ABE scheme yields a Ciphertext-Policy ABE scheme for bounded-size circuits.

The setup algorithm will take as inputs a maximum depth ℓ of all the circuits as well as the input size n for all ciphertexts. All circuits f in our system will be of depth ℓ (have the output gate at depth ℓ) and be layered as discussed in Section 2.4. Using layered circuits and having all circuits be of the same depth is primarily for ease of exposition, as we believe that our construction could directly be adapted to the general case. The fact that setup defines a maximum depth ℓ is more fundamental as the algorithm defines a $k = \ell + 1$ group sequence a k pairings.

We also use the convention here that (multi-bit) messages are be encoded as group elements. In Section 4 we will translate this construction to the GGH setting.

Setup($1^\lambda, n, \ell$). The setup algorithm takes as input a security parameter λ, the maximum depth ℓ of a circuit, and the number of boolean inputs n.

It then runs $\mathcal{G}(1^\lambda, k = \ell + 1)$ that produces groups $\mathbb{G} = (\mathbb{G}_1, \ldots, \mathbb{G}_k)$ of prime order p, with canonical generators $g_1, \ldots, g_k$. We let $g = g_1$. Next, it chooses random $\alpha \in \mathbb{Z}_p$ and $h_1, \ldots, h_n \in \mathbb{G}_1$.

The public parameters, PP, consist of the group sequence description plus:

$$g_k^\alpha, h_1, \ldots, h_n.$$

The master secret key MSK is $(g_{k-1})^\alpha$.

Encrypt(PP, $x \in \{0,1\}^n, M \in \mathbb{G}_k$). The encryption algorithm takes in the public parameters, an descriptor input $x \in \{0,1\}^n$, and a message bit $M \in \mathbb{G}_k$. We use the convention that M is a group element.

The encryption algorithm chooses a random $s \in \mathbb{Z}_p$. It then sets $C_M = M \cdot (g_k^\alpha)^s$. We let S be the set of i such that $x_i = 1$.

The ciphertext is created as

$$\text{CT} = (C_M, \ g^s, \ \forall i \in S \ \ C_i = h_i^s).$$

KeyGen(MSK, $f = (n, q, A, B, \texttt{GateType})$). The algorithm takes in the master secret key and a description f of a circuit. Recall that the circuit has $n + q$ wires with n input wires, q gates and the wire $n + q$ designated as the output wire.

The key generation algorithm chooses random $r_1, \ldots, r_{n+q} \in \mathbb{Z}_p$, where we think of randomness r_w as being associated with wire w. The algorithm produces a "header" component

$$K_H = (g_{k-1})^{\alpha - r_{n+q}}.$$

Next, the algorithm generates key components for every wire w. The structure of the key components depends upon whether w is an input wire, an OR gate, or an AND gate. We describe how it generates components for each case.

- *Input wire*

  By our convention if $w \in [1, n]$ then it corresponds to the w-th input. The key generation algorithm chooses random $z_w \in \mathbb{Z}_p$. The key components are:

  $$K_{w,1} = g^{r_w} h_w^{z_w}, \; K_{w,2} = g^{-z_w}.$$

- *OR gate*

  Suppose that wire $w \in$ Gates and that $\text{GateType}(w) = \text{OR}$. In addition, let $j = \text{depth}(w)$ be the depth of wire w. The algorithm will choose random $a_w, b_w \in \mathbb{Z}_p$. Then the algorithm creates key components:

  $$K_{w,1} = g^{a_w}, \; K_{w,2} = g^{b_w}, \; K_{w,3} = g_j^{r_w - a_w \cdot r_{A(w)}}, \; K_{w,4} = g_j^{r_w - b_w \cdot r_{B(w)}}.$$

- *AND gate*

  Suppose that wire $w \in$ Gates and that $\text{GateType}(w) = \text{AND}$. In addition, let $j = \text{depth}(w)$ be the depth of wire w. The algorithm will choose random $a_w, b_w \in \mathbb{Z}_p$. The components are:

  $$K_{w,1} = g^{a_w}, \; K_{w,2} = g^{b_w}, \; K_{w,3} = g_j^{r_w - a_w \cdot r_{A(w)} - b_w \cdot r_{B(w)}}.$$

We will sometimes refer to the $K_{w,3}, K_{w,4}$ of the AND and OR gates as the "shift" components. This terminology will take on more meaning when we see how they are used during decryption.

The secret key SK output consists of the description of f, the header component K_H and the key components for each wire w.

Decrypt(SK, CT). Suppose that we are evaluating decryption for a secret key associated with a circuit $f = (n, q, A, B, \text{GateType})$ and a cipherext with input x. We will be able to decrypt if $f(x) = 1$.

We begin by observing that the goal of decryption should be to compute $g_k^{\alpha s}$. One can then recover M by computing $M = C_M / g_k^{\alpha s}$. First, there is a header computation where we compute $E' = e(K_H), g^s) = e(g_{k-1}^{\alpha - r_{n+q}}, g^s) = g_k^{\alpha s} g_k^{-r_{n+q} \cdot s}$ Our goal is now reduced to computing $g_k^{r_{n+q} \cdot s}$.

Next, we will evaluate the circuit from the bottom up. Consider wire w at depth j; if $f_w(x) = 1$ then, our algorithm will compute $E_w = (g_{j+1})^{s r_w}$. (If $f_w(x) = 0$ nothing needs to be computed for that wire.) Our decryption algorithm proceeds iteratively starting with computing E_1 and proceeds in order to finally compute E_{n+q}. Computing these values in order ensures that the computation on a depth $j-1$ wire (that evaluates to 1) will be defined before computing for a depth j wire. We show how to compute E_w for all w where $f_w(x) = 1$, again breaking the cases according to whether the wire is an input, AND or OR gate.

– *Input wire*
  By our convention if $w \in [1, n]$ then it corresponds to the w-th input. Suppose that $x_w = f_w(x) = 1$. The algorithm computes:

$$E_w = e(K_{w,1}, g^s) \cdot e(K_{w,2}, C_w) = e(g^{r_w} h_w^{z_w}, g^s) \cdot e(g^{-z_w}, h_w^s) = g_2^{s r_w}.$$

  We observe that this mechanism is similar to many existing ABE schemes.
– *OR gate*
  Consider a wire $w \in$ Gates and that $\texttt{GateType}(w) = $ OR. In addition, let $j = \texttt{depth}(w)$ be the depth of wire w. Suppose that $f_w(x) = 1$. If $f_{A(w)}(x) = 1$ (the first input evaluated to 1) then we compute:

$$E_w = e(E_{A(w)}, K_{w,1}) \cdot e(K_{w,3}, g^s) = e(g_j^{s r_{A(w)}}, g^{a_w}) \cdot e(g_j^{r_w - a_w \cdot r_{A(w)}}, g^s) = (g_{j+1})^{s r_w}.$$

  Alternatively, if $f_{A(w)}(x) = 0$, but $f_{B(w)}(x) = 1$, then we compute:

$$E_w = e(E_{B(w)}, K_{w,2}) \cdot e(K_{w,4}, g^s) = e(g_j^{s r_{B(w)}}, g^{b_w}) \cdot e(g_j^{r_w - b_w \cdot r_{B(w)}}, g^s) = (g_{j+1})^{s r_w}.$$

  Let's examine this mechanism for the case where the first input is 1 ($f_{A(w)}(x) = 1$). In this case the algorithm "moves" the value $E_{A(w)}$ from group $\mathbb{G}_j$ to group $\mathbb{G}_{j+1}$ when pairing it with $K_{w,1}$. It then multiplies it by $e(K_{w,3}, g^s)$ which "shifts" that result to E_w.
  Suppose that $f_{A(w)}(x) = 1$, but $f_{B(w)}(x) = 0$. A critical feature of the mechanism is that an attacker cannot perform a "backtracking" attack to compute $E_{B(w)}$. The reason is that the pairing operation cannot be reverse to go from group $\mathbb{G}_{j+1}$ to group $\mathbb{G}_j$. If this were not the case, it would be debilitating for security as gate $B(w)$ might have fanout greater than 1. This type of backtracking attacking is why existing ABE constructions are limited to circuits with fanout of 1.
– *AND gate*
  Consider a wire $w \in$ Gates and that $\texttt{GateType}(w) = $ AND. In addition, let $j = \texttt{depth}(w)$ be the depth of wire w. Suppose that $f_w(x) = 1$. Then $f_{A(w)}(x) = f_{B(w)}(x) = 1$ and we compute:

$$E_w = e(E_{A(w)}, K_{w,1}) \cdot e(E_{B(w)}, K_{w,2}) \cdot e(K_{w,3}, g^s)$$

$$= e(g_j^{s r_{A(w)}}, g^{a_w}) \cdot e(g_j^{s r_{B(w)}}, g^{b_w}) \cdot e(g_j^{r_w - a_w \cdot r_{A(w)} - c_w \cdot r_{B(w)}}, g^s) = (g_{j+1})^{s r_w}.$$

If the $f(x) = f_{n+q}(x) = 1$, then the algorithm will compute $E_{n+q} = g_k^{r_{n+q} \cdot s}$. It finally computes $E' \cdot E_{n+q} = g_k^{\alpha s}$ and tests if this equals C_M, outputting $M = 1$ if so and $M = 0$ otherwise. Correctness holds with high probability.

A Few Remarks. Our OR and AND key components respectively have one and two "shift" components. It is conceivable to have a construction with one shift component for the OR and none for the AND. However, we designed it this way since it made the exposition of our proof provided in the full verison [GGH+13b](in particular the distribution of private keys) easier.

Finally, our construction uses a layered circuit, where a wire at depth j gets its inputs from depth $j' = j - 1$. We could imagine a small modification to our construction which allowed j' to be of any depth less than j. Suppose this were the case for the first input. Then instead of $K_{w,1} = g_1^{a_w}$ we might more generally let $K_{w,1} = (g_{j-j'})^{a_w}$. However, we stick to describing and proving the layered case for simplicity.

4 Our Construction: Based on GGH Graded Algebras

We now describe how to modify our construction to use the GGH [GGH12b] graded algebras analogue of multilinear maps. The translation of our scheme above is straightforward to the GGH setting. We start by providing background on Garg et al.'s lattice-based "approximate" multilinear maps (a.k.a. "graded encoding systems") [GGH12b].

4.1 Graded Encoding Systems: Definition

Garg, Gentry and Halevi (GGH) [GGH12b] defined an "approximate" version of a multilinear group family, which they call a *graded encoding system*. As a starting point, they view g_i^α in a multilinear group family as simply an *encoding* of α at "level-i". This encoding permits basic functionalities, such as equality testing (it is easy to check that two level-i encodings encode the same exponent), additive homomorphism (via the group operation in $\mathbb{G}_i$), and bounded multiplicative homomorphism (via the multilinear map e). They retain the notion of a somewhat homomorphic encoding with equality testing, but they use probabilistic encodings, and replace the multilinear group family with "less structured" sets of encodings related to lattices.

Abstractly, their n-graded encoding system for a ring R includes a system of sets $\mathcal{S} = \{S_i^{(\alpha)} \subset \{0,1\}^* : i \in [0,n], \alpha \in R\}$ such that, for every fixed $i \in [0,n]$, the sets $\{S_i^{(\alpha)} : \alpha \in R\}$ are disjoint (and thus form a partition of $S_i \stackrel{\text{def}}{=} \bigcup_\alpha S_i^{(\alpha)}$). The set $S_i^{(\alpha)}$ consists of the "level-i encodings of α". Moreover, the system comes equipped with efficient procedures, as follows:[7]

---

[7] Since GGH's realization of a graded encoding system uses "noisy" encodings over ideal lattices, the procedures incorporate information about the magnitude of the noise.

Instance Generation. The randomized $\mathsf{InstGen}(1^\lambda, 1^n)$ takes as input the security parameter λ and integer n. The procedure outputs $(\mathsf{params}, \mathbf{p}_{zt})$, where params is a description of an n-graded encoding system as above, and $\mathbf{p}_{zt}$ is a level-n "zero-test parameter".

Ring Sampler. The randomized $\mathsf{samp}(\mathsf{params})$ outputs a "level-zero encoding" $a \in S_0$, such that the induced distribution on α such that $a \in S_0^{(\alpha)}$ is statistically uniform.

Encoding. The (possibly randomized) $\mathsf{enc}(\mathsf{params}, i, a)$ takes $i \in [n]$ and a level-zero encoding $a \in S_0^{(\alpha)}$ for some $\alpha \in R$, and outputs a level-i encoding $u \in S_i^{(\alpha)}$ for the same α.

Re-Randomization. The randomized $\mathsf{reRand}(\mathsf{params}, i, u)$ re-randomizes encodings to the same level, as long as the initial encoding is under a given noise bound. Specifically, for a level $i \in [n]$ and encoding $u \in S_i^{(\alpha)}$, it outputs another encoding $u' \in S_i^{(\alpha)}$. Moreover for any two encodings $u_1, u_2 \in S_i^{(\alpha)}$ whose noise bound is at most some b, the output distributions of $\mathsf{reRand}(\mathsf{params}, i, u_1)$ and $\mathsf{reRand}(\mathsf{params}, i, u_2)$ are statistically the same.

Addition and negation. Given params and two encodings at the same level, $u_1 \in S_i^{(\alpha_1)}$ and $u_2 \in S_i^{(\alpha_2)}$, we have $\mathsf{add}(\mathsf{params}, u_1, u_2) \in S_i^{(\alpha_1 + \alpha_2)}$, and $\mathsf{neg}(\mathsf{params}, u_1) \in S_i^{(-\alpha_1)}$, subject to bounds on the noise.

Multiplication. For $u_1 \in S_{i_1}^{(\alpha_1)}$, $u_2 \in S_{i_2}^{(\alpha_2)}$, we have $\mathsf{mult}(\mathsf{params}, u_1, u_2) \in S_{i_1 + i_2}^{(\alpha_1 \cdot \alpha_2)}$.

Zero-test. The procedure $\mathsf{isZero}(\mathsf{params}, \mathbf{p}_{zt}, u)$ outputs 1 if $u \in S_n^{(0)}$ and 0 otherwise. Note that in conjunction with the procedure for subtracting encodings, this gives us an equality test.

Extraction. This procedure extracts a "canonical" and "random" representation of ring elements from their level-n encoding. Namely $\mathsf{ext}(\mathsf{params}, \mathbf{p}_{zt}, u)$ outputs (say) $K \in \{0, 1\}^\lambda$, such that:
(a) With overwhelming probability over the choice of $\alpha \in R$, for any two $u_1, u_2 \in S_n^{(\alpha)}$, $\mathsf{ext}(\mathsf{params}, \mathbf{p}_{zt}, u_1) = \mathsf{ext}(\mathsf{params}, \mathbf{p}_{zt}, u_2)$,
(b) The distribution $\{\mathsf{ext}(\mathsf{params}, \mathbf{p}_{zt}, u) : \alpha \in R, u \in S_n^{(\alpha)}\}$ is statistically uniform over $\{0, 1\}^\lambda$.

We can extend add and mult to handle more than two encodings as inputs, by applying the binary versions of add and mult iteratively. Also, we use the canonicalizing encoding algorithm (as defined in Remark 2 of [GGH12b]) $\mathsf{cenc}_\ell(\mathsf{params}, i, a)$ which takes as input encoding of a and generates another encoding according to a "nice" distribution. This parameter ℓ essentially captures the noise present in the encodings. In our scheme the maximum value ℓ takes will be a small constant.

Recall that the k-multilinear assumption for the graded encodings as follows:

Assumption 2 (k-GMDDH Assumption). *The k-Graded Multilinear Decisional Diffie-Hellman (k-GMDDH) assumption states the following: Given $\mathsf{cenc}_1($ $\mathsf{params}, 1, s), \mathsf{cenc}_1(\mathsf{params}, 1, c_1), \ldots, \mathsf{cenc}_1(\mathsf{params}, 1, c_k)$, it is hard to distinguish*

$T = cenc_1(\text{params}, k, s \prod_{j \in [1,k]} c_j)$ *from* $T = cenc_1(\text{params}, k, \text{samp}(\text{params}))$, *with better than negligible advantage (in security parameter λ), where* $(\text{params}, \mathbf{p}_{zt}) \leftarrow \text{InstGen}(1^\lambda, 1^k)$. *and* $s, c_1, \ldots, c_k \leftarrow \text{samp}(\text{params})$.

4.2 Graded Encoding Systems: Realization

Concretely, GGH's n-graded encoding system works as follows. (This is a whirl-wind overview; see [GGH12b] for details.) The system uses three rings. First, it uses the ring of integers $\mathcal{O}$ of the m-th cyclotomic field. This ring is typically represented as the ring of polynomials $\mathcal{O} = \mathbb{Z}[x]/(\Phi_m(x))$, where $\Phi_m(x)$ is the m-th cyclotomic polynomial, which has degree $N = \phi(m)$. Second, for some suitable integer modulus q, it uses the quotient ring $\mathcal{O}/(q) = \mathbb{Z}_q[x]/(\Phi_m(x))$, similar to the NTRU encryption scheme [HPS98]. The encodings live in $\mathcal{O}/(q)$. Finally, it uses the quotient ring $R = \mathcal{O}/\mathcal{I}$, where $\mathcal{I} = \langle g \rangle$ is a principal ideal of $\mathcal{O}$ that is generated by g and where $|\mathcal{O}/\mathcal{I}|$ is a large prime. This is the ring "R" referred to above; elements of R are what is encoded.

What does a GGH encoding look like? For a fixed random $z \in \mathcal{O}/(q)$, an element of $S_i^{(\alpha)}$ – that is, a level-i encoding of $\alpha \in R$ – has the form $e/z^i \in \mathcal{O}/(q)$, where $e \in \mathcal{O}$ is a "small" representative of the coset $\alpha + \mathcal{I}$ (it has coefficients that are very small compared to q). To add encodings $e_1/z^i \in S_i^{(\alpha_1)}$ and $e_2/z^i \in S_i^{(\alpha_2)}$, just add them in $\mathcal{O}/(q)$ to obtain $(e_1 + e_2)/z^i$, which is in $S_i^{(\alpha_1+\alpha_2)}$ if $e_1 + e_2$ is "small". To mult encodings $e_1/z^{i_1} \in S_{i_1}^{(\alpha_1)}$ and $e_2/z^{i_2} \in S_{i_2}^{(\alpha_2)}$, just multiply them in $\mathcal{O}/(q)$ to obtain $e_1 \cdot e_2/z^{i_1+i_2}$, which is in $S_{i_1+i_2}^{(\alpha_1 \cdot \alpha_2)}$ if $e_1 \cdot e_2$ is "small". This smallness condition limits the GGH encoding system to degree polynomial in the security parameter. Intuitively, dividing encodings does not "work", since the resulting denominator has a nontrivial term that is not z.

The GGH params allow everyone to generate encodings of random (known) values. The params include a level-1 encoding of 1 (from which one can generate encodings of 1 at other levels), and (for each $i \in [n]$) a sufficient number of level-i encodings of 0 to enable re-randomization. To encode (say at level-1), run samp(params) to sample a small element a from $\mathcal{O}$, e.g. according to a discrete Gaussian distribution. For a Gaussian with appropriate deviation, this will in-duce a statistically uniform distribution over the cosets of $\mathcal{I}$. Then, multiply a with the level-1 encoding of 1 to get a level-1 encoding u of $a \in R$. Finally, run reRand(params, 1, u), which involves adding a random Gaussian linear combina-tion of the level-1 encodings of 0, whose noisiness (i.e., numerator size) "drowns out" the initial encoding. The parameters for the GGH scheme can be instanti-ated such that the re-randomization procedure can be used for any pre-specified polynomial number of times.

To permit testing of whether a level-n encoding $u = e/z^n \in S_n$ encodes 0, GGH publishes a level-n zero-test parameter $\mathbf{p}_{zt} = hz^n/g$, where h is "somewhat small"[8] and g is the generator of $\mathcal{I}$. The procedure isZero(params, $\mathbf{p}_{zt}$, u) simply

---

[8] Its coefficients are on the order of (say) $q^{2/3}$, while other terms – such as a numerator e or the principal ideal generator g – are much, much smaller.

computes $\mathbf{p}_{zt} \cdot u$ and tests whether its coefficients are small modulo q. If u encodes 0, then $e \in \mathcal{I}$ and equals $g \cdot c$ for some (small) c, and thus $\mathbf{p}_{zt} \cdot u = h \cdot c$ has no denominator and is small modulo q. If u encodes something nonzero, $\mathbf{p}_{zt} \cdot u$ has g in the denominator and is not small modulo q. The $\mathsf{ext}(\mathsf{params}, \mathbf{p}_{zt}, u)$ procedure works by applying a strong extractor to the most significant bits of $\mathbf{p}_{zt} \cdot u$. For any two $u_1, u_2 \in S_n^{(\alpha)}$, we have (subject to noise issues) $u_1 - u_2 \in S_n^{(0)}$, which implies $\mathbf{p}_{zt}(u_1 - u_2)$ is small, and hence $\mathbf{p}_{zt} \cdot u_1$ and $\mathbf{p}_{zt} \cdot u_2$ have the same most significant bits (for an overwhelming fraction of α's).

4.3   Our Construction

Now we provide our construction in GGH's n-graded encoding system. **For ease of notation on the reader, we suppress repeated** params **arguments that are provided to every algorithm..** Thus, for instance, we will write $\alpha \leftarrow \mathsf{samp}()$ instead of $\alpha \leftarrow \mathsf{samp}(\mathsf{params})$. Note that in our scheme, there will only ever be a single uniquely chosen value for params throughout the scheme, so there is no cause for confusion.

Setup$(1^\lambda, n, \ell)$. The setup algorithm takes as input, a security parameter λ, the maximum depth ℓ of a circuit, and the number of boolean inputs n.

It then runs $(\mathbf{p}_{zt}) \leftarrow \mathsf{InstGen}(1^\lambda, 1^{k=\ell+1})$. Recall that params will be implicitly given as input to all GGH-related algorithms below. Next, it samples $\alpha, \hat{h}_1, \ldots, \hat{h}_n \leftarrow \mathsf{samp}()$.

The public parameters, PP, consist of $\mathbf{p}_{zt}$, plus:

$$H = \mathsf{cenc}_2(k, \alpha), h_1 = \mathsf{cenc}_2(1, \hat{h}_1), \ldots, h_n = \mathsf{cenc}_2(1, \hat{h}_n).$$

The master secret key MSK is α.

Encrypt$(\mathrm{PP}, x \in \{0,1\}^n, M \in \{0,1\})$. The encryption algorithm takes in the public parameters, an descriptor input $x \in \{0,1\}^n$, and a message bit $M \in \{0,1\}$.

The encryption algorithm chooses a random $s \leftarrow \mathsf{samp}()$. If $M = 0$ it sets C_M to be a random value:

$$C_M = \mathsf{cenc}_3(k, \mathsf{samp}())$$

otherwise it lets

$$C_M = \mathsf{cenc}_3(k, H \cdot s).$$

Next, let S be the set of i such that $x_i = 1$.

The ciphertext is created as

$$\mathrm{CT} = (C_M,\ \tilde{s} = \mathsf{cenc}_1(1, s),\ \forall i \in S\ \ C_i = \mathsf{cenc}_3(1, h_i \cdot s)).$$

KeyGen$(\mathrm{MSK} = \alpha, f = (n, q, A, B, \mathtt{GateType}))$. The algorithm takes in the master secret key and a description f of a circuit. Recall, that the circuit has $n + q$ wires with n input wires, q gates and the wire $n + q$ designated as the output wire.

The key generation algorithm chooses random $r_1, \ldots, r_{n+q} \leftarrow \mathsf{samp}()$, where we think of randomness r_w as being associated with wire w. The algorithm produces a "header" component

$$K_H = \mathsf{cenc}_3(k - 1, \alpha - r_{n+q}).$$

Next, the algorithm generates key components for every wire w. The structure of the key components depends upon if w is an input wire, an OR gate, or an AND gate. We describe how it generates components for each case.

- *Input wire*
  By our convention if $w \in [1, n]$ then it corresponds to the w-th input. The key generation algorithm chooses random $z_w \leftarrow \mathsf{samp}()$.
  The key components are:

$$K_{w,1} = \mathsf{cenc}_3(1, \mathsf{enc}(1, r_w) + h_w \cdot z_w), \quad K_{w,2} = \mathsf{cenc}_3(1, -z_w).$$

- *OR gate*
  Suppose that wire $w \in$ Gates and that $\mathsf{GateType}(w) = $ OR. In addition, let $j = \mathsf{depth}(w)$ be the depth of wire w. The algorithm will choose random $a_w, b_w \leftarrow \mathsf{samp}()$. Then the algorithm creates key components:

$$K_{w,1} = \mathsf{cenc}_3(1, a_w), \quad K_{w,2} = \mathsf{cenc}_3(1, b_w),$$

$$K_{w,3} = \mathsf{cenc}_3(j, r_w - a_w \cdot r_{A(w)}), \quad K_{w,4} = \mathsf{cenc}_3(j, r_w - b_w \cdot r_{B(w)}).$$

- *AND gate*
  Suppose that wire $w \in$ Gates and that $\mathsf{GateType}(w) = $ AND. In addition, let $j = \mathsf{depth}(w)$ be the depth of wire w. The algorithm will choose random $a_w, b_w \leftarrow \mathsf{samp}()$.

$$K_{w,1} = \mathsf{cenc}_3(1, a_w), \quad K_{w,2} = \mathsf{cenc}_3(1, b_w),$$

$$K_{w,3} = \mathsf{cenc}_3(j, r_w - a_w \cdot r_{A(w)} - b_w \cdot r_{B(w)}).$$

We will sometimes refer to the $K_{w,3}, K_{w,4}$ of the AND and OR gates as the "shift" components. This terminology will take on more meaning when we see how they are used during decryption.

The secret key SK output consists of the description of f, the header component K_H and the key components for each wire w.

Decrypt(SK, CT). Suppose that we are evaluating decryption for a secret key associated with a circuit $f = (n, q, A, B, \mathsf{GateType})$ and a cipherext with input x. We will be able to decrypt if $f(x) = 1$.

We begin by observing that the goal of decryption should be to compute a level k encoding of $\alpha \cdot s$ such that we can test if this is equal to C_M. First, there is a header computation where we compute $E' = K_H \cdot \tilde{s}$. Note that E' should thus be a level k encoding of $\alpha s - r_{n+q} \cdot s$. Our goal is now reduced to computing a level k encoding of $r_{n+q} \cdot s$.

Next, we will evaluate the circuit from the bottom up. Consider wire w at depth j; if $f_w(x) = 1$ then, our algorithm will compute E_w to be a level $j + 1$ encoding of sr_w. Note that if $f_w(x) = 0$ nothing needs to be computed for that wire, since we have a monotonic circuit. Our decryption algorithm proceeds iteratively starting with computing E_1 and proceeds in order to finally compute E_{n+q}. Computing these values in order ensures that the computation on a depth $j - 1$ wire (that evaluates to 1) will be defined before computing for a depth j wire. We show how to compute E_w for all w where $f_w(x) = 1$, again breaking the cases according to whether the wire is an input, AND or OR gate.

- *Input wire*
  By our convention if $w \in [1, n]$ then it corresponds to the w-th input. Suppose that $x_w = f_w(x) = 1$. The algorithm computes:

$$E_w = K_{w,1} \cdot \tilde{s} + K_{w,2} \cdot C_w.$$

  Thus, E_w computes a level 2 encoding of $(r_w + \hat{h}_w \cdot z_w) \cdot s + (-z_w) \cdot \hat{h}_w \cdot s = sr_w$.
- *OR gate*
  Consider a wire $w \in$ Gates and that $\texttt{GateType}(w) = $ OR. In addition, let $j = \texttt{depth}(w)$ be the depth of wire w. Suppose that $f_w(x) = 1$. If $f_{A(w)}(x) = 1$ (the first input evaluated to 1) then we compute:

$$E_w = E_{A(w)} \cdot K_{w,1} + K_{w,3} \cdot \tilde{s}.$$

Thus, E_w computes a level $j+1$ encoding of $sr_{A(w)} \cdot a_w + (r_w - a_w \cdot r_{A(w)}) \cdot s = sr_w$.

Alternatively, if $f_{A(w)}(x) = 0$, but $f_{B(w)}(x) = 1$, then we compute:

$$E_w = E_{B(w)} \cdot K_{w,2} + K_{w,4} \cdot \tilde{s}.$$

This similarly computes a level $j+1$ encoding of $sr_{B(w)} \cdot b_w + (r_w - b_w \cdot r_{B(w)}) \cdot s = sr_w$.

Let's examine this mechanism for the case where the first input is 1 ($f_{A(w)}(x) = 1$). In this case the algorithm "moves" the value $E_{A(w)}$ from level j to level $j + 1$ when multiplying it with $K_{w,1}$. It then adds it to $K_{w,3} \cdot \tilde{s}$ which "shifts" that result to E_w.

Suppose that $f_{A(w)}(x) = 1$, but $f_{B(w)}(x) = 0$. A critical feature of the mechanism is that an attacker cannot perform a "backtracking" attack to compute $E_{B(w)}$. The reason is that the GGH encoding cannot be reversed to go from level $j+1$ to level j. (See [GGH12b] for details on why this is the case.) If this were not the case, it would be debilitating for security as gate $B(w)$ might have fanout greater than 1. This type of backtracking attacking is why existing ABE constructions are limited to circuits with fanout of 1.

– *AND gate*

Consider a wire $w \in$ Gates and that GateType(w) = AND. In addition, let j = depth(w) be the depth of wire w. Suppose that $f_w(x) = 1$. Then $f_{A(w)}(x) = f_{B(w)}(x) = 1$ and we compute:

$$E_w = E_{A(w)} \cdot K_{w,1} + E_{B(w)} \cdot K_{w,2} + K_{w,3} \cdot \tilde{s}.$$

Note that this computes a level $j+1$ encoding of sr_w in a manner analogous to above.

If $f(x) = f_{n+q}(x) = 1$, then the algorithm will compute E_{n+q} to be a level k encoding of $r_{n+q} \cdot s$. It finally computes $E' + E_{n+q}$ which is a level k encoding of αs and tests if this equals C_M using isZero$(\mathbf{p}_{zt}, E' + E_{n+q} - C_M)$, outputting $M = 1$ if so and $M = 0$ otherwise. Correctness holds with high probability.

A Quick Remark about Message Length. Our encryption algorithm takes as input a single bit message. We can extend this to longer messages using the ext algorithm provided by the GGH encoding (see Section 4.1). We restrict ourselves to single bit messages for clarity of the scheme and proof of security. We postpone the proof itself to the full version [GGH$^+$13b].

References

[ABB10]   Agrawal, S., Boneh, D., Boyen, X.: Efficient lattice (H)IBE in the standard model. In: Gilbert, H. (ed.) EUROCRYPT 2010. LNCS, vol. 6110, pp. 553–572. Springer, Heidelberg (2010)

[ABV$^+$12] Agrawal, S., Boyen, X., Vaikuntanathan, V., Voulgaris, P., Wee, H.: Functional encryption for threshold functions (or fuzzy IBE) from lattices. In: Fischlin, M., Buchmann, J., Manulis, M. (eds.) PKC 2012. LNCS, vol. 7293, pp. 280–297. Springer, Heidelberg (2012)

[BGN05]   Boneh, D., Goh, E.-J., Nissim, K.: Evaluating 2-DNF formulas on ciphertexts. In: Kilian, J. (ed.) TCC 2005. LNCS, vol. 3378, pp. 325–341. Springer, Heidelberg (2005)

[BHR12]   Bellare, M., Hoang, V.T., Rogaway, P.: Foundations of garbled circuits. Cryptology ePrint Archive, Report 2012/265 (2012), http://eprint.iacr.org/

[Boy13]   Boyen, X.: Attribute-based functional encryption on lattices. In: Sahai, A. (ed.) TCC 2013. LNCS, vol. 7785, pp. 122–142. Springer, Heidelberg (2013)

[BS02]   Boneh, D., Silverberg, A.: Applications of multilinear forms to cryptography. IACR Cryptology ePrint Archive 2002:80 (2002)

[BSW06]   Boneh, D., Sahai, A., Waters, B.: Fully collusion resistant traitor tracing with short ciphertexts and private keys. In: Vaudenay, S. (ed.) EUROCRYPT 2006. LNCS, vol. 4004, pp. 573–592. Springer, Heidelberg (2006)

[BSW11]   Boneh, D., Sahai, A., Waters, B.: Functional encryption: Definitions and challenges. In: Ishai, Y. (ed.) TCC 2011. LNCS, vol. 6597, pp. 253–273. Springer, Heidelberg (2011)

[CC09]   Chase, M., Chow, S.S.M.: Improving privacy and security in multi-authority attribute-based encryption. In: ACM Conference on Computer and Communications Security, pp. 121–130 (2009)

[Cha07]   Chase, M.: Multi-authority attribute based encryption. In: Vadhan, S.P. (ed.) TCC 2007. LNCS, vol. 4392, pp. 515–534. Springer, Heidelberg (2007)

[CHKP10]   Cash, D., Hofheinz, D., Kiltz, E., Peikert, C.: Bonsai trees, or how to delegate a lattice basis. In: Gilbert, H. (ed.) EUROCRYPT 2010. LNCS, vol. 6110, pp. 523–552. Springer, Heidelberg (2010)

[GGH12a]   Garg, S., Gentry, C., Halevi, S.: Attribute-based encryption for circuits from multilinear maps (2012) (manuscript)

[GGH12b]   Garg, S., Gentry, C., Halevi, S.: Candidate multilinear maps from ideal lattices and applications. IACR Cryptology ePrint Archive, 2012:610 (2012)

[GGH13a]   Garg, S., Gentry, C., Halevi, S.: Candidate multilinear maps from ideal lattices. In: Johansson, T., Nguyen, P.Q. (eds.) EUROCRYPT 2013. LNCS, vol. 7881, pp. 1–17. Springer, Heidelberg (2013)

[GGH+13b]   Garg, S., Gentry, C., Halevi, S., Sahai, A., Waters, B.: Attribute-based encryption for circuits from multilinear maps. Cryptology ePrint Archive, Report 2013/128 (2013), http://eprint.iacr.org/

[GGSW13]   Garg, S., Gentry, C., Sahai, A., Waters, B.: Witness encryption and its applications. In: STOC, pp. 467–476 (2013)

[GKP+13]   Goldwasser, S., Kalai, Y., Popa, R.A., Vaikuntanathan, V., Zeldovich, N.: Succinct functional encryption and applications: Reusable garbled circuits and beyond. In: STOC, pp. 555–564 (2013)

[GPSW06]   Goyal, V., Pandey, O., Sahai, A., Waters, B.: Attribute-based encryption for fine-grained access control of encrypted data. In: ACM Conference on Computer and Communications Security, pp. 89–98 (2006)

[GVW12]   Gorbunov, S., Vaikuntanathan, V., Wee, H.: Functional encryption with bounded collusions via multi-party computation. In: Safavi-Naini, R., Canetti, R. (eds.) CRYPTO 2012. LNCS, vol. 7417, pp. 162–179. Springer, Heidelberg (2012)

[GVW13]   Gorbunov, S., Vaikuntanathan, V., Wee, H.: Attribute-based encryption for circuits. In: STOC, pp. 545–554 (2013)

[Ham11]   Hamburg, M.: Spatial encryption. IACR Cryptology ePrint Archive, 2011:389 (2011)

[HPS98]   Hoffstein, J., Pipher, J., Silverman, J.H.: NTRU: A ring-based public key cryptosystem. In: Buhler, J.P. (ed.) ANTS 1998. LNCS, vol. 1423, pp. 267–288. Springer, Heidelberg (1998)

[KSW08]   Katz, J., Sahai, A., Waters, B.: Predicate encryption supporting disjunctions, polynomial equations, and inner products. In: Smart, N.P. (ed.) EUROCRYPT 2008. LNCS, vol. 4965, pp. 146–162. Springer, Heidelberg (2008)

[LOS+10]   Lewko, A., Okamoto, T., Sahai, A., Takashima, K., Waters, B.: Fully secure functional encryption: Attribute-based encryption and (Hierarchical) inner product encryption. In: Gilbert, H. (ed.) EUROCRYPT 2010. LNCS, vol. 6110, pp. 62–91. Springer, Heidelberg (2010)

[LW11]   Lewko, A., Waters, B.: Decentralizing attribute-based encryption. In: Paterson, K.G. (ed.) EUROCRYPT 2011. LNCS, vol. 6632, pp. 568–588. Springer, Heidelberg (2011)

[LW12]   Lewko, A., Waters, B.: New proof methods for attribute-based encryption: Achieving full security through selective techniques. In: Safavi-Naini, R., Canetti, R. (eds.) CRYPTO 2012. LNCS, vol. 7417, pp. 180–198. Springer, Heidelberg (2012)

[OT10]   Okamoto, T., Takashima, K.: Fully secure functional encryption with general relations from the decisional linear assumption. In: Rabin, T. (ed.) CRYPTO 2010. LNCS, vol. 6223, pp. 191–208. Springer, Heidelberg (2010)

[OT12]   Okamoto, T., Takashima, K.: Adaptively attribute-hiding (Hierarchical) inner product encryption. In: Pointcheval, D., Johansson, T. (eds.) EUROCRYPT 2012. LNCS, vol. 7237, pp. 591–608. Springer, Heidelberg (2012)

[PRV12]   Parno, B., Raykova, M., Vaikuntanathan, V.: How to delegate and verify in public: Verifiable computation from attribute-based encryption. In: Cramer, R. (ed.) TCC 2012. LNCS, vol. 7194, pp. 422–439. Springer, Heidelberg (2012)

[Reg05]   Regev, O.: On lattices, learning with errors, random linear codes, and cryptography. In: STOC, pp. 84–93 (2005)

[Rot12]   Rothblum, R.: On the circular security of bit-encryption. Cryptology ePrint Archive, Report 2012/102 (2012), http://eprint.iacr.org/

[SS10]   Sahai, A., Seyalioglu, H.: Worry-free encryption: functional encryption with public keys. In: ACM Conference on Computer and Communications Security, pp. 463–472 (2010)

[SW05]   Sahai, A., Waters, B.: Fuzzy identity-based encryption. In: Cramer, R. (ed.) EUROCRYPT 2005. LNCS, vol. 3494, pp. 457–473. Springer, Heidelberg (2005)

[SW08]   Sahai, A., Waters, B.: Slides on functional encryption. PowerPoint presentation (2008), http://www.cs.utexas.edu/ bwaters/presentations/ files/functional.ppt

[SW12]   Sahai, A., Waters, B.: Attribute-based encryption for circuits from multilinear maps. IACR Cryptology ePrint Archive 2012:592 (2012)

[Wat12]   Waters, B.: Functional encryption for regular languages. In: Safavi-Naini, R., Canetti, R. (eds.) CRYPTO 2012. LNCS, vol. 7417, pp. 218–235. Springer, Heidelberg (2012)

Functional Encryption: New Perspectives and Lower Bounds

Shweta Agrawal[1,*], Sergey Gorbunov[2,**], Vinod Vaikuntanathan[2,***], and Hoeteck Wee[3,†]

[1] University of California, Los Angeles
[2] University of Toronto
[3] George Washington University

Abstract. Functional encryption is an emerging paradigm for public-key encryption that enables fine-grained control of access to encrypted data. In this work, we present new lower bounds and impossibility results on functional encryption, as well as new perspectives on security definitions. Our main contributions are as follows:

- We show that functional encryption schemes that satisfy even a weak (non-adaptive) simulation-based security notion are impossible to construct in general. This is the *first* impossibility result that exploits *unbounded* collusions in an essential way. In particular, we show that there are no such functional encryption schemes for the class of weak pseudo-random functions (and more generally, for any class of incompressible functions). More quantitatively, our technique also gives us a lower bound for functional encryption schemes secure against *bounded* collusions. To be secure against q collusions, we show that the ciphertext in any such scheme must have size $\Omega(q)$.
- We put forth and discuss a simulation-based notion of security for functional encryption, with an unbounded simulator (called USIM). We show that this notion interpolates indistinguishability and simulation-based security notions, and is inspired by results and barriers in the zero-knowledge and multi-party computation literature.

* Research supported in part from a DARPA/ONR PROCEED award, NSF grants 1136174, 1118096, 1065276, 0916574 and 0830803, a Xerox Faculty Research Award, a Google Faculty Research Award, an equipment grant from Intel, and an Okawa Foundation Research Grant (Amit Sahai). This material is based upon work supported by the Defense Advanced Research Projects Agency through the U.S. Office of Naval Research under Contract N00014-11-1-0389. The views expressed are those of the author and do not reflect the official policy or position of the Department of Defense, the National Science Foundation, or the U.S. Government.
** Supported by Ontario Graduate Scholarship (OGS).
*** Supported by an NSERC Discovery Grant and by DARPA under Agreement number FA8750-11-2-0225. The U.S. Government is authorized to reproduce and distribute reprints for Governmental purposes notwithstanding any copyright notation thereon. The views and conclusions contained herein are those of the author and should not be interpreted as necessarily representing the official policies or endorsements, either expressed or implied, of DARPA or the U.S. Government.
† Supported by NSF CAREER Award CNS-1237429.

R. Canetti and J.A. Garay (Eds.): CRYPTO 2013, Part II, LNCS 8043, pp. 500–518, 2013.

1   Introduction

Functional encryption [SW05,SW08] is a new paradigm for public-key encryption that enables fine-grained control of access to encrypted data. It extends several previous notions, most notably identity-based encryption [Sha84,BF01,Coc01], and provides, for instance, the ability to generate and release secret keys associated with a keyword that can decrypt only those documents that contain the keyword. More generally, functional encryption allows the owner of a "master" secret key to release restricted secret keys that reveal a specific function of encrypted data. This stands in stark contrast to traditional encryption, where access to the encrypted data is all or nothing: namely, given the secret key, one can decrypt and read the entire plaintext, but without it, nothing about the plaintext is revealed at all (other than its length).

Functional Encryption. A functional encryption scheme for a circuit family [BSW11,O'N10] $\mathcal{C}$, associates secret keys SK_C with every circuit $C \in \mathcal{C}$ and ciphertext CT with input messages x.[1]

In broad terms, functional encryption requires that the owner of a secret key SK_C and a ciphertext CT (corresponding to an input message x) be able to compute $C(x)$, but learn nothing else about x itself. (Typically, and throughout this work, we assume that the circuit family $\mathcal{C}$ as well as the circuit queries C are public, in the sense that they are not hidden from the key holders.)

Moreover, security should hold in the presence of collusions amongst "key holders", that is, malicious users should not be able to combine their secret keys to learn unauthorized information. More formally, a collusion of users that hold secret keys $\mathsf{SK}_{C_1}, \ldots, \mathsf{SK}_{C_q}$ and an encryption of x should learn nothing else about x apart from $C_1(x), \ldots, C_q(x)$, for any polynomial q.

An important subclass of functional encryption is that of public-index predicate encryption. Here, the input x is a pair (ind, μ) where ind is an index and μ the payload message. Let P be a Boolean predicate defined on indices, the circuit family $\mathcal{C}$ is given by:

$$C_P(\mathsf{ind}, \mu) = \begin{cases} (\mathsf{ind}, \mu) & \text{if } P(\mathsf{ind}) = 1 \\ (\mathsf{ind}, \bot) & \text{otherwise} \end{cases}$$

Even though public index predicate encryption seems like a weak object, it already captures identity-based encryption, and is also very useful in constructing protocols for verifiably delegating computation as shown recently by Parno, Raykova and Vaikuntanathan [PRV12].

Predicate encryption captures and generalizes a large number of previous constructions, including identity-based encryption (IBE) [Sha84,BF01,Coc01,BW06], fuzzy IBE [SW05,ABV+12], attribute-based encryption (ABE) [GPSW06,LOS+10], and inner product

---

[1] An alternative approach is associate secret keys to inputs and ciphertexts to circuits. This is equivalent to our approach by taking a new "universal" family U_x that on input C outputs $C(x)$.

encryption [KSW08,LOS$^+$10,AFV11]. Specifically, IBE corresponds to P encoding a point function. Moreover, essentially all known constructions are examples of public-index predicate encryption schemes or its variants, with a few exceptions – constructions in [BF01,BW06,KSW08] achieve a stronger private-index security notion in which the index ind also remains hidden from the adversary.

Security Notions. Boneh, Sahai and Waters [BSW11] and O'Neill [O'N10] were the first to put forth a general definitional framework for functional encryption. They considered two security notions for functional encryption, namely: *indistinguishability* (IND) based security and *simulation* (SIM) based security. The former stipulates that it is infeasible to distinguish encryptions of any two messages, without getting a secret key that decrypts the ciphertexts to distinct values; the latter stipulates the existence of an efficient simulator that given $C_1(x), \ldots, C_q(x)$, outputs the view of the colluders that are given an encryption of x as well as secret keys $\mathsf{SK}_{C_1}, \ldots, \mathsf{SK}_{C_q}$.

Both of these notions may be further refined in two ways:

- *adaptive* (AD) versus *non-adaptive* (NA) which capture whether the adversary's queries to the key derivation oracle may or may not depend on the challenge ciphertext; and
- *one* versus *many*, referring to whether the adversary receives a single or multiple challenge ciphertexts.

Together, these give rise to eight security notions xx-yy-zzz, where xx $\in \{1, \mathrm{many}\}$, yy $\in \{\mathsf{NA}, \mathsf{AD}\}$, and zzz $\in \{\mathsf{IND}, \mathsf{SIM}\}$.

Recent work. We briefly outline the known relationships amongst these eight notions. We note that in general, indistinguishability based security provides a weaker guarantee than simulation based security (that is, xx-yy-SIM implies xx-yy-IND and xx-yy-IND does not imply xx-yy-SIM in general); on the other hand, we have that 1-yy-IND implies many-yy-IND. Boneh, et al. [BSW11] pointed out that indistinguishability based security is vacuous and inadequate for certain circuit families, which indicate that we should opt for simulation-based security whenever possible.[2] O'Neill [O'N10] showed that NA-IND and NA-SIM are equivalent for some subclass of circuit families that are roughly speaking, "easy to invert".

All prior positive results achieve many-AD-IND security or relaxations thereof.[3] The only known impossibility result we have for general functional encryption is that of Boneh et al. [BSW11] for realizing the IBE functionality under many-AD-SIM security. In particular, in light of known results, it is entirely

---

[2] [BSW11, Section 5.3] presents an "equivalence" between many-AD-IND and many-AD-SIM in the *programmable* random oracle model for public-index predicate encryption. For this work, we consider only the standard model.

[3] A commonly used relaxation of AD-IND security for predicate encryption is that of "selective security" [CHK03].

	realizable for public-index	realizable for all circuits
xx-yy-IND	[GVW13,GGH+13][4]	open
xx-yy-SIM (xx = 1 OR yy = NA)	open	**no** (Section 4)
many-AD-SIM	no [BSW11]	no ←
xx-yy-USIM (xx = 1 OR yy = NA)	open	open
many-AD-USIM	**no** [BSW11] ♮	**no** ←

Fig. 1. Summary of results and open problems. Results from this work are marked with boldface. Results implicit in previous works are marked with ♮. Results that are trivially implied by results in a previous column are marked with ←. The second and third columns indicate whether the definition is realizable for all public-index predicate encryption schemes (e.g. IBE) and for all circuits respectively. USIM refers to the notion of unbounded simulation discussed in Section 1.2.

conceivable that we can realize functional encryption for all poly-size circuits under either 1-AD-SIM security (thus 1-AD-IND and many-AD-IND security) or many-NA-SIM security.

In this work, we narrow the gap between existing security definitions for functional encryption, as well as that between existing constructions and impossibility results. Our results are as follows.

1.1 New Lower Bound: Impossibility for Simulation-Based Definitions

Our main result rules out general functional encryption under the one message secure, non-adaptive simulation definition (1-NA-SIM). In particular, this rules out both of the scenarios presented at the end of the preceding section (i.e. 1-AD-SIM or many-NA-SIM for all circuits) in a strong sense. This is the *first* lower bound that exploits *unbounded* collusions in an essential way. We compare the impossibility result from [BSW11] with ours in the full version.

Theorem 1 (Informal). *There exists a circuit family C for which there is no 1-NA-SIM-secure functional encryption scheme.*

Specifically, assuming the existence of a family of weak pseudo-random function $\mathsf{wPRF}(\cdot, \cdot)$ (See Definition 3) that outputs one bit, we show that there does not exist a functional encryption scheme for the family:

$$C_d(x) = \mathsf{wPRF}(x, d), \text{where the input message } x \text{ is the PRF seed}$$

We show that the ciphertext size in a 1-NA-SIM-secure scheme realizing this circuit family must grow with the size of the collusion; this yields a contradiction, since the scheme must handle unbounded collusions. In fact, the result is unconditional since any non-trivial functional encryption scheme gives rise to a one-way function and thus pseudo-random functions.

The key observation is as follows. Suppose the adversary requests for q secret keys corresponding to random inputs $C_{d_1}, \ldots, C_{d_q}$ and then requests for an encryption of a random x. Then, the simulated ciphertext together with the q simulated secret keys constitute a description of the values $\mathsf{wPRF}(x, d_1), \ldots, \mathsf{wPRF}(x, d_q)$, which is computationally indistinguishable from a sequence of q truly random bits via pseudo-randomness. By a standard information-theoretic argument, this means that the length of the ciphertext plus the secret keys must grow with q. To obtain a lower bound on the ciphertext size, we carefully exploit the fact that the simulator has to generate the secret keys before it sees the output of $\mathsf{wPRF}(x, \cdot)$. Then, the simulator has to generate a small ciphertext that "explains" all these pseudorandom values which is impossible using a compressibility argument. More generally, we show that (1) weak pseudo-random family is "incompressible", and (2) NA-SIM-secure functional encryption only exists for "compressible" circuit families. (In particular, the circuit family for all *public-index* predicate encryption is compressible.)

This idea is reminiscent of the obfuscation impossibility result of Goldwasser and Kalai [GK05], although the precise settings are quite different (in particular, functional encryption and program obfuscation seem incomparable, although related, objects).

Implications. The basic idea described above can be extended to a lower bound for even weaker forms of the simulation-based definition, including (a non-adaptive variant of) the definition of Boneh, Sahai and Waters [BSW11]. Here, we mention yet another implication of this idea.

Gorbunov, Vaikuntanathan and Wee [GVW12] recently presented a 1-AD-SIM-secure functional encryption scheme for all circuits, assuming that the adversary can only corrupt an a-priori bounded number of users (and thus, get the corresponding secret keys). One of the shortcomings of their bounded-collusion security notion as well as their construction is that the parameters of the system, and especially the size of the ciphertext depends on the collusion bound q. A natural question is whether their ciphertexts can be made to have size independent of q (or, at the very least, $o(q)$).[5] Indeed, in light of the results of Dodis, Katz, Xu and Yung [DKXY02] and most recently, Goldwasser, Lewko and Wilson [GLW12] in the context of bounded-collusion IBE, one might expect that achieving "short" ciphertexts can actually be possible in general.

Unfortunately, our techniques result in a strong negative answer to this question.

---

[5] The previous lower bound for many-AD-SIM IBE in [BSW11] (which says that the secret key size must grow with the number of challenge ciphertexts) is not applicable here as the [GVW12] construction considers only a single challenge ciphertext.

Corollary 1. *There exists a family of circuits $\mathcal{C}$ such that for every $q = q(\kappa)$, there are no q-collusion resistant 1-NA-SIM-secure functional encryption schemes with ciphertexts of size $o(q)$.*

1.2 New Perspectives: Unbounded Simulation

The preceding lower bound together with those of Boneh, Sahai and Waters [BSW11] show that even fairly weak simulation-based definitions of functional encryption are unachievable for a large and natural class of circuits. This state of affairs begs the question:

> *What is a meaningful and generally realizable security notion for functional encryption?*

While we do not provide a definitive answer to this question in our work, we believe that the quest for the right definition should incorporate insights from secure computation and zero knowledge. Indeed, Sahai and Seyalioglu [SS10] used Yao's garbled circuits to construct a one-query secure functional encryption scheme for all circuits. Subsequently, Gorbunov et al. [GVW12] exploited more techniques and insights from secure computation [Yao86,BGW88,BMR90] to derive general feasibility results for functional encryption with bounded collusions.

We put forth USIM security, where the simulator has unbounded computational power. In particular, this would allow us to circumvent our lower bound in the previous section, since the lower bound crucially relies on the existence of an efficient simulator in order to break the weak pseudo-random function. Similar notions have been considered for zero knowledge and secure computation [Pas03,PS04,BS05].[6] In the more basic setting of public-key encryption, we know that IND and SIM are equivalent [GM82], and it follows readily that all of IND, USIM, and SIM are also equivalent.

We begin an intuitive interpretation of what USIM security buys us, via the real/ideal paradigm. Consider an efficient adversary $\mathcal{A}$ holding a secret key sk_C. Then, an encryption of x leaks no more information about x apart from what a computationally *unbounded* adversary can learn from $C(x)$. Specifically, in the case of public-index predicate encryption where the predicate is false, $C(x)$ hides the payload message μ completely, even against unbounded adversaries. Thus, USIM security for public-index predicate encryption offers very meaningful simulation-based security.[7] On the other hand, for circuits that only hide

---

[6] The works on zero knowledge and secure computation focus on quasi-polynomial-time simulators. We observe that our lower bound also rules out quasi-polynomial-time simulators assuming the existence of one-way functions with sub-exponential hardness.

[7] Prior work of O'Neill [O'N10, Section 4] implies that NA-IND, NA-USIM and NA-SIM are equivalent for public-index predicate encryption. This does not subsume the point we are making because our argument applies also to the adaptive setting, where AD-IND and AD-SIM are provably *not* equivalent for public-index predicate encryption.

information about x computationally, USIM security would be inadequate and SIM security remains the desirable notion.

We observe that USIM security is "sandwiched" between IND and SIM security, that is, for yy $\in$ {NA, AD}:

$$\text{yy-IND} \Leftarrow \text{yy-USIM} \Leftarrow \text{yy-SIM}$$

This result holds for both single and many message definitions. Then, we build upon the results in [BSW11] to obtain separations and impossibility results for USIM security:

- We present a counter-example separating SIM and USIM security. In fact, the example (which encodes a one-way permutation into the circuit family) is exactly that in [BSW11, Section 4.2] for separating SIM and IND security.
- We show that it is impossible to achieve many-AD-USIM security for the IBE functionality. This strengthens the many-AD-SIM lower bound for IBE in [BSW11, Section 5.2]. That is, the latter is fundamentally about the limitations of simulation-based security notion, and not about efficiency.
- A discussion in [BSW11] pointed that IND security is inadequate whenever "the output of the functionality is supposed to have some computational hiding properties"; however, there was no precise formalization of the latter. USIM security provides a way to make this statement precise. Recall that USIM security implies IND security, and therefore, if USIM security is inadequate for some functionality, then IND security must be inadequate for the same functionality. Thanks to the real/ideal paradigm, we have a simple "litmus test" for checking whether USIM security is adequate or not. Specifically, USIM security is inadequate if $C(x)$ reveals more information about x to an unbounded adversary than to an efficient adversary. (Indeed, this is trivially the case for the separation for USIM and SIM security since an unbounded adversary can invert the one-way permutation.)

We leave as an intriguing open problem the question of establishing either a separation or an equivalence between USIM and IND security. As a first step, we establish an equivalence between USIM and IND security in the "fully non-adaptive" setting, where all queries and messages are generated by the adversary before it sees the public parameters (See Remark ?? for details).

Organization. We refer the reader to Figure 1 for a survey of our results and open problems, and to Appendix ?? for results on the unbounded simulator definition.

1.3   Discussion

FunctoMania. Let's be wishful thinkers for a minute – suppose we can have whatever we hope for in functional encryption, call this world "Functomania".

What does Functomania look like? In light of the existing (im)possibilities, there will be two incomparable "dream results"[8]:

- 1-AD-SIM secure public index predicate encryption for all efficient predicates; such schemes also satisfy 1-AD-IND, 1-AD-USIM, and many-AD-IND security.
- 1-AD-USIM secure functional encryption for all poly-size circuits; such schemes also satisfy 1-AD-IND and many-AD-IND security.

The IND-(U)SIM *Conundrum.* From a definitional stand-point, SIM/USIM-based security notions are preferable to IND-based security notion, as they offer a stronger security guarantee that has a natural, intuitive and aesthetically pleasing interpretation via the real/ideal paradigm. On the other hand, IND-based security notion allows us to bypass the impossibility results given in [BSW11] and in this work; in addition, they guarantee *message composability* in that security with a single ciphertext implies security for multiple ciphertexts (and so does NA-SIM considered in [GVW12] and those considered in an independent work [BF13]). We do not offer a complete answer to this conundrum; instead, we point out that 1-AD-SIM and 1-AD-USIM appear to be an adequate compromise for predicate encryption and general functional encryption respectively. We also note that such a conundrum is not unique to functional encryption, and has indeed previously surfaced and widely studied in the context of zero knowledge [FS90,Pas03] and secure multi-party computation [PS04,BS05,MPR06]. One notable difference is that in zero knowledge and secure computation, super-polynomial time simulation offers concurrency; this is not the case for functional encryption. (The lower bound for many-AD-USIM-secure IBE indicates that even unbounded-time simulation does not help with message composability.)

Concurrent and Independent Work. In an independent work, Bellare and O'Neill [BO12] put forth simulation-based definitions for functional encryption with non-black-box simulators. In addition, they extended the [BSW11] lower bound for IBE to the setting of efficient, non-black-box simulators, assuming the existence of collision-resistant hash functions. At a high level, the work is similar in spirit to our results on USIM security in that both consider larger classes of simulators than that in [BSW11] The independent work of Barbosa and Farshim [BF13] takes a orthognal approach, namely to restrict the adversary's key queries via some "potential leakage relation".

As with [BSW11], the definitions we study in this work are "inherently black-box" since the simulator must explicitly provide the adversary with secret keys and ciphertexts. Moreover, our NA-SIM lower bound relies crucially on black-box simulation as the compression comes from the simulated ciphertext. This leaves as an open problem the question of realizing (or ruling out) many-AD-USIM IBE with a non-black-box simulator.

---

[8] Substantial progress were made recently on both of these problems in [SW12,GVW13,GKP+13].

2 Functional Encryption

Let $\mathcal{X} = \{\mathcal{X}_\kappa\}_{\kappa \in \mathbb{N}}$ and $\mathcal{Y} = \{\mathcal{Y}_\kappa\}_{\kappa \in \mathbb{N}}$ denote ensembles where each $\mathcal{X}_\kappa$ and $\mathcal{Y}_\kappa$ is a finite set. Let $\mathcal{C} = \{\mathcal{C}_\kappa\}_{\kappa \in \mathbb{N}}$ denote an ensemble where each $\mathcal{C}_\kappa$ is a finite collection of circuits, and each circuit $C \in \mathcal{C}_\kappa$ takes as input a string $x \in \mathcal{X}_\kappa$ and outputs $C(x) \in \mathcal{Y}_\kappa$.

A functional encryption scheme $\mathcal{FE}$ for $\mathcal{C}$ consists of four algorithms $\mathcal{FE} = $ (FE.Setup, FE.Keygen, FE.Enc, FE.Dec) defined as follows[9].

- **Setup** FE.Setup(1^κ) is a p.p.t. algorithm takes as input the unary representation of the security parameter and outputs the master public and secret keys (MPK, MSK).
- **Key Generation** FE.Keygen(MSK, C) is a p.p.t. algorithm that takes as input the master secret key MSK and a circuit $C \in \mathcal{C}_\kappa$ and outputs a corresponding secret key SK_C.
- **Encryption** FE.Enc(MPK, x) is a p.p.t. algorithm that takes as input the master public key MPK and an input message $x \in \mathcal{X}_\kappa$ and outputs a ciphertext CT.
- **Decryption** FE.Dec(SK_C, CT) is a deterministic algorithm that takes as input the secret key SK_C and a ciphertext CT and outputs $C(x)$.

Definition 1 (Correctness). *A functional encryption scheme $\mathcal{FE}$ is correct if for all $C \in \mathcal{C}_\kappa$ and all $x \in \mathcal{X}_\kappa$,*

$$\Pr\left[\begin{array}{l}(\mathsf{MPK}, \mathsf{MSK}) \leftarrow \mathsf{FE.Setup}(1^\kappa); \\ \mathsf{FE.Dec}(\mathsf{FE.Keygen}(\mathsf{MSK}, C), \mathsf{FE.Enc}(\mathsf{MPK}, x)) \neq C(x)\end{array}\right] = \mathrm{negl}(\kappa)$$

where the probability is taken over the coins of FE.Setup, FE.Keygen, *and* FE.Enc.

2.1 A Simulation-Based Definition of Security

In this section, we present a simulation-based definition of functional encryption, similar in spirit to the way one defines security for secure computation via the ideal/real paradigm. We define the security game for a single message since our lower bounds apply to this weaker setting. However, this definition can be easily extended to many messages setting (see Appendix **??**).

Definition 2 (1-NA-SIM- and 1-AD-SIM- Security). *Let $\mathcal{FE}$ be a functional encryption scheme for a circuit family $\mathcal{C}$. Consider a p.p.t. adversary $A = (A_1, A_2)$ and a stateful p.p.t. simulator* Sim.[10] *Let $U_x(\cdot)$ denote a universal oracle, such that $U_x(C) = C(x)$. Consider the following two experiments:*

---

[9] Unlike in [BSW11], we do not consider the "empty key".

[10] One can replace a stateful simulator can be replaced by a regular (stateless) simulator that outputs a state st_s upon each invocation which is carried over to its next invocation.

$\mathsf{Exp}^{real}_{\mathcal{FE},A}(1^\kappa)$:	$\mathsf{Exp}^{ideal}_{\mathcal{FE},\mathrm{Sim}}(1^\kappa)$:
1: $(\mathsf{MPK}, \mathsf{MSK}) \leftarrow \mathsf{FE.Setup}(1^\kappa)$	1: $\mathsf{MPK} \leftarrow \mathrm{Sim}(1^\kappa)$
2: $(x, st) \leftarrow A_1^{\mathsf{FE.Keygen}(\mathsf{MSK},\cdot)}(\mathsf{MPK})$	2: $(x, st) \leftarrow A_1^{\mathrm{Sim}(\cdot)}(\mathsf{MPK})$
3: $\mathsf{CT} \leftarrow \mathsf{FE.Enc}(\mathsf{MPK}, x)$	3: $\mathsf{CT} \leftarrow \mathrm{Sim}^{U_x(\cdot)}(1^\kappa, 1^{\|x\|})$
4: $\alpha \leftarrow A_2^{\mathcal{O}(\mathsf{MSK},\cdot)}(\mathsf{MPK}, \mathsf{CT}, st)$	4: $\alpha \leftarrow A_2^{\mathcal{O}'(\cdot)}(\mathsf{MPK}, \mathsf{CT}, st)$
5: $Output\ (x, \alpha)$	5: $Output\ (x, \alpha)$

We distinguish between two cases of the above experiment:

1. The adaptive experiment, *where:*
   - *the oracle* $\mathcal{O}(\mathsf{MSK}, \cdot) = \mathsf{FE.Keygen}(\mathsf{MSK}, \cdot)$ *and*
   - *the oracle* $\mathcal{O}'(\cdot)$ *is the simulator, namely* $\mathrm{Sim}^{U_x(\cdot)}(\cdot)$

   We call a stateful simulator algorithm Sim admissible *if, on each input C, Sim makes just a single query to its oracle $U_x(\cdot)$ on C itself.*
   The functional encryption scheme $\mathcal{FE}$ is then said to be simulation-secure *for one message against adaptive adversaries (1-AD-SIM-secure, for short) if there is an admissible stateful p.p.t. simulator* Sim *such that for every p.p.t. adversary $A = (A_1, A_2)$, the following two distributions are computationally indistinguishable:*

$$\left\{\mathsf{Exp}^{real}_{\mathcal{FE},A}(1^\kappa)\right\}_{\kappa \in \mathbb{N}} \overset{c}{\approx} \left\{\mathsf{Exp}^{ideal}_{\mathcal{FE},\mathrm{Sim}}(1^\kappa)\right\}_{\kappa \in \mathbb{N}}$$

2. The non-adaptive experiment, *where the oracles $\mathcal{O}(\mathsf{MSK}, \cdot)$ and $\mathcal{O}'(\cdot)$ are both the "empty oracles" that return nothing.*
   The functional encryption scheme $\mathcal{FE}$ is then said to be simulation-secure *for one message against non-adaptive adversaries (1-NA-SIM-secure, for short) if there is an admissible stateful p.p.t. simulator* Sim *such that for every p.p.t. adversary $A = (A_1, A_2)$, the two distributions above are computationally indistinguishable.*

Remarks on the Definition. Our definition is stronger than that in [BSW11] but weaker than that in [GVW12]; our lower bound in Section 4 holds for all three definitions. Amongst the three, the one in [GVW12] is the only for which we know a composition theorem where security for one message implies security for many messages, in the non-adaptive setting. Note that composition in the non-adaptive setting is the "best" we can hope for; composition in the adaptive setting is essentially impossible by many-AD-SIM lower bound for IBE [BSW11]. In more detail:

- In [BSW11], the simulator is given oracle access to A_2, which it can call on any ciphertext. Therefore, it can "rewind" the adversary A_2 and adaptively reconstruct the view, which is problematic for composition [PRS02,Lin04,BMQU07]. We call this a "rewinding" definition. In our "straight-line" definition, the simulator must commit to a ciphertext once and for all, which makes it stronger.

- Unlike our definition, the [GVW12] definition does not allow the simulator to fake or "program" the setup parameters and the secret keys. The difficulty in proving a composition theorem for our definition lies in that the simulator may use "trapdoor" information from faking the setup parameters and secret keys while simulating the ciphertext.

We note that in the equivalence of NA-IND and NA-SIM under pre-image sampleability in [O'N10, Section 4], the NA-SIM-simulator actually satisfies the stronger definition in [GVW12].

The Indistinguishability-based Definition of Security. We refer the reader to the full version for the non-adaptive NA-IND and the adaptive AD-IND notions of security.

3   Preliminaries

Notations. Let $\mathcal{D}$ denote a distribution over some finite set S. Then, $x \leftarrow \mathcal{D}$ is used to denote the fact that x is chosen from the distribution $\mathcal{D}$. When we say $x \leftarrow S$, we simply mean that x is chosen from the uniform distribution over S. Let κ denote the security parameter.

Definition 3 (wPRF). *Let* wPRF $= \{$wPRF$_\kappa\}_{\kappa \in \mathbb{N}}$ *denote a family of efficiently computable functions where* wPRF$_\kappa : \{0,1\}^{n(\kappa)} \times \{0,1\}^{m(\kappa)} \to \{0,1\}^{k(\kappa)}$, *the first argument of which is called the seed to the wPRF and the second argument is the input.*

For every probabilistic polynomial time oracle distinguisher Dist, *consider the following two experiments:*

- Real$_{\text{Dist}}(1^\kappa)$: *Choose* $x \xleftarrow{\$} \{0,1\}^{n(\kappa)}$ *and run* Dist *with access to a probabilistic oracle* $\mathcal{O}_{real}(x)$ *which, when invoked, chooses a uniformly random* $d \leftarrow \{0,1\}^{m(\kappa)}$ *and returns the pair* $(d, \text{wPRF}_\kappa(x,d))$. *This experiment outputs whatever* Dist *outputs.*
- Rand$_{\text{Dist}}(1^\kappa)$: *Choose a uniformly random function* $R : \{0,1\}^{m(\kappa)} \to \{0,1\}^{k(\kappa)}$ *and run* Dist *with access to a probabilistic oracle* $\mathcal{O}_{rand}(R)$ *which, when invoked, chooses a uniformly random* $d \leftarrow \{0,1\}^{m(\kappa)}$ *and returns the pair* $(d, R(d))$. *This experiment outputs whatever* Dist *outputs.*

We say wPRF *is a weak pseudo-random function if for all p.p.t. distinguishers* Dist,

$$\left| \Pr[\text{Real}_{\text{Dist}}(1^\kappa) = 1] - \Pr[\text{Rand}_{\text{Dist}}(1^\kappa) = 1] \right| = \text{negl}(\kappa)$$

where the probabilities are over the choice of x *and* R, *as well as the coin-tosses of* Dist *and the oracles* $\mathcal{O}_{real}$ *and* $\mathcal{O}_{rand}$.

This is in contrast to the stronger notion of (regular) pseudo-random functions where the distinguisher Dist gets query access to the function, namely it can query the function on inputs x of its choice and get either the output of the function (in the real world) or independent random bits (in the ideal world).

In our impossibility result, we will use a weak pseudo-random function with seed length $n(\kappa) = \kappa$ and output length $k(\kappa) = 1$.

4   Impossibility Results for Functional Encryption

In this section, we present our main lower bound for 1-NA-SIM-secure functional encryption. We begin with a notion of "incompressible" circuits. Then, we show that (1) weak pseudo-random functions are "incompressible", and (2) 1-NA-SIM-secure functional encryption only exists for "compressible" circuits. Putting the two together yields our lower bound.

4.1   Incompressible Circuits

We first define a family of compressible circuits. Informally, we say that a family of circuits $\{\mathcal{G}_\kappa\}$ is (ℓ, t)-compressible if for a list of uniformly random circuit descriptions $G_1, \ldots, G_\ell \in \mathcal{G}_\kappa$ and a uniformly chosen input x, there is some efficiently computable description of $G_1(x), \ldots, G_\ell(x)$ of size t.

Definition 4 (Incompressible Circuits). *Let $\ell = \ell(\kappa)$ and $t = t(\kappa)$ be functions of the security parameter κ. A family of circuits $\mathcal{G} = \{\mathcal{G}_\kappa\}_{\kappa \in \mathbb{N}}$ is (ℓ, t)-compressible if there exists a family of (deterministic) compressor circuits $\{\mathbf{C}_\kappa\}_{\kappa \in \mathbb{N}}$ and a family of decompressor circuits $\{\mathbf{D}_\kappa\}_{\kappa \in \mathbb{N}}$ such that:*

- *(polynomial size) the circuits $\mathbf{C}_\kappa$ and $\mathbf{D}_\kappa$ have size $\mathsf{poly}(\kappa, \ell)$.*
- *(mild compression) for sufficiently large κ and all x,*

$$\left| \mathbf{C}_\kappa(G_1, \ldots, G_\ell, y_1, \ldots, y_\ell) \right| = t$$

where $y_i = G_i(x)$.
- *(correctness) there is a polynomial $p = p(\kappa)$ such that*

$$\Pr[x \xleftarrow{\$} \{0,1\}^\kappa, G_1, \ldots, G_\ell \xleftarrow{\$} \mathcal{G}_\kappa, y_i = G_i(x):$$
$$\mathbf{D}_\kappa(G_1, \ldots, G_\ell, \mathbf{C}_\kappa(G_1, \ldots, G_\ell, y_1, \ldots, y_\ell)) = (y_1, \ldots, y_\ell)] \geq 1/p(\kappa)$$

where the probability is taken over the choice of x as well as the circuits $G_1, \ldots, G_\ell$.

The family $\mathcal{G}$ is (ℓ, t)-incompressible if it is not (ℓ, t)-compressible.

We now give examples of (in)compressible circuits. First, consider the notion of pre-image samplable family of circuits introduced by O'Neill [O'N10] which requires that given $G_1(x), \ldots, G_\ell(x)$, there is a polynomial-time algorithm that returns an arbitrary x' such that $G_i(x') = G_i(x)$ for all i. In our language, this says that the family $\mathcal{G}$ is $(\ell, |x'|)$-compressible; the compression algorithm simply outputs x'.

Next, consider an arbitrary public-index circuit family parametrized by predicates P and given by:

$$G_P(\mathsf{ind}, \mu) = \begin{cases} (\mathsf{ind}, \mu) & \text{if } P(\mathsf{ind}) = 1 \\ (\mathsf{ind}, \bot) & \text{otherwise} \end{cases}$$

It is easy to see that this circuit family is $(\ell, |(\mathsf{ind}, \mu)|)$-compressible. On input

$$G_{P_1}(\mathsf{ind}, \mu), \ldots, G_{P_\ell}(\mathsf{ind}, \mu)$$

the compression algorithm always learn ind. In addition, if $P_i(\mathsf{ind}) = 1$ for some i, then the compressor also learns μ and hence it outputs (ind, μ). If $P_i(\mathsf{ind}) = 0$ for all i, then the compressor outputs $(\mathsf{ind}, \perp)$. Given $\big(G_{P_1}, \ldots, G_{P_\ell}, (\mathsf{ind}, \mu)\big)$ the decoding algorithm outputs $y_i = (\mathsf{ind}, \mu)$ if $G_{P_i}(\mathsf{ind}) = 1$ and $y_i = (\mathsf{ind}, \perp)$ otherwise. Given $\big(G_{P_1}, \ldots, G_{P_\ell}, (\mathsf{ind}, \perp)\big)$ the decoder simply outputs $y_i = (\mathsf{ind}, \perp)$ for all i.

On the other hand, as we show below (see Lemma 1), any family of (weak) pseudo-random functions is incompressible in a strong sense. More precisely, consider a family of circuits $\mathcal{G} = \{G_{d_i}(\cdot) = \mathsf{wPRF}(\cdot, d_i)\}$ where d_i serves as the input to the pseudo-random function. Informally, the incompressibility is due to the fact that a sequence $(G_{d_1}(x), \ldots, G_{d_\ell}(x)) = (\mathsf{wPRF}(x, d_1), \ldots, \mathsf{wPRF}(x, d_\ell))$ is indistinguishable from a sequence of uniformly random bits, which are clearly incompressible.

Lemma 1 (weak PRFs are $(\ell, \ell-\kappa)$-incompressible). *Let* $\mathsf{wPRF} = \{\mathsf{wPRF}_\kappa : \{0,1\}^\kappa \times \{0,1\}^{m(\kappa)} \to \{0,1\}\}_{\kappa \in \mathbb{N}}$ *be a family of weak pseudo-random functions, where* $m(\kappa) = \omega(\log \kappa)$. *Define* $G_d(x) = \mathsf{wPRF}(x, d)$. *Consider a family* $\mathcal{G} = \{\mathcal{G}_\kappa\}_{\kappa \in \mathbb{N}}$ *defined as*

$$\mathcal{G}_\kappa = \big\{G_d(\cdot) : |d| = m(\kappa)\big\}$$

Then, $\mathcal{G}$ *is* $(\ell, \ell - \kappa)$-*incompressible.*

We refer the reader to the full version for the formal proof.

4.2   The Impossibility Result

We are now ready to state and prove our main theorem.

Theorem 2. *There exists a family of circuits* $\mathcal{G}$ *for which there are no 1-NA-SIM-secure functional encryption schemes.*

Proof. We consider two cases.

Case 1: Assume there exists a circuit family of weak pseudo-random functions

$$\mathsf{wPRF} = \{\mathsf{wPRF}_\kappa : \{0,1\}^\kappa \times \{0,1\}^{m(\kappa)} \to \{0,1\}\}_{\kappa \in \mathbb{N}}$$

where $m(\kappa) = \omega(\log \kappa)$. Let $G_d(x) = \mathsf{wPRF}(x, d)$ and consider a family $\mathcal{G} = \{\mathcal{G}_\kappa\}_{\kappa \in \mathbb{N}}$ defined as

$$\mathcal{G}_\kappa = \big\{G_d(\cdot) : |d| = m(\kappa)\big\}$$

Assume, for the sake of contradiction, there exist a 1-NA-SIM-secure function encryption scheme $\mathcal{FE}$ for $\mathcal{G}$, and let $|\mathsf{CT}|$ denote the length of a ciphertext in the scheme. Let $\ell = \ell(\kappa) = |\mathsf{CT}| + \kappa$.

From Lemma 1, we know that $\mathcal{G}$ is $(|\mathsf{CT}| + \kappa, |\mathsf{CT}|)$-incompressible. However, Lemma 2 below tells us that since there is a 1-NA-SIM secure scheme for $\mathcal{G}$, the family $\mathcal{G}$ is $(|\mathsf{CT}| + \kappa, |\mathsf{CT}|)$-compressible. This gives us the desired contradiction, and therefore, there cannot exist a 1-NA-SIM-secure functional encryption scheme for $\mathcal{G}$.

Case 2: Assume there does not exist a family of weak pseudo-random functions. Also, for the sake of contradiction, assume there exists a 1-NA-SIM-secure function encryption scheme for all families of circuits $\mathcal{G}$.

In particular, this means that there is a functional encryption scheme for the empty circuit family (namely, a family $\mathcal{G}$ that does not contain any circuits at all). A 1-NA-SIM-secure scheme $\mathcal{FE}$ for $\mathcal{G}$ is also a secure public-key encryption scheme. Since public-key encryption implies one-way functions, which in turn imply pseudo-random functions [GGM86,HILL99], we obtain the desired contradiction.

Lemma 2 (1-NA-SIM $\Rightarrow$ $(\ell, |\mathsf{CT}|)$-compressibility). *Let $\mathcal{G} = \{\mathcal{G}_\kappa\}_{\kappa \in \mathbb{N}}$ be a family of circuits. Suppose there exists a 1-NA-SIM-secure functional encryption scheme for the $\mathcal{G}$. Then, the family $\mathcal{G}$ is $(\ell, |\mathsf{CT}|)$-compressible for any polynomially bounded $\ell = \ell(\kappa)$, where $|\mathsf{CT}|$ denotes size of the encryption of input x.*

Informally, the compression algorithm works as follows: on input $G_1, \dots, G_\ell$ and $G_1(x), \dots, G_\ell(x)$, the output is the simulated ciphertext corresponding to an encryption of x. The decompression algorithm then evaluates the decryption algorithm, which is guaranteed to produce $G_1(x), \dots, G_\ell(x)$.

Proof. Let (FE.Setup, FE.Keygen, FE.Enc, FE.Dec) denote the encryption scheme for the family $\mathcal{G}$. Consider the adversary $A = (A_1, A_2)$ in the 1-NA-SIM security experiment that acts as follows:

- A_1 chooses $G_1, \dots, G_\ell \xleftarrow{\$} \mathcal{G}$ independently at random and requests for the corresponding secret keys $\mathsf{SK}_1, \dots, \mathsf{SK}_\ell$. In addition, it chooses $x \xleftarrow{\$} \{0,1\}^{m(\kappa)}$ and outputs x as the challenge message and state

$$(G_1, \dots, G_\ell, \mathsf{SK}_1, \dots, \mathsf{SK}_\ell)$$

- A_2 outputs α composed of the challenge ciphertext and the state

$$(G_1, \dots, G_\ell, \mathsf{SK}_1, \dots, \mathsf{SK}_\ell)$$

Let Sim denote the (admissible) stateful p.p.t. simulator guaranteed by 1-NA-SIM security. We show how to use the simulator to construct a family of (deterministic) compressor and decompressor circuits $\mathbf{C}_\rho$ and $\mathbf{D}_\rho$, indexed by a random string ρ corresponding to the random tape for the simulator:

- The compressor $\mathbf{C}_\rho$, on input $G_1, \dots, G_\ell$ and $y_1, \dots, y_\ell$ works as follows: first, compute $\mathsf{MPK} \leftarrow \mathsf{Sim}(1^\kappa; \rho)$ and secret keys $\{\mathsf{SK}_i : \mathsf{SK}_i \leftarrow \mathsf{Sim}(G_i; \rho)\}_{i \in [\ell]}$. Then compute and output CT as the compressed string, where queries $G_i(x)$ are answered with y_i:

$$\mathsf{CT} \leftarrow \mathsf{Sim}^{U_x(\cdot)}(1^{|m(\kappa)|})$$

– The decompressor $\mathbf{D}_\rho$, on input $G_1, \ldots, G_\ell$ and CT first reconstructs the master public key MPK $\leftarrow \mathrm{Sim}(1^\kappa; \rho)$ and the set of secret keys:

$$\{\mathsf{SK}_i : \mathsf{SK}_i \leftarrow \mathrm{Sim}(G_i; \rho)\}_{i \in [\ell]}$$

Note that $\mathbf{D}_\rho$ has the same randomness ρ hard-wired, and so the secret keys SK_i are exactly the same as those used by $\mathbf{C}_\rho$. Finally, it computes and outputs:

$$\{y_i \leftarrow \mathsf{FE.Dec}(\mathsf{SK}_i, \mathsf{CT})\}_{i \in [\ell]}$$

Formally, we output $(\mathbf{C}_\rho, \mathbf{D}_\rho)$ for a random ρ, which is a pair of polynomial-size circuits. Clearly, we achieve *mild compression* (where $|\mathsf{CT}|$ is the compressor's output size), since the size of CT is determined by the functional encryption scheme and independent of ℓ. To establish correctness, it suffices to show that:

$$\Pr_{\rho, x, G_1, \ldots, G_\ell}[\mathbf{D}_\rho(G_1, \ldots, G_\ell, \mathbf{C}_\rho(G_1, \ldots, G_\ell, G_1(x), \ldots, G_\ell(x))) =$$

$$(G_1(x), \ldots, G_\ell(x))] \geq 1 - \mathrm{negl}(\kappa)$$

Here, we will rely on the correctness of the functional encryption scheme as well as 1-NA-SIM-security. First, consider the distinguisher Dist that given the output $(x, \mathsf{CT}, G_1, \ldots, G_\ell, \mathsf{SK}_1, \ldots, \mathsf{SK}_\ell)$ of the adversary A_2 proceeds as follows:

Output 1 iff for all $i \in [\ell]$, $\mathsf{FE.Dec}(\mathsf{SK}_i, \mathsf{CT}) = G_i(x)$.

Observe that by correctness of the encryption scheme, Dist outputs 1 with probability $1 - \mathrm{negl}(\kappa)$ given the output of the adversary A_2 in the 1-NA-SIM experiment. Therefore, by 1-NA-SIM-security, Dist also outputs 1 with probability $1 - \mathrm{negl}(\kappa)$ given the output of the (admissible) simulator, where the randomness is taken over the coin tosses ρ of the simulator, along with the random choices of $x, G_1, \ldots, G_\ell$.

This shows that the pair of circuits $(\mathbf{C}_\rho, \mathbf{D}_\rho)$ for a uniformly random ρ is a correct compressor-decompressor pair. Therefore, we obtain a $(\ell, |\mathsf{CT}|)$-compressor and a decompressor, thus establishing the lemma.

We point out here that our lower bound extends to the setting where the simulator is not required to be admissible, by using a family of (standard) pseudo-random functions.

Finally, the argument here generalizes to showing that functional encryption secure against an a-priori bounded number $q = q(\kappa)$ of collusions is impossible if one insists on small ciphertexts (namely, ciphertexts with much fewer than q bits). This matches the recent result of [GVW12] who construct such functional encryption schemes with ciphertexts of size polynomial in q.

Corollary 2. *There exists a family of circuits $\mathcal{G}$ such that for every $q = q(\kappa)$, there are no q-collusion resistant 1-NA-SIM-secure functional encryption schemes with ciphertexts of size $o(q)$.*

4.3 Extensions: Impossibility of Weaker Simulation-Based Definitions

The idea behind our impossibility result is robust enough to apply to various relaxations of the simulation-based security definition. In this section, we describe a number of such extensions of our result.

Impossibility for the Selective and Random-Input Definitions. In the selective model, the adversary is required to commit to the secret key queries $G_1, \ldots, G_q$ as well as the challenge input x before the setup phase. In particular, this means that the adversary will not be able to pick up the circuits or the challenge input depending on the system parameters. Variants of the selective security model are frequently considered in the literature as a relaxations of regular security notions (see, e.g., [BB11,GPSW06,AFV11]). Another relaxation one can consider is one where the adversary is not allowed to choose the circuits or the challenge, but instead, they are chosen uniformly at random.

Our lower bound easily extends to these weaker notions, simply because the adversary we consider in the proof of Lemma 2 chooses the circuits and the challenge uniformly at random, and independent of the system parameters.

Impossibility for the Non-Adaptive BSW Definition (the "Rewinding Definition"). The main difference between the definition proposed by [BSW11] and our definition in Section 2 is that whereas our definition restricts the simulator to be "straight-line", the BSW definition allows the simulator to "rewind" the adversary and interact with it in order to generate the view. For more details, we direct the reader to the full version.

We also state the impossibility extension for secret-key functional encryption in the full version.

Acknowledgments. We thank Shafi Goldwasser, Yael Kalai, Raluca Ada Popa and Charles Rackoff for a number of insightful conversations that helped improve the presentation of the impossibility result. We would also like to thank Adam O'Neill and Amit Sahai for helpful pointers and discussions.

References

ABV⁺12. Agrawal, S., Boyen, X., Vaikuntanathan, V., Voulgaris, P., Wee, H.: Functional encryption for threshold functions (or fuzzy IBE) from lattices. In: Fischlin, M., Buchmann, J., Manulis, M. (eds.) PKC 2012. LNCS, vol. 7293, pp. 280–297. Springer, Heidelberg (2012)

AFV11. Agrawal, S., Freeman, D.M., Vaikuntanathan, V.: Functional encryption for inner product predicates from learning with errors. In: Lee, D.H., Wang, X. (eds.) ASIACRYPT 2011. LNCS, vol. 7073, pp. 21–40. Springer, Heidelberg (2011)

BB11. Boneh, D., Boyen, X.: Efficient selective identity-based encryption without random oracles. J. Cryptology 24(4), 659–693 (2011)

BF01.   Boneh, D., Franklin, M.: Identity-based encryption from the weil pairing.
           In: Kilian, J. (ed.) CRYPTO 2001. LNCS, vol. 2139, pp. 213–229. Springer,
           Heidelberg (2001)
BF13.   Barbosa, M., Farshim, P.: On the semantic security of functional encryption
           schemes. In: Kurosawa, K., Hanaoka, G. (eds.) PKC 2013. LNCS, vol. 7778,
           pp. 143–161. Springer, Heidelberg (2013)
BGW88.   Ben-Or, M., Goldwasser, S., Wigderson, A.: Completeness theorems for
           non-cryptographic fault-tolerant distributed computation. In: Proceedings
           of the Twentieth Annual ACM Symposium on Theory of Computing,
           STOC 1988, pp. 1–10. ACM, New York (1988)
BMQU07.   Backes, M., Müller-Quade, J., Unruh, D.: On the Necessity of Rewinding in
           Secure Multiparty Computation. In: Vadhan, S.P. (ed.) TCC 2007. LNCS,
           vol. 4392, pp. 157–173. Springer, Heidelberg (2007)
BMR90.   Beaver, D., Micali, S., Rogaway, P.: The round complexity of secure
           protocols (extended abstract). In: STOC, pp. 503–513 (1990)
BO12.   Bellare, M., O'Neill, A.: Semantically-secure functional encryption:
           Possibility results, impossibility results and the quest for a general
           definition. Cryptology ePrint Archive, Report 2012/515 (2012)
BS05.   Barak, B., Sahai, A.: How to play almost any mental game over the
           net - concurrent composition via super-polynomial simulation. In: FOCS,
           pp. 543–552 (2005)
BSW11.   Boneh, D., Sahai, A., Waters, B.: Functional encryption: Definitions and
           challenges. In: Ishai, Y. (ed.) TCC 2011. LNCS, vol. 6597, pp. 253–273.
           Springer, Heidelberg (2011)
BW06.   Boyen, X., Waters, B.: Anonymous hierarchical identity-based encryption
           (Without random oracles). In: Dwork, C. (ed.) CRYPTO 2006. LNCS,
           vol. 4117, pp. 290–307. Springer, Heidelberg (2006)
CHK03.   Canetti, R., Halevi, S., Katz, J.: A forward-secure public-key encryption
           scheme. In: Biham, E. (ed.) EUROCRYPT 2003. LNCS, vol. 2656,
           pp. 255–271. Springer, Heidelberg (2003)
Coc01.   Cocks, C.: An identity based encryption scheme based on quadratic
           residues. IMA Int. Conf., 360–363 (2001)
DKXY02.   Dodis, Y., Katz, J., Xu, S., Yung, M.: Key-insulated public key
           cryptosystems. In: Knudsen, L.R. (ed.) EUROCRYPT 2002. LNCS,
           vol. 2332, pp. 65–82. Springer, Heidelberg (2002)
FS90.   Feige, U., Shamir, A.: Witness indistinguishable and witness hiding
           protocols. In: STOC, pp. 416–426 (1990)
GGH+13.   Garg, S., Gentry, C., Halevi, S., Sahai, A., Waters, B.: Attribute-based
           encryption for circuits from multilinear maps. In: Canetti, R., Garay, J.A.
           (eds.) CRYPTO 2013, Part II. LNCS, vol. 8043, pp. 479–499. Springer,
           Heidelberg (2013)
GGM86.   Goldreich, O., Goldwasser, S., Micali, S.: How to construct random
           functions. J. ACM 33(4), 792–807 (1986)
GK05.   Goldwasser, S., Kalai, Y.T.: On the impossibility of obfuscation with
           auxiliary input. In: FOCS, pp. 553–562 (2005)
GKP+13.   Goldwasser, S., Kalai, Y.T., Popa, R.A., Vaikuntanathan, V., Zeldovich, N.:
           Succinct functional encryption and its power: Reusable garbled circuits and
           beyond. In: STOC (to appear, 2013)
GLW12.   Goldwasser, S., Lewko, A., Wilson, D.A.: Bounded-collusion IBE from
           key homomorphism. In: Cramer, R. (ed.) TCC 2012. LNCS, vol. 7194,
           pp. 564–581. Springer, Heidelberg (2012)

GM82.   Goldwasser, S., Micali, S.: Probabilistic encryption and how to play mental poker keeping secret all partial information. In: STOC, pp. 365–377 (1982)

GPSW06.   Goyal, V., Pandey, O., Sahai, A., Waters, B.: Attribute-based encryption for fine-grained access control of encrypted data. In: ACM Conference on Computer and Communications Security, pp. 89–98 (2006)

GVW12.   Gorbunov, S., Vaikuntanathan, V., Wee, H.: Functional encryption with bounded collusions via multi-party computation. In: Safavi-Naini, R., Canetti, R. (eds.) CRYPTO 2012. LNCS, vol. 7417, pp. 162–179. Springer, Heidelberg (2012)

GVW13.   Gorbunov, S., Vaikuntanathan, V., Wee, H.: Attribute-based encryption for circuits. In: Proceedings of the 45th Annual ACM Symposium on Symposium on Theory of Computing, STOC 2013, pp. 545–554. ACM, New York (2013)

HILL99.   Håstad, J., Impagliazzo, R., Levin, L.A., Luby, M.: A pseudorandom generator from any one-way function. SIAM J. Comput. 28(4), 1364–1396 (1999)

KSW08.   Katz, J., Sahai, A., Waters, B.: Predicate encryption supporting disjunctions, polynomial equations, and inner products. In: Smart, N.P. (ed.) EUROCRYPT 2008. LNCS, vol. 4965, pp. 146–162. Springer, Heidelberg (2008)

Lin04.   Lindell, Y.: Lower bounds and impossibility results for concurrent self composition. The Journal of Cryptology (2004)

LOS+10.   Lewko, A., Okamoto, T., Sahai, A., Takashima, K., Waters, B.: Fully secure functional encryption: Attribute-based encryption and (Hierarchical) inner product encryption. In: Gilbert, H. (ed.) EUROCRYPT 2010. LNCS, vol. 6110, pp. 62–91. Springer, Heidelberg (2010)

MPR06.   Micali, S., Pass, R., Rosen, A.: Input-indistinguishable computation. In: FOCS, pp. 367–378 (2006)

O'N10.   O'Neill, A.: Definitional issues in functional encryption. Cryptology ePrint Archive, Report 2010/556 (2010), http://eprint.iacr.org/

Pas03.   Pass, R.: Simulation in quasi-polynomial time and its application to protocol composition. In: Biham, E. (ed.) EUROCRYPT 2003. LNCS, vol. 2656, pp. 160–176. Springer, Heidelberg (2003)

PRS02.   Prabhakaran, M., Rosen, A., Sahai, A.: Concurrent zero knowledge with logarithmic round-complexity. In: 43rd FOCS, pp. 366–375 (2002)

PRV12.   Parno, B., Raykova, M., Vaikuntanathan, V.: How to Delegate and Verify in Public: Verifiable Computation from Attribute-Based Encryption. In: Cramer, R. (ed.) TCC 2012. LNCS, vol. 7194, pp. 422–439. Springer, Heidelberg (2012)

PS04.   Prabhakaran, M., Sahai, A.: New notions of security: achieving universal composability without trusted setup. In: STOC, pp. 242–251 (2004)

Sha84.   Shamir, A.: Identity-based cryptosystems and signature schemes. In: Blakely, G.R., Chaum, D. (eds.) CRYPTO 1984. LNCS, vol. 196, pp. 47–53. Springer, Heidelberg (1985)

SS10.   Sahai, A., Seyalioglu, H.: Worry-free encryption: functional encryption with public keys. In: ACM Conference on Computer and Communications Security, pp. 463–472 (2010)

SW05.   Sahai, A., Waters, B.: Fuzzy identity-based encryption. In: Cramer, R. (ed.) EUROCRYPT 2005. LNCS, vol. 3494, pp. 457–473. Springer, Heidelberg (2005)

SW08.   Sahai, A., Waters, B.: Slides on functional encryption. power point presenta-
          tion (2008), `http://www.cs.utexas.edu/ bwaters/presentations/`
          `files/functional.ppt`
SW12.   Sahai, A., Waters, B.: Attribute-based encryption for circuits from
          multilinear maps. Cryptology ePrint Archive, Report 2012/592 (2012)
Yao86.   Yao, A.C.-C.: How to generate and exchange secrets (extended abstract).
          In: FOCS, pp. 162–167 (1986)

On the Achievability of Simulation-Based Security for Functional Encryption

Angelo De Caro[1], Vincenzo Iovino[2], Abhishek Jain[3,*], Adam O'Neill[4,**],
Omer Paneth[4,***], and Giuseppe Persiano[2]

[1] NTT Secure Platform Laboratories, Japan
Angelo.Decaro@lab.ntt.co.jp
[2] Dipartimento di Informatica, University of Salerno, Italy
{iovino,giuper}@dia.unisa.it
[3] MIT and Boston University
abhishek@csail.mit.edu
[4] Boston University
{amoneill,omer}@bu.edu

Abstract. This work attempts to clarify to what extent simulation-based security (SIM-security) is achievable for functional encryption (FE) and its relation to the weaker indistinguishability-based security (IND-security). Our main result is a compiler that transforms any FE scheme for the general circuit functionality (which we denote by Circuit-FE) meeting indistinguishability-based security (IND-security) to a Circuit-FE scheme meeting SIM-security, where:

- In the random oracle model, the resulting scheme is secure for an unbounded number of encryption and key queries, which is the strongest security level one can ask for.
- In the standard model, the resulting scheme is secure for a bounded number of encryption and non-adaptive key queries, but an *unbounded* number of adaptive key queries. This matches known impossibility results and improves upon Gorbunov et al. [CRYPTO'12] (which is only secure for *non-adaptive* key queries).

Our compiler is inspired by the celebrated Fiat-Lapidot-Shamir paradigm [FOCS'90] for obtaining zero-knowledge proof systems from witness-indistinguishable proof systems. As it is currently unknown whether Circuit-FE meeting IND-security exists, the purpose of this result is to establish that it remains a good target for future research despite known deficiencies of IND-security [Boneh et al. – TCC'11, O'Neill – ePrint '10]. We also give a tailored construction of SIM-secure hidden vector encryption (HVE) in composite-order bilinear groups. Finally, we revisit the known negative results for SIM-secure FE, extending them to natural

* Supported by NSF Contract CCF-1018064 and DARPA Contract Number: FA8750-11-2-0225. The author also thanks BU RISCS (Reliable Information Systems and Cyber Security) Institute.
** Supported by NSF grants CNS-1012910 and CNS-0546614. The author also thanks BU RISCS (Reliable Information Systems and Cyber Security) Institute.
*** Supported by the Simons award for graduate students in theoretical computer science and NSF award 1218461.

R. Canetti and J.A. Garay (Eds.): CRYPTO 2013, Part II, LNCS 8043, pp. 519–535, 2013.

weakenings of the security definition and thus providing essentially a full picture of the (in)achievability of SIM-secure FE.

Keywords: Functional Encryption, Hidden Vector Encryption, Simulation-Based Security.

1 Introduction

Let $F : K \times M \to \Sigma$ be a functionality, where K is the *key space* and M is the *message space* and Σ is the *output space*. Then a *functional encryption scheme for F* (or *F-FE scheme*) [23,7] is a special encryption scheme in which, for every *key* $k \in K$, the owner of the master secret key Msk associated with the public key Pk can generate a special key or "token" Tok_k that allows the computation of $F(k, m)$ from a ciphertext of m computed under public key Pk. In other words, whereas in traditional encryption schemes decryption is an all-or-nothing affair, in FE it is possible to finely control the amount of information that is revealed by a ciphertext. This opens up exciting applications to access control, searching on encrypted data [6], and secure delegation of computation (cf. [22]), among others.

Unlike in the case of classical cryptosystems, a general study of the security of FE did not appear initially. Instead, progressively more expressive forms of FE were constructed in a series of works (see, e.g., [23,16,8,17,18,20,24,15]) that adopted indistinguishability-based (IND) notions of security. The study of simulation-based (SIM) notions of security for FE were initiated only comparatively recently by Boneh, Sahai, and Waters [7] and O'Neill [21].[1] In particular, they show there exist clearly insecure FE schemes for certain functionalities that are nonetheless deemed secure by IND-security, whereas these schemes do not meet the stronger notion of SIM-security. On the other hand, negative results have also emerged showing SIM-security is not always achievable [7,3,1]. This leads to the main questions that we study in this work:

> To what extent is SIM-security for FE achievable? In particular, can schemes for IND-secure FE be "compiled" to ones meeting the stronger notion of SIM-security?

In order to make these questions more precise, let us call an F-FE scheme (q_1, ℓ, q_2)-SIM-secure (resp. (q_1, ℓ, q_2)-IND-secure) if it is secure under the respective security definition for adversaries making at most q_1 "non-adaptive" key queries (i.e., before seeing the challenge ciphertexts), q_2 "adaptive" key queries (i.e., after seeing the challenge ciphertexts), and at most ℓ encryption queries

---

[1] Very roughly, in both definitions the adversary makes key-derivation queries, then queries for challenge ciphertexts, then again makes key-derivation queries. IND-security asks that the adversary cannot distinguish between encryptions of messages that it cannot trivially distinguish using the keys. SIM-security asks that the "view" of the adversary can be simulated by simulator given neither ciphertexts nor keys but only the corresponding outputs of the functionality on the underlying plaintexts.

(i.e., the number of challenge ciphertexts). Note that these bounds are fixed *a priori* and do not vary per adversary. In the case that a parameter is unbounded we denote it by poly, so for example (poly, poly, poly)-SIM security means SIM-security where the number of encryption and key queries are all unbounded. Of particular interest is $F =$ Circuit, meaning the general circuit functionality.

A Compiler for General Functionalities. Our main result is a compiler that takes a IND-secure Circuit-FE scheme and produces a SIM-secure Circuit-FE scheme. More specifically, in the random oracle (RO) model [4], we show the existence of a (poly, poly, poly)-IND-secure Circuit-FE scheme implies the existence of a (poly, poly, poly)-SIM-secure Circuit-FE scheme. In the standard (random oracle devoid) model, we show the existence of a (poly, poly, poly)-IND-secure[2] Circuit-FE scheme implies the existence of a (q_1, ℓ, poly)-SIM-secure Circuit-FE scheme for any polynomials q_1, ℓ. The result in the standard model is optimal in that it matches recent impossibility results [7,1,3] discussed later.

We note that it is currently a central open question in FE to construct a (poly, poly, poly)-IND-secure Circuit-FE scheme.[3] If such a scheme is achieved, we will obtain interesting new results via our compiler. To compare, Boneh et al. [7] achieve (poly, poly, poly)-SIM-secure *identity-based encryption* (IBE) in the RO model; in fact, they explicitly raise the open question of constructing SIM-secure Circuit-FE in the RO model. Gorbunov et al. [14] construct Circuit-FE (in the standard model) which achieves only $(q_1, \ell, 0)$-SIM-security (instead of $q_2 = \text{poly}$). See Table 1.

Our Techniques. Our compiler is inspired by the construction of zero-knowledge proof systems from witness indistinguishable proof systems, as studied in the celebrated work of Feige, Lapidot and Shamir [12]. Recall that in the FLS paradigm, the simulator operates the proof system in a "trapdoor" mode which is indistinguishable from the behavior of the honest party to the adversary. Adopting this paradigm to FE, our compiler produces "trapdoor circuits" which have additional "slots" in plaintext and keys that are used only by the simulator, not by the real system. To illustrate our techniques, consider the simpler case of a single challenge ciphertext and only *adaptive* key queries. Then, at a very high-level, instead of F we use a trapdoor circuit F_{trap} with an additional slot flag in the plaintext and an additional slot value in the key, namely:

$$F_{trap}((k, \text{value}), (m, \text{flag})) = \begin{cases} F(k, m) & \text{if flag} = 0 \\ \text{value} & \text{if flag} = 1 \end{cases}$$

(This is not actually sufficient because a key may reveal value, but we are just trying to get a rough idea across; see Section 3 for the full constructions.) An honest encryptor will always set flag $= 0$, but the simulator will set flag $= 1$

---

[2] This can be relaxed to (q_1, ℓ, poly)-IND-security.

[3] We emphasize that, since our transformation matches the known impossibility results, we do not obtain any impossibility result for (poly, poly, poly)-IND-secure Circuit-FE. Indeed, we believe (poly, poly, poly)-IND-secure Circuit-FE is possible.

and can then set value in the tokens it gives out to program the output F_{trap} appropriately. The proof of SIM-security is by reduction to IND-security of the underlying scheme, since the output of F_{trap} in the flag $= 0$ and flag $= 1$ cases will be the same.

Why is IND-security Enough? The above shows that, surprisingly, despite the weakness of IND-security for certain functionalities shown in [7,21], an IND-secure FE scheme for general circuits is enough to go "all the way" to a SIM-secure one. To see how this can be possible, let us look at the counter-example functionality of [21], which is an f (which we think of here as a circuit) for which there is another function g such that g is "hard to compute" from f but is "isomorphic" to g, meaning f and g have the same equality pattern across the domain. In this case, despite IND-security, a token for f may also allow computing g. However, the corresponding trapdoor circuit produced by our compiler can be programmed to agree on f (via the additional slots) on all the challenge messages but *no longer agree on g*, because g is computed on a "dummy" message in the first plaintext slot. This means that if a token for computing f in the compiled scheme allowed computing g an adversary could indeed violate IND-security.

Simulation-Secure Hidden Vector Encryption. By using a similar high-level approach as in our general compiler, we also give a tailored construction of (poly, ℓ, poly)-SIM-secure HVE-FE, where "HVE" denotes *hidden vector encryption*, a generalization of anonymous IBE introduced by [8]. Again, these parameters are optimal in that they match known impossibility results [7,3] discussed later. (Note that in this case we are able to achieve security for an unbounded number of non-adaptive key queries, which is impossible for the general Circuit functionality considered above [1].) The scheme is set in composite order bilinear groups and proven secure under the general subgroup decision assumption of [5]. In some sense, we compile existing IND-secure constructions of HVE-FE to a SIM-secure one in a "non-blackbox" way. Namely, our scheme mirrors existing IND-secure constructions of HVE-FE [20,11] except in some additional subgroups. The presence of an additional subgroup component in a simulated ciphertext acts as the "flag" that triggers an interaction with this additional subgroup component in the simulated keys.

Stronger Impossibility Results. As we mentioned, the positive results we obtain match the recent impossibility results for SIM-secure FE. Namely, Boneh et al. [7] show that $(0, \text{poly}, 2)$-SIM-secure IBE is impossible (though in the "non-programmable" RO model; this was recently extended to the standard model in [3]). Agrawal et al. [1] show impossibility of $(\text{poly}, 1, 0)$-SIM-security for a functionality that computes a weak pseudorandom function (wPRF-FE) and hence for Circuit-FE.

One mays wonder if there are weaker formulations of SIM-security under which these results might be circumvented. We identify two:

1. *Fully Non-Adaptivity Adversaries.* The above results crucially rely on the fact that the experiment proceeds in distinct phases of non-adaptive key

queries, encryption queries, then adaptive key queries. (In particular, the result of [1] for non-adaptive key queries crucially uses adaptivity of the encryption queries.) We thus ask whether these results can be circumvented for *fully non-adaptive adversaries* that must choose their encryption and key queries *simultaneously*.

2. *Non-Blackbox Simulation.* The result of [1] requires that the simulator use the adversary as a black-box. We ask whether this result can be circumvented by using *non-blackbox simulation*.

Unfortunately, by building on the techniques of [3,1], we go on to resolve these possibilities in the negative. See Table 2.

As a final contribution of independent interest, we show that in Circuit-FE the ciphertext length must grow with the output length of the functionality. Namely, we show impossibility of $(1, 1, 0)$-SIM-security for a functionality that computes a pseudorandom generator (PRG); that is, the ciphertext length must be as long as the output length of the PRG. To the best of our knowledge, this is the first impossibility result for non-adaptive key queries and bounded collusion. Note that Goldwasser et al. [13] recently give a construction of FE for *boolean* functionalities with "succinct" ciphertexts, but for functionalities with longer output the ciphertext length in all existing constructions grows linearly with the output length of the functionality. Our result shows this is inherent.

Table 1. Positive results for SIM-secure FE. UB and B denote unbounded and bounded, respectively. Single-boxed entries are inherent (matching impossibility results below).

Work	Func	#Non-Adaptive Key Queries	#Encryption Queries	#Adaptive Key Queries	Assumptions
[BSW11]	IBE	UB	UB	UB	RO
Ours	HVE	UB	B	UB	Standard
[GVW12]	Circuit	B	UB	0	Standard
Ours	Circuit	UB	UB	UB	RO, IND-security
Ours	Circuit	B	B	UB	IND-security

Table 2. Negative results for SIM-secure FE. Boxed entries are new to our work. UB denote unbounded. Pre-challenge and post-challenge key queries refer to non-adaptive and adaptive queries , respectively, while simultaneous queries are queried together with the challenge. Results using pre-challenge and post-challenge queries are incomparable, while results using simultaneous queries are stronger. PRG* refers to a pseudorandom generator functionality whose output is longer than the ciphertext size.

Work	Func	#Key Queries	#Encryption Queries	Key Query Time	Non-Black-Box Simulation?
[BSW11,BO12]	IBE	2	UB	Post-Challenge	YES
[AGVW12]	wPRF	UB	1	Pre-Challenge	NO
Ours	wPRF	UB	1	Pre-Challenge	YES
Ours	wPRF	UB	UB	Simultaneous	YES
Ours	PRG*	1	1	Pre-Challenge	NO

Organization. In Section 2 we give the basic definitions for FE. In Section 3 we describe the transformations from IND-secure to SIM-secure FE, both in the random oracle model and in the plain model. Section 4 describes the construction of SIM-secure FE for the hidden vector encryption functionality. In Section 5 we present our negative results.

2   Definitions

A *negligible* function $\mathsf{negl}(k)$ is a function that is smaller than the inverse of any polynomial in k. If D is a probability distribution, "$x \leftarrow D$" denotes that x is chosen according to D. If D is a finite set, "$x \leftarrow D$" denotes that x is chosen according to uniform probability on D. If $q > 0$ is an integer then $[q]$ denotes the set $\{1, \ldots, q\}$. "PPT" stands for "probabilistic polynomial time" and "PT" stands for "polynomial time." Algorithms are PPT unless explicitly noted otherwise. If A and B are algorithms, we denote by "$y \leftarrow A^{B(\cdot)}(x)$" that y is assigned the output of A when run on input x with oracle access to B. If A or B are randomized this is done using fresh random coins. If a and b are strings, then $a|b$ denotes the string representing their delimited concatenation. We will make use of standard primitives such as pseudorandom functions and symmetric-key encryption; definitions are in the full version of this paper [10].

2.1   Functional Encryption

Functional encryption (FE) schemes [7] are encryption schemes for which the owner of the master secret can compute restricted keys (also called "tokens") that allow to compute a *functionality* on the plaintext associated with a ciphertext. A formal definition follows.

Syntax. A *functionality* $F = \{F_n\}_{n>0}$ is a family of PT functions $F_n : K_n \times M_n \to \Sigma$ where K_n is the *key space* for parameter n, M_n is the *message space* for parameter n and Σ is the *output space*. Sometimes we will refer to functionality F as a function from $F : K \times M \to \Sigma$ with $K = \cup_n K_n$ and $M = \cup_n M_n$.

A *functional encryption scheme for F* (an F-FE scheme) is a tuple $\mathsf{FE} = (\mathsf{Setup}, \mathsf{KeyGen}, \mathsf{Enc}, \mathsf{Eval})$ of four algorithms with the following syntax: Algorithm $\mathsf{Setup}(1^\lambda, 1^n)$ outputs *public* and *master secret* keys $(\mathsf{Pk}, \mathsf{Msk})$ for *security parameter* λ and *length parameter* n that are polynomially related. Algorithm $\mathsf{KeyGen}(\mathsf{Msk}, k)$, on input a master secret key Msk outputs *secret key* (or *token*) Tok. Algorithm $\mathsf{Enc}(\mathsf{Pk}, m)$, on input public key Pk and *plaintext* $m \in M_n$ outputs *ciphertext* Ct. PT algorithm $\mathsf{Eval}(\mathsf{Pk}, \mathsf{Ct}, \mathsf{Tok})$ outputs $y \in \Sigma \cup \{\bot\}$.

For *correctness* we require for all $(\mathsf{Pk}, \mathsf{Msk}) \leftarrow \mathsf{Setup}(1^\lambda, 1^n)$, all $k \in K_n$ and $m \in M_n$, for $\mathsf{Tok} \leftarrow \mathsf{KeyGen}(\mathsf{Msk}, k)$ and $\mathsf{Ct} \leftarrow \mathsf{Enc}(\mathsf{Pk}, m)$, we have that $\mathsf{Eval}(\mathsf{Pk}, \mathsf{Ct}, \mathsf{Tok}) = F(k, m)$ whenever $F(k, m) \neq \bot$, except with negligible probability. (See [3] for a discussion about this condition.)

Functionalities of Interest. In this paper, we will mainly be concerned with two specific functionalities.

First, the Circuit functionality has key space K_n equals to the set of all n-input Boolean circuits and message space M_n the set $\{0,1\}^n$ of n-bit strings. For $C \in K_n$ and $m \in M_n$, we have $\mathsf{Circuit}(C, m) = C(m)$. In the random oracle (RO) model [4] we *allow* the circuits in the Circuit functionality to have RO gates. This is because, in practice, we replace the random oracle invocations with computation of a cryptographic hash function having an explicit circuit description.

Second, the HVE functionality [8] has message space M_n equal to the set of length n Boolean vectors $\boldsymbol{x} = \langle x_1, \ldots, x_n \rangle$ and key space K_n equal to the set length n Boolean vectors $\boldsymbol{y} = \langle y_1, \ldots, y_n \rangle$ with $\star$'s ("don't-care" entries). $\mathsf{HVE}(\boldsymbol{x}, \boldsymbol{y})$ is equal to 1 iff, for all $1 \le i \le n$, $x_i = y_i$ or $y_i = \star$.

Security. We next define indistinguishability-based and simulation-based security for FE based on [7,21]. Some remarks about the definitions follow them.

Definition 1. [Indistinguishability-based security.] We say that an F-FE scheme is (q_1, ℓ, q_2)-*IND-secure* if for every PPT adversary $\mathcal{A} = (\mathcal{A}_0, \mathcal{A}_1)$ where $\mathcal{A}_0$ makes at most q_1 oracle queries and $\mathcal{A}_1$ makes at most q_2 oracle queries, the advantage of $\mathcal{A}$ defined as

$$\mathsf{Adv}_{\mathcal{A}}^{\mathsf{FE},\mathsf{IND}}(1^\lambda, 1^n) = \left| \mathrm{Prob}[\mathsf{IND}_{\mathcal{A}}^{\mathsf{FE}}(1^\lambda, 1^n) = 1] - 1/2 \right|$$

is negligible, where:

Experiment $\mathsf{IND}_{\mathcal{A}}^{\mathsf{FE}}(1^\lambda, 1^n)$:
  $(\mathsf{Pk}, \mathsf{Msk}) \leftarrow \mathsf{Setup}(1^\lambda, 1^n)$
  $(\boldsymbol{m}_0, \boldsymbol{m}_1, \mathsf{st}) \leftarrow \mathcal{A}_0^{\mathsf{FE.KeyGen}(\mathsf{Msk}, \cdot)}(\mathsf{Pk})$ where $\boldsymbol{m}_0, \boldsymbol{m}_1 \in M_n^\ell$
  $b \leftarrow \{0,1\}$; $\mathsf{Ct}[i] \leftarrow \mathsf{FE.Enc}(\mathsf{Pk}, \boldsymbol{m}_b[i])$ for $i \in [\ell]$
  $b' \leftarrow \mathcal{A}_1^{\mathsf{FE.KeyGen}(\mathsf{Msk}, \cdot)}(\mathsf{st}, \mathsf{Ct})$
Output: $(b = b')$

Above we require that $F(k, \boldsymbol{m}_0[i]) = F(k, \boldsymbol{m}_1[i])$ for every $i \in [\ell]$ and every oracle query k made by either $\mathcal{A}_0$ or $\mathcal{A}_1$.

Definition 2. [Simulation-Based security.] We say that an F-FE scheme is (q_1, ℓ, q_2)-*SIM-secure* if for every PPT adversary $\mathcal{A} = (\mathcal{A}_0, \mathcal{A}_1)$ where $\mathcal{A}_0$ makes at most q_1 oracle queries and $\mathcal{A}_1$ makes at most q_2 oracle queries, there exists a PPT simulator $\mathsf{Sim} = (\mathsf{Sim}_0, \mathsf{Sim}_1)$ such that the outputs of the following two experiments are computationally indistinguishable:

Experiment $\mathsf{RealExp}^{\mathsf{FE}, \mathcal{A}}(1^\lambda, 1^n)$:
  $(\mathsf{Pk}, \mathsf{Msk}) \leftarrow \mathsf{FE.Setup}(1^\lambda, 1^n)$
  $(\boldsymbol{m}, \mathsf{st}) \leftarrow \mathcal{A}_0^{\mathsf{FE.KeyGen}(\mathsf{Msk}, \cdot)}(\mathsf{Pk})$
  $\mathsf{Ct} \leftarrow \mathsf{Enc}(\mathsf{Pk}, \boldsymbol{m})$
  $\alpha \leftarrow \mathcal{A}_1^{\mathsf{FE.KeyGen}(\mathsf{Msk}, \cdot)}(\mathsf{Pk}, \mathsf{Ct}, \mathsf{st})$
  Output: $(\mathsf{Pk}, \boldsymbol{m}, \{k_i\}, \alpha)$

Experiment $\mathsf{IdealExp}_{\mathsf{Sim}}^{\mathsf{FE}, \mathcal{A}}(1^\lambda, 1^n)$:
  $(\mathsf{Pk}, \mathsf{Msk}) \leftarrow \mathsf{FE.Setup}(1^\lambda, 1^n)$
  $(\boldsymbol{m}, \mathsf{st}) \leftarrow \mathcal{A}_0^{\mathsf{FE.KeyGen}(\mathsf{Msk}, \cdot)}(\mathsf{Pk})$
  $(\mathsf{Ct}, \mathsf{st}') \leftarrow \mathsf{Sim}_0(\mathsf{Pk}, |\boldsymbol{m}|, \{k_i, \mathsf{Tok}_{k_i}, F(k_i, \boldsymbol{m})\})$
  $\alpha \leftarrow \mathcal{A}_1^{\mathcal{O}(\cdot)}(\mathsf{Pk}, \mathsf{Ct}, \mathsf{st})$
  Output: $(\mathsf{Pk}, \boldsymbol{m}, \{k_i\}, \alpha)$

Above we require $|m| \leq \ell$. In the output of the experiments, $\{k_i\}$ contains the token queries of the adversary (i.e., the queries of $\mathcal{A}_0$ and $\mathcal{A}_1$ combined). The oracle $\mathcal{O}(\cdot)$ is the second stage of the simulator, namely algorithm $\mathsf{Sim}_1(\mathsf{Msk}, \mathsf{st}', \cdot, \cdot)$, which receives as its third argument a key k_j for which the adversary queries a token and as its fourth argument the output value $F(k_j, m)$. Further, note that the simulator algorithm Sim_1 is stateful in that after each invocation, it updates the state st' which is carried over to its next invocation.

We also note that above follows the security definitions of [21,14] in that in the ideal experiment, the setup and non-adaptive token queries are handled honestly (not by the simulator). This is just for simplicity. Additionally, the challenge messages are selected by $\mathcal{A}_0$ in "one-shot" and not adaptively depending on previous challenge ciphertexts as in [3]. Again, this is just for simplicity.

Random Oracle Model. To lift our definition to the random oracle (RO model [4], the output of the real experiment *includes* the queries made by any algorithm (i.e., either those of the scheme or the adversary) to the RO and the responses. In the ideal experiment, the simulator provides responses to the queries made by any algorithms to the RO and the output of the experiment again *includes* all these queries and responses. This is analogous to the "explicitly" programmable RO model formalized by Wee [25] for zero-knowledge and seems to us to be the most natural formalization of security in the RO model in our context.

3   From Indistinguishability to Simulation-Based Security

In this section, we show that from an IND-secure Circuit-FE scheme one can construct a SIM-secure Circuit-FE scheme. We give two constructions: one in the RO model [4] and one in the standard model, which (necessarily) achieve security for different parameters.

3.1   Trapdoor Circuits

The idea of our transformations is to replace the original circuit with a "trapdoor" one that a simulator can use to program the output in some way. This will be done via interaction of additional "slots" in the plaintext and key that interact when a flag is set in the plaintext. This approach is inspired by the FLS paradigm introduced by Feige, Lapidot and Shamir [12] to obtain zero-knowledge proof systems from witness indistinguishable proof systems.

Random Oracle Model Construction. Here, a plaintext will have four slots and a key will have two. In the plaintext, the slots will be: (1) actual message m (2) a bit flag to indicate trapdoor mode, (3) a random string x, and (4) a seed r for a pseudorandom function (PRF). In the key, the slots will be (1) the actual circuit C and (2) a random string y. For evaluation, in non-trapdoor mode we simply evaluate the original circuit C on m. in trapdoor mode, the output is

---

Circuit $\mathsf{Trap}_1[C, \mathsf{Hash}, \mathcal{F}]^y(m')$
  $(m, \mathsf{flag}, x, r) \leftarrow m'$
  If $\mathsf{flag} = 1$
    Then return $\mathsf{Hash}(x, y) \oplus f_r(y)$
    Else return $C(m)$

Circuit $\mathsf{Trap}_2[C, \mathsf{SE}]^{k'}(m')$
  $(r, c_{\mathsf{SE}}) \leftarrow k'$
  $(m, \mathsf{flag}, k_{\mathsf{SE}}, (r_1, y_1), \ldots, (r_q, y_q)) \leftarrow m'$
  If $\mathsf{flag} = 1$ then
    If there exists i such that $r = r_i$
      Then return y_i
      Else $y \leftarrow \mathsf{SE.Dec}(k_{\mathsf{SE}}, c_{\mathsf{SE}})$; Return y
    Else return $C(m)$

Fig. 1. Construction of trapdoor circuit in the RO model (left) and standard model (right) from a given circuit C

instead "programmed" as $\mathsf{Hash}(x, y) \oplus f_r(y)$, where Hash is a RO and f_r is the PRF keyed by r.

Formally, let C be a circuit on n-bits. Let $\mathsf{Hash} \colon \{0,1\}^n \times \{0,1\}^n \to \{0,1\}^n$ be a hash function and $\mathcal{F} = \{f_s : s \in \{0,1\}^k\}_{k \in \mathbb{N}}$ be a PRF. For $y \in \{0,1\}^n$ define the corresponding *RO-based trapdoor circuit* $\mathsf{Trap}_1[C, \mathsf{Hash}, \mathcal{F}]^y$ on $(3n+1)$-bits as in Figure 1.

Standard Model Construction. Here, a plaintext will have $3 + 2q$ slots (for a polynomial q) and a key will three two. In the plaintext, the slots will be: (1) the actual message m, (2) a bit flag to indicate trapdoor mode, (3) a key $\mathsf{sk}_{\mathsf{SE}}$ for a symmetric-key encryption scheme SE, and finally the last $2q$ slots will be pairs (r_i, z_i), where r_i is a random string and z_i is a desired output value. (Looking ahead, the third slot will be used to handle adaptive key queries and the last $2q$ slots will be used to handle up to q non-adaptive key queries.) On the other hand, in the key the slots will be: (1) the actual circuit C, (2) a random string r, and (3) a ciphertext c_{SE} under SE. For evaluation, in non-trapdoor mode we simply evaluate the original circuit C on m. In in trapdoor mode, if $r = r_i$ for some $i \in [q]$ then the output is "programmed"' as z_i, and otherwise as $\mathsf{SE.Dec}(\mathsf{sk}_{\mathsf{SE}}, c)$ where $\mathsf{SE.Dec}$ is the decryption algorithm of SE.

Formally, let C be a circuit with n-bit inputs and n-bit outputs, and let $\mathsf{SE} = (\mathsf{SE.Enc}, \mathsf{SE.Dec})$ be a symmetric-key encryption scheme with key-space $\{0,1\}^s$, message-space $\{0,1\}^n$, and ciphertext-space $\{0,1\}^\nu$. For $k' \in \{0,1\}^{n+\nu}$ define the corresponding *standard-model trapdoor circuit* $\mathsf{Trap}_2[C, \mathsf{SE}]^{k'}$ with $((2q + 1)n + 1 + s)$-bit inputs and n-bit outputs as in Figure 1.

3.2 Random Oracle Model Transformation

Let $\mathsf{IndFE} = (\mathsf{IndFE.Setup}, \mathsf{IndFE.Enc}, \mathsf{IndFE.KeyGen}, \mathsf{IndFE.Eval})$ be a functional encryption scheme for the functionality Circuit. Let $\mathsf{Hash} \colon \{0,1\}^n \times \{0,1\}^n \to \{0,1\}^n$ be a hash function (which will be modeled as a random oracle) and $\mathcal{F} =$

$\{f_s : s \in \{0,1\}^k\}_{k \in \mathbb{N}}$ be a PRF. We define a new FE scheme $\mathsf{SimFE}_1[\mathsf{Hash}, \mathcal{F}] = (\mathsf{Setup}, \mathsf{KeyGen}, \mathsf{Enc}, \mathsf{Eval})$ for Circuit as follows:

- $\mathsf{Setup}(1^\lambda, 1^n)$: returns the output of $\mathsf{IndFE.Setup}(1^\lambda, 1^{3n+1})$ as its own output.
- $\mathsf{Enc}(\mathsf{Pk}, m)$: on input Pk and $m \in \{0,1\}^n$, the algorithm chooses x at random from $\{0,1\}^n$, sets $m' = (m, 0, x, 0^n)$ and returns $\mathsf{IndFE.Enc}(\mathsf{Pk}, m')$ as its own output.
- $\mathsf{KeyGen}(\mathsf{Msk}, C)$: on input Msk and a n-input Boolean circuit C, the algorithm chooses random $y \in \{0,1\}^n$ and returns (y, Tok) where $\mathsf{Tok} \leftarrow \mathsf{IndFE.KeyGen}(\mathsf{Msk}, \mathsf{Trap}_1[C, \mathsf{Hash}, \mathcal{F}]^y)$.
- $\mathsf{Eval}(\mathsf{Pk}, \mathsf{Ct}, \mathsf{Tok})$: on input Pk, Ct and Tok, returns the output $\mathsf{IndFE.Eval}(\mathsf{Pk}, \mathsf{Ct}, \mathsf{Tok})$.

Theorem 3. Suppose IndFE is $(\mathsf{poly}, \mathsf{poly}, \mathsf{poly})$-IND-Secure. Then SimFE_1 is $(\mathsf{poly}, \mathsf{poly}, \mathsf{poly})$-SIM-secure in the random oracle model.

We defer the proof to the full version of this paper [10] and give some intuition here. it is instructive to consider a simpler system where $f_r(y)$ in the evaluation is simply replaced by r. In this case, the fourth slot in the plaintext acts as an encryption under Nielsen's RO-based non-committing encryption scheme [19], whose decryption can be adaptively programmed. However, this approach does not work for multiple tokens, since then the simulator would need to program two hash outputs to $r \oplus C_1(m)$ and $r \oplus C_2(m)$, which would not look independently random to the distinguisher. Since the number of tokens is unbounded, we need to generate more randomness than can be contained in the plaintext slot, and thus we use a PRF to generate a "fresh" ciphertext for each token.

A Note on Uninstantiability. We notice that, due to the result of Bellare and O'Neill [3], our construction in the RO model cannot be proven SIM-secure when implemented with any function ensemble in place of the RO. However, we stress that unlike other some other "uninstantiable" schemes (e.g., those of Canetti et al. [9]) which are *clearly* insecure (in the standard model) when implemented with any function ensemble, our construction does not seem to suffer any real-world attack. In this sense, we still view it as a good heuristic for our scheme to have a proof of security in the RO model.

3.3 Standard Model Transformation

Let $\mathsf{IndFE} = (\mathsf{IndFE.Setup}, \mathsf{IndFE.Enc}, \mathsf{IndFE.KeyGen}, \mathsf{IndFE.Eval})$ be a functional encryption scheme for the functionality Circuit. Let $\mathsf{SE} = (\mathsf{SE.Enc}, \mathsf{SE.Dec})$ be a symmetric-key encryption scheme with key-space $\{0,1\}^s$, message-space $\{0,1\}^n$, and ciphertext-space $\{0,1\}^\nu$. We define a new FE scheme $\mathsf{SimFE}_2[\mathsf{SE}] = (\mathsf{Setup}, \mathsf{KeyGen}, \mathsf{Enc}, \mathsf{Eval})$ for Circuit as follows:

- $\mathsf{Setup}(1^\lambda, 1^n)$: returns the output of $\mathsf{IndFE.Setup}(1^\lambda, 1^{n(2q+1)+s+1})$ as its own output. In addition the algorithm picks a random key $\mathsf{sk}_{\mathsf{SE}} \in \{0,1\}^s$ and keeps it in the master secret key Msk.

- Enc(Pk, m): on input Pk and $m \in \{0,1\}^n$, the algorithm sets $m' \leftarrow (m, 0, 0^s, (0^n, 0^n), \ldots, (0^n, 0^n))$ and returns the output of IndFE.Enc(Pk, m') as its own output.
- KeyGen(Msk, C): on input Msk and a n-input Boolean circuit C, the algorithm chooses random $r \in \{0,1\}^\lambda$ and $c \in \{0,1\}^\nu$, and returns (r, c, Tok) where it computes Tok $\leftarrow$ IndFE.KeyGen(Msk, Trap$_2[C, \text{SE}]^{k'}$) and sets $k' \leftarrow r\|c$.
- Eval(Pk, Ct, Tok): on input Pk, ciphertext Ct and token (r, c, Tok), returns the output of IndFE.Eval(Pk, Ct, Tok).

Theorem 4. Suppose IndFE is $(q_1, 1, \text{poly})$-IND-secure, $\mathcal{F}$ is a PRF, and SE has pseudorandom ciphertexts. Then SimFE$_2$ is $(q_1, 1, \text{poly})$-SIM-secure.

Again, we defer the proof to the full version [10] and give some intuition here. The intuition is very similar in spirit to that of Theorem 3. First, consider a simpler system where the $2q$ pairs (r_i, z_i) are replaced by a single pair $(\tilde{r}, \tilde{z})$. This approach does not work for multiple non-adaptive tokens, since then the simulator would need to program $\tilde{z}$ to be $C_1(m)$ and $C_2(m)$ at the same time. To solve this problem, we add additional pairs in the ciphertext, one for each non-adaptive query. This is also the reason why we need an *a priori* bound on the number of non-adaptive key queries. For adaptive key queries, the simulator can instead program c in the token to be an encryption of the desired output.

We note that it is straightforward to extend our construction to achieve (q_1, ℓ, poly)-SIM-security for any polynomial ℓ (where now we need to assume the starting scheme is (q_1, ℓ, poly)-IND-secure). Note that by [7,3], the restriction to a bounded ℓ is necessary in the standard model. Moreover, by [1], for the Circuit functionality the restriction to a bounded number of non-adaptive key queries q_1 is also necessary.

An Instantiation for Polynomial Evaluation. In the full version of this paper [10], we show how to adapt our standard model transformation to the polynomial evaluation functionality [17], for which which we have efficient constructions from bilinear maps and lattices.

4 Simulation-Secure Hidden Vector Encryption

In the section we present a SIM-secure HVE-FE scheme whose whose security can be proved under static assumptions in the bilinear pairing setting in the standard model. We use composite order bilinear groups whose order is the product of five distinct primes (see the full version for the standard definition of such groups [10]).

The Scheme. We now describe our HVE scheme. To make our description and proofs simpler, we add to all vectors $\boldsymbol{x}$ and $\boldsymbol{y}$ two dummy components and set both of them equal to 0. We can thus assume that all vectors have at least two non-star positions.

- Setup($1^\lambda, 1^\ell$): The setup algorithm chooses a description of a bilinear group $\mathcal{I} = (N = p_1 p_2 p_3 p_4 p_5, \mathbb{G}, \mathbb{G}_T, e) \leftarrow \mathcal{G}(1^\lambda)$ with known factorization, and random $g_1 \in \mathbb{G}_{p_1}$, $g_2 \in \mathbb{G}_{p_2}$, $g_3 \in \mathbb{G}_{p_3}$, $g_4 \in \mathbb{G}_{p_4}$, and, for $i \in [\ell]$ and $b \in \{0,1\}$, random $t_{i,b} \in Z_N$ and random $R_{i,b} \in \mathbb{G}_{p_3}$ and sets $T_{i,b} = g_1^{t_{i,b}} \cdot R_{i,b}$. The public key is $\mathsf{Pk} = [\mathcal{I}, g_3, (T_{i,b})_{i\in[\ell], b\in\{0,1\}}]$, and the master secret key is $\mathsf{Msk} = [g_{12}, g_4, (t_{i,b})_{i\in[\ell], b\in\{0,1\}}]$, where $g_{12} = g_1 \cdot g_2$. The algorithm returns $(\mathsf{Pk}, \mathsf{Msk})$.

- KeyGen($\mathsf{Msk}, \boldsymbol{y}$): Let $S_{\boldsymbol{y}}$ be the set of indices i such that $y_i \neq \star$. The key generation algorithm chooses random $a_i \in \mathbb{Z}_N$ for $i \in S_{\boldsymbol{y}}$ under the constraint that $\sum_{i\in S_{\boldsymbol{y}}} a_i = 0$. For $i \in S_{\boldsymbol{y}}$, the algorithm chooses random $W_i \in \mathbb{G}_{p_4}$ and sets $Y_i = g_{12}^{a_i/t_{i,y_i}} \cdot W_i$. The algorithm returns ciphertext $\mathsf{Ct} = (Y_i)_{i\in S_{\boldsymbol{y}}}$. Here we use the fact that $S_{\boldsymbol{y}}$ has size at least 2.

- Enc($\mathsf{Pk}, \boldsymbol{x}$): The encryption algorithm chooses random $s \in \mathbb{Z}_N$. For $i \in [\ell]$, the algorithm chooses random $Z_i \in \mathbb{G}_{p_3}$ and sets $X_i = T_{i,x_i}^s \cdot Z_i$, and returns the token $\mathsf{Tok}_{\boldsymbol{y}} = (X_i)_{i\in[\ell]}$.

- Eval($\mathsf{Pk}, \mathsf{Ct}, \mathsf{Tok}_{\boldsymbol{y}}$): The test algorithm computes $T = \prod_{i\in S_{\boldsymbol{y}}} e(X_i, Y_i)$. It returns TRUE if $T = 1$, FALSE otherwise.

It easy to see that the scheme is correct. Regarding security, we show:

Theorem 5. Under the General Subgroup Decision Assumption [5] the HVE scheme described is ($\mathsf{poly}, 1, \mathsf{poly}$)-SIM-secure.

The proof is in the full version of this paper [10]. Informally, we simulate the flag used in the trapdoor circuits by means of the presence of the $\mathbb{G}_{p_5}$ subgroup. Specifically, if the $\mathbb{G}_{p_5}$ part is absent the ciphertext is in normal mode, otherwise it acts in trapdoor mode. The simulator then modifies the distributions of the adaptive queries, adding a $\mathbb{G}_{p_5}$ part, to interact with the trapdoor mechanism of the ciphertext when needed.

We note that one can easily extend our construction to meet ($\mathsf{poly}, \ell, \mathsf{poly}$)-SIM security for polynomial ℓ. The idea is simply to use a different subgroup for each message in the "trapdoor" mode. By [7,3], the restriction to a bounded number of challenge ciphertexts ℓ is necessary. On the other hand, the impossibility result of [1] does not apply to HVE, so there is no contradiction with the fact that our result here has $q_1 = \mathsf{poly}$ (instead of bounded q_1 as for our standard-model construction of Circuit-FE in Section 3, which is necessary).

5   Impossibility Results

In the section we present new negative results for simulation-based secure FE. We refer the reader to Section 1 for a background on the previously known impossibility results. All of our negative results build on ideas from the impossibility result of [1] for ($\mathsf{poly}, 1, 0$)-SIM-secure wPRF-FE w.r.t. black-box simulation. Below, we first describe the weak PRF functionality (that will be used in our negative results as well) and recall the impossibility result of [1]. We then proceed to discuss our new results.

Impossibility of Agrawal et al. [1]. Let $\{F\}$ be a weak pseudo-random function family on domain K and key space M. The wPRF functionality on key $k \in K$ and input $m \in M$ outputs $F_m(k)$. Let $l-1$ be an upper bound on the ciphertext size of the wPRF-FE scheme. The adversary asks tokens for l random inputs $x_1, \ldots, x_l$ in the domain of F, and for an encryption of a random k from the key space of F. The simulator needs to produce tokens $\{\mathsf{Tok}_i\}_{i \in [l]}$, and then it is given the functionality's outputs $\{F_k(x_i)\}_{i \in [l]}$. Now the simulator has to produce a ciphertext Ct such that for every $i \in [l]$, $F_k(x_i) = \mathsf{Eval}(\mathsf{Pk}, \mathsf{Ct}, \mathsf{Tok}_i)$. Now, on the one hand, the simulator needs to "encode" all of the functionality's outputs into Ct. On the other hand, the functionality's outputs are l pseudo-random bits, while $|\mathsf{Ct}| < l-1$. Since a pseudo-random string cannot be efficiently compressed we get a contradiction. (Note that a *black-box* simulator cannot encode the functionality's outputs into the tokens $\{\mathsf{Tok}_i\}$ since these are fixed before the simulator learns the outputs.)

5.1 Fully Non-adaptive Adversaries

In this section we give an impossibility result for a natural relaxation of the simulation-security considering only adversaries that are *fully non-adaptive*. In particular, we consider adversaries who make *simultaneous* token and ciphertext queries in the SIM-security game for FE.

Below, we formally define security against fully non-adaptive adversaries. Our definition allows for non-black-box simulation.

Definition 6. [Fully Non-Adaptive Security] We say that an F-FE scheme is (q, ℓ)-*fully non-adaptively* SIM-secure if every PPT adversary $\mathcal{A} = (\mathcal{A}_0, \mathcal{A}_1)$ there exists a PPT simulator Sim such that the outputs of the following two experiments are computationally indistinguishable:

Experiment $\mathsf{RealExp}^{\mathsf{FE}, \mathcal{A}}(1^\lambda)$:
  $(\mathsf{Pk}, \mathsf{Msk}) \leftarrow \mathsf{Setup}(1^\lambda)$;
  $(\{m_i\}_{i=1}^q, \{k_j\}_{j=1}^\ell, \mathsf{st}) \leftarrow \mathcal{A}_0(\mathsf{Pk})$;
  $\mathsf{Tok}_{k_j} \leftarrow \mathsf{KeyGen}(\mathsf{Msk}, k_j)$;
  $\mathsf{Ct}_i \leftarrow \mathsf{Enc}(\mathsf{Pk}, m_i)$;
  $\alpha \leftarrow \mathcal{A}_1(\mathsf{Pk}, \{\mathsf{Tok}_{k_j}\}, \{\mathsf{Ct}_i\}, \mathsf{st})$;
Output: (Pk, α)

Experiment $\mathsf{IdealExp}^{\mathsf{FE}, \mathcal{A}}_{\mathsf{Sim}}(1^\lambda)$:
  $(\mathsf{Pk}, \mathsf{Msk}) \leftarrow \mathsf{Setup}(1^\lambda)$;
  $\{m_i\}_{i=1}^q, \{k_j\}_{j=1}^\ell, \mathsf{st}) \leftarrow \mathcal{A}_0(\mathsf{Pk})$;
  $\alpha \leftarrow \mathsf{Sim}(\mathsf{Pk}, \mathsf{Msk}, \mathsf{st}, \{k_j, F(k_j, m)\})$;
Output: (Pk, α)

Theorem 7. Assuming the existence of a collision-resistant hash function family, there does not exist a $(\mathsf{poly}, \mathsf{poly})$-fully-non-adaptively SIM-secure wPRF-FE.

We prove the above theorem by extending the impossibility of [1]. Roughly speaking, the central idea is to use many ciphertext queries instead of one. The intuition is that in the non-adaptive case, the simulator can encode information about the function outputs in the tokens that might be long; however, by making many ciphertext queries, the same tokens can be used to decrypt many ciphertexts, making the length of the tokens insignificant. Indeed, the same idea can be used in the impossibility for given in the next subsections (at the cost of an increase in a number of ciphertext queries). We defer details to the full version [10].

5.2 Non-Black-Box Simulation

The impossibility of [1] rules out SIM-security against adversaries who make an unbounded number of non-adaptive token queries assuming the simulator is using the code of the adversary as black-box. In this section we extend their result to non-black-box simulators using the techniques from [3].

Below, we give a non-black-box definition of SIM-security, which is similar to that of [7] except that we only consider an unbounded number of non-adaptive token queries and one ciphertext query (corresponding to $(\mathsf{poly}, 1, 0)$-SIM-security). Further, following [3] we let the adversary and the simulator use an auxiliary input sampled from some distribution. In our negative result, we use this auxiliary input to store a random key of a collision-resistant hash function.[4]

Definition 8. [Non-Black-Box Simulation] We say that an F-FE scheme SIM-secure with *non-black-box* simulator if for every distribution on auxiliary input Z, and every PPT adversary $\mathcal{A} = (\mathcal{A}_0, \mathcal{A}_1)$, there exists a PPT simulator $\mathsf{Sim} = (\mathsf{Sim}_0, \mathsf{Sim}_1)$ such that the outputs of the following two experiments are computationally indistinguishable:

Experiment $\mathsf{RealExp}^{\mathsf{FE}, \mathcal{A}}(1^\lambda)$:	**Experiment** $\mathsf{IdealExp}^{\mathsf{FE}}_{\mathsf{Sim}}(1^\lambda)$:
$(\mathsf{Pk}, \mathsf{Msk}) \leftarrow \mathsf{Setup}(1^\lambda); z \leftarrow Z;$	$z \leftarrow Z;$
$(M, \mathbf{st}) \leftarrow \mathcal{A}_0^{\mathsf{KeyGen}(\mathsf{Msk}, \cdot)}(\mathsf{Pk}, z);$	$(M, \mathbf{st}) \leftarrow \mathsf{Sim}_0(z);$
$m \leftarrow M, \mathsf{Ct} \leftarrow \mathsf{Enc}(\mathsf{Pk}, m);$	$m \leftarrow M;$
$\alpha \leftarrow \mathcal{A}_1(\mathsf{Ct}, \mathbf{st});$	$\alpha \leftarrow \mathsf{Sim}_1^{F(\cdot, m)}(\mathbf{st});$
Let $\{k_i\}$ be the queries of $\mathcal{A}_0$ to $\mathsf{KeyGen};$	Let $\{k_i\}$ be the queries of Sim_1 to $F;$
Output: $(z, M, m, \alpha, \{k_i\})$	**Output:** $(z, M, m, \alpha, \{k_i\})$

where the output of $\mathcal{A}_0$ and Sim_0 consists of an arbitrary state $\mathbf{st}$ and a description of a distribution over messages M.

We now state our result:

Theorem 9. Assuming collision-resistant hash functions, there does not exist a SIM-secure wPRF-FE with non-black-box simulator.

In the non-black-box simulation definition the real and the simulated outputs may contain the generated tokens and ciphertext. However, the simulator is only required to produce the simulated tokens and ciphertext after receiving the functionality's outputs. Since the tokens may encode a lot of information (at least

---

[4] A stronger variant of FE with non-black-box simulation is defined in [3] and our negative result holds also for their definition.

as much as the functionality's outputs), the impossibility of [1] is not applicable here. To commit the simulator to the tokens before learning the functionality's outputs we use technique of [2]. This technique was recently used by [3] to extend the impossibility of [7] to hold without a non-programmable random oracle. The main idea is to consider an adversary that computes a collision-resistant hash of the tokens, and selects the message distribution based on the hash value. Intuitively, this commits the simulator to the tokens before it learns the functionality's outputs. We defer details to the full version [10].

5.3   FE for Multi-bit Outputs with Succinct Ciphertexts

Finally, we show that it is impossible to construct FE schemes where the ciphertext length is *independent* of the output length of the functionality. Recently, Goldwasser et al. [13] construct a SIM-secure FE scheme with "succinct" ciphertexts, improving on Gorbunov et al. [14] (in which the ciphertext size depends on the size of the circuit computing the functionality). However, [13] is only for functionalities with *boolean* output; for functionalities with longer output, the ciphertexts in both of these constructions grows linearly with the output length of the functionality. Our result shows this dependency is *inherent*.

To prove this result we consider the functionality that computes a pseudorandom generator, and we set the output length of the generator to be longer then the size of the FE ciphertext. The proof uses an incompressibility argument similar to the one used in [1]. However, unlike in [1] we consider only one token query and one ciphertext query and do not rely on an unbounded collusion. Due to space constraints, formal definitions and details are deferred to the full version [10].

Acknowledgements. We thank the anonymous reviewers of Crypto 2013 for very helpful comments regarding the presentation.

References

1. Agrawal, S., Gorbunov, S., Vaikuntanathan, V., Wee, H.: Functional encryption: New perspectives and lower bounds. Cryptology ePrint Archive, Report 2012/468 (2012), http://eprint.iacr.org/
2. Bellare, M., Dowsley, R., Waters, B., Yilek, S.: Standard security does not imply security against selective-opening. In: Pointcheval, D., Johansson, T. (eds.) EUROCRYPT 2012. LNCS, vol. 7237, pp. 645–662. Springer, Heidelberg (2012)
3. Bellare, M., O'Neill, A.: Semantically-secure functional encryption: Possibility results, impossibility results and the quest for a general definition. Cryptology ePrint Archive, Report 2012/515 (2012), http://eprint.iacr.org/
4. Bellare, M., Rogaway, P.: Random oracles are practical: A paradigm for designing efficient protocols. In: ACM CCS 1993, pp. 62–73 (1993)
5. Bellare, M., Waters, B., Yilek, S.: Identity-based encryption secure against selective opening attack. In: Ishai, Y. (ed.) TCC 2011. LNCS, vol. 6597, pp. 235–252. Springer, Heidelberg (2011)

6. Boneh, D., Di Crescenzo, G., Ostrovsky, R., Persiano, G.: Public key encryption with keyword search. In: Cachin, C., Camenisch, J.L. (eds.) EUROCRYPT 2004. LNCS, vol. 3027, pp. 506–522. Springer, Heidelberg (2004)
7. Boneh, D., Sahai, A., Waters, B.: Functional encryption: Definitions and challenges. In: Ishai, Y. (ed.) TCC 2011. LNCS, vol. 6597, pp. 253–273. Springer, Heidelberg (2011)
8. Boneh, D., Waters, B.: Conjunctive, subset, and range queries on encrypted data. In: Vadhan, S.P. (ed.) TCC 2007. LNCS, vol. 4392, pp. 535–554. Springer, Heidelberg (2007)
9. Canetti, R., Goldreich, O., Halevi, S.: The random oracle methodology, revisited (preliminary version). In: 30th ACM STOC, pp. 209–218, Full version avaiable at Cryptology ePrint Archive, Report 1998/011
10. Caro, A.D., Iovino, V., Jain, A., O'Neill, A., Paneth, O., Persiano, G.: On the achievability of simulation-based security for functional encryption. IACR Cryptology ePrint Archive (2013)
11. De Caro, A., Iovino, V., Persiano, G.: Fully secure hidden vector encryption. In: Abdalla, M., Lange, T. (eds.) Pairing 2012. LNCS, vol. 7708, pp. 102–121. Springer, Heidelberg (2013)
12. Feige, U., Lapidot, D., Shamir, A.: Multiple non-interactive zero knowledge proofs based on a single random string (extended abstract). In: 31st Annual Symposium on Foundations of Computer Science, St. Louis, Missouri, USA, October 22-24, vol. I, pp. 308–317 (1990)
13. Goldwasser, S., Kalai, Y., Popa, R.A., Vaikuntanathan, V., Zeldovich, N.: Reusable garbled circuits and succinct functional encryption. In: 45th ACM STOC, pp. 555–564
14. Gorbunov, S., Vaikuntanathan, V., Wee, H.: Functional encryption with bounded collusions via multi-party computation. In: Safavi-Naini, R., Canetti, R. (eds.) CRYPTO 2012. LNCS, vol. 7417, pp. 162–179. Springer, Heidelberg (2012)
15. Gorbunov, S., Vaikuntanathan, V., Wee, H.: Attribute-based encryption for circuits. In: STOC
16. Goyal, V., Pandey, O., Sahai, A., Waters, B.: Attribute-based encryption for fine-grained access control of encrypted data. In: ACM CCS 2006, pp. 89–98 (2006), Available as Cryptology ePrint Archive Report 2006/309
17. Katz, J., Sahai, A., Waters, B.: Predicate encryption supporting disjunctions, polynomial equations, and inner products. In: Smart, N.P. (ed.) EUROCRYPT 2008. LNCS, vol. 4965, pp. 146–162. Springer, Heidelberg (2008)
18. Lewko, A., Okamoto, T., Sahai, A., Takashima, K., Waters, B.: Fully secure functional encryption: Attribute-based encryption and (Hierarchical) inner product encryption. In: Gilbert, H. (ed.) EUROCRYPT 2010. LNCS, vol. 6110, pp. 62–91. Springer, Heidelberg (2010)
19. Nielsen, J.B.: Separating random oracle proofs from complexity theoretic proofs: The non-committing encryption case. In: Yung, M. (ed.) CRYPTO 2002. LNCS, vol. 2442, pp. 111–126. Springer, Heidelberg (2002)
20. Okamoto, T., Takashima, K.: Adaptively attribute-hiding (Hierarchical) inner product encryption. In: Pointcheval, D., Johansson, T. (eds.) EUROCRYPT 2012. LNCS, vol. 7237, pp. 591–608. Springer, Heidelberg (2012)
21. O'Neill, A.: Definitional issues in functional encryption. Cryptology ePrint Archive, Report 2010/556 (2010), http://eprint.iacr.org/
22. Parno, B., Raykova, M., Vaikuntanathan, V.: How to delegate and verify in public: Verifiable computation from attribute-based encryption. In: Cramer, R. (ed.) TCC 2012. LNCS, vol. 7194, pp. 422–439. Springer, Heidelberg (2012)

23. Sahai, A., Waters, B.: Fuzzy identity-based encryption. In: Cramer, R. (ed.) EUROCRYPT 2005. LNCS, vol. 3494, pp. 457–473. Springer, Heidelberg (2005)
24. Waters, B.: Functional encryption for regular languages. In: Safavi-Naini, R., Canetti, R. (eds.) CRYPTO 2012. LNCS, vol. 7417, pp. 218–235. Springer, Heidelberg (2012)
25. Wee, H.: Zero knowledge in the random oracle model, revisited. In: Matsui, M. (ed.) ASIACRYPT 2009. LNCS, vol. 5912, pp. 417–434. Springer, Heidelberg (2009)

How to Run Turing Machines on Encrypted Data

Shafi Goldwasser[1], Yael Tauman Kalai[2], Raluca Ada Popa[1],
Vinod Vaikuntanathan[3], and Nickolai Zeldovich[1]

[1] MIT CSAIL
[2] Microsoft Research
[3] University of Toronto

Abstract. Cryptographic schemes for computing on encrypted data promise to be a fundamental building block of cryptography. The way one models such algorithms has a crucial effect on the efficiency and usefulness of the resulting cryptographic schemes. As of today, almost all known schemes for fully homomorphic encryption, functional encryption, and garbling schemes work by modeling algorithms as circuits rather than as Turing machines.

As a consequence of this modeling, evaluating an algorithm over encrypted data is as slow as the worst-case running time of that algorithm, a dire fact for many tasks. In addition, in settings where an evaluator needs a description of the algorithm itself in some "encoded" form, the cost of computing and communicating such encoding is as large as the worst-case running time of this algorithm.

In this work, we construct cryptographic schemes for computing Turing machines on encrypted data that avoid the worst-case problem. Specifically, we show:

- An attribute-based encryption scheme for any polynomial-time Turing machine and Random Access Machine (RAM).
- A (single-key and succinct) functional encryption scheme for any polynomial-time Turing machine.
- A reusable garbling scheme for any polynomial-time Turing machine.

  These three schemes have the property that the size of a key or of a garbling for a Turing machine is very short: it depends only on the description of the Turing machine and not on its running time.

  Previously, the only existing constructions of such schemes were for depth-d circuits, where all the parameters grow with d. Our constructions remove this depth d restriction, have short keys, and moreover, avoid the worst-case running time.

- A variant of fully homomorphic encryption scheme for Turing machines, where one can evaluate a Turing machine M on an encrypted input x in time that is dependent on the running time of M *on input* x as opposed to the worst-case runtime of M. Previously, such a result was known only for a restricted class of Turing machines and it required an expensive preprocessing phase (with worst-case runtime); our constructions remove both restrictions.

Our results are obtained via a reduction from SNARKs (Bitanski et al) and an "extractable" variant of witness encryption, a scheme introduced by Garg *et al.*. We prove that the new assumption is secure in the generic group model. We also point out the connection between (the variant of) witness encryption and the obfuscation of point filter functions as defined by Goldwasser and Kalai in 2005.

Keywords: Computing on encrypted data, Functional encryption, Fully homomorphic encryption, Turing machines, Input-specific running time.

R. Canetti and J.A. Garay (Eds.): CRYPTO 2013, Part II, LNCS 8043, pp. 536–553, 2013.

1   Introduction

Cryptographic schemes for computing on encrypted data promise to be a major focus of cryptographic research for years to come. We now have early constructions of fully homomorphic encryption, functional encryption, and attribute-based encryption, as well as more established constructions for garbling schemes. An important question for the practicality and usability of these schemes is *how to model* an algorithm that computes on encrypted data in cryptographic constructions.

Modeling algorithms as circuits instead of Turing machines has efficiency and usability disadvantages. Indeed, almost all known[1] cryptographic constructions of fully homomorphic encryption, attribute-based encryption, functional encryption and garbling schemes for general algorithms model these algorithms as Boolean or arithmetic circuits. As a consequence, these constructions suffer from the following two disadvantages.

The first disadvantage is that evaluating an algorithm A modeled as a circuit on encrypted data is at least as slow as the worst-case running time of algorithm A on all inputs of a certain size. Ideally, the runtime of A on input x should be the time A takes to run on x. The reason for this slowdown is that all the known transformations from Turing machines to circuits essentially work by unrolling loops to their worst-case runtime, and by considering all branches of a computation. Even if the cryptographic overhead of these schemes were zero, such worst-case runtime can still make the computation prohibitively slow: for example, the simplex algorithm for linear programming runs in polynomial time on most instances one encounters in practice, but in exponential time on rare inputs.

The second disadvantage arises for schemes that require an evaluator to obtain an encoded description of an algorithm A (called a *token*) in order to run A on the encrypted data. For example, in functional encryption, the token is a key for the algorithm A and in garbling schemes, the token is the garbling of the algorithm. In these settings, modeling algorithms as circuits makes the size of the token as large as the running time of the algorithm, instead of having the token size depend only on the description of the algorithm, which can be much shorter.

The earliest example of using circuits for computing on encrypted data is Yao's secure function evaluation protocol [Yao86] which takes as input any polynomial-time computable function f – specified by a circuit – and outputs a "garbled circuit" with the same input-output functionality. Such worst-case runtime also affects known two-party and multi-party protocols for general secure function evaluation [Yao86,GMW87,BGW88,CCD88].

More recent constructions for computing on encrypted data also use circuits to model computation and thus suffer from the worst-case slowdown: fully homomorphic encryption schemes (FHE) [Gen09,BV11a,BV11b,BGV12,Bra12], attribute-based encryption (ABE) schemes [GVW13,GGH+13b,GGH13a], and functional encryption (FE) schemes for general functions [SS10,GVW12,GKP+13b].

In this work, we present cryptographic schemes for Turing machines, thus removing the two major limitations of circuits discussed above. We construct attribute-based

---

[1] An exception is the garbling scheme of [LO12] for RAMs, but this scheme also suffers from the worst-case running time problem we address in this paper (see Sec. 1.1).

encryption, (succinct and single-key) functional encryption, reusable garbling schemes, and a version of FHE for polynomial-time Turing machines. For each of these schemes, we show that the time to evaluate a Turing machine M on an input x is *input specific*: it depends on the runtime of M on x and not on the worst-case runtime of M on all inputs of length n where $n = |x|$. Moreover, we show that the token for a Turing machine M is short: its size depends on the size of *the description of the Turing machine M* and not on M's runtime. Our schemes are for both uniform and non-uniform Turing machines (so in particular, they can compute circuits).

Our schemes are based on extractable witness encryption, a variant of the witness encryption notion of Garg *et al.* [GGSW13]. We show how to obtain such an extractable witness encryption scheme using the construction of Garg *et al.* [GGSW13], by strengthening their assumption with a knowledge property. We prove the new assumption secure in the generic group model. Interestingly, we show that extractable witness encryption is closely related to (weakly) obfuscatable point-filter functions [GK05].

1.1   Our Results

We now explain our results in detail.

Attribute-Based Encryption (ABE) for Turing Machines and RAMs. Attribute-based encryption schemes, originally defined by Sahai and Waters [SW05], allow a user holding the master secret key msk to generate a function key sk_f for any predicate f of his choice, where sk_f does not hide f. Using the master public key mpk, anyone can encrypt a message m with respect to an "attribute" x: such a ciphertext is denoted by $\mathsf{Enc}(x; m)$. The ciphertext $\mathsf{Enc}(x; m)$ does not hide x, and hides only m. Given a function key sk_f and a ciphertext $\mathsf{Enc}(x; m)$, one can compute m if $f(x) = 1$. On the other hand, if $f(x) = 0$, ABE leaks no information about m and provides semantic security.

Attribute-based encryption is a powerful primitive and has thus received significant attention [GPSW06,LOS+10,LW12,GVW13]. The state-of-the-art is the scheme of Gorbunov *et al.* [GVW13]: based on the LWE assumption, they construct an ABE for the class of all circuits of depth at most d, where the efficiency of the scheme (such as the size of the ciphertexts) decreases polynomially with d. In concurrent work, Garg *et al.* constructed ABE schemes with similar properties [GGH+13b], and an ABE scheme with large ciphertexts [GGSW13], both from candidate multi-linear maps.

In this work, we construct an attribute-based encryption scheme for all circuits, with no restriction on the depth. More importantly, we model functions as Turing machines (with possibly non-uniform advice), as opposed to circuits as in previous work. Computing a function key sk_M, corresponding to a Turing machine M, takes roughly linear time in the *size of the description* of M, *independent of the runtime* of M. Moreover, given sk_M and $\mathsf{Enc}(x; m)$ where $f(x) = 1$, one can compute m in time that depends only on the time it takes to compute M *on input x* as opposed to the worst-case running time of M. We prove the security of our scheme with respect to a non-adaptive simulation-based definition (we refer the reader to Sec. 3 for details). We then show that a modification of our construction provides ABE for RAMs.

Theorem 1 (Informal). *There exists an attribute-based encryption scheme (as defined in Defs. 3, 4) for (uniform or non-uniform) polynomial-time Turing machines and RAMs from the assumptions in Sec. 1.2.*

Interestingly, we show how to extend our ABE scheme beyond Turing machines and RAMs: for example, an evaluator can choose by himself which Turing machines to run on the ciphertexts, as long as they satisfy some property expressed in a function key.

Functional Encryption (FE) for Turing Machines. Functional encryption, formalized by Boneh, Sahai and Waters [BSW11], is a generalization of attribute-based encryption. In functional encryption, a user holding the master secret key msk can generate a function key sk_f corresponding to a function f; then, anyone having a ciphertext $Enc(x)$ and a function key sk_f can compute $f(x)$, but learns nothing else about the input x.

So far, the only many-keys FE schemes known (schemes in which the secret key owner can securely release an unbounded number of function keys) are for the inner-product predicates [KSW08,SSW09]. For general functions, Agrawal *et al.* [AGVW13] showed that there does not exist a many-keys FE scheme if one wants to achieve a natural simulation-based security definition[2], so the natural question was to construct a single-key functional encryption scheme for general functions. Sahai and Seyalioglu [SS10], Gorbunov *et al.* [GVW12], and Goldwasser *et al.* [GKP+13b] constructed such schemes for circuits. The work of Goldwasser *et al.* [GKP+13b] is the first to provide succinct ciphertexts: the ciphertext size is much smaller than the circuit size; they constructed a *succinct* single-key FE scheme for any depth d circuit, where the parameters of the scheme grow with d (but are independent of the circuit size).

In this work, we not only remove this depth-d restriction, but we model functions as (possibly non-uniform) Turing machines, as opposed to circuits as in prior work. Our schemes have short function keys: computing the function key of a Turing machine M depends only on the size of M and does not depend on the runtime of M. We note that in all previous schemes for general functions the size of a function key for a function f grows (at least linearly) with the worst-case runtime of f. We note however, that as opposed to our ABE scheme, in a functional encryption scheme, given $Enc(x)$ and sk_M, the time it takes to compute $M(x)$ must be proportional to the worst-case runtime of M, since the runtime of M on input x may leak sensitive information about x. However, if one is willing to slightly relax security and allow leaking the runtime of M on the secret input x, then we provide a second functional encryption scheme for which the decryption algorithm has input-specific runtime (i.e., it runs in time polynomial in the runtime of M on input x) – we denote this by *input-specific runtime functional encryption*.

Theorem 2 (Informal). *There exists a single-key (succinct) functional encryption scheme and input-specific runtime functional encryption scheme for (uniform or non-uniform) polynomial-time Turing machines from the assumptions in Sec. 1.2.*

Variant of FHE for Turing Machines. We construct a variant of FHE where one can evaluate a Turing machine M on a ciphertext $Enc(x)$ in time that depends on the runtime

---

[2] Their lower bound does not apply to weaker security definitions.

of *P* on the specific input x. We naturally call this scheme *input-specific FHE*. At first glance, this may seem impossible, since revealing the runtime of *P* on input x may reveal secret information about x. However, for many Turing machines M, revealing only the runtime of M is not harmful, and it can provide significant efficiency gains.

Our construction is an improvement of Goldwasser *et al.* [GKP+13b] who showed how to construct input-specific runtime FHE from single-key functional encryption. As in Goldwasser *et al.* [GKP+13b], we also encrypt a Turing machine M and x together into a token $\mathsf{tk}_{M,x}$. Producing such a token depends only on the size of x and M, and not on the running time of M. The evaluator can use $\mathsf{tk}_{M,x}$ and public information to compute $M(x)$ in input-specific time. The reason we provide a token for M at all is for security: the FHE evaluator must no longer be able to evaluate TMs of its choice on the encrypted inputs because the running time of those TMs can leak the input entirely. We combine M and x in $\mathsf{tk}_{M,x}$ for a technical reason stemming from the fact that the FE scheme we use in the construction is single-key – we elaborate in our full paper.

Comparing to [GKP+13b], we make the following improvements:

- *Remove costly preprocessing.* [GKP+13b] had an expensive preprocessing phase taking as long as the worst-case runtime. With our scheme, the preprocessing is cheap: polynomial in the size of the TMs and independent of the worst-case runtime (so in fact it can be performed in the online phase).

- *Works for any polynomial-time Turing machine.* Because the ciphertext size in [GKP+13b] depended on the depth of the worst-case circuit representation of the class of Turing machines, [GKP+13b] only allowed a restricted class of Turing machines: the class of TMs that can be expressed by shallow-depth circuits (e.g., log-space Turing machines). Our result does not have the depth restriction and thus applies to any class of Turing machines with runtime upper-bounded by a polynomial.

Theorem 3 (Informal). *There exists an input-specific-runtime fully homomorphic encryption scheme for (uniform or non-uniform) polynomial-time Turing machines based on the assumptions in Sec. 1.2.*

Reusable Garbling Scheme for Turing Machines. Garbling schemes, introduced in the seminal work of Yao [Yao86], have found many applications in cryptography. In such schemes, a user can "garble" a function f and then encode an input x in a token tk_x. Given a garbling of f and a token tk_x, one can compute $f(x)$, but learns nothing else about f or x. Some works also considered an authenticity property [BHR12,GVW13], on which we do not dwell. Traditional garbling schemes are one-time: they are secure only if an adversary gets a token for at most one input. A reusable garbling scheme is secure when the adversary gets an unbounded number of tokens.

In known garbling schemes (even non-reusable ones), the size of the garbling is as large as the worst-case runtime of f. Often, the reason is that programs are modeled as circuits, and the size of the garbling is at least the size of the corresponding circuit. In this work, we construct a (reusable) garbling scheme for (uniform or non-uniform) Turing machines, where the size of the garbling depends only on the size of the Turing machine, and is *independent of its runtime*. The work of [LO12] is an exception from the circuit model: they model computation as RAM, but their scheme still has large garbling size, at least as large as the worst-case running time.

As in our FHE and FE schemes, if one allows leaking the runtime of M on input x, we can additionally avoid worst-case evaluation time and obtain an input-specific reusable garbling scheme: given a garbling for a Turing machine M and a token tk_x, the time to compute $M(x)$ is polynomial in the runtime of M on the specific input x.

Goldwasser *et al.* [GKP$^+$13b] provide a reusable garbling scheme only for depth bounded circuits; our schemes remove the depth dependency, provide short garbling size, and can additionally avoid worst-case running time.

Theorem 4 (Informal). *There exists a reusable garbling scheme and an input-specific reusable garbling scheme for (uniform or non-uniform) polynomial-time Turing machines from the assumptions in Sec. 1.2.*

In summary, our work models computation on encrypted data as Turing machines and thus avoids the worst-case "curse" for a set of well-known cryptographic notions.

Remark 1. Interestingly, we can easily overcome the worst-case curse for interactive tasks such as two-party and multi-party protocols as follows. To securely evaluate a Turing machine M, we evaluate the Turing machines $M_1, \ldots, M_{\omega(\log n)}$ sequentially, where M_i runs the Turing machine M for 2^i steps and outputs M's answer if M halted in 2^i steps, otherwise $\perp$. To evaluate M_i, we simply use existing multi-party protocols. Note that the circuit size for M_i is $\mathrm{poly}(2^i)$, and since we halt the computation as soon as we get a non-$\perp$ answer, the protocol runs in input-specific time. The reason we can overcome the worst-case curse in this manner is that interaction is allowed. In this work, we focus on non-interactive tasks, which are more challenging.

1.2 Our Assumptions

Our schemes rely on two assumptions: extractable witness encryption and the existence of SNARKs.

Extractable Witness Encryption. The recent work of Garg *et al.* [GGSW13] constructs a new primitive called witness encryption (WE). Such a scheme is associated with some **NP** complete language L. Given an instance x and a message m, any user can encrypt m with respect to x; this is denoted by $\mathsf{Enc}_x(m)$. Given $\mathsf{Enc}_x(m)$ and a valid witness w of x, any user can decrypt x efficiently. On the other hand, if x is not in the language, the scheme provides semantic security.

In our work, we additionally assume that the [GGSW13] scheme is extractable: if an adversary can break semantic security for an instance x, an extractor can extract the witness for x. Such an extractable scheme can be constructed from an extractable version of the [GGSW13] assumption (called extractable DGE No-Exact-Cover assumption) so we strengthen their assumption. While we state our assumption in a decisional form for simplicity, the search version of the assumption suffices for our schemes because we can use hard-core predicates to mask the one bit we care to hide (m).

We validate our assumption in the generic group model: we prove that no polynomial-time adversary can break the assumption in the generic group model where adversaries can only use multilinear map operations as a black-box. We refer the reader to our full paper for more details on the assumption, and emphasize that we view our result as a reduction from any extractable witness encryption scheme, as opposed to a result that is tied to the specific computational assumption.

We show that, interestingly, extractable witness encryption is highly related to another task that was already well-known in the cryptographic literature: (weakly) obfuscating point-filter functions, defined by Goldwasser and Kalai [GK05]. Informally, point-filter functions for a language $L \in \mathbf{NP}$ with witness relation R_L are a class of functions $\{\delta_{x,b}\}$, indexed by a string $x \in \{0,1\}^n$ and a bit $b \in \{0,1\}$ that behave as follows:

$$\delta_{x,b}(w) = \begin{cases} (x,b), & \text{if } (x,w) \in R_L, \\ (x, \perp), & \text{otherwise.} \end{cases}$$

It can be shown that extractable witness encryption is indeed equivalent to (weakly) obfuscating point filter function. Thus, the former implies the consequences of the later regarding the impossibility of obfuscation for a wide range of natural tasks based on [GK05]. See our full paper for more details.

The Existence of SNARKs (Succinct Non-Interactive Arguments of Knowledge). Bitansky *et al.* [BCCT13] construct SNARKs in a generic way (via a reduction from weaker SNARKs). Their work is based on "knowledge of exponent assumptions", and the existence of collision resistant hash functions.

If we remove SNARKs from our constructions, we still obtain novel schemes over prior work because the sizes of the function keys and of the garbling remain short, linear in the size of the Turing machine. Without SNARKs, though, the loss is that the ciphertext size grows with the running time of the Turing machines.

Our FE, FHE, and reusable garbling schemes additionally rely on the existence of a fully homomorphic encryption scheme, which can be obtained from the LWE assumption with circular security [BGV12].

1.3 Techniques Overview

ABE for Turing Machines. The main technical challenge in this work is constructing an ABE scheme for Turing machines.

Our construction starts with witness encryption and a signature scheme. The function key for a Turing machine M is simply a signature of M. The master secret and public keys generated during setup are the secret and verification keys (SigSK, VK) for the signature scheme. To encrypt a bit b with respect to a (public) attribute x, we compute a witness encryption $\mathsf{Enc}_{x^*}(b)$, where $x^* = (x, \mathsf{VK})$ and where a valid witness for x^* is a tuple (M, σ, π), where M is a Turing machine, σ is a signature of M using SigSK, and π the tableau of the computation, which can be interpreted as a "proof" that $M(x) = 1$.

Loosely speaking, the security proof proceeds as follows. Suppose there exists a successful adversary $\mathcal{A}$ for our ABE scheme. Then, given $\mathsf{Enc}_{x^*}(b)$, the ABE encryption of a random bit b, and several secret keys $\mathsf{sk}_{M_i} = \sigma_i$ such that $M_i(x) = 0$, $\mathcal{A}$ succeeds in guessing b with non-negligible advantage. The security of the extractable witness encryption implies that there exists a poly-time extractor that extracts a valid witness from $\mathcal{A}$ with non-negligible probability. Recall that a valid witness is a triplet of the form (M^*, σ^*, π^*) where σ^* is a valid signature of the Turing machine M^* and

π^* is a proof that $M^*(x) = 1$. Note that since $M_i(x) = 0$ for every i, it must be the case that $M^* \neq M$, which contradicts the unforgeability of the signature scheme.

Unfortunately, this idea does not quite give us the results we want. The reason is that the time to check a witness for an instance $x^* = (x, \mathsf{VK})$ is very long because it involves checking the tableau π of M on input x. In this case, the witness encryption of Garg et al. [GGSW13] is not "succinct": the size of the ciphertext $\mathsf{Enc}_{x^*}(b)$ grows with the time to check the witness. Thus, the approach above gives us a non-succinct ABE scheme, where the size of a ciphertext depends on the worst-case runtime of any (allowed) Turing machine.

To obtain succinctness, we use a SNARG scheme [BCCT13]. A SNARG has a common reference string crs, which is assumed to be securely generated. Any user can prove any **NP** statement by computing a proof π. The length of the crs, the length of the proofs, and the time to verify a proof are all *short*: depending only on the security parameter, and not on the time to verify the **NP** witness.

$\mathsf{Enc}_{x^*}(b)$ now proceeds as follows. It generates a crs corresponding the underlying SNARG scheme. To encrypt a bit b w.r.t. a public attribute x, it simply computes $\mathsf{Enc}_{x^*}(b)$, where x^* is now $(x, \mathsf{crs}, \mathsf{VK})$. A valid witness for x^* is a tuple of the form (M, σ, π) where σ is a valid signature of the Turing machine M, and π is a *succinct* SNARG proof that $M(x) = 1$. The fact that π can be verified in a short time makes the WE ciphertext succinct, as desired.

This gives us an ABE for Turing machines. Because SNARKs are for NP, our resulting ABE scheme is for any class of Turing machines for which there exists a polynomial that upper bounds the runtime of all machines in the class.

There scheme still has a slight drawback: it is succinct only for uniform Turing machines. If the Turing machines have non-uniform advice as large as the runtime, the resulting ABE ciphertexts are non-succinct. We would like our ABE scheme to be a generalization of previous work on circuits, and in particular to be succinct for any non-uniform Turing machine. To this end, we replace the SNARG scheme with a SNARK scheme (succinct non-interactive argument of knowledge) scheme. SNARKs have the additional property that if an adversary $\mathcal{A}$ succeeds in proving that $x \in L$, an extractor can extract a corresponding witness w from $\mathcal{A}$.

The final ABE scheme is as before, except that now a valid witness for $x^* = (x, \mathsf{crs}, \mathsf{VK})$ is a pair (π, t) (without the Turing machine and the signature), where π is a proof-of-knowledge of a Turing machine M and a signature σ such that σ is a valid signature of M and $M(x) = 1$. Now the witness size and the verification time is efficient (independent of the size of the Turing machine or its runtime). We refer the reader to Sec. 3 for more details on our ABE scheme and the security proof.

Functional Encryption for Turing Machines. We use the reduction of Goldwasser et al. [GKP+13b] to construct a (single-key and succinct) FE scheme from FHE and ABE. Their reduction is for circuits so we need to adapt it to Turing machines. The main technical issue is that we need to perform the FHE evaluation of a Turing machine M. To achieve this goal, we construct a new Turing machine M_{FHE} that evaluates homomorphically the transition function of M for a t number of times. The problem is that M_{FHE} needs to know what inputs to read from M's tape to feed into the FHE evaluation, but the movement of the head in M is an output of the transition function, so it is encrypted with FHE and unavailable to M_{FHE}. To solve this issue, we transform M into an *oblivious* Turing machine using Pippenger-Fischer [PF79]: now the movement of the head follows a fixed and known pattern independent of the input to M.

If one allows the runtime of M on x to leak, we can provide a second FE scheme FE^* whose decryption algorithm runs in input-specific time. We construct FE^* as a reduction from our FE scheme above using the idea of [GKP+13b]: instead of generating a function key sk_M for a Turing machine M, we generate many function keys $\mathsf{sk}_{M_1}, \ldots, \mathsf{sk}_{M_{\log B_n}}$, where M_i is the Turing machine that runs M for 2^i time steps, and either outputs the output of M or $\bot$ if M did not halt in 2^i steps; the parameter B_n is a global bound on the runtime of the Turing machines we consider. To generate $\log B_n$ function keys, we use $\log B_n$ instances of our single-key functional encryption scheme above, by generating fresh keys for every instance of it. Moreover, since the underlying functional encryption scheme is for Turing machines, generating sk_{M_i} can be done very efficiently, in time polynomial in the *size* of M_i, independent on the runtime of M_i.

On input a ciphertext $\mathsf{Enc}(x)$ and a function key $(\mathsf{sk}_{M_1}, \ldots, \mathsf{sk}_{M_{\log B}})$ for the Turing machine M, the decryption algorithm first tries to decrypt with sk_{M_1}, then tries with sk_{M_2}, and so on. The first time that it succeeds it stops. Note that the runtime of this decryption algorithm depends on the runtime of M on the *specific input x*, denoted by t_x. This is the case since it runs the original decryption algorithm (which runs in the worst-case) only with the secret keys $\mathsf{sk}_{M_1}, \ldots, \mathsf{sk}_{M_{\log t_x}}$, and all the Turing machines $M_1, \ldots, M_{\log t_x}$ run in time at most t_x.

Reusable Garbling and a Variant of FHE for Turing Machines. In our full version, we show how to construct these schemes from our FE scheme using a similar reduction to [GKP+13b].

Other Related Work. We discuss other related work in the full version of our paper.

1.4 Paper Roadmap

The rest of this paper is organized as follows. We provide definitions for extractable witness encryption and ABE in Sec. 2, and refer the reader to our full paper [GKP+13a] for other relevant preliminaries. Next, Sec. 3 presents our ABE scheme for Turing machines, which we prove formally in our full paper. Finally, Sections 4 and 4.2 show how to construct functional encryption for Turing machines. Due to space constraints, in our full paper [GKP+13a], we present the construction of extractable witness encryption and prove the new assumption in the generic group model, we show that extractable witness encryption implies (weakly) obfuscatable point filter functions and deduce implications to obfuscation, and we present the construction of FHE for Turing machines.

2 Preliminaries

In this section, we define extractable witness encryption and ABE for Turing machines, and refer the reader to our full paper for definitions of FE for Turing machines, SNARKs, and other relevant preliminaries.

2.1 Notation

We let κ denote the security parameter throughout this paper. For a distribution $\mathcal{D}$, we say $x \leftarrow \mathcal{D}$ when x is sampled from the distribution $\mathcal{D}$. If S is a finite set, by $x \leftarrow S$, we mean x is sampled from the uniform distribution over the set S.

We say that a function f is negligible in an input parameter κ, if for all $d > 0$, there exists K such that for all $\kappa > K$, $f(\kappa) < \kappa^{-d}$. For brevity, we write: for all sufficiently large κ, $f(\kappa) = \mathrm{negl}(\kappa)$.

2.2 Witness Encryption (WE)

The syntax of WE is as defined by Garg et al. [GGSW13], but the security definition has an additional extractability property.

Definition 1 (Witness Encryption). *A witness encryption for a language $L \in$ NP with corresponding witness relation R_L consists of two polynomial-time algorithms* (WE.Enc, WE.Dec) *such that*

- *Encryption* WE.Enc$(1^\kappa, x, b)$: *takes as input a security parameter κ, $x \in \{0,1\}^*$ and a bit b and outputs a ciphertext* ct.
- *Decryption* WE.Dec(w, ct): *takes as input $w \in \{0,1\}^*$ and a ciphertext* ct *and outputs a bit b or the symbol $\perp$.*

Correctness: For all $(x, w) \in R_L$, for all bits b, for every sufficiently large security parameter κ:

$$\Pr[\mathsf{ct} \leftarrow \mathsf{WE.Enc}(1^\kappa, x, b) : \mathsf{WE.Dec}(w, \mathsf{ct}) = b] = 1 - \mathrm{negl}(\kappa).$$

Definition 2 (Extractable security). *A witness encryption scheme for a language $L \in$ NP is secure if for all p.p.t. adversaries A, and all poly q, there exists a p.p.t. extractor E and a poly p, such that for all auxiliary inputs z and for all $x \in \{0,1\}^*$, the following holds:*

$$\Pr[b \leftarrow \{0,1\}; \mathsf{ct} \leftarrow \mathsf{WE.Enc}(1^\kappa, x, b) : A(x, \mathsf{ct}, z) = b] \geq 1/2 + 1/q(|x|)$$
$$\Rightarrow \quad \Pr[E(x, z) = w : (x, w) \in R_L] \geq 1/p(|x|).$$

2.3 Attribute-Based Encryption (ABE) for Turing Machines

We define the syntax and security of ABE for Turing machines.

Definition 3 (ABE for Turing machines). *An attribute-based encryption scheme* ABE *for a class of Turing machines $\mathcal{T}$ is a tuple of four algorithms* (ABE.Setup, ABE.KeyGen, ABE.Enc, ABE.Dec), *the first three of which are p.p.t., such that:*

- ABE.Setup(1^κ) *takes as input the security parameter 1^κ and outputs a master public key* mpk *and a master secret key* msk.
- ABE.KeyGen(msk, M) *takes as input the master secret key* msk, *a Turing machine $M \in \mathcal{T}$, and outputs a function key* sk$_M$.

- ABE.Enc(mpk, x, b) *takes as input the master public key* mpk, *an attribute* $x \in \{0, 1\}^*$, *and a bit b and outputs a ciphertext* ct.
- ABE.Dec(sk$_M$, ct) *takes as input a key* sk$_M$ *and a ciphertext c and outputs a bit.*

Correctness. *For all Turing machines* $M \in \mathcal{T}$, *for all attributes* $x \in \{0, 1\}^*$, *for all bits b, for κ sufficiently large,*

$$\Pr[(\mathsf{mpk}, \mathsf{msk}) \leftarrow \mathsf{ABE.Setup}(1^\kappa); \mathsf{fsk}_f \leftarrow \mathsf{ABE.KeyGen}(\mathsf{fmsk}, f);$$
$$c \leftarrow \mathsf{ABE.Enc}(\mathsf{fmpk}, x) : \mathsf{ABE.Dec}(\mathsf{fsk}_f, 1^t, c) = f(x)]$$
$$= 1 - \mathsf{negl}(\kappa).$$

Efficiency. *There exists a polynomial p such that the running time of* ABE.Dec(sk$_M$, ct) *is at most* $p(\kappa, \mathsf{runtime}(M, x))$.

The efficiency property states that the work of the decryption depends on the run time of a Turing machine on the attribute. Since ABE.Setup, ABE.KeyGen and ABE.Enc are p.p.t.-s, their running time depends only on the security parameter and not on the running time of the Turing machines (except for a logarithmic dependency on it).

Our security definition is full (the adversary can choose the challenge attribute based on the public key) and non-adaptive (the adversary chooses the Turing machines before getting the challenge ciphertext).

Definition 4 (Attribute-based encryption security). *Let* ABE *be an attribute-based encryption scheme for a class of Turing machines* $\mathcal{T}$ *and let* $A = (A_1, A_2)$ *be an adversary. Consider the following experiment.*

$$\mathsf{Exp}_{\mathsf{ABE}}(1^\kappa)\mathbf{:}$$

1: (mpk, msk) $\leftarrow$ ABE.Setup(1^κ)
2: $(x, \mathsf{state}) \leftarrow A_1^{\mathsf{ABE.KeyGen}(\mathsf{msk}, \cdot)}(\mathsf{mpk})$
3: Choose a bit b at random and let ct $\leftarrow$ ABE.Enc(mpk, x, b).
4: $b' \leftarrow A_2(\mathsf{state}, \mathsf{ct})$.
5: If, $b = b'$ and for all Turing machines M that A requests to oracle ABE.KeyGen(msk, $\cdot$), we have $M(x) = 0$, output 1, else output 0.

We say that the scheme is a secure attribute-based encryption for Turing machines if for all p.p.t. adversaries A, and for all sufficiently large κ:

$$\mathsf{Adv}_{\mathsf{ABE}, A} := |\Pr[\mathsf{Exp}_{\mathsf{ABE}, A}(1^\kappa) = 1] - 1/2| = \mathsf{negl}(\kappa).$$

3   Attribute-Based Encryption for Turing Machines and RAMs

We construct an ABE scheme for Turing machines based on three ingredients:

1. an extractable witness encryption scheme WE = (WE.Enc, WE.Dec) based on the work of [GGSW13], on which we elaborate in Sec. 2.2,

2. a succinct argument of knowledge scheme, SNARK = (SNARK.Gen, SNARK.Prover, SNARK.Verify), based on the work of [BCCT13],

3. an existentially unforgeable signature scheme secure against adaptive chosen message attacks SIG = (SIG.KeyGen,SIG.Sign, SIG.Verify) [GMR88].

Theorem 5. *Assuming the above three primitives, there exists a secure attribute-based encryption scheme (as per Def. 4) for any class of (uniform or non-uniform) Turing machines $\mathcal{T}$, for which there exists a polynomial p such that the runtime of every machine in $\mathcal{T}$ is upper-bounded by p.*

The p restriction comes from the fact that SNARKs are for **NP**. From now on, for brevity, we will refer to such a class by "a class of Turing machines with runtime upper-bounded by some polynomial".

Corollary 1. *There exists a secure attribute-based encryption scheme for any class of (uniform or non-uniform) Turing machines whose runtime is upper-bounded by some polynomial under the extractable DGE No-Exact-Cover assumption, "knowledge of exponent assumption", and the existence of collision-resistant hash functions (Sec. 1.2).*

3.1 Construction preliminaries

We advise the reader to recall the intuition we provided in technique overview, Sec. 1.3.

The Language L for SNARK. We define L by defining its relation, R_L. Let R_L be the following instance-witness relation: the instance is of the form $y = (\mathsf{VK}, x, t)$ (a verification key VK for a signature scheme, an input x, and a time bound t) and the witness is of the form $w = (M, \sigma)$, for M a Turing machine and σ a signature. Then, $(y, w) \in R_L$ iff SIG.Verify$(\mathsf{VK}, M, \sigma) = 1$ and M halts on x in at most t steps and outputs one. Moreover, $t < p(|x|)$, where p is a polynomial upper-bound on the runtime of every Turing machine in the class of interest. Let (SNARK.Gen, SNARK.Prover, SNARK.Verify) be a SNARK system for L.

The Language L^* for WE. Based on the above language L and the SNARK system (SNARK.Gen, SNARK.Prover, SNARK.Verify) for L, we define a language L^* for the witness encryption scheme using the witness relation R_{L^*} as follows:

$$R_{L^*}\big[x^* = (x, \mathsf{crs}, \mathsf{VK}), w^* = (\pi, t)\big] = 1 \text{ iff SNARK.Verify}(\mathsf{crs}, (\mathsf{VK}, x, t), \pi) = 1.$$

Let WE = (WE.Enc, WE.Dec) be an extractable witness encryption scheme for the witness relation R_{L^*}.

3.2 Construction of ABE for Turing Machines

Our construction of ABE = (ABE.Setup, ABE.KeyGen, ABE.Enc, ABE.Dec) for Turing machines proceeds as follows. Let $\mathcal{T}$ be the class of (uniform or non-uniform) polynomial time Turing machines for the ABE scheme.

Setup ABE.Setup(1^κ) where κ is the security parameter:

1. Sample a verification key / signing key pair $(\mathsf{VK}, \mathsf{SigSK}) \leftarrow \mathsf{SIG.KeyGen}(1^\kappa)$, and output $\mathsf{mpk} := \mathsf{VK}$ and $\mathsf{msk} := \mathsf{SigSK}$.

Encryption $\mathsf{ABE.Enc}(\mathsf{mpk}, x, b)$ where $\mathsf{mpk} = \mathsf{VK}$, $x \in \{0,1\}^*$ and $b \in \{0,1\}$:

1. Run the SNARK generator $\mathsf{SNARK.Gen}$ to get $\mathsf{crs} \leftarrow \mathsf{SNARK.Gen}(1^\kappa)$.
2. Let $x^* = (x, \mathsf{crs}, \mathsf{VK})$. Compute $\mathsf{ct}_{\mathsf{WE}} \leftarrow \mathsf{WE.Enc}(1^\kappa, x^*, b)$.
3. Output $\mathsf{ct} := (x^*, \mathsf{ct}_{\mathsf{WE}})$.

Key generation $\mathsf{ABE.KeyGen}(\mathsf{msk}, M)$ where M is a Turing machine:

1. Compute $\sigma \leftarrow \mathsf{SIG.Sign}(\mathsf{SigSK}, M)$ and output $\mathsf{sk}_M := (M, \sigma)$.

Decryption $\mathsf{ABE.Dec}(\mathsf{sk}_M, \mathsf{ct})$ where $\mathsf{sk}_M = (M, \sigma)$ and $\mathsf{ct} = (x^* = (x, \mathsf{crs}, \mathsf{VK}), \mathsf{ct}_{\mathsf{WE}})$:

1. Run M on x and let t be the number of steps after which M halts (note that M is a polynomial time Turing machine so it must halt within a polynomial number of steps).
2. If $M(x) = 0$, output $\perp$ and exit.
3. Otherwise, let $w := (M, \sigma)$ and note that $\big((\mathsf{VK}, x, t), w\big) \in R_L$.
4. Run $\mathsf{SNARK.Prover}$ to obtain a proof $\pi \leftarrow \mathsf{SNARK.Prover}(\mathsf{crs}, (\mathsf{VK}, x, t), w)$.
5. Let $w^* = (\pi, t)$. Compute and output $\mathsf{WE.Dec}(w^*, \mathsf{ct}_{\mathsf{WE}})$.

Proof Intuition. We prove Th. 5 formally in our full version, and we only provide intuition here for the security proof. We start by assuming the ABE scheme is not secure, and reach a contradiction by showing that one can forge signatures using the extractability properties of the WE and SNARK schemes. Therefore, assume there is an adversary for ABE, $A_{\mathsf{ABE}} = (A_{\mathsf{ABE},1}, A_{\mathsf{ABE},2})$. We will show how to construct an adversary A_{WE} for the WE scheme: A_{WE} simply embeds its challenge ciphertext into the ciphertext for A_{ABE} and lets A_{ABE} decide.

Once we have the adversary A_{WE}, by the security definition of WE, we also have an extractor E_{WE} which on input x^*, outputs a valid witness $w^* = (\pi, t)$ of $(x^*, w^*) \in R_{L^*}$. Using E_{WE}, we construct a prover P^* for the SNARK system that is able to construct an instance $y = (\mathsf{VK}, x, t)$ and a proof π for which the SNARK verifier accepts. By the proof of knowledge property of the SNARK, there exists an extractor E_{SNARK} that outputs a witness for the SNARK language L, namely $w = (M, \sigma)$, such that $(y, w) \in R_L$. This means that $M(x) = 1$ and that σ is a correct signature on M; but A_{ABE} only asked for signatures of Turing machines M_i for which $M_i(x) = 0$. Therefore, (M, σ) are a new signature pair and thus we used P^* and E_{SNARK} to forge a signature and reach a contradiction.

3.3   ABE for RAMs

In this section, we discuss how to construct ABE for RAMs. This construction is similar to our construction for Turing machines, so we only mention the main differences here: the language L for the SNARK and $\mathsf{ABE.KeyGen}$. See our full paper for more details. Let (M, D) be a RAM pair: a RAM machine M and memory D.

The Language L for SNARK. Let R_L be the following instance-witness relation: the instance is of the form $y = (\mathsf{VK}, x, t)$ (a verification key VK for a signature scheme, an input x, and a time bound t) and the witness is of the form $w = (r, M, \sigma_{(r,M)}, S, \{i, D_i, \sigma_{(r,i,D_i)}\}_{i \in S})$, where r is a nonce, M a machine, $\sigma_{(M,r)}$ is a signature on the description of the machine M and the nonce r, S is a set of integers that represent memory addresses (the memory accesses M makes to D), D_i is the value in the i-th slot of memory and σ_{r,i,D_i} is a signature on r and D_i. Then, $(y, w) \in R_L$ iff

1. $\mathsf{SIG.Verify}(\mathsf{VK}, (r, M), \sigma_{(r,M)}) = 1$,
2. $\mathsf{SIG.Verify}(\mathsf{VK}, (r, i, D_i), \sigma_{(r,i,D_i)}) = 1$ for all $i \in S$,
3. M halts on x in at most t, all of its memory queries are in S, and outputs one.

Key generation $\mathsf{ABE.KeyGen}(\mathsf{msk}, M, D)$ where M is a RAM and D its memory:

1. Choose $r \leftarrow \{0, 1\}^{\mathrm{poly}(\kappa)}$.
2. Compute $\sigma_{(r,M)} \leftarrow \mathsf{SIG.Sign}(\mathsf{SigSK}, (r, M))$.
3. For every $i \in 1 \dots |D|$, compute $\sigma_{(r,i,D_i)} \leftarrow \mathsf{SIG.Sign}(\mathsf{SigSK}, (r, i, D_i))$.
4. Output $(r, M, \sigma_{(r,M)}, \{D_i, \sigma_{(r,i,D_i)}\}_{i=1}^{|D|})$.

Key generation runtime and the function key size are polynomial in the description of the RAM and the size of $|D|$, but they do not depend on the runtime of the RAM. (As a remark, to obtain a slightly shorter key size, one can sign a Merkle tree over the entries in D.) The time to decrypt also only depends on the time to run the RAM and not on its worst case running time or on the memory size.

3.4   Beyond ABE for Turing Machines and RAMs

Interestingly, it turns out the expressivity of our ABE construction goes beyond that of Turing machines and RAMs. The ABE construction can be easily changed to allow the evaluator to provide *an additional input* α to the computation. That is, given a function key sk_M, a ciphertext $\mathsf{ct}_{x,m}$, an evaluator can choose an input α by himself; then if $M(x, \alpha) = 1$, $\mathsf{ABE.Dec}$ outputs m, otherwise it outputs $\bot$. To construct such an ABE, one only has to change the SNARK language L such that an instance has the form (VK, x, t) and a witness is (M, σ, α) with $M(x, \alpha) = 1$ and σ verifies M.

This extra input α makes the scheme significantly more expressive. We illustrate on two examples. The first example allows the secret key owner to delegate the choice of Turing machines to another user, say Alice, by issuing a function key for Alice; then Alice can choose Turing machines of her choice to run on the ciphertexts, without contacting the secret key owner. To construct this example, the secret key owner generates $\mathsf{sk}_{U_{\mathsf{Alice}}}$ where U_{Alice} is a universal circuit containing Alice's public key. U_{Alice} takes as input $\alpha = (\mathsf{TM}, \sigma(\mathsf{TM}))$ and x: it first checks that $\sigma(\mathsf{TM})$ verifies with Alice's public key as being a signature of TM, and if so, it runs $\mathsf{TM}(x)$. Now Alice can choose any Turing machine TM she wishes, and as long as she signs it, she will be able to evaluate it on the ciphertext. In fact, the secret key owner can delegate the choice of Turing machines to any group of people, and he can even express complex policies, e.g. "allow any Turing machine that is signed by (Alice and Bob) or Chris".

The second example is to run any approved RAM on any approved database, where approved means that it was signed by the secret key owner. We do not elaborate further on this construction and its applications in this short paper version.

4   Functional Encryption for Turing Machines

In this section we construct a (single-key and succinct) functional encryption scheme for Turing machines. We refer the reader to our full paper for a definition of FE for Turing machines.

Theorem 6. *Assuming we have:*

- *an attribute-based encryption scheme for any class of (uniform or non-uniform) Turing machines with running time upper-bounded by a polynomial, and*

- *a fully homomorphic encryption scheme,*

there is a (single-key and succinct) functional encryption scheme for any class of (uniform or non-uniform) Turing machines with running time upper-bounded by a polynomial.

Theorem 7. *Assuming there exists a (single-key and succinct) functional encryption scheme for any class of (uniform or non-uniform) Turing machines with running time bounded by a polynomial, there is a (single-key and succinct) input-specific runtime functional encryption scheme for any class of (uniform or non-uniform) Turing machines with running time bounded by a polynomial.*

Corollary 2. *There exists a secure (single-key and succinct) functional encryption scheme* FE *and a (single-key) input-specific runtime functional encryption scheme* FE* *for any class of (uniform or non-uniform) Turing machines with runtime bounded by a polynomial under the extractable DGE No-Exact-Cover assumption, "knowledge of exponent assumption", and the LWE assumption with circular security (Sec. 1.2).*

4.1   FE for Turing Machines Construction (FE)

Recall the construction overview provided in Sec. 1.3. We follow the reduction of Goldwasser *et al.* [GKP+13b] who showed how to construct a (single-key and succinct) functional encryption scheme from any ABE and FHE scheme, where functions were modeled as circuits.

Our construction of FE = (FE.Setup, FE.KeyGen, FE.Enc, FE.Dec) proceeds similarly to the [GKP+13b] construction, with the main difference being that we work with Turing machines instead of circuits. There are two places in the reduction where the treatment of circuits is different from the treatment of Turing machines: in the use of the ABE and FHE schemes. To adapt the reduction to Turing machines, we first use our ABE for Turing machines scheme. Second, we need to construct a Turing machine M_{FHE} that performs the FHE evaluation of another Turing machine M. We only present here the construction of M_{FHE} and delegate the full FE construction to our full paper.

Based on the intuition provided in Sec. 1.3, we describe a compiler $\mathsf{Compile}_{\mathsf{FHE}}$ that takes as input a Turing machine M and a number of steps t and produces a Turing machine M_{FHE} that computes the FHE evaluation of M for t steps. In the following, let $\hat{x}$ denote the FHE encryption of x.

Algorithm 1. ($\mathsf{Compile}_{\mathsf{FHE}}(M, t)$)

1. Use the Pippenger-Fischer transformation [PF79] for time bound t to transform M into an oblivious Turing machine M_O with head movement function next. next is a function that takes as input i, the current step in the computation, and outputs whether the head of M_O should move left or right on the tape. The Turing machine M_O has a transition function δ: δ takes as input a tape input bit b, a state state and outputs a new state state$'$, and the new content b' for the new tape location which is indicated by next.

2. Based on (M_O, next), construct a new Turing machine M_{FHE} that takes as input an FHE public key hpk and an input encryption $\hat{x}$. M_{FHE} evaluates homomorphically the transition function δ of M_O for t steps. Each cell of the tape of M_O corresponds to the FHE encryption of the cell value for M_{FHE}. At step i, M_{FHE} maintains the FHE encryption of the state of M_O at time i: $\widehat{\text{state}}_i$. At step i, M_{FHE} takes as input the encrypted bit from the input tape $\hat{b}$ that the head currently points at, the current encrypted state $\widehat{\text{state}}_i$, and outputs an encrypted new state $\widehat{\text{state}}_{i+1}$ and a new content $\hat{b}'$. M_{FHE} updates the current cell with $\hat{b}'$ and then computes next(i) to determine whether to move left or right.

3. Output the description of M_{FHE}.

Note that the running time of $\mathsf{Compile}_{\mathsf{FHE}}$ and M_{FHE} is polynomial in t.

4.2 Input-Specific Runtime Functional Encryption for Turing Machines (FE*)

In what follows we show how to convert a (single-key) functional encryption scheme for Turing machines FE into one where the decryption algorithm, on input a function key for M denoted fsk_M and FE.Enc(MPK, x), runs in time that depends on the runtime of M on input x. Denote by FE* such a functional encryption scheme. We refer the reader to Sec. 1.3 for the construction overview and to our full paper for the definition of input-specific runtime functional encryption.

Setup FE*.Setup(1^κ):

1. Generate $\tau := \log B_n$ independent pair of keys for the FE scheme: $(\mathsf{msk}_i, \mathsf{mpk}_i) \leftarrow$ FE.Setup(1^κ).

2. Output MPK $:= (\mathsf{mpk}_1, \ldots, \mathsf{mpk}_\tau)$ and MSK $:= (\mathsf{msk}_1, \ldots, \mathsf{msk}_\tau)$.

Key Generation FE*.KeyGen(MSK, M): with MSK $= (\mathsf{msk}_1, \ldots, \mathsf{msk}_\tau)$.

1. Let M_i be the Turing machine that runs M for 2^i steps and outputs $M(x)$ if M finishes in that number of steps, otherwise, it outputs $\bot$. Let t_i be the number of steps M_i runs for.[3]

2. Let $\mathsf{fsk}_{M_i} \leftarrow$ FE.KeyGen(msk_i, M_i, t_i), for $i = 1 \ldots \tau$.

3. Output $\mathsf{fsk}_M := (\mathsf{fsk}_{M_1}, \ldots, \mathsf{fsk}_{M_\tau})$.

Encryption FE*.Enc(MPK, x) with MPK $= (\mathsf{mpk}_1, \ldots, \mathsf{mpk}_\tau)$

1. Compute $\mathsf{ct}_i \leftarrow$ FE.Enc(mpk_i, x) for $i = 1 \ldots \tau$.

---

[3] Note that t_i may be slightly larger than 2^i, since t_i is the number of steps it takes to simulate a Turing machine that runs for 2^i steps.

2. Output $\mathsf{ct} := (\mathsf{ct}_1, \ldots, \mathsf{ct}_\tau)$.

Decryption $\mathsf{FE}^*.\mathsf{Dec}(\mathsf{fsk}_M, \mathsf{ct})$: for $\mathsf{fsk}_M = (\mathsf{fsk}_{M_1}, \ldots, \mathsf{fsk}_{M_\tau})$, $\mathsf{ct} = (\mathsf{ct}_1, \ldots, \mathsf{ct}_\tau)$.

1. Starting with $i = 1$, repeat until $v \neq \bot$:

   (a) $v \leftarrow \mathsf{FE}.\mathsf{Dec}(\mathsf{fsk}_{M_i}, \mathsf{ct}_i)$

   (b) $i \leftarrow i + 1$

2. Output v.

   Based on this construction, we prove Th. 7 in our full paper.

Acknowledgments. We would like to thank the authors of the witness encryption paper [GGH13a] and Zvika Brakerski for useful discussions. This work was supported by an NSERC Discovery Grant, by DARPA awards FA8750-11-2-0225 and N66001-10-2-4089, by NSF awards CNS-1053143 and IIS-1065219, and by Google. The U.S. Government is authorized to reproduce and distribute reprints for governmental purposes notwithstanding any copyright notation thereon. The views and conclusions contained herein are those of the author and should not be interpreted as necessarily representing the official policies or endorsements, either expressed or implied, of DARPA or the U.S. Government.

References

AGVW13. Agrawal, S., Gorbunov, S., Vaikuntanathan, V., Wee, H.: Functional encryption: New perspectives and lower bounds. In: Canetti, R., Garay, J.A. (eds.) CRYPTO 2013, Part II. LNCS, vol. 8043, pp. 500–518. Springer, Heidelberg (2013)

BCCT13. Bitansky, N., Canetti, R., Chiesa, A., Tromer, E.: Recursive composition and bootstrapping for SNARKs and proof-carrying data. In: STOC (2013)

BGV12. Brakerski, Z., Gentry, C., Vaikuntanathan, V.: (Leveled) fully homomorphic encryption without bootstrapping. In: ITCS, pp. 309–325 (2012)

BGW88. Ben-Or, M., Goldwasser, S., Wigderson, A.: Completeness theorems for non-cryptographic fault-tolerant distributed computation. In: STOC, pp. 1–10 (1988)

BHR12. Bellare, M., Hoang, V.T., Rogaway, P.: Foundations of garbled circuits. In: ACM CCS, pp. 784–796 (2012)

Bra12. Brakerski, Z.: Fully homomorphic encryption without modulus switching from classical gapSVP. In: Safavi-Naini, R., Canetti, R. (eds.) CRYPTO 2012. LNCS, vol. 7417, pp. 868–886. Springer, Heidelberg (2012)

BSW11. Boneh, D., Sahai, A., Waters, B.: Functional encryption: Definitions and challenges. In: Ishai, Y. (ed.) TCC 2011. LNCS, vol. 6597, pp. 253–273. Springer, Heidelberg (2011)

BV11a. Brakerski, Z., Vaikuntanathan, V.: Efficient fully homomorphic encryption from (standard) LWE. In: FOCS, pp. 97–106 (2011)

BV11b. Brakerski, Z., Vaikuntanathan, V.: Fully homomorphic encryption from ring-LWE and security for key dependent messages. In: Rogaway, P. (ed.) CRYPTO 2011. LNCS, vol. 6841, pp. 505–524. Springer, Heidelberg (2011)

CCD88. Chaum, D., Crépeau, C., Damgård, I.: Multiparty unconditionally secure protocols (extended abstract). In: STOC, pp. 11–19 (1988)

Gen09.   Gentry, C.: Fully homomorphic encryption using ideal lattices. In: STOC, pp. 169–178 (2009)

GGH13a.   Garg, S., Gentry, C., Halevi, S.: Candidate multilinear maps from ideal lattices. In: Johansson, T., Nguyen, P.Q. (eds.) EUROCRYPT 2013. LNCS, vol. 7881, pp. 1–17. Springer, Heidelberg (2013)

GGH+13b.   Garg, S., Gentry, C., Halevi, S., Sahai, A., Waters, B.: Attribute-based encryption for circuits from multilinear maps (2013)

GGSW13.   Garg, S., Gentry, C., Sahai, A., Waters, B.: Witness encryption and its applications. In: STOC (2013)

GK05.   Goldwasser, S., Kalai, Y.T.: On the impossibility of obfuscation with auxiliary input. In: FOCS, pp. 553–562 (2005)

GKP+13a.   Goldwasser, S., Kalai, Y.T., Popa, R.A., Vaikuntanathan, V., Zeldovich, N.: How to run Turing machines on encrypted data. Cryptology ePrint Archive, Report 2013/229 (2013), http://eprint.iacr.org/

GKP+13b.   Goldwasser, S., Kalai, Y.T., Popa, R.A., Vaikuntanathan, V., Zeldovich, N.: Reusable garbled circuits and succinct functional encryption. In: STOC (2013)

GMR88.   Goldwasser, S., Micali, S., Rivest, R.L.: A digital signature scheme secure against adaptive chosen-message attacks. SIAM J. Comput., 281–308 (1988)

GMW87.   Goldreich, O., Micali, S., Wigderson, A.: How to play any mental game. In: STOC, pp. 218–229 (1987)

GPSW06.   Goyal, V., Pandey, O., Sahai, A., Waters, B.: Attribute-based encryption for fine-grained access control of encrypted data. In: ACM CCS, pp. 89–98 (2006)

GVW12.   Gorbunov, S., Vaikuntanathan, V., Wee, H.: Functional encryption with bounded collusions via multi-party computation. In: Safavi-Naini, R., Canetti, R. (eds.) CRYPTO 2012. LNCS, vol. 7417, pp. 162–179. Springer, Heidelberg (2012)

GVW13.   Gorbunov, S., Vaikuntanathan, V., Wee, H.: Attribute-based encryption for circuits. In: STOC (2013)

KSW08.   Katz, J., Sahai, A., Waters, B.: Predicate encryption supporting disjunctions, polynomial equations, and inner products. In: Smart, N.P. (ed.) EUROCRYPT 2008. LNCS, vol. 4965, pp. 146–162. Springer, Heidelberg (2008)

LO12.   Lu, S., Ostrovsky, R.: How to garble RAM programs? In: Johansson, T., Nguyen, P.Q. (eds.) EUROCRYPT 2013. LNCS, vol. 7881, pp. 719–734. Springer, Heidelberg (2013)

LOS+10.   Lewko, A., Okamoto, T., Sahai, A., Takashima, K., Waters, B.: Fully secure functional encryption: Attribute-based encryption and (Hierarchical) inner product encryption. In: Gilbert, H. (ed.) EUROCRYPT 2010. LNCS, vol. 6110, pp. 62–91. Springer, Heidelberg (2010)

LW12.   Lewko, A., Waters, B.: New proof methods for attribute-based encryption: Achieving full security through selective techniques. In: Safavi-Naini, R., Canetti, R. (eds.) CRYPTO 2012. LNCS, vol. 7417, pp. 180–198. Springer, Heidelberg (2012)

PF79.   Pippenger, N., Fischer, M.J.: Relations among complexity measures. J. ACM 26(2), 361–381 (1979)

SS10.   Sahai, A., Seyalioglu, H.: Worry-free encryption: functional encryption with public keys. In: ACM CCS, pp. 463–472 (2010)

SSW09.   Shen, E., Shi, E., Waters, B.: Predicate privacy in encryption systems. In: Reingold, O. (ed.) TCC 2009. LNCS, vol. 5444, pp. 457–473. Springer, Heidelberg (2009)

SW05.   Sahai, A., Waters, B.: Fuzzy identity-based encryption. In: Cramer, R. (ed.) EUROCRYPT 2005. LNCS, vol. 3494, pp. 457–473. Springer, Heidelberg (2005)

Yao86.   Yao, A.C.: How to generate and exchange secrets (extended abstract). In: FOCS, pp. 162–167 (1986)

Author Index